Schrödinger Equations in Nonlinear Systems

Wu-Ming Liu · Emmanuel Kengne

Schrödinger Equations in Nonlinear Systems

Springer

Wu-Ming Liu
Institute of Physics
Chinese Academy of Sciences
Beijing, China

Emmanuel Kengne
Department of Computer Science
and Engineering
University of Quebec at Outaouais
Gatineau, QC, Canada

ISBN 978-981-13-6580-5 ISBN 978-981-13-6581-2 (eBook)
https://doi.org/10.1007/978-981-13-6581-2

Library of Congress Control Number: 2019931988

This Springer imprint is published by the registered company Springer Nature Singapore Pte Ltd.
The registered company address is: 152 Beach Road, #21-01/04 Gateway East, Singapore 189721, Singapore

Preface

The nonlinear Schrödinger equation is a very versatile equation used in many branches of physics, dictating the behavior of wave packets in weakly nonlinear media. From the mathematical point of view, this equation has been studied on the Euclidean space since the seventies. During the past several decades, there have been a great deal of both theoretical and experimental investigations in models based on the nonlinear Schrödinger equations with or without external potential, cubic, and/or cubic–quintic nonlinearities. The physical models of nonlinear Schrödinger type emerge in various nonlinear physical phenomena, such as pulse propagation in optical fibers and waveguides, light propagation in chalcogenide glasses, the envelope of wave packets in ocean waves in an infinitely deep ocean, solitary wave propagation in nonlinear transmission networks, dynamics of Bose–Einstein condensates, convection in pure and binary fluids, mode-locked lasers, plasma–laser interaction, pattern formation, the spin–orbit coupled Bose–Einstein condensates, and some organic materials. Appearing in diverse branches of nonlinear systems such as in nonlinear optics, the nonlinear transmission networks, and the Bose–Einstein condensation, the cubic and/or cubic–quintic nonlinear Schrödinger equations with nonlinearity management present practical interest. This book surveys the advancements and applications of Schrödinger equations in nonlinear phenomena in both theoretical and experimental contexts, and from fundamental concepts to real applications in modeling nonlinear phenomena; it is mainly focused on the applications of the theory of nonlinear transmission networks and of the theory of Bose–Einstein condensates.

This book serves as a handbook for physicists, applied mathematicians, and engineers working on nonlinear phenomena. It is organized in a parallel structure to when and how the Schrödinger equations can be applied for modeling and studying nonlinear phenomena of the theories of nonlinear transmission networks and Bose–Einstein condensates in trapping external potential. Chapter 1 is devoted to the basic knowledge of Schrödinger models, and Chap. 2 to the well-posedness of nonlocal boundary value problems of the Schrödinger models. The well-posedness of problems is useful for modeling and in solving both analytically and numerically of nonlocal boundary value problems. Chapters 3–5 focus on the applications of Schrödinger equations in modeling and investigating nonlinear phenomena in nonlinear transmission networks. Chapters 6–8 study the exact and numerical solutions of inhomogeneous nonlinear Schrödinger equations, allowing to investigate analytically, numerically and experimentally the dynamics of nonlinear matter waves on Bose–Einstein condensates. Chapters 9–11 focus on the spin–orbit coupling for multicomponents Bose–Einstein condensates.

The authors would like to share with the readers their happiness in undertaking the scientific collaboration, which started in Autumn 2005. At that time little was known about the derivative nonlinear Schrödinger equation in nonlinear transmission networks. From September 2005, E. Kengne's first time to visit the Institute of Physics Chinese Academy of Sciences and to cooperate with W. M. Liu, the two authors kept longtime cooperation and published over 14 papers; this book is the gain of their longtime cooperation. Without the ensuing teamwork, it would have been impossible to achieve the main original results contained in this book! We apologize if any important references are omitted. Any such error is definitely due to the limits of the authors' knowledge of the literature.

Acknowledgements

Nobody has been more important to one of the authors, E. Kengne, in the pursuit of the multiple projects leading to this book than the members of his family. E. Kengne would like to thank his mother, maman Marie Djomou at Bandjoun-Cameroon, his second mother, maman Lydie Takotue at Douala-Cameroon, and his second dad, Ta'a Moukam Gakgoun Innocent Takotue at Douala-Cameroon, whose love and guidance are with him in whatever he pursue. They are the ultimate role models. Most importantly, E. Kengne wishes to thank his loving and supportive wife, Eleonore Nkuojip Sado, and his three wonderful children, Kengneson Delma Djomo, Kengneson Weierstrass Owan Wambo, and Kengneson Cris-Carelle Djike all at Gatineau-Quebec-Canada, who provide unending inspiration.

One of the authors, Emmanuel Kengne, would like to acknowledge the Chinese Academy of Sciences (CAS) for granting him three time the President's International Fellowship (PIF) for Visiting Scientists (2012, 2018, and 2019). These PIFs have enhanced his research cooperation with Wu-Ming Liu at the Institute of Physics CAS, and the present book is one of the gains of these Fellowships. The authors acknowledge all the reviewers for their constructive comments, critics, remarks, and opinions on some of the authors' original papers.

This book was supported by The Chinese Academy of Sciences President's International Fellowship for Visiting Scientists under Grants No. 2018VMA0002, the National Key R&D Program of China under grants No. 2016YFA0301500, NSFC under grants Nos. 11434015, 61227902.

Beijing, China Wu-Ming Liu
Gatineau, Canada Emmanuel Kengne

Contents

Chapter 1
Overview of Nonlinear Schrödinger Equations

Abstract The present chapter deals with an overview of nonlinear Schrödinger equations that are mainly used to investigate the dynamics of nonlinear matter waves of either the nonlinear transmission networks or the Bose–Einstein condensates: The one-dimensional cubic nonlinear Schröodinger equations including the standard and the dissipative cases, the derivative nonlinear Schrödinger equations, the inhomogeneous nonlinear Schrödinger equations, and the multicomponent nonlinear Schrödinger equations

The nonlinear Schrödinger equation is an equation that can be interpreted as the differential law of the development in time of a system. It is a very versatile model applicable to many disciplines in engineering and applied science, such as dynamical systems, nonlinear optics, fluid dynamics, materials science, statistical physics, particle physics, astrophysics, nonlinear transmission networks [5, 13, 23, 27, 33, 37, 43, 54, 55, 60]. It dictates the behavior of wave packets in weakly nonlinear media. The NLS equation represents the evolution of optical waves in a nonlinear fiber, the envelope of wave packets in ocean waves in an infinitely deep ocean, various biological systems, and the price of options in economics.

The one-dimensional (1D) generalized nonlinear Schrödinger equation that describes the slow evolution of a weakly nonlinear wave packet in a given medium reads [10, 13, 21]

$$i\frac{\partial\psi}{\partial t} = P\frac{\partial^2\psi}{\partial x^2} + Q\,|\psi|^2\,\psi + [V(x,t) + i\gamma]\,\psi + \chi\,|\psi|^4\,\psi + i\alpha\,|\psi|^2\,\frac{\partial\psi}{\partial x} + i\beta\psi^2\frac{\partial\psi^*}{\partial x}. \quad (1.1)$$

Here, ψ is the slowly varying complex envelope field, star * stands for the complex conjugate, $x \in \mathbf{R}$, $t \geq 0$, P, Q, χ, α, β, and γ are real-valued functions of t, and $V(x,t)$ is a real-valued function of x and t. In Eq. (1.1), P describes the dispersion, Q the nonlinearity, γ the dissipation; $V(x,t)$ accounts for the external potential, χ is the strength of the quintic nonlinearities; the derivative terms $|\psi|^2\,\frac{\partial\psi}{\partial x}$ and $\psi^2\frac{\partial\psi^*}{\partial x}$ appear in the asymptotic derivation and are commonly associated with self- steepening. These two derivative terms also appear when higher order physical effects are included in the dynamics of the wave packet [9]. In an optical fiber setting (in particular, as optical waveguides and waveguides are concerned), coordinates t and x denote,

W.-M. Liu and E. Kengne, *Schrödinger Equations in Nonlinear Systems*,
https://doi.org/10.1007/978-981-13-6581-2_1

respectively, the propagation distance (usually denoted z) and the retarded time, while for spatial waveguide, they denote the propagation coordinate and diffraction, respectively. In hydrodynamics, these variables t and x denote typically slow time and spatial coordinate moving with the group velocity. The cubic–quintic nonlinear Schrödinger models ($Q\chi \neq 0$) have been considered with various additional physical effects [50] such as fourth-order dispersion [16], non-Kerr media [59], and presence of energy gain/loss [24].

It is important to note that when either the dissipative parameter $\gamma \neq 0$ or at least one of the coefficients P, Q, χ, α, and β is complex valued, Eq. (1.1) is called the **complex Giznburg–Landau** (CGL) equations, well known as the dissipative generalizations of NLS equations. Throughout this book, Eq. (1.1) under the condition $\gamma \neq 0$ will be called the dissipative NLS equation.

It is important to mention that Eq. (1.1) is invariant with respect to a Galilean transformation, which, in contrast to the usual NLS equation, involves a change of coefficients of the equation. Indeed, if $\psi(x,t)$ is a solution of Eq. (1.1), then its boosted counterpart, moving at arbitrary velocity v, $\widetilde{\psi}(x-v,t)\exp\left[i\left(-\frac{c}{2P}x+\frac{c^2}{4P}t\right)\right]$ is a solution to Eq. (1.1) with the shifted cubic coefficient $Q \rightarrow Q + \frac{c(\alpha-\beta)}{2P}$.

One of the most important dynamical features of the NLS equation (1.1) is the modulational instability (MI) of its plane wave (continuous wave (CW)) solutions. The MI refers to a general phenomenon in nonlinear wave equations. It indicates that due to the interplay between nonlinearity and the dispersive effects, a small perturbation on the envelope of a plane wave may induce an exponential growth of the wave amplitude, resulting in the carrier-wave breakup into a train of localized waves [4, 44, 49]. The onset of MI leads to the formation of solitary waves [53].

This chapter is a brief introduction of the 1D nonlinear Schrödinger equation. Here, we enumerate some classes of NLS equations having mathematical and physical meaning. For these classes of NLS equations, we will take an overlook on the question of MI leading to the formation of solitary waves.

1.1 One-dimensional Cubic Nonlinear Schrödinger Equations: Standard and Dissipative Cases

The standard one-dimensional cubic nonlinear Schrödinger equation [2, 29]

$$i\frac{\partial\psi}{\partial t} = P\frac{\partial^2\psi}{\partial x^2} + Q\,|\psi|^2\,\psi \tag{1.2}$$

describes the self-modulation of one-dimensional monochromatic plane waves in dispersive media, when the dissipative effects are weak. When the dissipative effects are taken into account, i.e., when the medium is dissipative as well as dispersive, the modulation of the first-order amplitude modulation can be described by the one-dimensional cubic dissipative NLS equation [12, 30, 42]

$$i\frac{\partial \psi}{\partial t} = P\frac{\partial^2 \psi}{\partial x^2} + Q\left|\psi\right|^2 \psi + i\gamma\psi, \tag{1.3}$$

which is the simplest case of the complex Ginzburg–Landau equation. Therefore, the dissipative NLS equation can be used to model the dissipative self-modulating monochromatic waves with dispersion. Equations (1.2) and (1.3) have been extensively used to investigate the propagation of modulated matter-wave solitons in nonlinear transmission networks [29, 30, 35, 42].

Equations (1.2) and (1.3) arise frequently in mathematical physics, and occur, for example, in the modeling of deep water waves in hydrodynamics and the propagation of pulses in nonlinear optics [2], as well as the propagation of matter-wave solitons in nonlinear transmission networks [29, 35, 42]. In the context of nonlinear optics, these two equations govern the propagation of pulses inside single-mode fiber, where ψ represents the slowly varying amplitude of the pulse envelope and x is measured in a frame of reference moving with the pulse at the group velocity. Their linear term $P\frac{\partial^2 \psi}{\partial x^2}$ denotes, the second-order dispersion ($P \neq 0$), while their nonlinear term $Q\left|\psi\right|^2 \psi$ accounts for the self-phase-modulation effect. Generally, the dissipative term $\gamma\psi$ of Eq. (1.3) appears when dissipation effects (fiber loss) of the media are taken into account. Depending on the initial width and the peak power of the incident pulse, either dispersive or nonlinear effects may dominate along the fiber. In the context of BECs, ψ is the mean-field wave function, P is a constant, Q is proportional to the s-wave scattering length and represents the two-body interatomic interactions, positive for repulsive interatomic interactions (or defocusing nonlinearities) and negative for attractive ones (focusing nonlinearities), and γ is a small parameter related to the feeding ($\gamma > 0$) or loss ($\gamma < 0$) of atoms in the condensate resulting from the contact with the thermal cloud and three-body recombination [11, 34, 44, 56].

Analyzing the question of the MI of the plane wave solution

$\psi_0(x,t) = A_0 \exp\left[i\left(k_0 x - \left(Q\left|A_0\right|^2 - k_0^2 P\right)t\right)\right]$ of Eq. (1.2) with constant coefficients, it has been found that the plane wave will be unstable (stable) under modulation if product PQ is positive, leading to envelope solitons of Eq. (1.2) (negative, leading to hole solitons of Eq. (1.2)) [42]. Under the localized traveling wave assumption, Eq. (1.2) admits the following soliton solution:

$$\psi(x,t) = \frac{A}{\cosh\left[A\left(\frac{Q}{2P}x - kQt\right)\right]}\exp\left[i\left(kx - \left(Pk^2 - \frac{QA}{2}\right)t\right)\right] \tag{1.4a}$$

for $PQ > 0$, and

$$\psi(x,t) = A\tanh\left[A\left(-\frac{Q}{2P}x + kQt\right)\right]\exp\left[i\left(kx - \left(Pk^2 - QA^2\right)t\right)\right] \tag{1.4b}$$

for $PQ < 0$, where A is the soliton amplitude and k its wavenumber. The width of solitons (1.4a) and (1.4b) is inversely proportional to $\frac{Q}{2P}$ and $-\frac{Q}{2P}$, respectively, meaning that quantity $Q/2P$ can be used to manage the compression of the solitons of the standard NLS equation (1.2).

In order to investigate the MI of the dissipative NLS Eq. (1.3) in the special case of constant coefficients, we look for the solution in the form

$$\psi(x,t) = (q_0 + \delta\psi)\exp\left[\gamma t + i\frac{q_0^2 Q}{2\gamma}\left(1 - \exp[2\gamma t]\right)\right], \tag{1.5}$$

where q_0 is a real constant and $\delta\psi$ is a small perturbation on the wave amplitude. It is important to note that Eq. (1.5) leads to a solution of Eq. (1.3) when $\delta\psi = 0$. If we denote by κ the wavenumber of the modulation waves and follow the idea of Ref. [35], we find that the decaying/growing wave solution $\psi(x,t) = q_0\exp\left[\gamma t + i\frac{q_0^2 Q}{2\gamma}\left(1 - \exp[2\gamma t]\right)\right]$ of Eq. (1.3) will be unstable under modulation (stable under modulation) if the wavenumber κ satisfies condition $K^2 - \frac{2q_0^2 Q}{a^2 P}\exp[2\gamma t] < 0$ ($K^2 - \frac{2q_0^2 Q}{a^2 P}\exp[2\gamma t] > 0$), $a \neq 0$ being any real number. For condition $K^2 - \frac{2q_0^2 Q}{a^2 P}\exp[2\gamma t] < 0$ to be satisfied, it is necessary that $PQ > 0$.

Under the near-integrable condition $|\gamma| << |Q|$, the dissipative term $i\gamma\psi$ in Eq. (1.3) can be treated as a small perturbation. Using the adiabatic perturbation theory developed in Refs. [1, 15], one can derive the approximative solitary wave solution for Eq. (1.3). For example, in the case of modulational instability ($PQ > 0$), we have derived the following bright solitary wave solution:

$$\psi(x,t) = \frac{A\exp[2\gamma t]}{\cosh\left[A\exp[2\gamma t]\left(\sqrt{\frac{Q}{2P}}x + kQt + x_0\right)\right]}\exp\left[i\left\{k\xi + \phi(t)\right\}\right], \tag{1.6}$$

where $\phi(t) = \frac{1}{2}\left[Qk^2 t - \frac{A^2 Q}{4\gamma}\exp[4\gamma t] + \phi_0\right]$, $A \neq 0$, k, x_0, and ϕ_0 being arbitrary real constants. It follows from solution (1.6) that (i) the amplitude of the bright solitary wave in the near-integrable condition $|\gamma| << |Q|$ is proportional to $\exp[2\gamma t]$. (ii) The width of a bright solitary wave in the near-integrable condition $|\gamma| << |Q|$ is inversely proportional to $\sqrt{\frac{Q}{2P}}\exp[2\gamma t]$. This means that Eq. (1.6) can be used to describe the compression of a bright solitary waves when $\gamma < 0$. (iii) The centre of the bright solitary wave is given by $\xi(t) = -\sqrt{\frac{2P}{Q}}\left(kQt + x_0\right)$.

1.2 Derivative Nonlinear Schrödinger Equation

As it is known, the mathematical model for nonlinear pulse propagation in many physical systems as the picosecond timescale with the anomalous dispersion regime in an isotropic, homogeneous, lossless, non-amplifying, polarization-preserving single-mode optical fiber is the nonlinear NLS equation. However, it has been shown experimentally and theoretically that for some other physical systems, the NLS equation is no longer valid; for example, the NLS equation does not support and govern the propagation of high-power ultrashort pulses in long monomode optical fibers in the subpicosecond–femtosecond timescale. In this case, some additional nonlinear terms and/or higher order linear dispersion should be added to the standard (dissipative) NLS equation leading to the so-called derivative NLS equation [15, 33, 58]. In the case of the monomode optical fibers in the subpicosecond–femtosecond timescale, for example, the model can be described by the following nonlinear evolution equations of NLS type [36]

$$i\frac{\partial\psi}{\partial t} = P\frac{\partial^2\psi}{\partial x^2} + Q\left|\psi\right|^2\psi + i\gamma\psi + i\alpha\frac{\partial\left|\psi\right|^2\psi}{\partial x} + i\beta\psi\frac{\partial\left|\psi\right|^2}{\partial x} + i\delta\frac{\partial^3\psi}{\partial x^3} + \chi\left|\psi\right|^4\psi,$$
$$i\alpha\frac{\partial\left|\psi\right|^2\psi}{\partial x} + i\beta\psi\frac{\partial\left|\psi\right|^2}{\partial x} = i\left(2\alpha+\beta\right)\left|\psi\right|^2\frac{\partial\psi}{\partial x} + i\left(\alpha+\beta\right)\psi^2\frac{\partial\psi^*}{\partial x} \tag{1.7}$$

where ψ is the slowly varying amplitude of the complex field envelope, t is the propagation distance along the fiber length, t is the time measured in a frame of reference moving with the pulse at the group velocity, α governs the effects due to the intensity dependence of the group velocity, γ is the intrinsic fiber loss, δ governs the effects of the third-order linear dispersion, β governs the soliton self-frequency shift effect [3], and χ represents the quintic (non-Kerr) nonlinear effect. It should be noted that the two derivative cubic terms can be rewritten as $i\alpha\frac{\partial|\psi|^2\psi}{\partial x} + i\beta\psi\frac{\partial|\psi|^2}{\partial x} = i\left(2\alpha+\beta\right)\left|\psi\right|^2\frac{\partial\psi}{\partial x} + i\left(\alpha+\beta\right)\psi^2\frac{\partial\psi^*}{\partial x}$. In hydrodynamics, the coordinates t and x are typically slow time and spatial coordinate moving with the group velocity.

By considering different spectral problems, there are three kinds of celebrated derivative NLS equations, that can be obtained from Eq. (1.7). They are the Kaup–Newell (KN) equation [26]

$$i\frac{\partial\psi}{\partial t} = P\frac{\partial^2\psi}{\partial x^2} + i\alpha\frac{\partial\left|\psi\right|^2\psi}{\partial x}, \tag{1.8a}$$

the Chen–Lee–Liu (CLL) equation [9]

$$i\frac{\partial\psi}{\partial t} = P\frac{\partial^2\psi}{\partial x^2} + i\tilde{\alpha}\left|\psi\right|^2\frac{\partial\psi}{\partial x}, \tag{1.8b}$$

and the Gerdjikov–Ivanov(GI) equation [20]

$$i\frac{\partial\psi}{\partial t} = P\frac{\partial^2\psi}{\partial x^2} + i\tilde{\beta}\psi^2\frac{\partial\psi^*}{\partial x} + \chi\left|\psi\right|^4\psi. \tag{1.8c}$$

Equations (1.8a)–(1.8c) are used as generic models in plasma physics, in optical waveguides with the self-steepening effect, and in fluid mechanics [8, 25, 51]. The derivative nonlinear terms $\partial\left(\left|\psi\right|^2\psi\right)/\partial x$ and $\left|\psi\right|^2\partial\psi/\partial x$ appearing, respectively, in Eqs. (1.8a) and (1.8b) adequately describe the self-steepening effect [46]. The KN Eq. (1.8a) appears in plasma physics as a model for the propagation of Alfvén waves in magnetized plasma [54]; the same equation was also found to describe the behavior of large-amplitude magnetohydrodynamic waves, propagating in an arbitrary direction with respect to the magnetic field, in a high β-plasma [52]. The CLL Eq. (1.8b) appears as a model for ultrashort optical pulses [45]. The GI Eq. (1.8c) can be transformed into the KN or the CLL equation by corresponding gauge transformations [14].

Many other derivative NLS equations have been derived in the context of nonlinear transmission networks. To investigate the transmission of solitary wave signal through a one-dimensional transmission network, Kengne and Abdourahman [28] recently derived the cubic derivative NLS equation

$$i\frac{\partial\psi}{\partial t}+P\frac{\partial^2\psi}{\partial x^2}+Q\,|\psi|^2\,\psi+i\widetilde{\alpha}\,|\psi|^2\,\frac{\partial\psi}{\partial x}=0 \tag{1.9}$$

which generalizes the CLL equation. For the Stokes wave solution

$$\psi_s(x,t)=r\exp\left[i\left(lx-\left(Pl^2-Q+\widetilde{\alpha}l\right)t\right)\right]$$

of Eq. (1.9) (with real constants l and r), they established the MI criterion $PQr^2-2l^2P^2-3P\widetilde{\alpha}r^2l>0$ which generalizes the MI criterion of the standard NLS equation. Exact soliton-like solutions of Eq. (1.9) are presented in Ref. [28] and prove the possible transmission of solitary signal through the network under consideration.

The cubic–quintic derivative NLS equations

$$i\frac{\partial\psi}{\partial t}+P\frac{\partial^2\psi}{\partial x^2}=Q_1\,|\psi|^2\,\psi+Q_2\,|\psi|^4\,\psi+iQ_3\psi^2\frac{\partial\psi^*}{\partial x}+iQ_4\,|\psi|^2\,\frac{\partial\psi}{\partial x}, \tag{1.10a}$$

$$i\frac{\partial\psi}{\partial t}=P\frac{\partial^2\psi}{\partial x^2}+\gamma\psi+Q_1\,|\psi|^4\,\psi+iQ_2\,|\psi|^2\,\frac{\partial\psi}{\partial x}+iQ_3\psi^2\frac{\partial\psi^*}{\partial x} \tag{1.10b}$$

were derived [33, 48] to investigate the dynamics of modulated waves in a one-dimensional discrete nonlinear electrical transmission network. Using Eq. (1.10a), the authors generalized the Benjamin–Feir instability criterion in the network and showed that the discrete nonlinear transmission line under consideration can support solitary waves. Exact periodic and solitary wave solutions of Eq. (1.10b) have been presented in Refs. [33, 47]. With the help of Eq. (1.10b), Marklund and and Shukla [41] investigated the MI of partially coherent signals in electrical transmission lines; their results should help to understand the nonlinear propagation of broadband pulses in electrical transmission lines. Starting from the derivative cubic–quintic NLS equation (1.10b), Kengne et al. [32] used the modified Hirota's bilinear method to prove the possible transmission of higher order bright solitons and shock signals in nonlinear transmission networks.

1.3 Inhomogeneous Nonlinear Schrödinger Equations (Gross–Pitaevskii Equations)

The discoveries of the past decades in both Bose–Einstein condensates (BECs) and in optical fiber, as well as in many other branches of physical sciences have involved much interest in the study of nonlinear partial differential equations of Schrödinger type. These evolutive equations of NLS-type govern the propagation of optical pulses inside a fiber optics, of a waves in plasma, of waves in nonlinear transmission networks, or of wave functions of the condensates in a BECs. In the context of BECs, these NLS-type equations are generally referred to as the inhomogeneous nonlinear Schrödinger equations and are usually called Gross–Pitaevskii (GP) equations. Here, the inhomogeneity can be attributed to the trapping potential and the nonlinear interaction between the atoms in the condensates. In the standard physical literature, the GP equation is usually obtained in the framework of the second quantization formalism, which in many cases goes beyond the material covered in many advanced undergraduate courses.

At ultralow temperatures, the time-dependent wavefunction of the condensate $\Psi(\mathbf{r}, \tau)$ at position $\mathbf{r}$ and time τ in the presence of both the two- and three-bodies interatomic interactions can be described by the following cubic–quintic dissipative NLS equation with external potential [17]

$$i\hbar\frac{\partial\Psi}{\partial\tau} = -\frac{\hbar^2}{2m}\Delta\Psi + [V_{ext}(\mathbf{r}, \tau) + i\widetilde{\gamma}(\tau)]\,\Psi + \widetilde{g}(\tau)N\,|\Psi|^2\,\Psi + \widetilde{\chi}(\tau)\,|\Psi|^4\,\Psi, \tag{1.11a}$$

where N is the total number of atoms in the condensate, $\hbar$ is Planck's constant, m is the mass of the single bosonic atom, $V_{ext}(\mathbf{r}, \tau)$ is the external macroscopic potential, $\widetilde{\gamma}(\tau)$ is a time-dependent small parameter related to the feeding ($\widetilde{\gamma} > 0$) or loss ($\widetilde{\gamma} < 0$) of atoms in the condensate resulting from the contact with the thermal cloud and three-body recombination, $\widetilde{g}(\tau) = 4\pi\hbar^2 a_s(\tau)/m$ and $\widetilde{\chi}(\tau)$ are, respectively, the strengths of time-dependent two- and three-bodies interatomic interactions (they measure the atomic interactions). Here, $a_s(\tau)$ is the time-dependent s-wave scattering length which can be tuned to any desired value by using the Feshbach resonance technique; its sign indicates whether the two-body interactions are attractive ($a_s < 0$) or repulsive ($a_s > 0$). The scattering length a_s measures the intensity of the two-body interactions between the atoms. Mostly used as an envelope equation in optics, the cubic–quintic nonlinearities (i.e., $\widetilde{g}(\tau)N\,|\Psi|^2\,\Psi$ and $\widetilde{\chi}(\tau)\,|\Psi|^4\,\Psi$) is also considered in BEC for alkalime gases, in which case $\widetilde{\chi}(\tau) < 0$ [17]; here, the cubic term corresponds to a negative scattering length, and the quintic term to a repulsive three-body elastic interaction.

In Eq. (1.11a), $|\Psi|^2$ represents the atomic (probability) density; given the normalization condition $\int_{\mathbf{R}^n} |\Psi|^2\, d\mathbf{r} = 1$ (n being the spatial dimension), this quantity is the probability density of finding a boson. In Eq. (1.11a), the external potential V_{ext} is generally used to model the action of the external world on the condensate, in particular of a trap. By either optical or magnetic means, the bosons are most of the time confined in a small region of space. This situation is simulated with the help of a trapping potential with a minimum in the central part. Generally, potential with a minimum can be locally approximated by a harmonic potential, and such that the condensate remains in the part where the quadratic term is dominant, as for example, the ellipsoidal (spherically) symmetric trap

$$V_{ext}(\mathbf{r},\tau) = \frac{1}{2}m\left(\omega_x^2 x^2 + \omega_y^2 y^2 + \omega_z^2 z^2\right), \tag{1.12a}$$

where m is the mass of the bosons, and ω_x, ω_y, and ω_z are the trap frequencies in x, y, and z direction, respectively. In the special case where $\omega_x = \omega_y = \omega_z$, the trap potential is isotropic.

It is more convenient to use the GP equation in a dimensionless form. The 1D GP equation can be easily obtained with the use of the transformation of variables $r = \sqrt{2}\,|\mathbf{r}|^2 / \sqrt{\hbar/(m\omega)}$, $t = \omega\tau$, $\psi(r,t) = \Psi(\mathbf{r},\tau)\left(\frac{(\hbar/(m\omega))^{\frac{3}{2}}}{2\sqrt{2}}\right)^{\frac{1}{2}}$. Inserting this transformation of variables into Eqs. (1.11a) and (1.12a), we obtain the following radial part of the GP. equation (1.11a)

$$i\frac{\partial\psi}{\partial t} = -\frac{\partial^2\psi}{\partial r^2} + \frac{1-D}{r}\frac{\partial\psi}{\partial r} + \left[\frac{\widetilde{\kappa}(t)}{4}r^2 + i\gamma(t)\right]\psi + g(t)\,|\psi|^2\,\psi + \chi(t)\,|\psi|^4\,\psi, \tag{1.13}$$

where D is a spatial dimension, and the parameter $\widetilde{\kappa}(t)$ is the strength of the external trap which is to be reduced from 1 to 0 when the trap is switched off. In the case of the 1D GP equation (1.13), the normalization condition becomes $4\pi\int_0^{+\infty}|\psi(r,t)|\,dr = 1$.

In the physically important case of the cigar-shaped BECs with harmonic potential, the 1D GP equation takes the form

$$i\frac{\partial\psi}{\partial t} = -\frac{\partial^2\psi}{\partial\xi^2} + \left[\kappa(t)\xi^2 + i\gamma(t)\right]\psi + g(t)\,|\psi|^2\,\psi + \chi(t)\,|\psi|^4\,\psi. \tag{1.14}$$

In Eq. (1.14), time t and the spatial coordinate x are, respectively, measured in units $2/\omega_\perp$ and $a_\perp$, where $a_\perp = \sqrt{\frac{\hbar}{m\omega_\perp}}$ and $a_0 = \sqrt{\frac{\hbar}{m\omega_0}}$ are respectively the linear oscillator lengths in the transverse and cigar-axis directions, $\omega_\perp$ and ω_0 are corresponding harmonic oscillator frequencies. In the present situation, the normalized macroscopic wave function $\psi(x,t)$ is connected to the original order parameter $\Psi(\mathbf{r},\tau)$ as follows [39]:

$$\psi\left(\xi = \frac{x}{a_\perp}, t = \frac{\omega_\perp}{2}\tau\right) = \sqrt{2\pi a_B}a_\perp\Psi(\mathbf{r},\tau)\exp\left[i\omega_\perp\tau + \frac{y^2+z^2}{2a_\perp}\right].$$

The inhomogeneous NLS equation (1.14) arises in various physical contexts from the description of nonlinear waves such as propagation of a laser beam, plasma waves, the nonlinear transmission networks, the BECs, or fiber transmission lines with in-line phase modulators. In the absence of the external trap potential, it describes the beam propagation in an inhomogeneous medium, where ψ is the electric field in laser optics and $g(t)$ is proportional to the electric density. From BEC point of view, a basic question to Eq. (1.14) is the following one: when may the condensate be unstable to collapse or exist for all time? As well as we know, the ground-state solutions, i.e., the localized solutions $\psi(x, t)$ of Eq. (1.14) are known to play a distinguished role in the long-time evolution of the initial disturbance.

It has been shown that near the surface of a Bose–Einstein condensate, the quadratic trapping potential can be approximated by a linear potential and the surface of the condensate can be regarded as an infinite plane. In such a situation, the GP equation takes the form [6, 7]

$$i\hbar\frac{\partial\psi}{\partial t} = \left[-\frac{\hbar^2}{2m}\frac{\partial^2}{\partial x^2} + Fx + \frac{4\pi a_s(t)\hbar^2}{m}|\psi|^2\right]\psi, \tag{1.15}$$

where x is the coordinate normal to the surface of the condensate such that the bulk of the condensate exists in the region $x < 0$ and $F = m\omega^2 R$ is a constant force arisen from linearizing the harmonic potential near the surface; here, ω and R are, respectively, the characteristic frequency of a spherically symmetric harmonic trapping potential and the radius of the condensate. In the context of nonlinear transmission network, Kengne et al. [27] used the reductive perturbation method in the continuum limit to show that the propagation of matter-wave solitons in a modified Noguchi electrical network can be modeled by a GP equation of type (1.15).

1.4 Multicomponent Nonlinear Schrödinger Equation

The multicomponent nonlinear Schrödinger equation is the vectorial version of the NLS and appears in many fields of physics whenever polarization effects are to be considered. The most famous multicomponent generalization of the nonlinear Schrödinger is the so-called the standard coupled NLS equation

$$\frac{\partial\psi_1}{\partial t} + c_1\frac{\partial^2\psi_1}{\partial x^2} + \left(\varkappa_{11}|\psi_1|^2 + \varkappa_{12}|\psi_2|^2\right)\psi_1 = 0, \tag{1.16a}$$

$$\frac{\partial\psi_2}{\partial t} + c_2\frac{\partial^2\psi_2}{\partial x^2} + \left(\varkappa\,|_{21}\psi_1|^2 + \varkappa_{22}|\psi_2|^2\right)\psi_2 = 0, \tag{1.16b}$$

originally introduced by Manakov [40] to describe the propagation of intense electromagnetic waves. Recently, Kengne et al. showed that a similar system governs solitary wave signals in nonlinear transmission network [31]. By means of the Darboux transformation, the Hirota bilinear method, the Backlund-gauge or transformation, many kinds of solutions of system (1.16a), (1.16b) including bright soliton, dark solitons and rogue waves have been derived. Another most important generalization form of the NLS equation which has attracted extensive attention of many scholars is the coupled derivative NLS equation

$$\frac{\partial \psi_1}{\partial t} + \frac{\partial^2 \psi_1}{\partial x^2} + i\varkappa \frac{\partial}{\partial x}\left(|\psi_1|^2 \psi_1 + |\psi_2|^2 \psi_1\right) = 0, \tag{1.17a}$$

$$\frac{\partial \psi_2}{\partial t} + \frac{\partial^2 \psi_2}{\partial x^2} + i\varkappa \frac{\partial}{\partial x}\left(|\psi_1|^2 \psi_2 + |\psi_2|^2 \psi_2\right) = 0. \tag{1.17b}$$

With the use of many methods such as the inverse scattering transformation, the Hirota bilinear techniques, the Darboux transformation, N-soliton solutions and the complete integrability of system (1.17a), (1.17b) have been obtained.

In the theory of BECs, the three components ψ_0, ψ_1, and ψ_{-1} of a three-component normalized spinor wave vector $\psi = (\psi_1, \psi_0, \psi_{-1})^T$ of the BEC of alkali atoms in the $F = 1$ hyperfine state, elongated in x direction and confined in the transverse directions y, z by purely optical means, are described by the three-component NLS [22, 38]

$$\begin{aligned}
\frac{\partial \psi_1}{\partial t} + \frac{\partial^2 \psi_1}{\partial x^2} + 2\left(|\psi_1|^2 + |\psi_0|^2\right)\psi_1 + 2\psi_{-1}^*\psi_0^2 &= 0,\\
\frac{\partial \psi_0}{\partial t} + \frac{\partial^2 \psi_0}{\partial x^2} + 2\left(|\psi_{-1}|^2 + |\psi_0|^2 + |\psi_1|^2\right)\psi_0 + 2\psi_0\psi_1\psi_{-1} &= 0,\\
\frac{\partial \psi_{-1}}{\partial t} + \frac{\partial^2 \psi_{-1}}{\partial x^2} + 2\left(|\psi_{-1}|^2 + |\psi_0|^2\right)\psi_{-1} + 2\psi_1^*\psi_0^2 &= 0,
\end{aligned}$$

when considering dimensionless units and using special choices for the scattering lengths. Similarly, spinor BEC with $F = 2$ is described by a five-component normalized spinor wave vector $\psi = (\psi_1, \psi_2, \psi_0, \psi_{-1}, \psi_{-2})^T$. As shown in Ref. [57], for specific choices of the scattering lengths in dimensionless coordinates, components $\psi_{\pm 1}$, $\psi_{\pm 2}$, and ψ_0 satisfy the following five-component NLS equation:

$$\begin{aligned}
\frac{\partial \psi_{\pm 2}}{\partial t} + \frac{\partial^2 \psi_{\pm 2}}{\partial x^2} + 2\left(\psi \cdot \psi^*\right)\psi_{\pm 2} - \left(2\psi_2\psi_{-2} - 2\psi_1\psi_{-1} + \psi_0^2\right)\psi_{\pm 2}^* &= 0,\\
\frac{\partial \psi_{\pm 1}}{\partial t} + \frac{\partial^2 \psi_{\pm 1}}{\partial x^2} + 2\left(\psi \cdot \psi^*\right)\psi_{\pm 1} - \left(2\psi_2\psi_{-2} - 2\psi_1\psi_{-1} + \psi_0^2\right)\psi_{\pm 1}^* &= 0,\\
\frac{\partial \psi_0}{\partial t} + \frac{\partial^2 \psi_0}{\partial x^2} + 2\left(\psi \cdot \psi^*\right)\psi_0 - \left(2\psi_2\psi_{-2} - 2\psi_1\psi_{-1} + \psi_0^2\right)\psi_0^* &= 0.
\end{aligned}$$

These three- and five-component NLS equations are integrable by means of inverse scattering transform method [19]. Using a modification of the Zakharov–Shabat "dressing method", Gerdjikov [18] described soliton solutions and the effects of the reductions on them.

References

1. Abdullaev, F.K., Caputo, J.G., Kraenkel, R.A., Malomed, B.A.: Controlling collapse in Bose-Einstein condensates by temporal modulation of the scattering length. Phys. Rev. A **67**, 013605 (2003)
2. Ablowitz, M.J., Clarkson, P.A.: Solitons, Nonlinear Evolution Equations and Inverse Scattering. Cambridge University Press, UK (1991)
3. Agrawal, G.P.: Nonlinear Fiber Optics. Academic, San Diego (2007)
4. Agrawal, G.P.: Nonlinear Fiber Optics. Academic, San Diego (2001)
5. Akhmediev, N.N., Ankiewicz, A.: Solitons: Nonlinear Pulses and Beams. Kluwer, Boston (1997)
6. Al, Khawaja U.: Exact solitonic solutions of the Gross-Pitaevskii equation with a linear potential. Phys. Rev. E **75**, 066607 (2007)
7. Al, Khawaja U., Pethick, C.J., Smith, H.: Surface of a Bose-Einstein condensed atomic cloud. Phys. Rev. A **60**, 1507 (1999)
8. Anderson, A., Lisak, M.: Nonlinear asymmetric self-phase modulation and self-steepening of pulses in long optical waveguides. Phys. Rev. A **27**, 1393 (1983)
9. Chen, H.H., Lee, Y.C., Liu, C.S.: Integrability of nonlinear hamiltonian systems by inverse scattering method. Phys. Scripta **20**, 490 (1979)
10. Craig, W., Sulem, C., Sulem, P.L.: Nonlinear modulation of gravity waves: a rigorous approach. Nonlinearity **5**, 497–522 (1992)
11. Dalfovo, F., Giorgini, S., Pitaevskii, L.P., Stringari, S.: Theory of Bose-Einstein condensation in trapped gases. Rev. Mod. Phys. **71**, 463 (1999)
12. Demiray, H.: Modulation of nonlinear waves in a thin elastic tube filled with a viscous fluid. Int. J. Eng. Sci. **37**, 1877–1891 (1999)
13. Doelman, A.: On the nonlinear evolution of patterns (modulation equations and their solutions). Ph.D. thesis, University of Utrecht, The Netherlands (1990)
14. Fan, E.G.: Darboux transformation and soliton-like solutions for the Gerdjikov-Ivanov equation. J. Phys. A **33**, 6925 (2000)
15. FordyA, P.: Derivative nonlinear Schrödinger equations and Hermitian symmetric spaces. J. Phys. A **17**, 1235 (1984)
16. Fujioka, J., Espinosa, A.: Soliton-Like Solution of an extended NLS equation existing in resonance with linear dispersive waves. J. Phys. Soc. Jpn. **66**, 2601 (1997)
17. Gammal, A., Frederico, T., Tomio, L., Chomaz, P.: Atomic Bose-Einstein condensation with three-body interactions and collective excitations. Phys. B: At. Mol. Opt. Phys. **33**, 4053 (2000)
18. Gerdjikov, V.S.: Bose-Einsten condensate and spectral properties on multicomponent nonlinear Schrödinger equation. Discret. Contin. Dyn. Syst. Ser. S **4**(5), 1181 (2011)
19. Gerdjikov, V.S., Kostov, N.A., Valchev, T.I.: Solutions of multi-component NLS models and spinor Bose-Einstein condensates. Phys. D **238**, 1306 (2009)
20. Gerdzhikov, V.S., Invanov, M.I., Kulish, P.P.: Quadratic bundle and nonlinear equations. Theor. Math. Phys. **44**, 784–795 (1980)
21. Hasimoto, H., Ono, H.: Nonlinear modulation of gravity waves. J. Phys. Soc. Jpn. **33**, 805–811 (1972)
22. Ieda, J., Miyakawa, T., Wadati, M.: Matter-wave solitons in an F = 1 spinor Bose-Einstein condensate. J. Phys. Soc. Jpn. **73**, 2996 (2004)
23. Duan, Jinqiao, Holmes, Philip: Fronts, domain walls and pulses in a generalized Ginzburg-Landau equation. Proc. Edinb. Math. Soc. **38**(1), 77–97 (1995)
24. Jiotsa, A.K., Kofané, T.C.: Dynamics of pattern formation and modulational instabilities in the quintic complex Ginzburg-Landau equation. J. Phys. Soc. Jpn. **72**, 1800 (2003)
25. Johnson, R.S.: On the modulation of water waves in the neighbourhood of kh $\approx$ 1.363. Proc. R. Soc. Lond. A **357**, 131 (1977)
26. Kaup, D.J., Newell, A.C.: An exact solution for a derivative nonlinear Schrödinger equation. J. Math. Phys. **19**, 798 (1978)

27. Kengne, E., Lakhssassi, A., Liu, W.M.: Modeling of matter-wave solitons in a nonlinear inductor-capacitor network through a Gross-Pitaevskii equation with time-dependent linear potential. Phys. Rev E. **96**, 022221 (2017)
28. Kengne, E., Abdourahman.: Generation of nonlinear modulated waves in a modified Noguchi electrical transmission network. Chaos, Solitons and Fractals **92**, 1–8 (2016)
29. Kengne, E., Lakhssassi, A.: Analytical study of dynamics of matter-wave solitons in lossless nonlinear discrete bi-inductance transmission lines. Phys. Rev. E **91**, 032907 (2015)
30. Kengne, E., Lakhssassi, A., Liu, W.M.: Dynamics of modulated waves in a lossy modified Noguchi electrical transmission line. Phys. Rev. E **91**, 062915 (2015)
31. Kengne, E., Vaillancourt, R., Malomed, B.A.: Coupled nonlinear Schrödinger equations for solitary-wave and kink signals propagating in discret nonlinear dispersive transmission lines. Int. J. Mod. Phys. B **23**, 133 (2009)
32. Kengne, E., Tadmon, C., Nguyen-Ba, T., Vaillancourt, R.: Higher order bright solitons and shock signals in nonlinear transmission lines. Chin. J. Phys. **47**(5), 698703 (2009)
33. Kengne, E., Liu, W.M.: Exact solutions of the derivative nonlinear Schrödinger equation for a nonlinear transmission line. Phys. Rev. E **73**, 026603 (2006)
34. Kengne, E., Talla, P.K.: Dynamics of bright matter wave solitons in Bose-Einstein condensates in an expulsive parabolic and complex potential. J. Phys. B **39**, 3679 (2006)
35. Kengne, E.: Envelope modulational instability in the nonlinear dissipative transmission line. J. Nonlinear Oscil. **5**(1), 23–31 (2002)
36. Kitaev, A.V., Vartanian, A.H.: Higher order asymptotics of the modified non-linear Schrödinger equation. Commun. Partial Differ. Equ. **25**, 1043–1098 (2000)
37. Kivshar, Y.S., Agrawal, G.P.: Optical Solitons: From Fibers to Photonic Crystals. Academic, New York (2003)
38. Li, L., Li, Z., Malomed, B.A., Mihalache, D., Liu, W.M.: Exact soliton solutions and nonlinear modulation instability in spinor Bose-Einstein condensates. Phys. Rev. A **72** (2005)
39. Liang, Z.X., Zhang, Z.D., Liu, W.M.: Dynamics of a bright soliton in Bose-Einstein condensates with time-dependent atomic scattering length in an expulsive parabolic potential. Phys. Rev. Lett. **94**, 050402 (2005)
40. Manakov, S.V.: On the theory of two-dimensional stationary self-focusing of electromagnetic waves. Zh. Eksp. Teor. Fiz. **65**, 505 (1973)
41. Marklund, M., Shukla, P.K.: Modulational instability of partially coherent signals in electrical transmission lines. Phys. Rev. E. **73**, 057601 (2006)
42. Marquié, R., Bilbault, J.M., Remoissenet, M.: Nonlinear Schrödinger models and modulational instability in real electrical lattices. Phys. D **87**, 371–374 (1995)
43. Mei, C.C.: The Applied Dynamics of Ocean Waves. World Scientific, Singapore (1989)
44. Mohamadou, A., Wamba, E., Doka, S.Y., Ekogo, T.B., Kofane, T.C.: Generation of matter-wave solitons of the Gross-Pitaevskii equation with a time-dependent complicated potential. Phys. Rev. A **84**, 023602 (2011)
45. Moses, J., Malomed, B., Wise, F.: Self-steepening of ultrashort optical pulses without self-phase-modulation. Phys. Rev. A **76**, 1–4 (2007)
46. Moses, J., Malomed, B., Wise, F.: Self-steepening of ultrashort optical pulses without self-phase-modulation. Phys. Rev. A **76**, 021802(R) (2007)
47. Nickel, J., Schürmann, H.W.: Comment on "Exact solutions of the derivative nonlinear Schrödinger equation for a nonlinear transmission line?. Phys. Rev. E **75**, 038601 (2007)
48. Pelap, F.B., Faye, M.M.: Solitonlike excitations in a one-dimensional electrical transmission line. JMP **46**, 033502 (2005)
49. Rapti, Z., Kevrekidis, P.G., Frantzeskakis, D.J., Malomed, B.A.: On the modulational instability of the nonlinear Schrödinger equation with dissipation. Phys. Scr. **74**(T113) (2004)
50. Rogers C., Malomed B., LI, H.J., Chow, K.W.: Propagating wave patterns in a derivative nonlinear Schrödinger system with quintic nonlinearity. J. Phys. Soc. Jpn. **81**, 094005 (2012)
51. Rogister, A.: Parallel propagation of nonlinear low-frequency waves in high-β plasma. Phys. Fluids **14**, 2733 (1971)

52. Ruderman, M.S.: Propagation of solitons of the derivative nonlinear Schrödinger equation in a plasma with fluctuating density. Phys. Plasmas **9**, 2940 (2002)
53. Segur, H., Henderson, D.M., Hammack, J.L., Li, C.-M., Pheiff, D., Socha, K.: Stabilizing the Benjamin-Feir Instability. J. Fluid Mech. **539**, 229–271 (2005)
54. Sulem, C., Sulem, P.-L.: The Nonlinear Schrödinger Equation: Self Focusing and Wave Collapse. Springer, New York (1999)
55. Takhtajan, L.A., Faddeev, L.D.: Hamiltonian Methods in the Theory of Solitons. Nauka, Moscow (1986)
56. Theocharis, G., Rapti, Z., Kevrekidis, P.G., Frantzeskakis, D.J., Konotop, V.V.: Modulational instability of Gross-Pitaevskii-type equations in $1+1$ dimensions. Phys. Rev. A **67**, 063610 (2003)
57. Uchiyama, M., Ieda, J., Wadati, M.: Multicomponent bright solitons in $F = 2$ spinor Bose-Einstein condensates. J. Phys. Soc. Jpn. **76**, 74005 (2007)
58. Xu, J., Fan, E., Chen, Y.: Long-time asymptotic for the derivative Nonlinear Schrödinger equation with step-like initial value. Math. Phys. Anal. Geom. **16**, 253–288 (2013)
59. Yan, Z.: Optical solitary wave solutions to nonlinear Schrödinger equation with cubic-quintic nonlinearity in non-Kerr media. J. Phys. Soc. Jpn. **73**, 2397 (2004)
60. Zakharov, V.E., Manakov, S.V., Novikov, S.P., Pitaevskii, L.P.: The Theory of Solitons: The Inverse-Problem Method. Nauka, Moscow (1980)

Chapter 2
Well-Posedness of Nonlocal Boundary-Value Problems and Schrödinger Equations

Abstract Questions related to the well-posedness of differential problems in the mathematical modeling of real objects, phenomena, and processes form are essential and very important part of problems of the description of real objects by mathematical means. Mathematically, a given boundary-value problem is said to be well-posed (in Hadamard sense) if (a) a solution exists, (b) the solution is unique, and (c) the solution is stable, that is, continuously depends on the data (initial and/or boundary conditions) of the problem. If at least one of these three conditions is violated, the problem is said to be ill-posed. In this chapter, we introduce the notion of well-posedness of nonlocal boundary-value problems for equations of Schrödinger type.

In the course of investigation of real processes and phenomena governed by differential equations, there arises not only the necessity of constructing analytical or numerical solutions, but also the necessity of studying various properties of these solutions. In the case where the desired solutions are constructed analytically, it is easy to study their properties. Unfortunately, such cases are very rare. Therefore, it is necessary to establish the properties of solutions of differential problems by indirect methods, with respect to the properties of differential equations. An important role here is played by the well-posedness of problems. Problems related to the well-posedness of differential problems in the mathematical simulation of real objects, phenomena, and processes form are essential and very important part of problems of description of real objects by mathematical means. Referring to Hadamard [12], the classical definition of the well-posedness of a given boundary-value problem in a given space of function can be formulated as follows.

Definition: A given problem is said to be **well-posed** if the following conditions are satisfied simultaneously: (i) the solution to the problem exists; (ii) the solution to the problem is unique; and (iii) the solution continuously depend on the data for the problem. If at least one of these three conditions is violated, the problem is said to be **ill-posed** (according to Hadamard).

The problems with nonlocal boundary conditions in the time variable for partial differential equations are, in general, Hadamard ill-posed [12], in the scales of Sobolev spaces of functions periodic in spatial variables. Uniqueness of a solution to such problems in some cases depends on the Diophantine properties of their

W.-M. Liu and E. Kengne, *Schrödinger Equations in Nonlinear Systems*,
https://doi.org/10.1007/978-981-13-6581-2_2

coefficients and parameters, while the solvability and the smoothness of the solutions are connected with the problem of small denominators [15, 29]. In recent decades, problems with nonlocal boundary conditions for partial differential equations and differential operator equations have been studied by many authors. In particular, regular problems without small denominators or with conditions under which small denominators are not equal to zero, which guarantees the solvability of the problem, were mainly investigated [7, 8, 16, 17, 19, 20, 23, 24, 30]. However, in the general case, nonlocal boundary-value problems for general differential operators with partial derivatives are ill-posed, and the problem of their solvability is often connected with the problem of small denominators. It is important to note that many meaningful and important problems in physics and engineering are ill-posed. It is expected that if a continuous problem is well-posed, such should be its discrete approximation.

The present chapter deals with the well-posedness of one class of one-dimensional time-dependent boundary-value problems with global boundary conditions on the entire time interval. Here, we pay a special attention on nonlocal boundary value problems for (nonlinear) Schrödinger equations.

2.1 Boundary-Value Problem from the Viewpoint of the General Theory of Partial Differential Equations

In this section, we formulate the criterion of the well-posedness of nonlocal boundary-value problem for a linear evolution partial differential equation (PDE). From the viewpoint of the general theory of partial differential equations (PDEs), let us consider the following two-point boundary-value problem [4]:

$$\frac{\partial \psi}{\partial t} = P\left(\partial/\partial x\right)\psi(x,t), \quad (x,t) \in \Pi_T = \mathbf{R} \times [0,\ T]\,, T > 0, \tag{2.1a}$$

$$A\psi(x,0) + B\psi(x,T) + C\int_0^T \psi(x,\tau)d\tau = \psi_0(x),\ \ x \in \mathbf{R}, \tag{2.1b}$$

where $P(s)$ is an arbitrary polynomial with constant (complex) coefficients, A, B, and C are three complex parameters satisfying condition $|A| + |B| + |C| > 0$, $\psi_0(x) : \mathbf{R} \to \mathbf{C}$ is a given complex—valued function of spatial variable x, and $\psi(x,t) : \Pi_T \to \mathbf{C}$ is the unknown function of $(x,t) \in \Pi_T$. When $AB = C = 0$, problem (2.1a), (2.1b) is just a one-point problem and is sometimes called Cauchy problem. Under condition $|AB| + |C| > 0$, problem (2.1a), (2.1b) will be a two-point boundary-value problem, also called nonlocal boundary-value problem. In what follows, we denote by H_n the following space of all nth continuously differentiable functions φ defined on $\mathbf{R}$ with norm $\|\varphi\|_n$

$$H_n = \left\{\varphi \in C^n(\mathbf{R}) : \|\varphi\|_n = \max_{0\le j\le n} \mathop{Sup}_{\mathbf{R}} \left|\frac{\partial^j\varphi(x)}{\partial x^j}\right| < +\infty\right\}.$$

Following Petrovskii [28], we introduce the following definition.

Definition 1: We say that problem (2.1a), (2.1b) is well-posed in Π_T (according to Petrovskii) if for every positive integer m there exist a positive integer n and a positive real constant C_0 such that for every function $\psi_0(x) \in H_n$, problem (2.1a), (2.1b) possesses a unique solution $\psi(x, t) \in H_m$ (for every $t \in [0,\ T]$) which satisfies condition

$$\sup_{[0,\ T]} \|\psi(x,t)\|_m \le C_0 \|\psi_0(x)\|_n .$$

The criterion of the well-posedness of Cauchy problem for Eq. (2.1a), i.e., when $AB = C = 0$, is given by the condition (A) of Petrovskii [28] $|A| \sup_{\mathbf{R}} \mathrm{Re}[P(i\sigma)] - B \inf_{\mathbf{R}} \mathrm{Re}[P(i\sigma)] < +\infty$. Also, the well-posedness of problem (2.1a), (2.1b) in the special cases $C = 0$ and $AB \neq 0$ and $A = B = 0$ and $C \neq 0$ has been investigated in Refs. [3] and [9] respectively. In the general case when $ABC \neq 0$, the proof of the well-posedness of problem (2.1a), (2.1b) is based on a series of lemmas [12]. Before formulating these lemmas and the criterion of the well-posedness of problem (2.1a), (2.1b), we first introduce the following notations:

$$\begin{aligned}
\Delta_T(\sigma) &= h\left(TP(i\sigma)\right), \sigma \in \mathbf{R},\\
h(z) &= \begin{cases} A + B\exp[z] - CT\frac{1-\exp[z]}{z}, & \text{if } z \neq 0\\ A + B + CT, & \text{if } z = 0,\end{cases}\\
\delta(s) &= As + Bs\exp[s] - C\left(1 - \exp[s]\right),\\
P(i\sigma) &= P_1(\sigma) + iP_2(\sigma),\\
\mathbf{R}_{\sigma_0} &= \{\sigma \in \mathbf{R} : |\sigma| \ge \sigma_0\}, \text{ with } \sigma_0 > 0,\\
\mathbf{R}^{\lambda}_{\sigma_0} &= \left\{\sigma \in \mathbf{R}_{\sigma_0} : \lambda\sigma \ge \sigma_0\right\}, \ \lambda = \pm 1,\\
N_h &= \{\sigma \in \mathbf{R} : h(\sigma) = 0\}, \quad N_h^C = \{s \in \mathbf{C} : h(s) = 0\}.
\end{aligned}$$

The notation $f(\sigma) \le g(\sigma)$ means that there exist $\sigma_0 > 0$ and $\lambda \in \{-1,\ 1\}$ such that $f(\sigma) \le g(\sigma)$ for every $\sigma \in \mathbf{R}^{\lambda}_{\sigma_0}$; moreover, we designate by $\rho(E_1, E_2)$ the distance between sets E_1 and E_2.

Lemma 1: If $AB \neq 0$, then sufficiently large (in modulus) zeros z_k of the function $\delta(z)$ can be expressed in form

$$z_k = \ln(-A/B) + \sum_{j=1}^{+\infty} \zeta_j k^{-j},\ \zeta_j \in \mathbf{C}. \tag{2.2}$$

▲ Let $\{\widetilde{s}_k\}$ be the set of zeros of function $\widetilde{\delta}(z) = s\left(A + B\exp[s]\right)$. Then, with a suitable numeration, we get $\widetilde{s}_k = s_k + o(1)$ as $|k| \to +\infty$, $k \in \mathbf{Z}$ [1]. Setting

$$\begin{aligned}
\gamma_0 &= \arg\left(-A/B\right), \mu_k = s_k - \ln|A/B| + i\left(\gamma_0 + 2k\pi\right), \\
\delta(k,\mu) &= -\frac{A}{B}\left[B\ln\left|\frac{A}{B}\right| + iB\left(\gamma_0 + 2k\pi\right) + B_\mu + C\right]\exp\left[\mu\right] \\
&\quad + A\left[\ln\left|\frac{A}{B}\right| + \mu + i\left(\gamma_0 + 2k\pi\right)\right] - C, \\
\delta_1(k,\mu) &= k\delta(1/k,\mu)
\end{aligned}$$

lead to

$$\delta(k,\mu_k) = \delta(k_s) = 0,\ \delta_1(0,0) = 0,\ \left.\frac{\partial\delta_1(k,\mu)}{\partial\mu}\right|_{(k,\mu)=(0,0)} = 2i\pi A \neq 0.$$

Thus, for sufficiently small $|k|$, equation $\delta_1(k,\mu) = 0$ determines an analytic function

$$\mu = \mu(k) = \sum_{j=1}^{+\infty} \zeta_j k^j,\ \zeta_j \in \mathbf{C}.$$

This means that for large k (in modulus), the roots μ_k of equation $\delta(k,\mu) = 0$ can be represented as

$$\mu_k = \sum_{j=1}^{+\infty} \zeta_j k^{-j},\ \zeta_j \in \mathbf{C},$$

whence the representation (2.2) for the zeros of $\delta(s)$, and Lemma 1 is proved. ■

Lemma 2: If $\Delta_T(\sigma) \neq 0$ for every real σ, then there exist $M > 0$ and $\mu \in \mathbf{R}$ such that

$$|\Delta_T(\sigma)| \geq M\left(1 + |\sigma|\right)^\mu, \forall\sigma \in \mathbf{R}. \tag{2.3}$$

▲ If $P(i\sigma)$ is a constant, then inequality (2.3) is obvious. Now, let $P(i\sigma)$ be a non-constant function. First, let $AB \neq 0$ and $\inf\rho\left(TP(i\sigma), N_h^C\right) > 0$, Then estimation (2.3) is true with $\mu = 0$ [1]. If $\inf\rho\left(TP(i\sigma), N_h^C\right) = 0$, then $TP_1(\sigma) = \ln|A/B|$. If we assume that $p = \deg P_2(\sigma) > 0$, Eq. (2.2) implies that $\mathrm{Re}\left[\sum_{j=1}^{+\infty}\zeta_j k^{-j}\right]$ is not a trivial (zero) function. Hence, for some $j_0 \geq 1$, we have $\mathrm{Re}\left[\zeta_{j_0}\right] \neq 0$ and $\mathrm{Re}\left[\zeta_j\right] = 0$ for $1 \leq j < j_0$. If $k(\sigma) \in \mathbf{Z}$ such that $\inf_{\mathbf{Z}}\rho\left(TP(i\sigma), N_h^C\right) = \left|TP(i\sigma) - s_{k(\sigma)}\right|$, we get $TP_2(\sigma) - \gamma_0 - 2k(\sigma)\pi = o(1)$ $(|\sigma| \to +\infty)$ and, consequently, $|k(\sigma)| \leq a_0|\sigma|^p$ as $|\sigma| \to +\infty$, for some $a_0 > 0$. On the other hand, for some $a_1 > 0$, we have, for sufficiently large $|\sigma|$,

$$\left|TP(i\sigma) - s_{k(\sigma)}\right| \geq \left|TP(i\sigma) - \mathrm{Re}\left[s_{k(\sigma)}\right]\right| \geq \mathrm{Re}\left[\zeta_{j_0}\right]|k(\sigma)|^{-j_0}/2 \geq a_1|\sigma|^{-pj_0}.$$

Writing $\Delta_T(\sigma)$ in the form

$$\Delta_T(\sigma) = (s - s_k)\left[B\exp[s_k]\,\frac{\exp[s-s_k]-1}{s-s_k} + \frac{CT}{ss_k}\left(1 - \exp[s_k] + s_k\exp[s_k]\,\frac{\exp[s-s_k]-1}{s-s_k}\right)\right]$$

with $s = TP(i\sigma)$ and $s_k = s_{k(\sigma)}$, and taking into account that $\exp[s_k] \to -A/B$ and $\frac{\exp[s-s_k]-1}{s-s_k} \to 1$ as $|\sigma| \to +\infty$, we conclude that $|\Delta_T(\sigma)| \geq |A|\left|TP(i\sigma) - s_{k(\sigma)}\right|/2$, for all $\sigma \in \mathbf{R}_{\sigma_0}$, whence inequality (2.3).

If $A = 0$ and $B \neq 0$, then it follows from Ref. [1] that $s_k = x_k + iy_k$, where $x_k = \ln\left|\frac{CT}{B}\right| - \ln|2k\pi + \gamma_0 \mp \pi/2| + o(1)$ and $y_k = 2k\pi + \gamma_0 \mp \pi/2 + o(1)$ as $|k| \to +\infty$, where $\gamma_0 = \arg(C/B)$. This implies that $\inf \rho\left(TP(i\sigma), N_h^C\right) > 0$, meaning that inequality (2.3) is satisfied with $\mu = 0$.

Now, let $A \neq 0$ and $B = 0$. Then, the large (in modulus) zeros s_k of the function $h(s)$ are asymptotically close to the zeros $\widetilde{s}_k$ of function $\widetilde{h}(s) = As + CT\exp[Ts]$, and $\widetilde{s}_k = \xi_k + i\eta_k$, where ξ_k and η_k are real solutions of the system

$$\begin{cases} \exp[\xi]\cos\eta - a\xi + b\eta = 0, \\ \exp[\xi]\sin\eta - b\xi - a\eta = 0 \end{cases}$$

with large $|\xi + i\eta|$ (here, $a = \mathrm{Re}[-CT/A]$ and $b = \mathrm{Im}[-CT/A]$). But then $\xi_k \to +\infty$ and $\xi_k = \ln|k|\,(1 + o(1))$ and $\eta_k = k\pi\,(1 + o(1))$, as $|k| \to +\infty$. These relations imply that inequality (2.3) holds with $\mu = 0$, and this completes the proof of Lemma 2. ■

Lemma 3: Let $\Delta_T(\sigma) \neq 0$ for every real σ. Then function $R(\sigma, t) = \exp[tP(i\sigma)]/\Delta_T(\sigma)$ possesses the following property:

$$\forall q \in \mathbf{N}_0 = \mathbf{N} \cup \{\mathbf{0}\},\ \exists C_q > 0 \text{ and } \exists m_q \in \mathbf{R}:$$
$$\left|\frac{\partial^q R(\sigma, t)}{\partial t^q}\right| \leq C_q\,(1 + |\sigma|)^{m_q},\quad \forall(\sigma, t) \in \Pi_T. \tag{2.4}$$

▲ Clearly, it is sufficient to prove the inequality (2.4) when $\sigma \in \mathbf{R}_{\sigma_0}$ under the condition $\deg P > 0$, σ_0 being a positive real number. If $P(i\sigma) \neq 0$ on $\mathbf{R}_{\sigma_0}$, one can easily establish that

$$\frac{\partial^q R(\sigma, t)}{\partial t^q} = \frac{1}{(P(i\sigma)\Delta_T(\sigma))^{1+q}} \sum_{j=0}^{q} H_{qj}(\sigma, t)\exp[(t + jT)\,P(i\sigma)],$$

$H_{qj}(\sigma, t)$ being a polynomial with respect with σ. This leads to Eq. (2.4) if $\deg P_1(\sigma) > 0$. In the case of constant $P_1(\sigma)$, we apply the inequality (2.3) and obtain Eq. (2.4), and Lemma 3 is proved. ■

Theorem (Criterion of the well-posedness of problem (2.1a), (2.1b)): For problem (2.1a), (2.1b) to be well- posed (according to Petrovskii), it is necessary and sufficient that (i)

$$\Delta_T(\sigma) \neq 0 \text{ for every real } \sigma,\ \text{ if }\ |AB| + |C| > 0, \tag{2.5a}$$

(ii)

$$|A|\,\underset{\mathbf{R}}{\sup}\mathrm{Re}\,[P(i\sigma)] - |B|\,\underset{\mathbf{R}}{\inf}\mathrm{Re}\,[P(i\sigma)] < +\infty, \quad \text{if} \quad AB = C = 0. \tag{2.5b}$$

▲ If $AB = C = 0$, problem (2.1a), (2.1b) is Cauchy problem which is well-posed under condition (2.5b) [32]. Now, let $|AB| + |C| > 0$. Let us associate with problem (2.1a), (2.1b) the following problem (with a parameter $\sigma \in \mathbf{R}$):

$$\begin{cases} \frac{dv(\sigma,t)}{dt} = P(i\sigma)v(\sigma,t), \\ Av(\sigma,0) + Bv(\sigma,T) + C\int_0^T v(\sigma,\tau)d\tau = v_0(\sigma). \end{cases} \tag{2.6}$$

Let $\psi(x,t)$ be a solution of problem (2.1a), (2.1b) with the given $\psi_0(x)$ and let $\mathfrak{F}(\psi(x,t)) = v(\sigma,t)$ and $\mathfrak{F}(\psi_0(x)) = v_0(\sigma)$ be the Fourier transformation of respectively, $\psi(x,t)$ and $\psi_0(x)$. Then, $v(\sigma,t)$ is a solution of problem (2.6). For $\sigma \notin N_{\Delta_T}$, a solution of problem (2.6) can be expressed in the form

$$v(\sigma,t) = R(\sigma,t)v_0(\sigma), \tag{2.7}$$

where $R(\sigma,t)$ is the function defined in Lemma 3. By virtue of relation (2.7) and estimate (2.4), the well-posedness of problem (2.1a), (2.1b) can be established by using the Petrovskii's method by using a resolution of the identity [32].

If for some real σ_0 condition $\Delta_T(\sigma_0) = 0$, is satisfied, then the function $\psi(x,t) = \exp[tP(i\sigma_0) + ix\sigma_0]$ will be a solution of (2.1a) and satisfies (2.1b) with $\psi_0(x) = 0$. Then problem (2.1a), (2.1b) is ill-posed and the theorem is proved. ■

Example 1: Consider the following nonlocal boundary-value problem for the thermal conduction equation:

$$\frac{\partial\psi}{\partial t} = \frac{\partial^2\psi}{\partial x^2}, \quad (x,t) \in \Pi_T, \tag{2.8a}$$

$$(2+i)\,[\psi(x,0) + 2\psi(x,T)] + (2-i)\int_0^T \psi(x,\tau)d\tau = \psi_0(x), \; x \in \mathbf{R}. \tag{2.8b}$$

For the boundary-value problem (2.8a), (2.8b), $P(i\sigma) = -\sigma^2$, and

$$\Delta_T(\sigma) = \begin{cases} (2+i)\left(1 + 2\exp\left[-T\sigma^2\right]\right) + (2-i)\frac{1-\exp\left[-T\sigma^2\right]}{\sigma^2}, & \text{if } \sigma \neq 0 \\ 3 + 2T + i\,(3-T), & \text{if } \sigma = 0. \end{cases}$$

It is obvious that for $\forall T > 0$, $\forall\sigma \in \mathbf{R}$, $\Delta_T(\sigma) \neq 0$. Therefore, problem (2.8a), (2.8b) is well-posed in Π_T.

Example 2: Consider the following nonlocal boundary-value problem for the linear Schrödinger equation:

$$\frac{\partial \psi}{\partial t} = i\frac{\hbar}{2\mu}\frac{\partial^2 \psi}{\partial x^2}, \quad (x, t) \in \Pi_T \tag{2.9a}$$

$$\frac{\mu}{2\hbar}\psi(x, 0) + \frac{2\mu}{\hbar}\psi(x, T) - i\int_0^T \psi(x, \tau)d\tau = \psi_0(x), \; x \in \mathbf{R}. \tag{2.9b}$$

For problem (2.9a), (2.9b), we have $P\,(i\sigma) = -i\frac{\hbar}{2\mu}\sigma^2$, and

$$\Delta_T(\sigma) = \begin{cases} \frac{\mu}{\hbar}\left(\frac{1}{2} + 2\exp\left[-i\frac{T\hbar}{2\mu}\sigma^2\right] - \frac{2\left(1-\exp\left[-i\frac{T\hbar}{2\mu}\sigma^2\right]\right)}{\sigma^2}\right), & \text{if } \sigma \neq 0 \\ \frac{5\mu}{2\hbar} - iT, & \text{if } \sigma = 0. \end{cases}$$

Therefore, $\forall T > 0$, $\forall \sigma \in \mathbf{R}$, $\Delta_T(\sigma) \neq 0$, meaning that problem (2.9a), (2.9b) is well-posed in Π_T.

2.2 Perturbation of a Two-Point Boundary-Value Problem

In this section, we investigate the effects of the integral term of the boundary condition (2.1a) on the well-posedness of problem (2.1a), (2.1b). In other words, we investigate the problem of well-posedness of problem (2.1a), (2.1b) for small values of $|C|$. We will use the notation $\Delta(\sigma; C) = \Delta_T(\sigma)$, so that $\Delta(\sigma; 0) = A + B\exp[TP(i\sigma)]$. Let us associate with the boundary condition (2.1b) the following boundary condition in the absence of the integral term

$$A\psi(x, 0) + B\psi(x, T) = \psi_0(x), \; x \in \mathbf{R}. \tag{2.10}$$

We also denote $H = \bigcup_{m\geq 0} H_m$. We focus ourselves to the situation when $|AB| + |C| > 0$, that is when problem (2.1a), (2.1b) is not a Cauchy problem. As we have shown in the previous section, problem (2.1a), (2.1b) under condition $|AB| + |C| > 0$ will be well-posed (in H) if and only, if $\Delta(\sigma; C) \neq 0$ for every real σ. In what follows, we answer the following two questions:

I. Let problem (2.1a), (2.10) be ill-posed, i.e., equation $\Delta(\sigma; 0) = 0$ admits at least one solution in $\mathbf{R}$. It is true that for every $\varepsilon > 0$, there exists $C_\varepsilon \in \mathbf{C}$ so that $0 < |C_\varepsilon| < \varepsilon \Rightarrow \Delta(\sigma; C_\varepsilon) \neq 0$ for every real σ, i.e., problem (2.1a), (2.1b) with $C = C_\varepsilon$ is well-posed (in H)?

II. Let problem (2.1a), (2.10) be well-posed, i.e., $\Delta(\sigma; 0) \neq 0$ for every real σ. Is it true that there exists $\varepsilon_0 > 0$ such that for every complex number C, $0 < |C| < \varepsilon_0 \Rightarrow \Delta(\sigma; C) \neq 0$ for every real σ, i.e., problem (2.1a), (2.1b) will be well-posed (in H)?

2.2.1 Perturbation of Ill-Posed Boundary-Value Problems by an Integral Term in the Boundary Condition

Let us assume that problem (2.1a), (2.10) is ill-posed (in H), i.e., equation $\Delta(\sigma; 0) = 0$ admits at least one solution in **R**, meaning that $\{\sigma \in R : A + B\exp[TP(i\sigma)] = 0\} \neq \varnothing$.

Theorem: Let $\{\sigma \in \mathbf{R} : \Delta(\sigma; 0) = 0\} \neq \varnothing$ and $AB \neq 0$. In order that, for any $\varepsilon > 0$ there exists $C_\varepsilon \in \mathbf{C} : 0 < |C_\varepsilon| < \varepsilon \Rightarrow \Delta(\sigma; C_\varepsilon) \neq 0$ for all real σ, it is necessary and sufficient that either $A + B \neq 0$ or $A + B = 0$ and $\{\sigma \in \mathbf{R} : \Delta(\sigma; 0) = 0\} = \{\sigma \in \mathbf{R} : TP(i\sigma) = 0\}$.

▲ *Necessity*: Assume that $A + B = 0$ and $\{\sigma \in \mathbf{R} : \Delta(\sigma; 0) = 0\} \setminus \{\sigma \in \mathbf{R} : TP(i\sigma) = 0\} \neq \varnothing$. Let $\sigma_0 \in \mathbf{R} : P(i\sigma_0) \neq 0$ and $\Delta(\sigma_0; 0) = A + B\exp[TP(i\sigma_0)] = 0$. Then $\exp[TP(i\sigma_0)] = -A/B = 1$ and $\forall C \in \mathbf{C}$, $\Delta(\sigma_0; C) = 0$. This means that problem (2.1a), (2.1b) will be ill-posed for every complex number C.

Sufficiency: Let us assume that one of the conditions $A + B \neq 0$ or $A + B = 0$ and $\{\sigma \in \mathbf{R} : \Delta(\sigma; 0) = 0\} = \{\sigma \in \mathbf{R} : TP(i\sigma) = 0\}$ is satisfied. If we denote by

$$\varphi(\sigma) = \frac{\exp[TP(i\sigma)] - 1}{P(i\sigma)}, \text{ if } P(i\sigma) \neq 0,$$
$$\varphi(\sigma) = T, \text{ if } P(i\sigma) = 0,$$

then it is evident that

$$\Delta(\sigma; C) = A + B\exp[TP(i\sigma)] + C\varphi(\sigma). \tag{2.11}$$

(i) Let $A + B \neq 0$. In this situation, for every $\sigma \in N_\varphi = \{\sigma \in \mathbf{R} : \varphi(\sigma) = 0\}$, we have $\Delta(\sigma; C) = A + B \neq 0$. If $\sigma \notin N_\varphi$, then it follows from Eq. (2.22) that $\Delta(\sigma; C) = \varphi(\sigma)\left[C + \frac{A+B\exp[TP(i\sigma)]}{\varphi(\sigma}\right] \neq 0$ for all $C \notin \left\{-\frac{A+B\exp[TP(i\sigma)]}{\varphi(\sigma}, \ \sigma \in \mathbf{R}\right\}$. It is obvious that for every $\varepsilon > 0$, there exists $C_\varepsilon \in \mathbf{C} : 0 < |C_\varepsilon| < \varepsilon$ and $C_\varepsilon \neq -\frac{A+B\exp[TP(i\sigma)]}{\varphi(\sigma}$, $\forall \sigma \in \mathbf{R}$. Therefore, $\Delta(\sigma; C_\varepsilon) \neq 0$ and problem (2.1a), (2.1b) with $C = C_\varepsilon$ is well-posed.

(ii) Let $A + B = 0$ and $\{\sigma \in \mathbf{R} : \Delta(\sigma; 0) = 0\} = \{\sigma \in \mathbf{R} : TP(i\sigma) = 0\}$. In this case,

$$\Delta(\sigma; C) = \begin{cases} (AP(i\sigma) - CT)\,\frac{(1-\exp[TP(i\sigma)])}{P(i\sigma)}, & \text{if } P(i\sigma) \neq 0 \\ CT, \ \text{il } P(i\sigma) = 0 \end{cases}$$

and $\exp[TP(i\sigma)] = 1$ only when $P(i\sigma) = 1$. Therefore, $\Delta(\sigma; C) \neq 0$ if and only, if $C \neq \frac{AP(i\sigma)}{T}$. Thus, for every $\varepsilon > 0$, for every $C_\varepsilon \in \mathbf{C} : 0 < |C_\varepsilon| < \varepsilon$ and $C_\varepsilon \neq \frac{AP(i\sigma)}{T}$ for every real σ, we will have $\Delta(\sigma; C_\varepsilon) \neq 0$, meaning that problem (2.1a), (2.1b) in which $C = C_\varepsilon$ is well-posed, and the theorem if proved. ■

Example: The boundary-value problem for the linear Schrödinger equation

$$\frac{\partial \psi}{\partial t} = i\frac{\hbar}{2\mu}\frac{\partial^2 \psi}{\partial x^2}, \quad (x,t) \in \Pi_T = \mathbf{R} \times [0,\ T],$$
$$\psi(x,0) + \psi(x,T) = \psi_0(x), \quad x \in \mathbf{R}$$

is ill-posed (in H). Indeed, for this problem, $P(i\sigma) = -i\frac{\hbar}{2\mu}\sigma^2$, $A = 1$, $B = 1$ so that $A + B \neq 0$. Therefore $\Delta(\sigma;0) = 1 + \exp\left[-i\frac{T\hbar}{2\mu}\sigma^2\right]$ so that $N_{\Delta(\sigma;0)} \neq \varnothing$, and according to the well-posedness criterion for problem (2.1a), (2.1b) under condition $|AB| + |C| > 0$, the above boundary-value problem for the linear Schrödinger equation is ill-posed (in H).

For every $C \in \mathbf{C}$, we have $\Delta(\sigma;C) = 1 + \exp\left[-i\frac{T\hbar}{2\mu}\sigma^2\right] + C\frac{2i\mu\left(\exp\left[-i\frac{T\hbar}{2\mu}\sigma^2\right]-1\right)}{\hbar\sigma^2}$. $\forall\sigma \in \mathbf{R} : \exp\left[-i\frac{T\hbar}{2\mu}\sigma^2\right] - 1 = 0$, we have $\Delta(\sigma;C) = 2 \neq 0$. Let $\sigma \in \mathbf{R}$: $\exp\left[-i\frac{T\hbar}{2\mu}\sigma^2\right] - 1 \neq 0$. Then

$$\Delta(\sigma;C) = \left(\frac{2i\mu\left(\exp\left[-i\frac{T\hbar}{2\mu}\sigma^2\right]-1\right)}{\hbar\sigma^2}\right)\left(C - i\frac{\hbar\sigma^2\left(1+\exp\left[-i\frac{T\hbar}{2\mu}\sigma^2\right]\right)}{2\mu\left(\exp\left[-i\frac{T\hbar}{2\mu}\sigma^2\right]-1\right)}\right) \neq 0,$$

for every $C \neq i\frac{\hbar\sigma^2\left(1+\exp\left[-i\frac{T\hbar}{2\mu}\sigma^2\right]\right)}{2\mu\left(\exp\left[-i\frac{T\hbar}{2\mu}\sigma^2\right]-1\right)}$. For such C, we have $\Delta(\sigma;C) \neq 0$, for every real σ, and the corresponding problem (2.1a), (2.1b) is well-posed (in H).

2.2.2 *Perturbation of Well-Posed Boundary-Value Problems by Integral Terms in the Boundary Condition*

Here, we assume that the boundary-value problem (2.1a), (2.10) is well-posed (in H) and show that every perturbation with a small integral term in the boundary condition leads to a well-posed problem.

Theorem: If problem (2.1a), (2.10) with $AB \neq 0$ is well-posed, then there exists an $\varepsilon_0 > 0$ so that for every complex number C satisfying the condition $0 < |C| < \varepsilon$, the corresponding boundary-value problem (2.1a), (2.1b) will be well-posed.

▲ First, let $A + B = 0$. According to the criterion of the well-posedness of problem (2.1a), (2.1b), the well-posedness of (2.1a), (2.10) means that $N_{\Delta(\sigma;0)} = \{\sigma \in \mathbf{R} : \Delta(\sigma;0) = 0\} = \varnothing$. Hence, $N_P = \{\sigma \in \mathbf{R} : P(i\sigma) = 0\} = \varnothing$ and thus, $\inf_{\mathbf{R}} |P_1(\sigma)| = M_0 > 0$. Consequently, $\Delta(\sigma;C) = (A + BP(i\sigma))\,\varphi(\sigma)$ and, moreover, $N_\varphi = \{\sigma \in \mathbf{R} : \varphi(\sigma) = 0\} = \varnothing$. For every $C \in \mathbf{C}$ satisfying condition $0 < |C| < \varepsilon_0 = |B|\,M_0$, we have $\Delta(\sigma;C) \neq 0$, $\forall\sigma \in \mathbf{R}$, and problem (2.1a), (2.1b) with such C is well-posed (in H).

Now let $A + B \neq 0$. To prove the theorem in this situation, we distinguish three cases.

(i) Let $\mu_1 = \sup_{\mathbf{R}} |\varphi(\sigma)|$ and $\lambda_1 = \inf_{\mathbf{R}} |A + B\exp[TP(i\sigma)]|$. It follows from the well-posedness of problem (2.1a), (2.10) that $N_{\Delta(\sigma;0)} = \varnothing$, i.e., $A + B\exp[TP(i\sigma)] \neq 0$, $\forall\sigma \in \mathbf{R}$. Moreover, if $P_1(\sigma) \to -\infty$ as $|\sigma| \to +\infty$, we have $A \neq 0$ because, otherwise, the Petrovskii condition (A) is violated and the Cauchy problem for Eq. (2.1a) under condition (2.10) will be ill-posed. Hence $\lambda_1 > 0$, leading to $|\Delta(\sigma; C)| \geq |A + B\exp[TP(i\sigma)]| - |C|\,|\varphi(\sigma)| \geq \lambda_1 - |C|\,\mu_1 > 0$, $\forall C \in \mathbf{C} : 0 < |C| < \varepsilon_0 = \lambda_1/\mu_1$. Hence problem (2.1a), (2.1b) for such C is well-posed (in H).

(ii) Let $\mu_1 = \sup_{\mathbf{R}} |\varphi(\sigma)|\exp[-TP_1(\sigma)] < +\infty$, $\lambda_1 = \inf_{\mathbf{R}} |A + B\exp[-TP_1(\sigma)]|$. Then

$$\begin{aligned}|\Delta(\sigma; C)| &= \exp[TP_1(\sigma)]\,|B + C\varphi(\sigma)\exp[-TP(i\sigma)] + A\exp[-TP(i\sigma)]| \\ &\geq \exp[TP_1(\sigma)]\,(\lambda_1 - |C|\,\mu_1) > 0,\ \forall C \in \mathbf{C} : 0 < |C| < \varepsilon_0 = \lambda_1/\mu_1.\end{aligned}$$

Hence, the corresponding C problem (2.1a), (2.1b) is well-posed (in H).

(iii) Let us assume that $sign\,[P_1(\sigma)] = sign\,[\sigma]$ as $|\sigma| \to +\infty$ and set $\mu_1' = \sup_{\sigma\geq 0} |\varphi(\sigma)|\exp[-TP_1(\sigma)]$, $\mu_1'' = \sup_{\sigma\leq 0} |\varphi(\sigma)|$, $\lambda_1' = \inf_{\sigma\geq 0} |B + A\exp[-TP(i\sigma)]|$, $\lambda_1'' = \inf_{\sigma\leq 0} |A + B\exp[TP(i\sigma)]|$. Then $|\Delta(\sigma; C)| \geq \lambda_1'' - |C|\,\mu_1''$, $\forall\sigma \leq 0$, and $|\Delta(\sigma; C)| \geq \exp[TP_1(\sigma)]\left(\lambda_1' - |C|\,\mu_1'\right)$, $\forall\sigma \geq 0$. Hence, $\Delta(\sigma; C) \neq 0$, $\forall\sigma \in \mathbf{R}$ and $\forall C \in \mathbf{C} : 0 < |C| < \varepsilon_0 = \min\left(\lambda_1'/\mu_1',\ \lambda_1''/\mu_1''\right)$, and the corresponding problem (2.1a), (2.1b) is well-posed (in H). The theorem is completely proved. ■

2.3 Well-Posedness of a Two-Point Boundary-Value Problem in a Layer for a System of Evolution Equations

This section deals with the well-posedness of a boundary-value problem for a system of equations containing pseudo differential operators, whose symbols are defined and continuous in a certain domain $H \subseteq \mathbf{R}^m$. More precisely, we consider the following two-point boundary-value problem for a system of evolution equations in the infinite layer $\Pi = \{(x, t) : x \in \mathbf{R}^m \text{ and } t \in [0,\ T],\ T > 0\}$ [18]

$$L\psi = \frac{\partial\psi}{\partial t} + P\left(-i\frac{\partial}{\partial x}\right)\psi = 0, \tag{2.12a}$$

$$A\left(-i\frac{\partial}{\partial x}\right)\psi(x, 0) + B\left(-i\frac{\partial}{\partial x}\right)\psi(x, T) = \varphi(x),\ \ x \in \mathbf{R}^m, \tag{2.12b}$$

where $\partial/\partial x = (\partial/\partial x_1, \partial/\partial x_2, \ldots, \partial/\partial x_m)$, $\psi = \mathrm{col}(\psi_1, \psi_2, \ldots, \psi_m)$, $\varphi = \mathrm{col}\,(\varphi_1, \varphi_2, \ldots, \varphi_m)$, and $P\left(-i\frac{\partial}{\partial x}\right) = \left[P_{jk}\left(-i\frac{\partial}{\partial x}\right)\right]_{jk}$, $A\left(-i\frac{\partial}{\partial x}\right) = \left[A_{jk}\left(-i\frac{\partial}{\partial x}\right)\right]_{jk}$, and $B\left(-i\frac{\partial}{\partial x}\right) = \left[B_{jk}\left(-i\frac{\partial}{\partial x}\right)\right]_{jk}$ are $l \times l$-matrices whose elements are pseudodifferential operators with symbols $P(\sigma)$, $A(\sigma)$, and $B(\sigma)$, respectively, continuous in a certain domain $H \subseteq \mathbf{R}^m_\sigma$.

In this section, we use the following notations. We denote by W^∞_Ω the space of vector functions $u = \text{col}\,(u_1, u_2, \ldots, u_m)$ with components $u_j \in L_2(\mathbf{R}^m)$ so that the Fourier transform $\widehat{u}_j(\sigma) = \mathfrak{F}\{u_j\}$ is compactly supported in $\Omega \subset \mathbf{R}^\infty_\sigma$. If we denote by $\widetilde{A}(-i\partial/\partial x)$ the pseudodifferential operator defined by

$$\widetilde{A}\left(-i\frac{\partial}{\partial x}\right)u(x) = \frac{1}{(2\pi)^m}\int_\Omega \widehat{A}(\sigma)\widehat{u}(\sigma)\exp[ix\sigma]\,d\sigma,\quad u(x) \in W^\infty_\Omega$$

with a continuous in Ω matrix $\widehat{A}(\sigma)$, then one can easily verify that space W^∞_Ω is invariant with respect to operator $\widetilde{A}(-i\partial/\partial x)$. Also, $\widetilde{A}\left(-i\frac{\partial}{\partial x}\right)$ is a continuous operator from space W^∞_Ω to W^∞_Ω. The space $\left(W^\infty_\Omega\right)'$ of the space of generalized vector functions on W^∞_Ω is invariant with respect to $\widehat{A}(i\partial/\partial x)$. Let $\left(W^{+\infty}_\Omega\right)' = \left(W^\infty_\Omega\right)'$ and $\left(W^{-\infty}_\Omega\right)' = \left(W^\infty_{-\Omega}\right)'$, where $-\Omega = \{\sigma \in \mathbf{R}^m : -\sigma \in -\Omega\}$, and let $C^k\left([0,\ T],\ W^{\pm\infty}_\Omega\right)$ be the spaces of vector functions that, for every $t \in [0,\ T]$ are functions of the spaces $W^{\pm\infty}_\Omega$, that together with all their derivatives up to the kth order continuously depend on t. We will denote by $\lambda_j = \lambda_j(\sigma)$ any root of the characteristic equation $\det(I\lambda + P(\sigma)) = 0$ with multiplicity χ_j, where I is the $l \times l$ identical matrix.

Let us introduce the following definition [28].

Definition: Problem (2.12a), (2.12b) is said to be well-posed in space $C^k\left([0,\ T],\ W^{\pm\infty}_\Omega\right)$ if, for every boundary function $\varphi(x) \in W^{\pm\infty}_\Omega$, there exists one and only one solution $\psi(x,t)$ in $C^k\left([0,\ T],\ W^{\pm\infty}_\Omega\right)$, which depends continuously on $\varphi(x)$ and on operators A and B.

If we denote by $\widehat{\psi}(\sigma,t) = \mathfrak{F}[\psi(x,t)]$ the x—Fourier transforms of the solution $\psi(x,t)$ of problem (2.12a), (2.12b) and by $\widehat{\varphi}(\sigma) = \mathfrak{F}[\varphi(x)]$ the Fourier transforms of the boundary function $\varphi(x)$ appearing in the boundary condition (2.12b), then it is easy to verify that $\widehat{\psi}(\sigma,t)$ is the solution of the boundary-value problem

$$L\left(\frac{d}{dt},\sigma\right)\widehat{\psi}(\sigma,t) = \frac{d}{dt}\widehat{\psi}(\sigma,t) + P(\sigma)\widehat{\psi}(\sigma,t) = 0, \tag{2.13a}$$

$$M(\sigma)\widehat{\psi}(\sigma,t) = A(\sigma)\widehat{\psi}(\sigma,0) + B(\sigma)\widehat{\psi}(\sigma,T) = \widehat{\varphi}(\sigma). \tag{2.13b}$$

Because $P(\sigma)$ is a matrix continuous of σ in H, we conclude that any root $\lambda_j(\sigma)$ $(j = 1, 2, \ldots, \nu)$ with the multiplicity $\chi_j(\sigma)$ of the characteristic equation $\det(\lambda I + P(\sigma)) = 0$ is a continuous function of the variable σ in H. According to Ref. [10], there exist two polynomial $l \times l$ matrices $M_1(\lambda,\sigma) = \left[M_{jk}(\lambda,\sigma)\right]$ and $N(\lambda,\sigma) = \left[N_{jk}(\lambda,\sigma)\right]$ such that

$$M_1(\lambda,\sigma) \times L(\lambda,\sigma) \times N(\lambda,\sigma) = Q(\lambda,\sigma) = \left[h_j(\lambda,\sigma)\delta_{jk}\right],\ \det N(\lambda,\sigma) = N(\sigma) \neq 0,$$

where

$$h_j(\lambda, \sigma) = \Pi_{k=1}^{\nu} (\lambda - \lambda_k)^{q_{jk}(\sigma)}, \; j = 1, 2, \ldots, l,$$

$$\sum_{j=1}^{l} q_{jk}(\sigma) = \chi_k(\sigma), \; k = 1, 2, \ldots, l,$$

δ_{jk} being the Kronecker symbols. Therefore, $M_1(\lambda, \sigma) \times L(\lambda, \sigma) = Q(\lambda, \sigma) \times N^{-1}(\lambda, \sigma)$. With the use of ansatz $y(\sigma, t) = N^{-1}(d/dt, \sigma)\,\widehat{\psi}(\sigma, t)$, system (2.13a) becomes $Q(d/dt, \sigma)\, y(\sigma, t) = 0$, i.e.,

$$h_j\left(\frac{d}{dt}, \sigma\right) y_j(\sigma, t) = 0, \; j = 1, 2, \ldots, l. \tag{2.14a}$$

Solving the system of ordinary differential equations (2.14a) leads to the following fundamental system of solutions:

$$y_{j,k\alpha}(\sigma, t) = \left(\frac{d}{d\lambda}\right)^{\alpha-1} \exp[\lambda t]\Big|_{\lambda=\lambda_k(\sigma)}, \; \alpha = 1, 2, \ldots, q_{jk}(\sigma), \; k = 1, 2, \ldots, \nu.$$

Therefore, we can express the fundamental matrix of solutions of system (2.13a) as follows:

$$F(\sigma, t) = \left[\exp[\lambda_k t] \times (\alpha - 1)! \times \sum_{s=0}^{\alpha-1} \frac{t^s}{s!} \frac{1}{(\alpha - 1 - s)!} \left(\frac{d}{d\lambda}\right)^{\alpha-1-s} N_{j\beta}(\sigma, t)\Big|_{\lambda=\lambda_k}\right]_{\substack{j,\,\beta=1,\,2,\,\ldots,\,l\\ \alpha=1,\,2,\,\ldots,\,q_{\beta k}\\ k=1,\,2,\ldots,\,\nu}},$$

which is a continuous function with respect to σ [27]. With the help of $F(\sigma, t)$, we introduce the notation

$$\Delta(\sigma) = \det D(\sigma), \;\; \text{with} \;\; D(\sigma) = A(\sigma)F(\sigma, 0) + B(\sigma)F(\sigma, T).$$

To formulate the criterion of the well-posedness of problem (2.12a), (2.12b), we define the following set and pseudodifferential operator. Let denote by $\Omega = H \backslash N_\Delta$, where $N_\Delta = \{\sigma \in \mathbf{R}^m : \Delta(\sigma) = 0\}$ is the set of zeros of $\Delta(\sigma)$ on $\mathbf{R}^m$. Let us associate the matrix $W(\sigma, t)$ with the pseudodifferential operator $W(-i\partial/\partial x, t) : W_\Omega^{\pm\infty} \to C^n\left([0, T], W_\Omega^{\pm\infty}\right)$. Then $W(-i\partial/\partial x, t)\,\varphi(x)$ is a continuous application from $W_\Omega^{\pm\infty}$ to $C^n\left([0, T], W_\Omega^{\pm\infty}\right)$.

Theorem: For the well-posedness of the nonlocal boundary-value problem (2.12a), (2.12b) in space $C^n\left([0, T], W_\Omega^{\pm\infty}\right)$, it is necessary and sufficient that $\Omega = H \backslash N_\Delta$.

▲ It is obvious that $\Omega = H \backslash N_\Delta \subset H$. Without loss of generality, we prove the theorem in space $C^n\left([0, T], W_\Omega^{+\infty}\right)$.

Necessity: If $\Omega \cap N_\Delta \neq \varnothing$ and $\sigma_0 \in \Omega \cap N_\Delta$, then the homogeneous problem

$$L\left(\frac{d}{dt},\sigma_0\right)\widehat{\psi}(\sigma_0,t)=\frac{d}{dt}\widehat{\psi}(\sigma_0,t)+P(\sigma_0)\widehat{\psi}(\sigma_o,t)=0,$$
$$M(\sigma_0)\widehat{\psi}(\sigma_0,t)=A(\sigma)\widehat{\psi}(\sigma_0,0)+B(\sigma_0)\widehat{\psi}(\sigma_0,T)=0$$

will have more than one solution. Therefore, the solution of the homogeneous problem

$$L\left(\frac{\partial}{\partial t},-i\frac{\partial}{\partial x}\right)\psi(x,t)=0,$$
$$A\left(-i\frac{\partial}{\partial x}\right)\psi(x,0)+B\left(-i\frac{\partial}{\partial x}\right)\psi(x,T)=0,$$

if it exists, will not be unique in space $C^n\left([0,\ T],W_{\Omega}^{+\infty}\right)$; this leads to the nonuniqueness of solutions of problem (2.12a), (2.12b) in space $C^n\left([0,\ T],W_{\Omega}^{+\infty}\right)$, whence the ill-posedness of problem (2.12a), (2.12b) in space $C^n\left([0,\ T],W_{\Omega}^{+\infty}\right)$.

Sufficiency: Now, let us suppose that $\Omega=H\backslash N_{\Delta}\subset H$ and prove that problem (2.12a), (2.12b) will be well-posed in space $C^n\left([0,\ T],W_{\Omega}^{+\infty}\right)$. For every $\varphi(x)\in W_{\Omega}^{+\infty}$, function $\psi(x,t)=W\left(-i\partial/\partial x,t\right)\varphi(x)$ will be a solution of problem (2.12a), (2.12b) in space $C^n\left([0,\ T],W_{\Omega}^{+\infty}\right)$ and will continuously depend on $\varphi(x)$. Thus, for every $\varphi(x)\in W_{\Omega}^{+\infty}$, problem (2.12a), (2.12b) admit in space $C^n\left([0,\ T],W_{\Omega}^{+\infty}\right)$ a solution which continuously depends on $\varphi(x)$. Let us now prove the uniqueness of solutions in $C^n\left([0,\ T],W_{\Omega}^{+\infty}\right)$. It is evident that if $\psi(x,t)\in C^n\left([0,\ T],W_{\Omega}^{+\infty}\right)$ is a solution of problem (2.12a), (2.12b), its x—Fourier transform $\widehat{\psi}(\sigma,t)$ will be a solution of problem (2.13a), (2.13b). Under the conditions of the theorem, problem (2.13a), (2.13b) admits a unique solution for every $\sigma\in\Omega$. If $\sigma\in\mathbf{R}^m\backslash\Omega$, then $\widetilde{\psi}(\sigma,t)=0$. This implies the uniqueness of a solution of problem (2.12a), (2.12b) in space $C^n\left([0,\ T],W_{\Omega}^{+\infty}\right)$, and the theorem is proved. ■

Example: Let

$$\frac{d^2v_n(t)}{dt^2}=\frac{a^2}{4}\left[v_{n+1}(t)-2v_n(t)+v_{n-1}(t)\right],\ 0<t<T,\ n=0,\ \pm1,\ \pm2,\ \ldots \tag{2.16}$$

be the fundamental law of dynamics of an infinite set of point particles located on a string at the same distance l from each other, where $a=2\sqrt{F/(ml)}$. Here, m and F are, respectively, the mass of each particle and the tension of the string which are supposed to be constant everywhere and time independent. Equation (2.16) is obtained under the condition that particles have only one degree of freedom and the motion of the nth particle is completely defined in terms of the position of its $(n-1)$ and $(n+1)$th adjacent particles. Let us associate with Eq. (2.16) the nonlocal boundary conditions

$$\begin{cases} v_n(0)+v_n(T)=\varphi_n, \\ \frac{dv_n}{dt}\Big|_{t=0}+\frac{dv_n}{dt}\Big|_{t=T}=\psi_n, \end{cases} \tag{2.17}$$

where φ_n and ψ_n are two constants. We aim to prove that, by taking $H = \mathbf{R}$, problem (2.16)–(2.17) will be well-posed in space $C^n\left([0,\ T],\ W_{\Omega}^{\pm\infty}\right)$ with Ω to be determined.

Let $v(x,t)$ be any differentiable function which coincides with $v_j(t)$ at the nodes of the lattice, i.e., $v(jl,t) = v_j(t)$. In terms of $v(x,t)$, Eq. (2.16) becomes

$$\frac{\partial^2 v(x,t)}{\partial t^2} = \frac{a^2}{4}\left[v(x+l,t) - 2v(x,t) + v(x-l,t)\right],\ 0 < t < T. \tag{2.18}$$

With the help of the identity $v(x+l,t) = \exp\left[i\left(-2i\frac{\partial}{\partial x}\right)\right]v(x,t)$, Eq. (2.18) takes the form of a pseudodifferential equation

$$\frac{\partial^2 v(x,t)}{\partial t^2} = -\frac{a^2}{4}\sin^2\left(-i\frac{\partial}{\partial x}\right)v(x,t),\ 0 < t < T. \tag{2.19}$$

If we introduce the vector function $\psi = \mathrm{col}(\psi_1, \psi_2) = (v, \partial v/\partial t)$, Eq. (2.19) becomes

$$\begin{cases} \frac{\partial \psi_1(x,t)}{\partial t} = \frac{\partial v(x,t)}{\partial t} = \psi_2, \\ \frac{\partial \psi_2(x,t)}{\partial t} = \frac{\partial^2 v(x,t)}{\partial t^2} = -\frac{a^2}{4}\sin^2\left(-i\frac{\partial}{\partial x}\right)\psi_1, \end{cases}$$

i.e.,

$$\frac{\partial \psi}{\partial t} = \frac{\partial}{\partial t}\begin{pmatrix}\psi_1(x,t)\\ \psi_2(x,t)\end{pmatrix} + \begin{pmatrix} 0 & -1 \\ \frac{a^2}{4}\sin^2\left(-i\frac{\partial}{\partial x}\right) & 0\end{pmatrix}\begin{pmatrix}\psi_1\\ \psi_2\end{pmatrix} = 0.$$

Hence,

$$\frac{\partial \psi}{\partial t} + P\left(-i\frac{\partial}{\partial x}\right)\psi = 0, \tag{2.20}$$

where $P\left(-i\frac{\partial}{\partial x}\right) = \begin{pmatrix} 0 & -1 \\ \frac{a^2}{4}\sin^2\left(-i\frac{\partial}{\partial x}\right) & 0\end{pmatrix}$. The boundary conditions (2.17) become

$$\begin{pmatrix}1 & 0\\ 0 & 1\end{pmatrix}\begin{pmatrix}\psi_1(x,0)\\ \psi_2(x,0)\end{pmatrix} + \begin{pmatrix}1 & 0\\ 0 & 1\end{pmatrix}\begin{pmatrix}\psi_1(x,T)\\ \psi_2(x,T)\end{pmatrix} = \begin{pmatrix}\theta_1(x)\\ \theta_2(x)\end{pmatrix} \tag{2.21}$$

where $\theta_1(x)$ and $\theta_2(x)$ are two differentiable functions that take the values of φ_n and ψ_n, respectively, at the nodes of the lattice. Here, $A = B = I$. For the problem (2.20)–(2.21), matrix $F(\sigma,t)$ reads

$$F(\sigma,t) = \begin{pmatrix} \cos\left[at\sin[\sigma]\right] & -\sin\left[at\sin[\sigma]\right] \\ -a\sin[\sigma]\sin\left[at\sin[\sigma]\right] & -a\sin[\sigma]\cos\left[at\sin[\sigma]\right]\end{pmatrix},$$

leading to

$$\Delta(\sigma) = \det\left[A(\sigma)F(\sigma,0) + B(\sigma)F(\sigma,T)\right] = -2a\sin[\sigma]\left[1+\cos[aT\sin[\sigma]]\right].$$

Therefore,

$$N_\Delta = \{k\pi\}_{k\in\mathbf{Z}} \cup \left\{\arcsin\frac{(1+2k)\,\pi}{aT} : k\in\mathbf{Z} \text{ and } -\frac{\pi+aT}{2\pi} \le k \le \frac{\pi+aT}{2\pi}\right\}.$$

Thus, if we take $H = \mathbf{R}$, then it follows from the above theorem that problem (2.20)–(2.21) is well-posed in $C^n\left([0,\ T],\, W_\Omega^{\pm\infty}\right)$ with $\Omega = \mathbf{R}\backslash N_\Delta$. Therefore, problem (2.16)–(2.17) is well-posed in space $C^n\left([0,\ T],\, W_\Omega^{\pm\infty}\right)$ with $\Omega = \mathbf{R}\backslash N_\Delta$.

2.4 Boundary-Value Problem for Factorized Operators with Dirichlet-Type Boundary Conditions

In the section, we extend the investigations carried out in previous sections to PDEs with a variable with respect to t and x coefficients. We study a problem with nonlocal conditions with respect to the variable t and Dirichlet-type boundary conditions with respect to x for factorized operators with partial derivatives.

2.4.1 Formulation of Problem and Notations

In the domain $\prod = (0,l)\times(0,T)$ of the variables (x,t), we consider the partial differential equation

$$\frac{\partial}{\partial t}\left[\frac{\partial}{\partial t} - P\left(t,\frac{d}{dx}\right)\right]u = f(x,t), \tag{2.22}$$

where $P\left(t,\frac{d}{dx}\right) = a(t)\left[q(x) - \frac{d}{dx}\left(p(x)\frac{d}{dx}\right)\right] + b(t)$. Here, $a(t)$ and $b(t)$ are complex-valued functions, $a, b \in C^1([0,T])$, $p, q \in C^1([0,l])$, $p(x) \ge p_0 = const > 0$ and $q(x) \ge 0$, for every $x \in [0,l]$. We need to find a solution to Eq. (2.22) satisfying the conditions

$$\begin{pmatrix} 1 & \alpha_2 & -\mu & \alpha_4 \\ \beta_1 & 1 & -\mu\beta_1 & -\mu \end{pmatrix}\begin{pmatrix} u(x,0) \\ u_t(x,0) \\ u(x,T) \\ u_t(x,T) \end{pmatrix} = \begin{pmatrix} \varphi_1(x) \\ \varphi_2(x) \end{pmatrix}, \quad x\in[0,l], \tag{2.23}$$

and

$$u(0,t) = u(l,t) = 0, \tag{2.24}$$

where μ, α_2, α_4, and β_1 are four complex numbers such that $\mu \neq 0$, $1 - \mu - \beta_1\alpha_2 - \beta_1\alpha_4 \neq 0$, and $rank \begin{pmatrix} 1 & \alpha_2 & -\mu & \alpha_4 \\ \beta_1 & 1 & -\mu\beta_1 & -\mu \end{pmatrix} = 2$, and φ_1 and φ_2 are two given complex-valued functions.

The eigenvalue problem corresponding to problem (2.22)–(2.24) reads

$$\frac{d}{dx}\left(p(x)\frac{d}{dx}\right)X + (\lambda - q(x))\,X = 0,\ \ X(0) = X(l) = 0. \tag{2.25}$$

Let $\Lambda = \{\lambda_n,\ n \in \mathbb{N}\}$ and $X = \{X_n(x),\ n \in \mathbb{N}\}$ be the system of eigenvalues and the system of eigenfunctions of problem (2.25), respectively. According to [27], the system of eigenfunctions of problem (2.25) is complete and orthonormal in $L_2\,(0, l)$ (in what follows, we assume that it is normalized), $\lambda_n > 0$, and $\lambda_i \neq \lambda_j$ when $i \neq j$. Furthermore, $X_n \in C^2\,([0, l])$, $n \in \mathbb{N}$ and the following estimates are true [25, 27]:

$$c_1 n^2 \leq \lambda_n \leq c_2 n^2,\ 0 < c_1 \leq c_2,\ n \in \mathbb{N}, \tag{2.26}$$

$$\max_{x\in[0,l]} \left|X_n^{(j)}(x)\right| \leq c_{3j}\lambda_n^{j/2},\ j = 0,\ 1,\ 2,\ n \in \mathbb{N}. \tag{2.27}$$

It follows from (2.26) that $\lim\limits_{n\to+\infty} \lambda_n = +\infty$.

In what follows, we denote by $C^{(1,2)}\left(\overline{\prod}\right)$ the Banach space of functions $w(x, t)$ with the norm

$$\|w\|_{C^{(1,2)}\left(\overline{\prod}\right)} = \sum_{2k+m\leq 2} \max_{(x,t)\in\overline{\prod}} \left|\frac{\partial^{k+m} w(x, t)}{\partial t^k \partial x^m}\right|,$$

B_k, $k > 0$, is the space of functions $\varphi \in L_2\,(0, l)$,

$$\varphi(x) = \sum_{n=1}^{+\infty} \varphi_n X_n(x)$$

with the norm

$$\|\varphi\|_{B_k} = \sum_{n=1}^{+\infty} |\varphi_n| \exp\left(k\lambda_n\right),$$

and $C\,(B_k, [0,\ T])$ is the space of functions

$$\psi(x, t) = \sum_{n=1}^{+\infty} \psi_n(t) X_n(x)$$

defined in $\overline{\prod}$, continuous in t, and belonging to the space B_k for every fixed $t \in [0, T]$ with the norm

$$\|\psi\|_{C(B_k,\,[0,\,T])} = \sum_{n=1}^{+\infty} \max_{t\in[0,\,T]} |\psi_n(t)| \exp(k\lambda_n) .$$

For every complex function or constant α, we write $\alpha = \alpha^r + i\alpha^i$, which means that $\alpha^r = Re(\alpha)$ and $\alpha^i = Im(\alpha)$.

2.4.2 Uniqueness of Solutions to Problem (2.22)–(2.24)

We seek a solution of problem (2.22)–(2.24) in the form

$$u(x,t) = \sum_{n=1}^{+\infty} Y_n(t) X_n(x). \tag{2.28}$$

If series (2.28) and the series obtained by the termwise differentiation of series (2.28) with respect to x up to the order two inclusive converge uniformly in the domain $\overline{\prod}$, then function $u(x,t)$ defined by (2.22) satisfies the boundary conditions (2.24). In order to determine $Y_n(t)$, we assume that the boundary functions $\varphi_1(x)$ and $\varphi_2(x)$ and the free term $f(x,t)$ can be expanded into Fourier series:

$$\varphi_s(x) = \sum_{n=1}^{+\infty} \varphi_{s,n} X_n(x),\ s = 1,\ 2,\ f(x,t) = \sum_{n=1}^{+\infty} f_n(t) X_n(x), \tag{2.29}$$

where

$$\varphi_{s,n} = \int_0^l \varphi_s(x) X_n(x) dx,\ f_n(t) = \int_0^l f(x,t) X_n(x) dx,\ n \in \mathbb{N},\ s = 1,\ 2. \tag{2.30}$$

If we insert (2.22) and (2.23) into (2.28) and (2.29) we obtain for determination of $Y_n(t)$ the following Cauchy problem for an ordinary differential equation:

$$\frac{d}{dt}\left(\frac{d}{dt} - \lambda_n a - b\right) Y_n = f_n(t), \tag{2.31}$$

$$\begin{pmatrix} 1 & \alpha_2 & -\mu & \alpha_4 \\ \beta_1 & 1 & -\mu\beta_1 & -\mu \end{pmatrix} \begin{pmatrix} Y_n(0) \\ Y_n'(0) \\ Y_n(T) \\ Y_n'(T) \end{pmatrix} = \begin{pmatrix} \varphi_{1,n} \\ \varphi_{2,n} \end{pmatrix}. \tag{2.32}$$

A fundamental system of solutions of the homogeneous equation corresponding to equation (2.31) reads

$$\left\{Y_{n1}(t) = \exp\left(\int_0^t (\lambda_n a(\tau) + b(\tau))\, d\tau\right),\right.$$

$$\left. Y_{n2}(t) = \int_0^t \exp\left(\int_0^t (\lambda_n a(\tau) + b(\tau))\, d\tau - \int_0^y (\lambda_n a(\tau) + b(\tau))\, d\tau\right) dy\right\}.$$

In what follows, we restrict ourselves to the case where the coefficients of equation (2.28) satisfy the conditions

$$a(0) = a(T) = 0;\; b(0) = b(T) = 0. \tag{2.33}$$

Under conditions (2.33), the characteristic determinant [25] of the homogeneous problem

$$\frac{d}{dt}\left(\frac{d}{dt} - \lambda_n a - b\right) Y_n = 0,\; \begin{pmatrix} 1 & \alpha_2 & -\mu & \alpha_4 \\ \beta_1 & 1 & -\mu\beta_1 & -\mu \end{pmatrix} \begin{pmatrix} Y_n(0) \\ Y_n'(0) \\ Y_n(T) \\ Y_n'(T) \end{pmatrix} = \begin{pmatrix} 0 \\ 0 \end{pmatrix} \tag{2.34}$$

reads

$$\Delta(\lambda_n) = (1 - \mu - \beta_1\alpha_2 - \beta_1\alpha_4)\,(1 - \mu \exp(\lambda_n A + B)), \tag{2.35}$$

where

$$A = \int_0^T a(\tau) d\tau,\; B = \int_0^T b(\tau) d\tau.$$

since $1 - \mu - \beta_1\alpha_2 - \beta_1\alpha_4 \neq 0$, $\Delta(\lambda_n) = 0$ if and only, if $1 - \mu \exp(\lambda_n A + B) = 0$.

$$\begin{aligned} 1 - \mu \exp(\lambda_n A + B) &= 0 \Leftrightarrow \mu^{-1} = \exp(\lambda_n A + B) = 0 \\ &\Leftrightarrow -\ln|\mu| - i \arg\mu = \lambda_n A + B + 2\pi i m,\; m \in \mathbb{Z} \\ &\Leftrightarrow \ln|\mu| + \lambda_n A^r + B^r = 0 \\ \text{and } 2\pi \arg\mu + \lambda_n A^i + B^i &= 2\pi m, m \in .\mathbb{Z} \end{aligned}$$

Thus, in order that $\Delta(\lambda_n) \neq 0$ under condition $1 - \mu - \beta_1\alpha_2 - \beta_1\alpha_4 \neq 0$, it is necessary and sufficient that at least one of the following conditions holds:

$$(1)\;\ln|\mu| + \lambda_n A^r + B^r \neq 0$$

$$(2)\;\arg\mu + \lambda_n A^i + B^i \neq 2\pi m,\; m \in \mathbb{Z}.$$

Theorem: *For the uniqueness of solutions of problem* (2.22)–(2.24), (2.28) *in space* $C^{(1,2)}\left(\overline{\Pi}\right)$, *it is necessary and sufficient that*

$$1 - \mu \exp(\lambda_n A + B) \neq 0, \textit{for every } \lambda_n \in \Lambda. \tag{2.36}$$

▲ *Necessity.* Assume that for certain $\lambda_{n_0} \in \Lambda$, we have $1 - \mu \exp\left(\lambda_{n_0} A + B\right) = 0$. Then $\Delta(\lambda_{n_0}) = 0$ and there exist nontrivial solutions

$$Y_{n_0}(t) = C_1 \exp\left(\int_0^t \left(\lambda_{n_0} a(\tau) + b(\tau)\right) d\tau\right), \ C_1 \neq 0$$

of homogeneous problem (2.34). In this case, the homogeneous problem associated to problem (2.22)–(2.24) ($f(x,t) \equiv 0$, $\varphi_1(x) \equiv 0$, and $\varphi_2(x) \equiv 0$) admits nontrivial solutions of the form

$$u(x,t) = Y_{n_0}(t) X_{n_0}(x)$$

and a solution of the inhomogeneous problem (2.22)–(2.24), if it exists, is not unique.

Sufficiency. Let condition (2.36) be satisfied. Let $u_1(x,t)$ and $u_2(x,t)$ be two solutions of problem (2.22)–(2.24) in space $C^{(1,2)}\left(\overline{\prod}\right)$. Then function $\widetilde{u}(x,t) = u_1(x,t) - u_2(x,t)$ belongs to space $C^{(1,2)}\left(\overline{\prod}\right)$ and is a solution of homogeneous problem associated to problem (2.22)–(2.24) and can be expanded in Fourier series

$$\widetilde{u}(x,t) = \sum_{n=1}^{+\infty} \widetilde{u}_n(t) X_n(x). \tag{2.37}$$

We can also expand

$$\frac{\partial}{\partial t}\left[\frac{\partial}{\partial t} - P\left(t, \frac{d}{dx}\right)\right]\widetilde{u}$$

and

$$\begin{pmatrix} 1 & \alpha_2 & -\mu & \alpha_4 \\ \beta_1 & 1 & -\mu\beta_1 & -\mu \end{pmatrix} \begin{pmatrix} \widetilde{u}(x,0) \\ \widetilde{u}_t(x,0) \\ \widetilde{u}(x,T) \\ \widetilde{u}_t(x,T) \end{pmatrix}$$

into Fourier series in the system of functions $X_n(x)$, and these series coincide with

$$\frac{\partial}{\partial t}\left[\frac{\partial}{\partial t} - P\left(t, \frac{d}{dx}\right)\right]\sum_{n=1}^{+\infty} \widetilde{u}_n(t) X_n(x)$$

and

$$\begin{pmatrix} 1 & \alpha_2 & -\mu & \alpha_4 \\ \beta_1 & 1 & -\mu\beta_1 & -\mu \end{pmatrix} \begin{pmatrix} \sum_{n=1}^{+\infty} \widetilde{u}_n(0) X_n(x) \\ \sum_{n=1}^{+\infty} \widetilde{u}_n'(0) X_n(x) \\ \sum_{n=1}^{+\infty} \widetilde{u}_n(T) X_n(x) \\ \sum_{n=1}^{+\infty} \widetilde{u}_n'(T) X_n(x) \end{pmatrix},$$

respectively. Using Parseval equality for the functions $\frac{\partial}{\partial t}\left[\frac{\partial}{\partial t} - P\left(t, \frac{d}{dx}\right)\right]\widetilde{u}$ and $\begin{pmatrix} 1 & \alpha_2 & -\mu & \alpha_4 \\ \beta_1 & 1 & -\mu\beta_1 & -\mu \end{pmatrix} \begin{pmatrix} \widetilde{u}(x,0) \\ \widetilde{u}_t(x,0) \\ \widetilde{u}(x,T) \\ \widetilde{u}_t(x,T) \end{pmatrix}$, we obtain that for every $n \in \mathbb{N}$, $\widetilde{u}_n(t)$ is a solution of problem (2.34) under conditions (2.33). This solution can be expressed in the form

$$\widetilde{u}_n(t) = C_{n1}Y_{n1}(t) + C_{n2}Y_{n2}(t),$$

where (C_{n1}, C_{n2}) is a solution of a linear homogeneous algebraic system with determinant $\Delta(\lambda_n) \neq 0$. Therefore, $C_{n1} = C_{n2} = 0$, and $\widetilde{u}_n(t) \equiv 0$, for every $n \in \mathbb{N}$. From the Parseval equality for the function $\widetilde{u}(x,t)$ and continuity of this function in $\widetilde{\prod}$, we conclude that $\widetilde{u}(x,t) \equiv 0$, and this means that $u_1(x,t) \equiv u_2(x,t)$. We have then proved that the solution of problem (2.22)–(2.24) in space $C^{(1,2)}\left(\overline{\prod}\right)$, if it exists, is unique. Theorem is proven. ■

2.4.3 Existence of Solutions of Problem (2.22)–(2.24)

In this section, we assume that the solution of problem (2.22)–(2.24), (2.33) in space $C^{(1,2)}\left(\overline{\prod}\right)$, if it exists, is unique. According to the above theorem, condition (2.36) is satisfied. Then for every $n \in \mathbf{N}$, the solution of problem (2.31)–(2.33) reads

$$Y_n(t) = C_{n1}Y_{n1}(t) + C_{n2}Y_{n2}(t) + Y_n^*(t), \tag{2.38}$$

where

$$\begin{aligned} Y_n^*(t) &= \exp\left(\int_0^t (\lambda_n a(\tau) + b(\tau))\,d\tau\right) \int_0^t f_n(z) \\ &\left[\int_0^z \exp\left(-\int_0^y (\lambda_n a(\tau) + b(\tau))\,d\tau - 2\int_0^z (\lambda_n a(\tau) + b(\tau))\,d\tau\right) dy \right. \\ &\left. - \int_0^t \exp\left(-\int_0^y (\lambda_n a(\tau) + b(\tau))\,d\tau - 2\int_0^z (\lambda_n a(\tau) + b(\tau))\,d\tau\right) dy\right] dz, \end{aligned} \tag{2.39}$$

$Y_{n1}(t)$ and $Y_{n2}(t)$ are the above-defined functions, and coefficients C_{n1} and C_{n2} satisfy the system

$$\begin{pmatrix} 1 & \alpha_2 & -\mu & \alpha_4 \\ \beta_1 & 1 & -\mu\beta_1 & -\mu \end{pmatrix} \begin{pmatrix} C_{n1}Y_{n1}(0) + C_{n2}Y_{n2}(0) + Y_n^*(0) \\ C_{n1}Y'_{n1}(0) + C_{n2}Y'_{n2}(0) + Y_n^{*\prime}(0) \\ C_{n1}Y_{n1}(T) + C_{n2}Y_{n2}(T) + Y_n^*(T) \\ C_{n1}Y'_{n1}(T) + C_{n2}Y'_{n2}(T) + Y_n^{*\prime}(T) \end{pmatrix} = \begin{pmatrix} \varphi_{1,n} \\ \varphi_{2,n} \end{pmatrix},$$

that is,

$$\begin{cases} \left[1-\mu \exp\left(\lambda_n A+B\right)\right] C_{n1}-\varphi_{1,n}-\mu Y_n^*(T)+\alpha_4 Y_n^{*\prime}(T) \\ \quad +\left[\alpha_2+\alpha_4-\mu \exp\left(\lambda_n A+B\right)\right) \\ \quad \times \int_0^T \exp\left(-\int_0^y\left(\lambda_n a(\tau)+b(\tau)\right) d\tau\right) dy\Big] C_{n2}=0, \\ \\ \left[\beta_1+\exp\left(\lambda_n A+B\right)\right] C_{n1}-\varphi_{2,n}-\mu \beta_1 Y_n^*(T)-\mu Y_n^{*\prime}(T) \\ \quad +\left[1-\mu-\mu \beta_1 \exp\left(\lambda_n A+B\right)\right) \\ \quad \times \int_0^T \exp\left(-\int_0^y\left(\lambda_n a(\tau)+b(\tau)\right) d\tau\right) dy\Big] C_{n2}=0. \end{cases} \tag{2.40}$$

The determinant of system (2.40) coincides with $\Delta(\lambda_n)$. Therefore

$$\begin{aligned} C_{n1} = & \\ & \tfrac{1}{\Delta(\lambda_n)}\Big[\Big(\varphi_{1,n}+\mu \exp\left(\lambda_n A+B\right) \int_0^T f_n(z)\delta_T(z)dz+\alpha_4 T \exp\left(-\left(\lambda_n A+B\right)\right) \\ & \times \int_0^T f_n(z) \exp\left(2C(z)\right) dz\Big) \times \Big(1-\mu-\mu \beta_1 \exp\left(\lambda_n A+B\right)\Big) \int_0^T \exp\left(C(y)\right) dy\Big) \\ & -\Big(\alpha_2+\alpha_4-\mu \exp\left(\lambda_n A+B\right)\Big) \int_0^T \exp\left(C(y)\right) dy\Big) \\ & \times\Big(\varphi_{2,n}+\mu \beta_1 \exp\left(\lambda_n A+B\right) \int_0^T f_n(z)\delta_T(z)dz-\mu T \exp\left(-\left(\lambda_n A+B\right)\right) \\ & \times \int_0^T f_n(z) \exp\left(2C(z)\right) dz\Big)\Big] \end{aligned}$$

$$\begin{aligned} C_{n2} = & \\ & \tfrac{1}{\Delta(\lambda_n)}\Big[\Big(\varphi_{2,n}+\mu \beta_1 \exp\left(\lambda_n A+B\right) \int_0^T f_n(z)\delta_T(z)dz-\mu T \exp\left(-\left(\lambda_n A+B\right)\right) \\ & \times \int_0^T f_n(z) \exp\left(2C(z)\right) dz\Big) \times\left(1-\mu \exp\left(\lambda_n A+B\right)\right)-\left(\beta_1+\exp\left(\lambda_n A+B\right)\right) \\ & \Big(\varphi_{1,n}+\mu \exp\left(\lambda_n A+B\right) \int_0^T f_n(z)\delta_T(z)dz)+\alpha_4 T \exp\left(-\left(\lambda_n A+B\right)\right) \\ & \times \int_0^T f_n(z) \exp\left(2C(z)\right) dz\Big)\Big], \end{aligned} \tag{2.41}$$

where

$$\begin{aligned} C(\xi) &= -\int_0^\xi\left(\lambda_n a(\tau)+b(\tau)\right) d\tau, \\ \delta_T(z) &= \int_0^z \exp\left(C(y)+2C(z)\right) dy-\int_0^T \exp\left(C(y)+2C(z)\right) dy. \end{aligned} \tag{2.42}$$

Inserting (2.38) into (2.28), a formal solution of problem (2.22)–(2.24), (2.33) reads

$$u(x,t)=\sum_{n=1}^{+\infty}\left(C_{n1}Y_{n1}(t)+C_{n2}Y_{n2}(t)+Y_n^*(t)\right) X_n(x). \tag{2.43}$$

In general, series (2.43) is divergent because the modula of the nonzero expression (see (2.35) and (2.41)) $\widetilde{\Delta}(\lambda_n)=1-\mu \exp\left(\lambda_n A+B\right)$ can be arbitrarily small for infinitely many numbers $\lambda_n \in \Lambda$. Therefore, the question of the existence of a solution of problem (2.22)–(2.24), (2.33) is connected with the problem of small denominator.

Lemma: *Let $F(\lambda_n)$be a bounded sequence of real numbers. Then, for almost all (with respect to the Lebesgue measure in $\overline{\mathbf{R}}$) numbers $a \in \mathbf{R}$, the following inequality holds for all (except finitely many) pairs (λ_n, m) $\lambda_n \in \Lambda$, $m \in \mathbf{Z}$:*

$$\left| F(\lambda_k) - \frac{am}{\lambda_n} \right| \geq \lambda_n^{-3/2-\varepsilon}, 0 < \varepsilon < 1. \tag{2.44}$$

▲ To establish inequality (2.44), it is sufficient to prove that for almost all (with respect to the Lebesgue measure in $\overline{\mathbf{R}}$) numbers $a \in \mathbf{R}$, the inequality

$$\left| F(\lambda_k) - \frac{am}{\lambda_n} \right| < \lambda_n^{-3/2-\varepsilon}, 0 < \varepsilon < 1 \tag{2.45}$$

admits no more that a finite number of solutions $(\lambda_n, \ m)$, with $m \in \mathbf{Z}$ and $\lambda_n \in \Lambda$. To prove the inequality (2.45) we use the procedure of the proof of Lemma 2.4 in [Ref. [30], Chap. 1]. Let us denote by K the set of real numbers $a \in [d, \ 2d]$ $(d \in \mathbf{R})$ for which the inequality (2.45) has infinitely many solutions $(\lambda_n, \ m)$, $\lambda_n \in \Lambda$ and $m \in \mathbf{Z}$. Let us fix $\lambda_n = \lambda_{n^*} \in \Lambda$, which corresponds to $n^* \in \mathbf{N}$. The corresponding values of m for which inequality (2.45) holds belong to the interval

$$\frac{\lambda_{n^*}}{a}\left(F(\lambda_k) - \lambda_{n^*}^{-3/2-\varepsilon}\right) < m < \frac{\lambda_{n^*}}{a}\left(F(\lambda_k) + \lambda_{n^*}^{-3/2-\varepsilon}\right). \tag{2.46}$$

Let $m = m^0 \in \mathbf{Z}$ satisfying inequality (2.46). It is obvious that for some positive constant C_a that depends on a and does not depend on n^* we have $\left|m^0\right| < C_a \lambda_{n^*}$. Then the measure of the set $K(n^*, m^0)$ of real numbers a for which inequality (4.8) holds for fixed $n = n^*$ and $m = m^0$ satisfies the inequality

$$mes\ K(\lambda_{n^*}, m^0) < \frac{2}{\left|m^0\right| \lambda_{n^*}^{1/2+\varepsilon}}. \tag{2.47}$$

It follows from (2.46) and (2.47) that for the measure of the set $K(\lambda_{n^*})$ of real numbers a for which the inequality (2.45) for a fixed $\lambda_n = \lambda_{n^*}$ admits solutions in integers $m \neq 0$, the following estimate is satisfied:

$$mes\ K(\lambda_{n^*}) < \sum_{m^0 = -E(C_a|\lambda_{n^*}|)-1}^{E(C_a|\lambda_{n^*}|)+1} \frac{2}{\left|m^0\right| |\lambda_{n^*}|^{1/2+\varepsilon}} \leq \frac{C_{1a} \ln \lambda_{n^*}}{\lambda_{n^*}^{1/2+\varepsilon}}.$$

If we sum this last estimate for all λ_n we obtain

$$mes\ K < C_{1a} \sum_{n=1}^{+\infty} \frac{\ln \lambda_n}{\lambda_n^{1/2+\varepsilon}}.$$

From the convergence of the series at the right-hand side of the last inequality and Lemma 2.1 in [Ref. [30], Chap. 1], we conclude that the measure of set $K = [d, 2d]$ of real numbers a for which inequality (2.45) admits infinitely many solutions in integers m is equal to zero. Therefore, we establish that for almost all $a \in [d, 2d]$, estimate (2.44) is true for all (except finitely many) $\lambda_n \in \Lambda$. Because the real axis $(-\infty, +\infty)$ can be covered by finitely many segments of the form $[d, 2d]$ (e.g.,

$$(-\infty, +\infty) = \left(\bigcup_{l=-\infty}^{+\infty} [-2^l, -2^{l-1}] \right) \cup \left(\bigcup_{l=-\infty}^{+\infty} [2^{l-1}, 2^l] \right)),$$

we complete the proof of the lemma. ■

Theorem: *If condition (2.36) is satisfied then there exist $\alpha \geq 0$ and $C > 0$ so that the following estimate holds for all (except finitely many) $n \in \mathbf{N}$:*

$$|1 - \mu \exp(\lambda_n A + B)| \geq C\lambda_n^{-\alpha}. \tag{2.48}$$

▲ Estimate (2.48) is evident if $A = 0$. Suppose $A \neq 0$ and let $\mu = \exp(\rho + i\varphi)$, $\rho \in \mathbb{R}$ and $\varphi = \arg \mu \in [0, 2\pi)$. Then $1 - \mu \exp(\lambda_n A + B) = 1 - \exp((A^r\lambda_n + B^r + \rho) + i\left(A^i\lambda_n + B^i + \varphi\right))$. It is proven in [3] that for every polynomial $P(i\sigma)$ with complex coefficients, if $1 - \exp(P(i\sigma)) = 0$ admits no more than a finite number of solutions, then there exist $\widetilde{C} > 0$ and $r \in \mathbb{R}$ so that the following estimate is true:

$$|1 - \exp(P(i\sigma))| \geq \widetilde{C}(1 + |\sigma|)^r, \forall \sigma \in \mathbb{R} \text{ so that } |\sigma| \geq \sigma_0. \tag{2.49}$$

In (2.49) σ_0 is a positive number so that $1 - \exp(P(i\sigma)) \neq 0$ for every $|\sigma| \geq \sigma_0$. In other words, all the zeros of $1 - \exp(P(i\sigma)) = 0$ are in the interval $(-\sigma_0, \sigma_0)$. In particular for the first-degree polynomial $P(i\sigma) = (A^r\sigma + B^r + \rho) + i\left(A^i\sigma + B^i + \varphi\right) = A\sigma + B + \rho + i\varphi$, (2.45) is satisfied as soon as $|A^r| + |B^r + \rho| > 0$ or $A^r = A^i = 0$. Therefore from (2.49) we obtain for some constant $C > 0$

$$|1 - \mu \exp(\lambda_n A + B)| = \left|1 - \exp\left((A^r\lambda_n + B^r + \rho) + i\left(A^i\lambda_n + B^i + \varphi\right)\right)\right| \geq C,$$

and estimate (2.48) is established.

Let now

$$\left|A^r\right| + \left|B^r + \rho\right| = 0 \text{ and } A^i \neq 0. \tag{2.50}$$

In this case, equation $1 - \exp(P(i\sigma)) = 0$ with $P(i\sigma) = A\sigma + B + \rho + i\varphi$ admits an infinite number of solutions. Now we have

$$1 - \mu \exp(\lambda_n A + B) = 1 - \exp\left(i\left(A^i\lambda_n + B^i + \varphi\right)\right),$$

and $1 - \mu \exp(\lambda_n A + B) \neq 0$ if and only, if

$$A^i\lambda_n + B^i + \varphi \notin 2\pi\mathbf{Z}, \text{ for every } n \in \mathbf{N}.$$

$$\left|1-\exp\left(i\left(A^i\lambda_n+B^i+\varphi\right)\right)\right| = 2\sqrt{\sin^2\frac{\left(A^i\lambda_n+B^i+\varphi\right)}{2}}$$
$$= 2\left|\sin\left(\frac{\varphi}{2}+\frac{A^i\lambda_n+B^i}{2}\right)\right|.$$

For every $n \in \mathbf{N}$ there exists an integer $m = m(n)$ so that

$$\varphi + A^i\lambda_n + B^i - 2m\pi \in (-\pi, \pi)$$

and

$$\left|\sin\left(\frac{\varphi}{2}+\frac{A^i\lambda_n+B^i}{2}\right)\right| = \left|\sin\left(\frac{\varphi}{2}+\frac{A^i\lambda_n+B^i}{2}-m\pi\right)\right|$$
$$\geq 2\left|\frac{\varphi}{2\pi}+\frac{A^i\lambda_n+B^i}{2\pi}-m\right|$$
$$= \frac{1}{\pi}\left|B^i+A^i\lambda_n+\varphi-2\pi m\right|$$

Thus

$$\left|1-\exp\left(i\left(A^i\lambda_n+B^i+\varphi\right)\right)\right| \geq 4\lambda_n\left|A^i\right|\left|\frac{A^i\lambda_n+B^i+\varphi}{2\pi A^i\lambda_n}-\frac{1}{A^i}\frac{m}{\lambda_n}\right|.$$

Because $\Lambda = \{\lambda_n\}_{n=1}^{+\infty}$ is a sequence satisfying $\lim\limits_{n\to+\infty}\lambda_n = +\infty$, $F(\lambda_n) = \frac{A^i\lambda_n+B^i+\varphi}{2\pi A^i\lambda_n}$ is a bounded sequence. According to the above lemma in which we take $a = 1/A^i$, we have

$$\left|1-\exp\left(i\left(A^i\lambda_n+B^i+\varphi\right)\right)\right| \geq 4\lambda_n\left|A^i\right|\left|\frac{A^i\lambda_n+B^i+\varphi}{2\pi A^i\lambda_n}-\frac{1}{A^i}\frac{m}{\lambda_n}\right|$$
$$\geq 4\lambda_n\left|A^i\right|\lambda_n^{-3/2-\varepsilon}$$
$$= 4\left|A^i\right|\lambda_n^{-1/2-\varepsilon} = C\lambda_n^{-\alpha},$$

with

$$\alpha = 1/2+\varepsilon,\ 0<\varepsilon<1,$$

for almost all (with respect to the Lebesgue measure in $\mathbf{R}$) numbers A^i and all (except finitely many) $\lambda_n \in \Lambda$. The theorem is completely proven. ■

Theorem: *Suppose that condition* (2.36) *is satisfied and at least one of conditions* (2.50) *is not satisfied. Then if* $\varphi_q \in B_{\delta_1}$, $q = 1, 2$, $f \in C\left(B_{\delta_2}, [0, T]\right)$, *with* $\delta_1 > \varepsilon_1$ *and* $\delta_2 > \varepsilon_2$, *there always exists a unique solution of problem* (2.22)–(2.24), (2.33) *that belongs to the space* $C^{(1,2)}\left(\overline{\prod}\right)$ *and continuously depends on* $f(x,t)$ *and* φ_q, $q = 1, 2$.

Theorem: *Suppose that condition (2.36) is satisfied and each of conditions (2.50) is satisfied. Then if* $\varphi_q \in B_{\delta_1}$, $q = 1, 2$, $f \in C\left(B_{\delta_2}, [0, T]\right)$, *with* $\delta_1 > \alpha_1$ *and* $\delta_2 > \alpha_2$, *for almost all (with respect to the Lebesgue measure in* **R**) *values of* A^i, *there exists a unique solution of problem* (2.22)–(2.24), (2.33) *that belongs to the space* $C^{(1,2)}\left(\overline{\prod}\right)$ *and continuously depends on* $f(x,t)$ *and* φ_q, $q = 1, 2$.

▲ If the conditions of these two theorems are satisfied then according to the previous theorem, there exist $\alpha \geq 0$ and $C > 0$ so that the following estimate holds for all (except finitely many) integer $n \in \mathbf{N}$:

$$|1 - \mu \exp(\lambda_n A + B)| \geq C\lambda_n^{-\alpha}. \tag{2.51}$$

Under the conditions of the above theorems, by using estimates (2.26) and (2.27), the expressions for $Y_{n1}(t)$ and $Y_{n2}(t)$, and relations (2.38)–(2.41), (2.43), and (2.51), we have

$$\begin{aligned} \|u\|_{C^{(1,2)}\left(\overline{\prod}\right)} &\leq \textstyle\sum_{n=1}^{+\infty} \left|C_{n1}Y_{n1}(t) + C_{n2}Y_{n2}(t) + Y_n^*(t)\right| |X_n(x)| \\ &\leq C_1' \textstyle\sum_{n=1}^{+\infty} \lambda_n^{3\alpha-1} \left(|\varphi_{1n}| + |\varphi_{2n}|\right) \exp(\varepsilon_1 \lambda_n) \\ &\quad + C_2' \textstyle\sum_{n=1}^{+\infty} \lambda^{2+\alpha} \max_{t\in[0,\,T]} |f_n(t)| \exp(\varepsilon_2 \lambda_n), \end{aligned} \tag{2.52}$$

where C_1' and C_2' are two positive constants. Knowing that for every $p \geq 0$ and $\varepsilon > 0$, there exists a positive constant $C_3' = C(p, \varepsilon)$ and a real $h \geq 0$ so that $h^p \leq C_3' \exp(\varepsilon h)$, we can reduce (2.52) to the form

$$\begin{aligned} \|u\|_{C^{(1,2)}\left(\overline{\prod}\right)} &\leq C_4' \left\{ \sum_{n=1}^{+\infty} |\varphi_{1n}| \exp\left((\varepsilon_1 + \epsilon_1)\lambda_n\right) + \sum_{n=1}^{+\infty} |\varphi_{2n}| \exp\left((\varepsilon_1 + \epsilon_1)\lambda_n\right) \right\} \\ &\quad + C_5' \sum_{n=1}^{+\infty} \max_{t\in[0,\,T]} |f_n(t)| \exp\left((\varepsilon_2 + \epsilon_2)\lambda_n\right) \\ &= C_6' \left[\|\varphi_1\|_{B_{\varepsilon_1+\epsilon_1}} + \|\varphi_2\|_{B_{\varepsilon_1+\epsilon_1}} \right] + C_7' \|f\|_{C\left(B_{\varepsilon_2+\epsilon_2},\,[0,\,T]\right)} < +\infty, \end{aligned}$$

where C_6' and C_7' are positive constants, $\epsilon_j = \delta_j - \varepsilon_j$, $j = 1, 2$. This last estimate completes the proof of the two theorems. ■

2.5 Review of the Well-Posedness of the Cauchy Problem for the NLS Equation

In the present section, we present a review of the well-posedness of the Cauchy problem for a one-dimensional NLS equation. More precisely, we review the well-posedness of the Cauchy problem

$$i\frac{\partial\psi}{\partial t} = -\frac{\partial^2\psi}{\partial x^2} + F(\psi, \psi^*, \partial\psi/\partial x, \partial\psi^*/\partial x) = 0, \quad (2.53a)$$

$$\psi(x, 0) = \psi_0(x), \ x \in \mathbf{R}, \quad (2.53b)$$

where ψ is a complex function of variable $(x, t) \in \Pi_\infty = \mathbf{R} \times [0, +\infty[$, nonlinearity F is sufficiently regular in $\mathbf{C}^4$. When $F \equiv 0$, Eq. (2.53a) becomes a linear Schrödinger equation and condition (2.5b) of the well-posedness of the Cauchy problem (2.53a), (2.53b) is satisfied. Henceforth, we focus ourselves on Eq. (2.53a) when the nonlinear term is present.

We denote by

$$H^s(\mathbf{R}) = \left\{ f \in S' : \int_{\mathbf{R}} \left(1 + |\xi|^2\right)^s \left|\widehat{f}(\xi)\right|^2 d\xi \right\} < 0$$

a Sobolev space of regularity $s \in \mathbf{R}$, where $\widehat{f}(\xi) = \int_{\mathbf{R}} \exp[-ix\xi] f(x)dx$ is the Fourier transform of function f and S' being a space of tempered distributions in $\mathbf{R}$. $C^\infty(\mathbf{R})$ will denote the space of functions with continuous derivatives of all orders in $\mathbf{R}$; we use $\|\cdot\|_{L^p}$ to denote the $L^p(\mathbf{R})$ norm with $p \geq 1$. we denote by $int\,(B)$ the interior of a given set B. In Eq. (2.53b) $\psi_0(x) \in H^s(\mathbf{R})$. If the solutions found of problem (2.53a), (2.53b) are guaranteed for only some time interval $[-T, T]$, they are called *local*, while if they exist for all times, the are called *global*. Other natural questions to ask are whether the solutions are unique and whether small changes in the initial datum produce small changes in the solution. Together, such results are known as well-posedness theories for an equation.

Definitions: The Cauchy problem (2.53a), (2.53b) is said to be (locally) **well-posed** on $H^s(\mathbf{R})$ if, for every bounded subset $B \subset H^s(\mathbf{R})$, there exist a time $T > 0$ and a closed subspace B_T continuously contained in $C([-T, T], H^s(\mathbf{R}))$ such that
(i) for every $\psi_0 \in B$, Eq. (2.53a) admits a unique solution ψ in B_T satisfying the initial condition (2.53b);
(ii) if $\psi_0 \in H^\sigma(\mathbf{R})$ for $\sigma > s$, then $\psi \in C([-T, T], H^\sigma(\mathbf{R}))$;
(iii) the map $\begin{array}{c} B \to B_T \\ \psi_0 \longmapsto \psi \end{array}$ is continuous.
If conditions (i)–(iii) are satisfied for every time $T > 0$, problem (2.53a), (2.53b) is said to be globally well-posed on $H^s(\mathbf{R})$.
If problem (2.53a), (2.53b) is locally (globally) well-posed on $H^s(\mathbf{R})$ and the map shown in (iii) is uniformly continuous in B, we say that this problem is locally (globally) **uniformly well-posed** on $H^s(\mathbf{R})$.
If problem (2.53a), (2.53b) is locally (globally) well-posed on $H^s(\mathbf{R})$ and the map shown in (iii) is smooth in $int\,(B)$, we will say that (2.53a), (2.53b) is locally (globally) **regularly well-posed** in $H^s(\mathbf{R})$.

The problem of the well-posedness of problem (2.53a), (2.53b) has been investigated by many authors. It appears that the well-posedness or the ill-posedness of this problem depends on the regularity s and the properties of the nonlinear functional term F. It is proved in Ref. [11] that if $F(\psi, \psi^*, \partial\psi/\partial x, \partial\psi^*/\partial x) = F(\psi, \psi^*)$

derives from a potential (i.e., $F(\psi, \psi^*) = \partial V/\partial z^*$), is gauge-invariant, and satisfies conditions $F(0, 0) = 0$ and $\left|D^k F(z)\right| \le C_k \left(1 + |z|^2\right)^{1+\alpha-k}$, $k = 0, 1, 2, \ldots$, then problem (2.53a), (2.53b) will be regularly well-posed on $H^s(\mathbf{R})$ for $s > 1/2$. In the same work, it is proved that in the situation when $F(0, 0) = 0$ and $D^k F(0, 0) \ne 0$ for some $k \ge 2$, then if the Cauchy problem (2.53a), (2.53b) is regularly well-posed $H^s(\mathbf{R})$, then $s \ge 1/2 - 2/(k-1)$. This last situation means that if $F(0, 0) = 0$ and F is real analytic and is not polynomial, then the Cauchy (2.53a), (2.53b) is not regularly well-posed on $H^s(\mathbf{R})$ for $s < 1/2$.

Investigating the Cauchy problem (2.53a), (2.53b) under the condition that $F(\psi, \psi^*, \partial\psi/\partial x, \partial\psi^*/\partial x) = F(\psi, \psi^*)$ satisfies condition $F(0, 0) = 0$, Kenig et al. [22] applied the energy method to establish the local well-posedness of the problem on $H^s(\mathbf{R})$ for $s > 1/2$. They also used the energy method to prove the local well-posedness of the problem on $H^s(\mathbf{R})$ for $s > 3/2$ when the nonlinearity term $F(\psi, \psi^*, \partial\psi/\partial x, \partial\psi^*/\partial x)$ contains the derivatives.

In [21], Kenig et al. made a great progress on the well-posedness of problem (2.53a), (2.53b) and proved its local well-posedness for small data in the Sobolev space $H^s(\mathbf{R})$ and in its weighted version by pushing forward the linear estimates associated with the Schrödinger group $\left\{\exp\left[it\frac{\partial}{\partial x}\right]\right\}$ and by introducing suitable function spaces where these estimates act naturally. The smallness assumption on the size of the data was removed by Hayashi and Ozawa [14] by using a change of variable to obtain an equivalent system with a nonlinear term independent of the derivative $\partial\psi/\partial x$, where the new system could be treated by the standard energy method. Using an invertible classical pseudodifferential operator of order zero, Chihara [6] was able to remove the size restriction on the data.

Based on the nonlinear term F of Eq. (2.53a), Bienaimé and Boulkhemair [2] formulated the following theorem on the well-posedness of problem (2.53a), (2.53b).

Theorem: If function F vanishes to the third order at $(0, 0, 0, 0)$, i.e., F and its partial derivatives up to the second order vanishes at $(0, 0, 0, 0)$, then $\forall s > 5/2$, $\forall \psi_0 \in H^s(\mathbf{R})$, $\exists T > 0$ such that Eq. (2.53a) admits a unique solution ψ which is defined on the time interval $[0, T]$ and satisfying condition $\psi \in B_T = C([0, T], H^s(\mathbf{R}))$. Moreover, for every bounded subset $B \in H^s(\mathbf{R})$, $\exists T > 0 : \forall\psi_0 \in B$, the associated solution ψ of problem (2.53a), (2.53b) exists on the interval $[0, T]$ and the map
$$\begin{aligned} B &\to B_T = C([0, T], H^s(\mathbf{R})) \\ \psi_0 &\longmapsto \psi \end{aligned}$$
is continuous.

Takaoka [31] investigated the global well-posedness of problem (2.53a), (2.53b) with data in $H^s(\mathbf{R})$ in the case when Eq. (2.53a) models the propagation of circularly polarized Alfvèn waves in magnetized plasma with a constant external magnetic field, i.e., when the nonlinear term is of the form $F(\psi, \psi^*, \partial\psi/\partial x, \partial\psi^*/\partial x) = i\lambda\frac{\partial}{\partial x}|\psi|^2\psi$, λ being a real parameter. For this form of the nonlinear term, Cauchy problem (2.53a), (2.53b) is L^2 critical and completely integrable. Its H^1—solution obeys the mass, energy, and momentum conservation laws, and its local well-posedness in the energy space $H^1(\mathbf{R})$ was proved (see, for example, Ref. [13]). With the use of the mass and energy conservation law, and the gauge transformation, Ozawa [26] proved that problem (2.53a), (2.53b) with the above nonlinear term is globally well-posed in the energy space $H^1(\mathbf{R})$ if the data ψ_0 satisfies the condition $\|\psi\|_{L^2} \le \sqrt{2\pi/|\lambda|}$. Under some restrictions on the data $\psi_0 \in H^s(\mathbf{R})$, Takaoka proved the local well-posedness and the global well-posedness of problem (2.53a), (2.53b) when $\frac{32}{33} < s < 1$ and $s > \frac{32}{33}$, respectively. With these conditions on s, one may expect the local solution to be global by making the iteration process of local well-posedness. But, as showed by Takaoka, iteration method can not by itself yield the global well-posedness indeed, the proof of global well-posedness relies on providing the a priori estimate of solution, besides the proof of the local well-posedness [5, 22].

References

1. Bellman, R., Cooke, K.L.: Differential-Difference Equations. Academic, New York (1963)
2. Bienaimé, P.-Y., Boulkhemair, A.: Well-posedness and smoothing effect for generalized nonlinear Schrödinger equations (2017). https://hal.archives-ouvertes.fr/hal-01515658v2
3. Borok, V.M., Fardigola, L.V.: Nonlocal boundary-value problems in a layer. Mat. Zametki **48**(1), 20 (1990)
4. Borok, V.M., Kengne, E.: Classification of nonlocal boundary-value problems on a narrow strip. Ukr. Math. J. **46**(4), 352 (1994)
5. Bourgain, J.: Refinement of Strichartz inequality and applications to 2D NLS with critical nonlinearity. Int. Math. Res. Notices **5**, 253 (1998)
6. Chihara, H.: Local existence for semilinear Schrödinger equations. Math. Jpn. **42**, 35 (1995)
7. Dezin, A.A.: Simplest solvable extensions for ultrahyperbolic and pseudoparabolic operators. Dokl. Akad. Nauk SSSR **148**(5), 1013 (1963)
8. Dezin, A.A.: General Questions of the Theory of Boundary Value-Problems. Nauka, Moscow (1980)
9. Fardigola, L.V.: A well-posedness criterion in a layer for a boundary-value problem with an integral condition. Ukr. Mat. Zh. **42**(11), 1546 (1990)
10. Gantmakher, F.R.: Theory of Matrices. Nauka, Moscow (1988). [in Russian]
11. Gérard, P.: Nonlinear Schrödinger equations in inhomogeneous media: wellposedness and illposedness of the Cauchy problem. In: Proceedings of the International Congress of Mathematicians. Madrid, Spain (2006)
12. Hadamard, J.: Le probleme de Cauchy et les equations aux derivees partielles lineaires hyperboliques. Hermann, Paris (1932)
13. Hayashi, N.: The initial value problem for the derivative nonlinear Schrödinger equation in the energy space. Nonlinear Anal. **20**(7), 833 (1993)
14. Hayashi, N., Ozawa, T.: Remarks on nonlinear Schrödinger equations in one space dimension. Differ. Integr. Equ. **2**, 453 (1994)
15. Ilkiv, V.S.: A nonlocal boundary value-problem for normal anisotropic systems of partial differential equations with constant coefficients. Visn. L'viv. Univ. Ser. Mech.-Math. **54**, 84 (1999)
16. Kalenyuk, P.I., Nytrebych, Z.M.: Construction of solutions of certain boundary-value problems for linear partial differential equations admitting separation of variables. In: Boundary Value-Problems with Various Degenerations and Singularities, pp. 62–71. Ruta, Chernivtsi (1990)
17. Kengne, E.: Asymptotically well-posed boundary value-problems. Ukr. Math. J. **56**(2), 169 (2004)
18. Kengne, E.: On the well-posedness of a two-point boundary-value problem for a system with pseudodifferential operators. Ukr. Math. J. **57**(8), 1334 (2005)
19. Kengne, E.: Nonlocal boundary value-problems for partial differential equations with variable coefficients. Afr. Diaspora J. Math. **5**(1), 11 (2006)

20. Kengne, E., Pelap, F.B.: Regularity of two-point boundary value problems. Afr. Mat. **3**(12), 61 (2001)
21. Kenig, C.E., Ponce, G., Vega, L.: Small solutions to nonlinear Schrödinger equations. Ann. Inst. H. Poincaré Anal. Non Linéaire **10**, 255 (1993)
22. Kenig, C.E., Ponce, G., Vega, L.: The cauchy problem for the quasilinear Schrödinger equation. Invent. Math. **158**, 343 (2004)
23. Kiguradze, T.I.: On solutions of quasilinear hyperbolic systems bounded and periodic in a strip. Differentsial'nye Uravneniya **30**(10), 1760 (1994)
24. Mitropolskii, Y.A., Urmancheva, L.B.: On a two-point problem for a system of hyperbolic equations. Ukr. Math. J. **42**(2), 1657 (1990)
25. Naimark, M.A.: Linear Differential Operators. Nauka, Moscow (1969)
26. Ozawa, T.: On the nonlinear Schrödinger equations of derivative type. Indiana Univ. Math. J. **45**(1), 137 (1996)
27. Petrovskii, I.G.: Lectures on the Theory of Ordinary Differential Equations. Nauka, Moscow (1970). [in Russian]
28. Petrovskii, I.G.: On the Cauchy problem for a system of linear partial differential equations in a region of nonanalytic functions. Bull. Mosk. Univ. Sect. A **1**(7), 1 (1938)
29. Ptashnik, B.I.: Ill-Posed Boundary-Value Problems for Partial Differential Equations. Naukova Dumka, Kiev (1984)
30. Romanko, V.K.: On systems of operator differential equations. Differentsial'nye Uravneniya **23**(9), 1574 (1987)
31. Takaoka, Hideo: Global well-posedness for Schrödinger equations with derivative in a nonlinear term and data in low-order Sobolev spaces. Electron. J. Differ. Equ. **42**, 1 (2001)
32. Takhtajan, L.A., Faddeev, L.D.: Hamiltonian Methods in the Theory of Solitons. Nauka, Moscow (1986)

Chapter 3
Nonlinear Schrödinger Equations of Single Transmission Lines

Abstract The present chapter deals with the investigation of the dynamics of modulated waves propagating through a modified single nonlinear transmission network in both cases when the second-neighbor interactions are neglected and when the second-neighbor interactions are taken into account. Using the reductive perturbation approach in the semidiscrete approximation, we show that the dynamics of modulated wave in the two systems are governed by equations of Schrödinger type. With the help of these equations, we investigate quantitatively and qualitatively the properties of modulated waves including bright solitons, dark solitons, kink, and anti-kink solitonic pulses in the networks under consideration.

3.1 Introduction

High-impedance transmission lines, periodically loaded with shunt-connected varactor diodes, are structures that exhibit nonlinearity and dispersion [7]. Because of their periodicity, these circuits, known as nonlinear transmission lines (NLTL), also exhibit passband–stopband characteristics. The transition between these two regions, which occurs at the Bragg frequency (defined as the cutoff of the first passband), is a key parameter in NLTL design [36]. Since a pioneering work of Hirota and Suzuki [8] on a simulation line of the integrable Toda lattice, many theoretical and numerical studies have been carried out on the effective consideration of dynamics of modulated waves in nonlinear transmission lines [3, 12, 13, 15, 17, 18, 22, 23, 34, 38, 41, 43]. A NLTL is a system of repeating inductors and capacitors, where the inductors, the capacitors, or both are nonlinear in their response to current and voltage, respectively. They are very convenient tools to study wave propagation in nonlinear dispersive media and provide a useful way to check how the nonlinear excitations behave inside the nonlinear medium and to model the exotic properties of new systems. In electronics, for example, they serve as nonlinear dispersive media where electrical solitons can propagate in the form of voltage waves [29]. Solitons form in a transmission line by feeding a signal into a transmission line using voltage-dependent capacitors elements. The semiconductor devices such as Schottky diodes and heterostructure barrier varactors can also be incorporated into a soliton generating NLTL; these devices are used to generate oscillations with frequencies up to

W.-M. Liu and E. Kengne, *Schrödinger Equations in Nonlinear Systems*,
https://doi.org/10.1007/978-981-13-6581-2_3

the terahertz range and pulses with subpicosecond durations [30, 35]. In the linear regime, the NLTLs can be used as phase shifters in phased antenna arrays, where time delay can be controlled by means of a DC bias applied to Schottky diodes acting as variable reactance [45, 48].

Using accumulation mode metal–oxide–semiconductor varactors and the gradual scaling lines, Afshari and Hajimiri [3] have introduced and analyzed pulse narrowing and edge sharpening passive NLTLs, showing simultaneous edge sharpening for both rising and falling edges in silicon. Their experimental results show considerable improvement in the rise and fall times of the pulses. These NLTLs can have applications in ultra-wideband systems, broadband signal generations, and high-speed serial communications [2, 4]. Also, the NLTLs have been used for the experimental study of nonlinearity; they allows investigations of new exotic designs [20, 21]. Transient analysis of coupled networks with nonlinear loads is important for the study and design of high-speed electronic circuits. Owing to the increasing frequencies and bit rates, one has to take the transmission line effects on the interconnections, such as the signal-delay, reflexion, attenuation, and cross talk, into account. The performance of high-speed systems is mainly determined and limited by the ability to transmit signals undisturbed, undistorted, and with the desired speed; the interconnections are quite often terminated with nonlinear loads such as diodes or transistors [40].

Nonlinear transmission lines, like every nonlinear system, can exhibit an instability that leads to a self-induced modulation of input plane wave with the subsequence generation of localized pulses [17, 44]. Known as a Benjamin–Feir modulational instability [19], this phenomenon is responsible for many physically interesting effects such as the generation and propagation of solitonic waves, the domain formation, and quantum phase transition [6, 9, 12, 18, 19, 37, 42]. In nonlinear dispersive systems, the MI indicates that due to the interplay between nonlinearity and the dispersive effects, a small perturbation on the envelope of a Stokes wave may induce an exponential growth of the wave amplitude, resulting in the carrier-wave breakup into a train of localized pulses. In homogeneous nonlinear systems, MI may be considered as the leading mechanism for energy localization as well as the formation of traveling intrinsic localized modes [31, 32]. The corresponding mathematical model is the Korteweg–de Vries (KdV) equation, the nonlinear Schrödinger (NLS) equation, or the complex Ginzburg–Landau (CGL) equation with periodically varying dispersion and nonlinear coefficients. Scott [34] showed that the KdV equation describes weakly nonlinear waves in a nonlinear LC transmission line containing a finite number of cells, which consist of a linear inductor in the series branch and a nonlinear capacitor in the shunt branch. If the nonlinearity is moved from the capacitor parallel to the shunt branch of the line to a capacitor parallel to the series branch, the NLS equation is obtained instead [24, 25].

Using the reductive perturbation approach in the semidiscrete approximation, several authors have shown that the dynamics of modulated waves in NLTLs can be described, to a good approximation, with the well-known NLS equations [15, 17, 21–23]. Some years ago, we presented a model for wave propagation on a discrete NLTL based on the cubic–quintic derivative NLS equation, derived in the small amplitude and long wavelength limit using the standard reductive perturbation technique and

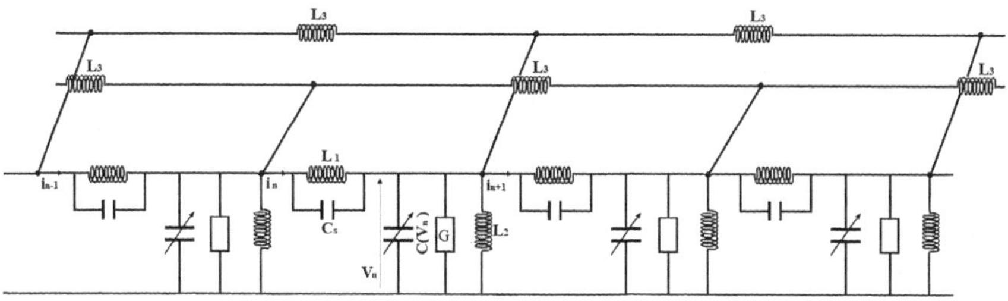

Fig. 3.1 Schematic representation of parts of a dissipative 1D modified discrete Noguchi electrical transmission line with second-neighbor interactions

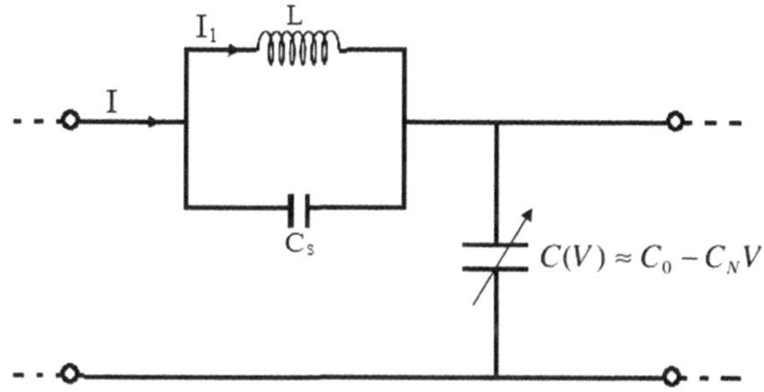

Fig. 3.2 Schematic representation of one elementary cell of a nonlinear dispersive transmission line

complex expansion on the governing nonlinear equations [15]. In our recent works on the NLTLs, we showed that the dynamics of modulated wave of the network may be governed by either an NLS equation with linear external potential [12] or a generalized Chen–Lee–Liu equation [1], which appears as a cubic NLS equation with an additional cubic derivative term.

The present chapter deals with the dynamics of modulated waves governed by NLS equations and propagating through the two models of NLTLs as shown in Figs. 3.1 and 3.2. Figure 3.1 presents a modified version of the Noguchi electrical transmission line, built study experimentally the propagation of solitons [15, 26]. This network contains a finite number of cells N and each cell contains a linear inductor of inductance L_1 in parallel with a capacitance C_s (which may be either linear or nonlinear) in the series branch and a linear inductor of inductance L_2 in parallel with a nonlinear capacitor C (throughout this chapter, we work with a constant C_s). This capacitor consists of a reverse-biased diode with differential capacitance function of the voltage V_n across the nth capacitor. It is biased by a constant voltage V_b and depends on the voltage V_n, for low voltage, at cell n. The presence of a linear inductor of inductance L_3 brings the contribution of the second-neighbor interactions. In order to take into account the dissipation of the network, a conductance G is connected in parallel with inductor L_2; the conductance G accounts for the dissipation of the inductor L_2 in addition with the loss of the nonlinear capacitor C. The dissipation in the inductor L_1 is neglected. The model shown in Fig. 3.2 [10, 11, 17] is a discrete nonlinear mono-inductance LC transmission line.

3.2 Cubic NLS Equation for the Network of Fig. 3.1

In this section, we investigate, through an NLS equation with linear external potential, the dynamics of modulated waves propagating through the NLTL of Fig. 3.1 for a linear capacitor C_s.

3.2.1 *Model Description and Basic Equations*

The model shown in Fig. 3.1 is a one-dimensional dissipative modified Noguchi electrical transmission line with second-neighbor interactions. This line contains a finite number of N identical cells. Each contains a linear inductor of inductance L_1 in parallel with a linear capacitance C_S in the series branch and a linear inductor of inductance L_2 in parallel with a nonlinear capacitor $C(V_n)$. The introduction of a third linear inductor of inductance L_3 brings the contribution of the second-neighbor interactions. In order to take into account the dissipations of the network, a conductance G is connected in parallel with inductor L_2. The conductance G accounts for the dissipation of the inductor L_2 in addition with the loss of the nonlinear capacitor $C(V_n)$. The dissipation of the inductors L_1 and L_3 is neglected. This line of Fig. 3.1 is an assemblage of band-pass filter in which nonlinearity is introduced by a varicap diode admitting that the capacitance varies with the applied voltage. Therefore, the nonlinear capacitor C of the network is voltage dependent and consists of a reverse-biased diode with differential capacitance function of the voltage V_n across the nth capacitor. It is biased by a constant voltage V_b as $C(V_b + V_n) = \frac{dQ_n}{dt} \approx C_0\left(1 - \alpha V_n + \beta V_n^2\right)$, with the characteristic capacitance $C_0 = C(V_b)$; here, α and β are the nonlinear positive coefficients of the electrical charge Q_n stored in the nth capacitor [13, 27]. Using the Kirchhoff's laws, we obtain the following differential system governing the wave propagation in the network:

$$\frac{d^2V_n}{dt^2} + \left(u_0^2 + \gamma\frac{d^2}{dt^2}\right)(2V_n - V_{n+1} - V_{n-1}) + \Omega_0^2\,(2V_n - V_{n+2} - V_{n-2})$$
$$= -\frac{G}{C_0}\frac{dV_n}{dt} - \omega_0^2 V_n + \frac{d^2}{dt^2}\left(\alpha V_n^2 - \beta V_n^3\right),\ n = 1,\ 2,\ ...,\ N, \tag{3.1}$$

where $u_0^2 = \frac{1}{L_1C_0}$ and $\omega_0^2 = \frac{1}{L_2C_0}$ are the characteristic frequencies of the network, $\Omega_0^2 = \frac{1}{L_3C_0}$, and $\gamma = C_S/C_0$. Here, Ω_0^2 is the coupling frequency representing the contribution of the second-neighbor interactions. The situation when the second-neighbor interactions are not taken into account ($\Omega_0^2 = 0$) corresponds to $L_3 = +\infty$. When using the differential system (3.1) to experience the transmission of solitary signals in the network of Fig. 3.1, one excites the left extremity of the network and chooses the total number N of cells of the network so as not to encounter the wave reflection at the end of the network. In the numerical simulations, we will use the following typical line parameters [13]:

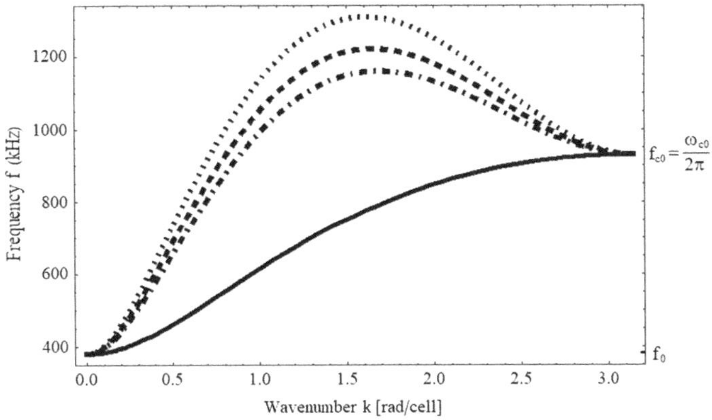

Fig. 3.3 Linear dispersive curve showing the frequency $f = \omega/2\pi$ as function of the wave number k for different values of the coupling linear inductor L_3 for the network parameters given in Eq. (3.2). The solid line is obtained when the second-neighbor interactions in the network are neglected, i.e., when $L_3 = +\infty$; dotted line, dashed line, and dot-dashed line are obtained with $L_3 = 0.4L_2$, $L_3 = 0.5L_2$, and $L_3 = 0.6L_2$, respectively. Reprint from Ref. [14], Copyright 2018, with permission from American Physical Society

$$\begin{aligned} L_1 &= 220\,\mu\text{H},\ L_2 = 470\,\mu\text{H},\ V_b = 2\,\text{V},\ C_0 = C(V_b) = 370\,\text{pF}, \\ C_S &= 56\,\text{pF}, \alpha = 0.21\,\text{V}^{-1},\ \beta = 0.0197\text{V}^{-2}. \end{aligned} \tag{3.2}$$

To study the linear properties of our model, we assume a sinusoidal wave of the form

$$V_n(t) = V \exp\left[i\,(kn - \omega t)\right] + c.c., \tag{3.3}$$

where V is a constant amplitude, k and ω are the wave number and the angular frequency, respectively, and $c.c.$ stands for complex conjugate. Inserting Eq. (3.2) into Eq. (3.1) yields the following linear dispersion relation:

$$\omega^2 = \frac{\omega_0^2 + 4u_0^2 \sin^2 \frac{k}{2}}{1 + 4\gamma \sin^2 \frac{k}{2}} + \frac{4\Omega_0^2 \sin^2 k}{1 + 4\gamma \sin^2 \frac{k}{2}} \tag{3.4}$$

in which the wave number k is taken in the first Brillouin zone ($0 \leq k \leq \pi$). The second term in the right-hand side of Eq. (3.4) is the second-neighbors' contribution to the angular frequency. It is seen from the linear dispersion relation (3.4) that the introduction of the second-neighbors in the network increases the angular frequency of the propagating wave, meaning that the second-neighbors add (increase) the number of frequencies that the network can support and this allows to transmit several information in the network. Moreover, the angular frequency ω increases (decreases) as inductance L_3 decreases (increases). These behaviors can be seen from plots of Fig. 3.3 showing the dependence of the frequency $f = \omega/(2\pi)$ as function of wave number k for different values coupling linear inductor L_3 and represent a band-pass filter. It is seen from these plots that, independently of the choice of the coupling linear inductor L_3, the corresponding linear spectrum has a gap $f_0 = \omega_0/2\pi$ which represents the lower cutoff frequency introduced by the parallel inductance L_2 and it is limited by the cutoff frequency $f_{\max} = \omega(k_c)/2\pi \geq \omega_{c0} = \omega(\pi)/2\pi = \frac{1}{2\pi}\sqrt{\frac{\omega_0^2+4u_0^2}{1+4\gamma}}$, where either $k_c = \pi$ or $k_c < \pi$, depending on the value of the coupling linear inductor L_3. Plots of Fig. 3.3 show that the cutoff frequency $f_{\max}$ decreases with the growth of L_3 which means that the coupling linear inductor L_3 contributes to reduce the network effects on the wave during its propagation. We can then conclude that the frequency band-pass of the modified Noguchi's filter decreases with values of L_3 and it becomes more selective in terms of frequency. It is important to notice the following important feature: When the second-neighbor interactions are taken into account, two different wave numbers may be associated to the same frequency. Therefore, when the second-neighbor interactions in the network are taken into account, two different waves may simultaneously propagate at the same frequency.

3.2.2 Nonlinear Schrödinger Equations of the Dynamics

To derive the NLS equations of the dynamics of the NLTL of Fig. 3.1, we focus on waves with a slowly varying envelope in time and space with regard to a given carrier wave with angular frequency $\omega = \omega_p = 2\pi f_p$ and wave number $k = k_p$. In order to apply the reductive perturbation method [37, 39] allowing us to treat properly the carrier with its discrete character and to describe the envelope in the continuum approximation, we introduce the slow envelope variables $x = \varepsilon\left(n - v_g t\right)$ and $\tau = \varepsilon^2 t$, where n is the cell number, ε is a small parameter that measures the smallness of the modulation frequency and the amplitude of the input waves, and $v_g = \frac{d\omega}{dk} = \frac{\sin k}{\omega}\frac{u_0^2-\gamma\omega^2+4\Omega_0^2\cos k}{1+4\gamma\sin^2\frac{k}{2}}$ is the group velocity of the linear wave packets. Following the idea of Ref. [21], the solution of (3.1) is assumed to have the following general form:

$$V_n(t) = \varepsilon\psi(x,\tau)\exp[i\theta] + \varepsilon^2\left[\psi_0(x,\tau) + \psi_1(x,\tau)\exp[2i\theta]\right] + c.c., \qquad (3.5)$$

where $\theta = kn - \omega t$ is the rapidly varying phase. The DC term ψ_0 and the second-harmonic term ψ_1 are added to the fundamental term ψ in order to take into account the asymmetry of the charge–voltage relation given by Eq. (3.1). We order the damping coefficient so that the effect of damping and nonlinearity appear in the same perturbation equation; in other words, we set $G/C_0 = 2u_0\varepsilon^2\sigma$. Substituting Eq. (3.5) into Eq. (3.1) yields a series of inhomogeneous equations at different orders of $(\varepsilon, \exp[i\theta])$. The terms of proportional to $(\varepsilon, \exp[i\theta])$ lead to the linear dispersion relation (3.4). At order $\left(\varepsilon^2, \exp[i\theta]\right)$, we obtain equation $\partial\psi/\partial\tau + \frac{d\omega}{dk}\partial\psi/\partial x = 0$, showing that $\psi(x,\tau)$ represents a traveling wave moving with group velocity v_g. Equations at orders $\left(\varepsilon^4, \exp[0i\theta]\right)$ and $\left(\varepsilon^2, \exp[2i\theta]\right)$ lead, respectively, to the DC term ψ_0 and the second-harmonic term ψ_1

$$\psi_0(x,\tau) = \frac{2\alpha v_g^2}{v_g^2 - u_0^2 - 2\Omega_0^2}\,|\psi|^2 + \lambda(\tau)x, \tag{3.6a}$$

$$\psi_1(x,\tau) = \frac{\alpha\omega^2}{\omega^2 + \left(4\gamma\omega^2 - u_0^2\right)\sin^2 k - \Omega_0^2 \sin^2 2k - \frac{1}{4}\omega_0^2}\psi^2, \tag{3.6b}$$

where $\lambda(\tau)$ is an arbitrary complex function of τ. With the use of Eqs. (3.6a) and (3.6b), we obtain at order $\left(\varepsilon^3, \exp[i\theta]\right)$ the following parameterized nonlinear Schrödinger equation with an external linear potential:

$$i\frac{\partial\psi}{\partial\tau} + P\frac{\partial^2\psi}{\partial x^2} + Q\,|\psi|^2\,\psi + \lambda(\tau)x\psi + i\Gamma\psi = 0, \tag{3.7}$$

where $P = P(k)$, $Q = Q(k)$ and $\Gamma = \Gamma(k)$ are given as follows:

$$\begin{cases} P = -\frac{v_g^2}{2\omega}\left(1 + 4\gamma\sin^2\frac{k}{2}\right) + \left(\frac{u_0^2}{2\omega} - \frac{\gamma\omega}{2}\right)\cos k - 2\gamma v_g \sin k + \frac{2\Omega_0^2}{\omega\left(1+4\gamma\sin^2\frac{k}{2}\right)}\cos k, \\ Q = \frac{\alpha^2\omega}{1+4\gamma\sin^2\frac{k}{2}}\left(\frac{3\beta}{2\alpha^2} - \frac{2v_g^2}{v_g^2-u_0^2-2\Omega_0^2} - \frac{\omega^2}{\omega^2+\left(4\gamma\omega^2-u_0^2\right)\sin^2 k-\Omega_0^2\sin^2 2k-\frac{1}{4}\omega_0^2}\right), \\ \Gamma = \frac{u_0\sigma}{1+4\gamma\sin^2\frac{k}{2}}. \end{cases} \tag{3.8}$$

Under condition $\Gamma\lambda \neq 0$, Eq. (3.7) with coefficients (3.8) is referred to as a dissipative NLS equation with an external linear potential; it has been derived in Ref. [12] for a modified Noguchi lossless electrical network when the second-neighbor interactions are neglected. In the context of Bose-Einstein condensates, function $\psi(x,\tau)$ represents the condensate wave function in which x is the coordinate normal to the surface of the condensate such that the bulk of the condensate exists in the region $x < 0$, and $\lambda(\tau)$ that arises from linearizing the harmonic potential near the surface represents the potential strength; in the context of nonlinear optics, $2P$ and Q is the group velocity dispersion and the nonlinear parameter, respectively.

3.2.3 *Effects of the Coupling Second-Neighbor Interactions on the MI Phenomena*

We proved in work [12] that for the NLS Eq. (3.7) with $\Gamma = 0$, the modulational instability of the plane wave and the possible bright/dark solitary wave solutions depend only on the sign of product PQ: The positivity of product PQ leads to the modulational instability of a plane wave and envelope solitary wave solutions, while its negativity leads to the modulational stability of a plane wave and dark solitary wave solutions. With the network parameters given in Eq. (3.3) and for different values of the coupling linear inductance L_3, we report in Fig. 3.4 the product $P(k)Q(k)$ (top panels) and the frequency $f(k)$ (bottom panels) showing regions of positive and negative dispersion, and regions of positive and negative nonlinearity. It is seen from Fig. 3.4 that for a given wave number k, the sign of each of

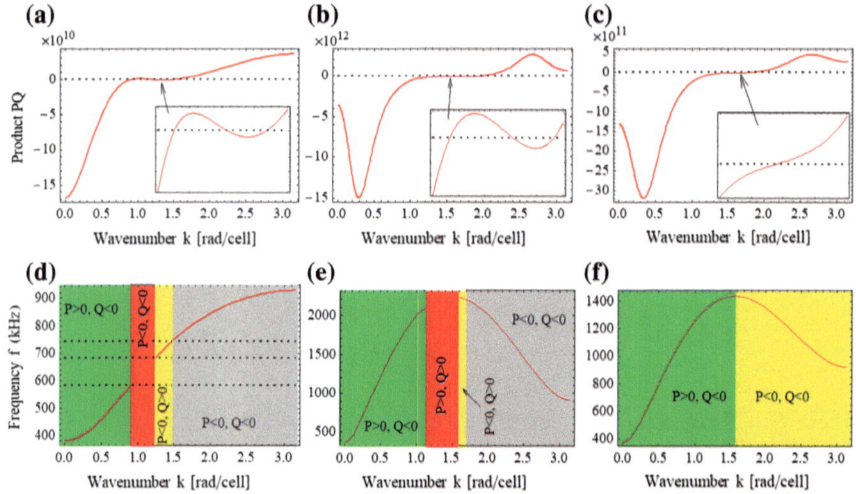

Fig. 3.4 Plots of product $P(k)Q(k)$ (top plots) and linear dispersive curve (bottom plots) showing the evolution of the frequency $f = \omega/2\pi$ for the network parameters (3.2) and different values of the coupling linear inductance L_3: **a**, **d**: $L_3 = +\infty$ (second-neighbor interactions in the network are neglected); **b**, **e**: $L_3 = 01.L_2$; and **c**, **f**: $L_3 = 0.3L_2$. Reprint from Ref. [14], Copyright 2018, with permission from American Physical Society

P and Q may change with the change in the value of coupling linear inductance L_3. These plots show that in the case where the second-neighbors are neglected ($L_3 = +\infty$) as well as in the case of small coupling linear inductance L_3 (Fig. 3.4d and e), there exist four domains (with respect with the wave number k) in which the model can exhibit a stability or an instability: $(P > 0, Q < 0)$, $(P < 0, Q < 0)$, $(P < 0, Q > 0)$, and $(P < 0, Q < 0)$ for plot (d) and $(P > 0, Q < 0)$, $(P > 0, Q > 0)$, $(P < 0, Q > 0)$, and $(P < 0, Q < 0)$ for plot (e); in the case of large coupling linear inductance L_3 (Fig. 3.4c and f), there exist only two domains (with respect with the wave number k) in which the model can exhibits either a stability or an instability, $(P > 0, Q < 0)$ and $(P < 0, \ Q < 0)$. Our investigations showed that for the set of the network parameters given in (3.3), only one of the three scenarios reported in Fig. 3.4 may occur for any value of the coupling linear inductance L_3 (Fig. 3.4d, e, or Fig. 3.4f). Figure 3.4d shows that when the second-neighbor interactions are neglected, every frequency associated with a wave number taken from the first Brillouin zone corresponds to the propagation of one and only one soliton (either envelope or hole soliton) in the network. Figure 3.4e and f inform us that in the presence of the second-neighbor interactions, both envelope and hole solitons may propagate with the same frequency (associated with different wave numbers k) in the network. Moreover, as it is well seen from Fig. 3.4e, two bright solitary wave may propagate in the network at the same frequency (see the red band) when the second-neighbor interactions are taken into account. It is important to notice that the situation presented in Fig. 3.4d corresponding to the case when the second-neighbor interactions are neglected may also occur when the second-neighbor interactions are taken into account. From the above analysis, we conclude that the introduction of second-neighbors in the modified Noguchi network shown in Fig. 3.1 may promote the simultaneous propagation of either two envelope solitons or both envelope and hole solitons with the same frequency.

3.2.4 Transmission of Soliton-Like Pulses Through the Lossless Network of Fig. 3.1 ($\Gamma = 0$): Analytical Studies

In this section, we investigate analytically the transmission of soliton-like pulses through the lossless NLTL shown in Fig. 3.1 when the dissipations in the network are neglected. We focus ourselves on the search and analysis of solitary wave solutions of Eq. (3.7), which will allow us to investigate analytically the transmission of soliton-like pulses through the network of Fig. 3.1. We start by finding exact solitary wave solutions of Eq. (3.7) under the condition $\Gamma = 0$. For simplicity, we restrict ourselves to the case where $\lambda(\tau) = \lambda$ is a real constant. To investigate the dynamics of solitary signal voltage in the network of Fig. 3.1, we employ the method used in Ref. [16].

Exact Solitary Wave Solutions

Following Kengne and Talla [16] and seeking a seed solution of equation (3.7) in the form $\psi_c(x, \tau) = A_0 \exp[i\varphi_c(x, \tau)]$, we obtain the following continuous-wave background solution:

$$\psi_c(x,\tau) = A_c \sqrt[6]{p^2 \frac{\lambda^2}{Q^2}\frac{P}{Q}} \exp\left[i\sqrt[3]{\lambda^2 P}\left(p^2 A_c^2 - \left(\frac{\lambda}{P}\right)^{\frac{1}{3}} \left\{ x + \frac{P\lambda}{3}\tau^2 \right\} \right)\tau \right], \tag{3.9}$$

where A_c is an arbitrary real constant, and $p^2 = 1$ if $PQ > 0$ and $p^2 = -1$ if $PQ < 0$.

Using the seed solution (3.9) and the method proposed in Ref. [12] leads to the following families of, respectively, envelope and hole solitary wave solutions of Eq. (3.7):

$$\psi(x,\tau) = \sqrt[6]{p^2 \frac{\lambda^2}{Q^2}\frac{P}{Q}} \exp[i\varphi_c] \tag{3.10}$$

$$\times \left[A_c + A_s \frac{(|\mu|^2 - 1)\sin\phi - 2\mu_I \sinh\vartheta}{(|\mu|^2 + 1)\cosh\vartheta + 2\mu_R\cos\phi} - i A_s \frac{2\mu_R \cosh\vartheta + (1 + |\mu|^2)\cos\phi}{(|\mu|^2 + 1)\cosh\vartheta + 2\mu_R \cos\phi} \right]$$

for $p^2 = 1$, and

$$\psi(x,\tau) = \sqrt[6]{p^2 \frac{\lambda^2}{Q^2}\frac{P}{Q}} \exp\left[i\varphi_c\right] \times \left[A_c \pm A_s \frac{\left(1-|\mu|^2\right)\sin\phi + 2\mu_I \sinh\vartheta}{\left(|\mu|^2-1\right)\sinh\vartheta + 2\mu_I\sin\phi} \pm i A_s \frac{2\mu_R\cosh\vartheta + \left(1+|\mu|^2\right)\cos\phi}{\left(|\mu|^2-1\right)\sinh\vartheta + 2\mu_I\sin\phi}\right], \tag{3.11}$$

for $p^2 = -1$, where

$$\varphi_c = \varphi_c(x,\tau) = \sqrt[3]{\lambda^2 P}\left(p^2 A_c^2 - \left(\frac{\lambda}{P}\right)^{\frac{1}{3}}\left\{x + \frac{P\lambda}{3}\tau^2\right\}\right)\tau,$$

$$\vartheta = \vartheta(x,\tau) = \sqrt{2}\left[\sqrt[3]{\frac{\lambda}{P}} M_R\left(x + \lambda P\tau^2\right) + \lambda\sqrt[3]{\frac{P}{\lambda}}\frac{2M_R\Lambda - M_I A_s}{\sqrt{2}}\tau\right] + \vartheta_0,$$

$$\phi = \phi(x,\tau) = -\sqrt{2}\left[\sqrt[3]{\frac{\lambda}{P}} M_I\left(x + \lambda P\tau^2\right) + \lambda\sqrt[3]{\frac{P}{\lambda}}\frac{2M_I\Lambda + M_R A_s}{\sqrt{2}}\tau\right] + \phi_0,$$

$$\mu = \frac{2\sqrt{2}pA_c}{M_R + \sqrt{2}A_s + i\left(M_I - 2\sqrt{2}\Lambda\right)} = \mu_R + i\mu_I,$$

$$M = \sqrt{\left(\frac{A_s}{2} - i\Lambda\right)^2 - A_c^2} = M_R + iM_I,$$

and ϑ_0, φ_0, A_s, and Λ being four arbitrary real constants. Here, the subscripts R and I denote, respectively, real and imaginary parts. Because A_c is an arbitrary real constant, we conclude that each of the families (3.10) and (3.11) of solitary wave solutions of Eq. (3.7) contains six free real parameters, namely $\lambda \neq 0$, A_c, ϑ_0, ϕ_0, A_s, and Λ. Parameters ϑ_0 and ϕ_0 have the trivial effect of shifting the solutions in the $\xi-$ and $T-$ coordinates. If necessary, we will, without loss of generality, take $\vartheta_0 = \phi_0 = 0$. Three of the above parameters, namely A_c, A_s, and Λ define the "sign" of M; M will be (i) a real number if $A_s\Lambda = 0$ and $\frac{A_s^2}{4} - \Lambda^2 - p^2A_c^2 > 0$, (ii) a pure imaginary number if $A_s\Lambda = 0$ and $\frac{A_s^2}{4} - \Lambda^2 - p^2A_c^2 < 0$, (iii) and complex if $A_s\Lambda \neq 0$.

Solutions (3.10) and (3.11) of Eq. (3.7) confirm the propagation of both bright and dark solitary waves in the modified Noguchi transmission network with second-neighbor interactions as shown in Fig. 3.1. As we will see below, free solution parameters A_c, A_s, and Λ play an important role in the type of solitary wave propagating in our network.

Analysis of Solitary Wave Solutions (3.10) **and** (3.11)

In this subsection, we study, in terms of parameters A_c, A_s, and Λ some properties of solutions (3.10) and (3.11). First of all, it is important to note that a necessary condition for dark solitary wave solutions (3.11) to be bounded is that $A_s \neq 0$. In what follows, we analyze some special cases of solutions (3.10) and (3.11).

Case $A_c A_s = 0$ When A_s vanishes, each of the solutions (3.10) and (3.11) turns into the CW solution $\psi_c(x, \tau)$ given in Eq. (3.9). On the other hand, when A_c vanishes (possible only for solution (3.10)), which means there is no CW background, we will have $M = \frac{A_s}{2} - i\Lambda$ and solution (3.10) turns into the bright single-solitary wave solution

$$\psi(x,\tau) = \frac{A_s\sqrt[6]{\frac{\lambda^2}{Q^2}\frac{P}{Q}}}{\cosh\left[A_s\left(\frac{\left(\frac{\lambda}{P}\right)^{\frac{1}{3}}}{\sqrt{2}}x + 2\sqrt[3]{\lambda^2 P}\Lambda\tau + \frac{(\lambda^4P^2)^{\frac{1}{3}}}{\sqrt{2}}\tau^2\right)\right]}\exp\left[i\phi_s(x,\tau)\right],$$

$$\phi_s(x,\tau) = 4\sqrt[3]{\lambda^2 P}\left(\frac{A_s^2}{8} - \Lambda^2\right)\tau - \sqrt{2}\Lambda\left(\frac{\lambda}{P}\right)^{\frac{1}{3}}\left(x + \lambda P\tau^2\right) + \phi_0. \qquad (3.12)$$

Therefore in general, the bright solitary wave solution (3.10) represents a bright solitary wave which propagates on a CW background, and when $A_c << A_s$, the background is small within the existence of a bright solitary wave. From the bright solitary wave solution (3.12) we can conclude that (i) if $\Lambda = 0$, $\phi_s(x, \tau)$ will have no x -dependence and the solution will be non-oscillatory; (ii) the amplitude and the width of bright solitary wave are proportional to $\sqrt[6]{\frac{\lambda^2}{Q^2}\frac{P}{Q}}$ and to $\sqrt{2}\left(\frac{P}{\lambda}\right)^{\frac{1}{3}}$, respectively. As it is seen from Fig. 3.5, Eq. (3.12) can be used to to describe the compression of a bright solitary waves when P/λ decreases with L_3; (iii) the centre (the trajectory) of bright solitary wave is $\xi(\tau) = -\left(2\sqrt{2}\sqrt[3]{\lambda P^2}\Lambda\tau + \lambda P\tau^2\right)$; the trajectory of the bright solitary wave is thus parabolic in time with an acceleration of $-2\lambda P$. With the use of network parameters given in Eq. (3.3), we show, for different values of L_3, the amplitude $|\psi(x, \tau)|$ of bright solitary waves at time $t = 5$. Here, we took $\vartheta_0 = \phi_0 = 0$.

Case $A_s A_c \neq 0$, $\Lambda = 0$, **and** $\left(\frac{A_s}{2}\right)^2 - A_c^2 < 0$. In this special case, $M = i\frac{\sqrt{4A_c^2 - A_s^2}}{2}$ is a pure imaginary number, leading to $\vartheta(x,\tau) = -\frac{A_s\lambda\sqrt{4A_c^2 - A_s^2}}{2}\sqrt[3]{\frac{P}{\lambda}}\tau + \vartheta_0$, and $\phi(x,\tau) = -\frac{\sqrt{4A_c^2 - A_s^2}}{\sqrt{2}}\sqrt[3]{\frac{\lambda}{P}}\left(x + \lambda P\tau^2\right) + \phi_0$. From the expressions of $\vartheta(x, \tau)$ and $\phi(x, \tau)$, we conclude that each of the solutions (3.10) and (3.11) in the special case where $A_s A_c \neq 0$, $\Lambda = 0$, and $\left(\frac{A_s}{2}\right)^2 - A_c^2 < 0$ is aperiodic in the temporal coordinate and periodic with the period $-2\pi\sqrt[3]{\frac{P}{\lambda}}\sqrt{\frac{2}{4A_c^2 - A_s^2}}$ in the spatial coordinate.

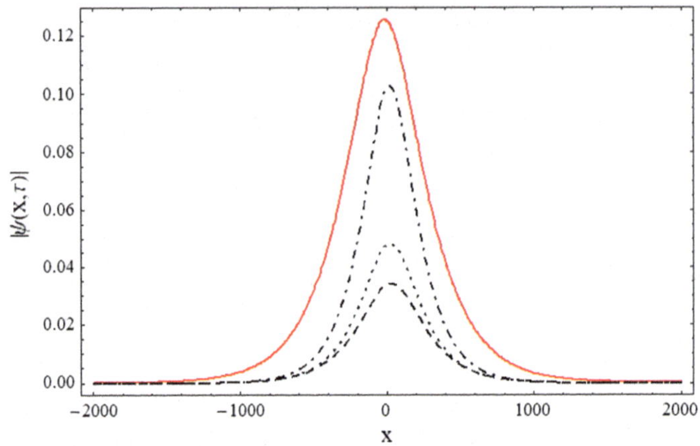

Fig. 3.5 Amplitude $|\psi(x,\tau)|$ of bright solitary waves propagating with frequency $f = f(1.55)$ at time $\tau = 5$ for the network parameters (3.2) and different values of the coupling linear inductance L_3: Solid line: $L_3 = +\infty$, corresponding to the case when second-neighbor interactions in the network are neglected; Dashed lines: $L_3 = 0.1L_2$; Dotted lines: $L_3 = 0.14L_2$; Dash-dotted lines: $L_3 = 0.18L_2$. Here, we use the equation and solution parameters $\lambda = 2 \times 10^{-6}$, $A_s = 40$, and $\Lambda = 10^{-2}$. These plots show that the amplitude and the width of the bright solitary wave increase when the parameter L_3 increases. Reprint from Ref. [14], Copyright 2018, with permission from American Physical Society

Case $A_s A_c \neq 0$, $\Lambda = 0$, **and** $\left(\frac{A_s}{2}\right)^2 - A_c^2 > 0$. In the special case $A_s A_c \neq 0$, $\Lambda = 0$, and $\left(\frac{A_s}{2}\right)^2 - A_c^2 > 0$, we have $M = \frac{\sqrt{A_s^2 - 4A_c^2}}{2}$, leading to $\vartheta(x,\tau) = \sqrt[3]{\frac{\lambda}{P}}\sqrt{\frac{A_s^2-4A_c^2}{2}}$ $\left(x + \lambda P \tau^2\right) + \vartheta_0$ and $\phi(x,\tau) = -\lambda A_s \sqrt[3]{\frac{P}{\lambda}}\sqrt{\frac{A_s^2}{4} - A_c^2}\tau + \phi_0$. Because $\phi(x,\tau)$ has no x -dependence, the solitary wave solutions (3.10) and (3.11) will be non-oscillatory in x-direction.

Case $A_s A_c \Lambda \neq 0$. In the special $A_s A_c \Lambda \neq 0$, we have $M_R M_I \neq 0$ and each of the functions $\psi(x,\tau)$ and $\vartheta(x,\tau)$ depends on both x and τ. Therefore, each of the solitary wave solutions (3.10) and (3.11) will be a combination of both non-oscillatory and oscillatory with respect to x but with a localized envelope/hole which has, in the case of solution (3.10), a density profile that is similar to the of gap solitons as shown in Fig. 3.1c of Ref. [5]. For the line parameters (3.3), we show in Fig. 3.6 the spatial profile at time $\tau = 0.01$ of bright solitary wave for different values of L_3 in the special case where $A_s A_c \Lambda \neq 0$; here, we have chosen $\vartheta_0 = \phi_0 = 0$.

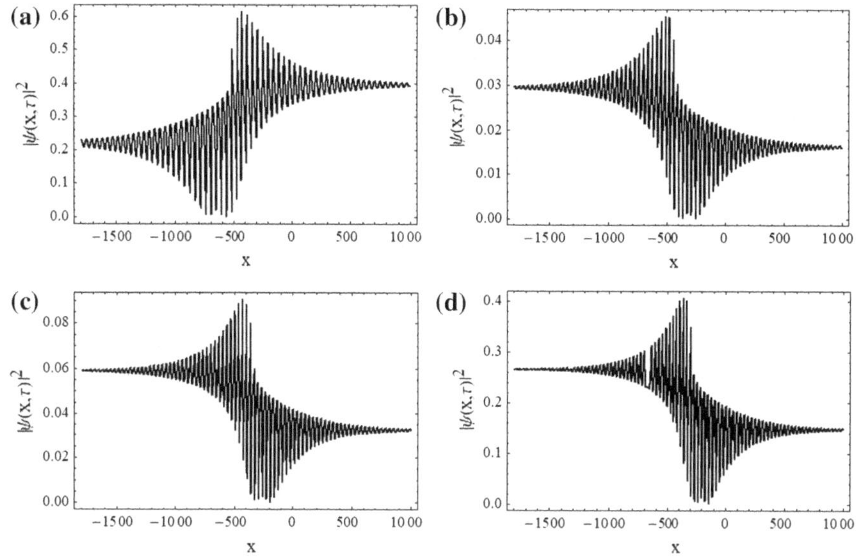

Fig. 3.6 Density $|\psi(x,\tau)|^2$ of bright solitary waves at time $\tau = 0.01$ for the network parameters (3.2) and different values of L_3. **a**: $L_3 = +\infty$ corresponding to the situation when second-neighbor interactions in the network are neglected; **b**: $L_3 = 0.1\ L_2$; **c**: $L_3 = 0.14\ L_2$. Different plots are obtained with $k = 1.55$, $A_c = 10$, $A_s = 15$, $\lambda = 0.002$, and $\Lambda = 0.1$. Plots of this figure show that once the second-neighbors are introduced in the network, the amplitude of the bright solitary wave increases when the second-neighbor parameter L_3 increases. Reprint from Ref. [14], Copyright 2018, with permission from American Physical Society

3.2.5 *Propagation of Solitonic Pulses Through the Lossless Network of Fig. 3.1*

Now, we come to the electrical envelope/hole soliton-like pulses propagating through the network of Fig. 3.1. For this purpose, we use the decomposition (3.5) of the signal voltage $V_n(t)$ and return to dimensioned parameters. The analytical technique used here is to inject the analytical solitary wave solutions (3.10) and (3.11) of Eq. (3.7), together with ψ_0 and ψ_1 form Eqs. (3.6a) and (3.6b) into Eq. (3.5). This operation leads to an expression for $V_n(t)$ expressed in terms of the network parameters, dimensionless parameter ε, time t and cell number n. In our study, we analyze many possible situations associated with the coupling second-neighbor parameter L_3, solution parameters Λ, A_c, A_s, as well as with parameter λ of Eq. (3.7). In our simulations, we will use the network parameters (3.3) and $\varepsilon = 0.01$. Without loss of generality, we limit ourselves to the transmission of bright soliton-like pulses associated with the bright solitary wave solution (3.10) in the network of Fig. 3.1. When applying the solitary wave solution (3.10), we will distinguish two main situations, namely, the situation where $A_s^2 - 4A_c^2 > 0$ and the situation where $A_s^2 - 4A_c^2 < 0$. We will focus ourselves on the transmission of solitary signals through a given cell n of our network.

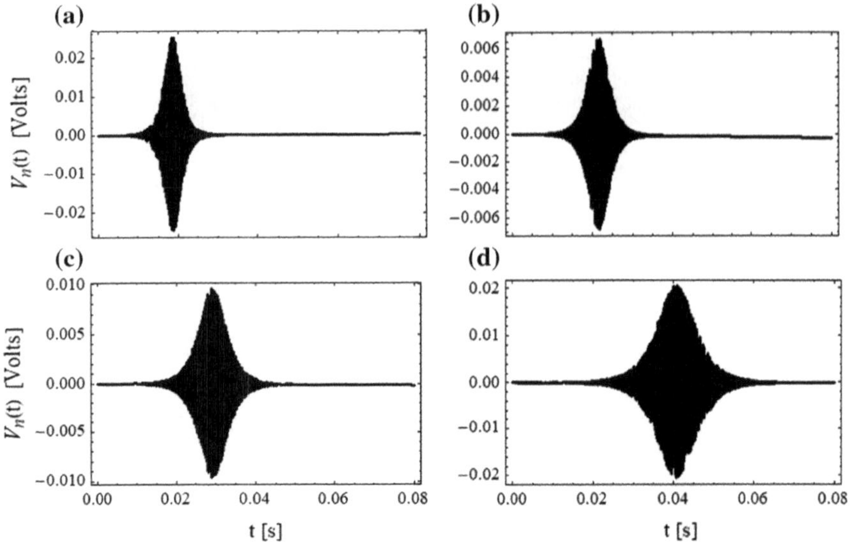

Fig. 3.7 Propagation through the network at the cell $n = 500$ of bright solitary signals without a CW background for the network parameters (3.2) and different values of the coupling second-neighbor parameter L_3 leading to different frequencies f_p associated with the wave number $k = 1.55$. **a**: $L_3 = +\infty$ (second-neighbor interactions are neglected) leading to $f_p = f(1.55) = 763.18$ kHz; **b**: $L_3 = 0.1L_2$ leading to $f_p = f(1.55) = 2252.72$ kHz; **c**: $L_3 = 0.14L_2$ leading to $f_p = f(1.55) = 1947.11$ kHz; **d**: $L_3 = 0.18L_2$ leading to $f_p = f(1.55) = 1754.47$ kHz. Other parameters used in these plots are $A_c = 0$, $A_s = 40$, $\vartheta_0 = 10$, $\phi_0 = 0$, $\lambda = -0.002$, and $\Lambda = 0.01$. Reprint from Ref. [14], Copyright 2018, with permission from American Physical Society

Special Case Where $A_s^2 - 4A_c^2 > 0$

We start our analytical investigations of the transmission of bright solitary signal through the network of Fig. 3.1 with the special situation parameters A_s and A_c appearing in solution (3.10) satisfy condition $A_s^2 - 4A_c^2 > 0$. Our numerical simulations will be carried out on both cases when $A_c = 0$ and when $A_c \neq 0$.

Figures 3.7 and 3.8 show the propagation through cell $n = 500$ of, respectively, bright solitary signals when there is no CW background ($A_c = 0$) and bright solitary signals embedded on a CW background ($A_c \neq 0$) for different values of the coupling second-neighbor parameter L_3. From plots of Figs. 3.7 and 3.8, it can be seen that, with the increasing of the coupling second-neighbor parameter L_3, bright solitary waves have an increase in the peak value and a broadening in its width. These plots also inform us that the wave velocity increases with parameter L_3.

Figures 3.9 and 3.10 show the propagation of bright solitary signals embedded on a CW background through cell $n = 500$ of the network with, respectively, negative A_s and A_s positive for $A_c = 4$ and different values of the coupling second-neighbor parameter L_3 with $\Lambda = 0$. Figure 3.9, associated with $\Lambda = 0$ and $A_s < 0$ show solitary signals with sharp edges, while Fig. 3.10, associated with $\Lambda = 0$ and $A_s > 0$

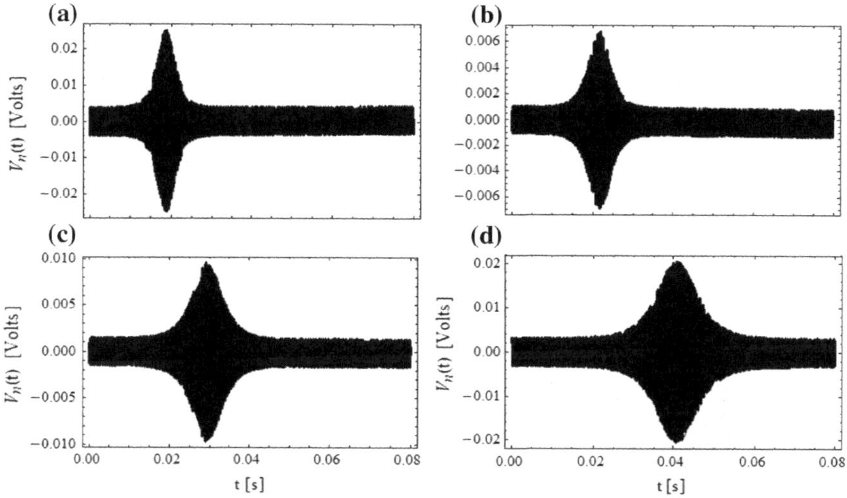

Fig. 3.8 Propagation through the network at the cell $n = 500$ of bright solitary signals embedded on a CW background for the network parameters (3.2) and different values of the coupling second-neighbor parameter L_3 leading to different frequencies f_p associated with the wave number $k = 1.55$. **a**: $L_3 = +\infty$ (second-neighbor interactions are neglected) leading to $f_p = f(1.55) = 763.18$ kHz; **b**: $L_3 = 0.1L_2$ leading to $f_p = f(1.55) = 2252.72$ kHz; **c**: $L_3 = 0.14L_2$ leading to $f_p = f(1.55) = 1947.11$ kHz; **d**: $L_3 = 0.18L_2$ leading to $f_p = f(1.55) = 1754.47$ kHz. Other parameters used in these plots are $A_c = 2$, $A_s = 40$, $\vartheta_0 = 10$, $\phi_0 = 0$, $\lambda = -0.002$, and $\Lambda = 0.01$. Reprint from Ref. [14], Copyright 2018, with permission from American Physical Society

show solitary signal with broad edges. As in Figs. 3.7, 3.8, 3.9, and 3.10 show that the introduction of the second-neighbor interactions in the network of Fig. 3.1 can be used to manage the amplitude, the width, as well as the speed of the solitary signals propagating through a given cell of the network.

As we have mentioned above, two different waves may propagate through a given network at the same frequency. Figure 3.11 shows two pairs of bright solitary waves propagating at the same frequency through cell $n = 500$ of the network of Fig. 3.1 with a coupling second-neighbor parameter $L_3 = 0.18L_2$. Waves showed in Fig. 3.11a and b propagate at frequency $f_p = 1754.96$ kHz corresponding to the wave numbers $k \simeq 1.516$ for Fig. 3.11a and $k = 1.530$ for Fig. 3.11b, while waves of Fig. 3.11c and d propagate at frequency $f_p = 1754.47$ kHz corresponding to the wave numbers $k \simeq 1.496$ for Fig. 3.11c and $k = 1.550$ for Fig. 3.11d. Figure 3.11a and c are obtained with wave numbers taken from the region of increasing frequency f (as a function of wave number k), while Fig. 3.11b and d are generated with wave numbers taken from the region of frequency growth (as a function of wave number k). From plots of Fig. 3.11a and c, it can be seen that when the propagating frequency f is taken from its region of growth (as function of wave number k), the amplitude and the width of the solitary wave increase when the frequency decreases; the same, from plots of Fig. 3.11b and d, we conclude that when the propagating frequency is taken from its decreasing region (as function of wave number k) the amplitude and

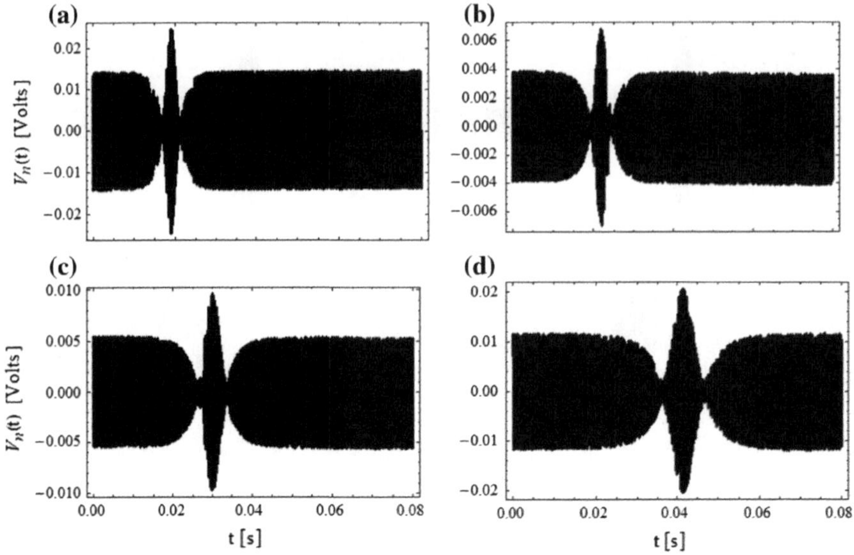

Fig. 3.9 Propagation through the network at the cell $n = 500$ of bright solitary signals embedded on a CW background with $\Lambda = 0$ and negative A_s for the network parameters (3.2) and different values of the coupling second-neighbor parameter L_3 leading to different frequencies f_p associated with the wave number $k = 1.55$. **a**: $L_3 = +\infty$ (second-neighbor interactions are neglected) leading to $f_p = f(1.55) = 763.18$ kHz; **b**: $L_3 = 0.1L_2$ leading to $f_p = f(1.55) = 2252.72$ kHz; **c**: $L_3 = 0.14L_2$ leading to $f_p = f(1.55) = 1947.11$ kHz; **d**: $L_3 = 0.18L_2$ leading to $f_p = f(1.55) = 1754.47$ kHz. Other parameters used in these plots are $A_c = 4$, $A_s = -40$, $\vartheta_0 = 10$, $\phi_0 = 0$, $\lambda = -0.002$. Reprint from Ref. [14], Copyright 2018, with permission from American Physical Society

the width of the solitary signal decrease with the frequency. These facts are clearly seen in Fig. 3.12.

Now, we study the effects of parameter λ on the transmission of solitary signals through the network of Fig. 3.1. In order to show how parameter λ influences on the transmission of solitary signals through our network, we have presented in Figs. 3.13 and 3.14 the propagation of bright solitary waves through the network for different negative values of parameter λ. Plots of Fig. 3.13 are generated by considering the presence of second-neighbor interactions in the network, while Fig. 3.14 are obtained when the second-neighbor interactions are not taking into account. Our computations show that the wave profile depends on the sign of the solution parameters A_s and Λ: Plots (a)–(c) and plots (d)–(f) of Figs. 3.13 and 3.14 are obtained with $A_s = 20$ and $\Lambda = 0.01$ (in general for $A_s\Lambda > 0$) and $A_s = -20$ and $\Lambda = 0.01$ (in general for $A_s\Lambda < 0$), respectively. The two main conclusions are that (i) the wave amplitude increases as parameter λ decreases and (ii) the center (the trajectory) of bright solitary wave is an oblique line, and the introduction of the second-neighbor interactions into the network change the wave trajectory [in the case when the second-neighbors are neglected, the trajectory is an oblique increasing line (see Fig. 3.14), while in the presence of second-neighbors, it is an oblique decreasing line (see Fig. 3.13)].

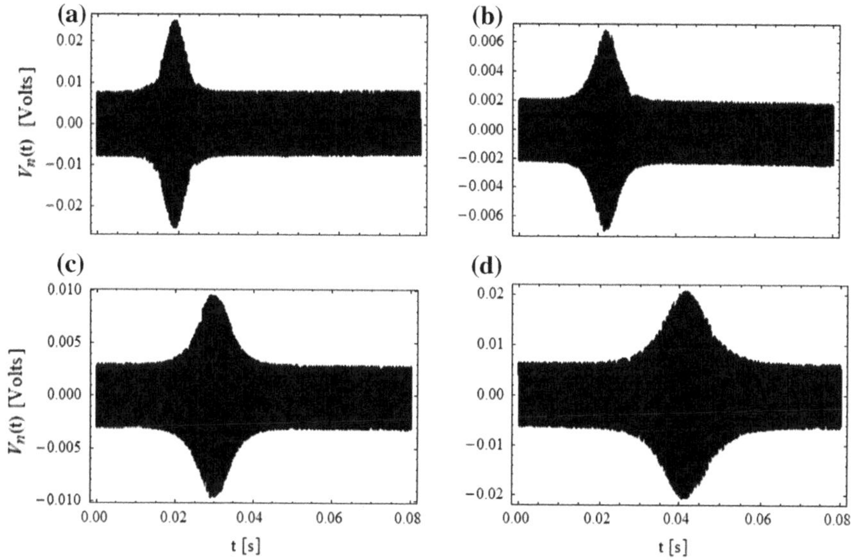

Fig. 3.10 Propagation through the network at the cell $n = 500$ of bright solitary signals embedded on a CW background with $\Lambda = 0$ and positive A_s for the network parameters (3.2) and different values of the coupling second-neighbor parameter L_3 leading to different frequencies f_p associated with the wave number $k = 1.55$. **a**: $L_3 = +\infty$ (second-neighbor interactions are neglected) leading to $f_p = f(1.55) = 763.18$ kHz; **b**: $L_3 = 0.1L_2$ leading to $f_p = f(1.55) = 2252.72$ kHz; **c**: $L_3 = 0.14L_2$ leading to $f_p = f(1.55) = 1947.11$ kHz; **d**: $L_3 = 0.18L_2$ leading to $f_p = f(1.55) = 1754.47$ kHz. Other parameters used in these plots are $A_c = 4$, $A_s = 40$, $\vartheta_0 = 10$, $\phi_0 = 0$, $\lambda = -0.002$. Reprint from Ref. [14], Copyright 2018, with permission from American Physical Society

Special Case Where $A_s^2 - 4A_c^2 < 0$

Now, we turn to the situation of solution (3.10) when parameters A_s and A_c satisfy condition $A_s^2 - 4A_c^2 < 0$. It is obvious that in the present situation, $A_c \neq 0$. With the equation parameter $|\lambda| = 0.2$ and solution parameters $\phi_0 = \vartheta_0 = 0$, $|A_s| = 15$, $A_c = 11$, and $\Lambda = 0.1$, we have presented in Fig. 3.15. Plots of the top panels show periodic solitary waves propagating at frequency $f_p = 923.56$ kHz in the case when the second-neighbor interactions are neglected ($L_3 = \infty$), while plots of bottom panels show periodic solitary waves propagating at frequency $f_p = 1071.01$ kHz in the presence of second-neighbor interactions with coupling parameter $L_3 = 0.14L_2$. The first observation is that the introduction of second-neighbor interactions in the network increases the period of propagating solitary waves through the network (see the top plots obtained with $L_3 = \infty$ and the bottom plots generated with $L_3 \neq \infty$). Second, the change in the sign of parameter λ changes the trajectory of solitary waves propagating through the network [see, for example, plots (a) and (f) generated with $A_s = 15$ and $\lambda = 0.2$ and plots (b) and (f), obtained with $A_s = 15$ and $\lambda = -0.2$]. Lastly, the change in the sign of A_s maintains the wave trajectory but, changes the direction of the solitary wave propagation (as we can see for example from Fig. 3.15a

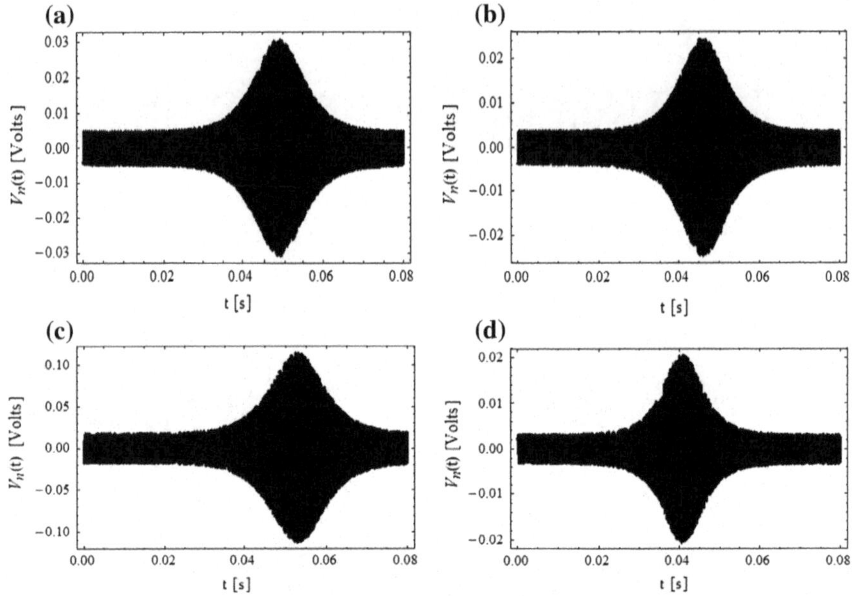

Fig. 3.11 Pairs of bright solitary signals propagating at the same frequency f_p through cell $n = 500$: **a**, **b**: Two bright solitary signals propagating at the same frequency $f_p = 1754.96$ kHz; **c**, **d**: Two bright solitary signals propagating at the same frequency $f_p = 1754.47$ kHz. The curves of Fig. 3.11a and b are generated with wave numbers $k \simeq 1.516$ and $k = 1.530$, respectively; to plot curves Fig. 3.11c and d, we have used the wave numbers $k \simeq 1.496$ and $k = 1.55$, respectively. Other parameters used in these plots are $A_c = 2$, $A_s = 40$, $\vartheta_0 = 10$, $\phi_0 = 0$, $\lambda = -0.002$, and $\Lambda = 0.01$. Reprint from Ref. [14], Copyright 2018, with permission from American Physical Society

and e obtained with $A_s = 15$ and $\lambda = 0.2$ and Fig. 3.15c and g generated with $A_s = -15$ and $\lambda = 0.2$, the solitary waves have the same trajectory, but propagate in opposite directions).

Figure 3.16 shows the propagation of periodic solitary signals through cell $n = 450$ of the network either for different values of the coupling second-neighbor parameter L_3. The plots of this figure show that the period of wave propagation decreases as parameter L_3 increases. It is also seen from these plots that the solitary wave magnitude increases with parameter L_3.

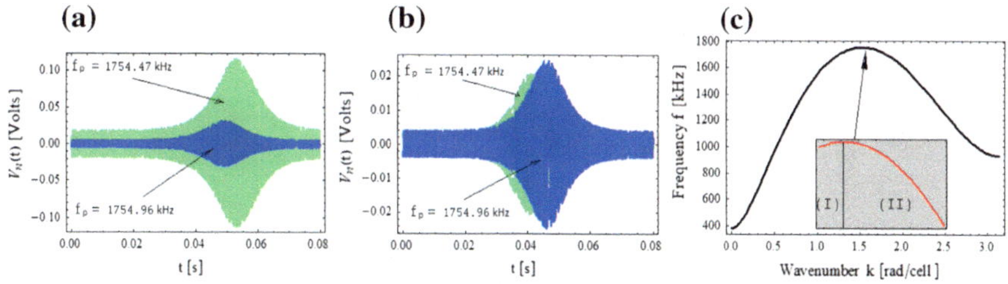

Fig. 3.12 Propagation of bright solitary waves through cell $n = 500$ of the network of Fig. 3.1 with the use of the same parameters as in Fig. 3.11. **a**: The two frequencies showed in plot **a** are taken in the region where frequency f increases as a function of wave number k; **b**: The two frequencies showed in plot **b** are taken in the region where frequency f decreases as a function of wave number k; **c**: Frequency plot associated with the coupling second-neighbor parameter $L_3 = 0.18L_2$. Plots of figure **a** show that the amplitude and the width of the solitary waves increase when frequency f decreases. It is seen from the plots **a** that the amplitude and the width of the solitary waves increase when frequency f increases. In plot **c**, we have zoomed the wave number region where the frequencies used in plots **a** and **b** have been taken; in terms of the wave number k, this region is defined as $1.4944863318959838 < k < 1.596560380167552$. In this region of course, $PQ > 0$ (here, the product PQ has the same behavior as shown in Fig. 3.3b). (I) and (II) show the region of frequency growth and the frequency decay region, respectively. Reprint from Ref. [14], Copyright 2018, with permission from American Physical Society

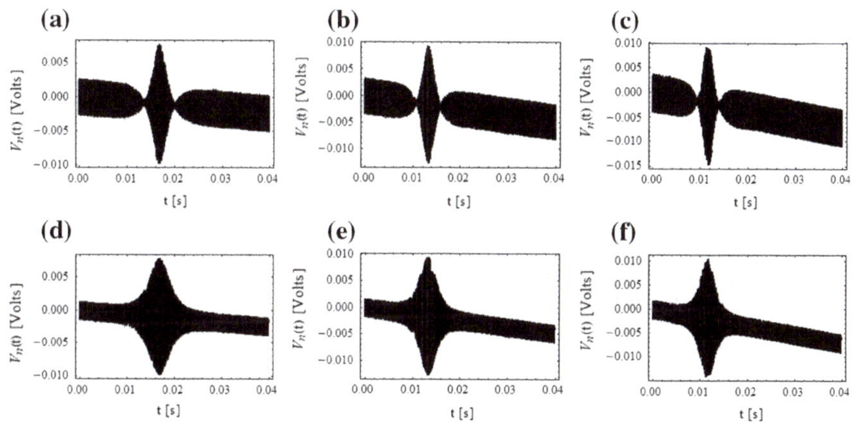

Fig. 3.13 Propagation of bright solitary waves through cell $n = 350$ of the network at frequency $f_p = 2252.72$ kHz for different values of parameter λ with $\Lambda = 0.01$, $A_c = 1$, $\phi_0 = 0$, $\vartheta_0 = 10$, and the coupling second-neighbor parameter $L_3 = 0.1L_2$. **a**, **d**: $\lambda = -0.034$; **b**, **e**: $\lambda = -0.068$; **c**, **f**: $\lambda = -0.1$. on the dynamics of modulated matter waves in the network of Fig. 3.1. Plots of the top and bottom panels are generated with $A_s = 20$ and $A_s = -20$, respectively. Wave amplitude increases as parameter λ (< 0) decreases, and the trajectory of bright solitary waves is a decreasing oblique line. Reprint from Ref. [14], Copyright 2018, with permission from American Physical Society

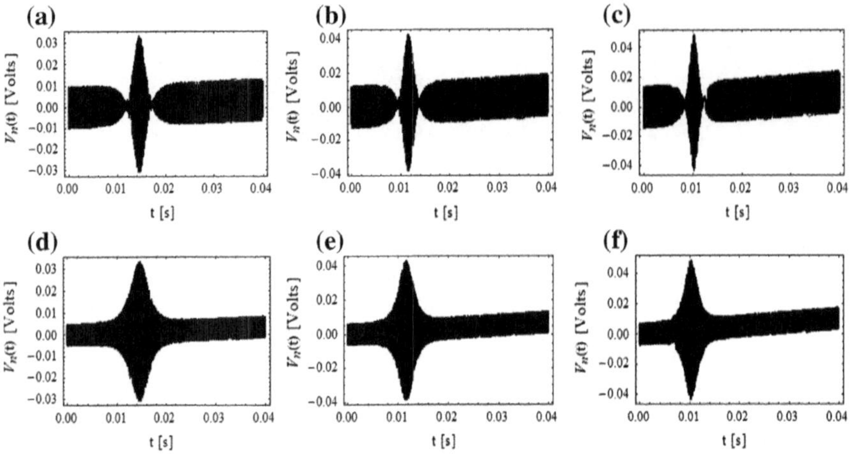

Fig. 3.14 Propagation of bright solitary waves through cell $n = 350$ of the network at frequency $f_p = 2252.72$ kHz for different values of parameter λ when second–neighbor interactions are neglected ($L_3 = +\infty$). Different plots are generated with the same parameters as in Fig. 3.12. Wave amplitude increases as parameter $\lambda(< 0)$ decreases, and the trajectory of bright solitary waves is an increasing oblique line. Reprint from Ref. [14], Copyright 2018, with permission from American Physical Society

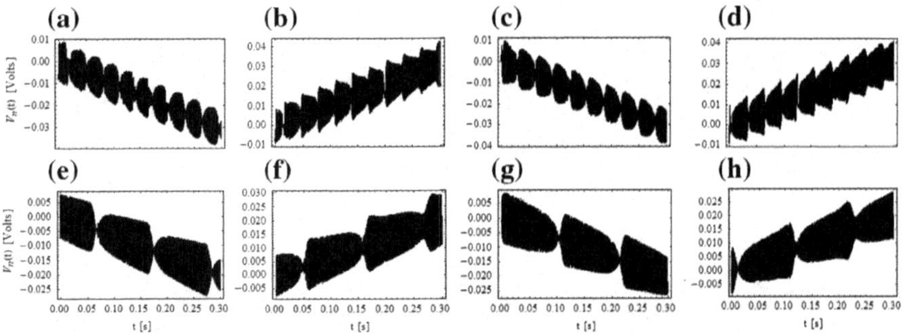

Fig. 3.15 Plot of voltage $V_n(t)$ as a function of the time t showing the propagation of a periodic soliton-like pulses propagating through cell $n = 450$ of the network for network parameters (3.2). Plots **a**–**d** show solitary waves propagating at frequency $f_p = 923.56$ kHz for $L_3 = +\infty$ (second-neighbors in the network are neglected), while plots **e**–**h** show solitary waves propagating at frequency $f_p = 1071.01$ kHz when the second-neighbors are taken into account. **a**, **e**: $A_s = 15$, $\lambda = 0.2$; **b**, **f**: $A_s = 15$, $\lambda = -0.2$; **c**, **g**: $A_s = -15$, $\lambda = 0.2$; **d**, **h**: $A_s = -15$, $\lambda = -0.2$. Other parameters are shown in the text. Reprint from Ref. [14], Copyright 2018, with permission from American Physical Society

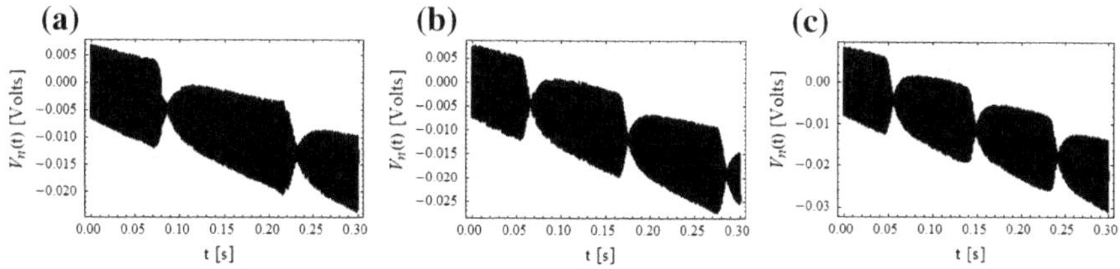

Fig. 3.16 Plot of voltage $V_n(t)$ as a function of the time t showing the propagation of a periodic solitary signals propagating through cell $n = 450$ of the network for network parameters (3.2), different values of the coupling second-neighbor parameter L_3, and solution parameters $A_s = 15$, $A_c = 11$, $\lambda = 0.2$, and $\Lambda = 0.1$. **a**: solitary wave propagating at frequency $f_p = 1124.59$ kHz associated with $L_3 = 0.1L_2$; **b** solitary wave propagating at frequency $f_p = 1071.01$ kHz associated with $L_3 = 0.14L_2$; **c** solitary wave propagating at frequency $f_p = 1040.05$ kHz associated with $L_3 = 0.18L_2$. Reprint from Ref. [14], Copyright 2018, with permission from American Physical Society

3.2.6 Conclusion

Using the analytical solutions of the NLS Eq. (3.7) for $\Gamma = 0$, we have investigated the transmission of soliton-like pulses through the network of Fig. 3.1 when the dissipative effects in the network are neglected. Our simulations have shown that the choice of the three main parameters A_s, A_c, and Λ of the analytical solitary wave solution (3.10) influences the form of the pulses propagating through the network. Investigating the effects of the coupling second-neighbor interactions on the transmission of solitary signals through the network, we have obtained the following main results: (i) The inclusion of second-order couplings in the network influences on the trajectory of the bright solitary wave propagating through the network. (ii) The magnitude of solitary signals increases with the coupling second-neighbor parameter L_3 and the system might be more robust to weak external perturbations. (iii) By considering the second-neighbor interactions, two different solitaries (either both bright or one bright and one dark) signals can simultaneously propagate at the same frequency. It is important to notice that our study can be useful in better choice of parameters of the network and in better understanding of the results of experiments of the propagating modulated wave in network with second-neighbor interactions.

3.2.7 Transmission of Solitonic Pulses Through the Dissipative Network of Fig. 3.1 ($\Gamma \neq 0$): Analytical Studies

In this section, we use the dissipative NLS Eq. (3.7) to investigate analytically the propagation of soliton-like pulses through the dissipative network with second-neighbors of Fig. 3.1. For simplicity, we set $\lambda(\tau) = 0$. We start by finding analytical solitonic solutions of Eq. (3.7). For this purpose,, we introduce ansatz

$$\psi(x,\tau) = a(\tau)u(X)\exp\left[i\left(Kx - \varphi(\tau)\right)\right],\ X = X(x,\tau) = \alpha_0(\tau)x + \beta_0(\tau), \tag{3.13}$$

where K is a real constant, $a(\tau)$, $\alpha_0(\tau)$, $\beta_0(\tau)$, and $\varphi(\tau)$ are real function of variable τ to be determined, and u is a real function of variable X. Inserting ansatz (3.13) into Eq. (3.7) and separating real and imaginary parts yield

$$P\alpha_0^2\left(\frac{du}{dX}\right)^2 + 2Qa^2u^4 + \left(\frac{d\varphi(\tau)}{d\tau} - PK^2\right)u^2 - A_0 = 0, \tag{3.14a}$$

$$\left(\frac{\frac{da}{d\tau}}{a} + \Gamma\right)u + \left(\frac{d\beta_0}{d\tau} - \frac{1}{\alpha_0}\frac{d\alpha_0}{d\tau}\beta_0 + 2PK\alpha_0 + \frac{\frac{d\alpha_0}{d\tau}}{\alpha_0}X\right)\frac{du}{dX} = 0, \tag{3.14b}$$

where $A_0(\tau)$ is an arbitrary function of τ (constant of integration with respect with X). Demanding that α_0 and β_0 satisfy equation

$$\frac{d\beta_0}{d\tau} - \frac{1}{\alpha_0}\frac{d\alpha_0}{d\tau}\beta_0(\tau) + 2PK\alpha_0 = 0$$

yields

$$\beta_0(\tau) = -2PK\tau\alpha_0(\tau) + K_0\alpha_0(\tau),\ X = \alpha_0(\tau)\left[x - 2PK\tau + K_0\right], \tag{3.15}$$

where K_0 is an arbitrary real parameter. Inserting Eq. (3.15) into system (3.14a), (3.14b) yields

$$P\alpha_0^2(\tau)\left(\frac{du}{dX}\right)^2 + 2Qa^2u^4 + \left(\frac{d\varphi(\tau)}{d\tau} - PK^2\right)u^2 - A_0 = 0, \tag{3.16a}$$

$$\left(\frac{1}{a}\frac{da}{d\tau} + \Gamma\right)u + \left(\frac{X}{\alpha_0}\frac{d\alpha_0}{d\tau}\right)\frac{du}{dX} = 0. \tag{3.16b}$$

Equation (3.16a) is an elliptic ordinary differential equation (EODE). The simplest functions $\alpha_0(\tau)$ and $a(\tau)$ for which Eq. (3.16b) is satisfies for every $u(X)$ are obtained by asking that the coefficients of u and du/dX in this equation vanish, i.e., $\frac{1}{a}\frac{da}{d\tau} + \Gamma = 0$ and $\frac{X}{\alpha_0}\frac{d\alpha_0}{d\tau} = 0$. This leads to

$$a(\tau) = a_0\exp\left[-\Gamma\tau\right],\ \alpha_0(\tau) = \alpha_{00}, \tag{3.17}$$

where $a_0 \neq 0$ and $\alpha_{00} \neq 0$ are two real constants. Thus, if $a(\tau)$ and $\alpha_0(\tau)$ are defined by Eq. (3.17), system (3.16a), (3.16b) becomes

$$\left(\frac{du}{dX}\right)^2 = -\frac{2Qa^2}{P\alpha_0^2}u^4 + \frac{PK^2 - \frac{d\varphi}{d\tau}}{P\alpha_0^2}u^2 + \frac{A_0}{P\alpha_0^2}, \tag{3.18}$$

that we rewrite in the form

$$\left(\frac{d\rho}{dX}\right)^2 = -\frac{8Qa^2}{6P\alpha_0^2}\rho^3 + \frac{4\left(PK^2 - \frac{d\varphi(\tau)}{d\tau}\right)}{P\alpha_0^2}\rho^2 + 4\frac{A_0}{P\alpha_0^2}\rho, \ u^2 = \rho. \tag{3.19}$$

Equations (3.18) and (3.19) are elliptic ordinary differential equations (EODEs). Each of these two equations is a special case of equation

$$\left(\frac{dy}{dX}\right)^2 = \alpha_1 y^4 + 4\beta_1 y^3 + 6\gamma_1 y^2 + 4\delta_1 y + \varepsilon_1 \stackrel{\text{def.}}{=} \Lambda(y), \tag{3.20}$$

where α_1, β_1, γ_1, δ_1, and ε_1 are functions of variable τ defined as follows:

$$\alpha_1 = \begin{cases} -\frac{2Qa^2}{P\alpha_0^2} \text{ for Eq. (3.18)}, \\ 0 \text{ for Eq. (3.19)}; \end{cases} \quad \beta_1 = \begin{cases} 0 \text{ for Eq. (3.18)}, \\ -\frac{4Qa^2}{3P\alpha_0^2} \text{ for Eq. (3.19)}; \end{cases}$$

$$\gamma_1 = \begin{cases} \frac{2\left(PK^2 - \frac{d\varphi(\tau)}{d\tau}\right)}{3P\alpha_0^2} \text{ for Eq. (3.18)}, \\ 0 \text{ for Eq. (3.19)}; \end{cases} \quad \delta_1 = \begin{cases} 0 \text{ for Eq. (3.18)}, \\ \frac{A_0}{P\alpha_0^2} \text{ for Eq. (3.19)}; \end{cases}$$

$$\varepsilon_1 = \begin{cases} \frac{A_0}{P\alpha_0^2} \text{ for Eq. (3.18)}, \\ 0 \text{ for Eq. (3.19)}. \end{cases}$$

In some special cases, we can express solitary wavelike solutions of Eq. (3.20) in terms of P, Q, α_0, a, φ, K, and A_0. In our analysis, we mainly use Eq. (3.18) (Eq. (3.19) will be used only occasionally).

To find analytical solitary wavelike solutions of equation (3.20) in explicit form, we use either the direct method or the Weierstrass' elliptic function method [46, 47]. It is well known that a large class of solutions of Eq. (3.20) is given by expression [46, 47]

$$y(X) = y_0 + \frac{\sqrt{\Lambda(y_0)}\frac{d\wp(X;g_2,g_3)}{dX} + \frac{1}{2}\Lambda'(y_0)\left[\wp(X;g_2,g_3) - \frac{1}{24}\Lambda''(y_0)\right] + \frac{1}{24}h(y_0)\Lambda'''(y_0)}{2\left[\wp(X;g_2,g_3) - \frac{1}{24}\Lambda''(y_0)\right]^2 - \frac{1}{48}\Lambda(y_0)\Lambda''''(y_0)}, \tag{3.21}$$

where y_0 is any real function of τ (not necessary a root of polynomial $\Lambda(y)$) and $\wp(X; g_2, g_3)$ is the Weierstrass' elliptic function with invariants $g_2 = \alpha_1\varepsilon_1 - 4\beta_1\delta_1 + 3\gamma_1^2$ and $g_3 = \alpha_1\gamma_1\varepsilon_1 + 2\beta\gamma\delta - \alpha\delta^2 - \gamma^3 - \varepsilon_1\beta_1$. Equation (3.21) leads to solitary wavelike solutions of Eq. (3.20) only when the following three conditions are satisfied simultaneously [33]:

$$\Delta = g_2^3 - 27g_3^2 = 0, \ g_2 \geq 0, \ g_3 \leq 0. \tag{3.22}$$

When y_0 is a simple root of polynomial $\Lambda(y)$, Eq. (3.21) leads to the solitary wavelike solution having the form [33]

$$y(X) = y_0 + \frac{\Lambda'(y_0)}{4\left[e_1 - \frac{\Lambda''(y_0)}{24} + 3e_1\cosh^2\left[\sqrt{3e_1}X\right]\right]}, \; e_1 = \frac{1}{2}\sqrt[3]{-g_3}. \tag{3.23}$$

It is important to note that expression (3.23) is used for solitary wavelike solution of Eq. (3.20) only if y_0 is a simple root of polynomial $\Lambda(y)$.

We notice that any solitary wavelike solution of the EODE (3.20) leading to a solitary wavelike solution $u(X)$ of Eq. (3.18) gives the following solitary wavelike solution of the dissipative NLS Eq. (3.7)

$$\psi(x,\tau) = a_0 \exp\left[-\Gamma\tau\right] u(\alpha_{00}\left[x - 2PK\tau + K_0\right]) \exp\left[i\left(Kx - \varphi(\tau)\right)\right]. \tag{3.24}$$

In what follows, we distinguish the case of bright, dark, and kink solitonic solutions.

Analytical Bright Solitonic Solutions of the EODE

As we have mentioned above, solitonic solutions of Eq. (3.18) will be sought by either the direct method or the Weierstrass' elliptic function method.

Direct method for exact bright soliton solutions to the EODE As examples of the use of the direct method, we seek soliton solutions of Eq. (3.18) having either the form $u(X) = 1/\cosh[mX]$ or $u(X) = 1/(A_1 + B_1\cosh^2[mX])$. Imposing to $u(X)$ to satisfy Eq. (3.18) yields

$$u(X) = \frac{1}{\cosh\left[\frac{a}{\alpha_0}\sqrt{\frac{2Q}{P}}X\right]}, \; \varphi(\tau) = \begin{cases} \left(K^2P - 2Qa_0^2\right)\tau - \varphi_0, \text{ if } \Gamma = 0 \\ K^2P\tau + \frac{Qa_0^2}{\Gamma}\exp\left[-2\Gamma\tau\right] - \varphi_0, \text{ if } \Gamma \neq 0, \end{cases} \tag{3.25a}$$

$$u(X) = \pm\frac{1}{a}\sqrt{\frac{PK^2 - \frac{d\varphi(\tau)}{d\tau}}{2Q}} \frac{1}{1 - 2\cosh^2\left[\frac{1}{2\alpha_0}\sqrt{\frac{PK^2 - \frac{d\varphi(\tau)}{d\tau}}{P}}X\right]}, \tag{3.25b}$$

respectively. In solution (3.25b), $\varphi(\tau)$ is an arbitrary function that satisfies conditions $P\left(PK^2 - \frac{d\varphi(\tau)}{d\tau}\right) > 0$ and $Q\left(PK^2 - \frac{d\varphi(\tau)}{d\tau}\right) > 0$. Both Eqs. (3.25a) and (3.25b) give soliton solutions of Eq. (3.18) with vanishing boundary conditions $u(\pm\infty) = 0$.

Weierstrass' elliptic function method for exact solitonic solutions to the EODE. Now, we apply the Weierstrass' elliptic function method to find exact bright solitary wavelike solutions of either Eq. (3.18) or (3.19). We start with Eq. (3.18). For Eq. (3.18), the conditions (3.22) for solitary wavelike solutions read

$$\begin{cases} A_0\left[8A_0Qa^2 + \left(PK^2 - \frac{d\varphi(\tau)}{d\tau}\right)^2\right] = 0, \\ \left(PK^2 - \frac{d\varphi(\tau)}{d\tau}\right)^2 - 24A_0Qa^2 \geq 0, \\ P\left(PK^2 - \frac{d\varphi(\tau)}{d\tau}\right)\left[72A_0Qa^2 + \left(PK^2 - \frac{d\varphi(\tau)}{d\tau}\right)^2\right] \geq 0. \end{cases} \tag{3.26}$$

If $A_0 = 0$, the first two conditions are satisfied, while the last one is satisfied only if $P\left(PK^2 - \frac{d\varphi(\tau)}{d\tau}\right) \geq 0$. In this case, polynomial Λ admits two simple roots, namely $u_0 = \pm\sqrt{\frac{PK^2 - \frac{d\varphi}{d\tau}}{2Qa^2}}$ if and only, if $Q\left(PK^2 - \frac{d\varphi(\tau)}{d\tau}\right) > 0$. Using expression (3.23) for each of these simple roots yields

$$u(X) = \pm\frac{1}{a}\sqrt{\frac{PK^2 - \frac{d\varphi(\tau)}{d\tau}}{2Q}}\left(1 - \frac{12}{1 + 6\cosh^2\left[\frac{1}{2\alpha_0}\sqrt{P^{-1}\left(PK^2 - \frac{d\varphi}{d\tau}\right)}X\right]}\right). \tag{3.27}$$

It is obvious that (3.27) is a solitary wavelike solution of Eq. (3.18) with non-vanishing boundary conditions.

If $A_0 \neq 0$, the first of conditions (3.26) for solitary wavelike solutions of Eq. (3.18) yields $A_0 = -\frac{1}{8Qa^2}\left(PK^2 - \frac{d\varphi(\tau)}{d\tau}\right)^2$, leading to $\Lambda(u) = -\frac{2Qa^2}{P\alpha_0^2}\left(u^2 - \frac{PK^2 - \frac{d\varphi}{d\tau}}{4Qa^2}\right)^2$. Evaluating e_1 in this case yields $e_1 = -\frac{\left(PK^2 - \frac{d\varphi(\tau)}{d\tau}\right)}{6P\alpha_0^2}$. It follows from the expression for $\Lambda(u)$ that polynomial $\Lambda(u)$ does not have any simple root.

Analytical Expression for Dark (Kink) Soliton Solutions of the EODE

We now apply direct method and Weierstrass' elliptic function method to find analytical dark soliton solutions $u(X)$ of Eq. (3.18).

Direct method for exact dark solutions to the EODE. Here, we find an exact dark soliton solution of Eq. (3.18) by substituting $u(X) = \tanh[mX]$ in Eq. (3.18). As a result, we obtain the dark soliton solution

$$u(X) = \tanh\left[\frac{a}{\alpha_0}\sqrt{-\frac{2Q}{P}}X\right], \ \varphi(\tau) = K^2P\tau - \frac{2Qa_0^2}{\Gamma}\exp[-2\Gamma\tau] - \varphi_0, \tag{3.28}$$

where φ_0 is an arbitrary real constant.

Weierstrass' elliptic function method for exact dark and/or kink soliton solutions to the EODE. Now, we apply the Weierstrass' elliptic function method to find dark and/or kink soliton solutions of Eq. (3.18). These solutions are associated with nonzero multiple roots of polynomial Λ for Eq. (3.18) and/or Eq. (3.19) [33]. For a multiple root of polynomial Λ, we use Eq. (3.21) instead of Eq. (3.23) to evaluate the corresponding soliton solutions.

We start with Eq. (3.18). When $A_0 \neq 0$, we found, under the conditions of solitary wavelike solutions (3.26), that polynomial Λ for Eq. (3.18) admits two double roots $u_0 = \pm\frac{1}{2}\sqrt{\frac{PK^2 - \frac{d\varphi}{d\tau}}{Qa^2}}$, if $\varphi(\tau)$ is chosen from condition $Q\left(PK^2 - \frac{d\varphi}{d\tau}\right) > 0$. Let us, as an example, consider the double root $u_0 = \frac{1}{2}\sqrt{\frac{PK^2-\frac{d\varphi}{d\tau}}{Qa^2}}$. Noting that $e_1 = -\frac{PK^2-\frac{d\varphi(\tau)}{d\tau}}{6P\alpha_0^2}$ and choosing $u_0 = \frac{1}{4a}\sqrt{\frac{PK^2-\frac{d\varphi(\tau)}{d\tau}}{Q}}$ [33], Eq. (3.21) for $\wp = e_1 + \frac{3e_1}{\sinh^2(\sqrt{3e_1}X)}$ gives the kink soliton solution

$$u(X) = \frac{1}{2a}\sqrt{\frac{PK^2 - \frac{d\varphi(\tau)}{d\tau}}{Q}} \frac{\left(4 + 3\sinh\left[\frac{2}{\alpha_0}\sqrt{-\frac{PK^2-\frac{d\varphi(\tau)}{d\tau}}{2P}}X\right]\right)}{5 + 3\cosh\left[\frac{2}{\alpha_0}\sqrt{-\frac{PK^2-\frac{d\varphi(\tau)}{d\tau}}{2P}}X\right]}, \tag{3.29}$$

where $Q\left(PK^2 - \frac{d\varphi}{d\tau}\right) > 0$, and consequently, $P\left(PK^2 - \frac{d\varphi}{d\tau}\right) < 0$.

We now use Eq. (3.19) to generate another dark soliton solution of Eq. (3.18). For this equation, the conditions (3.22) for soliton solutions give $8Qa^2A_0 + \left(\frac{d\varphi(\tau)}{d\tau} - PK^2\right)^2 = 0$ and $P\left(PK^2 - \frac{d\varphi(\tau)}{d\tau}\right) \leq 0$. Solving the first equation in A_0 and inserting the result in $\Lambda(\rho)$ yield $\Lambda(\rho) = -\frac{8Qa^2}{P\alpha_0^2}\rho\left(\rho - \frac{PK^2-\frac{d\varphi(\tau)}{d\tau}}{4Qa^2}\right)^2$. Therefore, $\Lambda(\rho)$ admits one double root, $\rho_0 = \frac{PK^2-\frac{d\varphi(\tau)}{d\tau}}{4Qa^2}$. Noting that $e_1 = -\frac{PK^2-\frac{d\varphi(\tau)}{d\tau}}{6P\alpha_0^2}$ and choosing $\rho_0 = \frac{PK^2-\frac{d\varphi(\tau)}{d\tau}}{8Qa^2}$ [33], we evaluate Eq. (3.21) for $\wp = e_1 + \frac{3e_1}{\sinh^2(\sqrt{3e_1}X)}$ and use the relationship $u^2 = \rho$ to get the following dark soliton solution:

$$u(X) = \pm\frac{1}{2a}\sqrt{\frac{PK^2 - \frac{d\varphi(\tau)}{d\tau}}{Q}} \times \frac{\sqrt{2 + \sinh^2\left(\sqrt{3e_1}X\right) + \sinh^4\left(\sqrt{3e_1}X\right) - 2\cosh\left(\sqrt{3e_1}X\right)\sinh\left(\sqrt{3e_1}X\right)}}{2 + \sinh^2\left(\sqrt{3e_1}X\right)}. \tag{3.30}$$

Here, $e_1 = -\frac{PK^2-\frac{d\varphi(\tau)}{d\tau}}{6P\alpha_0^2}$. Note that under condition $P\left(PK^2 - \frac{d\varphi(\tau)}{d\tau}\right) < 0$, we automatically have $Q\left(PK^2 - \frac{d\varphi(\tau)}{d\tau}\right) > 0$ since $PQ < 0$.

3.2.8 Propagation of Bright, Dark, and Kink Solitonic Pulses Through the Network of Fig. 3.1: Case of $L_3 = +\infty$

In this section, we use the exact solutions of the dissipative NLS Eq. (3.7) with $\lambda(\tau) \equiv 0$ to investigate analytically the propagation of bright, dark, and kink solitary wavelike through the network of Fig. 3.1 when the second-neighbor interactions are neglected, i.e., when $L_3 = +\infty$. We focus on the case of $\Gamma \neq 0$ and show how the dissipative elements of the network can be used to manipulate the motion of bright, kink, and dark solitonic signals through the network.

Propagation of Bright Solitonic Pulses Through the Network of Fig. 3.1

Each of the bright soliton solutions (3.25a) and (3.25b), and the solitary wavelike solution (3.27) of the EODE (3.18) leads, respectively, to bright soliton solutions and a bright solitary wavelike solution of Eq. (3.7), and consequently, to the propagation of bright solitons and bright solitary wavelike in the network of Fig. 3.1. It is important to notice that each of the solutions (3.25b) and (3.27) of the EODE (3.18) contains the free functional parameter $\varphi(\tau)$ which will be chosen under some conditions on P and/or Q. This free functional parameter will allow us to manipulate the motion of the solitary wavelike on the network. We will limit ourselves to the use of solutions (3.25a) and (3.27) [one solution without the free parameter $\varphi(\tau)$ and one solution that contains $\varphi(\tau)$].

We start with the solution (3.25a) that we write the corresponding solution of Eq. (3.7) for, respectively, $\Gamma \neq 0$ and $\Gamma = 0$ as follows:

$$\psi(x,\tau) = \frac{a_0 \exp[-\Gamma\tau] \exp\left[i\left(Kx - K^2 P\tau - \frac{Qa_0^2}{\Gamma}\exp[-2\Gamma\tau] + \varphi_0\right)\right]}{\cosh\left[a_0\sqrt{\frac{2Q}{P}}\exp[-\Gamma\tau][x - 2PK\tau + K_0]\right]} \tag{3.31a}$$

$$\psi(x,\tau) = \frac{a_0 \exp\left[i\left(Kx - \left(K^2 P - 2Qa_0^2\right)\tau + \varphi_0\right)\right]}{\cosh\left[a_0\sqrt{\frac{2Q}{P}}[x - 2PK\tau + K_0]\right]}. \tag{3.31b}$$

It is seen from Eq. (3.31a) that the soliton amplitude a_0 is proportional to function $\exp[-\Gamma\tau]$, while its width is inversely proportional to function $\sqrt{\frac{2Q}{P}}\exp[-\Gamma\tau]$. Because Γ is a positive parameter, function $\exp[-\Gamma\tau]$ decreases when Γ increases (we remember that $\tau = \epsilon^2 t > 0$). In other words, the soliton width increases when the dissipative element G of the network increases. Meantime, the soliton amplitude decreases when the dissipative element of the network increases. The centre of bright soliton is $\xi(\tau) = \sqrt{\frac{P}{2Q}}\,(2PK\tau + K_0)\exp[\Gamma\tau]$, and satisfies equation $\frac{d^2\xi(\tau)}{d\tau^2} - \frac{\Gamma[4PK + \Gamma(2PK\tau + K_0)]}{2PK\tau + K_0}\xi(\tau) = 0$. This equation means that the centre of mass of the macroscopic wave packet behaves like a classical particle, and allows us to manipulate the motion of bright soliton in the network by controlling the dissipative element G.

Figure 3.17 shows the propagation of solitary signal at frequency $f = 406.881$ kHz for the above network parameters and solution parameters $a_0 = 50$, $\varphi_0 = -20$, $K = -0.1$, and $\epsilon = 10^{-3}$. It is seen from Fig. 3.17a and c that the cell number n does not affect neither the peak value of wave amplitude nor the wave width. Figure 3.17b shows that the wave amplitude decreases while the wave width increases when the dissipative parameter σ increases.

Now, we consider the solitary wavelike solution (3.27) of the EODE (3.18). This solution of the EODE (3.18) leads to the following bright solitary wavelike solution of the dissipative NLS Eq. (3.7):

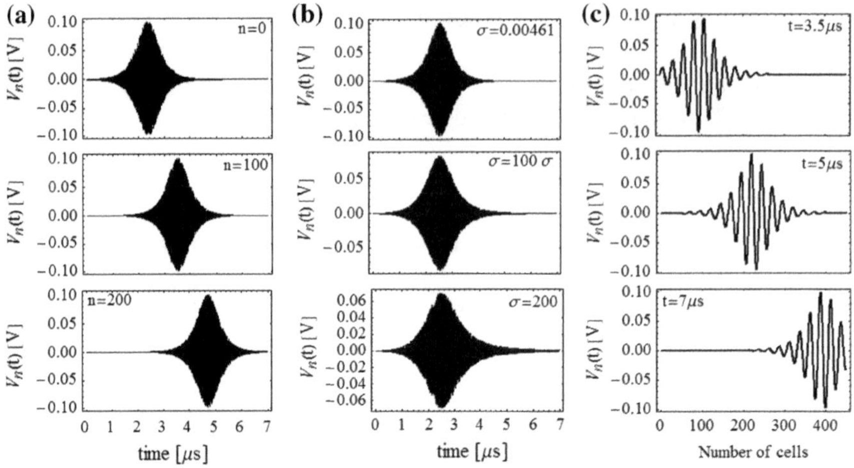

Fig. 3.17 Signal voltage obtained with the exact bright soliton solution (3.31a) of the dissipative NLS equation (1.5) for the network parameters shown in the text with $L_3 = +\infty$ and for frequency $f = 406.881$ kHz. **a**: Propagation of bright soliton signal at different cells of the network; **b**: Propagation of bright soliton signal at cell $n = 15$ of the network for different dissipative parameter σ; **c**: Spatial propagation of bright soliton through the network at different times for $\sigma = 0.00461$. Reprint from Ref. [13], Copyright 2018, with permission from American Physical Society

$$\psi(x,\tau) = \pm\sqrt{\frac{PK^2 - \frac{d\varphi(\tau)}{d\tau}}{2Q}}\exp\left[i\left(Kx - \varphi(\tau)\right)\right] \times \left(1 - \frac{12}{1 + 6\cosh^2\left[\frac{1}{2}\sqrt{\frac{PK^2 - \frac{d\varphi}{d\tau}}{P}}\left(x - 2PK\tau + K_0\right)\right]}\right), \tag{3.32}$$

where the functional parameter $\varphi(\tau)$ and constant K are to be taken from conditions $P\left(PK^2 - \frac{d\varphi(\tau)}{d\tau}\right) > 0$ and $Q\left(PK^2 - \frac{d\varphi(\tau)}{d\tau}\right) > 0$. Although the solitary wavelike solution (3.32) of the dissipative NLS Eq. (3.7) does not contain (explicitly) the dissipative coefficient Γ, parameters K and $\varphi(\tau)$ appearing in this solution result from the existence of the dissipative term in Eq. (3.7).

For the above network parameters and a given set of solution parameters, we show in Fig. 3.18 the propagation of bright solitary wavelike through cell $n = 25$ of the network for different expressions for $d\varphi(\tau)/\,d\tau$. This figure shows how we can manage the bright solitary wavelike motion in the line just by manipulating the functional parameter $\varphi(\tau)$.

Propagation of Kink Solitons in the Network of Fig. 3.1

Based on exact kink soliton solution of Eq. (3.18), we investigate analytically the propagation of kink soliton in the network of Fig. 3.1. The kink soliton solution

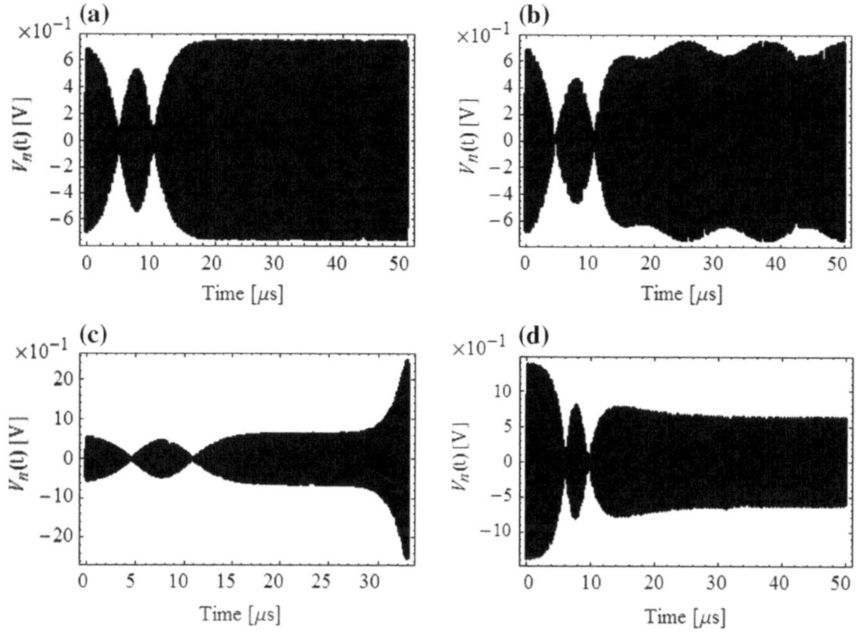

Fig. 3.18 Propagation of bright solitary wavelike in the network at frequency $f = 849.872$ kHz obtained with the exact bright solitary wavelike solution (3.32) of the dissipative NLS equation (3.7) for the network parameters shown in the text with $L_3 = +\infty$ and the solution parameters $\varepsilon = 10^{-2}$, $K = 0.1$, $K_0 = 40$ and **a**: $\frac{d\varphi(\tau)}{d\tau} = -10^9$; **b**: $\frac{d\varphi(\tau)}{d\tau} = -10^9 \cos^2\left[250000\varepsilon^{-2}\tau\right]$; **c**: $\frac{d\varphi}{d\tau} = -2 \times 10^{-4} \exp\left[10^6\varepsilon^{-2}\tau\right]$; **d**: $\frac{d\varphi}{d\tau} = -10^{10} \cosh^{-2}\left[10^5\varepsilon^{-2}\tau\right]$. Reprint from Ref. [13], Copyright 2018, with permission from American Physical Society

(3.28) of the EODE (3.18) is associated with the following kink soliton solution of the dissipative NLS Eq. (3.7)

$$\psi(x,\tau) = a_0 \exp\left[-\Gamma\tau\right] \tanh\left[a_0\sqrt{-\frac{2Q}{P}} \exp\left[-\Gamma\tau\right]\left[x - 2PK\tau + K_0\right]\right]$$
$$\times \exp\left[i\left(Kx - K^2P\tau + \frac{2Qa_0^2}{\Gamma}\exp\left[-2\Gamma\tau\right] + \varphi_0\right)\right]. \qquad (3.33)$$

It is seen from solution (3.33) that the wave amplitude a_0 is proportional to function $\exp\left[-\Gamma\tau\right]$ while its width is inversely proportional to $\sqrt{-\frac{2Q}{P}}\exp\left[-\Gamma\tau\right]$. Therefore, solution (3.33) can be used to describe the compression of kink solitons of the network of Fig. 3.1. With the help of the exact kink soliton solution (3.33) of the dissipative NLS Eq. (3.7), we show in figure 3.19 the dynamics of kink solitons propagating through the network for the above line parameters. Plots of Fig. 3.19a and c show that the wave amplitude does not vary with the cell number n. Plots of Fig. 3.19b show that the wave amplitude decreases while its width increases when the dissipative

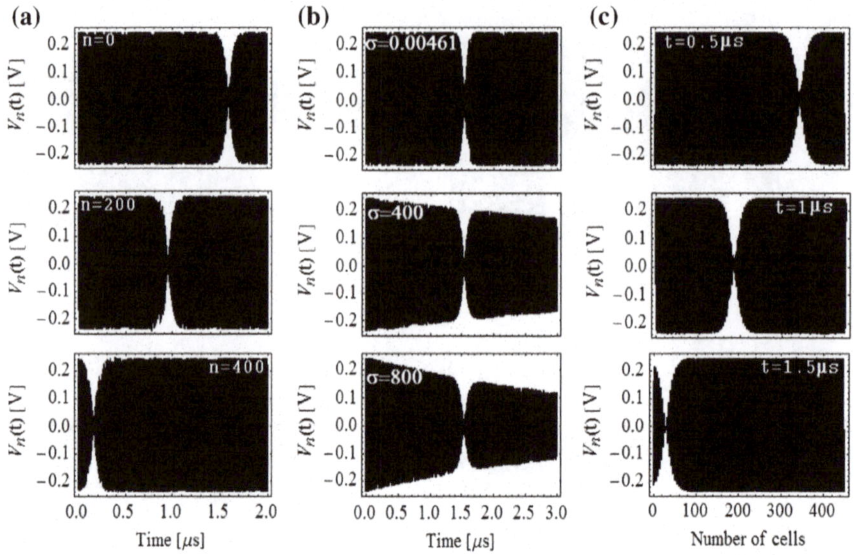

Fig. 3.19 Propagation of kink soliton in the network at frequency $f = 777.275$ kHz for the above line parameters and the solution parameters $a_0 = 12$, $\varphi_0 = 0.1$, $K_0 = -5$, $K = 0.1$, and $\varepsilon = 10^{-2}$. **a**: Evolution of kink soliton at three different cells; **b**: Evolution of kink soliton at cell $n = 25$ for three values of the dissipative parameter σ; **c** Profile of the kink soliton at three different times. Reprint from Ref. [13], Copyright 2018, with permission from American Physical Society

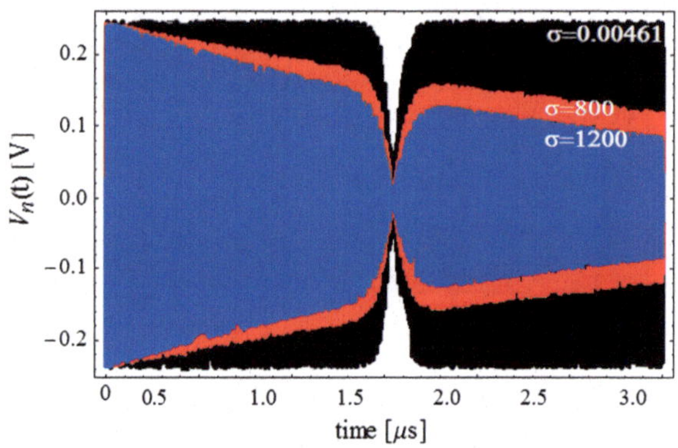

Fig. 3.20 Effects of the dissipative element of the network on the kink soliton propagating through cell $n = 15$ at frequency $f = 777.275$ kHz. Reprint from Ref. [13], Copyright 2018, with permission from American Physical Society

parameter σ increases. For a better understanding, we show in Fig. 3.20 the time evolution of kink solitons through cell $n = 15$ at frequency $f = 777.275$ kHz for three different values of the dissipative parameter σ.

Propagation of Dark Solitons in the Network of Fig. 3.1

We end this section with the analytical investigation of the dynamics of dark solitons propagating through the network. For this purpose, we use the exact solution (3.30)

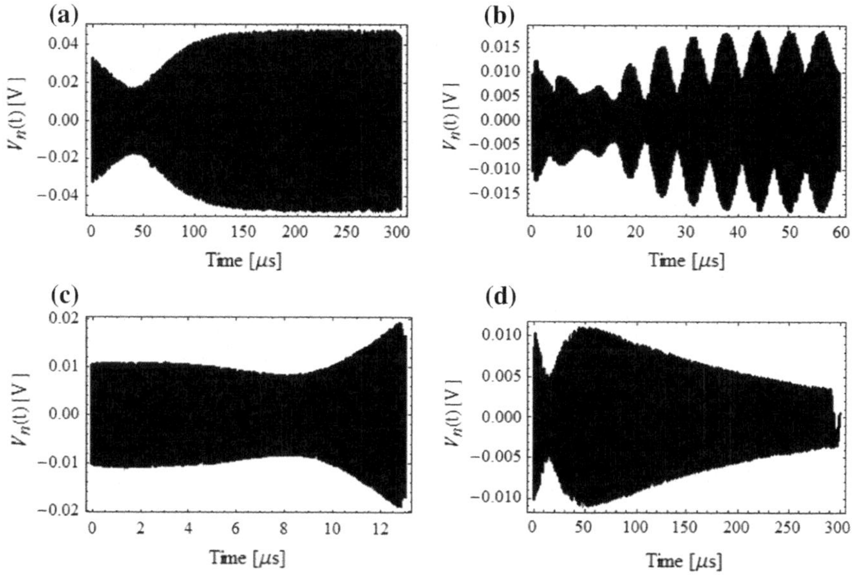

Fig. 3.21 Propagation of dark soliton in the network at frequency $f = 935.515$ kHz with the network parameters shown in the text for the solution parameters $\varepsilon = 10^{-3}$, $K = -0.1$, $K_0 = 500$, and different functional parameters $\varphi(\tau)$ defined from equation. **a**: $\frac{d\varphi(\tau)}{d\tau} = PK^2 + a_0$; **b**: $\frac{d\varphi(\tau)}{d\tau} = PK^2 + a_0\left(1 + 0.5\cos\left[\frac{10^6\tau}{\varepsilon^2}\right]\right)$; **c**: $\frac{d\varphi(\tau)}{d\tau} = PK^2 + a_0\exp\left[\frac{10^5}{\varepsilon^2}\tau\right]$; **d**: $\frac{d\varphi(\tau)}{d\tau} = PK^2 + a_0\exp\left[-\frac{10^5}{\varepsilon^2}\tau\right]$, where $a_0 = 2000$. Reprint from Ref. [13], Copyright 2018, with permission from American Physical Society

of the EODE (3.18) which leads to the following exact dark soliton solution of the dissipative NLS Eq. (3.7):

$$\psi(x,\tau) = \frac{\pm\sqrt{\frac{PK^2-\frac{d\varphi(\tau)}{d\tau}}{Q}}\exp[i(Kx-\varphi(\tau))]}{2\left(2+\sinh^2\left[\sqrt{-\frac{PK^2-\frac{d\varphi(\tau)}{d\tau}}{2P}}(x-2PK\tau+K_0)\right]\right)}\left(2-\sinh\left[2\sqrt{-\frac{PK^2-\frac{d\varphi(\tau)}{d\tau}}{2P}}(x-2PK\tau+K_0)\right]\right.$$
$$\left.+\sinh^2\left[\sqrt{-\frac{PK^2-\frac{d\varphi(\tau)}{d\tau}}{2P}}(x-2PK\tau+K_0)\right]+\sinh^4\left[\sqrt{-\frac{PK^2-\frac{d\varphi(\tau)}{d\tau}}{2P}}(x-2PK\tau+K_0)\right]\right)^{\frac{1}{2}}, \tag{3.34}$$

where constant K and the functional parameter $\varphi(\tau)$ are to be chosen from conditions $P\left(PK^2 - \frac{d\varphi(\tau)}{d\tau}\right) < 0$ and $Q\left(PK^2 - \frac{d\varphi(\tau)}{d\tau}\right) > 0$. It is obvious that the dissipative parameter Γ does not appear explicitly in (3.34). Meanwhile, parameters K and K_0 and the functional parameter $\varphi(\tau)$ result from the presence of Γ in Eq. (3.7). For a given set of solution parameters and the above network parameters, we show in Fig. 3.21 the time evolution of dark soliton through cell $n = 250$ of the network for different functional parameter $\varphi(\tau)$. Plots of this figure show how we may use parameter $\varphi(\tau)$ to manipulate the motion of dark solitons through the network.

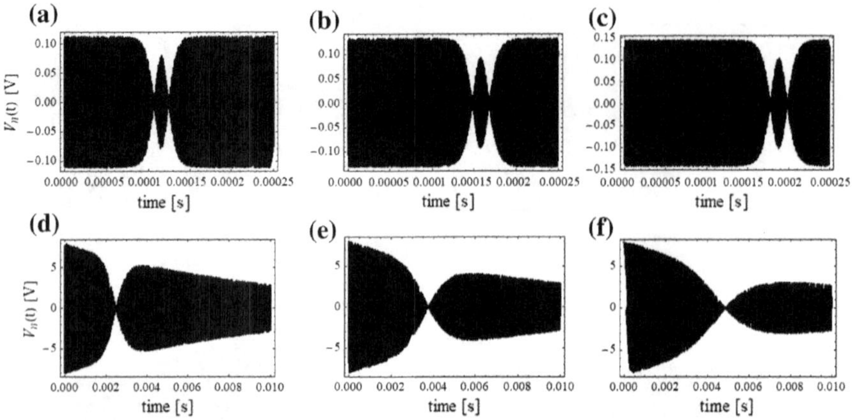

Fig. 3.22 Solitonic pulses propagating through cell $n = 250$ the network of Fig. 3.1 when the second-neighbor interactions are taken into account for the line parameters (3.2) and $\varepsilon = 10^{-3}$. **a**, **b**, **c**: Bright solitonic pulses propagating at frequencies $f = 3858.47$ kH, $f = 2773.48$kH, and $f = 2300.8$ kH, respectively, corresponding to the three different values of line parameter L_3, namely, $L_3 = 0.3L_2$ (**a**), $L_3 = 0.6L_2$ (**b**), and $L_3 = 0.9L_2$ (**c**). **d**, **e**, **f**: Dark solitonic pulses propagating at frequencies $f = 3121.64$ kH, $f = 2302.97$ kH, and $f = 1955.35$ kH, respectively, corresponding to the three different values of line parameter L_3, namely, $L_3 = 0.3L_2$ (**d**), $L_3 = 0.6L_2$ (**e**), and $L_3 = 0.9L_2$ (**f**). Plots **a**, **b**, and **c** are obtained with the use of the exact solution (3.32) for $K_0 = 0.5542$, $K = 0.1$, and $\frac{d\varphi}{d\tau} = -10^9 \cos^2(250000\varepsilon^2\tau)$, while plots **d**, **e**, and **f** are generated with the help of the exact solution (3.33) for $K_0 = -35.4688$, $\varphi_0 = 0.1$, $a_0 = 4000$, $K = 0.1$, and $\sigma = 40$. Reprint from Ref. [13], Copyright 2018, with permission from American Physical Society

3.2.9 *Effects of the Second-Neighbor Interactions on Solitonic Pulses Propagating Through the Lossy Network of Fig. 3.1*

With the help of the exact solutions of the dissipative NLS Eq. (3.7) with $\lambda(\tau) \equiv 0$, we investigate analytically the emission solitonic pulses through the network of Fig. 3.1 when the second-neighbor interactions are taken into account, i.e., when $L_3 \neq +\infty$. We focus on the case of $\Gamma \neq 0$ and show how the second-neighbors affect the motion of solitonic pulses during their propagation through the network of Fig. 3.1. As an example of bright and dark solitonic pulses, we use the exact solutions (3.32) and (3.33), respectively. For different values of the network parameter L_3, we show in Fig. 3.22 the propagation of bright solitonic pulses (top plots) and bark solitonic pulses (bottom plots) generated with the use of solutions (3.32) and (3.33), respectively. Both the top and the bottom plots show that solitonic speed increases when the second-neighbor parameter L_3 increases. It is seen from the top plots that the amplitude of the bright solitonic pulse increases with parameter L_3. It is seen from the plots of the bottom panels that the amplitude of dark solitonic pulse decreases when parameter L_3 increases; this situation is well seen in Fig. 3.23.

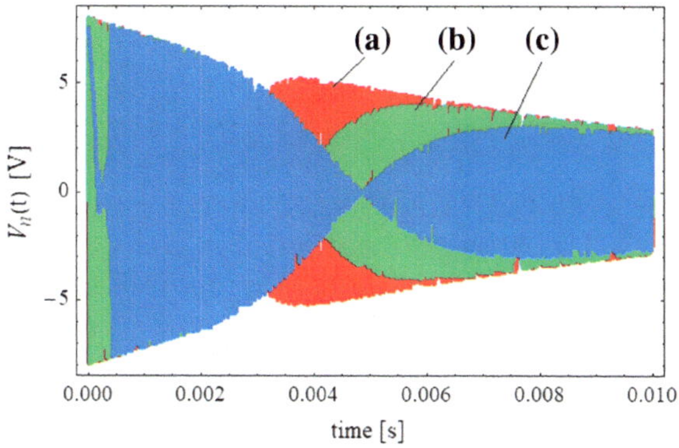

Fig. 3.23 Dark solitonic pulses propagating through cell $n = 250$ of the network of Fig. 3.1 under the influence of second-neighbor interactions for the line parameters (3.2) with $\varepsilon = 10^{-3}$. Plots **a**, **b**, **c** show dark solitonic pulses propagating at frequencies $f = 3121.64$ kH, $f = 2302.97$ kH, and $f = 1955.35$ kH, respectively, corresponding to three different values of line parameter L_3, namely, $L_3 = 0.3L_2$ (**a**), $L_3 = 0.6L_2$ (**b**), and $L_3 = 0.9L_2$ (**c**). The exact solution (3.33) with parameters $K = 0.1$, $K_0 = -35.4688$, $\varphi_0 = 0.1$, $a_0 = 4000$, and $\sigma = 40$ have been used to generate these plots. Reprint from Ref. [13], Copyright 2018, with permission from American Physical Society

3.3 Stability of the Modulated Amplitude Waves (MAWs) Propagating Through the Network of Fig. 3.2

In this section, we consider the nonlinear mono-inductance network shown in Fig. 3.2. In this transmission line C_n is a nonlinear capacitor such as a "VARICAP" or a reverse-biased $p - n$ junction diode, the capacitance of which depends on the voltage applied across it.

Model Description

We assume that if a given section of the line is located at position x, the next section will be located at position $x + \Delta x$, where $|\Delta x| << 1$. Picking one section and applying Kirchhoff's current theorem and voltage theorem yield

$$\frac{\partial I}{\partial x} + \frac{\partial \widetilde{Q}}{\partial t} = 0, \; \widetilde{L}\frac{\partial I_1}{\partial t} + \frac{\partial V}{\partial x} = 0, \; \frac{\partial^2 V}{\partial x \partial t} + \frac{1}{\widetilde{C}_s}(I - I_1) = 0, \tag{3.35}$$

where the current through the nonlinear capacitor is given by $\partial Q/\partial t$, ($Q = Q(V)$ being the charge density), $\widetilde{Q} = Q/\Delta x$, $\widetilde{L} = L/\Delta x$, and $\widetilde{C}_s = \Delta x C_s$. In what follows, we omit the '~' on Q, L, and C_s. Eliminating currents I and I_1 from Eq. (3.35) yields the following partial differential equation for the voltage V and charge Q

$$C_s \frac{\partial^4 V}{\partial x^2 \partial t^2} + \frac{1}{L}\frac{\partial^2 V}{\partial x^2} - \frac{\partial^2 Q(V)}{\partial t^2} = 0. \tag{3.36}$$

Expanding $Q(V)$ into the Taylor series $Q(V) = C_0 V - C_N V^2 + O(V^3)$ reduces Eq. (3.36) to

$$\frac{1}{L}\frac{\partial^2 V}{\partial x^2} - C_0\frac{\partial^2 V}{\partial t^2} + C_s\frac{\partial^4 V}{\partial x^2 \partial t^2} + C_N\frac{\partial^2 V^2}{\partial t^2}. \tag{3.37}$$

The first two terms in Eq. (3.37) are identical to the linear wave equation. The third term accounts for the dispersion introduced by the capacitor CS, while the last term is the nonlinear term. If we neglect the nonlinear term and seek the solution of equation (3.37) in the form $V(x,t) \sim \exp[i\,(kx - \omega t)]$, we get the dispersion relation

$$C_0\omega^2 - k^2\frac{1}{L} + C_s k^2\omega^2 = 0. \tag{3.38}$$

In our numerical simulations, we will use the line parameters

$$C_0 = 540\,\text{pF},\ C_s = 270\,\text{pF},\ C_N = 135\,\text{pF},\ L = 28\,\mu\text{H}. \tag{3.39}$$

3.3.1 NLS Equation of the MAWs of Network of Fig. 3.2

Using Eq. (3.37) and applying the perturbation theory, we wish to show that the MAWs of the network of Fig. 3.2 can be governed by a coupled NLS equations. Let us consider the interaction of any two wavepackets centered at (k_A, ω_A) and (k_B, ω_B). In order to resolve the weakly nonlinear equation (3.37), we introduce two different slow timescales $T_1 = \epsilon t$ and $T_2 = \epsilon^2 t$ in addition to the original timescale t; moreover we introduce one large scale $X_1 = \epsilon x$ in addition to the original space scale x, where $0 << \epsilon << 1$. With these three independent timescales T_0, T_1, and T_2, we may carry out third-order expansion [28] of $V = V(x,t)$

$$V(x,t) = \epsilon V_1 + \epsilon^2 V_2 + \epsilon^3 V_3 + O(\epsilon^4), \tag{3.40}$$

where $V_n = V_n(x, t, X_1, T_1, T_2)$. In order to analyze the propagation of any two wavepackets centered at (k_A, ω_A) and (k_B, ω_B), we take the lowest order term V_1 in the form

$$V_1 = A\exp[i\theta_A] + B\exp[i\theta_B] + c.c., \tag{3.41}$$

where $\theta_A = k_A x - \omega_A t$, $\theta_B = k_B x - \omega_B t$, A and B are the complex modulation amplitudes, functions of slow and space coordinates, i.e., $A = A(X_1, T_1, T_2)$, $B = B(X_1, T_1, T_2)$, and $c.c.$ stands for the complex conjugate.

Inserting the perturbation expansions (3.40) into the nonlinear equation (3.37) yield a series of non-homogeneous equations in different orders of ϵ:

$$D_0 V_1 = \left(\frac{1}{L} \frac{\partial^2}{\partial x^2} - C_0 \frac{\partial^2}{\partial t^2} + C_s \frac{\partial^4}{\partial x^2 \partial t^2} \right) V_1 = 0, \tag{3.42a}$$

$$D_0 V_2 = 2C_0 \frac{\partial^2 V_1}{\partial t \partial T_1} - C_N \frac{\partial^2 V_1^2}{\partial t^2} - \frac{2}{L} \frac{\partial^2 V_1}{\partial x \partial X_1} - 2C_s \left(\frac{\partial^4}{\partial x^2 \partial t \partial T_1} + \frac{\partial^4}{\partial t^2 \partial x \partial X_1} \right) V_1, \tag{3.42b}$$

$$\begin{aligned} D_0 V_3 = {} & C_0 \left(2 \frac{\partial^2 V_1}{\partial t \partial T_2} + \frac{\partial^2 V_1}{\partial T_1^2} + 2 \frac{\partial^2 V_1}{\partial t \partial T_1} \right) - 2C_N \left(\frac{\partial^2 V_1 V_2}{\partial t^2} + \frac{\partial V_1^2}{\partial t \partial T_1} \right) \\ & - \frac{1}{L} \left(\frac{\partial^2 V_1}{\partial X_1^2} + 2 \frac{\partial^2 V_2}{\partial x \partial X_1} \right) - C_s \left(2 \frac{\partial^4 V_1}{\partial x^2 \partial t \partial T_2} + \frac{\partial^4 V_1}{\partial x^2 \partial T_1^2} + 4 \frac{\partial^4 V_1}{\partial x \partial X_1 \partial t \partial T_1} \right. \\ & \left. + 2 \frac{\partial^4 V_1}{\partial x \partial X_2 \partial t^2} + \frac{\partial^4 V_1}{\partial X_1^2 \partial t^2} + 2 \frac{\partial^4 V_2}{\partial x^2 \partial t \partial T_1} + 2 \frac{\partial^4 V_2}{\partial x \partial X_1 \partial t^2} \right). \end{aligned} \tag{3.42c}$$

Inserting Eq. (3.41) into Eq. (3.42a) yields the dispersion relations

$$\left(C_0 + C_s k_A^2\right) \omega_A^2 - k_A^2 \frac{1}{L} = 0, \ \left(C_0 + C_s k_B^2\right) \omega_B^2 - k_B^2 \frac{1}{L} = 0. \tag{3.43}$$

From Eq. (3.42b), we obtain the solvability conditions

$$\frac{\partial A}{\partial T_1} + v_{gA} \frac{\partial A}{\partial X_1} = 0, \ \frac{\partial B}{\partial T_1} + v_{gB} \frac{\partial B}{\partial X_1} = 0, \tag{3.44}$$

where $v_{gA} = -d\omega_A/dk_A$ and $v_{gB} = -d\omega_B/dk_B$ are the group velocities. Equation (3.44) mean that $A = A(X_A, T_2)$ and $B = B(X_B, T_2)$, where $X_A = X_1 - v_{gA} T_1$ and $X_B = X_1 - v_{gB} T_1$ are the shifted coordinates. Under the solvability conditions (3.44), Eq. (3.42b) for V_2 becomes $D_0 V_2 = -C_N \partial^2 V_1^2 / \partial t^2$. Therefore,

$$\begin{aligned} V_2 = {} & \frac{C_N}{3C_0 k_A^2} A^2 \exp\left[2i\theta_A\right] + \frac{C_N}{3C_0 k_B^2} B^2 \exp\left[2i\theta_B\right] \\ & + \frac{8LC_N \left(\omega_A + \omega_B\right)^2 AB \exp\left[i\left(k_A + k_B\right)x - i\left(\omega_A + \omega_B\right)t\right]}{L\left(\omega_A + \omega_B\right)^2 \left[C_0 + C_N\left(k_A + k_B\right)^2\right] - \left(k_A + k_B\right)^2} \\ & + \frac{4LC_N \left(\omega_A + \omega_B\right)^2 AB^* \exp\left[i\left(k_A - k_B\right)x - i\left(\omega_A - \omega_B\right)t\right]}{L\left(\omega_A + \omega_B\right)^2 \left[C_0 + C_N\left(k_A + k_B\right)^2\right] - \left(k_A + k_B\right)^2} + c.c., \end{aligned} \tag{3.45}$$

where the asterisk stands for the complex conjugate. If we insert (3.41) and (3.45) into (3.42c) and eliminate the terms that produce secular terms in (3.42c), we obtain the following system:

$$\frac{\partial A}{\partial T_2}=\frac{1}{2\omega_A}\frac{\partial^2 A}{\partial T_1^2}+\frac{1}{C_0+C_s k_A^2}\left(\frac{C_s\omega_A-\frac{1}{L}}{2\omega_A}\frac{\partial^2 A}{\partial X_1^2}-2C_s k_A\frac{\partial^2 A}{\partial T_1\partial X_1}\right)$$
$$+\frac{C_N\omega_A}{C_0+C_s k_A^2}\left(\widetilde{a}_A\left|A\right|^2+\left(\widetilde{c}+\widetilde{d}\right)\left|B\right|^2\right)A$$
$$\frac{\partial B}{\partial T_2}=\frac{1}{2\omega_B}\frac{\partial^2 B}{\partial T_1^2}+\frac{1}{C_0+C_s k_B^2}\left(\frac{C_s\omega_B-\frac{1}{L}}{2\omega_B}\frac{\partial^2 B}{\partial X_1^2}-2C_s k_B\frac{\partial^2 B}{\partial T_1\partial X_1}\right)$$
$$+\frac{C_N\omega_B}{C_0+C_s k_B^2}\left(\widetilde{a}_B\left|B\right|^2+\left(\widetilde{c}+\widetilde{d}\right)\left|A\right|^2\right)B,$$

where

$$\widetilde{a}_j=\frac{C_N}{3C_0k_j^2},\ \widetilde{c}=\frac{8LC_N\left(\omega_A+\omega_B\right)^2}{L\left(\omega_A+\omega_B\right)^2\left[C_0+C_N\left(k_A+k_B\right)^2\right]-\left(k_A+k_B\right)^2},$$
$$\widetilde{d}=\frac{4LC_N\left(\omega_A-\omega_B\right)^2}{L\left(\omega_A-\omega_B\right)^2\left[C_0+C_N\left(k_A-k_B\right)^2\right]-\left(k_A+k_B\right)^2},\ j=A,B. \quad (3.46)$$

Using relations (3.44) and coming back to the original coordinates x and t, the above system for A and B becomes the couple NLS equations

$$i\frac{\partial A}{\partial t}+P_A\frac{\partial^2 A}{\partial x^2}-\left(Q_{AA}\left|A\right|^2+Q_{AB}\left|B\right|^2\right)A=0, \quad (3.47a)$$
$$i\frac{\partial B}{\partial t}+P_B\frac{\partial^2 B}{\partial x^2}-\left(Q_{BB}\left|B\right|^2+Q_{BA}\left|A\right|^2\right)B=0, \quad (3.47b)$$

where

$$\begin{cases} P_j=\frac{3}{2}\frac{LC_0C_s\omega_j^3}{C_0+C_sk_j^2},\ Q_{jj}=\widetilde{a}_j\frac{C_N\omega_j\epsilon^2}{C_0+C_sk_j^2},\ j=A,\ B,\\ Q_{AB}=\frac{(\widetilde{c}+\widetilde{d})C_N\omega_A\epsilon^2}{C_0+C_sk_A^2},\ Q_{BA}=\frac{(\widetilde{c}+\widetilde{d})C_N\omega_B\epsilon^2}{C_0+C_sk_B^2} \end{cases} \quad (3.48)$$

$\widetilde{a}_j$, $\widetilde{c}$, and $\widetilde{d}$ being given by Eq. (3.46).

Definition: Any solution (A, B) of coupled NLS (CNLS) Eq. (3.47a), (3.47b) is called phase (winding) solution if A and B are of the form $A(x,t)=a(t)\exp\left[i\left(q_Ax\right.\right.$ $\left.\left.-\Omega_A(t)\right)\right]$ and $B(x,t)=b(t)\exp\left[i\left(q_Bx-\Omega_B(t)\right)\right]$, where a and b and Ω_A and Ω_B are real functional amplitudes and phases, respectively, and q_A and q_b are the phase winding numbers.

The simplest nontrivial solutions to the CNLS (3.47a), (3.47b) are the phase winding solutions (also called the plane wave solutions) of the form

$$\begin{cases} A(x,t)=a_0\exp\left[iq_Ax-i\left(P_Aq_A^2+Q_{AA}a_0^2+b_0^2Q_{AB}\right)t+\Omega_A^0\right],\\ B(x,t)=b_0\exp\left[iq_Bx-i\left(P_Bq_B+Q_{BB}b_0^2+a_0^2Q_{BA}\right)t+\Omega_B^0\right],\end{cases}$$

where a_0, b_0, Ω_A^0, and Ω_B^0 are real constants. With the line parameters (3.39) and $\epsilon = 0.1$, we compute for $k_A \approx 0.585\,07$ and $k_B \approx 0.514\,62$, $\omega_A = 8 \times 10^6$ and $\omega = 6 \times 10^6$, and the coefficients of system (3.47a), (3.47b)

$$\begin{cases} P_A = -2996866.2528, \ P_B = -1835667.6375, \ Q_{AA} = 458.8657, \\ Q_{BB} = 1327.1282, \ Q_{AB} = 20575.8688, \ Q_{BA} = 14937.26503. \end{cases} \tag{3.49}$$

3.3.2 Amplitude Dynamics of Phase Winding Solutions

In order to investigate the existence and stability of MAWs, we first express A and B in their polar form

$$A(x,t) = a(x,t) \exp\left[i\varphi_A(x,t)\right], \ B(x,t) = b(x,t) \exp\left[i\varphi_B(x,t)\right], \tag{3.50}$$

where a, b, φ_A, and φ_B are real functions of variables x and t. Inserting Eq. (3.50) into system (3.47a), (3.47b) and separating real and imaginary parts yield

$$P_A \frac{\partial^2 a}{\partial x^2} + P_A a \left(\frac{\partial \varphi_A}{\partial x}\right)^2 + a\frac{\partial \varphi_A}{\partial t} + \left(Q_{AA} a^2 + Q_{AB} b^2\right) a = 0, \tag{3.51a}$$

$$P_B \frac{\partial^2 b}{\partial x^2} + P_B b \left(\frac{\partial \varphi_B}{\partial x}\right)^2 + b\frac{\partial \varphi B}{\partial t} + \left(Q_{BB} b^2 + Q_{BA} a^2\right) b = 0, \tag{3.51b}$$

$$\frac{\partial a}{\partial t} + P_A \left(2\frac{\partial a}{\partial x}\frac{\partial \varphi_A}{\partial x} + a\frac{\partial^2 \varphi_A}{\partial x^2}\right) = 0, \tag{3.51c}$$

$$\frac{\partial b}{\partial t} + P_B \left(2\frac{\partial b}{\partial x}\frac{\partial \varphi_B}{\partial x} + b\frac{\partial^2 \varphi_B}{\partial x^2}\right) = 0. \tag{3.51d}$$

The temporal evolution of coherent structures in the CNLS equations amounts to a uniform propagation with velocities v_A and v_B and an overall phase oscillation with frequencies ω_A and ω_B:

$$a = a(z_a), \ b = b(z_b), \ \varphi_A = \phi_A(z_A) - \omega_A t, \ \varphi_B = \phi_B(z_B) - \omega_B t, \tag{3.52}$$

where $z_a = x - v_A t$ and $z_b = x - v_B t$. In Eq. (3.52), $a(z_a)$, $b(z_b)$, $\phi_A(z_A)$, and $\phi_B(z_B)$ represent coherent structures. In what follows, we focus ourselves on the uniformly moving solutions. Substitution of ansatz (3.52) into system (3.51a), (3.51d) leads to the following system of nonlinear ordinary differential equations (N-ODEs):

$$-P_A \frac{d^2 a}{dz_A^2} + P_A a \left(\frac{d\phi_A}{dz_A}\right)^2 - v_A a \frac{d\phi_A}{dz_A} - \omega_A a + \left(Q_{AA} a^2 + Q_{AB} b^2\right) a = 0, \tag{3.53a}$$

$$-P_B\frac{d^2b}{dz_B^2}+P_Bb\left(\frac{d\phi_B}{dz_B}\right)^2-\upsilon_Bb\frac{d\phi_B}{dz_B}-\omega_Bb+\left(Q_{BB}b^2+Q_{BA}a^2\right)b=0, \tag{3.53b}$$

$$-\upsilon_A\frac{da}{dz_A}+P_A\left(2\frac{da}{dz_A}\frac{d\phi_A}{dz_A}+a\frac{d^2\phi_A}{dz_A^2}\right)=0, \tag{3.53c}$$

$$-\upsilon_B\frac{db}{dz_B}+P_B\left(2\frac{db}{dz_B}\frac{d\phi_B}{dz_B}+b\frac{d^2\phi_B}{dz_B^2}\right)=0. \tag{3.53d}$$

Multiplying (3.53c) by a and (3.53d) by b and integrating the resulting equations yield, respectively

$$\left(P_A\frac{d\phi_A}{dz_A}-\frac{\upsilon_A}{2}\right)a^2=c_A,\quad \left(P_B\frac{d\phi_B}{dz_B}-\frac{\upsilon_B}{2}\right)b^2=c_B,$$

where c_A and c_B are two constants of integration. Taking the constant of integration as $c_A=c_B=0$ yields

$$\frac{d\phi_j}{dz_j}=\frac{\upsilon_j}{2P_j},\quad j=A,\ B. \tag{3.54}$$

Inserting Eq. (3.54) into Eqs. (3.53a) and (3.53b) leads to the system

$$a''=-\frac{\upsilon_A^2+4\omega_A}{4P_A}+\frac{Q_{AA}}{P_A}a^3+\frac{Q_{AB}}{P_A}b^2a, \tag{3.55a}$$

$$b''=-\frac{\upsilon_B^2+4\omega_B}{4P_B}+\frac{Q_{BB}}{P_B}b^3+\frac{Q_{BA}}{P_B}a^2b. \tag{3.55b}$$

Introducing $X=a'$ and $Y=b'$, system (3.55a), (3.55b) becomes

$$\begin{cases} a'=X,\\ b'=Y\\ X'=-\frac{\upsilon_A^2+4\omega_A}{4P_A}+\frac{Q_{AA}}{P_A}a^3+\frac{Q_{AB}}{P_A}b^2a,\\ Y'=-\frac{\upsilon_B^2+4\omega_B}{4P_B}+\frac{Q_{BB}}{P_B}b^3+\frac{Q_{BA}}{P_B}a^2b. \end{cases} \tag{3.56a}$$

Let us denote by $G(\upsilon_A,\upsilon_B,\omega_A,\omega_B)$ the following function which will play a crucial role in the stability analysis of system (3.56a):

$$G(\upsilon_A,\upsilon_B,\omega_A,\omega_B)=\frac{\upsilon_B^2+4\omega_B}{\upsilon_A^2+4\omega_A}. \tag{3.57}$$

Hereafter, we will work under the conditions

$$\omega_j > 0,\ Q_{ij} > 0,\ D = Q_{AA}Q_{BB} - Q_{AB}Q_{BA} < 0,\ \ j = A,\ B, \tag{3.58}$$

leading to $G(\upsilon_A, \upsilon_B, \omega_A, \omega_B) > 0$.

Seeking the stationary solutions of system (3.56a) yields $X = Y = 0$, and

$$a_p = \frac{1}{2}\sqrt{\frac{\upsilon_A^2 + 4\omega_A}{Q_{AA}}},\ a_m = \sqrt{\frac{Q_{BB}\left(\upsilon_A^2 + 4\omega_A\right) - Q_{AB}\left(\upsilon_B^2 + 4\omega_B\right)}{4D}},$$

$$b_p = \frac{1}{2}\sqrt{\frac{\upsilon_B^2 + 4\omega_B}{Q_{BB}}},\ b_m = \sqrt{\frac{Q_{AA}\left(\upsilon_B^2 + 4\omega_B\right) - Q_{BA}\left(\upsilon_A^2 + 4\omega_A\right)}{4D}}.$$

Depending on the value of the function G generically, three cases can occur

(i) $G < \frac{Q_{BB}}{Q_{AB}}$, leading to three nonnegative stationary solutions: the zero solution $(0, 0, 0, 0)$ and two pure modes $\left(a_p, 0, 0, 0\right)$ and $\left(0, b_p, 0, 0\right)$. The zero solution $(0, 0, 0, 0)$ and the pure mode $\left(0, b_p, 0, 0\right)$ are unstable, and $\left(a_p, 0, 0, 0\right)$ is a stable fixed point.

(ii) $G > \frac{Q_{BA}}{Q_{AA}}$, leading to three nonnegative fixed points: the zero solution $(0, 0, 0, 0)$ and the pure modes $\left(a_p, 0, 0, 0,\right)$ and $\left(0, b_p, 0, 0\right)$. The zero solution $(0, 0, 0, 0)$ and the fixed point $\left(a_p, 0, 0, 0\right)$ are unstable, while $\left(0, b_p, 0, 0\right)$ is a stable fixed point.

(iii) $\frac{Q_{BB}}{Q_{AB}} < G < \frac{Q_{BA}}{Q_{AA}}$, leading to four fixed points: the zero fixed point $(0, 0, 0, 0)$, the pure modes $\left(a_p, 0, 0, 0\right)$ and $\left(0, b_p, 0, 0\right)$, and the mixed mode $(a_m, b_m, 0, 0)$. The pure modes $\left(a_p, 0, 0, 0\right)$ and $\left(0, b_p, 0, 0\right)$ are stable, while the zero fixed point $(0, 0, 0, 0)$ and the mixed point $(a_m, b_m, 0, 0)$ are unstable.

Passing through the critical value $G_{c_1} = \frac{Q_{BB}}{Q_{AB}}$, the stable point $\left(a_p, 0, 0, 0\right)$ becomes unstable and passing through the second critical value $G_{c_2} = \frac{Q_{BA}}{Q_{AA}}$, the stable fixed point $\left(0, b_p, 0, 0\right)$ becomes unstable. Figure 3.24 shows the phase portraits for the planar system of ordinary differential equations (3.56a) corresponding to the above cases (i), (ii), and (iii).

3.4 Interacting Signal Packets in the Lossless Network of Fig. 3.1 in Absence of Second-Neighbor Interactions

The goal in the present section is to derive the context of lossless network of Fig. 3.1 in the absence of second-neighbor interactions, the coupled nonlinear Schrödinger (CNLS) equations for the voltage, and examine if the scenario of the modulational instability can occur in the context of NLTLs and, especially for long waves. Because

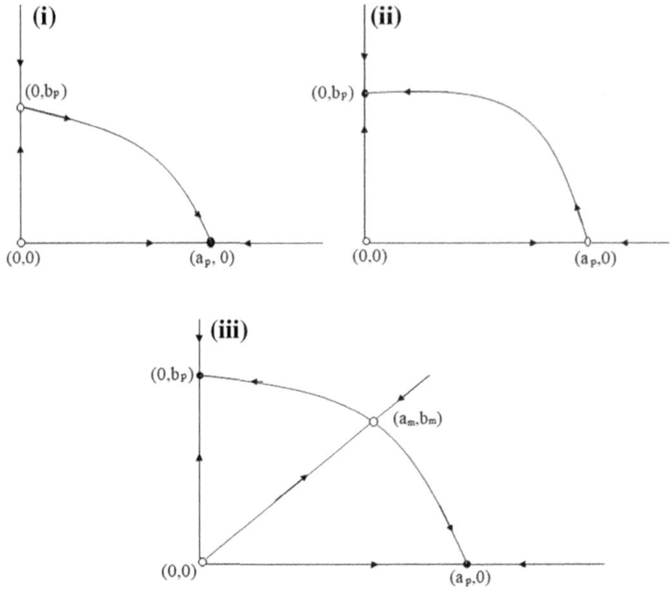

Fig. 3.24 Phase portraits for the planar system of ordinary differential equations (3.56a) corresponding to the above cases (*i*), (*ii*), and (*iii*)

long waves are often governed by the well-known Korteweg-de Vries (KdV) equation, whose wave packet solutions are always stable under modulation, it is appropriate to explore the prospect of the MI arising from the interaction of two wave packets, in the long-wave context.

3.4.1 Coupled NLS Equations for the Lossless Discrete Nonlinear Electrical Network of Fig. 3.1 for $L_3 = \infty$

In the absence of dissipations in the line of Fig. 3.1 and for $L_3 = \infty$, Eq. (3.1) of the network becomes

$$\frac{d^2V_n}{dt^2} + \left(u_0^2 + \gamma\frac{d^2}{dt^2}\right)(2V_n - V_{n+1} - V_{n-1}) + \omega_0^2 V_n = \frac{d^2}{dt^2}\left(\alpha V_n^2 - \beta V_n^3\right), \tag{3.59}$$

$n = 1, 2, \ldots, N$. The expression $2V_n - V_{n-1} - V_{n+1}$ can be approximated with partial derivatives with respect to distance x, from the beginning of the line, assuming that the spacing between two adjacent sections is δ_0 (i.e., $x_n = n\,\delta_0$). Let $V(x,t)$ be a continuous function of the variables x and t so that $V(n,t) = V_n(t)$ and $V_{n\pm1}(t) = V(x \pm \delta_0, t)$. Inserting the Taylor expansion of $V_{n\pm1}(t) = V(x \pm \delta_0, t)$ into Eq. (3.59) and by eliminating terms of higher order than δ_0^3 yield

$$\frac{\partial^2 V}{\partial t^2} - \delta_0^2 u_0^2 \frac{\partial^2 V}{\partial x^2} + \omega_0^2 V - \delta_0^2 \lambda \frac{\partial^4 V}{\partial t^2 \partial x^2} = \alpha \frac{d^2 V^2}{dt^2} - \beta \frac{d^2 V^3}{dt^2}. \quad (3.60)$$

Instead of the line parameters (3.2), we use in this section the network parameters

$$\begin{aligned} &L_1 = 200\,\mu H,\ L_2 = 470\,\mu H,\ V_b = 2\ \text{V},\ C_0 = 320\ \text{pF}, \\ &C_S \le 56\ \text{pF},\ \alpha = 0.21\ \text{V}^{-1},\ \beta = 0.0197\ \text{V}^{-2}. \end{aligned} \quad (3.61)$$

To derive a well-known CNLS equations that may admit analytical solutions, we consider the interaction of any two wave packets centered at (k_1, ω_1) and (k_2, ω_2) [15], where k_j and ω_j are the wave vectors and the frequencies, respectively, $j = 1, 2$. To resolve the weakly nonlinear equation, we introduce two different low timescales $T_1 = \varepsilon t$ and $T_2 = \varepsilon^2 t$ in addition to the original timescale $T_0 = t$. Moreover, we introduce the large spatial scale $X_1 = \varepsilon x$ in addition to the original $X_0 = x$, where ε is a small parameter. Hence, the solution of equation (3.60) is assumed to have the following general form:

$$V = \varepsilon U_1 + \varepsilon^2 U_2 + \varepsilon^3 U_3 + O(\varepsilon^4), \quad (3.62)$$

where $U_j = U_j(X_0, X_1, T_0, T_1, T_2)$; in Eq. (3.62) the small parameter ε measures the size of the amplitudes of the perturbation. To derive a CNLS equations in the small amplitude and long wavelength limit using a standard reductive perturbation method and complex expansion, the lowest order term U_1 of expansion (3.62) is taken in the form

$$U_1 = u_1 \exp\left[i\,(k_1 X_0 - \omega_1 T_0)\right] + u_2 \exp\left[i\,(k_2 X_0 - \omega_2 T_0)\right] + cc, \quad (3.63)$$

where $u_j = u_j(X_1, T_1, T_2)$ are two complex amplitudes and $c.c$ stands for the complex conjugate of the preceding terms.

Inserting the perturbation expansions (3.62) into the nonlinear partial differential equation (3.60) and equating coefficients of like powers of ε lead to a system of three PDEs in U_1, U_2, U_3. Equation of order ε leads to the linear dispersion relation

$$\omega_0^2 + \delta_0^2 u_0^2 k_j^2 - \left(1 + \delta_0^2 \lambda k_j^2\right) \omega_j^2 = 0,\ j = 1,\ 2. \quad (3.64)$$

It follows from Eq. (3.64) that $\frac{d\omega_j}{dk_j} = \left(u_0^2 - \lambda\omega_0^2\right) \delta_0^2 k_j / \left[\left(1 + \lambda\delta_0^2 k_j^2\right)^2 \omega_j\right]$ and $\lim_{k_j \to +\infty} \omega_j / 2\pi = u_0/(2\pi\sqrt{\lambda})$. Therefore, for small values of the perturbation element C_S, each of the frequencies $f_j = \omega_j/2\pi$ will be an increasing function of the wave number k_j. The corresponding linear spectrum has a gap $f_{j0} = \omega_0/2\pi$ which is the lower cutoff frequency introduced by the parallel inductance L_2 and it is limited by the upper cutoff frequency $f_{j\,\max} = u_0/(2\pi\sqrt{\lambda})$. It is evident that the upper cutoff frequency $f_{j\,\max}$ decreases when the perturbation parameter λ increases. This means that the dispersive element C_S contributes to reduce the network effects on the wave during its propagation.

Using Eq. (3.63) for U_1 and using the condition of the suppression of secular terms in either T_0 or X_0 or both from the equation of order ε^2 yield

$$\frac{\partial u_j}{\partial T_1} + \frac{d\omega_j}{dk_j}\frac{\partial u_j}{\partial X_1} = 0, \; j = 1, \; 2, \tag{3.65}$$

$$\begin{aligned} U_2 = {} & A_1 \exp\left[2i\left(k_1 X_0 - \omega_1 T_0\right)\right] + A_2 \exp\left[2i\left(k_2 X_0 - \omega_2 T_0\right)\right] \\ & + A_3 \exp\left[i\left(\left(k_1 + k_2\right) X_0 - \left(\omega_1 + \omega_2\right) T_0\right)\right] + A_4 \exp\left[i\left(\left(k_1 - k_2\right) X_0 - \left(\omega_1 - \omega_2\right) T_0\right)\right] \\ & + A_5 \exp\left[-i\left(\left(k_1 - k_2\right) X_0 - \left(\omega_1 - \omega_2\right) T_0\right)\right] + c.c., \end{aligned} \tag{3.66}$$

where

$$\begin{aligned} A_1 &= \frac{4\alpha\omega_1^2}{3\left(\omega_1^2 + 5\delta_0^2\lambda k_1^2\omega_1^2 - \delta_0^2 u_0^2 k_1^2\right)} u_1^2, \\ A_2 &= \frac{4\alpha\omega_2^2}{3\left(\omega_2^2 + 5\delta_0^2\lambda k_2^2\omega_2^2 - \delta_0^2 u_0^2 k_2^2\right)} u_2^2, \\ A_3 &= -\frac{4\alpha\left(\omega_1 + \omega_2\right)^2}{\omega_0^2 - \left(\omega_1 + \omega_2\right)^2 + \delta_0^2 u_0^2\left(k_1 + k_2\right)^2 - \delta_0^2\lambda\left(\omega_1 + \omega_2\right)^2\left(k_1 + k_2\right)^2} u_1 u_2, \\ A_4 &= -\frac{2\alpha\left(\omega_1 - \omega_2\right)^2}{\left(\omega_0^2 - \left(\omega_1 - \omega_2\right)^2 + \delta_0^2 u_0^2\left(k_1 - k_2\right)^2 - \delta_0^2\lambda\left(\omega_1 - \omega_2\right)^2\left(k_1 - k_2\right)^2\right)} u_1 u_2^*, \\ A_5 &= -\frac{2\alpha\left(\omega_1 - \omega_2\right)^2}{\left(\omega_0^2 - \left(\omega_1 - \omega_2\right)^2 + \delta_0^2 u_0^2\left(k_1 - k_2\right)^2 - \delta_0^2\lambda\left(\omega_1 - \omega_2\right)^2\left(k_1 - k_2\right)^2\right)} u_1^* u_2, \end{aligned} \tag{3.67}$$

where superscript $*$ denotes complex conjugation. Equation (3.65) means that $u_j(X_1, T_1, T_2)$ as functions of variable X_1 and T_1 represents a traveling wave moving with group velocity $V_{gj} = d\omega_j/dk_j$, that is,

$$u_j = \psi_j\left(\xi = X_1 - \frac{d\omega_j}{dk_j}T_1, \tau = T_2\right), \; j = 1, \; 2. \tag{3.68}$$

Inserting Eqs. (3.63) and (3.66) into the equation of order ε^3 and using again the condition of the suppression of secular terms leads to the desired coupled nonlinear Schrödinger equations

$$i\frac{\partial\psi_j}{\partial\tau} + P_j\frac{\partial^2\psi_j}{\partial\xi^2} + \left[Q_{jj}\left|\psi_j\right|^2 + Q_{j3-j}\left|\psi_{3-j}\right|^2\right]\psi_j = 0, \; j = 1, \; 2, \tag{3.69}$$

where coefficients are given by

$$P_j = -\frac{1}{2\omega_j\left(1+\delta_0^2\lambda k_j^2\right)}\left[\delta_0^2\left(\lambda\omega_j^2 - u_0^2\right) + 4\delta_0^2\lambda k_j\omega_j\frac{d\omega_j}{dk_j} + \left(1+\delta_0^2\lambda k_j^2\right)\left(\frac{d\omega_j}{dk_j}\right)^2\right],$$

$$Q_{jj} = \frac{\omega_j}{1+\delta_0^2\lambda k_j^2}\left[\frac{5\beta}{2} - \frac{4\alpha^2\omega_j^2}{3\left(\omega_j^2 + 5\delta_0^2\lambda k_j^2\omega_j^2 - \delta_0^2 u_0^2 k_j^2\right)}\right],$$

$$Q_{j3-j} = \frac{\omega_j}{1+\delta_0^2\lambda k_j^2}\left[6\beta \right. \tag{3.70}$$

$$+\frac{4\alpha^2\left(\omega_j-\omega_{3-j}\right)^2}{\left(\omega_0^2-\left(\omega_j-\omega_{3-j}\right)^2+\delta_0^2u_0^2\left(k_j-k_{3-j}\right)^2-\delta_0^2\lambda\left(\omega_j-\omega_{3-j}\right)^2\left(k_j-k_{3-j}\right)^2\right)}$$

$$\left.+\frac{4\alpha^2\left(\omega_j+\omega_{3-j}\right)^2}{\omega_0^2-\left(\omega_j+\omega_{3-j}\right)^2+\delta_0^2u_0^2\left(k_j+k_{3-j}\right)^2-\delta_0^2\lambda\left(\omega_j+\omega_{3-j}\right)^2\left(k_j+k_{3-j}\right)^2}\right].$$

It should be noted that at this order (i.e., at order ε^3), the expression for U_3 can be derived (we will not give this cumbersome expression here). In the CNLS equation (3.69), the dispersion term P_j corresponding to jth wave is related to the curvature of the dispersion curve via $P_j = d^2\omega_j/dk_j^2$. Q_{jj} and Q_{j3-j} model carrier self-phase modulation (SPM) and cross-phase modulation (XPM), respectively. It is also important to notice that $P_j = P_j(k_j)$, $Q_{jj} = Q_{jj}(k_j)$, and $Q_{j3-j} = Q_{j3-j}(k_j, k_{3-j})$; moreover, Q_{12} and Q_{21} have the same sign (for any (k_1, k_2), they are simultaneously either negative or positive). In the absence of the cross-phase modulation coefficients, i.e., if $Q_{12} = Q_{21} = 0$, then Eq. (3.69) is just a pair of two separated NLS equations. It is seen from Eq. (3.70) that $Q_{11}(k_1 = k) = Q_{22}(k_2 = k)$ and $Q_{12}(k_1, k_2) = Q_{21}(k_2, k_1)$. For the network parameters (3.61) with $C_S = 56$ pF and $\delta_0 = 0.0175$, In Fig. 3.25, we show (a) the linear dispersive curve showing evolution of the frequency $f_j = \omega_j/2\pi$ as function of the wave number k_j, (b) the dispersive coefficient $P_j(k_j)$, (c) the self-phase modulation coefficient $Q_{jj}(k_j)$, and (d) the cross-phase modulation coefficient $Q_{12}(k_1, k_2)$ for three values of either k_2 or k_1. The dashed and the dash-dotted lines of the linear dispersive curve (a) show the lower cutoff frequency $f_{j0} = 410.389$ kHz and $f_{j\,\max} = 1503.87$ kHz. It is seen from plots (b) and (c) that for the line parameters (3.61) with $C_S = 56$ pF, each of the dispersive coefficient $P_j(k_j)$ and self-phase modulation coefficient $Q_{jj}(k_j)$ changes its sign for positive wave number k_j. Curves of Fig. 3.25d show that the cross-phase modulation coefficient $Q_{j3-j}(k_1, k_2)$ also changes its sign.

For the network parameters (3.61) with $C_S = 56$ pF and $\delta_0 = 0.0175$, we show in Fig. 3.26 the curves of both the self-phase modulation coefficient Q_{jj} (solid line) and cross-phase modulation coefficient Q_{j3-j} (dashed line). Plots of Fig. 3.26 show that for a few values of k_1 and k_2, we have the equality $Q_{jj}(k_j) = Q_{j3-j}(k_j, k_{3-j})$. From Fig. 3.26a, we have either $Q_{11}(21.6376) = Q_{12}(21.6376, 0)$ or $Q_{22}(21.6376) = Q_{21}(0, 21.6376)$. From Fig. 3.26b, we either have $Q_{11}(k_1) = Q_{12}(k_1, 10)$ with $k_1 \in \{2.99098, 6.99262, 11.016\}$ or $Q_{22}(k_2) = Q_{12}(10, k_2)$ with $k_2 \in \{2.99098, 6.99262, 11.016\}$. It is seen from Fig. 3.26c that $Q_{jj}(k_j) = Q_{j3-j}(k_j, k_j)$ with $k_j = 10.5388$. Plots of Fig. 3.26 show that we may have $Q_{jj}(k_j) = Q_{j3-j}(k_j, k_{3-j})$ when either $k_1 \neq k_2$ or $k_1 = k_2$.

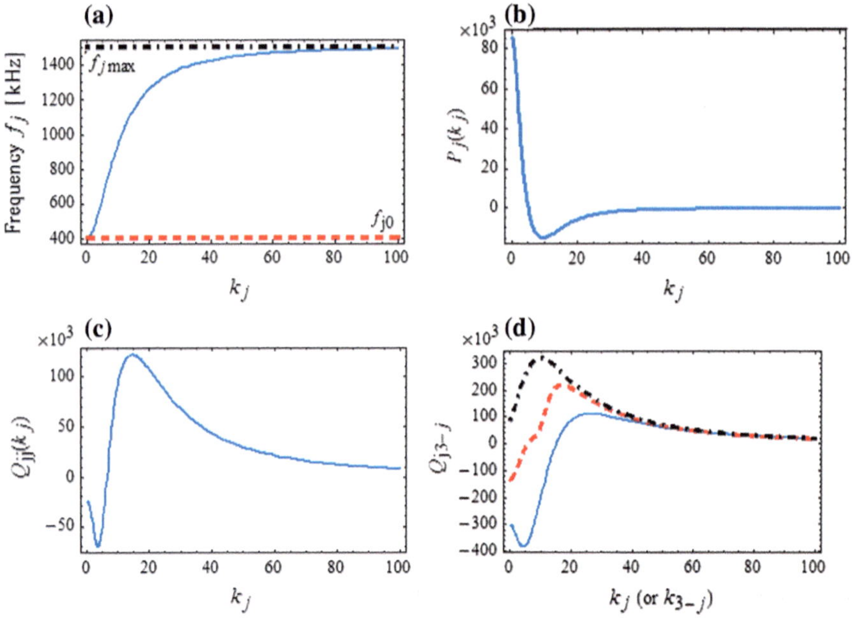

Fig. 3.25 Linear dispersive curve and plots of the divers coefficients of the CNLS equations for the network parameters (3.61) with $C_S = 56$ pF. **a**: Linear dispersive curve showing evolution of the frequency $f_j = \omega_j/2\pi$ as function of the wave number k; **b**: Behavior of the dispersive coefficient P_j as function of the wave number of the carrier k; **c**: Plot of the self-phase modulation coefficient Q_{jj} versus the wave number k_j; **d**: Behavior of the self-phase modulation coefficient Q_{12} or Q_{21} as function of the wave number of the carrier k_1 or k_2 for different values of k_2 or k_1 [$k_2 = 0$ (solid line), $k_2 = 10$ (dashed line), $k_2 = 20$ (dot-dashed line) or $k_1 = 0$ (solid line), $k_1 = 10$ (dashed line), $k_1 = 20$ (dot-dashed line)]

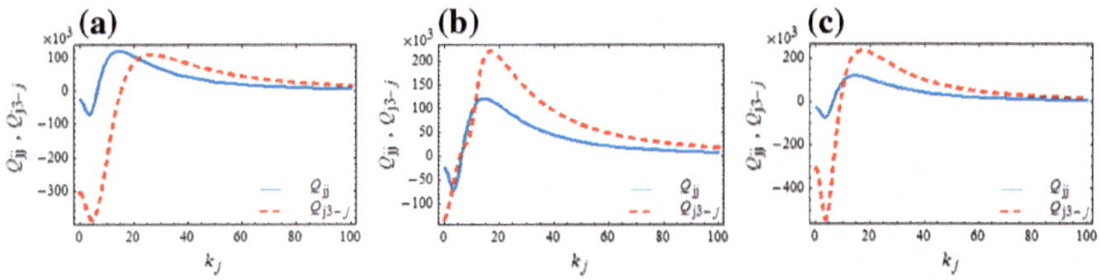

Fig. 3.26 Behavior of both the self-phase modulation coefficient Q_{jj} and cross-phase modulation coefficient $Q_{12}(k_1, k_{20})$ or $Q_{21}(k_{10}, k_2)$ for three values of either k_{10} or k_{20}. **a**: either $k_{10} = 0$ or $k_{20} = 0$; **b**: either $k_{10} = 10$ or $k_{10} = 10$; **c**: $k_1 = k_2$

3.4.2 Modulational Instability

In this section, we investigate the conditions under which a packet of uniform wave trains propagating along the network of Fig. 3.1 will become unstable to a small perturbation. We study the effect of the dispersive element C_S on the MI.

3.4.2.1 Conditions of the Modulational Instability of Stokes Waves

To find the conditions of the modulational instability of Stokes waves, we use the CNLS equations (3.69) derived from the exact Eq. (3.59) governing wave propagation in the network. The starting point is to obtain the Stokes wave, or continuous-wave solutions of system (3.69). The simplest Stokes wave solution of system (3.69) is

$$\begin{pmatrix} \psi_{01}(\xi,\tau) \\ \\ \psi_{02}(\xi,\tau) \end{pmatrix} = \begin{pmatrix} \rho_{01}\exp\left[i\left(Q_{11}\left|\rho_{01}\right|^2 + Q_{12}\left|\rho_{02}\right|^2\right)\tau\right] \\ \\ \rho_{02}\exp\left[i\left(Q_{21}\left|\rho_{01}\right|^2 + Q_{22}\left|\rho_{02}\right|^2\right)\tau\right] \end{pmatrix}, \tag{3.71}$$

where ρ_{01} and ρ_{02} constant complex amplitudes of the wave guides. It should be noted that the constant complex amplitudes ρ_{01} and ρ_{02} of the Stokes wave solution (3.71) define a wave having fixed polarization, i.e., a wave where the division of intensity between the two components ρ_1 and ρ_2 is uniform in both time and space.

The linear stability of solution (3.71) can be investigated by introducing the perturbations as follows:

$$\psi_1(\xi,\tau) = \rho_{01}\left(1+\epsilon A_1\right)\exp\left[i\omega_1\tau\right],\ \psi_2(\xi,\tau) = \rho_{02}\left(1+\epsilon A_2\right)\exp\left[i\omega_2\tau\right] \tag{3.72}$$

where $\omega_j = Q_{jj}\left|\rho_{0j}\right|^2 + Q_{j3-j}\left|\rho_{03-j}\right|^2$, ϵ is a small parameter, and $A_j = A_j(\xi,\tau)$, $j = 1, 2$, are the perturbation functions. Substituting Eq. (3.72) into the coupled NLS equations (3.69) and keeping only terms linear in ϵ lead to the linearized disturbance equations

$$i\frac{\partial A_1}{\partial\tau} + P_1\frac{\partial^2 A_1}{\partial\xi^2} + Q_{11}\left|\rho_{01}\right|^2\left(A_1 + A_1^*\right) + Q_{12}\left|\rho_{02}\right|^2\left(A_2 + A_2^*\right) = 0, \tag{3.73a}$$

$$i\frac{\partial A_2}{\partial\tau} + P_2\frac{\partial^2 A_2}{\partial\xi^2} + Q_{22}\left|\rho_{02}\right|^2\left(A_2 + A_2^*\right) + Q_{21}\left|\rho_{01}\right|^2\left(A_1 + A_1^*\right) = 0, \tag{3.73b}$$

Because of the presence of conjugates in system (3.73a), (3.73b), it is most convenient to express the perturbation functions A_1 and A_2 as linear combinations of pure Fourier modes

$$\begin{pmatrix} A_1(\xi,\tau) \\ \\ A_2(\xi,\tau) \end{pmatrix} = \begin{pmatrix} a_1\exp\left[i\left(K\xi - \mathrm{Re}\left[\Omega\right]\tau\right)\right] + a_2^*\exp\left[i(\mathrm{Re}\left[\Omega\right]\tau - \xi)\right] \\ \\ b_1\exp\left[i\left(K\xi - \mathrm{Re}\left[\Omega\right]\tau\right)\right] + b_2^*\exp\left[i\left(\mathrm{Re}\left[\Omega\right]\tau - \xi\right)\right] \end{pmatrix}\exp\left[\mathrm{Im}\left[\Omega\right]\tau\right] \tag{3.74}$$

where the real wave number $K \geq 0$ of the disturbance and the complex frequency Ω parameterize the two eigenmodes A_1 and A_2. As it is seen from Eq. (3.74), a plane wave with a given carrier-wave frequency will be unstable to any small modulation if the imaginary part $\mathrm{Im}[\Omega]$ of the modulation frequency Ω is positive.

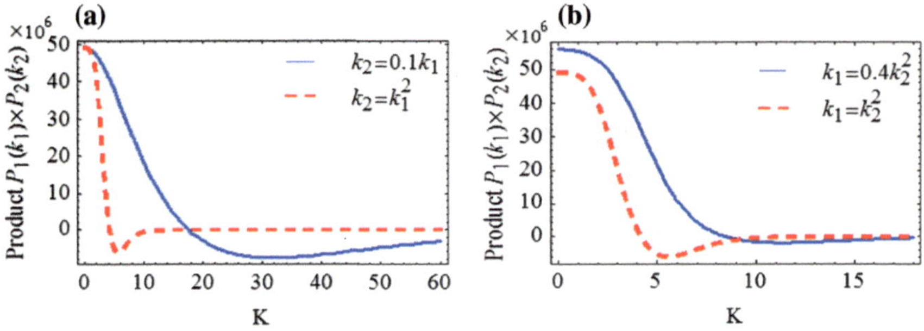

Fig. 3.27 Variations of the product $P_1(k_1)P_2(k_2)$ as function of wave number when either k_1 or k_2 is taken as a function k_2 or k_1 **a** Curves obtained with $C_S = 56$ pF; **b**: Curves obtained with $C_S = 9.6$ pF. The plot is obtained with line parameter (3.61) and $\delta_0 = 0.05$

Substituting Eq. (3.74) into the linearized system (3.73a), (3.73b) and using the condition of nontrivial solution, i.e., the condition that $(a_1, a_2, b_1, b_2) \neq (0, 0, 0, 0)$ lead to the following dispersion relation of the modulation:

$$\Omega = \pm K\sqrt{\frac{P_1^2 + P_2^2}{2}K^2 - P_2 Q_{22}|\rho_{02}|^2 - P_1 Q_{11}|\rho_{01}|^2 \pm \frac{1}{2}\sqrt{\Delta(K)}}, \qquad (3.75)$$

where

$$\begin{aligned}\Delta(K) = {} & 16 P_1 P_2 Q_{12} Q_{21}|\rho_{01}|^2|\rho_{02}|^2 \\ & + \left[\left(P_1^2 - P_2^2\right)K^2 + 2\left(P_2 Q_{22}|\rho_{02}|^2 - P_1 Q_{11}|\rho_{01}|^2\right)\right]^2.\end{aligned} \qquad (3.76)$$

Analyzing the sign of $\Delta(K)$ leads to

(1) If $P_1 P_2 Q_{12} Q_{21} \geq 0$, then $\Delta \geq 0$ for all wave number K;
(2) If $P_1 P_2 Q_{12} Q_{21} < 0$, then Δ will be either nonnegative or negative for all wave number K (this situation happens if $P_1^2 - P_2^2 = 0$), or nonnegative for some wave numbers K and negative for other wavenumebrs K (this situation occurs when $P_1^2 - P_2^2 \neq 0$). Thus, a necessary condition for $\Delta(K)$ to be negative is that $P_1 P_2 Q_{12} Q_{21} < 0$. Because Q_{12} and Q_{21} are of the same sign, we have $\text{sign}[P_1 P_2 Q_{12} Q_{21}] = \text{sign}[P_1 P_2]$. Figure 3.27 shows the behavior of the product $P_1(k_1)P_2(k_2)$ for two values of parameter λ when either k_1 is a function of k_2 or k_2 is a function of k_1. Plots of this figure show that the product $P_1 P_2$ can be positive, negative, or nil.

As it is seen from Eqs. (3.75) and (3.76), the frequency of modulation Ω will be a complex with a non-nil imaginary part if and only, if either

$$\Delta(K) < 0 \text{ or} \qquad (3.77a)$$

$$\Delta(K) \geq 0 \text{ and } \delta_\pm(K) = \frac{P_1^2 + P_2^2}{2}K^2 - P_1 Q_{11}|\rho_{01}|^2 - P_2 Q_{22}|\rho_{02}|^2 \mp \frac{1}{2}\sqrt{\Delta(K)} < 0. \qquad (3.77b)$$

It follows from expression (3.76) for discriminant Δ that condition (3.77a) can occur only in the presence of the XPM coefficients. In the first case, that is, when conditions (3.77b) are satisfied, the growth rate (gain) of modulational instability is given by the relation

$$|\mathrm{Im}\,[\Omega]| = \sqrt{\frac{-a+\sqrt{a^2+b^2}}{2}}, \tag{3.78a}$$

$$a = \frac{P_1^2+P_2^2}{2}K^2 - P_2Q_{22}\,|\rho_{02}|^2 - P_1Q_{11}\,|\rho_{01}|^2 \text{ and } b = \pm\frac{\sqrt{-\Delta(K)}}{2}. \tag{3.78b}$$

The second case, that is, when conditions (3.77b) are satisfied, the growth rate (gain) of modulational instability is given by the relation

$$|\mathrm{Im}\,[\Omega]|_{\mp} = \sqrt{P_1Q_{11}\,|\rho_{01}|^2 + P_2Q_{22}\,|\rho_{02}|^2 - \frac{P_1^2+P_2^2}{2}K^2 \pm \frac{1}{2}\sqrt{\Delta(K)}}. \tag{3.79}$$

Condition (3.79) may define either one branch of the growth rate of MI or two branches of the growth rate of MI, the upper branch associated with sign "+" corresponding to $\delta_-(K)$, and the lower branches associated with sign "− "corresponding to $\delta_+(K)$. It is obvious that two branches of growth rate of MI exist for only those wave numbers K for which both $\delta_-(K)$ and $\delta_+(K)$ are simultaneously negative.

Remark 1: Because $\lim_{K\to 0}\Delta(K) = 16P_1P_2Q_{12}Q_{21}\,|\rho_{01}|^2\,|\rho_{02}|^2$, $\lim_{K\to+\infty}\Delta(K) = +\infty$, and $\lim_{K\to+\infty}\delta_{\pm}(K) = +\infty$, we conclude that conditions (3.77a) and (3.77b) can occur only for "small" nonzero wave numbers K (here, "small" in comparison with infinity). Therefore, the phenomenon of the modulational instability can be observed only for "small" nonzero wave numbers; consequently, plane waves will be stable under modulation for high-wave numbers K. Such instability and stability are called **long-wave instability** and **short-wave stability**, respectively.

Remark 2: Let us examine the special case when $Q_{12} = Q_{21} = 0$. In the absence of the XPM coefficients, i.e., when $Q_{12} = Q_{21} = 0$, we have $\Delta(K) \geq 0$ and condition (3.77b) becomes $\delta_+(K) = P_1^2K^2 - P_1Q_{11}\,|\rho_{01}|^2 < 0$ and $\delta_-(K) = P_2^2K^2 - P_2Q_{22}\,|\rho_{02}|^2$. The necessary conditions to have either $\delta_+(K) < 0$ or $\delta_-(K) < 0$ are $P_1Q_{11} > 0$ or $P_2Q_{22} > 0$, and we recover the condition of the MI of plane waves for a single NLS equation. Thus, if $P_1Q_{11} > 0$ and $P_2Q_{22} > 0$, then the two components of the plane wave solution (3.71) will be simultaneously unstable under modulation for $0 < K < \min\left(\frac{Q_{11}}{P_1}, \frac{Q_{11}}{P_2}\right)$; for the modulation wave numbers K taken from interval $\min\left(\frac{Q_{11}}{P_1}, \frac{Q_{11}}{P_2}\right) < K < \max\left(\frac{Q_{11}}{P_1}, \frac{Q_{11}}{P_2}\right)$, only one of the components will be unstable.

Remark 3: It follows from Remarks 1 and 2 that we can simultaneously have $P_1Q_{11} < 0$, $P_2Q_2 < 0$, and $P_1Q_1 < 0$ (for example, these three conditions are satisfied for $C_S = 56\ \mu\mathrm{F}$, $k_1 = 250$, and $k_2 = 25$). Because $Q_{12}Q_{21} > 0$, we have

$P_1 P_2 Q_{12} Q_{21} < 0$; if one of conditions (3.77a) and (3.77b) is satisfied, then the plane waves will be unstable. This means that cross-phase modulations can induce instability in two coupled waveguides even if each participating mode is in the stable regime.

In order to conclude on the MI of a given Stokes wave solution (3.71), we follow the below algorithm.

(a) If either $P_1 P_2 Q_{12} Q_{21} \geq 0$ or $P_1 P_2 Q_{12} Q_{21} < 0$ and $\Delta(K) \geq 0$, then if

$$\frac{P_1^2 + P_2^2}{2} K^2 - P_2 Q_{22} |\rho_{02}|^2 - P_1 Q_{11} |\rho_{01}|^2 \pm \frac{1}{2}\sqrt{\Delta(K)} < 0,$$

the plane wave is unstable under modulation and the growth rate of MI is given by Eq. (3.79); if

$$\frac{P_1^2 + P_2^2}{2} K^2 - P_2 Q_{22} |\rho_{02}|^2 - P_1 Q_{11} |\rho_{01}|^2 \pm \frac{1}{2}\sqrt{\Delta(K)} \geq 0,$$

the plane wave is stable under modulation.

(b) If $P_1 P_2 Q_{12} Q_{21} < 0$ and $\Delta(K) < 0$, then the plane wave is unstable under modulation and the growth rate of MI is defined by Eq. (3.78a).

Figure 3.28 shows typical situations for which $\Delta(K)$ changes its sign. We have used the sign of functions $\Delta(K)$, $\delta_-(K)$, and $\delta_+(K)$ to divide the domain of variation of the modulation wave number K into sub-regions [see Fig. 3.28a and c]. Regions (I), (II), and (III) of Fig. 3.28a are associated with the sign of discriminant $\Delta(K)$: $\Delta(K)$ is positive in regions (I) and (III), and negative in region (II). In region (I), both $\delta_-(K)$ and $\delta_+(K)$ are negative and lead respectively to the upper and the lower branches of instability growth rate shown in Fig. 3.28b. In region (III), both $\delta_-(K)$ and $\delta_+(K)$ are positive, meaning that region (III) is associated with the stability domain of the plane wave. Region (II) where $\Delta(K)$ is negative is associated with the instability growth rate given by Eqs. (3.78a), (3.78b) and shown in Fig. 3.28b [solid line]. The curve of discriminant $\Delta(K)$ in Fig. 3.28c divides the region of variation of the modulation wave number K into two main regions, region (i) where $\Delta(K) < 0$ and regions (ii), (iii), and (iv) where $\Delta(K) > 0$. δ_+ is positive in regions (ii), (iii), and (iv), while $\delta_-(K)$ is positive in regions (ii) and (iv), and is negative in region (iii); this means that $\delta_-(K)$ and $\delta_+(K)$ are positive in regions (ii) and (iv). Therefore, regions (ii) and (iv) are regions of stable plane waves, while regions (i) and (iii) are regions of MI of the plane waves. The instability growth rates obtained from regions (i) and (iii) are respectively shown by the solid line and dash-dotted line in Fig. 3.28d.

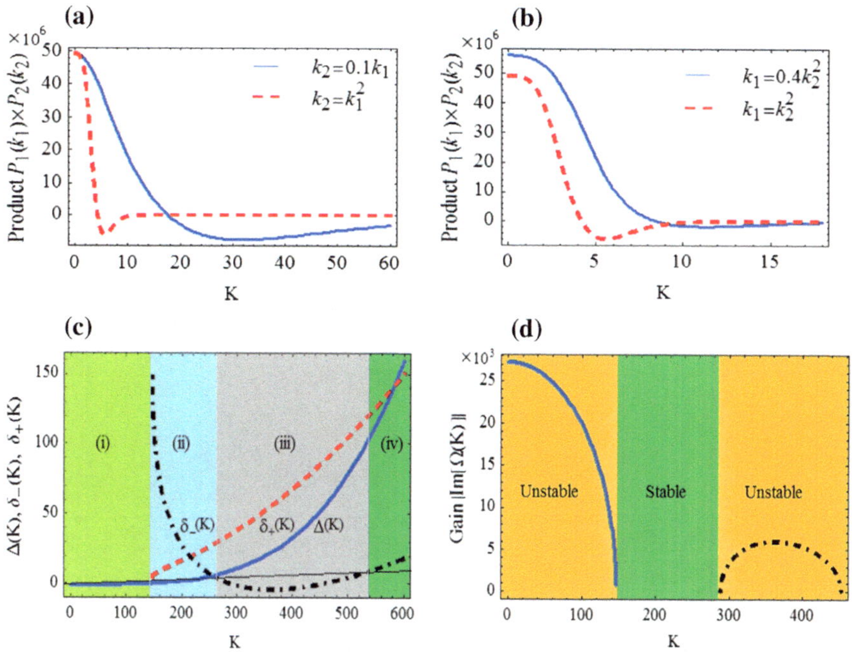

Fig. 3.28 Instability domain in terms of the modulation wave number K for the network parameter (3.61), $\delta_0 = 0.05$, $|\rho_{01}|^2 = 1$, and $|\rho_{02}|^2 = 0.412$. **a, c**: Sign of functions $\Delta(K)$, $\delta_-(K)$, and $\delta_+(K)$; **b, d**: Instability growth rate $|\text{Im}\,[\Omega]|$ according to either Eq. (3.79) or Eqs. (3.78a), (3.78b). Plots **a** and **b** are generated with $C_S = 5.6$ pF for the wave packets propagating at frequency $(f_1, f_2) = (1087.84$ kHz, 1934.78 kHz$)$. Plots **c** and **d** are generated with $C_S = 0.64$ pF for the wave packets propagating at frequency $(f_1, f_2) = (2087.61$ kHz, 1110.29 kHz$)$

3.4.2.2 Effects of the Frequency of Wave Packets and Element C_S of the Network on the MI

We now turn to the investigation of effects of the frequency $f_{12} = (\omega_1(k_1)/2\pi, \omega_2(k_2)/2\pi)$ and perturbation element C_S on the modulational instability. As we have seen in the previous section, the condition of the MI of the Stokes wave solutions of the CNLS equation (3.69) mainly depends on the sign of the product $P = P_1 P_2 Q_{12} Q_{21}$. In other hand, the sign of this product for any given set of line parameter with different values of perturbation element C_s depends on both C_S and the frequency $f_{12} = (\omega_1(k_1)/2\pi, \omega_2(k_2)/2\pi)$ of the wave packets. It is thus reasonable to investigate the effects of both frequency f_{12} and perturbation element C_S on the MI.

Effect of Frequency of the Wave Packets on the MI

Analyzing Eqs. (3.77a) and (3.77b), it is seen that for a given set of the network parameters, δ_0, and amplitudes ρ_{01} and ρ_{02}, the phenomenon of the MI can occur, depending on the frequency $f_{12} = (f_1, f_2)$, in each of the following situations:

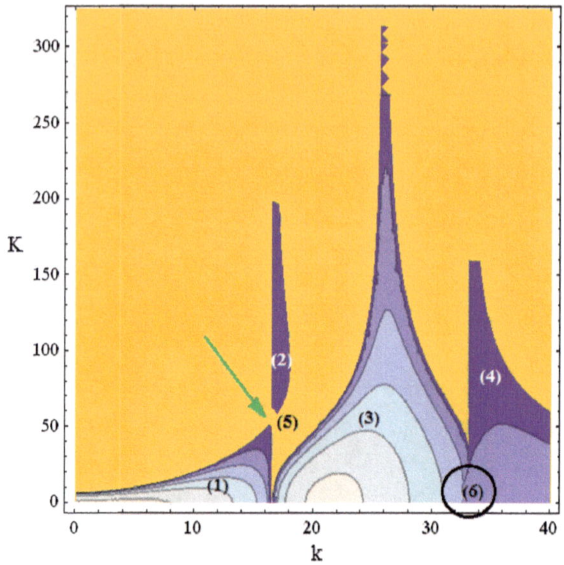

Fig. 3.29 Instability and stability domain in the (k, K) plane according to Eqs. (3.77a) and (3.77b), where $k_1/2 = k_2 = k$ and $k_2 = k$. The network parameters are given by Eq. (3.61) with $C_S = 5.44$ pF, while other parameters are $\delta_0 = 0.05$, $|\rho_{01}|^2 = 1$, $|\rho_{02}|^2 = 0.412$. Yellow color shows the stability region, while other colors show the instability region

(i) $\Delta(K)$ is positive for all modulation wave numbers K, $\delta_-(K)$ is negative for all K, and $\delta_+(K)$ is positive for all K. In this situation, the instability growth rate of MI is defined by Eq. (3.79).
(ii) $\Delta(K)$ changes its sign and both $\delta_-(K)$ and $\delta_+(K)$ are positive for all wave numbers K. In this situation, the instability growth rate of MI is defined by Eqs. (3.78a), (3.78b).
(iii) $\Delta(K)$ changes its sign and both $\delta_-(K)$ and $\delta_+(K)$ are negative for a set of wave numbers K satisfying condition $\Delta(K) > 0$. In this situation, Eqs. (3.78a), (3.78b) and (3.79) define the instability growth rate; moreover, Eq. (3.79) will define two branches of the instability growth rate.
(iv) $\Delta(K)$ changes its sign, $\delta_+(K)$ is positive for all wave numbers K, $\delta_-(K)$ changes its sign in the region of the positivity of $\Delta(K)$. In the present situation, the instability growth rates of MI are defined by Eqs. (3.78a), (3.78b) and (3.79); in this case, Eq. (3.79) defines only one branch of the instability growth rate.

The above four situations are summarized in Fig. 3.29 and are detailed in Fig. 3.30. Situation (i) occurs when the propagating frequency of the wave packets is associated with wave numbers $k_1 = 2k$ and $k_2 = k$, where k is taken from either region (1) or region (4) of Fig. 3.29. Case (ii) happens for wave packets propagating at frequency associated with wave numbers $k_1 = 2k$ and $k_2 = k$, where k is taken from region (3) of Fig. 3.29. Situation (iii) occurs for wave packets propagating at frequency corresponding to wave numbers $k_1 = 2k$ and $k_2 = k$, where k is taken from region (6) of Fig. 3.29. The (iv) and last situation happens for wave packets propagating at frequency associated with wave numbers $k_1 = 2k$ and $k_2 = k$, where k is taken from region (2) of Fig. 3.29. In Fig. 3.29, (5) shows a region between two regions where

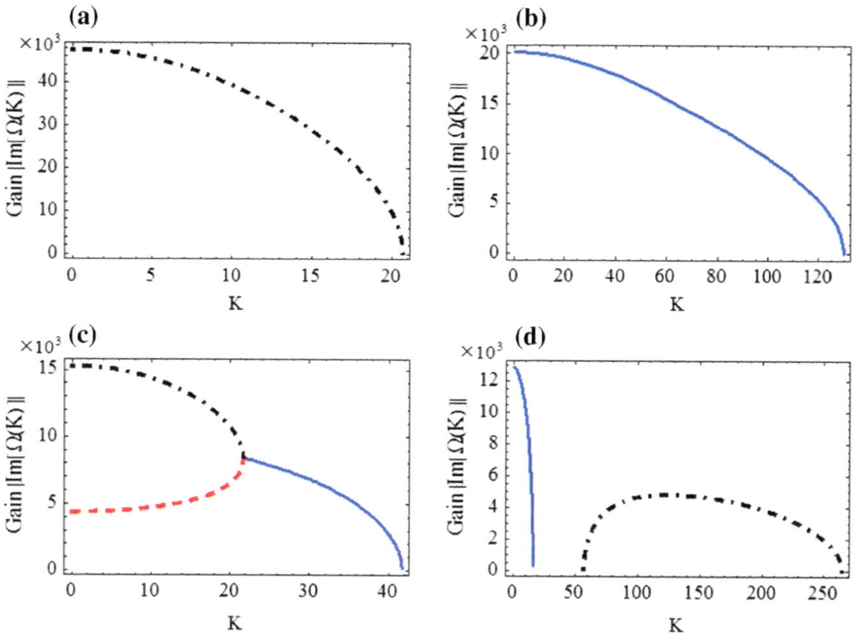

Fig. 3.30 Instability growth rate according to Eqs. (3.78a), (3.78b) and (3.79) for the same parameters as in Fig. 3.28. **a**: Instability growth rate of MI corresponding to situation (i) described in the text and obtained from wave packets propagating at frequency f_{12} = (744.832 kHz, 515.979 kHz). **b**: Instability growth rate of MI corresponding to situation (ii) described in the text and obtained from wave packets propagating at frequency f_{12} = (1699.01 kHz, 955.884 kHz). **c**: Instability growth rate of MI corresponding to situation (iii) described in the text and obtained from wave packets propagating at frequency f_{12} = (1943.95 kHz, 1091.25 kHz). The dashed line and the dash-dotted line show the lower and the upper branches of growth rate, respectively, obtained from Eq. (3.79); the solid line shows the instability growth rate according to Eqs. (3.78a), (3.78b). **d**: Instability growth rate of MI corresponding to situation (iv) described in the text and obtained from wave packets propagating at frequency f_{12} = (1112.96 kHz, 667.541 kHz). The solid line and the dash-dotted line are, respectively, obtained from Eqs. (3.78a), (3.78b), and (3.79)

the phenomena of the MI is observed (see Fig. 3.30d): it is a region where $\Delta(K)$, $\delta_-(K)$, and $\delta_+(K)$ are all positive.

Examples of situations (i), (ii), (iii), and (iv) are, respectively, shown in Fig. 3.30a, b, c and d. This four figures are generated with $k = 10$, $k = 28$, $k = 33$,, and $k = 16.9$, respectively.It is seen from Fig. 3.30d that plane waves are stable under modulation for wave numbers $K \in [16.225, 57.254] = [K_{01}, K_{02}]$; this region of stability of plane waves corresponds to region (5) in Fig. 3.29. For a better understanding of this situation, we show in Fig. 3.31 the domain of instability and stability associated to Fig. 3.30d; here, we repeat Fig. 3.30d. It is seen from Fig. 3.31 that $K_{01} \approx 16.225$ is a zero of $\Delta(K)$, while $K_{02} \approx 57.254$ is a zero of $\delta_-(K)$; moreover, all the three functions $\Delta(K)$, $\delta_-(K)$, and $\delta_+(K)$ are positive on $[K_{01}, K_{02}]$, meaning that none of conditions (3.77a) and (3.77b) is satisfied. Therefore, $[K_{01}, K_{02}]$ is a

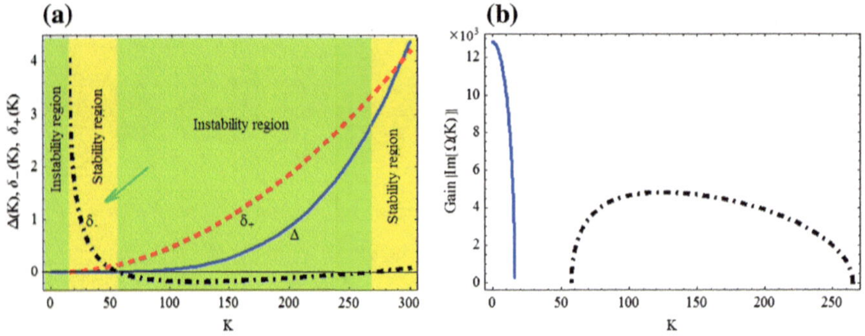

Fig. 3.31 **a** Instability domain according to Eqs. (3.77a) and (3.77b) and **b** instability growth rate of MI showed in Fig. 3.29d. The parameters are the same as in Fig. 3.29

region of modulational stability of plane waves. The green arrow in Figs. 3.29 and 3.31a shows the same stability region between two instability regions.

The instability and stability regions in K - axis are displayed in Fig. 3.32a and b for $f_{12} = (1088.54$ kHz, 1939.2 kHz) [left panel] and
$f_{12} = (1939.2$ kHz, 1088.54 kHz) [right panel],
that is, when the frequencies of the two waves of the wave packets are inverted. Figure 3.32c and d show the corresponding instability growth rate of MI according either to Eqs. (3.78a), (3.78b), or Eq. (3.79). Figure 3.31 shows how much the MI is affected when the frequencies of the two waves of the wave packets are inverted.

Effects Element C_S of the Network on the MI

Here, we investigate the effects of the dispersive element C_S on the MI. In order to investigate the effects of the dispersive element C_S on the MI, we maintain all line parameters (3.61) constant, except element C_S which varies between 0 and C_S. Therefore, parameter λ varies between 0 and 0.175. We have maintain δ_0, ρ_{01}, and ρ_{02} at $\delta_0 = 0.05$, $|\rho_{01}|^2 = 1$, and $|\rho_{02}|^2 = 0.412$. We also maintain the wave number of the wave packets at $(k_1, k_2) = (66, 33)$. With these parameters, we show in Fig. 3.33 the variation of the instability growth rate $|\mathrm{Im}\,[\Omega]|$ versus parameter λ and modulation wave number K. Based on the sign of Δ, δ_-, and δ_+, we who the instability growth rate in four plots. Plots (a), (c), and (d) are obtained for (λ, K) that satisfy condition $\Delta(\lambda, K) > 0$, while plot (b) corresponds to $\Delta(\lambda, K) < 0$. Plots (a) and (d) are obtained with $\delta_-(\lambda, K) < 0$, while plot (c) is obtained for (λ, K) that satisfy condition $\delta_-(\lambda, K) < 0$. Analyzing plots of Fig. 3.33, it is seen that (i) for either $0 \leq \lambda < 0.00119$ or $0.00119 \leq \lambda \leq 0.016199477$, or $0.01733112 \leq \lambda < 0.175$, there exists only one branch of the instability growth; (ii) for $0.016199477 < \lambda < 0.01733112$, there exist two branches of instability growth rate, one shown in Fig. 3.33c and another in Fig. 3.33d. Indeed, on region $0.016199477 < \lambda < 0.01733112$, discriminant Δ is negative while functions δ_- and δ_+ are positive for a set of modulation wave numbers K. Figure 3.33a shows that for small-wave number disturbances K,

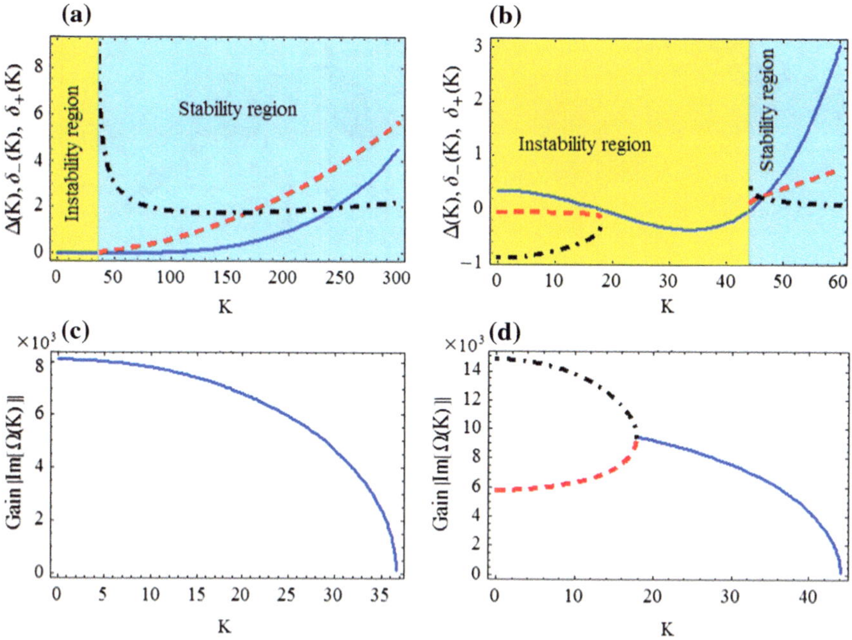

Fig. 3.32 Effects of the inversion of the wave frequencies of the wave packets on the MI. Plots of the top panels show the instability and stability regions in K-axis, while the plots of the bottom panels show the instability growth rate of MI according either to Eqs. (3.78a), (3.78b) or Eq. (3.79). In the top panels, the solid line, dashed line, and dash-dotted show the variation of function $\Delta(K)$, $\delta_+(K)$, and $\delta_-(K)$, respectively. The different plots are generated with the network parameters (3.61) with $C_S = 5.44$ pF, $\delta_0 = 0.05$, $|\rho_{01}|^2 = 1$, $|\rho_{02}|^2 = 0.412$ and correspond to wave packets propagating at frequency. **a**, **c**: $f_{12} =$ (1088.54 kHz, 1939.2 kHz) and **b**, **d**: $f_{12} =$ (1939.2 kHz, 1088.54 kHz)

the instability growth rate decreases as parameter λ increases; inversely, the instability growth rate increases as parameter λ increases for large-wave number disturbances K. It is also seen from Fig. 3.33a that the instability domain (in modulation wave numbers K) enhances when parameter λ increases. Moreover, the instability growth rate is a decreasing function of modulation wave number K. Figure 3.33b show that (i) for every parameter $\lambda \in [0.00119, 0.016199477]$, the instability growth rate decreases as a function of modulation wave number K; (ii) for some $\lambda_0 \in]0.00119, 0.016199477[$, the instability growth rate increase (respectively, decreases) as λ increases in interval $[0.00119, \lambda_0]$ (respectively, λ increases in interval $[\lambda_0, 0.016199477]$); (iii) the instability region (with respect to the modulation wave number K) enhances as $\lambda \to \lambda_0$. From Fig. 3.33c, it is clearly seen that (i) the instability growth rate increases as a function of modulation wave number K; (ii) the instability region (with respect to the modulation wave number K) enhances when parameter λ increases in interval $0.016199477 < \lambda < 0.01733112$. Lastly, Fig. 3.33d shows that (i) in region $0.01733112 < \lambda \leq 0.01733112$, the instability growth rate decreases as a function of wave number K, while the instability region

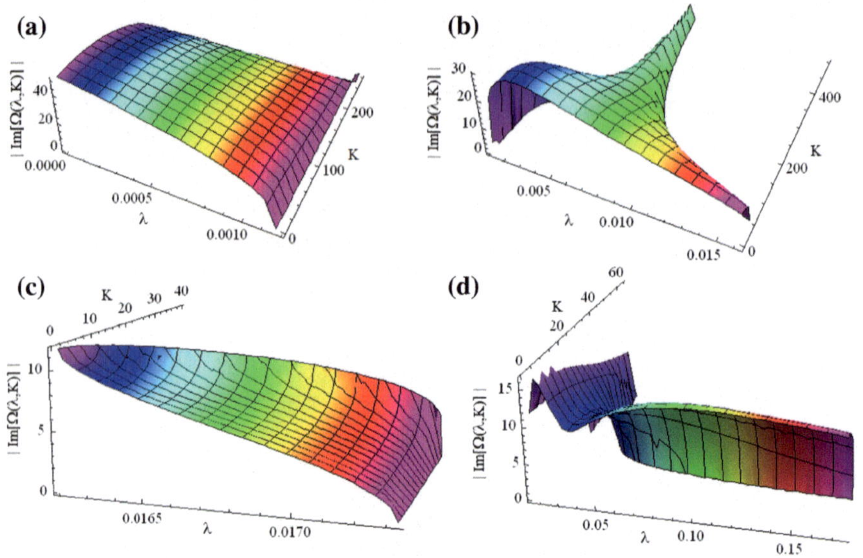

Fig. 3.33 Instability growth rate $|\mathrm{Im}\,[\Omega]|$ versus the dispersive parameter λ and modulation wave number K for parameters shown in the text. **a**: Instability growth rate associated with $0 \le \lambda < 0.00119$ according to Eq. (3.79) when $\Delta(\lambda, K) > 0$ and $\delta_-(\lambda, K) < 0$; **b**: Instability growth rate associated with $0.00119 \le \lambda \le 0.016199477$ according to Eqs. (3.78a), (3.78b) when $\Delta(\lambda, K) < 0$; **c**: Instability growth rate associated with $0.016199477 < \lambda < 0.01733112$ according to Eq. (3.79) when $\Delta(\lambda, K) > 0$ and $\delta_+(\lambda, K) < 0$; **d**: Instability growth rate associated with $0.016199477 < \lambda < 0.175$ according to Eq. (3.79) when $\Delta(\lambda, K) > 0$ and $\delta_-(\lambda, K) < 0$

(in K) enhances as parameter λ increases; (ii) in region $0.01733112 < \lambda \le 0.175$, the instability growth rate decrease (as a function of wave number K); the instability growth rate decreases as parameter λ increases; the instability domain (in K) shrinks as parameter λ increases.

3.4.3 Analytical Study of Matter-Wave Solitons in the Network

Based on the CNLS equation (3.69), we analytically investigate in this section the propagation of matter-solitary waves in the network of Fig. 3.1. We start by finding analytical solitary wave solutions of system (3.69). We seek traveling wave solutions of (3.69) in the form

$$\begin{pmatrix} \psi_1(\xi,\tau) \\ \\ \psi_2(\xi,\tau) \end{pmatrix} = \begin{pmatrix} \rho_1(z)\exp\left[iq_1\xi - i\lambda_1\tau\right] \\ \\ \rho_2(z)\exp\left[iq_2\xi - i\lambda_2\tau\right] \end{pmatrix},\ z = \xi - 2P_2q_2\tau, \tag{3.80}$$

where $\rho_1(z)$ and $\rho_2(z)$ are two real functions of variable z, and λ_j and q_j are real constants. Inserting Eq. (3.80) into system (3.69) yields

$$P_1 \frac{d^2\rho_1}{dz^2} + \left(\lambda_1 - P_1 q_1^2\right)\rho_1 + Q_{11}\rho_1^3 + Q_{12}\rho_2^2\rho_1 = 0, \quad (3.81a)$$

$$P_2 \frac{d^2\rho_2}{dz^2} + \left(\lambda_2 - P_2 q_2^2\right)\rho_2 + Q_{22}\rho_2^3 + Q_{21}\rho_1^2\rho_2 = 0, \quad (3.81b)$$

$$P_1 q_1 - P_2 q_2 = 0. \quad (3.81c)$$

Condition (3.81c) means that $q_1 = P_2 P_1^{-1} q_2$.

Bright–bright Solitary Wave Solutions

The bright–bright solitary wave solutions of system (3.81a), (3.81b) is sought in the form

$$(\rho_1, \rho_2) = \left(\frac{\eta_1}{a_1 \exp[\mu z + \mu_0] + b_1 \exp[-\mu z - \mu_0]}, \frac{\eta_2}{a_2 \exp[\mu z + \mu_0] + b_2 \exp[-\mu z - \mu_0]}\right), \quad (3.82)$$

where η_1, η_2, μ, μ_0, a_1, a_2, b_1, and b_2 are nonzero real constants to be determined. For ρ_1 and ρ_2 to be bounded, it is necessary and sufficient that $a_1 b_1 > 0$ and $a_2 b_2 > 0$. Inserting Eq. (3.82) into system (3.81a), (3.81b) yields

$$\begin{cases} P_j\mu^2 + \lambda_j - P_j q_j^2 = 0, \\ a_{3-j}^2 Q_{jj}\eta_j^2 + a_j^2 Q_{j3-j}\eta_{3-j}^2 - 8P_j a_j b_j a_{3-j}^2\mu^2 = 0, \\ b_{3-j}^2 Q_{jj}\eta_j^2 + b_j^2 Q_{j3-j}\eta_{3-j}^2 - 8P_j a_j b_j b_{3-j}^2\mu^2 = 0, \\ a_{3-j}b_{3-j} Q_{jj}\eta_j^2 + a_j b_j Q_{j3-j}\eta_{3-j}^2 - 8P_j a_j b_j a_{3-j} b_{3-j}\mu^2 = 0, \end{cases} \quad (3.83a)$$

where $j = 1, 2$. Solving the system (3.83a) for $Q_{3-jj}Q_{j3-j} - Q_{3-j3-j}Q_{jj} \neq 0$ yields,

$$a_j b_{3-j} - a_{3-j} b_j = 0, \; \lambda_j = P_j\left(q_j^2 - \mu^2\right), \; \eta_j^2 = \frac{8a_j b_j \left(P_{3-j}Q_{j3-j} - P_j Q_{3-j3-j}\right)}{Q_{3-jj}Q_{j3-j} - Q_{3-j3-j}Q_{jj}}\mu^2 > 0, \quad (3.84)$$

where $j = 1, 2$ and μ is any nonzero real constant. It follows from condition $a_j b_{3-j} - a_{3-j} b_j = 0$ that the possible bright–bright solution (3.82) is of the form $(\rho_1, \rho_2) = \left(\exp[\mu z] + b_1 a_1^{-1} \exp[-\mu z]\right)^{-1}\left(\frac{\eta_1}{a_1}, \frac{\eta_2}{a_2}\right)$.

Figure 3.34 presents the spatial profile (top panel) and the temporal profile (bottom panel) of bright soliton according to solution (3.82) with parameters (3.84). Plots of this figure show the first approximation $U_1(n, t)$ of the voltage V (see Eq. (3.63)). It is seen from plots of Fig. 3.34 that the wave amplitude increases when the dispersive parameter λ of the network increases. Plots of the top panel show that the soliton amplitude decreases as the wave propagate along the network. It is seen from plots

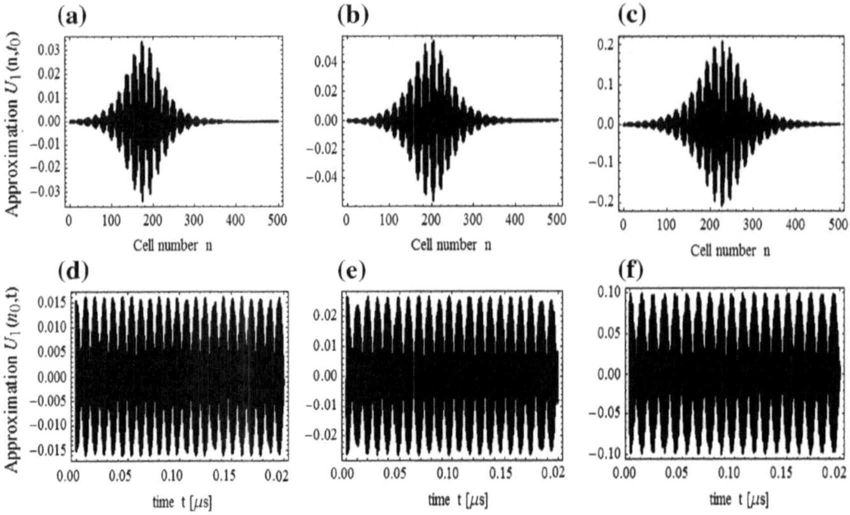

Fig. 3.34 Propagation of bright soliton associated with the bright–bright soliton solution (3.82) with parameters (3.84) for three values of dispersive parameter $\lambda = C_S/C_0$ with line parameters (3.61) and $\delta_0 = 0.01$, $\varepsilon = 0.01$, $\mu = 0.01$, $a_1 = 0.01$, $b_1 = 0.02$, $a_2 = 3$, $b_2 = 6$, $\mu_0 = 5.75$, and $q_2 = \pi/3$. The wave numbers of the wave packets are taken as $(k_1, k_2) = (473.385, 498.3)$. The top and bottom panels show respectively the spatial profile at time $t = 0.4$ ¯s and the temporal profile of cell $n = 1$ of the first term U_1 of voltage V according to Eq. (3.63). **a** and **d**: First approximation U_1 of the voltage for $\lambda = C_S/C_0 = 0.01$; **b** and **e**: First approximation U_1 of the voltage for $\lambda = C_S/C_0 = 0.013$; **c** and **f**: First approximation U_1 of the voltage for $\lambda = C_S/C_0 = 0.017$

of the bottom panel that the wave maintains a constant amplitude when propagating along a given cell of the network.

Kink–kink Solitary Wave Solutions

Here, we seek the exact kink–kink solitary wave solutions of system (3.69) in form (ρ_1, ρ_2) with

$$\begin{cases} \rho_1 = \eta_1 \frac{a_1 \exp[\mu z+\mu_0]-b_1 \exp[-\mu z-\mu_0]}{a_1 \exp[\mu z+\mu_0]+b_1 \exp[-\mu z-\mu_0]} \\ \rho_2 = \eta_2 \frac{a_2 \exp[\mu z+\mu_0]-b_2 \exp[-\mu z-\mu_0]}{a_2 \exp[\mu z+\mu_0]+b_2 \exp[-\mu z-\mu_0]} \end{cases} \tag{3.85}$$

where μ, μ_0, a_j, b_j,and η_j are real constants satisfying conditions $a_j b_j > 0$, $\eta_j \eta_{3-j} \neq 0$. Inserting Eq. (3.85) into system (3.81a), (3.81b) yields

$$\begin{cases} Q_{jj}\eta_j^2 + Q_{j3-j}\eta_{3-j}^2 + \left(\lambda_j - P_j q_j^2\right) = 0, \\ a_j b_{3-j} Q_{jj}\eta_j^2 + a_{3-j} b_j Q_{j3-j}\eta_{3-j}^2 + 2P_j a_j b_{3-j}\mu^2 = 0, \\ a_{3-j} b_j Q_{jj}\eta_j^2 + a_j b_{3-j} Q_{j3-j}\eta_{3-j}^2 + 2P_j a_{3-j} b_j \mu^2 = 0, \\ Q_{jj}\left(2a_j b_{3-j} - a_{3-j} b_j\right)\eta_j^2 + a_j b_{3-j} Q_{j3-j}\eta_{3-j}^2 + 2P_j\left(2a_j b_{3-j} - a_{3-j} b_j\right)\mu^2 = 0, \\ Q_{jj}\left(2a_{3-j} b_j - a_j b_{3-j}\right)\eta_j^2 + a_{3-j} b_j Q_{j3-j}\eta_{3-j}^2 + 2P_j\left(2a_{3-j} b_j - a_j b_{3-j}\right)\mu^2 = 0, \end{cases} \tag{3.86}$$

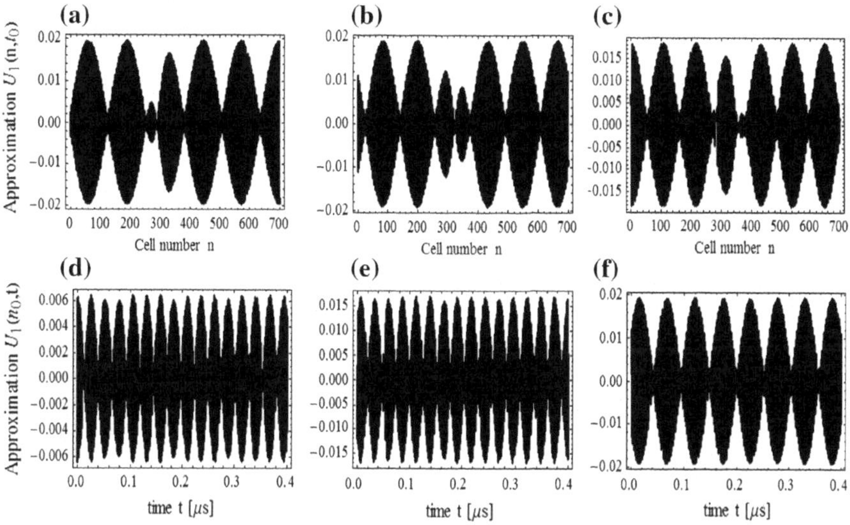

Fig. 3.35 Propagation of kink soliton associated with the kink–kink soliton solution (3.85) with parameters (3.87) for network parameters (3.61) and $\delta_0 = 0.01$, $\varepsilon = 0.01$, $\mu = 0.01$, $a_1 = 0.01$, $b_1 = 0.02$, $a_2 = 3$, $b_2 = 6$, $\mu_0 = 10.43$, and $q_2 = \pi/3$. In plots of top panel, $t_0 = 2$ ¯s. The wave numbers of the wave packets are taken as $(k_1, k_2) = (133, 140)$. The top and bottom panels show, respectively, the spatial profile and the temporal profile of the first term U_1 of voltage V according to Eq. (1.4b). **a**, **b**, **c**: First approximation U_1 of the voltage for $\lambda = C_S/C_0 = 0.009$, $\lambda = C_S/C_0 = 0.0135$, and $\lambda = C_S/C_0 = 0.0175$, respectively; **d**, **e**, **f**: First approximation U_1 of the voltage showing the voltage signal at cells $n = 0$, $n = 30$, and $n = 150$, respectively

$j = 1, 2$. Solving system (3.86) for $\left(Q_{j3-j}Q_{3-jj} - Q_{jj}Q_{3-j3-j}\right) \neq 0$ yields

$$a_j b_{3-j} - a_{3-j} b_j = 0,\ \lambda_j = P_j q_j^2 - \frac{2P_j\left(Q_{jj}Q_{3-j3-j} - Q_{j3-j}Q_{3-jj}\right)}{\left(Q_{j3-j}Q_{3-jj} - Q_{jj}Q_{3-j3-j}\right)}\mu^2, \quad (3.87)$$
$$\eta_j^2 = \frac{2\left(P_j Q_{3-j3-j} - P_{3-j}Q_{j3-j}\right)}{\left(Q_{j3-j}Q_{3-jj} - Q_{jj}Q_{3-j3-j}\right)}\mu^2,$$

where μ is any nonzero real constant, $\eta_j^2 > 0$, $j = 1, 2$. Condition $a_j b_{3-j} - a_{3-j} b_j = 0$ means that the possible the kink–kink solution (3.85) is of the form $\frac{\exp[\mu z+\mu_0] - b_1 a_1^{-1}\exp[-\mu z-\mu_0]}{\exp[\mu z+\mu_0] + b_1 a_1^{-1}\exp[-\mu z-\mu_0]}\,(\eta_1, \eta_2)$.

With the use of kink–kink soliton solution (3.85) with solution parameters (3.87), we show in Fig. 3.35 the propagation of kink soliton on the network of Fig. 3.1. Plots of this figure show the spatial (top) and temporal (bottom) profiles of kink soliton in the network for given wave numbers k_1 and k_2 of the wave packets. Figure 3.35 shows that the wave amplitude increases with the cell number n (this is clearly seen from the bottom plots; at the input cell $n = 0$, the amplitude is 0.006 V while at cells $n = 30$ and $n = 150$, it increases to 0.015 V and 0.02 V).

Remark: It follows from the above form of possible bright–bright and kink–kink solitary wave solutions that under the condition $(Q_{11} - Q_{21})(Q_{22} - Q_{12}) > 0$, then by taking $\lambda_2 = \lambda_1 + \frac{P_2(P_1-P_2)}{P_1}q_2^2$, other bright–bright and kink–kink solitary wave solutions of system (3.81a), (3.81b) can be sought in the form $(\rho_1, \rho_2) = \left(A, \pm A\sqrt{(Q_{11} - Q_{21})(Q_{22} - Q_{12})^{-1}}\right)\phi$, where ϕ is any solitary wave solution of the elliptic ordinary differential equation (EODE)

$$\left(\frac{d\phi}{dz}\right)^2 + \frac{A^2(Q_{11}Q_{22} - Q_{12}Q_{21})}{2P_2(Q_{22} - Q_{12})}\phi^4 + \frac{\lambda_1 - P_1q_1^2}{P_2}\phi^2 + A_0 = 0, \qquad (3.88)$$

A_0 being an arbitrary real parameter.

3.4.4 Bright-Kink and Kink-Bright Solitary Wave Solutions

Now, we assume that system (3.81a), (3.81b) admits the bright-kink solitary wave solutions (ρ_1, ρ_2) with

$$\begin{cases} \rho_1 = \frac{\eta_1}{a_1\exp[\mu z+\mu_0]+b_1\exp[-\mu z-\mu_0]}, \\ \rho_2 = \eta_2\frac{a_2\exp[\mu z+\mu_0]-b_2\exp[-\mu z-\mu_0]}{a_2\exp[\mu z+\mu_0]+b_2\exp[-\mu z-\mu_0]}, \end{cases} \qquad (3.89)$$

where a_j, b_j, η_j, μ_0, and μ are nonzero real constants to be determined, $a_jb_j > 0$, $j = 1, 2$. Inserting Eq. (3.89) into system (3.81a), (3.81b) and solving the resulting system yield

$$\begin{cases} a_2b_1 - a_1b_2 = 0, \\ \eta_1^2 = \frac{8a_1b_1(P_1Q_{22}-Q_{12}P_2)}{Q_{11}Q_{22}-Q_{21}Q_{12}}\mu^2, \\ \eta_2^2 = \frac{2(P_1Q_{21}-Q_{11}P_2)}{Q_{11}Q_{22}-Q_{21}Q_{12}}\mu^2, \\ \lambda_1 = P_1q_1^2 - \frac{P_1(Q_{12}Q_{21}+Q_{11}Q_{22})-2Q_{12}Q_{11}P_2}{Q_{11}Q_{22}-Q_{21}Q_{12}}\mu^2, \\ \lambda_2 = P_2q_2^2 - \frac{2Q_{22}(P_1Q_{21}-Q_{11}P_2)}{Q_{11}Q_{22}-Q_{21}Q_{12}}\mu^2, \end{cases} \qquad (3.90a)$$

where μ is any nonzero real parameter.

Assuming that system (3.81a), (3.81b) admits the kink-bright solitary wave solutions of form (ρ_1, ρ_2) with

$$\begin{cases} \rho_1 = \eta_1\frac{a_1\exp[\mu z+\mu_0]-b_1\exp[-\mu z-\mu_0]}{a_1\exp[\mu z+\mu_0]+b_1\exp[-\mu z-\mu_0]} \\ \rho_2 = \frac{\eta_2}{a_2\exp[\mu z+\mu_0]+b_2\exp[-\mu z-\mu_0]} \end{cases} \qquad (3.91a)$$

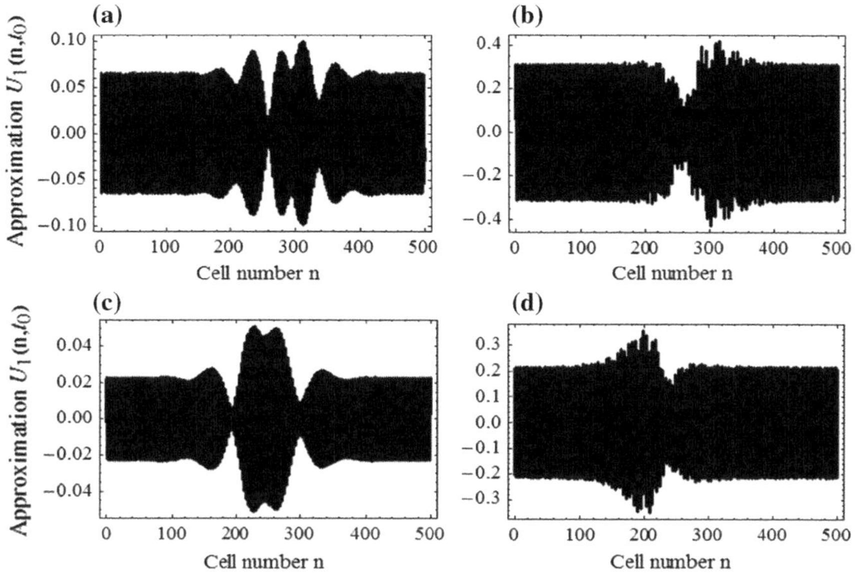

Fig. 3.36 Spatial propagation of bright-kink (top) and kink-bright (bottom) solitons associated with the bright-kink soliton solution (3.89)–(3.90a) and kink-bright soliton solution (3.91a)–(3.92), respectively, network parameters (3.61) for $C_S = 5.6$ pF, and $\delta_0 = 0.01$, $\varepsilon = 0.01$, $\mu = 0.01$, $a_1 = 0.01$, $b_1 = 0.02$, $a_2 = 3$, $b_2 = 6$, $q_2 = \pi/3$, $t_0 = 0.2$ ¯s, and $\mu_0 = 8.47$. Different plots are generated with wave numbers k_1 and k_2 of wave packets are taken as: **a**: $k_1 = 190$ and $k_2 = 200$; **b**: $k_1 = 90$ and $k_2 = 180$; **c**: $k_1 = 173$ and $k_2 = 164.35$ and **d**: $k_1 = 300$ and $k_2 = 150$

yields

$$\begin{cases} a_2 b_1 - a_1 b_2 = 0, \\ \eta_1^2 = \frac{2(P_1 Q_{22} - P_2 Q_{12})}{(Q_{21} Q_{12} - Q_{11} Q_{22})}\mu^2, \\ \eta_2^2 = \frac{8a_2 b_2 (P_1 Q_{21} - P_2 Q_{11})}{Q_{21} Q_{12} - Q_{11} Q_{22}}\mu^2, \\ \lambda_1 = P_1 q_1^2 - \frac{2Q_{11}(P_1 Q_{22} - P_2 Q_{12})}{Q_{21} Q_{12} - Q_{11} Q_{22}}\mu^2, \\ \lambda_2 = P_2 q_2^2 - \frac{2P_1 Q_{21} Q_{22} - P_2 (Q_{11} Q_{22} + Q_{12} Q_{21})}{Q_{21} Q_{12} - Q_{11} Q_{22}}\mu^2. \end{cases} \tag{3.92}$$

Figure 3.36 shows the spatial profile of the signal voltage according to the first approximation (3.63). Here, plots of top panel are obtained with the used of the bright-kink soliton solution (3.89)–(3.90a) while those of the bottom panel are associated with the kink-bright soliton solution (3.91a)–(3.92). Plots of Figs. 3.35, 3.36 show that bright-kink soliton solutions (3.89)–(3.90a) and kink-bright soliton solutions (3.91a)–(3.92) lead to the different behavior of solitons propagating on the network of Fig. 3.1.

References

1. Abdourahman and Emmanuel Kengne: Generation of nonlinear modulated waves in a modified Noguchi electrical transmission network. Chaos, Solitons Fractals **92**, 8 (2016)
2. Afshari, E., Bhat, H.S., Hajimiri, A., Marsden, J.E.: Extremely wideband signal shaping using one and two dimensional non-uniform nonlinear transmission lines. J. Appl. Phys. **99**(5), 054901 (2006)
3. Afshari, E., Hajimiri, A.: Nonlinear transmission lines for pulse shaping in silicon. IEEE J. Solid-State Circuits **40**, 744 (2005)
4. Bhat, H.S., Afshari, E.: Nonlinear constructive inter-ference in electrical lattices. Phys. Rev. E **77**(6), 066602 (2008)
5. Dabrowska, B.J., Ostrovskaya, E.A., Kivshar, Y.S.: Interaction of matter-wave gap solitons in optical lattices. J. Opt. B: Quantum Semiclass. Opt. **6**, 423 (2004)
6. Fukushima, K., Wadati, M., Narahara, Y.: Envelope solition in a new nonlinear transmission line. J. Phys. Soc. Jpn. **49**, 1593 (1980)
7. Hirota, R., Suzuki, K.: Theoretical and experimental studies of lattice solitons in nonlinear lumped networks. Proc IEEE **61**, 1483 (1973)
8. Hirota, R., Suzuki, K.: Studies on lattice solitons by using electrical networks. Phys. Soc. Jpn. **28**, 1366 (1970)
9. Kasamatsu, K., Tsubota, M.: Vortex molecules in coherently coupled two-component Bose-Einstein condensates. Phys. Rev. Lett. **93**, 100402 (2004)
10. Kengne, E.: Ginzburg-Landau system of complex modulation equations for a distributed nonlinear-dispersive transmission line. J. Phys. A: Math. Gen. **37**, 6053 (2004)
11. Kengne, E., Chui, S.T., Liu, W.M.: Modulational instability criteria for coupled nonlinear transmission lines with dispersive elements. Phys. Rev. E **74**, 036614 (2006)
12. Kengne, E., Lakhssassi, A., Liu, W.M.: Modeling of matter-wave solitons in a nonlinear inductor-capacitor network through a Gross-Pitaevskii equation with time-dependent linear potential. Phys. Rev. E **96**, 022221 (2017)
13. Kengne, E., Lakhssassi, A., Liu, W.M.: Dynamics of modulated waves in a lossy modified Noguchi electrical transmission line. Phys. Rev. E **91**, 062915 (2015)
14. Kengne, E., Liu, W.M.: Solitonlike pulses along a modified Noguchi nonlinear electrical network with second-neighbor interactions: analytical studies. Phys. Rev. E **97**, 052205 (2018)
15. Kengne, E., Liu, W.M.: Exact solutions of the derivative nonlinear Schrödinger equation for a nonlinear transmission line. Phys. Rev. E **73**, 026603 (2006)
16. Kengne, E., Talla, P.K.: Dynamics of bright matter wave solitons in Bose-Einstein condensates in an expulsive parabolic and complex potential. J. Phys. B: At. Mol. Opt. Phys. **39**, 3679 (2006)
17. Kengne, E., Vaillancourt, R., Malomed, B.A.: Coupled nonlinear Schrödinger equations for solitary-wave and kink signals propagating in discrete nonlinear dispersive transmission lines. Int. J. Mod. Phys. B **23**, 133 (2009)
18. Kofane, T.C., Michaux, B., Remoissenet, M.: Theoretical and experimental studies of diatomic lattice solitons using an electrical transmission line. J. Phys. C: Solid State Phys. **21**, 1395 (1988)
19. Lange, C.G., Newell, A.C.: A stability criterion for envelope equations. SIAM J. Appl. Math. **27**(3), 441 (1974)
20. Lonngren, K.: Observation of solitons on nonlinear dispersive transmission lines. In: Lonngren, K., Scott, A. (eds.) Solitons in Action. Academic, New York (1978)
21. Marquié, R., Bilbault, J.M., Remoissenet, M.: Nonlinear Schrödinger models and modulational instability in real electrical lattices. Phys. D **87**, 371 (1995)
22. Marquié, P., Bilbault, J.M., Remoissenet, M.: Observation of nonlinear localized modes in an electrical lattice. Phys. Rev. E **51**, 6127 (1995)
23. Marquié, P., Bilbault, J.M., Remoissenet, M.: Generation of envelope and hole solitons in an experimental transmission line. Phys. Rev. E **49**, 828 (1994)

24. Nejoh, Y.: Cusp solitons, shock waves and envelope solitons in a new non-linear transmission line. J. Phys. A **20**, 1733 (1987)
25. Nejoh, Y.: Envelope soliton of the electron plasma wave in a nonlinear transmission line. Phys. Scr. **31**, 415 (1985)
26. Noguchi, A.: Solitons in a nonlinear transmission line. Electron. Commun. Jpn. A **57**, 9 (1974)
27. Remoissenet, M.: Waves Called Solitons, 3rd edn. Springer, Berlin (1999)
28. Riecke, H., Kramer, L.: The stability of standing waves with small group velocity. Phys. D **137**, 124 (2000)
29. Rodwell, M.J., Allen, S.T., Yu, R.Y., Case, M.G., Bhattacharya, U., Reddy, M., Carman, E., Kamegawa, M., Konishi, Y., Pus, J., Pullela, R.: Active and nonlinear wave propagation devices in ultrafast electronics and optoelectronics. Proc. IEEE **82**, 1037 (1994)
30. Rodwell, M.J.W., Kamegawa, M., Yu, R., Case, M., Carman, E., Giboney, K.S.: GaAs nonlinear transmission lines for picosecond pulse generation and millimeter-wave sampling. IEEE Trans. Microw. Theory Tech. **39**, 1194 (1991)
31. Abdoulkary, S., Beda, T., Doka, S.Y., Ndzana, F.I., Kavitha, L., Mohamadou, A.: Dissipative discrete system with nearest-neighbor interaction for the nonlinear electrical lattice. J. Mod. Phys. **3**, 438 (2012)
32. Sato, M., Yasui, S., Kimura, M., Hikihara, T., Sievers, A.J.: Management of localized energy in discrete nonlinear transmission lines. Eur. Lett. **80**(3), 30002 (2007)
33. Schürmann, H.W.: Traveling-wave solutions of the cubic-quintic nonlinear Schrödinger equation. Phys. Rev. E **54**, 4312 (1996)
34. Scott, A.C.: Active and Nonlinear Wave Propagation in Electronics. Wiley-Interscience, New York (1970)
35. Sekulic, D.L., Zivanov, M.B.: Computational study on soliton–like pulses in the nonlinear RLC transmission lines. In: MIPRO, Proceedings of the 35th International Convention (2012)
36. Shi, H., Zhang, W.M., Domier, C.W., Luhmann, N.C., Sjogren, Jr. L.B., Liu, H.X.L.: Novel concepts for improved nonlinear transmission line performance. IEEE Trans. Microw. Theory Tech. MTT **43**, 780 (1995)
37. Smerzi, A., Trombettoni, A., Kevrekidis, P.G., Bishop, A.R.: Dynamical superfluid-insulator transition in a chain of weakly coupled Bose-Einstein condensates. Phys. Rev. Lett. **89**, 170402 (2002)
38. Yoshinaga, Takao, Kakutani, Tsunehiko: Solitary and shock waves on a coupled transmission line. J. Phys. Soc. Jpn. **49**, 2072 (1980)
39. Taniuti, T., Yajima, N.: Perturbation method for a nonlinear wave modulation I. J. Math. Phys. **10**, 1369 (1969)
40. Dhaene, Tom, De Zutter, Daniel: Selection of lumped models for coupled lossy transmission lines. IEEE Trans. Comput.-Aided Des. **11**(7), 805 (1992)
41. Yagi, Tsuneo: Solitons in a nonlinear inhomogeneous transmission line. J. Phys. Soc. Jpn. **50**, 2737 (1981)
42. Volyak, K.I., Marchenko, V.F., Strel'Tsov, A.M.: Envelope solitons in a transmission line with quadratic-cubic nonlinearity. Radiophys. Quantum Electron. **31**, 954 (1988)
43. Yagi, T., Noguchi, A.: Experimental studies on modulational instability by using nonlinear transmission lines. Trans. Inst. Electron. Commun. Eng. Jpn., Part E **59**, 17 (1976)
44. Yemele, D., Kenmogne, F.: Compact envelope dark solitary wave in a discrete nonlinear electrical transmission line. Phys. Lett. A **373**(42), 3801 (2009)
45. Yoneda, H., Tokuyama, K., Ueda, K., Yamamoto, H., Baba, K.: High power terahertz radiation with diamond photoconductive antenna array. In: Liu, S., Chen, X. (eds.), 25th International Conference on Infrared and Millimeter Waves Conference Digest, Beijing, p. 61 (2000)
46. Weierstrass, K.: Mathematische Werke V, pp. 4–16. Johnson, New York (1915)
47. Whittaker, E.T., Watson, G.N.: A Course of Modern Analysis, p. 454. Cambridge University Press, Cambridge (1927)
48. Zhang, W.M., Hsia, R.P., Liang, C., Song, G., Domier, C.W., Luhman, N.C.: Novel low-loss delay line for broadband phased antenna array applications. IEEE Microw. Guid. Wave Lett. **6**, 395 (1996)

Chapter 4
Derivative Nonlinear Schrödinger Equations for Single Transmission Lines

Abstract In the present chapter, we study the generation of nonlinear modulated waves in one-dimensional real electrical lattices. In the continuum limit, we use the semi-discrete approximation to show that the dynamics of modulated waves through the networks under considerations are governed by derivative equations of Schrödinger type including the generalized Chen–Lee–Liu equation and the cubic–quintic derivative NLS equations. Through the analytical exact solutions of the derived network equations, we investigate the transmission of modulated pulses through the networks. The effects of linear dispersive element on the voltage signals are investigated.

In the present chapter, we derive, for lines in Figs. 3.1 and 3.2, derivative NLS equation for the investigation of matter-wave solitons in these lines. For the network of Fig. 3.1, the second-neighbor interactions will be neglected; in other words, we work with the network of Fig. 3.1 under the condition that $L_3 = \infty$.

4.1 Generation of Nonlinear Modulated Waves in the Lossless Network of Fig. 3.1

In this section, we study the generation of nonlinear modulated waves in the lossless network of Fig. 3.1 when the second-neighbor interactions are neglected. Applying the reductive perturbation method in the semi-discrete approximation, we establish that the dynamics of modulated waves in this network is governed by a derivative NLS equation of the Chen–Lee–Liu (CLL) type (referred to as the generalized Chen–Lee–Liu (GCLL) equation). Then, we investigate the effects of the self-steepening on the modulation instabilities of Stokes waves. Through the obtained GCLL equation, we apply both the method of elliptic ordinary differential equation (EODE) and the Hirota bilinear operator method to investigate analytically the effects of the self-steepening on the propagation of soliton-like wave signals in the network.

W.-M. Liu and E. Kengne, *Schrödinger Equations in Nonlinear Systems*,
https://doi.org/10.1007/978-981-13-6581-2_4

4.1.1 Generalized Chen–Lee–Liu Equation for the Network of Fig. 3.1

As we have shown in Chap. 3, the voltage V_n across the nth capacitor, in the case when the dissipation and the second-neighbor interactions are neglected, satisfies the difference-differential system

$$\frac{d^2V_n}{dt^2}+\left(u_0^2+\lambda_S\frac{d^2}{dt^2}\right)(2V_n-V_{n+1}-V_{n-1})+\omega_0^2V_n=\frac{d^2}{dt^2}\left(\alpha V_n^2-\beta V_n^3\right), \tag{4.1}$$

where $n=1,2,...,N$ (N being the number of cells considered), $u_0^2=\frac{1}{L_1C_0}$ and $\omega_0^2=\frac{1}{L_2C_0}$ are the characteristic frequencies of the network, and $\lambda_S=C_S/C_0$. In the linear approximation ($V_n<<1$), Eq. (4.1) leads to the linear dispersion relation of a typical pass-band filter: $\omega^2=(\omega_0^2+4u_0^2\sin^2\frac{k}{2})/(1+4\lambda_S\sin^2\frac{k}{2})$, where the wave number k is taken in the Brillouin zone ($k\in[0,\ \pi]$ in rad/cell). The corresponding linear group velocity is $v_g=d\omega/dk=\frac{(u_0^2-\lambda_S\omega_0^2)\sin k}{\omega(1+4\lambda_S\sin^2\frac{k}{2})^2}$. Focusing ourselves on the situation when $C_s<L_2C_0/L_1$ (i.e., $u_0^2-\lambda_S\omega_0^2>0$), the dependence of the frequency $f=\omega/2\pi$ as function of k represents a band-pass filter bounded with the lower cut-off frequency $f_{\min}=\omega_{\min}/2\pi=\omega_0/2\pi$ introduced by the parallel inductance L_2, and the upper the cut-off frequency $f_{\max}=\omega_{\max}/2\pi=(2\pi)^{-1}\sqrt{\left(\omega_0^2+4u_0^2\right)/(1+4\lambda_S)}$ due to the intrinsic discrete character of the lattice. As a function of parameter λ_S, the upper cut-off frequency $f_{\max}$ decreases with the growth of $\lambda_S=C_S/C_0$; this means that the linear capacitor C_S contributes to reduce the network effects on the wave during its motion.

Now, we focus our attention on modulated waves with a slowly varying envelope in time and space with regard to a given carrier wave with angular frequency $\omega=\omega_p=2\pi f_p$ and wave vector $k=k_p$. In order to apply the reductive perturbation method in the semi-discrete approximation, we introduce the slow envelope variables $x=\varepsilon(n-v_gt)$ and $\tau=\varepsilon^2t$, where n is the cell number and ε is a small parameter that measures the smallness of the modulation frequency and the amplitude of the input waves. It is thus possible with the signal voltage V_n depending explicitly not only on n and t, but also on x and τ, to separate in a natural way its fast and slow variations, in both space and time. That is, $V_n(t)=V(n,x;t,\tau)$. To derive the appropriate NLS equation that governs the modulated waves in the lossless network of Fig. 3.1, we first assume the continuity of the analyzed system which means that the excited wavelengths in the transmission line are much longer than the typical length scale of its elementary cell (this continuity condition is satisfied as soon as the wave number k is chosen in the first Brillouin zone ($0\le k\le\pi$) and satisfied condition $k/\pi\ll 1$). Secondary, we opt for the rotating-wave approximation of the solution $V_n(t)$ of Eq. (4.1). In other words, we assume $V_n(t)$ to have the following general form: [15, 18]

$$V_n(t)=\varepsilon u(x,\tau)\exp\left[2i\Lambda(x,\tau)+i\theta\right]+c.c, \tag{4.2}$$

where $\theta = kn - \omega t$ is the rapidly varying phase, $\Lambda(x, \tau)$ is, in general, a real-valued function of variables x and τ, a phase-imprint (in the case of real-valued function) [15], and $c.c$ stands for the complex conjugation.

With the above assumptions, the substitution of solution (4.2) into Eq. (4.1) leads, for the terms proportional to $\left(\varepsilon^3, \exp[i\theta]\right)$, to the partial differential equation

$$i\frac{\partial u}{\partial \tau} + P\frac{\partial^2 u}{\partial x^2} + Q|u|^2 u + 2iP\left[\frac{\partial}{\partial x}\left(\frac{\partial \Lambda}{\partial x}u\right) + \frac{\partial \Lambda}{\partial x}\left(\frac{\partial u}{\partial x} + 2iu\frac{\partial \Lambda}{\partial x}\right)\right] - 2u\frac{\partial \Lambda}{\partial \tau} = 0, \tag{4.3}$$

where the dispersive coefficient P and the nonlinear coefficient Q are given by

$$P = P(k) = -\frac{v_g^2}{2\omega}\left(1 + 4\lambda_S \sin^2\frac{k}{2}\right) + \left(\frac{u_0^2}{2\omega} - \frac{\lambda_S}{2}\omega\right)\cos k - 2\lambda_S v_g \sin k, \tag{4.4a}$$

$$Q = Q(k) = \frac{3\beta}{2}\omega. \tag{4.4b}$$

It is important to mention that, due to our option for the rotating-wave approximation consistent with a network with band-pass filter properties, parameters P and Q of Eq. (4.3) are independent of the line parameter α, which is the symmetry breaker coefficient of diode nonlinearities [18]. In fact, parameter α usually introduces corrections induced by the presence of the second harmonics of the signal and less affects the dynamics of the network. Because of our option for the rotating-wave approximation (4.2), parameters P and Q appearing in Eq. (4.3) and defined by Eqs. (4.4a) and (4.4b) are independent of the line parameter α. Equation (4.3) governs the propagation of modulated waves in the NLTL of Fig. 4.1. As we well know, in the standard cubic NLS equation, parameter P describes the group velocity dispersion while the self-phase modulation parameter Q measures the strength of the standard nonlinearity. The last two terms of Eq. (4.3) result from the inclusion of the function $\Lambda(x, \tau)$. Different choices of function $\Lambda(x, \tau)$ (real or complex) lead to different modified NLS equations. For example, (i) when Λ is a real constant, Eq. (4.3) is just the standard cubic NLS equation; (ii) when $\Lambda = -\frac{i}{2}\int \Lambda_0(\tau)d\tau$ (with real-valued function $\Lambda_0(\tau)$), Eq. (4.3) becomes a dissipative NLS equation with a gain/loss functional parameter $\Lambda_0(\tau)$. Because the dissipative NLS equation describes the propagation of pulse envelopes in a nonlinear loss medium, we conclude that the choice of function $\Lambda(x, \tau) = -\frac{i}{2}\int \Lambda_0(\tau)d\tau$ can be made if we need to take into account the losses in the system of Fig. 3.1 (in general, losses of the network of Fig. 4.1 are neglected).

In order that the dynamics of modulated waves in the network of Fig. 3.1 be governed by a GCLL equation, we impose to the functional parameter $\Lambda(x, \tau)$ to be a solution of equation

$$2iP\left[\frac{\partial}{\partial x}\left(\frac{\partial \Lambda}{\partial x}u\right) + \frac{\partial \Lambda}{\partial x}\left(\frac{\partial u}{\partial x} + 2iu\frac{\partial \Lambda}{\partial x}\right)\right] - 2u\frac{\partial \Lambda}{\partial \tau} = i\widetilde{Q}|u|^2\frac{\partial u}{\partial x} \tag{4.5}$$

for any nonzero real coefficient $\widetilde{Q}$. One can easily verify that for any real parameter λ, function Λ chosen from conditions

$$\frac{\partial \Lambda}{\partial x} = \lambda |u|^2 \text{ and } \frac{\partial \Lambda}{\partial \tau} = iP\lambda \left(u \frac{\partial u^*}{\partial x} - u^* \frac{\partial u}{\partial x} \right) - 2P\lambda^2 |u|^4 \tag{4.6}$$

is a particular solution of Eq. (4.5); here, u^* denotes the complex conjugate of u. Inserting Eq. (4.6) [i.e., expressions $\frac{\partial \Lambda}{\partial x}$ and $\frac{\partial \Lambda}{\partial \tau}$ from Eq. (4.6)] into Eq. (4.3) and introducing the new variable $T = P\tau$ (under condition $P \neq 0$) lead to the generalized Chen–Lee–Liu equation

$$i\frac{\partial u}{\partial T} + \frac{\partial^2 u}{\partial x^2} + \frac{Q}{P} |u|^2 u + 8i\lambda |u|^2 \frac{\partial u}{\partial x} = 0 \tag{4.7}$$

with one free real parameter λ, referred to as the "self-steepening" parameter; here, $P = P(k)$ and $Q = Q(k)$ are given by Eq. (4.4a) and (4.4b), respectively. Equation (4.7) may be viewed, here and thereafter, as a modified NLS equation of the network of Fig. 4.1 since it reduces to the standard cubic NLS equation governing the dynamics of modulated waves in the network of Fig. 3.1. When parameter λ is neglected, this standard cubic NLS equation admits different types of solution depending on the sign of the product PQ. Namely, if $PQ > 0$, the solution of the standard cubic NLS equation is a bright-type while when $PQ < 0$, it is a hole or dark soliton. As we will see below, the equation parameter λ will affect the wave propagation through the network of Fig. 3.1.

4.1.2 *Effects of the Self-Steepening on the Modulational Instability*

Applying the modulational instability criterion for the cubic NLS equation [2], envelope solitons (hole solitons) of the network of Fig. 3.1 under the above conditions modeled by a cubic NLS equation (4.7) [without λ] with positive product PQ (with negative product PQ) occurs only when a Stokes wave exhibits modulation instability (modulation stability). Here, we aim to demonstrate that with the introduction of the derivative term in the standard NLS via the parameter λ, envelope solitons (hole solitons) can propagate in the network of Fig. 3.1 even if product PQ is negative (positive).

First of all, we point out that through the GCLL equation (4.7), the network of Fig. 3.1 supports the propagation of Stokes waves

$$u = r \exp\left[i\left(lx - \omega_0 T\right)\right], \omega_0 = l^2 + (8l\lambda - Q/P)\, r^2, \tag{4.8}$$

with real amplitude r and wave number l. Using Eq. (4.6), we obtain that Stokes waves (4.8) are associated with the phase-imprint $\Lambda(x, \tau) = \lambda r^2 \left[x + 2P\left(l - \lambda r^2\right)\tau\right]$. For MI investigations, we perturb the Stokes wave solution (4.8) of Eq. (4.7) as follows:

$$u(x, T) = (r + \delta B(x, T)) \exp\left[i\left(lx - \omega_0 T\right)\right], \tag{4.9}$$

where δB is a small perturbation on the wave amplitude. Substituting Eq. (4.9) into Eq. (1.7) and keeping only linear terms in δB and its complex conjugate δB^* leads to the following equation describing the dynamics of the perturbation:

$$i\frac{\partial \delta B}{\partial T}+\frac{\partial^2 \delta B}{\partial x^2}+i\left(2k+8\lambda r^2\right)\frac{\partial \delta B}{\partial x}+B_{01}\left(\delta B+\delta B^*\right)=0, \qquad (4.10)$$

where $B_{01}=Qr^2-8\lambda l r^2$. Because Eq. (4.10) contains both B and its conjugate B^*, it is reasonable to seek its nontrivial solutions in the form

$$\delta B=b_1\exp\left[iKx-i\Omega T\right]+b_2^*\exp\left[-iKx+i\Omega^* T\right], \qquad (4.11)$$

where K and Ω are, respectively, the real wave number and complex frequency of the modulation waves (for negative P, i.e., when $T\rightarrow-\infty$ as $t\rightarrow+\infty$, it is reasonable to replace Ω by $-\Omega$). Inserting Eq. (4.11) into Eq. (4.10) and asking that $|b_1|+|b_2|>0$ lead to the following dispersion relation of the modulation waves $\left[\Omega-K\left(2l+8\lambda r^2\right)\right]^2-K^2\left[K^2-K_0(r,l)\right]=0$, where $K_0(r,l)=\left(2\frac{Q}{P}r^2-4l^2\right)+r^2\left(64r^2\lambda^2-48l\lambda\right)$. It is well known that the MI may set in provided that the frequency of modulation is complex with a non-nil imaginary part, which is possible, as we can see from the above dispersion relation, if and only, if the wave number K of the modulation waves satisfies condition $K^2<K_0(r,l)$; this last inequality makes sense if and only, if $K_0(r,l)>0$. Because $K_0(r,l)>[2Q/P-48\, l\lambda)]r^2-4l^2$, condition $K_0(r,l)>0$ is satisfied as soon as

$$\frac{Q}{P}-2\frac{l^2}{r^2}-24\lambda l>0. \qquad (4.12)$$

Condition (4.12) represents the MI criterion associated with the GCLL Eq. (4.7). Under the MI criterion (4.12), the growth rate (gain) of modulational instability is given by the relation

$$|\mathrm{Im}\Omega(K)|=|K|\sqrt{K_0(r,l)-K^2}, \qquad (4.13)$$

imposing that $0<K<\sqrt{K_0(r,l)}$. In the special case when $\lambda=0$, that is, the case of the NLS equation, MI criterion (4.12) reads $Q/P-2l^2/r^2>0$, which can make sense only when $PQ>0$ (this is known as the criterion of the MI of the standard cubic NLS equation). For the network parameters

$$L_1=220\mu H,\ L_2=470\mu H,\ V_0=2\text{V},\ C_0=320\text{pF}, \qquad (4.14)$$
$$\alpha=0.21\text{V}^{-1},\ \beta=0.0197\text{V}^{-2},\ C_S=56\text{pF},$$

$Q(k)$ is positive while $P(k)$ changes its sign in the Brillouin zone $0\leq k\leq\pi$ (positive on $[0,k_z[$, negative on $]k_z,\pi]$, and $P(k_z)=0$, where $k_z=0.78164$). Therefore, PQ will be positive or negative on $[0,k_z[$ and $]k_z,\pi]$, respectively.

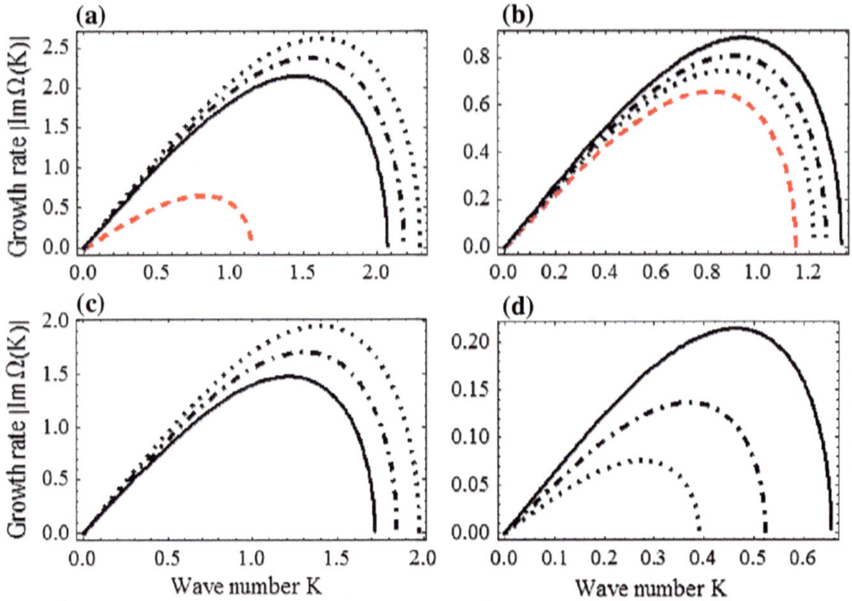

Fig. 4.1 Instability growth rate according to Eq. (4.13) for the Stokes wave parameters $l = 0.01$ and $r = 0.4$ and different values of equation parameter λ. Plots of the top and bottom panels are associated with $PQ = P(0.78)Q(0.78) > 0$ and $PQ = P(1.2)Q(1.2) < 0$, respectively. Curves of plots **a** and **c** are obtained with three positive values of parameter λ, namely, $\lambda = 1.37$ (solid lines), $\lambda = 1.47$ (dot-dashed lines), and $\lambda = 1.57$ (dotted lines); curves of plots **b** and **d** are obtained with three negative values of parameter λ, namely, $\lambda = -0.5$ (solid lines), $\lambda = -0.4$ (dot-dashed lines), and $\lambda = -0.3$ (dotted lines). Dashed lines of plots **a** and **b** correspond to $\lambda = 0$, i.e., the instability growth rate associated with the NLS equation. Reprint from Ref. [1], Copyright 2018, with permission from Elsevier

It is obvious that for given Stokes wave parameters l and r, one can always choose, independently of the sign of product PQ, the equation parameter λ so that K_0 $(r, l) > 0$, leading to the growth rate of modulational instability (4.13). This behavior is shown in Fig. 4.1 where we have plotted the instability growth rate of modulational instability according to Eq. (4.13) for the GCLL Eq. (4.7) associated with positive product PQ (top plots) and with negative product PQ (bottom plots) with different values of parameter λ. In plots (a) and (b), dashed lines are associated with λ=0 and correspond to the instability growth rate for the cubic NLS equation. The left and right panels are associated, respectively, with positive λ and with negative λ. Plots of Fig. 4.2 show that the growth rate of modulational instability associated with Eq. (4.7) increases with positive λ and decreases when negative λ increases. Therefore, the equation parameter λ can be adequately used either to enhance or to soften the instability of the NLS equation. Bottom plots, obtained for negative product PQ, show that the MI for the GCLL equation can occur in the frequency region where the cubic NLS equation is stable under modulation. Plots (a) and (b), obtained in the frequency region where the cubic NLS equation is modulational unstable ($PQ > 0$), show that the MI of the GCLL equation is more stronger than that of the cubic NLS equation.

4.1.3 Analytical Exact Soliton-like Wave Solutions of the GCLL Equation

In this section, we present new classes of analytical exact soliton-like solutions of the GCLL Eq. (4.7). For this purpose, we use the method of elliptic ordinary differential equation (EODE) and the Hirota bilinear transformation method. First of all, we mention that the GCLL Eq. (4.7) is invariant with respect to a Galilean transformation, which, in contrast to the usual cubic NLS equation, involves a change of coefficients of the equation. Indeed, if $u(x,T)$ is solution of Eq. (4.7), then its boosted counterpart, moving at arbitrary velocity v, $\widetilde{u} = u(x - vT, T)\exp\left[\left(i\frac{v}{2}x - i\frac{v^2}{4}T\right)\right]$ will be a solution to Eq. 4.7) with the shifted cubic coefficient $Q/P \rightarrow Q/P - 4v\lambda$.

Method of EODE for the GCLL Equation (4.7)

When using the cubic NLS equation associated with Eq. (4.7) (case $\lambda = 0$) to investigate the propagation of soliton-like wave signal in the network of Fig. 4.1, it is evident that bright soliton-like wave signals may propagate through the network only when the product PQ is positive (Cf. Ref. [12]). In this subsection, we aim to use the GCLL Eq. (4.7) to prove the propagation of bright soliton-like wave signals in the network of Fig. 4.1 in both situations when PQ is positive and when PQ is negative. We will limit ourselves to the pulselike solutions of Eq. (4.7) that have zero transverse velocity and look for solution of the form

$$u(x,T) = a(x)\exp\left[i\left(\phi(x) - \mu T\right)\right], \tag{4.15a}$$

where μ is a real constant and $a(x)$ and $\phi(x)$ are two real functions of variable x. Situation $\mu = 0$ is associated with the steady-state solution of Eq. (4.7). To ansatz (4.15a) corresponds the phase-imprint

$$\Lambda(x,\tau) = \lambda\int a^2(x)dx + \left(2P\lambda a^2\frac{d\phi}{dx} - 2P\lambda^2 a^4\right)\tau + \frac{\Lambda_{00}}{2}. \tag{4.15b}$$

Inserting Eqs. (4.15a) into Eq. (4.7) and separating imaginary and real terms yield

$$\frac{d}{dx}\left(a^2\frac{d\phi}{dx} + 2\lambda a^4\right) = 0, \tag{4.16a}$$

$$\mu a + \frac{d^2 a}{dx^2} - a\left(\frac{d\phi}{dx}\right)^2 + \frac{Q}{P}a^3 - 8\lambda a^3\frac{d\phi}{dx} = 0. \tag{4.16b}$$

Solving Eq. (4.16a) in $d\phi/dx$ and integrating the result yield

$$\phi(x) = \int\left(\frac{\lambda_0}{a^2(x)} - 2\lambda a^2(x)\right)dx + \lambda_{01}, \tag{4.17}$$

where λ_0 and λ_{01} are two constants of integration. For simplicity, we set the constant of integration $\lambda_0 = 0$. This choice of λ_0 is useful for the localized solution of the field equation, i.e., $a(x)$ and its various order derivatives vanish as $|x| \to +\infty$. However, we do not limit ourselves to such solutions. Inserting Eq. (4.17) with $\lambda_0 = 0$ into Eq. (4.16b) and integrating the result yield

$$\left(\frac{da}{dx}\right)^2 + 4\lambda^2 a^6 + \frac{Q}{2P}a^4 + \mu a^2 + \frac{1}{4}\varepsilon_0 = 0, \tag{4.18}$$

where ε_0 is a constant of integration equal to zero. Substitution $a^2 = \frac{1}{\rho}$ transforms Eq. (4.18) to the EODE

$$\left(\frac{d\rho}{dx}\right)^2 = -\varepsilon_0\rho^3 - 4\mu\rho^2 - \frac{2Q}{P}\rho - 16\lambda^2. \tag{4.19}$$

Soliton-Like Solutions $a(x)$ with Vanishing Boundary Conditions ($a(\pm\infty) = 0$)

We start with soliton-like solutions of Eq. (4.18) with vanishing boundary conditions. It follows from ansatz $a^2 = 1/\rho(x)$ and condition $a(x) \to 0$ as $|x| \to +\infty$ that $\rho(x) \to \infty$ as $|x| \to +\infty$. For simplicity, we set the constant of integration $\varepsilon_0 = 0$. Because we are interesting in soliton-like solutions $a(x)$, we seek $\rho(x)$ in either the form $\rho(x) = A + B\cosh^m[Cx]$ or the form $\rho(x) = A + Bx^n$ and find for $\mu \leq 0$ the following solutions:

$$\rho(x) = -\frac{\frac{Q}{P} + \sqrt{\frac{Q}{P^2}^2 - 64\lambda^2\mu}\cosh\left[2\sqrt{-\mu}x\right]}{4\mu}, \quad \mu < 0, \tag{4.20a}$$

$$\rho(x) = -\frac{\frac{Q}{P} + \sqrt{\frac{Q^2}{P^2} - 64\lambda^2\mu}\left(\pm 1 + 2\cosh^2\left[\sqrt{-\mu}x\right]\right)}{4\mu}\mu < 0, \tag{4.20b}$$

$$\rho(x) = -\frac{16P^2\lambda^2 + Q^2x^2}{2PQ}, \quad (\text{if } \mu = 0). \tag{4.20c}$$

For the positivity of $\rho(\xi)$ given by Eq. (4.20c), we must have $PQ < 0$. Equations (4.20a)–(4.20c) give the following soliton-like solutions of Eq. (4.18)

$$a^2(x) = \frac{-4\mu}{\frac{Q}{P} + \sqrt{\frac{Q^2}{P^2} - 64\mu\lambda^2}\cosh\left[2\sqrt{-\mu}x\right]}, \mu < 0 \tag{4.21a}$$

$$a^2(x) = \frac{-4\mu}{\frac{Q}{P} + \sqrt{\frac{Q^2}{P^2} - 64\lambda^2\mu}\left(\pm 1 + 2\cosh^2\left[\sqrt{-\mu}x\right]\right)}, \mu < 0 \tag{4.21b}$$

$$a^2(x) = -\frac{4PQ}{16\lambda^2P^2 + Q^2x^2} \; (\text{if } \mu = 0 \text{ and } PQ < 0), \tag{4.21c}$$

respectively. Inserting Eqs. (4.21a)–(4.21c) with Eq. (4.17) into ansatz (4.15a) leads to bright localized soliton-like solutions of the GCLL Eq. (4.7).

Soliton-like Solutions $a(x)$ with Nonvanishing Boundary Conditions ($a(\pm\infty) \neq 0$)

First of all, we remark that any positive soliton-like solution of Eq. (4.19) with nonvanishing boundary conditions leads to a bounded soliton-like solutions of Eq. (4.18). Then, we conclude from the theory of EODEs that bounded soliton-like solutions of Eq. (4.19) exist only if $\mu \leq 0$. In the situation when $\lambda \neq 0$, we set for simplicity $\varepsilon_0 = 0$. For $\mu < 0$, we write down nonnegative soliton-like solutions of Eq. (4.19) as follows:

$$\rho_{\pm}(x) = -\frac{1}{4\mu}\left\{\frac{Q}{P} \pm \sqrt{\frac{Q^2}{P^2} - 64\lambda^2\mu\left(3 - 2\tanh^2\left[\sqrt{-\mu}x\right]\right)}\right\}. \tag{4.22}$$

If $PQ < 0$, then only $\rho(x) = \rho_+(x)$ will be considered (i.e., solution with sign "+" in Eq. (4.22)). Equation (4.22) leads to the following soliton-like solution with nonvanishing boundary conditions of Eq. (4.18):

$$a_{\pm}^2(x) = -\frac{4\mu}{\frac{Q}{P} \pm \sqrt{\frac{Q^2}{P^2} - 64\lambda^2\mu\left(3 - 2\tanh^2\left[\sqrt{-\mu}x\right]\right)}}. \tag{4.23}$$

When $\mu = 0$, we follow Schürmann [20] and obtain the algebraic soliton-like solution with nonvanishing boundary conditions of Eq. (4.18)

$$a^2(x) = -\frac{Q}{8\lambda^2 P}\left(1 + \frac{17Q^2x^2}{6\left(16P^2\lambda^2 + Q^2x^2\right)}\right), \quad PQ < 0. \tag{4.24}$$

Inserting Eqs. (4.23) and (4.24) with Eq. (4.17) into ansatz (4.15a) leads to dark nonlocalized soliton-like solutions (solutions with nonvanishing boundary conditions) of Eq. (4.7).

Lastly, when $\lambda = 0$, we obtain that Eq. (4.18) admits the following soliton-like solution with nonvanishing boundary conditions [20]

$$a(x) = \pm\sqrt{-\frac{P\mu}{Q}}\tanh\left[\sqrt{\frac{\mu}{2}}x\right], \quad \text{if } PQ < 0,\ \mu > 0, \text{ with } \varepsilon_0 = \frac{2P\mu^2}{Q}, \tag{4.25}$$

$$a^2(x) = -\frac{2P\mu}{Q\left(2 - \tanh^2\left[\sqrt{-\mu}x\right]\right)}, \quad \text{if } PQ > 0,\ \mu < 0, \text{ with } \varepsilon_0 = 0. \tag{4.26}$$

It is important to notice that each solution $a(x)$ leads to solution $u(x,\tau) = a(x)\exp\left[i\left(\phi(x) - \mu P\tau\right)\right]$ of Eq. (4.3) under conditions (4.6), where $\phi(x)$ is defined by Eq. (4.17).

Hirota Bilinear Operator Method

In order to derive other classes of exact solitary wave solutions of Eq. (4.7), we use a generalized format of the bilinear transform [16] and seek the solution in the form

$$u(x, T) = \frac{G(x, T)}{F(x, T)}, \tag{4.27}$$

where G and F may be complex functions. Equation (4.7) for $u(x, T)$ can then be rewritten as three bilinear equations:

$$\left[iD_T + D_x^2 - K\right] G.F = 0, \tag{4.28a}$$

$$\left[D_x^2 - K\right] F.F^* = 4i\lambda D_x G.G^* + \frac{Q}{P} GG^*, \tag{4.28b}$$

$$D_x F.F^* = 4i\lambda GG^*, \tag{4.28c}$$

where * denotes the complex conjugate, D is the Hirota bilinear operator, and K is a constant. Solitary wave solutions $u(x, T)$ of Eq. (4.7) can be obtained by appropriate choice of functions F and G.

First Example of the Choice of Functions F and G Under Condition $\lambda \neq 0$

The first class of solitary wave solutions of Eq. (4.7) through the Hirota bilinear operator method is obtained with F and G defined as

$$G = \exp(xh_1 + h_0) \quad F = 1 + \eta \exp\left(x(h_1 + h_1^*) + h_0 + h_0^*\right), \tag{4.29a}$$

where h_0, h_1, and η are functions of variable T. Inserting Eq. (4.29a) into system (4.28a)–(4.28c) shows that $K = 0$, h_0 is a linear complex function of T defined by

$$h_0(T) = h_{00} + ih_1^2 T, \tag{4.29b}$$

and h_1 and η are two complex constants satisfying system

$$\begin{cases} \mathrm{Im}(\eta)\mathrm{Re}(h_1) = \lambda, \\ \lambda\mathrm{Im}(h_1) + \mathrm{Re}(\eta)\left[\mathrm{Re}(h_1)\right]^2 = \frac{Q}{8P}, \end{cases} \tag{4.29c}$$

h_{00} being an arbitrary complex constant. Inserting Eqs. (4.29a)– (4.29c) into Eq. (4.27) and replacing T by $P\tau$, we obtain

$$u(x, \tau) = \frac{1}{\exp\left[-xh_1 - iPh_1^2\tau - h_{00}\right] + \eta \exp\left[h_1^* x + iPh_1^{*2}\tau + h_{00}^*\right]}. \tag{4.30}$$

Solving system (4.29c) in η, that is, writing $\mathrm{Re}(\eta) = [\mathrm{Re}(h_1)]^{-2}\left[\frac{Q}{8P} - \lambda\mathrm{Im}(h_1)\right]$ and $\mathrm{Im}(\eta) = \lambda/\mathrm{Re}(h_1)$, and inserting results into Eq. (4.30) show that the obtained solution $u(x, T)$ contains two free complex constants, namely, h_1 and h_{00}.

Second Example of the Choice of Functions F and G Under Condition $\lambda \neq 0$

For the second example, we expand functions F and G in terms of ϵ as follows:

$$G(x,T) = \epsilon g(x,T), \quad F(x,T) = 1 + \epsilon^2 f(x,T), \tag{4.31a}$$

where g and f are two complex functions of variables x and T. Inserting (4.31a) into system (4.28a)–(4.28c) and equating to zero coefficients of ϵ^m $(m = 0, 1, 2, 3)$ yield

$$\left\{\begin{aligned} K &= 0,\\ i\frac{\partial g}{\partial T} + \frac{\partial^2 g}{\partial x^2} &= 0,\\ \frac{\partial f}{\partial x} - \frac{\partial f^*}{\partial x} - 4i\lambda g g^* &= 0,\\ f^*\frac{\partial f}{\partial x} - f\frac{\partial f^*}{\partial x} &= 0,\\ f\frac{\partial^2 g}{\partial x^2} - 2\frac{\partial g}{\partial x}\frac{\partial f}{\partial x} + g\frac{\partial^2 f}{\partial x^2} + i\left(f\frac{\partial g}{\partial T} - g\frac{\partial f}{\partial T}\right) &= 0,\\ f^*\frac{\partial^2 f}{\partial x^2} - 2\frac{\partial f}{\partial x}\frac{\partial f^*}{\partial x} + f\frac{\partial^2 f^*}{\partial x^2} &= 0,\\ 4i\lambda g\frac{\partial g^*}{\partial x} - 4i\lambda g^*\frac{\partial g}{\partial x} - \frac{Q}{P}gg^* + \frac{\partial^2 f}{\partial x^2} + \frac{\partial^2 f^*}{\partial x^2} &= 0. \end{aligned}\right. \tag{4.31b}$$

The simplest solution of system (4.31b) is found to be

$$\begin{aligned} g &= \exp\left[h_1 x + i h_1^2 T + h_0\right],\\ f &= \frac{Q + 8iP\lambda h_1}{2P\left(h_1 + h_1^*\right)^2}\exp\left[\left(h_1 + h_1^*\right)x + i\left(h_1^2 + h_1^{*2}\right)T + h_0 + h_0^*\right], \end{aligned} \tag{4.31c}$$

where h_0 and h_1 are two arbitrary complex constants satisfying condition $h_1 + h_1^* \neq 0$. Inserting Eqs. (4.31a) and (4.31c) into Eq. (4.27) and replacing T by $P\tau$, we obtain

$$u(x,\tau) = \frac{\epsilon}{\exp\left[-h_1 x - iPh_1^2\tau - h_0\right] + \epsilon^2\eta\exp\left[h_1^* x + iPh_1^{*2}\tau + h_0^*\right]}, \tag{4.32}$$

with $\eta = \frac{Q - 8P\lambda \mathrm{Im}(h_1)}{8P[\mathrm{Re}(h_1)]^2} + i\frac{\lambda}{\mathrm{Re}(h_1)}$. Solution (4.32) contains one free real parameter, ϵ, one free complex parameter, h_0, and one free complex parameter, h_1, with a nonzero real part ($\mathrm{Re}(h_1) \neq 0$). It is easily seen that solution (4.30) can be obtained from Eq. (4.32) by taking $\epsilon = 1$.

4.1.4 Effects of the Self-steepening on the Propagation of Soliton-Like Wave Signals in the Network

In this section, we use the exact solutions found in the previous section to present the effects of the self-steepening on the propagation of soliton-like wave signals in the network. For simplicity, we focus ourselves to the solutions obtained with the help of the method of EODE; indeed, these solutions are associated with the explicit expression of the phase-imprint parameter Λ given by Eq. (4.15b). For our numerical simulations, we will use the line parameters (4.14).

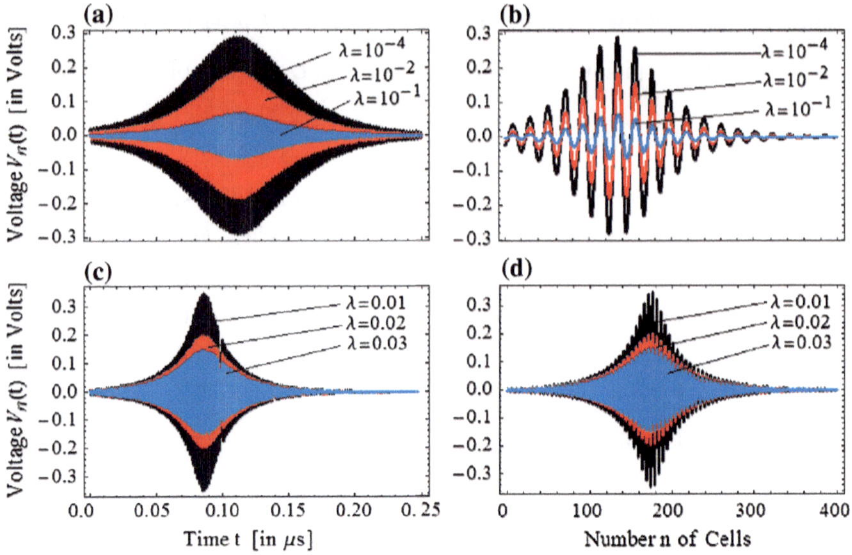

Fig. 4.2 Propagation of the bright soliton-like signal voltages in the network at frequencies $f = 444.379$ KHz (top plots) and $f = 733.981$ KHz (bottom plots) belonging to the regions of, respectively, Envelope Solitons ($PQ > 0$) and Hole Solitons ($PQ < 0$) of the NLS equation. **a** and **c** show the time evolution of the voltage signal propagating at cell $n = 150$ of the network, while **b** and **d** show the wave profile at time $t = 0.1$ μs. To generate these plots, we have used the exact solution associated with solution (4.21a) of Eq. (4.18) with $\mu = -5$ and the line parameters (4.14). Other used parameters are $\Lambda_{00} = \lambda_{01} = 0$, and $\varepsilon = 10^{-3}$. Reprint from Ref. [1], Copyright 2018, with permission from Elsevier

With the use of the exact solution $u(x, \tau)$ of the GCLL Eq. (4.7) associated with the wave amplitude function (4.21a), we show, respectively, in Figs. 4.2 and 4.3 the propagation of bright and dark soliton-like signal voltages in the network for different values of the "self-steepening" parameter λ. Plots of the top and bottom panels show soliton-like waves propagating at a frequency chosen from the region of Envelope Solitons and from the region of Hole Solitons of the NLS equation, respectively. Therefore, with the introduction of the "self-steepening" term in the NLS equation of the network of Fig. 4.1, bright and dark soliton-like signal voltages may propagate in both the regions of Envelope Solitons and Hole Solitons of the NLS equation. Plots of Figs. 4.2 and 4.3 show that the wave amplitude decreases when the "self-steepening" parameter λ increases.

Propagation of dark soliton-like signal voltages in the network is at frequencies $f = 444.379$ KHz (top plots) and $f = 733.981$ KHz (bottom plots) belonging to the regions of, respectively, positivity of product $PQ > 0$ and negativity of product $PQ < 0$. Plots (a) and (c) show the time evolution of the voltage signal propagating at cell $n = 150$ of the network, while plots (b) and (d) show the wave profile at time $t = 0.1$ μs. To generate these plots, we have used the exact solution associated with solution (4.23) (with sign "+") of Eq. (4.18) with $\mu = -5$ and the line parameters (4.14). Other used parameters are $\Lambda_{00} = \lambda_{01} = 0$, and $\varepsilon = 10^{-3}$.

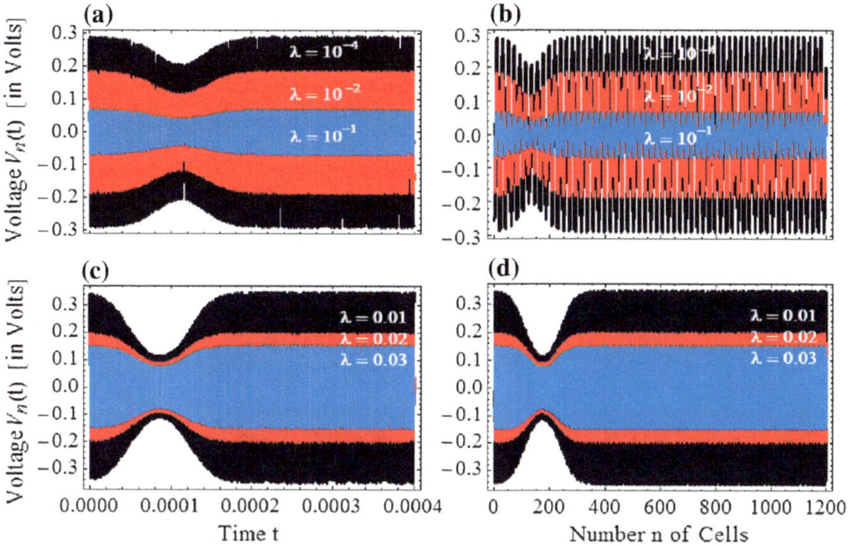

Fig. 4.3 Propagation of dark soliton-like signal voltages in the network at frequencies $f = 444.379$ KHz (top plots) and $f = 733.981$ KHz (bottom plots) belonging to the regions of, respectively, positivity of product $PQ > 0$ and negativity of product $PQ < 0$. Plots **a** and **c** show the time evolution of the voltage signal propagating at cell $n = 150$ of the network, while plots **b** and **d** show the wave profile at time $t = 0.1$ μs. To generate these plots, we have used the exact solution associated with solution (4.23) (with sign "+") of Eq. (4.18) with $\mu = -5$ and the line parameters (4.14). Other used parameters are $\Lambda_{00} = \lambda_{01} = 0$, and $\varepsilon = 10^{-3}$. Reprint from Ref. [1], Copyright 2018, with permission from Elsevier

4.2 Cubic–Quintic Derivative NLS (CQD-NLS) Equation for the Lossless Network of Fig. 3.1

In this section, we wish to show that the modulated waves of the lossless network of Fig. 3.1, in the case when the second-neighbor interactions are neglected, may be governed by a derivative cubic–quintic NLS equation. By applying the Kirchhoff laws, the circuit shown in Fig. 3.1 can be described with the differential-difference equation

$$\frac{\partial^2 Q_n}{\partial t^2} - C_S \frac{\partial^2}{\partial t^2}(V_{n-1} - 2V_n + V_{n+1}) - \frac{1}{L_1}(V_{n-1} - 2V_n + V_{n+1}) + \frac{1}{L_2} V_n = 0, \tag{4.33}$$

where Q_n is the charge stored in the nth capacitor by a DC bias voltage V_0 and the AC voltage V_n. The charge $Q_n(t)$ can be divided into two parts:

$$Q_n(t) = \int_0^{V_0} C(V)dV + \int_{V_0}^{V_0+V_n} C(V)dV.$$

Since the first part of $Q_n(t)$ is constant, it does not appear in the circuit equation (4.33). The nonlinear capacitance over the range V_0 to $(V_0 + V_n)$ is given by $C(V) = \frac{Q(V_0)}{F(V_0)-V_0+V}$, where $Q(V_0)$ and $F(V_0)$ are parameters depending on V_0. Therefore, the circuit Eq. (4.33) takes the form

$$\begin{aligned} &Q(V_0)\tfrac{\partial^2}{\partial t^2}\log\left[1+\tfrac{V_n}{F(V_0)}\right] - C_S\tfrac{\partial^2}{\partial t^2}\left(V_{n-1}-2V_n+V_{n+1}\right) \\ &\quad -\tfrac{1}{L_1}\left(V_{n-1}-2V_n+V_{n+1}\right)+\tfrac{1}{L_2}V_n=0. \end{aligned} \tag{4.34a}$$

Assuming that $V_n \ll F(V_0)$ leads to the expansion $\log\left[1+\frac{V_n}{F(V_0)}\right] = \frac{V_n}{F(V_0)} - \frac{1}{2}\left(\frac{V_n}{F(V_0)}\right)^2 + \ldots$Hence, expression $Q(V_0)\log\left[1+\frac{V_n}{F(V_0)}\right]$ can be then approximated by

$$Q(V_0)\log\left[1+\frac{V_n}{F(V_0)}\right] \approx C_0\left(V_n - aV_n^2\right), \tag{4.34b}$$

where $C_0 = Q(V_0)/F(V_0)$ and $a = 1/2F(V_0)$. Inserting Eq. (4.34b) into Eq. (4.34a) yields

$$\frac{\partial^2 V_n}{\partial t^2} - \left(\frac{C_S}{C_0}\frac{\partial^2}{\partial t^2}+\frac{1}{C_0L_1}\right)\left(V_{n-1}-2V_n+V_{n+1}\right)+\frac{1}{C_0L_2}V_n = a\frac{\partial^2 V_n^2}{\partial t^2}. \tag{4.34c}$$

Expression $V_{n-1} - 2V_n + V_{n+1}$ can be approximated with partial derivatives with respect to distance, x, from the beginning of the line, assuming that the spacing between two adjacent sections is δ (i.e., $x_n = n\delta$). We first, for long wavelengths, approximate V_{n-1} and V_{n+1} by the truncated series

$$V_{n\pm1} = V_n \pm \frac{\partial V_n}{\partial n} + \frac{1}{2}\frac{\partial^2 V_n}{\partial n^2} \pm \frac{1}{6}\frac{\partial^3 V_n}{\partial n^3} + \frac{1}{24}\frac{\partial^4 V_n}{\partial n^4} + \ldots$$

Assuming a small δ and eliminating terms of higher order than δ^4, we obtain the following wave equation valid in weakly dispersive and nonlinear regime:

$$\begin{aligned} &\tfrac{\partial^2 V}{\partial t^2} - \tfrac{\partial^2}{\partial x^2}\left(\tfrac{C_S\delta^2}{C_0}\tfrac{\partial^2 V}{\partial t^2} + \tfrac{\delta^2}{L_1C_0}V\right) \\ &\quad -\tfrac{\partial^4}{\partial x^4}\left(\tfrac{C_S\delta^4}{12C_0}\tfrac{\partial^2 V}{\partial t^2} + \tfrac{\delta^4}{12L_1C_0}V\right) - a\tfrac{\partial^2 V^2}{\partial t^2} + \tfrac{1}{L_2C_0}V = 0 \end{aligned} \tag{4.35}$$

where V is any function satisfying condition $V(n,t) = V_n(t)$.

4.2.1 Derivation of the CQD-NLS Equation for the Lossless Network of Fig. 3.1

The system of partial differential equations (4.35) is not an exactly solvable soliton equation, but it can be approximated by an equation of Schrödinger type. The approximation to an equation of Schrödinger type can be obtained with the perturbation reduction method [21, 23]. By assuming the wave propagation in one direction, we introduce the following two different slow time scales $T_1 = \epsilon t$ and $T_2 = \epsilon^2 t$ in addition to initial time t and one large length scale $X = \epsilon x$ in addition to the spatial variable x, where ϵ is an indicator of slowness. By assuming ϵ sufficiently small (i.e., the weak amplitude limit of voltage), we express the voltage V in the form [22]

$$\begin{aligned} V &= \epsilon^{1/2} u_{11} \exp[i\theta] + \epsilon\,(u_{20} + u_{22}\exp[2i\theta]) + \epsilon^{3/2}\,(u_{30} + u_{33}\exp[3i\theta]) \\ &\quad + \epsilon^2\,(u_{40} + u_{42}\exp[2i\theta] + u_{44}\exp[4i\theta]) + \epsilon^{5/2}\,(u_{50} + u_{53}\exp[3i\theta] \\ &\quad + u_{55}\exp[5i\theta]) + c.c. + O(\epsilon^3). \end{aligned} \tag{4.36}$$

Here $u_{ij} = u_{ij}(T_1, T_2, X_1)$, $\theta = kx - \omega t$, and $c.c.$ stands for the complex conjugate of the preceding terms. Substituting Eq. (4.36) into Eq. (4.35) yields a series of equations distinguished by the power of $\epsilon \times \exp[i\theta]$.

From the equation proportional to $\epsilon^{1/2} \times \exp[i\theta]$, we derive the dispersion equation

$$\frac{\delta^2}{L_1 C_0}k^2 - \frac{\delta^4}{12 L_1 C_0}k^4 + \frac{1}{L_2 C_0} - \left(1 + \frac{C_S \delta^2}{C_0}k^2 - \frac{C_S \delta^4}{12 C_0}k^4\right)\omega^2 = 0. \tag{4.37}$$

It follows from the dispersion relation (4.37) that ω is not defined for every real k.

From the terms proportional to $\epsilon \exp[0i\theta]$ and $\epsilon^2 \exp[0i\theta]$, we obtain $u_{20} = u_{40} = 0$, respectively. The terms proportional to $\epsilon \exp[2i\theta]$ and $\epsilon^{3/2} \exp[3i\theta]$ give, respectively, $u_{22} = A_{22} u_{11}^2$ and $u_{33} = A_{33} u_{11}^3$, where

$$A_{22} = \frac{4a\omega^2}{\frac{3}{L_2 C_0} + \frac{\delta^4}{L_1 C_0}k^4 + 4\left(\frac{3 C_S \delta^2}{C_0}k^2 - \frac{15 C_S \delta^4}{12 C_0}k^4\right)\omega^2}$$

and

$$A_{33} = \frac{18 a\omega^2 A_{22}}{9\left(1 + \frac{9 C_S \delta^2}{C_0}k^2\omega^2 - \frac{81 C_S \delta^4}{12 C_0}k^4\right)\omega^2 + \frac{81\delta^4}{12 L_1 C_0}k^4 - \frac{9\delta^2}{L_1 C_0}k^2 - \frac{1}{L_2 C_0}}.$$

From the term proportional to $\epsilon^2 \exp[2i\theta]$, we then we find

$$u_{42} = A_{42}^1 \left|u_{11}\right|^2 u_{11}^2 + A_{42}^2 u_{11}\frac{\partial u_{11}}{\partial T_1} + A_{42}^3 u_{11}\frac{\partial u_{11}}{\partial X_1},$$

where

$$A_{42}^1 = \frac{8a\omega^2 A_{33}}{4\left(1+\frac{4C_S\delta^2}{C_0}k^2 - \frac{16C_S\delta^4}{12C_0}k^4\right)\omega^2 + \frac{4\delta^4}{3L_1C_0}k^4 - \frac{4\delta^2}{L_1C_0}k^2 - \frac{1}{L_2C_0}},$$

$$A_{42}^2 = -i\frac{8\left[A_{22}\left(\omega + \frac{4C_S\delta^2}{C_0}k^2\omega - \frac{\delta^2}{L_1C_0}\omega - \frac{4C_S\delta^4}{3C_0}k^4\omega\right) - a\omega\right]}{4\left(1+\frac{4C_S\delta^2}{C_0}k^2 - \frac{16C_S\delta^4}{12C_0}k^4\right)\omega^2 + \frac{4\delta^4}{3L_1C_0}k^4 - \frac{4\delta^2}{L_1C_0}k^2 - \frac{1}{L_2C_0}},$$

$$A_{42}^3 = -i\frac{18A_{22}\left(\frac{2C_S\delta^4}{3C_0}k^3\omega^2 - \frac{2C_S\delta^2}{C_0}k\omega^2 - k^3\frac{\delta^4}{3L_1C_0}\right)}{4\left(1+\frac{4C_S\delta^2}{C_0}k^2 - \frac{16C_S\delta^4}{12C_0}k^4\right)\omega^2 + \frac{4\delta^4}{3L_1C_0}k^4 - \frac{4\delta^2}{L_1C_0}k^2 - \frac{1}{L_2C_0}}.$$

Using now the term proportional to $\epsilon^{3/2}\exp[i\theta]$, we obtain

$$i\left[\frac{\partial u_{11}}{\partial T_1} + V_g\frac{\partial u_{11}}{\partial X_1}\right] = A_{00}(k)\,|u_{11}|^2\,u_{11},$$

where

$$A_{00}(k) = \frac{a\omega A_{22}}{\left(1+\frac{C_S\delta^2}{C_0}k^2 - \frac{C_S\delta^4}{12C_0}k^4\right)}$$

and V_g is the group velocity defined as

$$V_g = \frac{d\omega}{dk} = \frac{\frac{\delta^4}{3L_1C_0}k^3 - \frac{2\delta^2}{L_1C_0}k + \left(\frac{2C_S\delta^2}{C_0}k - \frac{C_S\delta^4}{3C_0}k^3\right)\omega^2}{2\left(1+\frac{C_S\delta^2}{C_0}k^2 - \frac{C_S\delta^4}{12C_0}k^4\right)\omega}.$$

It is evident that $A_{00}(k) = 0$ if and only, if $\omega(k) = 0$, and from the dispersion relation (4.37) we get

$$k = k_c \in \left\{-\sqrt{\frac{6}{\delta^2} + \frac{2\sqrt{3}}{\delta^2}\sqrt{\left(\frac{L_1}{L_2}+3\right)}},\ \sqrt{\frac{6}{\delta^2} + \frac{2\sqrt{3}}{\delta^2}\sqrt{\left(\frac{L_1}{L_2}+3\right)}}\right\}.$$

Such $k = k_c$ is called the critical value of the wave number k of the signal. Following the idea of Kakutani et al. [8, 10], we assume that $\Delta k = k - k_c$ is of order $O(\epsilon)$ and write $A_{00}(k) = O(\epsilon)$ as $k \to k_c$, which means that $A_{00}(k) = \epsilon A_{00}^1(k)$, where $A_{00}^1(k) = O(1)$, as $k \to k_c$. Therefore, the term proportional to $\epsilon^{3/2}\exp[i\theta]$ gives

$$\frac{\partial u_{11}}{\partial T_1} = -V_g\frac{\partial u_{11}}{\partial X_1}. \tag{4.38}$$

Equation (4.38) means that in the reference frame moving with the group velocity V_g the complex amplitude u_{11} of the signal remains constant to the concerned scale.

Using the above expressions for different u_{ij} and setting $u_{11} = \epsilon^{-1/2} u \exp\left(i \frac{V_g}{2P} x\right)$, we obtain from the term proportional to $\epsilon^{5/2} \exp[i\theta]$ the desired CQD-NLS equation [14]

$$i\frac{\partial u}{\partial t} = P\frac{\partial^2 u}{\partial x^2} + \gamma u + Q_1 |u|^4 u + i Q_2 |u|^2 \frac{\partial u}{\partial x} + i Q_3 u^2 \frac{\partial u^*}{\partial x}, \tag{4.39}$$

where

$$\begin{aligned}
P &= \left(1 + \frac{C_S\delta^2}{C_0}k^2 - \frac{C_S\delta^4}{12C_0}k^4\right)V_g^2 + \left(\frac{\delta^4}{2L_1C_0} - \frac{C_S\delta^4}{2C_0}k^2 + \frac{C_S\delta^2}{C_0}\right)\omega^2 \\
&\quad - \left(\frac{4C_S\delta^2}{C_0}k^2 + \frac{2C_S\delta^4}{3C_0}k^3\omega\right)V_g - \frac{\delta^2}{L_1C_0}, \\
Q_1 &= 2a\omega^2\left(A_{42}^1 + A_{33}A_{22}\right), \\
Q_2 &= ia\omega^2\left(V_gA_{42}^2 - A_{42}^3\right) - 4a\omega A_{22}V_g, \\
Q_3 &= -2a\omega A_{22}V_g \\
\gamma &= \frac{V_g^2}{4P}.
\end{aligned}$$

Under the Galilean transformation

$$\psi(\xi = x - v_g t, \tau = t) = u(x,t)\exp\left[-i\left(\frac{v_g}{2P}x + \left(\gamma - \frac{v_g^2}{4P}\right)t\right)\right], \quad v_g \text{ - constant}, \tag{4.40}$$

equation (4.39) becomes

$$i\frac{\partial \psi}{\partial \tau} + P\frac{\partial^2\psi}{\partial \xi^2} = Q|\psi|^2\psi + i\chi_1\psi^2\frac{\partial\psi^*}{\partial\xi} + i\chi_2|\psi|^2\frac{\partial\psi}{\partial\xi} + \chi_3|\psi|^4\psi, \tag{4.41}$$

where $Q = \frac{v_g(Q_2 - Q_3)}{2P}$, $\chi_1 = -Q_3$, $\chi_2 = -Q_2$, $\chi_3 = -Q_1$. This means that a stationary wave is transformed into a traveling wave by transformation (4.40). We note that Eq. (4.41) is the most general equation to the lowest consistent order in an envelope expansion near the onset of perturbation and is thus generic in nature. In general, Eq. (4.41) is not integrable, and only a few exact solutions can be obtained under some restrictions on the coefficients of the equations. The CQD-NLS Eq. (4.39), as well as the GCQD-NLS Eq. (4.41), is a well-known dynamical model with numerous physical applications ranging from nonlinear optics to the theory of molecular vibrations. Its demonstrates a rich variety of dynamical behaviors, including bright and dark solitary waves and shock waves (SWs), that is, sharp expanding fronts followed by localized excitations and background oscillations. In what follows,

Eq. (4.39) will be called the CQD-NLS equation, while Eq. (4.41) will be called the generalized CQD-NLS (GCQD-NLS) equation.

4.2.2 Traveling Waves Solutions of the CQD-NLS Equation for the Lossless Network of Fig. 3.1

In this section, we use a complete discrimination system for polynomials to deal with respective ordinary differential equation (ODE) reduced by Eq. (4.39), and obtain the exact traveling wave solutions of the CQD-NLS Eq. (4.39). The obtained exact solutions may be useful to investigate analytically the propagation of modulated waves through the lossless network of Fig. 3.1 when the second-neighbor interactions are neglected.

We seek solutions of Eq. (4.39) in the form

$$u(x,t) = a(z = x - \upsilon t)\exp\left(i\varphi(z) - i\varsigma t\right),\ d\varphi dz = A + Ba^2(z), \tag{4.42}$$

where $a(z)$ and $\varphi(z)$ are two real function, υ and ς are two real numbers, and A and B are two real constants to be defined. Substituting Eq. (4.42) into Eq. (4.39) and separating imaginary and real terms yield

$$a'' = \frac{4P(\varsigma-\gamma)-\upsilon^2}{4P^2}a + \frac{5Q_3^2 - 3Q_2^2 + 2Q_2Q_3 - 16Q_1P}{16P^2}a^5 + \frac{(Q_3-Q_2)\upsilon}{2P^2}a^3, \tag{4.43a}$$

$$\varphi' = -\frac{\upsilon}{2P} - \frac{Q_2+Q_3}{4P}a^2. \tag{4.43b}$$

Integrating Eq. (4.43a) yields

$$a'^2 = \frac{4P(\varsigma-\gamma)-\upsilon^2}{4P^2}a^2 + \frac{Q_3\upsilon - Q_2\upsilon}{4P^2}a^4 + \frac{5Q_3^2 - 3Q_2^2 - 16Q_1P + 2Q_2Q_3}{48P^2}a^6 + \frac{C_1}{P^2}, \tag{4.43c}$$

where C_1 is a constant of integration.

Type I Solutions

For these solutions, $a^2(z)$ is a function of either $\mathrm{dn}^2(\mu z, m)$, $\mathrm{sn}^2(\mu z, m)$, or $\mathrm{cn}^2(\mu z, m)$, where $\mathrm{dn}(y, m)$, $\mathrm{sn}(y, m)$, and $\mathrm{cn}(y, m)$ denote, respectively, the Jacobian elliptic function of the third kind, the Jacobian elliptic sine function, and the Jacobian elliptic cosine function with modulus $0 \le m \le 1$. To obtain this family of solutions, we let $a^2(z) = \zeta^{-1}(z)$. Inserting this ansatz into Eq. (4.43c) yields

$$\left(\frac{d\zeta}{dz}\right)^2 = \frac{5Q_3^2 - 3Q_2^2 - 16Q_1P + 2Q_2Q_3}{12P^2} + \frac{(Q_3 - Q_2)\,\upsilon}{P^2}\zeta$$
$$+\frac{4P(\varsigma - \gamma) - \upsilon^2}{P^2}\zeta^2 + \frac{C_1}{P^2}\zeta^3. \tag{4.44}$$

Because of the positivity of $a^2(z) = \zeta^{-1}(z)$, we will be interested in positive solutions of Eq. (4.44).

Case $5Q_3^2 - 3Q_2^2 - 16Q_1P + 2Q_2Q_3 = 0$

In the case when $5Q_3^2 - 3Q_2^2 - 16Q_1P + 2Q_2Q_3 = 0$, Eq. (4.44) becomes

$$\zeta'^2 = 0\beta_0 + \beta_1\zeta + \beta_2\zeta^2 + \beta_3\zeta^3, \tag{4.45}$$

where

$$\beta_0 = \frac{5Q_3^2 - 3Q_2^2 - 16Q_1P + 2Q_2Q_3}{12P^2},\ \beta_1 = \frac{(Q_3 - Q_2)\,\upsilon}{P^2},$$
$$\beta_2 = \frac{4P(\varsigma - \gamma) - \upsilon^2}{P^2},\ \beta_3 = \frac{C_1}{P^2}. \tag{4.46}$$

The solutions of Eq. (4.45) have the following properties:

Proposition 1 *If $\zeta_1$is a solution of Eq. (4.45), then $w = \zeta_1^{-1}$ will be a solution of equation $w'^2 = \beta_3 w + \beta_2 w^2 + \beta_1 w^3$. It is obvious that this last equation takes the same form as Eq. (4.45) with β_1 exchanged with β_3.*

It should be noted that equation $w'^2 = \beta_3 w + \beta_2 w^2 + \beta_1 w^3$ is obtained from Eq. (4.43c) under the transformation $a^2(z) = w(z)$.

Proposition 2 *If $\Lambda\zeta_1$ (where Λ is a nonzero constant) is a solution of Eq. (4.45), then $w = \Lambda\zeta_1^{-1}$ will be a solution of equation $w'^2 = \beta_3\Lambda^2 w + \beta_2 w^2 + \beta_1\Lambda^{-2}w^3$. It is obvious that this last equation takes the same form as Eq. (4.45) with β_1 exchanged with β_3.*

Proposition 3 *For the equation*

$$\zeta'^2 = 4A\zeta - 4\left(1 + m^2\right)\zeta^2 + \frac{4m^2}{A}\zeta^3,$$

besides the solution $\zeta(z) = A\mathrm{sn}^2(z, k)$, there is another sister solution $\zeta_1(z) = \frac{A}{m^2}\mathrm{sn}^{-2}(z, m)$.

As corollary to this last proposition, we can have the following corollary.

Corollary 4 *For the equation*

$$\zeta'^2 = 4A\mu^2\zeta - 4\mu^2\left(1 + m^2\right)\zeta^2 + \frac{4\mu^2 m^2}{A}\zeta^3,$$

besides the solution $\zeta(z) = A\mathrm{sn}^2(\mu z, m)$, there is another sister solution $\zeta_1(z) = \frac{A}{m^2}\mathrm{sn}^{-2}(\mu z, m)$.

We first seek the solutions of Eq. (4.45) of the form $\zeta(z) = A + B\mathrm{dn}^2(z, k)$. Substituting this ansatz in Eq. (4.45) and equating the coefficients of equal powers of $\mathrm{dn}(x, m)$ results in relations among the solution parameters β_1, β_2, β_3, A, B, and m. They are $A = \frac{16-2\beta_3\beta_1-4\beta_2}{\beta_3(\beta_2+12)}$; $B = -\frac{4}{\beta_3}$, and $m^2 = \frac{48-\beta_2^2+8\beta_2+6\beta_3\beta_1}{4(\beta_2+12)}$. It follows from the positivity of m^2 and condition $m \leq 1$ that $0 \leq \frac{48-\beta_2^2+8\beta_2+6\beta_3\beta_1}{4(\beta_2+12)} \leq 1$. Then if the condition

$$0 \leq \frac{48 - \beta_2^2 + 8\beta_2 + 6\beta_3\beta_1}{4\,(\beta_2 + 12)} \leq 1 \tag{4.47a}$$

is satisfied, Eq. (4.45) will have the special solution

$$\zeta(z) = -\frac{4}{\beta_3}\left(\frac{2\beta_2 + \beta_3\beta_1 - 8}{2\,(\beta_2 + 12)} + \mathrm{dn}^2(z, m)\right), \tag{4.47b}$$

where

$$m^2 = \frac{48 - \beta_2^2 + 8\beta_2 + 6\beta_3\beta_1}{4\,(\beta_2 + 12)}. \tag{4.47c}$$

Because $\zeta^{-1}(z) = a^2(z)$ is positive, function $\zeta(z)$ given by Eq. (4.47b) must be positive for every real z. For the positivity of $\zeta(z)$ given by Eq. (4.47b), we must have either $(\beta_2 + 12)\,(2\beta_2 + \beta_3\beta_1 - 8) \geq 0$ and $\beta_3 < 0$ or $(\beta_2 + 12)\,(2\beta_2 + \beta_3\beta_1 - 8) < 0$ and either $\frac{2\beta_2+\beta_3\beta_1-8}{2(\beta_2+12)} + 1 - m^2 > 0$ and $\beta_3 < 0$ or $\frac{2\beta_2+\beta_3\beta_1-8}{2(\beta_2+12)} + 1 < 0$ and $\beta_3 > 0$. These conditions are consequence of the fact that $\mathrm{dn}(z, k) \in \left[\sqrt{1 - m^2},\ 1\right]$, for every real z. Using the expression of m^2 given by Eq. (4.47c), we obtain that for the positivity of $\zeta(z)$ given by Eq. (4.47b), it is necessary and sufficient that either

$$(\beta_2 + 12)\,(2\beta_2 + \beta_3\beta_1 - 8) \geq 0 \text{ and } \beta_3 < 0, \tag{4.48a}$$

or

$$\begin{gathered}(\beta_2 + 12)\,(2\beta_2 + \beta_3\beta_1 - 8) < 0,\\ (\beta_2 + 12)\left(\beta_2^2 - 4\beta_3\beta_1 - 16\right) > 0 \text{ and } \beta_3 < 0,\end{gathered} \tag{4.48b}$$

or

$$\begin{gathered}(\beta_2 + 12)\,(2\beta_2 + \beta_3\beta_1 - 8) < 0,\\ (\beta_2 + 12)\,(4\beta_2 + \beta_3\beta_1 + 16) < 0 \text{ and } \beta_3 > 0.\end{gathered} \tag{4.48c}$$

We have thus obtained the following result: If real parameters β_1, β_2, and β_3 defined by Eq. (4.46) satisfy condition (4.47a) and one of conditions (4.48a), (4.48b), or (4.48c), then Eq. (4.45) admits a particular positive solution given by Eq. (4.47b). To this solution of Eq. (4.45) corresponds

$$a^2(z) = -\frac{\beta_3}{4}\left(\frac{2\beta_2 + \beta_3\beta_1 - 8}{2\,(\beta_2 + 12)} + \mathrm{dn}^2(z, m)\right)^{-1}, \tag{4.49}$$

where the modulus m is given by Eq. (4.47c).

If $\beta_3 = \beta_1$, that is, if the constant of integration C_1 is taken as $C_1 = (a_2 - a_1)\,\upsilon$, then according to Proposition 1, function

$$\zeta(z) = -\frac{\beta_3}{4}\left(\frac{2\beta_2 + \beta_3\beta_1 - 8}{2\,(\beta_2 + 12)} + \mathrm{dn}^2(z, m)\right)^{-1} \tag{4.50a}$$

will also be a solution of Eq. (4.45). To this solution corresponds

$$a^2(z) = -\frac{4}{\beta_3}\left(\frac{2\beta_2 + \beta_3\beta_1 - 8}{2\,(\beta_2 + 12)} + \mathrm{dn}^2(z, m)\right). \tag{4.50b}$$

At the limit $m \to 1$, that is, when $\beta_2^2 - 4\beta_2 - 6\beta_3\beta_1 = 0$, Eqs. (4.49) and (4.50b) give the following soliton solution:

$$a^2(z) = -\frac{\beta_3}{4}\left(\frac{2\beta_2 + \beta_3\beta_1 - 8}{2\,(\beta_2 + 12)} + \frac{1}{\cosh^2 z}\right)^{-1} \tag{4.51}$$

and

$$a^2(z) = -\frac{4}{\beta_3}\left(\frac{2\beta_2 + \beta_3\beta_1 - 8}{2\,(\beta_2 + 12)} + \frac{1}{\cosh^2 z}\right), \tag{4.52}$$

respectively.

Now, we can use Proposition 2 to construct from solution (4.47b) another sister solution to Eq. (4.45). One can easily verify that for $\Lambda = -4/\beta_3$ and $C_1 = \frac{16P^4}{(a_2-a_1)\upsilon - 2PQ}$, $\Lambda\zeta_1(z) = \Lambda\left(\frac{2\beta_2+\beta_3\beta_1-8}{2(\beta_2+12)} + \mathrm{dn}^2(z, m)\right)^{-1}$ will be a solution of Eq. (4.45), i.e.,

$$\zeta(z) = -\frac{4}{\beta_3}\left(\frac{2\beta_2 + \beta_3\beta_1 - 8}{2\,(\beta_2 + 12)} + \mathrm{dn}^2(z, k)\right)^{-1} \tag{4.53a}$$

will be a solution of Eq. (4.45). To function (4.53a) corresponds solution

$$a^2(z) = -\frac{\beta_3}{4}\left(\frac{2\beta_2 + \beta_3\beta_1 - 8}{2\,(\beta_2 + 12)} + \mathrm{dn}^2(z, m)\right). \tag{4.53b}$$

of Eq. (4.45). At the limit $m \to 1$ (i.e., $\beta_2^2 - 4\beta_2 - 6\beta_3\beta_1 = 0$), solution (4.53b) gives the soliton solution

$$a^2(z) = -\frac{\beta_3}{4}\left(\frac{2\beta_2 + \beta_3\beta_1 - 8}{2\,(\beta_2 + 12)} + \frac{1}{\cosh^2 z}\right). \tag{4.53c}$$

Seeking a solution of Eq. (4.45) in the form $\zeta(z) = A\mathrm{sn}^2(z, m)$, we find that A, β_1, β_2, β_3, and m must satisfy conditions $A = \frac{\beta_1}{4}$, $m^2 = \frac{\beta_3\beta_1}{16}$, $\beta_3\beta_1 + 4\beta_2 + 16 = 0$. Thus under conditions

$$0 < \frac{\beta_3\beta_1}{16} \leq 1, \ \beta_3\beta_1 + 4\beta_2 + 16 = 0, \ \text{and} \ \beta_1 > 0, \tag{4.54}$$

a nonnegative solution of Eq. (4.45) reads $\zeta(z) = \frac{\beta_1}{4}\mathrm{sn}^2(z, m)$, with $m = \frac{\sqrt{\beta_3\beta_1}}{4}$. If we impose to β_1 and β_3 to verify condition $\beta_1 = \beta_3$, that is, $C_1 = (Q_3 - Q_2)\,\upsilon$, then according to Proposition 1, $\zeta(z) = \frac{4}{\beta_1 \mathrm{sn}^2(z,k)}$ will be a positive solution of Eq. (4.45) and corresponds to $a^2(z) = \frac{\beta_1}{4}\mathrm{sn}^2(z, mk)$. If $\beta_1\beta_3 = 16$, then according to Proposition 2, $\zeta(z) = \frac{\beta_1}{4}\frac{1}{\mathrm{sn}^2(z,m)}$ will be a positive solution of Eq. (4.45) and corresponds to $a^2(z) = \frac{4}{\beta_1}\tanh^2 z$. Thus if conditions (4.54) are satisfied and either $\beta_1 = \beta_3$ or $\beta_1\beta_3 = 16$, then the Eq. (4.45) admits a particular solution

$$a^2(z) = \frac{(Q_3 - Q_2)\,\upsilon}{4P^2}\mathrm{sn}^2(z, m), \tag{4.55a}$$

or

$$a^2(z) = \frac{4}{\beta_1}\tanh^2 z, \tag{4.55b}$$

where $m = \frac{(Q_3 - Q_2)\upsilon}{4P^2}$. Amplitude (4.55a) corresponds to the constant of integration $C_1 = (Q_3 - Q_2)\,\upsilon$, while amplitude (4.55b) corresponds to the constant of integration $C_1 = \frac{16P^4}{(Q_3 - Q_2)\upsilon}$.

Seeking now a nonnegative solution of Eq. (4.45) in the form $\zeta(z) = A\mathrm{cn}^2(z, m)$, we find that $A = -\frac{\beta_2+4}{2\beta_3}$, $m^2 = \frac{\beta_2+4}{8}$, $4\beta_1\beta_3 - \beta_2^2 + 16 = 0$. Thus if conditions

$$\beta_3\,(\beta_2 + 4) < 0, \quad 0 \leq \frac{\beta_2 + 4}{8} \leq 1, \ \text{and} \ 4\beta_1\beta_3 - \beta_2^2 + 16 = 0 \tag{4.56}$$

are satisfied, then $\zeta(z) = -\frac{\beta_2+4}{2\beta_3}\mathrm{cn}^2(z, m)$ with $m = \sqrt{(\beta_2 + 4)/8}$ will be a nonnegative solution of Eq. (4.45). According to Propositions 1 and 2, $\zeta(z) = -\frac{2\beta_3}{\beta_2+4}\mathrm{cn}^{-2}(z, m)$ and $\zeta(z) = -\frac{\beta_2+4}{2\beta_3}\mathrm{cn}^{-2}(z, m)$ will be positive solutions of Eq. (4.45) if $C_1 = (Q_3 - Q_2)\,\upsilon$ and $4\beta_3\beta_1 - (\beta_2 + 4)^2 = 0$, respectively. To these positive solutions correspond $a^2(z) = -\frac{\beta_2+4}{2\beta_3}\mathrm{cn}^2(z, m)$ and $a^2(x) = -\frac{2\beta_3}{\beta_2+4}\mathrm{cn}^2(z, m)$. We then conclude that if conditions (4.56) are satisfied, then the CQD-NLS Eq. (4.39) admits special solutions of form (4.42) with amplitude $a(z)$ given by

$$a^2(z) = \frac{4P\,(\omega + P - \gamma) - \upsilon^2}{2\,(Q_2 - Q_3)\,\upsilon}\mathrm{cn}^2(z, m), \tag{4.57}$$

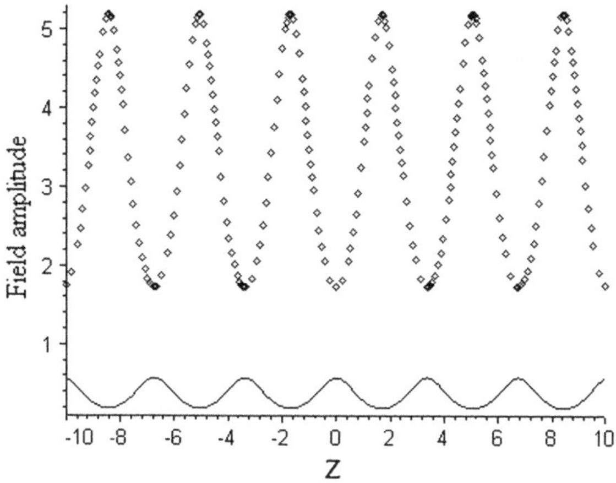

Fig. 4.4 Field amplitude of type I $\mathrm{dn}(z, m)$ and $1/\mathrm{dn}(z, m)$ solution. The values of different parameters used in these plots are given in the text

with $m = \sqrt{(\beta_2 + 4)/8}$. Amplitude (4.56) corresponds to the constant of integration $C_1 \in \left\{ (Q_3 - Q_2)\, \upsilon,\ \frac{\left(4P(\omega+P-\gamma)-\upsilon^2\right)^2}{4(Q_3-Q_2)\upsilon} \right\}$. Now, let A, m, and μ be three numbers defined by $A = \frac{\beta_1}{4\mu^2}$, $m^2 = \frac{\beta_1\beta_3}{16\mu^4}$, $\mu^2 = \frac{1}{8}\left(-2\beta_2 \pm \sqrt{\beta_2^2 - 4\beta_1\beta_3}\right)$. Then Eq. (4.45) takes the form $\zeta'^2 = 4A\mu^2\zeta - 4\mu^2\left(1 + m^2\right)\zeta^2 + \frac{4\mu^2 m^2}{A}\zeta^3$. According to corollary 4, this last equation has a positive solution $\zeta(z) = \frac{A}{k^2}\mathrm{sn}^{-2}(\mu z, m)$. Thus, if $\beta_3 > 0$, $\beta_2^2 - 4\beta_1\beta_3 \geq 0$, $\beta_2 < 0$, $0 < 4\beta_1\beta_3 \leq \left(-2\beta_2 \pm \sqrt{\beta_2^2 - 4\beta_1\beta_3}\right)^2$ then $\zeta(z) = \frac{-2\beta_2 \pm \sqrt{\beta_2^2 - 4\beta_1\beta_3}}{2\beta_3}\mathrm{sn}^{-2}(\mu z, m)$ will be a positive solution of Eq. (4.45). To this solution correspond amplitudes a defined by

$$a^2(z) = \frac{2\beta_3}{-2\beta_2 \pm \sqrt{\beta_2^2 - 4\beta_1\beta_3}}\mathrm{sn}^2(\mu z, m), \tag{4.58}$$

with

$$m^2 = \frac{2\sqrt{\beta_1\beta_3}}{-2\beta_2 \pm \sqrt{\beta_2^2 - 4\beta_1\beta_3}},\ \mu^2 = \frac{-2\beta_2 \pm \sqrt{\beta_2^2 - 4\beta_1\beta_3}}{8}.$$

For $Q_2 = Q_3 = 0.1$, $Q_1 = 4.9505 \times 10^{-3}$, $P = 5.05 \times 10^{-3}$, $\varsigma = \gamma$, $m = 1/2$, $C_1 = (Q_3 - Q_2)\,\upsilon$, and $\upsilon = 1.7494$, we have $\beta_1 = \beta_3$. For these values, we plot in Fig. 4.4 the curves of $|u(x, t)| = a(x, t) = a(z = x - \upsilon t)$ for $t = 0$. The solid line corresponds to amplitude a given Eq. (4.49), while the dotted line corresponds to amplitude a given by Eq. (4.50b). It is obvious that the graphical presentations have mirror-symmetry, have the same period and different amplitudes, but inverse phases. In Fig. 4.5, we plot the amplitude curves corresponding to parameters $Q_2 = Q_3 =$

Fig. 4.5 Field amplitude of type I soliton solutions. The values of different parameters used in these plots are given in the text

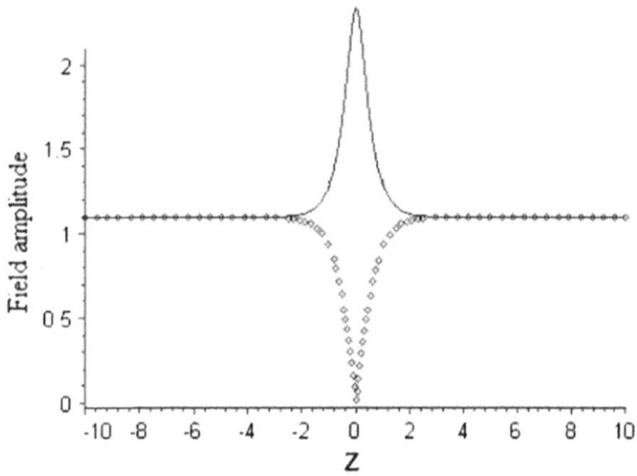

0.1, $Q_1 = 4.9505 \times 10^{-3}$, $P = 5.05 \times 10^{-3}$, $\varsigma = \gamma$, $m = 1$, $\upsilon = 1.5$, and $C_1 = (Q_3 - Q_2)\,\upsilon$, and for given time $t = 0$. For these values of parameters, $\beta_1 = \beta_3$. The solid line corresponds to amplitude (4.51), while the dotted line corresponds to Eq. (4.53c). From Fig. 4.5, one observes that the graphical presentations have mirror-symmetry and different amplitudes.

Remark 5 *We note that if one uses the properties* $\mathrm{dn}^2(z, m) = 1 - m^2\mathrm{sn}^2(z, m)$ *and* $\mathrm{dn}^2(z, k) = 1 - k^2 + k^2\mathrm{cn}^2(z, k)$, *one can write* $a^2(z)$ *given by Eqs.* (4.49), (4.50b), *and* (4.53b) *in terms of either* $\mathrm{sn}^2(z, k)$ *or* $\mathrm{cn}^2(z, k)$, *where* $\mathrm{sn}(z, k)$ *and* $\mathrm{cn}(z, m)$ *are the Jacobian sine and cosine elliptic functions, respectively.*

Remark 6 *One can find many other special solutions of Eq.* (4.45) *from which one has* $a(z)$ *defined either by* $a^2(z) = A^2\mathrm{sn}^2(z, m)$ *or* $a^2(z) = B^2\mathrm{cn}^2(z, m)$.

Case $5Q_3^2 - 3Q_2^2 + 2Q_2Q_3 - 16Q_1P \neq 0$

In the present case, we write Eq. (4.44) as

$$\zeta'^2 = \beta_0 + \beta_1\zeta + \beta_2\zeta^2 + \beta_3\zeta^3, \tag{4.59}$$

where β_0, β_1, β_2, and β_3 are given by Eq. (4.46).

We first formulate the following properties of solutions of Eq. (4.59).

Proposition 7 *If Eq.* (4.59) *admits a solution* $\zeta(z) = A_0 + A_1\mathrm{sn}^2(\mu z, m)$, *then there exists another sister solution* $\zeta(z) = A_0 + \frac{A_1}{m^2}\mathrm{sn}^{-2}(\mu z, m)$.

Proposition 8 *If Eq.* (4.59) *admits a solution* $\zeta(z) = A_0 + A_1\mathrm{cn}^2(\mu z, m)$, *then there exists another sister solution* $\zeta(z) = A_0 + \frac{(m^2-1)A_1}{m^2}\mathrm{cn}^{-2}(\mu z, m)$.

Applying the Jacobian elliptic function expansion method, let us prove that Eq. (4.59) admits solutions

$$\zeta(z) = A_0 + A_1\mathrm{sn}^2(\mu z, m) \tag{4.60a}$$

and

$$\zeta(z) = A_0 + A_1 \mathrm{cn}^2(\mu z, m). \tag{4.60b}$$

Substituting ansatz (4.60a) into Eq. (4.59) and equating the coefficients of equal powers of $\mathrm{sn}^2(\mu z, m)$ results in the following relations among the solution parameters and equation parameters

$$A_1 = \frac{4\mu^2 m^2}{\beta_3},\ A_0 = -\frac{\beta_2 + 4\mu^2\left(m^2+1\right)}{3\beta_3}, \tag{4.61a}$$

$$m^2 = \frac{1}{2} \pm \frac{\sqrt{\beta_2^2 - 3\beta_1\beta_3 - 12\mu^4}}{4\mu^2}, \tag{4.61b}$$

$$\begin{aligned} 0 = {} & 27\beta_0\beta_3^2 - \left(6\mu^2 + \beta_2 \pm \sqrt{\beta_2^2 - 3\beta_1\beta_3 - 12\mu^4}\right)^3 \\ & + 3\beta_2\left(6\mu^2 + \beta_2 \pm \sqrt{\beta_2^2 - 3\beta_1\beta_3 - 12\mu^4}\right)^2 \\ & - 9\beta_1\beta_3\left(6\mu^2 + \beta_2 \pm \sqrt{\beta_2^2 - 3\beta_1\beta_3 - 12\mu^4}\right). \end{aligned} \tag{4.61c}$$

The requirements of the positivity of $\zeta(z)$ and $\beta_2^2 - 3\beta_1\beta_3 - 12\mu^4$ and the condition $0 \le m \le 1$ impose some restrictions on the domain of coefficients of Eq. (4.59) and parameters μ and m : For the square root to make sense, we must have $\beta_2^2 - 3\beta_1\beta_3 - 12\mu^4 \ge 0$. In order to have $0 \le m \le 1$, it is necessary and sufficient that $\beta_2^2 - 3\beta_1\beta_3 - 16\mu^4 < 0$. For the positivity of $\zeta(z)$, we must have either $\beta_2 + 4\mu^2\left(m^2+1\right) < 0$ and $\beta_3 > 0$ or $\beta_2 + 4\mu^2\left(1 - 2m^2\right) > 0$ and $\beta_3 < 0$. Therefore, if for any μ, real solution of Eq. (4.61c) and for any m given by Eq. (4.61b) the following conditions

$$\beta_2^2 - 3\beta_1\beta_3 - 12\mu^4 \ge 0,\ \beta_2^2 - 3\beta_1\beta_3 - 16\mu^4 < 0$$

and either

$$\beta_2 + 4\mu^2\left(m^2+1\right) < 0 \text{ and } \beta_3 > 0$$

or

$$\beta_2 + 4\mu^2\left(1 - 2m^2\right) > 0 \text{ and } \beta_3 < 0$$

are satisfied, then Eq. (4.60a) gives the positive solution

$$\zeta(z) = \frac{4\mu^2 k^2}{\beta_3}\left(-\frac{\beta_2 + 4\mu^2\left(k^2+1\right)}{12\mu^2 k^2} + sn^2(\mu z, k)\right) \tag{4.62a}$$

of Eq. (4.59) where m and μ are given by Eqs. (4.61b) and (4.61c), respectively.

According to Remark 5,

$$\zeta(z) = \frac{4\mu^2}{\beta_3}\left(-\frac{\beta_2 + 4\mu^2\left(m^2+1\right)}{12\mu^2} + \frac{1}{\mathrm{sn}^2(\mu z, m)}\right) \tag{4.62b}$$

is also a solution of Eq. (4.59). For the positivity of this solution, we must have either $\beta_2 + 4\mu^2\left(m^2+1\right) < 0$ and $\beta_3 > 0$ or $-4\mu^2\left(m^2+1\right) < \beta_2 < 4\mu^2\left(2-m^2\right)$ and $\beta_3 < 0$. To solutions (4.62a) and (4.62b) correspond amplitudes

$$a^2(z) = \frac{3\beta_3}{12\mu^2 m^2 \mathrm{sn}^2(\mu z, m) - \left(\beta_2 + 4\mu^2\left(m^2+1\right)\right)} \tag{4.63a}$$

and

$$a^2(z) = \frac{3\beta_3 \mathrm{sn}^2(\mu z, m)}{12\mu^2 - \left(\beta_2 + 4\mu^2\left(m^2+1\right)\right)\mathrm{sn}^2(\mu z, m)}, \tag{4.63b}$$

respectively.

Now, if we substitute ansatz (4.60b) into Eq. (4.59) and equating the coefficients of equal powers of $\mathrm{cn}^2(\mu z, m)$, we obtain relations among the solution parameters and equation parameters

$$A_0 = \frac{4\mu^2\left(2m^2-1\right) - \beta_2}{3\beta_3},\; A_1 = -\frac{4m^2\mu^2}{\beta_3}, \tag{4.64}$$

$$m^2 = \frac{1}{2} \pm \frac{\sqrt{\beta_2^2 - 3\beta_1\beta_3 - 12\mu^4}}{4\mu^2}, \tag{4.65}$$

$$\begin{aligned} 0 = {} & 27\beta_0\beta_3^2 + 9\beta_1\beta_3\left(-\beta_2 \pm 2\sqrt{\beta_2^2 - 3\beta_1\beta_3 - 12\mu^4}\right) \\ & +3\beta_2\left(-\beta_2 \pm 2\sqrt{\beta_2^2 - 3\beta_1\beta_3 - 12\mu^4}\right)^2 \\ & +\left(-\beta_2 \pm 2\sqrt{\beta_2^2 - 3\beta_1\beta_3 - 12\mu^4}\right)^3. \end{aligned} \tag{4.66}$$

The following conditions are imposed on the above parameters (4.64)–(4.66): $\beta_2^2 - 3\beta_1\beta_3 - 12\mu^4 \geq 0$ and $\beta_2^2 - 3\beta_1\beta_3 - 16\mu^4 < 0$. Under these conditions, $0 \leq m \leq 1$ and the square roots in Eqs. (4.65) and (4.66) make sense. Inserting the expression of A_0 and A_1 from Eq. (4.64) in Eq. (4.60b) yields

$$\zeta(z) = -\frac{\beta_2 + 4\mu^2\left(1-2k^2\right) + 12k^2\mu^2\mathrm{cn}^2(\mu z, m)}{3\beta_3}, \tag{4.67}$$

which will be positive if and only if either $\beta_2 + 4\mu^2\left(1 + m^2\right) < 0$ and $\beta_3 > 0$ or $\beta_2 + 4\mu^2\left(1 - 2m^2\right) \geq 0$ and $\beta_3 < 0$. By using Remark 6, we obtain that under one of conditions

$$\beta_2 + 4\mu^2\left(1 - 2m^2\right) \leq 0,\ \beta_3 > 0$$

and

$$\beta_2 + 4\mu^2\left(k^2 - 2\right) > 0,\ \beta_3 < 0,$$

the function

$$\zeta(z) = \frac{12\mu^2\left(1 - k^2\right) - \left(\beta_2 + 4\mu^2\left(1 - 2m^2\right)\right)\mathrm{cn}^2(\mu z, m)}{3\beta_3\mathrm{cn}^2(\mu z, m)} \tag{4.68}$$

will be another sister positive solution. To functions $\zeta(z)$ given by Eqs. (4.67) and (4.68) correspond amplitudes

$$a^2(z) = \frac{-3\beta_3}{\beta_2 + 4\mu^2\left(1 - 2k^2\right) + 12m^2\mu^2\mathrm{cn}^2(\mu z, m)} \tag{4.69}$$

and

$$a^2(z) = \frac{3\beta_3\mathrm{cn}^2(\mu z, m)}{12\mu^2\left(1 - m^2\right) - \left(\beta_2 + 4\mu^2\left(1 - 2m^2\right)\right)\mathrm{cn}^2(\mu z, m)}, \tag{4.70}$$

respectively.

Now, if we pass in the limit as $m \to 1$ in Eqs. (4.63a), (4.63b), and (4.69), we obtain

$$a^2(z) = \frac{\beta_3}{4\mu^2}\left(\frac{12\mu^2}{12\mu^2\tanh^2(\mu z) - \left(\beta_2 + 8\mu^2\right)}\right), \tag{4.71a}$$

$$a^2(z) = \frac{3\beta_3\tanh^2(\mu z)}{12\mu^2 - \left(\beta_2 + 8\mu^2\right)\tanh^2(\mu z)}, \tag{4.71b}$$

$$a^2(z) = \frac{-3\beta_3\cosh^2(\mu z)}{12\mu^2 + \left(\beta_2 - 4\mu^2\right)\cosh^2(\mu z)}, \tag{4.71c}$$

respectively. Passing in the limit $k \to 0$ in Eqs. (4.63b) and (4.70) leads, respectively, to

$$a^2(z) = \frac{3\beta_3\sin^2(\mu z)}{12\mu^2 - \left(\beta_2 + 4\mu^2\right)\sin^2(\mu z)},$$

$$a^2(z) = \frac{3\beta_3\cos^2(\mu z)}{12\mu^2 - \left(\beta_2 + 4\mu^2\right)\cos^2(\mu z)}.$$

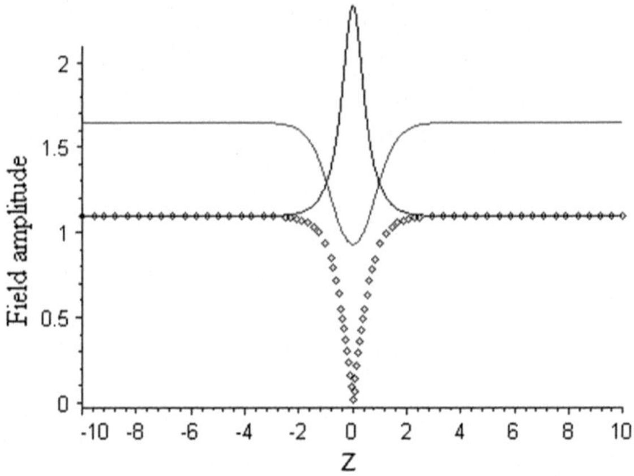

Fig. 4.6 Field amplitude of exact soliton solutions. For these curves, the black curves denote the hyperbolic tangent secant solutions (the solid curve corresponds to the positive amplitude, while the dotted curve corresponds to the nonnegative amplitude) for $Q_2 = Q_3 = -P = -0.505$, $Q_1 = 0.505$, and $\upsilon^2 = 2.02\,(\varsigma - \gamma) + 3.6994$; the solid blue curve denotes the hyperbolic cosine secant solutions for $Q_3 = Q_2 = -P = -0.505$, $Q_1 = 0.505$, $\upsilon^2 = 2.02\,(\varsigma - \gamma) - 3.8177$

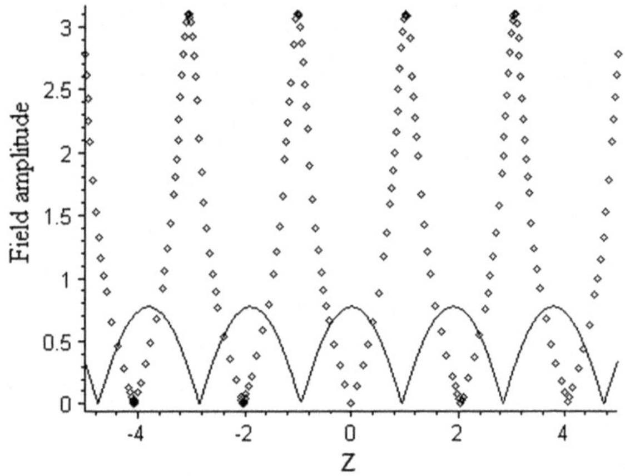

Fig. 4.7 Field amplitude of trigonometric solutions. The solid line denotes the cosine amplitude for $Q_3 = Q_2 = -P = -0.505$, $Q_1 = 0.505$, and $\upsilon^2 = 2.02\,(\varsigma - \gamma) + 4.4504$; the dotted line denotes the sine amplitude for $Q_3 = Q_2 = -P = -0.505$, $Q_1 = 0.505$, and $\upsilon^2 = 2.02\,(\varsigma - \gamma) - 4.2288$

The soliton amplitudes (case when $m \to 1$) as function of z are shown in Fig. 4.6, while the amplitudes of the trigonometric solutions (case when $m \to 0$) are shown in Fig. 4.7.

Type II Solutions

For these solutions, $a^2(z)$ is a function of either $\mathrm{dn}(\mu z, m)$, or $\mathrm{sn}(\mu z, m)$, or $\mathrm{cn}(\mu z, m)$. For this family of solutions, we make the ansatz $a^2 = \zeta$, which reduces Eq. (4.44) in the form

$$\zeta'^2 = \beta_3\zeta + \beta_2\zeta^2 + \beta_1\zeta^3 + \beta_0\zeta^4, \tag{4.72}$$

where β_0, β_1, β_2, and β_3 are given by Eqs. (4.46). In the case when $\beta_0 = 0$, Eq. (4.72) takes the form $w'^2 = \beta_3 w + \beta_2 w^2 + \beta_1 w^3$, which has the same form as Eq. (4.45) and has been discussed in Proposition 1. Thus for type II solutions, we are interested ourselves in the case when $\beta_0 \neq 0$.

In what follows, we show that under some conditions on β_j, i.e., on the coefficients of the CQD-NLS equation and solution parameter v and ς, Eq. (4.72) admits special bounded solutions of the form

$$\zeta(z) = A_0 + A_1 \mathrm{dn}(\mu z, m), \tag{4.73a}$$

$$\zeta(z) = B_0 + B_1 \mathrm{sn}(\mu z, m), \tag{4.73b}$$

$$\zeta(z) = C_0 + C_1 \mathrm{cn}(\mu z, m). \tag{4.73c}$$

First, we formulate the following properties of solutions of Eq. (4.72):

Proposition 11 *For Eq.* (4.72), *besides the bounded solutions* (4.73a), *there exist another sister solution*

$$\zeta(z) = A_0 + \frac{A_1\sqrt{1-m^2}}{\mathrm{dn}(\mu z, m)}. \tag{4.73d}$$

We first discuss the case where $\beta_3 = 0$. This corresponds to the case when the integration constant is zero. Equation (4.72) under condition $\beta_3 = 0$ becomes $\zeta'^2 = \beta_2\zeta^2 + \beta_1\zeta^3 + \beta_0\zeta^4$ and admits the positive soliton solution

$$\zeta(z) = \frac{2\beta_2}{-\beta_1 \pm \sqrt{\beta_1^2 - 4\beta_0\beta_2}\cosh\left(\sqrt{\beta_2}x\right)}, \tag{4.74a}$$

if $\beta_2 > 0$ and $\beta_1 < 0$. Equation (4.74a) defines two positive solutions if $\beta_0 > 0$. In the case when $\beta_0 < 0$, it defines one positive solution corresponding to sign "+". To the soliton solutions (4.74a) correspond amplitudes

$$a^2(z) = \frac{2\beta_2}{-\beta_1 \pm \sqrt{\beta_1^2 - 4\beta_0\beta_2}\cosh\left(\sqrt{\beta_2}x\right)}. \tag{4.74b}$$

We now consider the case where $\beta_3 \neq 0$. If we seek the bounded positive solution of Eq. (4.72) in form (4.73a), then substituting ansatz (4.73a) into Eq. (4.72) and equating the coefficients of equal powers of $\mathrm{dn}(\mu z, m)$ results in relations among the solution parameters A_0, A_1, m, and μ, and the equation coefficients. These are

$$A_0 = -\frac{\beta_1}{4\beta_0},\quad A_1^2 = -\frac{\mu^2}{\beta_0},\quad m^2 = 2 + \frac{3\beta_1^2 - 8\beta_2\beta_0}{8\beta_0\mu^2}$$

$$\mu^2 = \frac{1}{4}\left[\frac{8\beta_2\beta_0 - 3\beta_1^2}{4\beta_0} + \sqrt{\frac{4\beta_0\beta_2^2 - \beta_1^2\beta_2 - 2\beta_1\beta_3\beta_0}{\beta_0}}\right],$$

$$8\beta_3\beta_0^2 - 4\beta_2\beta_1\beta_0 + \beta_1^3 = 0.$$

The requirements that both $a^2(z) = \zeta(z)$ and A_1^2 are positive impose the following conditions on the domain of parameters: $\beta_0 < 0$, and either $16\beta_0\mu^2 + 16\beta_2\beta_0 - 7\beta_1^2 \geq 0$ for $A_1 > 0$ or $\beta_1 > 0$ and $0 \leq \beta_1^2 + 16\mu^2\beta_0$, for $A_1 < 0$. For $\beta_0 < 0$, the square root in the expression of μ^2 makes a sense only if $4\beta_0\beta_2^2 - \beta_1^2\beta_2 - 2\beta_0\beta_1\beta_3 \leq 0$. Condition $0 \leq k \leq 1$ imposes the condition $10\beta_0\beta_1\beta_3 - \beta_1^2\beta_2 \geq 0$ on coefficients β_0, β_1, and β_2 of Eq. (4.72).

According to Proposition 7, besides solution (4.73a), there exists another bounded sister solution (4.73d), $\zeta(z) = \frac{\frac{-\beta_1}{4\beta_0}\left(\mathrm{dn}(\mu z, m) \pm \frac{4}{\beta_1}\sqrt{-\beta_0\mu^2(1-m^2)}\right)}{\mathrm{dn}(\mu z, m)}$, which will be nonnegative for $A_1 > 0$ if either $\beta_1 > 0$ or $\beta_1 < 0$ and $\beta_1^2 - 4\beta_2\beta_0 \geq 0$, and for $A_1 < 0$ if $\beta_1 > 0$ and $\beta_1^2 - 4\beta_2\beta_0 \leq 0$. To solutions (4.73a) and (4.73d) correspond amplitudes

$$a^2(z) = A_0 + A_1 \mathrm{dn}\,(\mu z, m) \;\text{ and }\; a^2(z) = A_0 + \frac{A_1\sqrt{1-m^2}}{\mathrm{dn}\,(\mu z, m)}.$$

In order to obtain the solutions of form (4.73b), we insert ansatz (4.73b) into Eq. (4.72) and equate the coefficients of equal powers of $\mathrm{sn}(\mu z, m)$. This results in relations among the solution parameters B_0, B_1, m, and μ, and the equation coefficients. These are

$$\begin{aligned}
B_0 &= -\frac{\beta_1}{4\beta_0},\; B_1^2 = \frac{3\beta_1^2 - 8\beta_0\beta_2 - 8\beta_0\mu^2}{8\beta_0^2},\\
m^2 &= \frac{3\beta_1^2 - 8\beta_0\beta_2}{8\beta_0\mu^2} - 1,\; 8\beta_0^2\beta_3 - 4\beta_2\beta_0\beta_1 + \beta_1^3 = 0,\\
\mu^2 &= \frac{1}{4}\sqrt{\frac{2\beta_0\left(2\beta_0\beta_2^2 - \beta_1\beta_3\right) - 2\beta_1^2\beta_2 + \beta_1\beta_2^2}{\beta_0}} + \frac{\left(3\beta_1^2 - 8\beta_0\beta_2\right)}{16\beta_0},\\
\beta_1^3 &= 4\beta_2\beta_0\beta_1 - 8\beta_0^2\beta_3.
\end{aligned}$$

In order that $0 \leq m \leq 1$ and $B_1^2 > 0$, it is necessary and sufficient that $\beta_0 > 0$, $3\beta_1^2 - 8\beta_0\beta_2 > 0$, and $\beta_1\left(18\beta_0\beta_3 + 4\left(\beta_2^2 - 2\beta_0\beta_3\right)\right) - 5\beta_1^2\beta_2 \leq 0$. Under these conditions, μ^2 will be defined if and only, if $4\beta_0\beta_2^2 - 2\beta_1^2\beta_2 + \beta_1\beta_2^2 - 2\beta_0\beta_1\beta_3 \geq 0$. Under these conditions, nonnegative solutions will be obtained if $\beta_1 < 0$ and either $\beta_1^2 - 4\beta_0\beta_2 \leq 0$ or $\beta_1^2 - 4\beta_0\beta_2 > 0$ and $\beta_2^2 - \beta_2\beta_1 < 0$. To the two nonnegative solutions associated with $B_1 > 0$ and $B_1 < 0$ correspond two amplitudes

$$a^2(z) = B_0 + B_1 \mathrm{sn}(\mu z, m).$$

Now, we find solutions of form (4.73c). Inserting ansatz (4.73c) into Eq. (4.72) and equating the coefficients of equal powers of $\mathrm{cn}(\mu z, m)$ results in relations among the solution parameters C_0, C_1, m, and μ, and the equation coefficients. These are $C_0 = -\frac{\beta_1}{4\beta_0}$, $C_1^2 = \frac{-m^2\mu^2}{\beta_0}$, $m^2 = \frac{1}{2} - \frac{8\beta_0\beta_2 - 3\beta_1^2}{8\sqrt{4\beta_2^2\beta_0^2 - \beta_2\beta_1^2\beta_0 - 2\beta_3\beta_1\beta_0^2}}$, $\mu^2 = -\frac{\sqrt{4\beta_2^2\beta_0^2 - \beta_2\beta_1^2\beta_0 - 2\beta_3\beta_1\beta_0^2}}{2\beta_0}$, and $8\beta_0^2\beta_3 - 4\beta_2\beta_0\beta_1 + \beta_1^3 = 0$. In order that $C_1^2 > 0$, $0 \le m \le 1$ and μ^2 is defined and positive, it is necessary and sufficient that $\beta_0 < 0$, $4\beta_2^2\beta_0 - \beta_2\beta_1^2 - 2\beta_3\beta_1\beta_0 < 0$, and either $8\beta_0\beta_2 - 3\beta_1^2 > 0$ and $10\beta_0^2\beta_1\beta_3 - \beta_0\beta_1^2\beta_2 \ge 0$ or $8\beta_0\beta_2 - 3\beta_1^2 < 0$ and $10\beta_0\beta_1\beta_3 - \beta_1^2\beta_2 \le 0$. For the positivity solution (4.73c) with parameters above parameters, we must have $\beta_1 > 0$, $10\beta_0\beta_1\beta_3 - \beta_1^2\beta_2 < 0$, and $\beta_1^2 - 4\beta_0\beta_2 < 0$. Thus, if each of conditions $\beta_0 < 0$, $\beta_1 > 0$, $4\beta_2^2\beta_0 - \beta_2\beta_1^2 - 2\beta_3\beta_1\beta_0 < 0$, $10\beta_0\beta_1\beta_3 - \beta_1^2\beta_2 < 0$, and $\beta_1^2 - 4\beta_0\beta_2 < 0$ is satisfied, solution (4.73c) will give the amplitudes

$$a^2(z) = C_0 + C_1 \mathrm{cn}\,(\mu z, m)\,.$$

The following trigonometric amplitudes and soliton amplitudes can be obtained when passing at limit $m \to 0$ and $m \to 1$

$$a^2(z) = B_0 + B_1 \sin(\mu z) \text{ and } a^2(z) = C_0 + C_1 \cos(\mu z),$$

and

$$a^2(z) = A_0 + \frac{A_1}{\cosh(\mu z)}, \text{ and } a^2(z) = C_0 + \frac{C_1}{\cosh(\mu z)},$$

respectively.

4.3 Transmission of Dispersive Shock Signals Through the Lossless Network of Fig. 3.1 When $L_3 = \infty$

Dispersive shock waves (DSWs), also called unsteady undular bores, are nonlinear wavelike structures, which are generated in the breaking profiles of large-scale nonlinear waves propagating in dispersive media. The DSWs have been experimentally observed or their existence has been theoretically predicted in various nonlinear media—water, plasma, optical fibers, lattices, and in Bose–Einstein condensates. In contrast to usual dissipative shocks where combined action of nonlinear and dissipation effects leads to sharp jumps of the wave intensity, accompanied by abrupt changes of other wave characteristics, in dispersive shocks the viscosity effect is either absent or negligibly small compared with the dispersive one, and, instead of intensity jumps, the combined action of nonlinear and dispersion effects leads to formation oscillatory wave region [9]. Dispersive shock waves (DSWs) have yielded novel dynamics and interesting interaction behavior which has only recently begun to be studied theoretically (see for example [6, 8]). It is well known (see, e.g., [24]) that if the initial pulse is strong enough, so that the nonlinear term dominates over the dispersive one at the initial stage of the pulse evolution, then the dispersive shock wave develops after the wave breaking point. The modulational instability of partially coherent signals in electrical transmission lines that are governed by the CQD-NLS Eq. (4.39) has been investigated in Ref. [17], and the following condition on the modulational instability has been found

$$PQ_1 > 0. \tag{4.75}$$

The aim of this section is to show the possibility for shock wave to propagate in a system described by a CQD-NLS equation (4.39) without dissipative term; this leads to the transmission of dispersive shock signals through the lossless network of Fig. 3.1 in the absence of second-neighbors.

4.3.1 Dispersive Shock Wave Solutions of the CQD-NLS Equation

The shock wave-type solutions of the CQD-NLS equation (4.39) can be sought in the generic form

$$u(x,t) = a(\xi)\exp\left[i\left(\varphi(\xi) - \omega t\right)\right], \tag{4.76}$$

where $\xi = x - vt$ and the phase function $\varphi(\xi)$ is in general a function of amplitude $a(\xi)$, v being a real constant. We will distinguish three classes of solutions, a class of one-parameter solutions, a class of zero-parameter solutions, and a class of two-parameter solutions.

One-Parameter Solutions

Following Nozaki and Bekki [19], we seek the first class of the shock-like solution of Eq. (4.39) in the form

$$u(x,t) = \frac{a\exp\left[i\left(Kx - \Omega t\right)\right]}{\left[1 + \exp\left\{-2\mu\left(x - \eta t\right)\right\}\right]^{\frac{1}{2}+i\alpha}}, \quad \mu a \neq 0, \tag{4.77}$$

where a, K, Ω, μ, η, and α are real parameters. Inserting relation (4.77) into Eq. (4.39) results in the following set of equations:

$$\begin{cases} Q_1 a^4 - K(Q_2 + Q_3)a^2 - \left(\Omega + K^2 P\right) = 0, \\ \eta + 2P(K + 2\alpha\mu) = 0, \\ \eta + (Q_2 + 3Q_3)a^2 + 2P(K - 2\alpha\mu) = 0, \\ P\left(\mu^2 - K^2\right) - \Omega - 2\eta\alpha\mu - 4P\mu\alpha(K + \alpha\mu) = 0, \\ 2\Omega + 2P\left(K^2 + \mu^2\right) + 2\eta\alpha\mu + 4KP\alpha\mu - a^2 Q_3(K + 2\alpha\mu) + a^2(K + 2\alpha\mu)(Q_2 + 2Q_3) = 0. \end{cases} \tag{4.78}$$

It is important to notice that a necessary condition for system (4.78) under condition $\mu a \neq 0$ to admit nontrivial solution is that $|Q_3| + |Q_2 + 2Q_3| > 0$. This means that the derivative nonlinear terms in Eq. (4.39) are responsible for dispersive shock wave propagation in the network.

Solving system (4.78), we obtain

$$\begin{cases} K = \frac{8PQ_1 - Q_3(Q_2+3Q_3)}{6P(Q_2+Q_3)} a^2, \\ \eta = -\frac{16PQ_1+(Q_2+3Q_3)(Q_3+3Q_2)}{6(Q_2+Q_3)} a^2, \\ \mu^2 = -\frac{16PQ_1+(Q_2+3Q_3)(Q_3+3Q_2)}{48P^2} a^4, \\ \alpha = \frac{(Q_2+3Q_3)}{8P\mu} a^2, \\ \Omega = \frac{6(Q_3(Q_2+3Q_3)-2PQ_1)(Q_2+Q_3)^2-(8PQ_1-Q_3(Q_2+3Q_3))^2}{36P(Q_2+Q_3)^2} a^4, \end{cases} \tag{4.79}$$

a being a free real parameter. Since $\mu^2 > 0$, the transmission line parameters P and Q_j $(j = 1, 2, 3)$ must satisfy the relationship

$$16PQ_1 + (Q_2 + 3Q_3)(Q_3 + 3Q_2) < 0. \tag{4.80}$$

Inequality (4.80) is then the condition under which the propagation of shock waves is possible on the NLTLs. Hence, the SW exists only with both positive and negative velocity η, depending on the sign of $Q_2 + Q_3$ (see the second equation in Eq. (4.79)). It should be noted that condition $Q_2 + Q_3 \neq 0$ follows from our assumption $\mu \neq 0$ (see Eqs. (4.78)). It follows from conditions (4.75) and (1.4b) that the motion of shock waves is possible in the modulational instability domain only when the coefficients Q_2 and Q_3 of derivative terms satisfy the relationship

$$\frac{Q_2}{Q_3} < 0. \tag{4.81}$$

As we can see from Eqs. (4.79), the SWs in the CQD-NLS equation move at a constant velocity which explicitly depends on the dispersive coefficient P, on the higher order nonlinearity coefficient Q_1, and on the two derivative nonlinearity coefficients Q_2 and Q_3. Using Eq. (4.79), one obtains the following group velocity:

$$\frac{\partial\Omega}{\partial K} = -2KP - \frac{(Q_2 + Q_3)\,a^2}{4} = -\frac{32PQ_1 + 3Q_2^2 + 2Q_2Q_3 - 9Q_3^2}{12\,(Q_2 + Q_3)} a^2. \tag{4.82}$$

In Fig. 4.8, we plot the shock wave velocity (1) and the group velocity (2) as functions of derivative nonlinearity. Figure 4.8a shows the plot versus Q_3 for the equation parameters $P = 0.4$, $Q_1 = 0.2$, $Q_2 = 1.3$, while Fig. 4.8b corresponds to the plot versus Q_2 for the parameters $P = 0.4$, $Q_1 = 0.2$, $Q_3 = 0.7$. The plots of this figure show that the shock wave velocity as the group velocity decreases as the derivative nonlinearities increase. We can then conclude that the two derivative nonlinearity terms can significantly slow down the propagating speed of the shock wave. Figure 4.8a shows that the shock wave velocity is always greater than the group velocity and this suggests that there may exist many types of structures evolving behind the shock wave for the given equation parameters for $-1.29 \leq Q_3 \leq -0.78757$. As it is seen from Fig 4.8b, the group velocity is always greater that the front's speed, meaning that for the given set of equation parameters and for $-0.69 \leq Q_2 \leq -0.5$, there may exist many types of structures evolving in front of the shock wave.

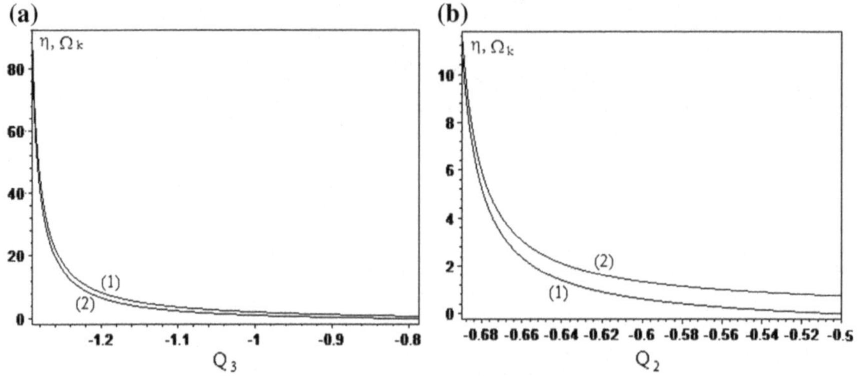

Fig. 4.8 Velocity η of the shock wave (*1*) and group velocity Ω_K (*2*) versus **a** the derivative nonlinearity Q_3 for $P = 0.4$, $Q_1 = 0.2$, $Q_2 = 1.3$ and **b** versus the derivative nonlinearity Q_2 for $P = 0.4$, $Q_1 = 0.2$, $Q_3 = 0.7$. Reprint from Ref. [13], Copyright 2018, with permission from Canadian Science Publishing

Zero-Parameter and Two-Parameter Solutions

For both classes of zero-parameter and two-parameter solutions, it is preferable to write solution (4.76) in the following generic form:

$$u(x,t) = R(x - Vt = z)\exp\left(i\int_0^z \Phi(z)\,dz - i\rho^2 t\right), \tag{4.83}$$

where V is the SWs velocity, $R(z)$ is a real amplitude function with boundary conditions $R(-\infty) = 0$, $R(+\infty) = \sqrt{2}\rho$, where ρ is a asymptotic continuous wave amplitude, and $\Phi(z)$ is a real function which takes the boundary values $\Phi(\pm\infty) = -\left[(Q_2 + 3Q_3)\rho^2 + V\right](2P)^{-1}$. We limit ourselves to the shock waves with positive velocity ($V > 0$) corresponding to a situation where the zero-state domain expands into the energy-carrying one. As in the previous section, we will see that dispersive shock wave propagation in the network is possible only if $|Q_3| + |Q_2 + 2Q_3| > 0$, that is, in the presence of at least one derivative nonlinear term in Eq. (4.39).

Inserting Eq. (4.83) into Eq. (4.39) and taking the boundary conditions into account lead to system

$$\begin{cases} R'' + \frac{(P+1)V-2\rho^2 P}{2P^2}R + \frac{(P+2V+1)Q_2+(3P+2V+3)Q_3}{4P^2}R^3 + \frac{4PQ_1+4Q_2Q_3+Q_2^2+3Q_3^2}{4P^2}R^5 = 0, \\ \Phi(z) = -\frac{V}{2P} - \frac{Q_2+3Q_3}{4P}R^2(z). \end{cases} \tag{4.84}$$

The first equation in (4.84) makes its presence felt in many contexts: it appears as a reduction equation of various nonlinear equations such as Rangawala–Rao equation, Ablowitz equation, and Gerdjikov and Ivanov equation. Many new solitary wave solutions are obtained by mapping this equation into the field equation of ϕ^6 like in field theory.

The shock wave solution of the first equation in (4.84) is sought in the form

$$R(z) = \rho\sqrt{1 + \tanh mz}, \tag{4.85}$$

where m is a real parameter. If we substitute Eq. (4.85) into the first equation in system (4.84), we can obtain, after long calculations, a system of equations in V, m, and ρ. The different cases when the equation is solvable are analyzed below.

Case $Q_2 + Q_3 = 0$

When

$$Q_2 + Q_3 = 0,\ \ PQ_1 < 0,\ \ (P+1)\,Q_3 > 0, \tag{4.86}$$

one obtains

$$m^2 = -\frac{3\,[Q_3\,(P+1)]^2}{64P^3Q_1},\ \ \rho^2 = -\frac{3\,(P+1)\,Q_3}{16PQ_1},\ \ V = \frac{3Q_3\,[4P - (P+1)\,Q_3]}{2\left(5Q_3^2 - 16PQ_1\right)}. \tag{4.87}$$

We then obtain that shock wave solutions exist under the condition $Q_2 + Q_3 = 0$ only in the modulational stability region $PQ_1 < 0$. Using Eq. (4.87), one can express the shock wave velocity in terms of the asymptotic continuous wave amplitude ρ as follows:

$$V(\rho) = -\frac{13\,(PQ_3 + Q_3 - 4P)\,\rho^2}{2\left(3 + 3P + 5Q_3\rho^2\right)}. \tag{4.88}$$

Case $4PQ_1 + Q_2^2 = Q_3 = 0$

In the present case, one obtain that

$$V = -\frac{P+1}{2}, \tag{4.89}$$

and solution (4.87) is governed by the two free parameters m and ρ. For the shock wave velocity to be positive, we must have $P + 1 < 0$. The resulting two-parameter shock wave solution holds only in the stability ($PQ_1 < 0$) domains. It should be noted that in the present case, the shock wave velocity does not depend on the asymptotic continuous wave amplitude ρ; moreover, it is free from Q_1, Q_2, and Q_3.

Case $\left(\left|4PQ_1 + Q_2^2\right| + |Q_3|\right)(Q_2 + Q_3) > 0$

In this case, we have $(Q_2 + Q_3)\left(4PQ_1 + 4Q_2Q_3 + Q_2^2 + 3Q_3^2\right) \neq 0$ and

$$m^2 = \frac{b}{2}\rho^2,\ \ \rho^2 = -\frac{6\,(Q_2 + Q_3)\,V + 3\,(P+1)\,(Q_2 + 3Q_3)}{8\left(4PQ_1 + 4Q_2Q_3 + Q_2^2 + 3Q_3^2\right)}, \tag{4.90}$$

where $b = ((P + 2V + 1)\,Q_2 + (3P + 2V + 3)\,Q_3)\left(4P^2\right)^{-1}$ and V is any positive solution of the quadratic equation

$$-12(Q_2+Q_3)^2V^2+3(P+1)(Q_2+3Q_3)[8P-(P+1)(Q_2+3Q_3)]$$
$$+4\left[(P+1)\left(32PQ_1+20Q_2Q_3+5Q_2^2+5Q_3^2\right)+12P(Q_2+Q_3)\right]V=0. \tag{4.91}$$

In order to obtain the conditions of validity of solution (4.85), one may first find the positive solution of Eq. (4.91), second insert the expression for V in b and into Eq. (4.90), and lastly, one may impose the condition of positivity of b and ρ^2.

It should be noted that it is also possible to express the shock wave velocity V in terms of the asymptotic continuous wave amplitude ρ

$$V(\rho)=-\frac{4\left(4PQ_1+4Q_2Q_3+Q_2^2+3Q_3^2\right)}{3(Q_2+Q_3)}\rho^2-\frac{(P+1)(Q_2+3Q_3)}{2(Q_2+Q_3)}. \tag{4.92}$$

4.3.2 *Some Characteristics of the Established DSW*

The most essential characteristics of the established DSW are the dependence of its velocity V on the parameter ρ and derivative nonlinearities. As it has been said above, the velocity V is independent of ρ and of the derivative nonlinearities in the case $4PQ_1+Q_2^2=Q_3=0$.

In the case of solution (4.87) with parameters (4.85), results demonstrating the (1) $V(Q_3)$ and (2) $\rho(Q_3)$ dependence are displayed in Fig. 4.9a for the parameter values $P=1$ and $Q_1=-0.5$. Figure 4.9b, obtained from Fig. 4.9a, shows the dependence of $V(\rho)$ for the above values of the equation parameters. An obvious inference suggested by Fig. 4.9 is that the dependence $V(\rho)$ or $V(Q_3)$ is neither linear nor monotone. This implies that the derivative nonlinear terms in Eq. (4.39) cannot be neglected. Another conclusion from Fig. 4.9 is that, depending on the parameter values in the equations, the derivative nonlinearity terms in Eq. (4.39) may either increase or decrease the speed of the shock wave. For the chosen parameter values $P=1$ and $Q_1=-0.5$, the asymptotic continuous wave amplitude increases with the derivative nonlinearity Q_3 and its maximal and minimal values are obtained for $V=0$, corresponding to two critical values of $Q_3=0.42265$ and $Q_3=1.5774$, respectively.

Plots of Fig. 4.10 correspond to the shock wave solution (4.87) with parameters (4.90) and (4.91). Curves (1) and (2) of Fig. 4.10a show the dependence of V and ρ on Q_3, respectively, while Fig. 4.10b shows the evolution of V versus ρ. The two plots are obtained with the equation parameters $P=1$ and $Q_1=-0.5$ when the derivative nonlinearities satisfy the relationship $Q_2+3Q_3=0$. As in the above case, Fig. 4.10b is obtained from Fig. 4.10a. Neither the shock velocity V nor the asymptotic continuous wave amplitude ρ is monotone as function of Q_3 for $-10\le Q_3\le 0$, so that for a specific choice of equation parameters, the derivative terms in Eq. (4.39) can either increase or decrease the propagating speed of the shock. The two plots correspond to a shock wave moving in the modulational stability domain ($PQ_1=-0.5<0$).

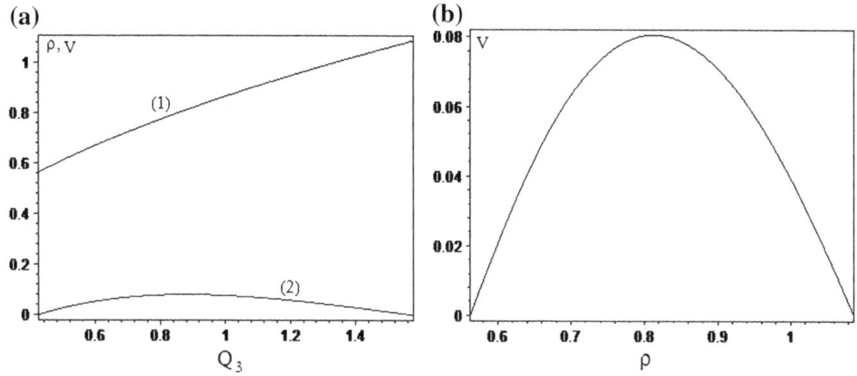

Fig. 4.9 Plots (*1*) of V and (*2*) ρ versus the derivative nonlinearity Q_3 (**a**), and dependence of V on ρ (**b**) for $P = 1$ and $Q_1 = -0.5$ (plotted quantities are dimensionless). Reprint from Ref. [13], Copyright 2018, with permission from Canadian Science Publishing

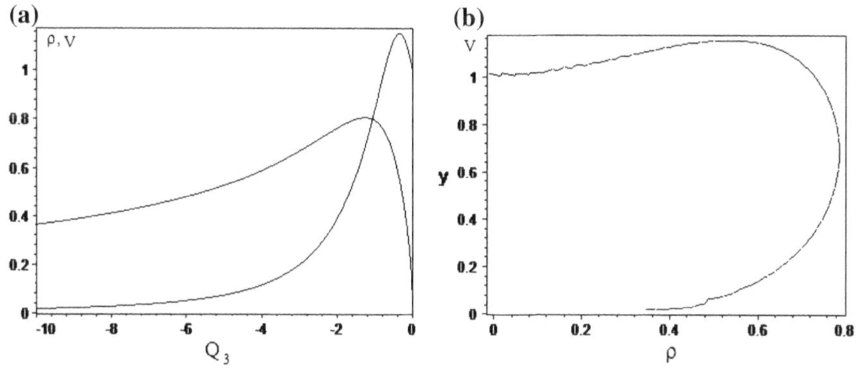

Fig. 4.10 Velocity V of the shock (*1*) and asymptotic continuous wave amplitude ρ (*2*) versus the parameter Q_3 (**a**) and dependence of V on ρ (**b**). The equation parameters are taken to be $P = 1$ and $Q_1 = -0.5$, while the derivative nonlinearities are related by $Q_2 + 3Q_3 = 0$ (plotted quantities are dimensionless). Reprint from Ref. [13], Copyright 2018, with permission from Canadian Science Publishing

4.4 Solitary Signals in the Network of Fig. 3.1

The present section deals with solitary wave solutions of the GCQD-NLS equation (4.41) which may be used to investigate analytically the transmission of solitary signals through the network of Fig. 3.1 when the dissipation and the second-neighbor interactions in the line are neglected. For simplicity, we limit ourselves to pulselike solutions that have zero transverse velocity. The linear stability of solitary wave solution is also investigated.

4.4.1 Analytical Pulselike Solutions of the GCQD-NLS Equation

As we have mentioned, we focus ourselves on pulselike solutions of Eq. (4.41) that have zero transverse velocity. Hence, we look for solution of the form

$$\psi(\xi, \tau) = r(\xi) \exp(-i\varsigma\tau), \tag{4.93}$$

where ς is a real constant and $r(\xi)$ is a complex function that can always be rewritten in an explicit form as

$$r(\xi) = a(\xi) \exp[i\phi(\xi)], \tag{4.94}$$

where a and ϕ are two real functions of variable ξ. Inserting Eqs. (4.93) and (4.94) into Eq. (4.41) and separating real and imaginary parts yield

$$\begin{aligned} a'^2 = & \frac{(3\chi_1 - \chi_2) C_1 - 2P\varsigma}{2P^2} a^2 + \frac{Q}{2P} a^4 \\ & + \frac{3\chi_2^2 - 5\chi_1^2 - 2\chi_1\chi_2 - 16P\chi_3}{48P^2} a^6 - \frac{C_1^2}{P^2 a^2} + \frac{C_2}{4}, \end{aligned} \tag{4.95}$$

$$\phi(x) = \int \left(\frac{C_1}{Pa^2} + \frac{\chi_1 + \chi_2}{4P} a^2 \right) dx + \phi_0, \tag{4.96}$$

where C_1 and C_2 are arbitrary parameters of integration, and ϕ_0 is an initial phase; we suppose $\phi_0 = 0$ for simplicity. Ansatz $a^2 = \zeta$ reduces Eq. (4.95) to

$$\begin{aligned} \zeta'^2 = & \frac{3\chi_2^2 - 5\chi_1^2 - 2\chi_1\chi_2 - 16P\chi_3}{12P^2} \zeta^4 + \frac{2Q}{P} \zeta^3 \\ & + \frac{2[(3\chi_1 - \chi_2) C_1 - 2P\varsigma]}{P^2} \zeta^2 + C_2\zeta - \frac{4C_1^2}{P^2}. \end{aligned} \tag{4.97}$$

Equation (4.97) is an elliptic equation, and its solutions can be found relatively easily. Bounded soliton-like solutions exist only if either $C_1 = C_2 = 0$ (this corresponds to solutions decreasing to zero at infinity) and $P\varsigma < 0$ or $5\chi_2^2 - 3\chi_1^2 + 2\chi_1\chi_2 - 16P\chi_3 = 0$ with some complementary conditions on C_1 and C_2.

In the case of solutions decreasing to zero at infinity ($C_1 = C_2 = 0$), Eq. (4.95) admits the bounded soliton-like solutions

$$a^2(\xi) = \frac{12P\varsigma}{3PQ - \sqrt{3\left(3P^2Q^2 + \varsigma P\left(3\chi_2^2 - 5\chi_1^2 - 2\chi_1\chi_2 - 16P\chi_3\right)\right)} \cosh 2\sqrt{-\varsigma P^{-1}} x} \tag{4.98}$$

under conditions

$P\varsigma < 0$ and either $PQ < 0$ or $PQ > 0$ and $5\chi_1^2 + 2\chi_1\chi_2 + 16P\chi_3 - 3\chi_2^2 > 0.$ (4.99)

In the special case when $3\chi_2^2 - 5\chi_1^2 - 2\chi_1\chi_2 - 16P\chi_3 = 0$, Eq. (4.95) admits the soliton-like solution

$$a^2(\xi) = A_0 + \frac{A_1}{\cosh^2 h\xi}, \qquad (4.100)$$

where

$$A_0 = \frac{2(\chi_2 - 3\chi_1)C_1 + 2P\varsigma + \sqrt{4[(3\chi_1 - \chi_2)C_1 - 2P\varsigma]^2 - 6P^3QC_2}}{6PQ},$$

$$A_1 = \frac{4(\chi_2 - 3\chi_1)C_1 + 8P\varsigma + \sqrt{4[(3\chi_1 - \chi_2)C_1 - 2P\varsigma]^2 - 6P^3QC_2}}{2PQ}, \qquad (4.101)$$

$$h = \frac{1}{2P}\left[4((\chi_2 - 3\chi_1)C_1 + 2P\varsigma)^2 - 6P^3QC_2\right]^{1/4},$$

C_1 being any solution of equation

$$6C_1^2 + \left((3\chi_1 - \chi_2)C_1 - 2P\varsigma + \sqrt{4((3\chi_1 - \chi_2)C_1 - 2P\varsigma)^2 - 6P^3QC_2}\right)A_0^2 = 0. \qquad (4.102)$$

Apart from the necessary condition $3\chi_2{}^2 - 5\chi_1^2 - 2\chi_1\chi_2 - 16P\chi_3 = 0$, the soliton solution (4.100) exists if $4[(\chi_2 - 3\chi_1)C_1 + 2P\varsigma]^2 - 6P^3QC_2 \geq 0$, where C_1 is a solution of Eq. (4.102) and C_2 and ς are two free parameters. The requirement that $a^2(\xi)$ is nonnegative imposes some conditions on the domain of these parameters, namely,

$$A_0 \geq 0 \text{ and either } A_1 > 0 \text{ or } A_1 < 0 \text{ and } A_0 + A_1 \geq 0. \qquad (4.103)$$

4.4.2 Linear Stability Analysis of Found Solutions

To analyze the stability of any of the solutions found in the previous section, we seek a perturbed solution of the form

$$\psi(\xi, \tau) = [a(\xi) + g(\xi, \tau)] \exp(i\phi(\xi) - i\varsigma\tau), \qquad (4.104)$$

where $a(\xi)\exp(i\phi(\xi) - i\varsigma\tau)$ is the stationary wave solution under study, and $g(\xi, \tau)$ is a perturbation function. Inserting Eq. (4.104) into Eq. (4.41) and collecting terms linear to g yield the linearized equation. Its real and imaginary parts are $\mathbf{U} = (U_1, U_2)^T = (\mathrm{Re}[g], \mathrm{Im}[g])^T$:

$$\frac{\partial \mathbf{U}}{\partial \tau} = JL\mathbf{U} = \begin{pmatrix} 0 & 1 \\ -1 & 0 \end{pmatrix} \begin{pmatrix} L_+ & S \\ -S & L_- \end{pmatrix} \mathbf{U}, \qquad (4.105)$$

$$L_+ = -P\frac{\partial^2}{\partial\xi^2} + \frac{C_1^2}{Pa^4} + 3Qa^2 - \left(\varsigma + \frac{(5\chi_2 - 7\chi_1)\,C_1}{2P}\right)$$
$$- \left(\frac{11\chi_2^2 - 13\chi_1^2 - 2\chi_1\chi_2}{16P} - 5\chi_3\right)a^4,$$
$$L_- = -P\frac{\partial^2}{\partial\xi^2} + \frac{C_1^2}{Pa^4} + Qa^2 + \left(\frac{(3\chi_1 - \chi_2)\,C_1}{2P} - \varsigma\right) \tag{4.106}$$
$$+ \frac{5\chi_1^2 - 3\chi_2^2 + 2\chi_1\chi_2 + 12P\chi_3}{16P}a^4,$$
$$S = -\frac{1}{a}\frac{\partial}{\partial\xi}\left(\frac{(\chi_1 + \chi_2)}{2}a^3 - 2C_1\frac{1}{a}\right),$$

$$J = \begin{pmatrix} 0 & 1 \\ -1 & 0 \end{pmatrix} \text{ and } L = \begin{pmatrix} L_+ & S \\ -S & L_- \end{pmatrix}.$$

Operator L is Hermitian as are L_+ and L_-, while S is anti-Hermitian. Equation (4.105) for the perturbation function g has potentially many possible types of solutions. Moreover, because the linear operator JL is not Hermitian, its eigenvalues are complex in general.

Seeking nontrivial solution of Eq. (4.105) of the form $U(\xi, \tau) = (\xi)\exp(\lambda\tau)$ yields the eigenvalue problem

$$(JL - E\lambda) = 0, \tag{4.107}$$

where E is the second-order identity matrix, λ is a complex, and $= \left(\gamma_1(\xi), \gamma_2(\xi)^T\right)$ is a complex vector function. If all λ are imaginary, then linear stability is established. In contrast, if there is at least one eigenvalue λ with a positive real part, then instability results. From the results of the previous section, one easily verifies that $L_-a(\xi) = 0$. Thus $\lambda = 0$ is in the spectrum of operator L_-. For general solutions of form (4.93), 4.94), (4.96), determining the spectrum of the associated linearized eigenvalue problem (4.97) is beyond the scope of current methods. So we restrict ourselves to the stability analysis of trivial-phase solutions $\phi(\xi) = 0$, which correspond to $C_1 = 0$ and $\chi_1 + \chi_2 = 0$ (see Eq. (4.96) with $\phi_0 = 0$).

Under the conditions $C_1 = \chi_1 + \chi_2 = 0$, the spectra of the Hermitian operator L_+ and L_- are real and consist of bands of continuous spectrum contained in $[\lambda_\pm, +\infty)$ [5]. Here $\lambda_\pm$ denotes the ground-state eigenvalues of operators $L_\pm$ given by Eqs. (1.13). In the case under consideration ($C_1 = \chi_1 + \chi_2 = 0$), we have $L_+ = L_- + 2a^2\left(Q + 2\chi_3a^2\right)$ so that $\lambda_+ < \lambda_-$ if $\text{sign}(Q) + \text{sign}(\chi_3) \leq -1$, and $\lambda_+ > \lambda_-$, if $\text{sign}(Q_1) + \text{sign}(\chi_3) \geq +1$. Because $\lambda = 0$ is in the spectrum of operator L_-, we also have $\lambda_- \leq 0$.

Reasoning similar to the arguments in [3, 4] leads to the following conclusions **criterion**: of the linear stability of the above solitary wave solutions of Eq. (4.41)

Criterion of the linear (in)stability of the above solitary wave solutions of Eq. (4.41)

1. If $a(\xi) > 0$, then $a(\xi)$ is the ground state of L_- and $\psi(\xi, \tau) = a(\xi) \exp[-i\varsigma t]$ is linearly stable (linearly unstable) if $\text{sign}(Q_1) + \text{sign}(\chi_3) \geq +1$ (if $\text{sign}(Q_1) + \text{sign}(\chi_3) \leq -1$).
2. If $a(\xi)$ has any zeros, it is not the ground state of L_- and $\lambda_- < 0$. If $\text{sign}(Q_1) + \text{sign}(\chi_3) \geq +1$ and in addition L_+ is a positive operator, then $\psi(\xi, \tau) = a(\xi) \exp[-i\varsigma\tau]$ is linearly unstable.
3. In all other cases, no conclusion is obtained from this stability analysis.

This analysis immediately establishes the linear (un)stability of a solution $\psi(\xi, \tau) = a(\xi) \exp[-i\varsigma\tau]$, where $a(\xi)$ is strictly positive. From Eqs. (4.98) and (4.99), we obtain for trivial-phase solution

$$a^2(\xi) = \frac{-12\varsigma}{-3Q_1 + \text{sign}(P)\sqrt{3\left(3Q_1^2 - 16\varsigma\chi_3\right)} \cosh\left[2\sqrt{-\varsigma P^{-1}}\xi\right]}$$

with $P\varsigma < 0$ and either $PQ < 0$ or $PQ > 0$ and $P\chi_3 > 0$, and criterion 1 applies on $\psi(\xi, \tau)$ for the linear stability. For the solutions with amplitude given by Eq. (4.100) with conditions (4.101), (4.102), the necessary condition gives $\chi_3 = 0$ for the trivial-phase solution, and the (in)stability criterion 1 applies if $A_0 \geq 0$ and either $A_1 > 0$ or $A_1 < 0$ and $A_0 + A_1 > 0$. We note that $a^2(\xi)$ given by Eq. (4.100) has zeros only if $A_0 \geq 0$, $A_1 < 0$ and $A_0 + A_1 \leq 0$. Thus, if $A_0 \geq 0$, $A_1 < 0$, and $A_0 + A_1 \leq 0$, then either the instability criterion 2 or the indeterminate stability criterion 3 occurs.

4.4.3 Higher Order Soliton-Like Solutions of the GCQD-NLS Equation

Now, we construct the higher order soliton-like solution of Eq. (4.41). If we denote by $u_1 = \psi$ and by $u_2 = \psi^*$ (i.e., the complex conjugate of ψ) and then, we take the complex conjugate of Eq. (4.41), we obtain that u_1 and u_2 are solutions of system

$$i\frac{\partial u_j}{\partial \tau} + P^j \frac{\partial^2 u_j}{\partial \xi^2} = u_j \left(Q^j u_j u_{3-j} + i\chi_1^j u_j \frac{\partial u_{3-j}}{\partial \xi} + i\chi_2^j u_{3-j} \frac{\partial u_j}{\partial \xi} + \chi_3^j u_j^2 u_{3-j}^2\right), \quad j = 1,\ 2, \tag{4.108}$$

where $P^1 = P^2 = P$, $Q^1 = Q^2 = Q$, $\chi_1^1 = \chi_1$, $\chi_1^2 = -\chi_1$, $\chi_2^1 = \chi_2$, $\chi_2^2 = -\chi_2$, $\chi_3^1 = \chi_3^2 = \chi_3$. Following Hirota [7], we introduce the ansatz

$$u_j = \frac{U_j}{\Phi}, \quad j = 1,\ 2, \tag{4.109}$$

where U_j, $j = 1, 2$ are complex functions and Φ is a real function. Using the fact that $u_2 = u_1^*$ and Φ is a real function, we must have $U_2 = U_1^*$. Inserting ansatz (4.109) into Eq. (4.108) yields

$$i\left(\Phi\frac{\partial U_j}{\partial\tau}-U_j\frac{\partial\Phi}{\partial\tau}\right)+P^j\left(\Phi\frac{\partial^2U_j}{\partial\xi^2}-2\frac{\partial U_j}{\partial\xi}\frac{\partial\Phi}{\partial\xi}+U_j\frac{\partial^2\Phi}{\partial\xi^2}\right)=0,\ j=1,\ 2,\tag{4.110}$$

$$\begin{aligned}&2P\Phi^2\left(\Phi\frac{\partial^2\Phi}{\partial\xi^2}-\left(\frac{\partial\Phi}{\partial\xi}\right)^2\right)+Q\Phi^2U_1U_2+\chi_3U_1^2U_2^2\\&=i\chi_1\Phi U_1\left(U_2\frac{\partial\Phi}{\partial\xi}-\Phi\frac{\partial U_2}{\partial\xi}\right)+i\chi_2\Phi U_2\left(U_1\frac{\partial\Phi}{\partial\xi}-\Phi\frac{\partial U_1}{\partial\xi}\right),\end{aligned}\tag{4.111}$$

$$\begin{aligned}&2P\Phi^2\left(\Phi\frac{\partial^2\Phi}{\partial\xi^2}-\left(\frac{\partial\Phi}{\partial\xi}\right)^2\right)+Q\Phi^2U_2U_1+\chi_3U_2^2U_1^2\\&=i\chi_1\Phi U_2\left(\Phi\frac{\partial U_1}{\partial\xi}-U_1\frac{\partial\Phi}{\partial\xi}\right)+i\chi_2\Phi U_1\left(\Phi\frac{\partial U_2}{\partial\xi}-U_2\frac{\partial\Phi}{\partial\xi}\right).\end{aligned}\tag{4.112}$$

So the left-hand sides of Eqs. (4.111) and (4.112) become equal. Hence, the right-hand sides of Eqs. (4.111) and (4.112) should also be equal which is true only when

$$\chi_1+\chi_2=0.\tag{4.113}$$

Having obtained condition (4.113), our next aim is to obtain the soliton solutions. Now to generate the soliton solutions, we expand the dependent variables in terms of ϵ and the solutions can be constructed as usual. For

$$U_j=\epsilon U_{j1},\ \Phi=1+\epsilon^2\Phi_2,\ j=1,2,\tag{4.114}$$

the solutions are found to be

$$\begin{aligned}U_{11}(\xi,\tau)&=\exp\left[k_1\xi+iPk_1^2\tau+k_{10}\right],\ U_{21}(\xi,\tau)=\exp\left[k_1^*\xi-iPk_1^{*2}\tau+k_{10}^*\right],\\ \Phi_2(\xi,\tau)&=\frac{\left[i\chi_1\left(k_1-k_1^*\right)-Q\right]}{2P\left(k_1+k_1^*\right)^2}\exp\left[\left(k_1+k_1^*\right)\xi+iP\left(k_1^2-k_1^{*2}\right)\tau+k_{10}+k_{10}^*\right],\end{aligned}$$

where k_1 and k_1 are two arbitrary complex numbers with $k_1\notin i\mathbb{R}$, that is, k_1 is not an imaginary number. Because $u_1(\xi,\tau)=\psi(\xi,\tau)$, we then obtain, under condition (4.113) and after using Eqs. (4.114) and (4.109), the following higher order soliton solutions of Eq. (4.41):

$$\psi(\xi,\tau)=\frac{\epsilon\exp\left(k_1\xi+iPk_1^2t+k_{10}\right)}{1+\epsilon^2\frac{\left[i\chi_1\left(k_1-k_1^*\right)-Q\right]}{2P\left(k_1+k_1^*\right)^2}\exp\left(\left(k_1+k_1^*\right)\xi+iP\left(k_1^2-k_1^{*2}\right)\tau+k_{10}+k_{10}^*\right)},\tag{4.115}$$

where ϵ is a real parameter. Parameters k_1, k_{10}, and ϵ are to be taken so that the denominator of the expression in the right-hand side of Eq. (4.115) does not vanish. For example, if k_1 is taken so that $P\left[i\chi_1\left(k_1-k_1^*\right)-Q\right]\geq 0$ (wit $P\neq 0$), then solution (4.115) will be well-defined. We note that the higher order soliton solution (4.115) does not depend on the quintic nonlinearity coefficient χ_3. This situation happens because for the determination of the mentioned solution, we have used only the lower order of ϵ, and namely, ϵ^0 in Eqs. (4.111) and (4.112). We also note that in the case of real k_1, the amplitude of soliton solution (4.115) will be a time-independent function. In this case, one can get from Eq. (4.115) the ordinary soliton solutions of Eq. (4.41) which is not included in those above. For example, if $PQ<0$ then by taking $k_1^2=-\epsilon^2\frac{Q}{8P}$ and $\epsilon>0$, we obtain the soliton solution with amplitude $|\psi(\xi,\tau)|=\frac{\epsilon}{2\cosh(k_1\xi+k_{10})}$.

4.5 Dissipative CQD-NLS Equation for Lossy Network of Fig. 3.1

In this section, we consider the dissipative nonlinear dispersive transmission line of Fig. 3.1 when the second-neighbor interactions are neglected. We show that the envelope modulation of the network in this case can be reduced to a cubic–quintic derivative nonlinear Schrödinger equation with a linear gain term.

4.5.1 *Circuit Equations*

To obtain the circuit equation, we follow the idea introduced in Sect. 3.2.1. Thus, by applying Kirchhoff's theorems to the dissipative circuit of Fig. 3.1 in the case when the second-neighbor interactions are neglected ($L_3=\infty$), we have the following differential-difference equation for the voltage $V_n(t)$

$$\frac{d^2V_n}{dt^2}-\left(u_0^2+\lambda\frac{d^2}{dt^2}\right)(V_{n+1}-2V_n+V_{n-1})+\left(2u_0\sigma\frac{d}{dt}+\omega_0^2\right)V_n$$
$$=\frac{d^2}{dt^2}\left(\alpha V_n^2-\beta V_n^3\right),\quad n=1,2,...,N, \qquad (4.116)$$

where $\omega_0=1/\sqrt{C_0L_2}$, $u_0=1/\sqrt{C_0L_1}$, $\lambda=C_S/C_0$, $\sigma=\frac{G}{2\sqrt{L_1^{-1}C_0}}$. Here, ω_0 and u_0 are the characteristic frequencies of the network. Now, we take the continuum limit of Eq. (4.116) in the n direction, along with the additional provisos that the amplitude of the voltage pulse is small and its wavelength is much greater than the spacing δ between two adjacent sections. Then $V_{n\pm1}$ can be approximated by partial derivatives with respect to distance, x, from the beginning of the line (i.e., $x_n=n\delta$). An approximate continuous PDE can be obtained by using the Taylor expansions of $V(x\pm\delta,t)$. Thus, we obtain

$$\frac{\partial^2 V}{\partial t^2} - u_0^2\delta^2 \frac{\partial^2 V}{\partial x^2} - \lambda\delta^2 \frac{\partial^4 V}{\partial t^2 \partial x^2} + 2u_0\sigma \frac{\partial V}{\partial t} + \omega_0^2 V = \frac{\partial^2}{\partial t^2}\left(\alpha\, V^2 - \beta V^3\right). \tag{4.117}$$

Looking for a solution of the form $\exp\left[-i(kx - \omega t)\right]$ of the linear Eq. (4.117) yields the complex dispersion equation

$$\omega^2 - \omega_0^2 + \delta^2\left(\lambda\omega^2 - u_0^2\right)k^2 - 2iu_0\sigma\omega = 0, \tag{4.118}$$

where $k = k_r(\omega) + ik_i(\omega)$ is the complex wave vector and ω is the angular frequency. Equation (4.118) defines the dispersion relation

$$\delta^4\left(u_0^2 - \lambda\omega^2\right)^2 k_r^4 + \delta^2\left(\omega^2 - \omega_0^2\right)\left(\lambda\omega^2 - u_0^2\right)k_r^2 - u_0^2\sigma^2\omega^2 = 0 \tag{4.119}$$

and the below linear dissipation relation for the definition of the linear attenuation coefficients $k_i = k_i^{\pm}$ as a function of k_r

$$\frac{1 + \delta^2\lambda\left(k_r^2 - k_i^2\right)}{\left(u_0\sigma \pm u_0\sqrt{\sigma^2 + 4k_r^2k_i^2\delta^4\lambda}\right)^{-1}} - \frac{2\lambda\delta^2 k_r k_i}{\left[\omega_0^2 + \delta^2 u_0^2\left(k_r^2 - k_i^2\right)\right]^{-1}} = 0. \tag{4.120}$$

If solving Eq. (4.119) with respect with the frequency the angular ω, one easily sees that the relation dispersion (4.119) provides two kinds of frequency modes: the Low-Frequency (LF) mode in the region $0 \le \omega \le \omega_c = u_0/\sqrt{\lambda}$ and the High-Frequency (HF) mode in the region $\omega \ge \omega_c$. It follows from Eq. (4.119) that $\omega(k_r = 0) = 0$, leading, when using Eq. (4.120) to $k_i^+(k_r = 0) = 1/(\delta\sqrt{\lambda})$ and an arbitrary $k_i^-(k_r = 0)$.

In a dissipative transmission line, the expression for propagating linear waves with the amplitude V_m at $x = 0$ and angular frequency ω is

$$V(x,t) = V_m \exp(k_i x)\cos\left(k_r x - \omega t\right). \tag{4.121}$$

For the dissipative transmission line, the linear waves must propagate and decay in the $+x$-direction. Hence, the attenuation coefficient k_i appearing in Eq. (4.121) must be negative. In other words, for the dissipative network of Fig. 3.1, we are interested in only negative solutions (with respect with k_i) of Eq. (4.120). Since the direction of the wave propagation is usually determined by the direction of the group velocity (GV) $v_g = d\omega/dk_r$, and because waves with positive GV in the presence of a loss decay in the $+x$ -direction, the suitable frequency for us is the one given by a positive GV.

In the absence of the dissipation in the network of Fig. 3.1, i.e., when $\sigma = 0$, the dispersion equation (4.119) coincides with the well-known dispersion relation $\omega^2 = \frac{\omega_0^2 + \delta^2 u_0^2 k_r^2}{1 + \lambda\delta^2 k_r^2}$, which admits a gap ω_c (minimal frequency) at $k_r = +\infty$ and attains the maximal frequency ω_0 at $k_r = 0$. Thus, Eq. (4.119) establishes the relation between the wave number and the angular frequency in the general case of dissipative network.

4.5.2 Dissipative CQD-NLS Equation

To show that the envelope modulation of the network when the dissipation in the network is taken into account and when the second-neighbor interactions are neglected, we use the standard reductive perturbation technique and complex expansion to obtain short wavelength envelope solitons. Assuming that waves propagate in one direction ($+x$-direction) and using this approach, we can introduce different slow time scales, let us say, for example, two time scales $T_1 = \epsilon t$ and $T_2 = \epsilon^2 t$ in addition to the time variable t, and one large length scale $X = \epsilon x$ in addition to the spatial variable x. Here, ϵ is a real parameter that measures the smallness of the modulation frequency and the amplitude of the input waves. With the signal V, depending not only on the original variables x and t, but also on X, T_1, and T_2, it is possible to separate, in a natural way, its fast and slow variations in both space and time. Thus $V(x,t) = V(x, X; t, T_1, T_2)$. Next, we order the damping coefficient σ so that the dissipation effect, as well as the nonlinearity effect, does not appear in all orders of ϵ. Thus, we set

$$\sigma = \epsilon^2 \sigma_0. \tag{4.122}$$

By assuming ϵ to be sufficiently small, we seek the solution V of Eq. (4.117) in the form

$$\begin{aligned} V &= \epsilon^{1/2} u_{11} \exp[i\theta] + \epsilon\left(u_{20} + u_{22}\exp[2i\theta]\right) + \epsilon^{3/2}\left(u_{30} + u_{33}\exp[3i\theta]\right) \\ &\quad + \epsilon^2\left(u_{40} + u_{42}\exp[2i\theta] + u_{44}\exp[4i\theta]\right) + \epsilon^{5/2}\left(u_{50} + u_{53}\exp[3i\theta]\right. \\ &\quad \left. + u_{55}\exp[5i\theta]\right) + cc + O(\epsilon^3) + \ldots, \end{aligned} \tag{4.123}$$

where $\theta = kx - \omega t$ is the rapidly varying phase, $u_{ij} = u_{ij}(T_1, T_2, X_1)$, and $c.c.$ stands for the complex conjugate of the preceding terms. Here k and ω are, respectively, the wave number and the angular frequency of the carrier. Under all the above assumptions and after inserting the perturbation expansion (4.123) into the nonlinear equation (4.117), we obtain a series of equations at different orders of $\epsilon^{1/2}$ for different harmonics.

Equations of order $O(\epsilon^{1/2})$ for the first harmonic leads to the linear dispersion relation

$$\omega^2 - \omega_0^2 + \delta^2\left(\lambda\omega^2 - u_0^2\right)k^2 = 0, \tag{4.124}$$

which is different from the dispersion equation (4.118) by the absence of the dissipation term (the last term in the right-hand side of Eq. (4.118)). At order $O(\epsilon^{3/2})$ for the first harmonic, we obtain

$$i\left[\frac{\partial u_{11}}{\partial T_1} + v_g \frac{\partial u_{11}}{\partial X}\right] = Q(k)\left|u_{11}\right|^2 u_{11}, \tag{4.125a}$$

$$Q(k) = -\frac{\omega\left[2\alpha\left(2A_{20} + A_{22}\right) - 3\beta\right]}{2\left(1 + \lambda\delta^2 k^2\right)}, \tag{4.125b}$$

where

$$A_{20} = \frac{2\alpha V_g^2}{\left(V_g^2 - u_0^2\delta^2\right)},\ A_{22} = \frac{4\alpha\omega^2}{3\delta^2\left(5\lambda\omega^2 - u_0^2\right)k^2 + 3\omega^2}, \tag{4.126}$$
$$v_g = \frac{d\omega}{dk} = \frac{k\delta^2\left(u_0^2 - \lambda\omega^2\right)}{\omega\left(1 + \lambda\delta^2k^2\right)}.$$

Following Kakutani et al. [10], we assume that $\Delta k = k - k_c = O(\epsilon)$ and write $Q(k) = \epsilon Q_1(k)$, where $Q_1(k)$ is approximately given by

$$Q_1(k) = \frac{1}{\epsilon}Q(k_c) + \frac{\Delta k}{\epsilon}\frac{dQ(k)}{dk}, \tag{4.127}$$

k_c being the critical value of the wave number k of the signal. We then obtain at order $O(\epsilon^{3/2})$ for the first harmonic, after shifting the right-hand side of Eq. (4.125a) to the equations of order $O(\epsilon^{5/2})$ for the first harmonic,

$$\frac{\partial u_{11}}{\partial T_1} + V_g\frac{\partial u_{11}}{\partial X} = 0, \tag{4.128}$$

implying that $u_{11} = u_{11}(X - v_gT_1, T_2) = u_{11}(\xi, \tau)$, where $\xi = X - v_gT_1 = \epsilon(x - v_gt)$ and $\tau = T_2 = \epsilon^2 t$. Hence, in a reference frame moving with the group velocity v_g, the complex amplitude u_{11} of the signal remains constant with respect to the concerned scale.

From the equations of order $O(\epsilon^{5/2})$ for the first harmonics where the right-hand side of Eq. (4.125a) has been shifted, we obtain the resulting equation that describes dynamics of a packet wave in the discrete dissipative nonlinear transmission line of Fig. 3.1 when $L_3 = \infty$

$$i\frac{\partial u}{\partial t} = -\frac{1}{2}P\frac{\partial^2 u}{\partial z^2} + Q|u|^2u + i\Gamma u + Q_1|u|^4u + iQ_2\frac{\partial}{\partial z}|u|^2u, \tag{4.129}$$

where

$$u = \epsilon^{1/2}u_{11},\ z = x - v_gt;$$
$$P = \frac{d^2\omega}{dk^2} = \frac{\delta^2\left(u_0^2 - \lambda\omega^2\right) - 4\lambda\delta^2k\omega v_g - \left(1 + \lambda\delta^2k^2\right)v_g^2}{\left(1 + \lambda\delta^2k^2\right)\omega},$$
$$Q = \frac{\omega\left[2\alpha\left(2A_{20} + A_{22}\right) - 3\beta\right]}{2\left(1 + \lambda\delta^2k^2\right)},\ \Gamma = \frac{u_0\sigma}{1 + \lambda\delta^2k^2}, \tag{4.130}$$
$$Q_1 = -\frac{\omega\left[\beta\left(11A_{20}^2 + 4A_{22}^2 + 8A_{20}A_{22} + 2A_{33}\right) - 2\alpha A_{22}A_{33}\right]}{2\left(1 + \lambda\delta^2k^2\right)},$$
$$Q_2 = -\frac{\left[3\beta\omega V_g - 2\alpha k\left(2A_{20} + A_{22}\right)\right]}{\left(1 + \lambda\delta^2k^2\right)\omega}.$$

In Eq. (4.130), A_{20}, A_{22} and v_g are defined by Eq. (4.126), and A_{33} is given by defined as $A_{33} = \frac{9\omega^2(\beta-\alpha A_{22})}{4(\delta^2(u_0^2-10\lambda\omega^2)k^2-\omega^2)}$. The dispersion coefficient P, the parameter of the cubic nonlinearity Q, the linear gain coefficient Γ, the parameter of the quintic nonlinearity Q_1, and the derivative term parameter Q_2 of Eq. (4.129) are all functions of the wave number k. It should be noted that only one coefficient of Eq. (4.130), namely, the linear gain coefficient Γ, is affected by the presence of the dissipation in the network. This is due to assumption (4.122), an assumption without which it was not possible to derive the nonlinear Schrödinger equation type.

The dissipative CQD-NLS equation (4.129) is different from the CQD-NLS (4.39), derived in the case of the lossless transmission line in Fig. 3.1 by the presence of the linear gain with parameter Γ and the cubic nonlinear term with coefficient Q. When observing the two equations (Eq. (4.129) and Eq. (4.39)), one easily sees that Eq. (4.39) possesses the Stokes wave solutions, while Eq. (4.129) does not. The absence of the Stokes wave solutions for Eq. (4.129) is due to the presence of the linear gain term. Thus the dissipative element in the nonlinear transmission line does not allow the propagation of Stokes wave in the network.

By change of variables

$$t = \tau,\ z = \xi + P\frac{Q}{Q_2}\tau, u = q(\xi,\tau)\exp\left[i\frac{Q}{Q_2}\left(z - \frac{PQ}{2Q_2}t\right)\right],$$

the dissipative CQD-NLS Eq. (4.129) becomes

$$i\frac{\partial q}{\partial \tau} = -\frac{1}{2}P\frac{\partial^2 q}{\partial \xi^2} + iQ_2\frac{\partial}{\partial \xi}|q|^2 q + Q_1|q|^4 q + i\Gamma u. \tag{4.131}$$

Ignoring the quintic and dissipative terms, Eq. (4.131) reduces to

$$i\frac{\partial q}{\partial \tau} = -\frac{1}{2}P\frac{\partial^2 q}{\partial \xi^2} + iQ_2\frac{\partial}{\partial \xi}|q|^2 q, \tag{4.132}$$

which is related to the Kaup–Newell (KN) equation [11]. Therefore, the dissipative CQD-NLS equation (4.131) is a generalized dissipative KN equation. The nonlinear term $\frac{\partial}{\partial \xi}|q|^2 q$ in the KN Eq. (4.132) adequately describe the self-steepening effect.

The dissipative CQD–NLS equation (4.129) possesses the plane wave solution with time-verifying amplitude

$$u(z,t) = \rho_0(t)\exp\left[ik_0 z - i\frac{PA_0k_0^2}{2}t\right] \tag{4.133}$$

where $\rho_0(t)$ is the complex amplitude defined as

$$\rho_0(t) = A_0\exp\left[\Gamma t + \frac{i}{4\Gamma}\left(2A_0^2\left(Q_2k_0 - Q\right)\exp\left(2\Gamma t\right)\right)\right.$$

$$-Q_1 A_0^4 \exp(4\Gamma t) + A_0^2 \left(Q_1 A_0^2 + 2Q - 2Q_2 k_0\right)\big)\big], \quad (4.134)$$

A_0 being an arbitrary constant. Because $\Gamma > 0$, function $|\rho_0(t)|$ increases and tends to $+\infty$ as $t \to +\infty$. In the limit $\Gamma \to +0$, the plane wave solution (4.133), (4.134) reduces to the periodic Stokes wave with constant amplitude

$$u(z,t) = a_0 \exp\left[ik_0 x - i\left(\frac{1}{2}Pk_0^2 + Qa_0^2 + Q_1 a_0^4 - Q_2 a_0^2 k_0\right)t\right],$$

plane wave solution of Eq. (4.129) in the absence of linear gain ($\Gamma = 0$).

References

1. Abdourahman., Kengne E.: Generation of nonlinear modulated waves in a modified Noguchi electrical transmission network. Chaos, Solitons Fractals **92**, 1–8 (2016)
2. Benjamin, T.B., Feir, J.E.: The disintegration of wave trains on deep water. Part 1. Theory. J. Fluid Mech. **27**, 417 (1967)
3. Bronski, J.C., Carr, L.D., Deconinck, B., Kutz, J.N.: Stability of attractive Bose-Einstein condensates in a periodic potential. Phys. Rev. Lett. **86**, 1401 (2001)
4. Bronski, J.C., Carr, L.D., Deconinck, B., Kutz, J.N., Promislow, K.: Stability of repulsive Bose-Einstein condensates in a periodic potential. Phys. Rev. E **63**, 036612 (2001)
5. Courant, R., Hilbert, D.: Method of mathematical physics. Wiley, New York (1989)
6. El, G.A., Grimshaw, R.H.J.: Generation of undular bores in the shelves of slowly-varying solitary waves. Chaos **12**, 1015 (2002)
7. Hirota, R.: Solitary signals in electrical nonlinear transmission line. J. Math. Phys. **14**, 805 (1978)
8. Hoefer, M.A., Ablowitz, M.J.: Interactions of dispersive shock waves. Phys. D **44** (2007)
9. Kamchatnov, A.M.: Nonlinear periodic waves and their modulations - an introductory course. World Scientific, Singapore (2000)
10. Kakutani, T., Michihiro, K.: Marginal state of modulational instability -note on Benjamin-Feir instability. J. Phys. Soc. Jpn. **52**, 4129 (1983)
11. Kaup, D.J., Newell, A.C.: An exact solution for a derivative nonlinear Schrödinger equation. J. Math. Phys. **19**, 798 (1978)
12. Kengne, E., Lakhssassi, A.: Dynamics of modulated waves in a lossy modified Noguchi electrical transmission line. Phys. Rev. E **91**, 032907 (2015)
13. Kengne, E., Lakhssassi, A., Nguyen-Ba, T., Vaillancourt, R.: Dispersive shock waves propagating in the cubic-quintic derivative nonlinear Schrödinger equation. Can. J. Phys. **88**, 55–66 (2010)
14. Kengne, E., Liu, W.M.: Exact solutions of the derivative nonlinear Schrödinger equation for a nonlinear transmission line. Phys. Rev. E **73**, 026603 (2006)
15. Kumar, V.R., Radha, R., Wadati, M.: Phase engineering and Solitons of Bose-Einstein condensates with two- and three-body interactions. J. Phys. Soc. Jpn. **79**, 074005 (2010)
16. Lai, D.W.C., Chow, K.W.: Coalescence of ripplons, breathers, dromions and dark solitons. J. Phys. Soc. Jpn. **70**, 666 (2001)
17. Marklund, M., Shukla, P.K.: Modulational instability of partially coherent signals in electrical transmission lines. Phys. Rev. E **73**, 057601 (2006)
18. Marquié, P., Bilbault, J.M., Remoissenet, M.: Generation of envelope and hole solitons in an experimental transmission line. Phys. Rev. E **49**, 828 (1994)
19. Nozaki, K., Bekki, N.: Pattern selection and spatiotemporal transition to Chaos in the Ginzburg-Landau equation. Phys. Rev. Lett. **51**, 2171 (1983)

20. Schürmann, H.W.: Traveling wave solutions of the cubic-quintic nonlinear Schrödinger equation. Phys. Rev. E **54**, 4312 (1996)
21. Taniuti, T., Wei, C.C.: Reductive perturbation method in nonlinear wave propagation II. Application to hydromagnetic waves in cold plasma. J. Phys. Soc. Jpn. **24,** 941 (1968)
22. Taniuti, T., Yajima, N.: Perturbation method for a nonlinear wave modulation I. J. Math. Phys. **10**, 1369 (1969)
23. Washimi, H., Taniuti, T.: Propagation of ion-acoustic solitary waves of small amplitude. Phys. Rev. Lett. **17**, 966 (1966)
24. Whitham, G.B.: Linear and Nonlinear Waves. Wiley-Interscience, New York (1974)

Chapter 5
NLS Equations for the System of the Nonlinear Transmission Lines Coupled by Capacitor C_2

Abstract This chapter investigates the dynamics of modulated waves through the system of the nonlinear transmission lines coupled by a linear capacitor. We prove that the dynamics of modulated waves through the system under consideration are governed by either a 2D NLS equation or a 2D NLS system. Based on these equations, we analytically investigate the transverse stability of plane solitary waves propagating through our system, the transmission of fundamental solitary pulses including the kink solitary pulses, the dark and the bright solitons through the system. The criterion of the modulational instability of the Stokes waves propagating through the system is also established.

5.1 Introduction

Distributed single electrical transmission lines that consist of a large number of identical sections have been used for the experimental study of the propagation of KdV solitons which satisfy the famous Korteweg–de Vries equation. This equation was originally derived in order to model the shallow-water wave experiments of John Scott Russel in the 19th century. Extending the transmission line model to higher dimensions has proven to be difficult both theoretically and experimentally. Experiments and resonances of a two-dimensional transmission line have been examined in [4, 6, 9] by assuming that the nonlinear capacitance C is of the form $C = \frac{C_0}{1+(V/V_0)^p}$. Duan et al. have studied the nonlinear waves on coupled nonlinear transmission line [4] where V is the voltage in the transmission line, C_0, V_0, and p are all constants; it was found that, in the continuum limit and in suitably scaled coordinates, the voltage for this system is governed by a modified Zakharov–Kuznetsov (ZK) equation. In the special case where $p = 1$ Kengne et al. [6] followed Duan et al. [4] and found that the dynamic of any two wave packets in the coupled nonlinear transmission line is described by a coupled nonlinear Schrödinger equations; they derived a set of explicit criteria of modulational (in)stability of the Stokes waves propagating in the lines. In this chapter, we assume that the capacitance C is voltage dependent and of a more general form, but only for small-amplitude perturbation voltage, as follows

W.-M. Liu and E. Kengne, *Schrödinger Equations in Nonlinear Systems*,
https://doi.org/10.1007/978-981-13-6581-2_5

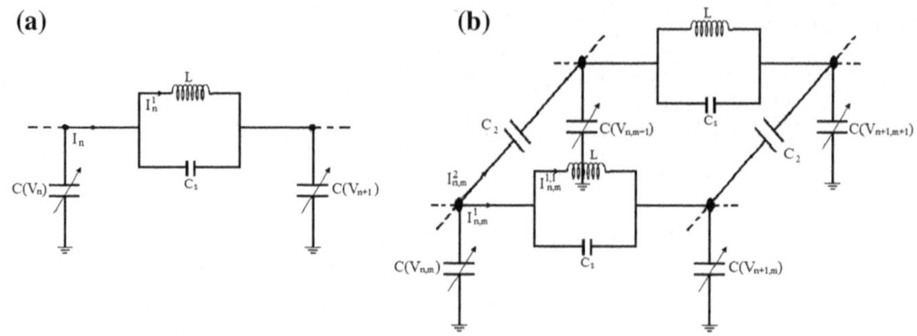

Fig. 5.1 **a**: Schematic representation of the elementary cell of the single lossless nonlinear transmission line. Each cell contains, in the series branch, a linear inductor L shunted by a linear capacitor C_s and in the shunted branch, a nonlinear capacitor $C(V)$. The nonlinear capacitor $C(V)$ introduces the standard nonlinearity in the network. The network possesses N identical unit cells; **b**: Part of the system of the nonlinear transmission line coupled by capacitor C_2. The system contains M lines and each line possesses N identical unit cells

$$C(V) = \frac{dQ}{dV} = C_0 \left(1 - 2\alpha V + 3\beta V^2 + \cdots\right), \tag{5.1}$$

where C_0 is a constants, V is the perturbation voltage in the transmission line shown in Fig. 5.1a, and α and β are the nonlinear constant coefficients of the electrical charge Q stored in the capacitor of the line. Throughout this chapter, we restrict ourselves to the case in which the perturbation voltage V is sufficiently small compared to the equilibrium voltage, so we neglect the higher-order terms in Eq. (5.1) [3]. It is important to note that only the first three terms of the expansion in Eq. (5.1) are used [3]. This is due to the following reasons: If only the first two terms of the expansion in Eq. (5.1) are used, then the ZK equation is obtained to describe this phenomenon, which is the special case with $C = 0$ in the work of Ref. [9]. For small voltage V, but not very small, we should consider the third term of the expansion in Eq. (5.1). For larger V, the higher-order terms of the expansion should be considered.

The present chapter deals with the NLS equations for the system of the nonlinear transmission lines coupled by capacitor C_2, as shown in Fig. 5.1b. This consists of a nonlinear network with many coupled nonlinear LC dispersive-transmission lines. We imagine that there are many identical dispersive lines of Fig. 5.1a which are coupled by means of a linear capacitors C_2 at each node. Each section of the system consists of a linear inductor L in parallel with a linear capacitor C_S (dispersive element) in the series branch and a nonlinear capacitor of capacitance $C = C(V)$ in the shunt branch. The nodes in the system are labelled with two discrete coordinates n and m, where n specifies the nodes in the direction of propagation of the pulse, and m labels the lines in the transverse direction.

5.2 Modulated Waves in the System of Network of Fig. 5.1b

In the continuum limit and suitably scaled of coordinates, we first show that the voltage for the coupled transmission system of Fig. 5.1b is described by a two-dimensional NLS equation. Then, with the use of this 2D NLS equation, we investigate the propagation of modulated waves in the system.

5.2.1 *Mathematical Formalism*

Here, we aim to derive the circuit equations for the system shown in Fig. 5.1b. For this purpose, we apply the Kirchhoff's laws in the orthogonal loops and obtain the circuit equations for this system as follows

$$L\frac{\partial I^1_{n,m}}{\partial t} = V_{n,m} - V_{n+1,m}, \tag{5.2a}$$

$$I^2_{n,m} = C_2\frac{d}{dt}\left(V_{n,m} - V_{n,m+1}\right), \quad I^1_{n,m} - I^{1,1}_{n,m} = C_S\frac{d}{dt}\left(V_{n-1,m} - V_{n,m}\right), \tag{5.2b}$$

$$\frac{\partial Q_{n,m}}{\partial t} = I^1_{n-1,m} - I^1_{n,m} + I^2_{n,m-1} - I^2_{n,m}. \tag{5.2c}$$

Eliminating currents in system (5.2a)–(5.2c) yields

$$\frac{\partial^2 Q_{n,m}}{\partial t^2} = \left(\frac{1}{L} + C_S\frac{\partial^2}{\partial t^2}\right)\left(V_{n-1,m} - 2V_{n,m} + V_{n+1,m}\right) + C_2\frac{\partial^2}{\partial t^2}\left(V_{n,m-1} - 2V_{n,m} + V_{n,m+1}\right). \tag{5.3}$$

Inserting the expansion (5.1) of the capacitance into Eq. (5.3) yields

$$\begin{aligned}\frac{\partial^2 V_{n,m}}{\partial t^2} + \left(\Omega_0^2 + d_0\frac{\partial^2}{\partial t^2}\right)\left(2V_{n,m} - V_{n-1,m} - V_{n+1,m}\right)\\ -\gamma_0\frac{\partial^2}{\partial t^2}\left(V_{n,m-1} - 2V_{n,m} + V_{n,m+1}\right) = \alpha\frac{\partial^2 V^2_{n,m}}{\partial t^2} - \beta\frac{\partial^2 V^3_{n,m}}{\partial t^2},\end{aligned} \tag{5.4}$$

where $n = 1, 2, \ldots, N$, $m = 1, 2, \ldots, M$, the linear wave speed Ω_0, the coupling coefficient γ_0, and the dispersion coefficient d_0 are given by

$$\Omega_0^2 = \frac{1}{C_0 L}, \quad \gamma_0 = \frac{C_2}{C_0}, \quad d_0 = \frac{C_S}{C_0}. \tag{5.5}$$

Equation (5.4) with parameters (5.5) is the circuit equations.

Now, we use the reductive perturbation method in the semi-discrete limit [10] assuming that the wave number is much greater than the lattice spacing, and that the perturbation voltage V is also small enough compared with equilibrium value. We assume that the dominant wave motion is in the n direction. Therefore, modulated waves in the network can be described by considering waves with a slowly varying envelope in time and space with respect to a given carrier with angular frequency ω and wavenumbers k and q. We thus define three stretched variables in the wave frame $x = \epsilon(n - v_g t)$, $y = \epsilon m$, $\tau = \epsilon^2 t$, where ϵ is a small parameter and v_g is a constant. In addition, the voltage in the system of Fig. 5.1b is expanded in the form [10]

$$V_{n,m} = \epsilon A(x, y, \tau) \exp(i\theta) + \epsilon^2 \left[\phi(x, y, \tau) + B(x, y, \tau) \exp(2i\theta)\right] + c.c., \quad (5.6)$$

with $\theta = kn + qm - \omega t$. Substituting Eq. (5.6) into Eq. (5.4) yields different equations as power series of ϵ and $\exp(i\theta)$. At order $\epsilon \times \exp(i\theta)$, we derive the dispersion relation

$$\omega^2(k, q) = \frac{4\Omega_0^2 \sin^2 \frac{k}{2}}{1 + 4d_0 \sin^2 \frac{k}{2} + 4\gamma_0 \sin^2 \frac{q}{2}} \quad (5.7)$$

showing that the angular frequency ω decreases as the coupling parameter $C_2 = C_0\gamma_0$ increases. At orders $\epsilon^2 \times \exp(2i\theta)$ and $\epsilon^4 \times \exp(0i\theta)$, we respectively obtain

$$B = \frac{\omega^2\alpha}{\left(d_0 - \Omega_0^2\right) \sin^2 k + \gamma_0 \sin^2 q - \omega^2} A^2, \quad \phi = \frac{2\alpha v_g^2}{v_g^2 - \Omega_0^2} |A|^2 .$$

At order $\epsilon^2 \times \exp(i\theta)$, we obtain
$\left[\omega\left(1 + 4d_0 \sin^2 \frac{k}{2} + 4\gamma_0 \sin^2 \frac{q}{2}\right) v_g + \left(d_0\omega^2 - \Omega_0^2\right) \sin k\right] \frac{\partial A}{\partial x} + \gamma_0\omega^2 \sin q \frac{\partial A}{\partial y} = 0$,
which is always satisfied if $q \in \pi\mathbf{Z}$ and

$$v_g = \frac{\left(\Omega_0^2 - d_0\omega^2\right) \sin k}{\omega\left(1 + 4d_0 \sin^2 \frac{k}{2} + 4\gamma_0 \sin^2 \frac{\ell'\pi}{2}\right)}, \quad \ell' = 0, \pm 1, \pm 2, \ldots \quad (5.8)$$

Condition $q \in \pi\mathbf{Z}$ leads to $\sin q = 0$, meaning that the group velocity $d\omega/dq$ in the m direction is zero, and this confirms that the dominant motion of the waves is in the n direction. Henceforth, we will distinguish two cases, the case of even ℓ' (i.e., $\ell' = 2\ell$) and the case of odd ℓ' (i.e., $\ell' = 2\ell + 1$), $\ell = 0, \pm 1, \pm 2, \ldots$

Case of even $\ell' : \ell' = 2\ell$
In the present case, Eq. (5.8) becomes

$$v_g = \frac{\Omega_0^2 \sin k}{\omega\left(1 + 4d_0 \sin^2 \frac{k}{2}\right)^2} = \frac{d\omega}{dk}, \quad (5.9)$$

meaning that v_g is the group velocity of the wave package in n direction. From the coefficient of ϵ^3, proportional to $\exp(i\theta)$, we obtain the following 2D nonlinear Schrödinger equation for A

$$i\frac{\partial A}{\partial \tau} + \frac{1}{2} P \left(\frac{\partial^2}{\partial x^2} + \frac{\partial^2}{\partial y^2}\right) A + Q |A|^2 A = 0, \quad (5.10)$$

with

$$P = -\frac{\omega\gamma_0}{1 + 4d_0 \sin^2 \frac{k}{2}}, \qquad Q = -\frac{\omega\left(3\beta + \frac{4\alpha^2 v_g^2}{\Omega_0^2 - v_g^2} + \frac{2\alpha^2\omega^2}{(\Omega_0^2 - d_0)\sin^2 k + \omega^2}\right)}{2\left(1 + 4d_0 \sin^2 \frac{k}{2}\right)}, \tag{5.11}$$

where v_g is given by Eq. (5.9) and $\omega^2 = \frac{4\Omega_0^2 \sin^2 \frac{k}{2}}{1+4d_0 \sin^2 \frac{k}{2}}$. The 2D NLS Eq. (5.10) possesses a y-independent (that is, plane or line) solitary wave solution. We are interested in the stability of this plane wave solution under transverse perturbations.

Case of odd $\ell' : \ell' = 2\ell + 1$

In the case of odd ℓ' ($\ell' = 2\ell + 1$), Eq. (5.8) leads to

$$v_g = \frac{(1 + 4\gamma_0)\,\Omega_0^2 \sin k}{\omega\left(1 + 4\gamma_0 + 4d_0 \sin^2 \frac{k}{2}\right)^2} = \frac{d\omega}{dk}, \tag{5.12}$$

meaning that v_g is the group velocity of the wave package in n direction. From the coefficient of ϵ^3, proportional to $\exp(i\theta)$, we obtain the following

$$i\frac{\partial A}{\partial \tau} + P_1 \frac{\partial^2 A}{\partial x^2} + P_2 \frac{\partial^2 A}{\partial y^2} + Q\,|A|^2 A = 0, \tag{5.13}$$

where

$$P_1 = -\frac{v_g^2\left(1 + 4\gamma_0 + 4d_0 \sin^2 \frac{k}{2}\right) + \left(d_0\omega^2 - \Omega_0^2\right)\cos k + 4d_0\omega v_g \sin k}{2\omega\left(1 + 4\gamma_0 + 4d_0 \sin^2 \frac{k}{2}\right)}, \tag{5.14a}$$

$$P_2 = \frac{\gamma_0\omega}{2\left(1 + 4\gamma_0 + 4d_0 \sin^2 \frac{k}{2}\right)}, \tag{5.14b}$$

$$Q = \frac{\omega\left(3\beta - 2\alpha^2\left[\frac{4v_g^2}{v_g^2 - \Omega_0^2} + \frac{\omega^2}{\omega^2 + (4d_0\omega^2 - \Omega_0^2)\sin^2 k}\right]\right)}{2\left(1 + 4\gamma_0 + 4d_0 \sin^2 \frac{k}{2}\right)}. \tag{5.14c}$$

Here, v_g is given by Eq. (5.12) and $\omega^2 = \frac{4\Omega_0^2 \sin^2 \frac{k}{2}}{1+4\gamma_0+4d_0 \sin^2 \frac{k}{2}}$. Equation (5.13) with coefficients (5.14a)–(5.14c) constitutes one set of two PDEs corresponding to the two different modes of propagation along the network. Although the two modes are independent of each other, we can simultaneously investigate the two modes of propagation. It is important to notice that $P_1 \neq P_2$, meaning that Eq. (5.13) is not a classical 2D NLS equation. Thus, when the wave vector $q = (2\ell + 1)\,\pi$, the dynamics of modulated waves propagating in the system shown in Fig. 5.1b is governed by a 2D PDE of Schrödinger type.

5.2.2 Transverse Stability of Plane Solitary Waves

As it is well-known, the Benjamin–Feir instabilities, exhibited by a dispersive nonlinear medium, constitute the proof of its capacity to support envelope solitons in certain propagation domains [1]. It follows from Eq. (5.10) that a continuous slowly modulated plane wave should be unstable if $PQ > 0$. This instability leads to the formation of envelope pulse solitons train, plane wave solution of Eq. (5.10)

$$A_0(x,\tau) = \frac{A_{\max}\exp\left[i\upsilon_e\left(2x - P\upsilon_p\tau\right)/4\right]}{\cosh\left[\left(2x - P\upsilon_e\tau\right)/2S_e\right]}, \tag{5.15}$$

where υ_e and υ_p are the envelope and phase velocities, respectively, $S_e = \sqrt{P/Q}/\varepsilon A_{\max}$ is the spatial soliton extension, and $A_{\max}$ is the nonlinear maximum amplitude of the soliton. In what follows, we restrict ourselves to the cases $\varepsilon^2 \in \{1,\ 3\}$.

Following Ody et al. [8], we impose on the line soliton (5.15) a sinusoidal perturbation in the y-direction and study the growth of this perturbation as the soliton propagates in the x-direction. Using the transformations

$$x' = \frac{x}{S_e} - \frac{P\upsilon_e}{2S_e}\tau,\quad \tau' = \tau,\quad y' = \frac{y}{S_e},\quad u = A\exp\left(-i\frac{S_e\upsilon_e}{2}x'\right), \tag{5.16}$$

and dropping primes, we find that Eq. (5.10) and the plane wave solution (5.15) become

$$i\frac{\partial u}{\partial \tau} + \frac{P}{2S_e^2}\Delta u + Q\left|u\right|^2 u + P\frac{\upsilon_e^2}{8}u = 0, \tag{5.17a}$$

$$u_0(x) = \frac{A_{\max}}{\cosh x}, \tag{5.17b}$$

respectively. To study the transverse stability of the soliton solution (5.17b), consider superimposing periodic modulations in y on it such that

$$u_p\left(x,y,\tau\right) = u_0(x) + \frac{1}{2}\delta\left[\Phi\left(x\right)\exp\left(iKy + \Omega\tau\right) + \Phi^*\left(x\right)\exp\left(-iKy + \Omega^*\tau\right)\right], \tag{5.18}$$

where δ is a small real parameter controlling the perturbation strength, $\Phi(x)$ is the perturbation complex function, K and Ω are the wavenumber of the perturbation and the complex frequency, respectively; "$*$" stands for the complex conjugate. Inserting Eq. (5.18) into Eq. (5.17a) and keeping only the linear terms in δ lead to the following equation for the complex function $\Phi\left(x\right)$

$$\Phi'' + \left(\frac{\upsilon_e^2 P + 8i\Omega}{4Q\varepsilon^2 A_{\max}^2} - K^2 + \frac{6}{\varepsilon^2}\mathrm{sech}^2 x\right)\Phi = 0. \tag{5.19}$$

Equation (5.19) in the limit $x \to \pm\infty$ becomes

$$\Phi'' + \left(\frac{v_e^2 P + 8i\Omega}{4Q\varepsilon^2 A_{\max}^2} - K^2 \right) \Phi = 0, \tag{5.20}$$

which is the homogeneous equation associated with Eq. (5.19). Seeking the solutions of Eq. (5.20) in the form $\Phi = \widetilde{\Phi} \exp(\lambda x)$ leads to satisfies the following quadratic equation for λ

$$\lambda^2 - \left(\frac{4Q\varepsilon^2 A_{\max}^2 K^2 - v_e^2 P}{4Q\varepsilon^2 A_{\max}^2} - i \frac{2\Omega}{Q\varepsilon^2 A_{\max}^2} \right) = 0. \tag{5.21}$$

Because the roots λ_j, $j = 1, 2$, of Eq. (5.21) are distinct, the corresponding solutions of Eq. (5.19) are given by

$$\Phi_j(x) = \exp\left(\lambda_j x\right) h_j(x),$$

where $h_j(x)$ is the solution of equation

$$\left(\frac{d^2}{dx^2} + 2\lambda_j \frac{d}{dx} + \frac{6}{\varepsilon^2} \operatorname{sech}^2 x \right) h_j = 0. \tag{5.22}$$

Case where $\varepsilon^2 = 3$

In the special case when $\varepsilon^2 = 3$, Eq. (5.22) admits special solutions $h_j(x) = \tanh x - \lambda_j$. Consequently,

$$\Phi_j(x) = \left(\tanh[x] - \lambda_j\right) \exp\left[\lambda_j x\right], \quad j = 1, 2. \tag{5.23}$$

Writing $\Omega = \Omega_r + i\Omega_i$, we find from Eq. (5.21)

$$\lambda_j = (-1)^{j-1} (a + ib), \tag{5.24}$$

where a and b are defined by

$$a^2 = \frac{\alpha_0 + \sqrt{\alpha_0^2 + \beta_0^2}}{2}, \quad b^2 = \frac{-\alpha_0 + \sqrt{\alpha_0^2 + \beta_0^2}}{2}, \tag{5.25}$$

with

$$\alpha_0 = \frac{4Q\varepsilon^2 A_{\max}^2 K^2 + 8\Omega_i - v_e^2 P}{4Q\varepsilon^2 A_{\max}^2}, \quad \beta_0 = -\frac{2\Omega_r}{Q\varepsilon^2 A_{\max}^2}. \tag{5.26}$$

It is evident that $\mathrm{Re}(\lambda_1)\,\mathrm{Re}(\lambda_2) \leq 0$. Without loss of generality, we consider that $\mathrm{Re}(\lambda_1) \geq 0$ and $\mathrm{Re}(\lambda_2) \leq 0$. In order that all the $\Phi_j(x)$, $j = 1, 2$, should be bounded, it is necessary and sufficient that $\mathrm{Re}\left(\lambda_j\right) = 0$, i.e., $a = 0$. Equation (5.25) then gives

$$\Omega_i \geq \frac{v_e^2 P}{8} - \frac{Q\varepsilon^2 A_{\max}^2 K^2}{2}, \quad \Omega_r = 0. \tag{5.27}$$

Conditions (5.27) are the conditions of the transverse stability.

If conditions (5.27) are violated, that is, either $\Omega_i < \frac{v_e^2 P}{8} - \frac{Q\varepsilon^2 A_{\max}^2 K^2}{2}$ or $\Omega_r \neq 0$, then $\mathrm{Re}(\lambda_1)$ will be positive and $\mathrm{Re}(\lambda_2)$ will be negative. If either $\Omega_r = 0$ or $\Omega_r < 0$, then the perturbed solution (5.18) that corresponds to the perturbation function $\Phi_1(x)$ will increase exponentially with $x > 0$. If $\Omega_r > 0$, then the perturbed solution (5.18) corresponding to the perturbation function $\Phi_1(x)$ will increase exponentially with both $x > 0$ and $\tau > 0$. With the perturbation function $\Phi_2(x)$, the perturbed solution (5.18) will remaind bounded in the both $x-$direction and $\tau-$direction for any $\Omega_r \leq 0$; this perturbed solution will grow exponentially in $\tau-$direction, if $\Omega_r > 0$.

Case where $\varepsilon^2 = 1$

In this special case, it is helpful to write Eq. (5.19) in the form

$$\left(-\frac{d^2}{dx^2} + 1 - \frac{6}{\cosh^2 x}\right)\Phi = \omega\Phi, \tag{5.28a}$$

$$\omega = 1 + \frac{v_e^2 P + 8i\Omega}{4Q\varepsilon^2 A_{\max}^2} - K^2. \tag{5.28b}$$

Seeking the solutions of Eq. (5.28a) in the form $\Phi(x) = h(x)\exp[qx]$, we find that either $h(x) = -\tanh^2 x - \frac{(\omega-3)}{6}$, $q = 0$ for $\omega \in \{-3, 1\}$ or $h(x) = \sinh x / \cosh^2 x$, $q = 0$, for $\omega = 0$. we thus obtain the following solutions of Eq. (5.19)

$$\Phi_3(x) = \frac{1}{\cosh^2 x}, \quad \Omega = i\frac{4Q\varepsilon^2 A_{\max}^2 (4 - K^2) + v_e^2 P}{8}, \tag{5.29a}$$

$$\Phi_4(x) = \frac{1}{\cosh^2 x} - \frac{2}{3}, \quad \Omega = i\frac{v_e^2 P - 4Q\varepsilon^2 A_{\max}^2 K^2}{8}, \tag{5.29b}$$

$$\Phi_5(x) = \frac{\sinh x}{\cosh^2 x}, \Omega = i\frac{4Q\varepsilon^2 A_{\max}^2 (1 - K^2) + v_e^2 P}{8} \tag{5.29c}$$

Thus, in the case where $\varepsilon^2 = 1$, the perturbation functions $\Phi_3(x)$, $\Phi_4(x)$, and $\Phi_5(x)$ are always bounded, and the complex frequency $\Omega \in i\mathbf{R}$. Therefore, the perturbed solutions (5.18) with the perturbation functions and complex frequencies given by Eqs. (5.29a), (5.29b), and (5.29c) are bounded in both the $x-$ and the $\tau-$direction. The soliton solution is then stable under the transverse perturbation in the y direction.

5.2.3 Propagation of Soliton-Like Signals Through the System of Fig. 5.1b

In this section, we present exact soliton-like solutions of the 2D PDE of Schrödinger type (5.13) with coefficients (5.14a)–(5.14c), which may allow one to investigate analytically the transmission of soliton-like signals through the coupled NLTL of Fig. 5.1b. In our investigations, we restrict ourselves to traveling-wave solutions of Eq. (5.13), that is, solutions of form

$$A(x, y, \tau) = \rho(x, y, \tau) \exp\left[i\left(k_p x + q_p y - \omega_p \tau\right)\right], \tag{5.30}$$

where k_p, q_p, and ω_p are three real parameter, and $\rho(x, y, \tau)$ is a real value function of its variables. Here, k_p and q_p are, respectively, the wave numbers in x-direction and y-direction, ω_p is the angular frequency, and $|\rho(x, y, \tau)|$ is the spatiotemporal wave amplitude. The substitution of ansatz (5.30) into Eq. (5.13) and the separation of the real and imaginary parts lead to

$$P_1 \frac{\partial^2 \rho}{\partial x^2} + P_2 \frac{\partial^2 \rho}{\partial y^2} + Q\rho^3 + \left[\omega_p - P_1 k_p^2 - P_2 q_p^2\right]\rho = 0, \tag{5.31a}$$

$$\frac{\partial \rho}{\partial \tau} + 2k_p P_1 \frac{\partial \rho}{\partial x} + 2q_p P_2 \frac{\partial \rho}{\partial y} = 0. \tag{5.31b}$$

Equation (5.31b) can be integrated to give

$$\rho(x, y, \tau) = F(X, Y, Z), \tag{5.32a}$$

$$\begin{cases} X = X(x, y, \tau) = 2q_p P_2^{l_0} x - 2k_p P_1^{l_0} y, \\ Y = Y(x, y, \tau) = 2k_p P_1^{l_0} \tau - x, \\ Z = Z(x, y, \tau) = 2q_p P_2^{l_0} \tau - y, \end{cases} \tag{5.32b}$$

where $F(X, Y, Z)$ is an arbitrary differentiable function of its variables. Because each of X, Y, and Z given by Eq. (5.32b) is a function of variables x, y, and τ, we conclude that $F(X, Y, Z)$ is a composite function of variables x, y, and τ. It is clearly seen from Eq. (5.32a) that the problem of finding $\rho(x, y, \tau)$ is reduced to the problem of finding $F(X, Y, Z)$. In order to find $F(X, Y, Z)$, we impose that $\rho(x, y, \tau)$ given by Eq. (5.32a) satisfies Eq. (5.31a).

Inserting Eq. (5.32a) into Eq. (5.31a) yields

$$P_0 \frac{\partial^2 F}{\partial X^2} + P_1 \frac{\partial^2 F}{\partial Y^2} + P_2 \frac{\partial^2 F}{\partial Z^2} + \Gamma_0 F + QF^3 = 0, \tag{5.33}$$

where

$$P_0 = 4\left[P_1\left(q_p P_2\right)^2 + P_2\left(k_p P_1\right)^2\right], \Gamma_0 = \omega_p - P_1 k_p^2 - P_2 q_p^2. \tag{5.34}$$

The problem of finding $\rho(x, y, \tau)$ is thus reduced to the problem of solving equation (5.33) under restriction (5.34). Henceforth, we focus our attention on the seeking of solutions of Eq. (5.33). In what follows, we use special and different mathematical techniques to find analytical solutions of Eq. (5.33).

Truncated Painlevé Expansion and Kink Soliton Solutions

To find the family of kink soliton solutions of Eq. (5.33), we apply the truncated Painlevé expansion

$$F(X, Y, Z) = \sum_{j=0}^{m_0} F_j(X, Y, Z)\Phi^{j-m_0}(X, Y, Z), \tag{5.35}$$

where $F_{m_0}(X, Y, Z)$ is an arbitrary solution of Eq. (5.33), and $F_0, F_1, \ldots, F_{m_0-1}$ are functions of the derivatives of $\Phi(X, Y, Z)$ with $F_0 \neq 0$. Substituting $F(X, Y, Z) \sim F_0\Phi^{-m_0}$ into Eq. (5.33) and balancing the items which diverge fastest yield $m_0 = 1$. Hence, the truncated Painlevé expansion reads

$$F(X, Y, Z) = F_1(X, Y, Z) + \frac{F_0(X, Y, Z)}{\Phi(X, Y, Z)}, \tag{5.36}$$

where F_1 is an arbitrary solution of Eq. (5.33), which is just a Bäcklund transformation of Eq. (5.33). For simplicity, we fix the seed solution F_1 as a constant. Substituting Eq. (5.36) into Eq. (5.33) yields

$$\begin{aligned}
0 = {} & F_1\left[\Gamma_0 + QF_1^2\right]\Phi^{-0} + \left[\Gamma_0 F_0 + 3QF_1^2F_0 + P_2\frac{\partial^2 F_0}{\partial Z^2} + P_1\frac{\partial^2 F_0}{\partial Y^2} + P_0\frac{\partial^2 F_0}{\partial X^2}\right]\Phi^{-1} \\
& + \left[3QF_1F_0^2 - 2P_2\frac{\partial F_0}{\partial Z}\frac{\partial \Phi}{\partial Z} - P_2F_0\frac{\partial^2\Phi}{\partial Z^2} - 2P_1\frac{\partial F_0}{\partial Y}\frac{\partial \Phi}{\partial Y} - 2P_0\frac{\partial F_0}{\partial X}\frac{\partial \Phi}{\partial X}\right. \\
& \left. - P_0F_0\frac{\partial^2\Phi}{\partial X^2} - P_1F_0\frac{\partial^2\Phi}{\partial Y^2}\right]\Phi^{-2} + F_0\left[QF_0^2 + 2P_0\left(\frac{\partial\Phi}{\partial X}\right)^2\right. \\
& \left. + 2P_1\left(\frac{\partial\Phi}{\partial Y}\right)^2 + 2P_2\left(\frac{\partial\Phi}{\partial Z}\right)^2\right]\Phi^{-3}.
\end{aligned} \tag{5.37}$$

Vanishing the coefficient of Φ^{-0} yield $F_1 = \varepsilon_1\sqrt{-\Gamma_0/Q}$, where $\varepsilon_1 = \pm 1$. Now, vanishing the coefficient of Φ^{-3} yields the meaningful solution of F_0 which is $F_0(X, Y, Z) = \varepsilon_2\sqrt{-\left[\frac{2P_0}{Q}\left(\frac{\partial\Phi}{\partial X}\right)^2 + \frac{2P_1}{Q}\left(\frac{\partial\Phi}{\partial Y}\right)^2 + \frac{2P_2}{Q}\left(\frac{\partial\Phi}{\partial Z}\right)^2\right]}$, $\varepsilon_2 = \pm 1$. To obtain an example of one-soliton solution of Eq. (5.33), we use the ansatz $\Phi(X, Y, Z) = 1 + \exp\left[k_{01}X + k_{02}Y - k_{03}Z\right]$, where k_{01}, k_{02}, k_{03} are arbitrary real parameters. Using this ansatz and imposing to the coefficient of Φ^{-1} in Eq. (5.37) to vanish lead to the following the dispersion relation between k_{01}, k_{02}, k_{03}

$$\omega_p = \frac{k_{01}^2 P_0 + \left(k_{02}^2 + 2k_p^2\right) P_1 + \left(k_{03}^2 + 2q_p^2\right) P_2}{2}. \tag{5.38}$$

Now, using the above ansatz and vanishing the coefficient leads to $1 + \varepsilon_1\varepsilon_2 = 0$. Therefore

$$\begin{array}{l} F_1 = \pm\sqrt{-\frac{\Gamma_0}{Q}},\ F_0 = \mp 2\sqrt{-\frac{\Gamma_0}{Q}} \exp\left[k_{01}X + k_{02}Y - k_{03}Z\right], \\ \Phi = 1 + \exp\left[k_{01}X + k_{02}Y - k_{03}Z\right]. \end{array} \tag{5.39}$$

Inserting Eq. (5.39) into ansatz (5.36) leads to the following one-soliton solution of Eq. (5.33)

$$F(X,Y,Z) = \sqrt{-\frac{k_{01}^2 P_0 + k_{02}^2 P_1 + k_{03}^2 P_2}{2Q^{l_0}}} \frac{\pm 1 \mp \exp\left[k_{01}X + k_{02}Y - k_{03}Z\right]}{1 + \exp\left[k_{01}X + k_{02}Y - k_{03}Z\right]}, \tag{5.40}$$

where k_{01}, k_{02}, and k_{03} are three arbitrary real parameters to be taken from condition

$$\left(k_{01}^2 P_0 + k_{02}^2 P_1 + k_{03}^2 P_2\right) Q = \left(k_{02}^2 P_1 + k_{03}^2 P_2 + 4k_{01}^2 P_1 P_2 \left[k_p^2 P_1 + q_p^2 P_2\right]\right) Q < 0. \tag{5.41}$$

Inserting Eq. (5.40) into Eq. (5.32a), we obtain $\rho(x, y, \tau)$ which together with Eq. (5.30) lead to the following soliton wave solutions of Eq. (5.13)

$$\begin{array}{l} A(x,y,\tau) = \sqrt{-\frac{k_{02}^2 P_1 + k_{03}^2 P_2 + 4k_{01}^2 P_1 P_2 \left(k_p^2 P_1 + q_p^2 P_2\right)}{2Q}} \times \exp\left[i\left(k_p x + q_p y - \omega_p \tau\right)\right] \\ \times \frac{\pm 1 \mp \exp\left[\left(2k_{01}q_p P_2 - k_{02}\right)x + \left(k_{03} - 2k_{01}k_p P_1\right)y + 2\left(k_{02}k_p P_1 - k_{03}q_p P_2\right)\tau\right]}{1 + \exp\left[\left(2k_{01}q_p P_2 - k_{02}\right)x + \left(k_{03} - 2k_{01}k_p P_1\right)y + 2\left(k_{02}k_p P_1 - k_{03}q_p P_2\right)\tau\right]}, \end{array} \tag{5.42}$$

where ω_p is given by Eq. (5.38) and k_{01}, k_{02}, k_{03}, k_p and q_p are five real parameters satisfying condition (5.41).

Method of Elliptic Ordinary Differential Equations (EODEs) for Kink and Dark Soliton Solutions

Here, we use the method of EODEs to find the classes of dark and kink soliton solutions of Eq. (5.33). For these two families, we employ the ansatz

$$F(X,Y,Z) = r(\xi = c_1 X + c_2 Y - c_3 Z) \tag{5.43}$$

which reduces Eq. (5.33) into the nonlinear ordinary differential equation (NODE)

$$\begin{cases} \frac{d^2 r}{d\xi^2} + \alpha_0 r + \beta_0 r^3 = 0, \\ \alpha_0 = \frac{\Gamma_0}{c_1^2 P_0 + c_2^2 P_1 + c_3^2 P_2},\ \beta_0 = \frac{Q^{l_0}}{c_1^2 P_0 + c_2^2 P_1 + c_3^2 P_2}. \end{cases} \tag{5.44}$$

Integrating the NODE (5.44) once leads to the following EODE of the first kind

$$\left(\frac{dr}{d\xi}\right)^2 = \widetilde{\alpha}r^4 + 6\widetilde{\gamma}r^2 + \varepsilon \stackrel{\text{def}}{=} R(r) \tag{5.45a}$$

$$\widetilde{\alpha} = -\frac{Q}{2\left(c_1^2 P_0 + c_2^2 P_1 + c_3^2 P_2\right)}, \widetilde{\gamma} = -\frac{\Gamma_0}{6\left(c_1^2 P_0 + c_2^2 P_1 + c_3^2 P_2\right)}, \tag{5.45b}$$

where ε is an arbitrary real parameter (constant of integration). Because $r(\xi)$ is a real function, polynomial $R(r)$ must be nonnegative.

Second family of exact kink soliton solutions By taking $\widetilde{\alpha} > 0$ and $\widetilde{\gamma} < 0$, we can obtain the arbitrary parameter ε by solving equation $9\widetilde{\gamma}^2 - \widetilde{\alpha}\varepsilon = 0$. In this case, Eq. (5.45a) can be rewritten as

$$\frac{dr}{\partial\xi} = \varepsilon_1\sqrt{\widetilde{\alpha}}\left(r^2 + \frac{3\widetilde{\gamma}}{\widetilde{\alpha}}\right),$$

with $\varepsilon_1 = \pm 1$. Solving this equation leads to the exact kink soliton solution

$$r(\xi) = \varepsilon_2\sqrt{-\frac{3\widetilde{\gamma}}{\widetilde{\alpha}}\frac{-1 + C_{00}\exp\left[-2\varepsilon_1\varepsilon_2\sqrt{-3\widetilde{\gamma}}\xi\right]}{1 + C_{00}\exp\left[-2\varepsilon_1\varepsilon_2\sqrt{-3\widetilde{\gamma}}\xi\right]}},$$

where $\varepsilon_2 = \pm 1$ and C_{00} is an arbitrary positive parameter. Going back to ansatz (5.43) yields the kink soliton solution of Eq. (5.33)

$$F(X, Y, Z) = \varepsilon_2\sqrt{-\frac{\Gamma_0}{Q}\frac{C_{00} - \exp\left[2\varepsilon_1\varepsilon_2\sqrt{-3\widetilde{\gamma}}(c_1 X + c_2 Y - c_3 Z)\right]}{C_{00} + \exp\left[2\varepsilon_1\varepsilon_2\sqrt{-3\widetilde{\gamma}}(c_1 X + c_2 Y - c_3 Z)\right]}}. \tag{5.46}$$

Class of exact dark soliton solutions In order to use (5.45a)–(5.45b) for seeking exact dark soliton solutions of Eq. (5.33), we consider only those $\varepsilon \neq 9\widetilde{\gamma}^2/\widetilde{\alpha}$. For the soliton solutions of Eq. (5.45a), one must have $g_2 = \widetilde{\alpha}\varepsilon + 3\widetilde{\gamma}^2 \geq 0$, $g_3 = \widetilde{\alpha}\widetilde{\gamma}\varepsilon - \widetilde{\gamma}^3 \leq 0$, $\Delta(\varepsilon) = g_2^3 - 27g_3^2 = \varepsilon\widetilde{\alpha}\left[\widetilde{\alpha}\varepsilon - 9\widetilde{\gamma}^2\right]^2 = 0$. Because $\varepsilon \neq 9\widetilde{\gamma}^2/\widetilde{\alpha}$ and $\widetilde{\alpha} \neq 0$, condition $\Delta(\varepsilon) = 0$ means that $\varepsilon = 0$. By taking c_1, c_2, and c_3 and other parameters appearing in the definition of Γ_0 and P_0 (see Eq. (5.34)) from conditions $\widetilde{\gamma} > 0$ and $\widetilde{\alpha} < 0$, a soliton solution of Eq. (5.45a) can be written as

$$r(\xi) = \varepsilon_1\sqrt{-\frac{6\widetilde{\gamma}}{\widetilde{\alpha}}\frac{\cosh^2\left[\sqrt{\frac{3}{2}\widetilde{\gamma}}\xi\right]}{2 + \cosh^2\left[\sqrt{\frac{3}{2}\widetilde{\gamma}}\xi\right]}}, \tag{5.47}$$

where $\varepsilon_1 = \pm 1$. To the soliton solution (5.47) corresponds the following dark soliton solution of Eq. (5.33)

$$F(X,Y,Z)=\varepsilon_1\sqrt{-\frac{2\Gamma_0}{Q}}\frac{\cosh^2\left[\frac{1}{2}\sqrt{-\frac{\Gamma_0}{c_1^2P_0+c_2^2P_1+c_3^2P_2}}(c_1X+c_2Y-c_3Z)\right]}{2+\cosh^2\left[\frac{1}{2}\sqrt{-\frac{\Gamma_0}{c_1^2P_0+c_2^2P_1+c_3^2P_2}}(c_1X+c_2Y-c_3Z)\right]}. \tag{5.48}$$

Riccati Equation Mapping Method for Kink Soliton Solutions

Now, we aim to apply the Riccati equation mapping method to find exact kink soliton solution of Eq. (5.33). Here, the Riccati equation mapping method is obtained as a combination of the truncated Painlevé expansion with the technique of EODEs. This is performed by applying the truncated Painlevé expansion for seeking exact solutions of the NODE (5.44). The idea is to reduce the seeking of exact solutions of Eq. (5.44) to that of a Riccati equation.

Let $r_1(\xi)$ be a particular solution of Eq. (5.44) and let us seek an exact solution of (5.44) in the form $r(\xi)=r_1(\xi)+r_0(\xi)/\Phi(\xi)$, where $r_0(\xi)\neq 0$ and $\Phi(\xi)\neq$ constant are two real value functions to be determined. Inserting this expression for $r(\xi)$ into Eq. (5.44) and equating to zero the coefficients of different powers of $1/\Phi$ lead to the differential system

$$\frac{d^2r_1}{d\xi^2}+\alpha_0r_1+\beta_0r_1^3=0, \tag{5.49a}$$

$$\frac{d^2r_0}{d\xi^2}+\left(\alpha_0+3\beta_0r_1^2\right)r_0=0, \tag{5.49b}$$

$$3r_1\beta_0r_0^2-2\frac{dr_0}{d\xi}\frac{d\Phi}{d\xi}-r_0\frac{d^2\Phi}{d\xi^2}=0, \tag{5.49c}$$

$$\beta_0r_0^2+2\left(\frac{d\Phi}{d\xi}\right)^2=0. \tag{5.49d}$$

It follows from Eq. (5.49d) that β_0 must be negative. Solving Eq. (5.49d) in r_0 yields $r_0(\xi)=\varepsilon_1\sqrt{-\frac{2}{\beta_0}\frac{d\Phi}{d\xi}}$, $\varepsilon_1=\pm1$. Inserting $r_0(\xi)$ into Eq. (5.49c) yields $2r_1\frac{d\Phi}{d\xi}+\varepsilon_1\sqrt{-\frac{2}{\beta_0}\frac{d^2\Phi}{d\xi^2}}=0$. Integrating this last equation once yields $\frac{d\Phi}{d\xi}=C_{01}\exp\left[-\varepsilon_1\sqrt{-2\beta_0}\int r_1(\xi)d\xi\right]$, where $C_{01}\neq 0$ is a constant of integration. Inserting $r_0(\xi)$ and its derivatives in Eq. (5.49b) and using the expression for $d\Phi/d\xi$ lead to the Riccati equation

$$\frac{dr_1}{d\xi}=\frac{\varepsilon_1\alpha_0}{\sqrt{-2\beta_0}}+\frac{\varepsilon_1\beta_0}{\sqrt{-2\beta_0}}r_1^2. \tag{5.50}$$

If $r_1(\xi)$ is a solution of Eq. (5.50), we will have $\frac{d^2r_1}{d\xi^2}=-\alpha_0r_1-\beta_0r_1^3$, which means that $r_1(\xi)$ satisfies Eq. (5.49a), and this confirms that $r_1(\xi)$ is a particular solution of Eq. (5.44). For soliton-like solution of the Riccati equation (5.50), we must have $\alpha_0/\beta_0=\Gamma_0/Q<0$. For $\Gamma_0/Q<0$, the general solution of Eq. (5.50) reads

$$r_1(\xi) = -\varepsilon_2 \sqrt{-\frac{\Gamma_0}{Q} \frac{1 - C_{02} \exp\left[\varepsilon_1 \varepsilon_2 \sqrt{2\alpha_0} \xi\right]}{1 + C_{02} \exp\left[\varepsilon_1 \varepsilon_2 \sqrt{2\alpha_0} \xi\right]}},$$

where C_{02} is the constant of integration. Inserting $r_1(\xi)$ into $d\Phi/d\xi$ and integrating the result yields

$$\Phi(\xi) = -\frac{\varepsilon_1 \varepsilon_2 C_{01} C_{02} \left(1 - C_{03} \left(1 + C_{02} \exp\left[\varepsilon_1 \varepsilon_2 \sqrt{2\alpha_0} \xi\right]\right)\right)}{\sqrt{2\alpha_0} \left(1 + C_{02} \exp\left[\varepsilon_1 \varepsilon_2 \sqrt{2\alpha_0} \xi\right]\right)}, \text{if} C_{02} \neq 0,$$

$$\Phi(\xi) = \frac{\varepsilon_1 \varepsilon_2 C_{01} \left(C_{03} + \exp\left[\varepsilon_1 \varepsilon_2 \sqrt{2\alpha_0} \xi\right]\right)}{\sqrt{2\alpha_0}}, \text{ if } C_{02} = 0,$$

C_{03} being a constant of integration. Inserting the expressions for $r_1(\xi)$, $r_0(\xi)$, and $\Phi(\xi)$ into the expression $r(\xi) = r_1(\xi) + r_0(\xi)/\Phi(\xi)$ leads the following kink soliton-like solution of Eq. (5.50)

$$r(\xi) = -\frac{\varepsilon_2 \sqrt{-\Gamma_0/Q}}{1 + C_{02} \exp\left[\varepsilon_1 \varepsilon_2 \sqrt{2\alpha_0} \xi\right]} \left[1 - C_{02} \exp\left[\frac{\sqrt{2\alpha_0}}{\varepsilon_1 \varepsilon_2} \xi\right] + \frac{2 C_{02} \exp\left[\varepsilon_1 \varepsilon_2 \sqrt{2\alpha_0} \xi\right]}{1 + C_{03} \left(1 + C_{02} \exp\left[\frac{\sqrt{2\alpha_0}}{\varepsilon_1 \varepsilon_2} \xi\right]\right)}\right], \text{ if } C_{20} \neq 0,$$

$$r(\xi) = -\varepsilon_2 \sqrt{-\frac{\Gamma_0}{Q}} \left(\frac{C_{03} - \exp\left[\varepsilon_1 \varepsilon_2 \sqrt{2\alpha_0} \xi\right]}{C_{03} + \exp\left[\varepsilon_1 \varepsilon_2 \sqrt{2\alpha_0} \xi\right]}\right), \text{ if } C_{02} = 0. \tag{5.51}$$

For solution $r(\xi)$ to be bounded, it is necessary that $C_{02} > 0$ and $C_{03} \geq 0$. Replacing ξ by $c_1 X + c_2 Y - c_3 Z$ in $r(\xi)$, we obtain $r(\xi) = r(c_1 X + c_2 Y - c_3 Z) = F(X, Y, Z)$, which is a traveling wave-like solution of Eq. (5.33). For $C_{02} \neq 0$ and $C_{02} = 0$, we obtain from Eq. (5.51)

$$F(X, Y, Z) = -\frac{\varepsilon_2 \sqrt{-\Gamma_0/Q}}{1 + C_{02} \exp\left[\varepsilon_1 \varepsilon_2 \sqrt{2\alpha_0} (c_1 X + c_2 Y - c_3 Z)\right]} \times \left[1 - C_{02} \exp\left[\varepsilon_1 \varepsilon_2 \sqrt{2\alpha_0} (c_1 X + c_2 Y - c_3 Z)\right] + \frac{2 C_{02} \exp\left[\varepsilon_1 \varepsilon_2 \sqrt{2\alpha_0} (c_1 X + c_2 Y - c_3 Z)\right]}{1 + C_{03} \left(1 + C_{02} \exp\left[\varepsilon_1 \varepsilon_2 \sqrt{2\alpha_0} (c_1 X + c_2 Y - c_3 Z)\right]\right)}\right], \tag{5.52}$$

$$F(X, Y, Z) = -\varepsilon_2 \sqrt{-\frac{\Gamma_0}{Q}} \left(\frac{C_{03} - \exp\left[\varepsilon_1 \varepsilon_2 \sqrt{2\alpha_0} (c_1 X + c_2 Y - c_3 Z)\right]}{C_{03} + \exp\left[\varepsilon_1 \varepsilon_2 \sqrt{2\alpha_0} (c_1 X + c_2 Y - c_3 Z)\right]}\right), \tag{5.53}$$

respectively. Here, $\varepsilon_1 = \pm 1$, $\varepsilon_2 = \pm 1$, c_1, c_2, and c_3 are three arbitrary parameters. For the validity of solutions (5.52) and (5.53), the solution parameters k_p, q_p, ω_p appearing in transformation (5.30), and the wave vector k appearing in the definition of Q (see Eq. (5.14c)) must be taken from condition $\Gamma_0 Q < 0$.

5.3 Modulational Instability for the Coupled NLTL of Fig. 5.1b

In this section, we start with the derivation of a 2D system of NLS equations that accurately model the effects of nonlinearity of the network of Fig. 5.1b. Through the traveling wave solution of the derived 2D system of NLS equations, we investigate analytically the modulational instability of the coupled nonlinear transmission lines of Fig. 5.1b.

5.3.1 2D Coupled NLS Equation for the System of Fig. 5.1b

We start with our analysis with the system of equations (5.3), obtained after applying the Kirchhoff's laws in the orthogonal loops. In our analysis, we take the nonlinear capacitance $C(V)$ to be of the form $C(V) = \frac{C_0}{1+\frac{V}{V_0}}$, where C_0 and V_0 are arbitrary capacitance and voltage scales, respectively. We focus ourselves to case when the perturbation voltage V is small enough compared to the equilibrium voltage. Then, we use the continuous limit assuming that the wave number is much greater than the lattice spacing, and that the perturbation voltage V is also small enough compared with equilibrium value. The continuum approximation has the merit that it is easier to treat analytically as well as numerically than the discrete lattice, and the results obtained can be related to those for the discrete version in most cases. To obtain the continuum approximation, we assume the validity of the Taylor expansion to rewrite the equation for nonlinear lattices. Moreover, we assume that the waves are smooth enough n and m are two continuous variables, the perturbation voltage $V_{n,m}$ is a function of the two continuous variables n and m. Based on these reasonable assumptions, we can expand $V_{n\pm1,m}$ and $V_{n,m\pm1}$ as follows

$$\begin{aligned} V_{n\pm1,m} &= V_{n,m} \pm \frac{\partial V_{n,m}}{\partial n} + \frac{1}{2}\frac{\partial^2 V_{n,m}}{\partial n^2} \pm \frac{1}{3!}\frac{\partial^3 V_{n,m}}{\partial n^3} + \frac{1}{4!}\frac{\partial^4 V_{n,m}}{\partial n^4} + \cdots, \\ V_{n,m\pm1} &= V_{n,m} \pm \frac{\partial V_{n,m}}{\partial m} + \frac{1}{2}\frac{\partial^2 V_{n,m}}{\partial m^2} \pm \frac{1}{3!}\frac{\partial^3 V_{n,m}}{\partial m^3} + \frac{1}{4!}\frac{\partial^4 V_{n,m}}{\partial m^4} + \cdots, \end{aligned}$$

Letting $V_{n,m}(t) = V(n, m, t)$ reduces Eq. (5.3) to

$$\begin{aligned} LC_0\frac{\partial^2}{\partial t^2}\left(V - bV^2\right) &= \frac{\partial^2}{\partial n^2}\left(V + \frac{1}{12}\frac{\partial^2 V}{\partial n^2}\right) + LC_2\frac{\partial^4}{\partial t^2\partial m^2}\left(V + \frac{1}{12}\frac{\partial^2 V}{\partial m^2}\right) \\ &\quad + LC_S\frac{\partial^4}{\partial t^2\partial n^2}\left(V + \frac{1}{12}\frac{\partial^2 V}{\partial n^2}\right), \end{aligned} \tag{5.54}$$

where $b = \frac{1}{2}V_0$. Then, we use the traditional perturbation technique which means that the dominant motion is in the $n-$ direction, and define four stretched variables in the wave frame as $\xi = \epsilon^{\frac{1}{2}}(n - vt)$, $\eta = \epsilon^{\frac{1}{2}}m$, $\tau = \epsilon^{\frac{3}{2}}t$, $V = \epsilon u(\xi, \eta, \tau)$, where $\epsilon << 1$ is a small parameter, and $v^2 = \frac{1}{LC_0}$ is the linear wave speed. In terms of these stretched variables, Eq. (5.54) becomes, to lowest order in ϵ,

$$12C_0 b \frac{\partial u^2}{\partial \xi} + \frac{24C_0}{v} \frac{\partial u}{\partial \tau} + (C_0 + 12C_S) \frac{\partial^3 u}{\partial \xi^3} + 12C_2 \frac{\partial^3 u}{\partial \xi \partial \eta^2} = 0. \quad (5.55)$$

Let us consider the interaction of any two wave-packets centered at $(\mathbf{k}_A, \omega_A)$ and $(\mathbf{k}_B, \omega_B)$; here, $\mathbf{k}_j$ and ω_j are the wave-vectors and the frequencies, respectively. In order to resolve the weakly nonlinear equation, two different low time scales $T_1 = \epsilon\tau$ and $T_2 = \epsilon^2\tau$ must be introduced in addition to the original time scale $T_0 = \tau$. Moreover we introduce the large spatial scale $X_1 = \epsilon\xi$ and $Y_1 = \epsilon\eta$ in addition to the original $X_0 = \xi$ and $Y_0 = \eta$. Then the voltage in the transmission line is expanded through function $u(\xi, \eta, \tau)$ as

$$u = \epsilon u_1 + \epsilon^2 u_2 + \epsilon^3 u_3 + O(\epsilon^4), \quad (5.56)$$

where $u_j = u_j(X_0, X_1, Y_0, Y_1, T_0, T_1, T_2)$; here, the small parameter ϵ measures the size of the amplitudes of the perturbation. To analyze the propagation of any two wave-packets centered around $(\mathbf{k}_A, \omega_A)$ and $(\mathbf{k}_B, \omega_B)$, we take the lowest order term u_1 in the form

$$u_1 = A \exp\left(i\left(\mathbf{k}_A.\mathbf{r} - \omega_A T_0\right)\right) + B \exp\left(i\left(\mathbf{k}_B.\mathbf{r} - \omega_B T_0\right)\right) + c.c., \quad (5.57)$$

where $A = A(X_1, Y_1, T_1, T_2)$, $B = B(X_1, Y_1, T_1, T_2)$ are two complex amplitudes, $\mathbf{r} = (X_0, Y_0)$, and $\mathbf{k}_j = \left(k_j^{X_0}, k_j^{Y_0}\right)$ so that

$$(C_0 + 12C_S)\left(k_B^{X_0}\right)^2 - 4C_2\left(k_B^{Y_0}\right)^2 > 0. \quad (5.58)$$

Equation (5.58) imposes some restriction on $\mathbf{k}_B = \left(k_B^{X_0}, k_B^{Y_0}\right)$. The carrier wave number $\mathbf{k}_j$ and the frequency ω_j of each wave are related by a dispersion relation function $\omega_j = \omega_j(\mathbf{k}_j)$. As we will show below, nonlinearity is manifested via a slow modulation of the wave amplitudes A and B, in time and space, say, along the $x-$axis a the $y-$ axis. Inserting the perturbation expansions (5.56) into the nonlinear equation (5.55) and equating coefficients of like powers of ϵ leads to a system of three PDEs in u_1, u_2, and u_3. The first of these equations contains only u_1, the second contains u_1 and u_2, and the third one contains u_1, u_2, and u_3 :

$$D_0 u_1 = \left(\frac{24C_0}{v}\frac{\partial}{\partial T_0} + (C_0 + 12C_S)\frac{\partial^3}{\partial X_0^3} + 12C_2\frac{\partial^3}{\partial X_0 \partial Y_0^2}\right) u_1 = 0, \quad (5.59a)$$

$$D_0 u_2 = -\frac{24C_0}{v}\frac{\partial u_1}{\partial T_1} - 3\left(C_0 + 12C_S\right)\frac{\partial^3 u_1}{\partial X_0^2 \partial X_1}$$
$$-12C_2\left(\frac{\partial^3 u_1}{\partial X_1 \partial Y_0^2} + 2\frac{\partial^3 u_1}{\partial X_0 \partial Y_0 \partial Y_1}\right) - 12C_0 b\frac{\partial u_1^2}{\partial X_0}, \quad (5.59b)$$

and

$$D_0 u_3 = -12C_0 b\left(2\frac{\partial u_1 u_2}{\partial X_0} + \frac{\partial u_1^2}{\partial X_1}\right) - \frac{24C_0}{v}\left(\frac{\partial u_2}{\partial T_1} + \frac{\partial u_1}{\partial T_2}\right)$$
$$-3\left(C_0 + 12C_S\right)\left(\frac{\partial^3 u_2}{\partial X_0^2 \partial X_1} + \frac{\partial^3 u_1}{\partial X_0 \partial X_1^2}\right) - 12C_2\left(\frac{\partial^3 u_1}{\partial X_0 \partial Y_1^2}\right. \quad (5.59c)$$
$$\left. +2\frac{\partial^3 u_1}{\partial Y_0 \partial X_1 \partial Y_1} + 2\frac{\partial^3 u_2}{\partial X_0 \partial Y_0 \partial Y_1} + \frac{\partial^3 u_2}{\partial Y_0^2 \partial X_1}\right).$$

Substituting Eq. (5.57) into Eq. (5.59a) leads to the dispersion relations

$$\frac{24C_0}{v}\omega_j + \left(C_0 + 12C_S\right)\left(k_j^{X_0}\right)^3 + 12C_2 k_j^{X_0}\left(k_j^{Y_0}\right)^2 = 0, \quad j = A,\ B. \quad (5.59d)$$

Because ω_j and $\mathbf{k}_j$ are related by the dispersion relation (5.59d), a special solution of Eq. (5.59b) contains secular terms in either X_0 or T_0 or both, leading to a nonuniform expansion for either long times or large distances or both. For a uniform expansion we require the vanishing of terms that produce secular terms. This solvability condition leads to the following equations for A and B

$$\begin{aligned} \frac{\partial A}{\partial T_1} &= -\frac{\partial \omega_A}{\partial k_A^{X_0}}\frac{\partial A}{\partial X_1} - \frac{\partial \omega_A}{\partial k_A^{Y_0}}\frac{\partial A}{\partial Y_1}, \\ \frac{\partial B}{\partial T_1} &= -\frac{\partial \omega_B}{\partial k_B^{X_0}}\frac{\partial B}{\partial X_1} - \frac{\partial \omega_B}{\partial k_B^{Y_0}}\frac{\partial B}{\partial Y_1}. \end{aligned} \quad (5.60)$$

Eliminating from Eq. (5.59c) the terms that produce secular terms and using Eqs. (5.58) and (5.60) lead to the following 2D coupled NLS equations the system of Fig. 5.1b

$$i\frac{\partial \psi_j}{\partial \tau} = M_j \nabla^2 \psi_j + \left(V_{jj}\left|\psi_j\right|^2 + V_{j3-j}\left|\psi_{3-j}\right|^2 + \mu_j\right)\psi_j, \quad j = 1, 2, \quad (5.61)$$

where

$$\psi_1(x, y, \tau) = \epsilon A(x, y, \tau)\exp\left[-i\left(\alpha_1 x + \beta_1 y\right)\right],$$
$$\psi_2(x, y, \tau) = \epsilon B(x, y, \tau)\exp\left[-i\left(\gamma_1 x + \delta_1 y\right)\right],$$

$$\alpha_1 = -\frac{1}{8}\frac{(C_0+12C_S)\,k_A^{X_0}\frac{\partial\omega_A}{\partial k_A^{Y_0}} - 4C_2 k_{Y_0}\frac{\partial\omega_A}{\partial k_A^{X_0}}}{C_2(C_0+12C_S)\left(k_A^{X_0}\right)^2 - 4C_2^2\left(k_A^{Y_0}\right)^2},$$

$$\beta_1 = \frac{1}{2\sqrt{(C_0+12C_S)\,C_2\left(k_B^{X_0}\right)^2 - 4C_2^2\left(k_B^{Y_0}\right)^2}}\frac{\partial\omega_A}{\partial k_A^{X_0}},$$

$$\gamma_1 = -\frac{1}{8}\frac{(C_0+12C_S)\,k_B^{X_0}\frac{\partial\omega_B}{\partial k_B^{Y_0}} - 4C_2 k_{Y_0}\frac{\partial\omega_B}{\partial k_B^{X_0}}}{C_2(C_0+12C_S)\left(k_B^{X_0}\right)^2 - 4C_2^2\left(k_B^{Y_0}\right)^2},$$

$$\delta_1 = \frac{1}{2\sqrt{(C_0+12C_S)\,C_2\left(k_B^{X_0}\right)^2 - 4C_2^2\left(k_B^{Y_0}\right)^2}}\frac{\partial\omega_B}{\partial k_B^{X_0}},$$

$$M_j = \frac{2\,(C_0+12C_S)\,C_2\left(k_j^{X_0}\right)^2 - 8C_2^2\left(k_j^{Y_0}\right)^2}{(C_0+12C_S)\,k_j^{X_0}},$$

$$V_{jj} = \frac{3C_0 \upsilon^2 b^2\left(k_j^{X_0}\right)^2}{12C_0\omega_j + 2\upsilon\left(k_j^{X_0}\right)^2(C_0+12C_S) + 24\upsilon C_2 k_j^{X_0}\left(k_j^{Y_0}\right)^2},$$

$$V_{j3-j} = \frac{\upsilon\gamma b k_j^{X_0}}{2},\ \text{with}\ k_1^{X_0} = k_A^{X_0},\ k_2^{X_0} = k_B^{X_0},$$

$$\begin{aligned}\gamma = {} & 24C_0 b\left(k_A^{X_0}+k_B^{X_0}\right)\Bigg(12\left(k_A^{X_0}+k_B^{X_0}\right)\left(k_A^{Y_0}+k_B^{Y_0}\right)^2 C_2 \\ & + (\omega_A+\omega_B)\frac{24C_0}{\upsilon} + \left(k_A^{X_0}+k_B^{X_0}\right)^3(C_0+12C_S)\Bigg)^{-1} \\ & + 24C_0 b\upsilon\left(k_A^{X_0}-k_B^{X_0}\right)\Bigg(\upsilon\,(C_0+12C_S)\left(k_A^{X_0}-k_B^{X_0}\right)^3 \\ & + 24\,(\omega_A-\omega_B)\,C_0 + 12C_2\upsilon\left(k_A^{X_0}-k_B^{X_0}\right)\left(k_A^{Y_0}-k_B^{Y_0}\right)^2\Bigg)^{-1},\end{aligned}$$

$$\begin{aligned}\mu_1 = {} & 2\frac{\sqrt{(C_0+12C_S)\,C_2\left(k_A^{X_0}\right)^2 - 4C_2^2\left(k_A^{Y_0}\right)^2}}{(C_0+12C_S)\,k_A^{X_0}}\frac{\partial\omega_A}{\partial k_A^{X_0}}\beta_1 \\ & - \frac{1}{2}\left(\frac{\partial\omega_A}{\partial k_A^{Y_0}} - \frac{\partial\omega_A}{\partial k_A^{X_0}}\frac{4C_2 k_A^{Y_0}}{(C_0+12C_S)\,k_A^{X_0}}\right)\alpha_1 - 2\frac{\left((C_0+12C_S)\,C_2\left(k_A^{X_0}\right)^2 - 4C_2^2\left(k_A^{Y_0}\right)^2\right)}{\left(\alpha_1^2+\beta_1^2\right)^{-1}(C_0+12C_S)\,k_A^{X_0}},\end{aligned}$$

$$\mu_2 = 2\frac{\sqrt{(C_0+12C_S)\,C_2\left(k_B^{X_0}\right)^2 - 4C_2^2\left(k_B^{Y_0}\right)^2}}{(C_0+12C_S)\,k_B^{X_0}}\frac{\partial\omega_B}{\partial k_B^{X_0}}\delta_1$$

$$-\frac{1}{2}\left(\frac{\partial\omega_B}{\partial k_B^{Y_0}} - \frac{\partial\omega_B}{\partial k_B^{X_0}}\frac{4C_2k_B^{Y_0}}{(C_0+12C_S)\,k_B^{X_0}}\right)\gamma_1 - 2\frac{\left((C_0+12C_S)\,C_2\left(k_B^{X_0}\right)^2 - 4C_2^2\left(k_B^{Y_0}\right)^2\right)}{\left(\gamma_1^2+\delta_1^2\right)^{-1}(C_0+12C_S)\,k_B^{X_0}},$$

$$x = \frac{4C_2k_B^{Y_0}}{(C_0+12C_S)\,k_B^{X_0}}\xi - \eta,\quad y = \frac{2\sqrt{(C_0+12C_S)\,C_2\left(k_B^{X_0}\right)^2 - 4C_2^2\left(k_B^{Y_0}\right)^2}}{(C_0+12C_S)\,k_B^{X_0}}\xi,$$

and $\nabla^2 = \frac{\partial^2}{\partial x^2} + \frac{\partial^2}{\partial y^2}$ is the Laplace operator (a 2D Cartesian geometry is considered for clarity).

We have thus established that the dynamics of any two wave-packets in the coupled discrete nonlinear transmission line shown in Fig. 5.1b is described by the coupled nonlinear Schrödinger equations (5.61). In this system, $\psi_1(x, y, \tau)$ and $\psi_2(x, y, \tau)$ are the complex amplitudes of the signal in the coupled nonlinear transmission lines of Fig. 5.1b. In system (5.61), the quantity M_j with sign "−", that is, $-M_j$, is the group-velocity-dispersion, V_{jj} and V_{j3-j} model carrier self-modulation and wave coupling, respectively, $\mu_j = \mu_j(\omega_j, \mathbf{k}_j)$ is the growing complex eigenvalue. In the above set of coefficients, the quantities $\partial\omega_j/\partial k_j^{X_0}$ and $\partial\omega_j/\partial k_j^{Y_0}$ are known as group velocities. If $\mu_1 = \mu_2 = 0$, then under the condition of equality of the nonlinear coefficients, Eq. (5.61) become the celebrated integrable Manakov model [7]. The (linear) last terms (terms with coefficients μ_j) in each equation of system (5.61) may readily be eliminated via a simple phase-shift transformation; this is however deliberately not done at this stage, for generality.

5.3.2 *Modulational Instability of the Coupled NLS Equations*

Here, we find the conditions under which a uniform wave-train moving along the coupled discrete nonlinear electrical transmission line of Fig. 5.1b will become stable/unstable to a small perturbation. We start with linear stability analysis.

Linear Stability Analysis

First of all, we note that system (5.61) possesses the Stokes wave solutions

$$\psi_j(x, y, \tau) = \psi_{j0}\exp\left[i\left(k_{0j}x + q_{0j}y + \Omega_{j0}\tau\right)\right],\quad j = 1,\ 2, \tag{5.62a}$$

where ψ_{j0} is a constant real amplitude, $\Omega_{j0} = \left[M\left(k_{0j}^2 + q_{0j}^2\right) - \left(V_{jj}\psi_{j0}^2 + V_{j3-j}\psi_{l0}^2 + \mu_j\right)\right]$ is the frequency of the carrier waves, and k_{0j} and q_{0j} are real wave-numbers of the carrier waves.

Let us consider a small perturbation around the stationary state $\psi_j = \psi_{j0} \exp\left[-i\left(V_{jj}\psi_{j0}^2 + V_{jl}\psi_{3-j0}^2 + \mu_j\right)\tau\right]$ by taking

$$\psi_j = (\psi_{j0} + \varepsilon\psi_{j1}) \exp\left[i\varphi_j(\tau)\right], \tag{5.62b}$$

where $\psi_{j1}(x, y, \tau)$ is a complex function denoting the small ($\varepsilon << 1$) perturbation of the slowly varying modulated complex amplitude (it includes both amplitude and phase corrections), and $\varphi_j(\tau) = \Omega_{j0}\tau$. Substituting Eq. (5.62b) into system (5.61) and separating into real and imaginary parts by writing $\psi_{j1} = a_j + ib_j$, the first order terms (in ε) yield $\frac{\partial b_j}{\partial \tau} = -M_j\Delta a_j - 2V_{jj}\psi_{j0}^2 a_j - 2V_{jl}\psi_{j0}\psi_{l0}a_l$, $\frac{\partial a_j}{\partial \tau} = M_j\Delta b_j$, where j and $l\,(\neq j) = 1, 2$ (this will be henceforth understood unless otherwise stated). Eliminating b_j yields

$$\frac{\partial^2 a_j}{\partial \tau^2} = -M_j\Delta\left(M_j\Delta a_j + 2V_{jj}\psi_{j0}^2 a_j\right) - 2V_{jl}\psi_{3-j0}\psi_{j0}\Delta a_{3-j},\ j = 1,\ 2. \tag{5.63a}$$

Now, we let $a_j = a_{j0} \exp\left[i\left(Kx + Qy - \Omega_k\tau\right)\right] + c.c.$, where $\mathbf{k} = (K, Q)$ and Ω_k are the wave number and the frequency of low-frequency perturbations modulating the carrier signal, respectively, and a_{j0} is a (constant) complex amplitude. Ω_k is the growth rate. For the determination of a_{j0}, we insert the expression of a_j into Eq. (5.63a) and obtain the linear algebraic system $\left(2M_jV_{11}\psi_{10}^2\mathbf{k}^2 - M_j^2\mathbf{k}^4 + \Omega_k^2\right)a_{10} + 2M_jV_{12}\psi_{10}\psi_{20}\mathbf{k}^2 a_{20} = 0,\ \ 2M_jV_{21}\psi_{20}\psi_{10}\mathbf{k}^2 a_{10} + \left(2M_jV_{22}\psi_{20}^2\mathbf{k}^2 - M_j^2\mathbf{k}^4 + \Omega_k^2\right)$ $a_{20} = 0$, where $\mathbf{k}^2 = \mathbf{k}.\mathbf{k} = K^2 + Q^2$. For a nontrivial solution the determinant of the coefficients matrix of this system must vanish. That is, frequency Ω_k and the wave number $\mathbf{k} = (K, Q)$ must be related by the dispersion relation

$$\left(\Omega_k^2 + \left(M_1V_{11}\psi_{10}^2 + M_2V_{22}\psi_{20}^2\right)\mathbf{k}^2 - \frac{M_1^2 + M_2^2}{2}\mathbf{k}^4\right)^2$$
$$= \mathbf{k}^4\left[\left(\frac{M_2^2 - M_1^2}{2}\mathbf{k}^2 + M_1V_{11}\psi_{10}^2 - M_2V_{22}\psi_{20}^2\right)^2 + 4V_{12}V_{21}M_1M_2\psi_{10}^2\psi_{20}^2\right] \tag{5.64}$$

Modulational Instability Criteria for Coupled Nonlinear Transmission Lines

The phenomenon of the MI is observed when the angular frequency $\Omega_k = \Omega_k(\mathbf{k})$ of the perturbation functions has a nonzero imaginary part leading to an exponential growth of the amplitude versus time. The dispersion relation (5.64) takes the form of a bi-quadratic polynomial equation and has the solutions

$$\Omega_{k\pm}^2 = \mathbf{k}^2\left[\frac{M_1^2+M_2^2}{2}\mathbf{k}^2 - \left(M_1V_{11}\psi_{10}^2 + M_2V_{22}\psi_{20}^2\right)\right.$$
$$\left.\pm\sqrt{\left(\frac{M_2^2-M_1^2}{2}\mathbf{k}^2 + \left(M_1V_{11}\psi_{10}^2 - M_2V_{22}\psi_{20}^2\right)\right)^2 + 4V_{12}V_{21}M_1M_2\psi_{10}^2\psi_{20}^2}\,\right]. \quad (5.65)$$

Equation (5.65) defines four branches of instability growth rate. We note that the right-handside of Eq. (5.64) is real and is either nonnegative or negative. The stability is ensured (for any wave-vector **k**) if (and only if) both solutions $\Omega_{k\pm}^2$ are nonnegative. In fact, for a given wave-vector **k**, the growth rate $\Omega_{k\pm}$ are real, and hence a_1 and a_2 are bounded if, and only if, the right-hand-side of Eq. (5.65) is a real quantity. We formulate the results in terms of theorem.

Theorem 5.1 *Let $M_1^2 - M_2^2 \neq 0$. In order that $\Omega_{k\pm}^2$ should be nonnegative for any wave-vector k, it is necessary that either $V_{12}V_{21}M_1M_2 \geq 0$ or $V_{12}V_{21}M_1M_2 < 0$, $\left(M_1V_{11}\psi_{10}^2 - M_2V_{22}\psi_{20}^2\right)^2 + 4V_{12}V_{21}M_1M_2\psi_{10}^2\psi_{20}^2 > 0$, $M_1V_{11}\psi_{10}^2 - M_2V_{22}\psi_{20}^2 > 0$, and $M_2^2 - M_1^2 > 0$, and sufficient that $M_1V_{11}\psi_{10}^2 + M_2V_{22}\psi_{20}^2 \leq 0$, and either $\left(M_2V_{11}\psi_{10}^2 - M_1V_{22}\psi_{20}^2\right)^2 + 4M_1M_2V_{12}V_{21}\psi_{10}^2\psi_{20}^2 < 0$ or $\left(M_2V_{11}\psi_{10}^2 - M_1V_{22}\psi_{20}^2\right)^2 + 4M_1M_2V_{12}V_{21}\psi_{10}^2\psi_{20}^2 \geq 0$,*
$3\left(M_1V_{22}\psi_{20}^2 + M_2V_{11}\psi_{10}^2\right)^2 + 4M_1M_2\left(V_{11}V_{22} - V_{12}V_{21}\right)\psi_{10}^2\psi_{20}^2 > 0$, and
$M_1M_2\left(M_1V_{22}\psi_{20}^2 + M_2V_{11}\psi_{10}^2\right) < 0$.

An analysis of Theorem 5.1 shows that its conditions are satisfied for any perturbation amplitude $|\psi_{j0}|$ as soon as $M_1M_2V_{11}V_{22} > M_1M_2V_{12}V_{21} > 0$, and $M_jV_{jj} < 0$. In particular, if $M_j < 0$, then the conditions of Theorem 5.1 are met as soon as $V_{11}V_{22} > V_{12}V_{21} > 0$ and $V_{jj} > 0$.

Theorem 5.2 *Let $M_1^2 - M_2^2 = 0$. In order that that $\Omega_{k\pm}^2$ should be nonnegative for any wave-vector **k**, it is necessary that $\left(M_1V_{11}\psi_{10}^2 - M_2V_{22}\psi_{20}^2\right)^2 + 4M_1M_2V_{12}V_{21}\psi_{10}^2\psi_{20}^2 \geq 0$ and sufficient that $M_1V_{11}\psi_{10}^2 + M_2V_{22}\psi_{20}^2 < 0$ and $M_1M_2\left(V_{12}V_{21} - V_{11}V_{22}\right) < 0$.*

After some analyzes of Theorem 5.2 we easily see that the conditions of this theorem are valid for any perturbation amplitude $|\psi_{j0}|$ as soon as one of the following conditions is satisfied:

(i) $M_1 = M_2 < 0$, $V_{12}V_{21} > 0$, $V_{jj} > 0$, $V_{12}V_{21} - V_{11}V_{22} < 0$;
(ii) $M_1 = M_2 > 0$, $V_{12}V_{21} > 0$, $V_{jj} < 0$, $V_{12}V_{21} - V_{11}V_{22} < 0$;
(iii) $M_1 = -M_2 > 0$, $V_{12}V_{21} < 0$, $V_{11} < 0$, $V_{22} > 0$, and $V_{11}V_{22} - V_{12}V_{21} < 0$;
(iv) $M_1 = -M_2 < 0$, $V_{12}V_{21} < 0$, $V_{11} > 0$, $V_{22} < 0$, and $V_{11}V_{22} - V_{12}V_{21} < 0$.

We note that if the necessary conditions in Theorem 5.1 is violated, then the right-hand-side of Eq. (5.64) will be negative for any wave-vector **k** satisfying either the condition $\mathbf{k}^2 < 2\frac{-\left(M_1V_{11}\psi_{10}^2 - M_2V_{22}\psi_{20}^2\right) \pm 2|\psi_{10}\psi_{20}|\sqrt{-V_{12}V_{21}M_1M_2}}{M_2^2 - M_1^2}$ or the condition $\mathbf{k}^2 > \frac{-2\left(M_1V_{11}\psi_{10}^2 - M_2V_{22}\psi_{20}^2\right) \pm 4|\psi_{10}\psi_{20}|\sqrt{-V_{12}V_{21}M_1M_2}}{M_2^2 - M_1^2}$. For these wave-vectors, $\Omega_{k\pm}^2$ will be complex number. In the case of the violation of the necessary conditions in Theorem 5.2, the right-hand-side of Eq. (5.64) will be negative for every wave-vector, of then $\Omega_{k\pm}^2$ will be always complex. In the other hand, if the necessary conditions in both Theorems 5.1 and 5.2 are fulfilled, then the right-hand-side of Eq. (5.65) is real and is either nonnegative or negative. In order that both $\Omega_{k\pm}^2$ should be nonnegative for any wave-vector $\mathbf{k}$, the sufficient conditions in Theorems 5.1 and 5.2 must be satisfied. Thus, if the sufficient conditions in either Theorem 5.1 or Theorem 5.2 are violated, then for some values of wave-vector, the right-hand-side of Eq. (5.65) will be negative, and then the corresponding $\Omega_{k\pm}$ will be complex, which leads to the modulational instability. It is also possible for $\Omega_{k\pm}^2$ to be complex for any wave-vector $\mathbf{k}$. For example, if $M_2^2 - M_1^2 < 0$, $V_{12}V_{21}M_1M_2 < 0$, $M_1V_{11}\psi_{10}^2 - M_2V_{22}\psi_{20}^2 < 0$, and $\left(M_1V_{11}\psi_{10}^2 - M_2V_{22}\psi_{20}^2\right)^2 + 4V_{12}V_{21}M_1M_2\psi_{10}^2\psi_{20}^2 > 0$, then the right-hand-side of Eq. (5.64) will be negative for any wave-vector $\mathbf{k}$. Hence, $\Omega_{k\pm}^2$ will be complex for any wave-vector $\mathbf{k}$. This situation also happens when $M_2^2 - M_1^2 = 0$ and $\left(M_1V_{11}\psi_{10}^2 - M_2V_{22}\psi_{20}^2\right)^2 + 4M_1M_2V_{12}V_{21}\psi_{10}^2\psi_{20}^2 < 0$, because under these conditions, the right-hand-side of Eq. (5.64) will negative for any wave-vector $\mathbf{k}$, and then $\Omega_{k\pm}^2$ given by Eq. (5.65) will be complex for every $\mathbf{k}$. We then obtain the following theorem:

Theorem 5.3 *If either* $M_2^2 - M_1^2 < 0$, $V_{12}V_{21}M_1M_2 < 0$, $M_1V_{11}\psi_{10}^2 - M_2V_{22}\psi_{20}^2 < 0$, *and* $\left(M_1V_{11}\psi_{10}^2 - M_2V_{22}\psi_{20}^2\right)^2 + 4V_{12}V_{21}M_1M_2\psi_{10}^2\psi_{20}^2 > 0$ *or* $M_2^2 - M_1^2 = 0$ *and* $\left(M_1V_{11}\psi_{10}^2 - M_2V_{22}\psi_{20}^2\right)^2 + 4M_1M_2V_{12}V_{21}\psi_{10}^2\psi_{20}^2 < 0$, *then all the* $\Omega_{k\pm}^2$ *given by Eq. (5.65) will be complex for any wave-vector* $\mathbf{k}$.

Analyzing Theorem 5.3, we find that if $M_2^2 - M_1^2 < 0$, $V_{12}V_{21}M_1M_2 < 0$, $M_1V_{11} < 0$, $M_2V_{22} > 0$, and $M_1M_2\left(2V_{12}V_{21} - V_{11}V_{22}\right) > 0$, then the conditions of Theorem 5.3 are satisfied for any perturbation amplitude $\left|\psi_{j0}\right|$. In particular, for $M_j < 0$, the conditions of Theorem 5.3 will be valid as soon as $M_2^2 - M_1^2 < 0$, $V_{12}V_{21} < 0$, $V_{11} > 0$, $V_{22} < 0$, and $2V_{12}V_{21} - V_{11}V_{22} > 0$

It is important to note that if the conditions of each of Theorems 5.1, 5.2, and 5.3 are violated, then at least one of Ω_{k+}^2 and Ω_{k-}^2 will be positive for some wave-vectors $\mathbf{k}$ and the other one will be negative for some other wave-vectors $\mathbf{k}$. For example, let $M_2^2 - M_1^2 = 0$, $\left(M_1V_{11}\psi_{10}^2 - M_2V_{22}\psi_{20}^2\right)^2 + 4M_1M_2V_{12}V_{21}\psi_{10}^2\psi_{20}^2 \geq 0$. Then if $M_1V_{11}\psi_{10}^2 + M_2V_{22}\psi_{20}^2 < 0$ and $M_1M_2\left(V_{12}V_{21} - V_{11}V_{22}\right) > 0$, we will have $\Omega_{k+}^2 \geq 0$ for any $\mathbf{k}$, and $\Omega_{k+}^2 < 0$ for every wave-vector $\mathbf{k}$ verifying the condition

$$\mathbf{k}^2 < \frac{2}{M_1^2 + M_2^2}\Big(M_1V_{11}\psi_{10}^2 + M_2V_{22}\psi_{20}^2 + \sqrt{\left(M_1V_{11}\psi_{10}^2 - M_2V_{22}\psi_{20}^2\right)^2 + 4V_{12}V_{21}M_1M_2\psi_{10}^2\psi_{20}^2}\Big);$$

if the wave-vector $\mathbf{k}$ is taking from the condition

$$\mathbf{k}^2 < \frac{2}{M_1^2 + M_2^2} \left(M_1 V_{11} \psi_{10}^2 + M_2 V_{22} \psi_{20}^2 \right.$$
$$\left. + \sqrt{\left(M_1 V_{11} \psi_{10}^2 - M_2 V_{22} \psi_{20}^2 \right)^2 + 4 V_{12} V_{21} M_1 M_2 \psi_{10}^2 \psi_{20}^2} \right),$$

then Ω_{k+}^2 will be positive. If $M_1 V_{11} \psi_{10}^2 + M_2 V_{22} \psi_{20}^2 > 0$ and $M_1 M_2 (V_{12} V_{21} - V_{11} V_{22}) < 0$, then both $\Omega_{k\pm}^2$ will be positive for any wave-vector **k** satisfying the condition

$$\mathbf{k}^2 > \frac{2}{M_1^2 + M_2^2} \left(M_1 V_{11} \psi_{10}^2 + M_2 V_{22} \psi_{20}^2 \right.$$
$$\left. + \sqrt{\left(M_1 V_{11} \psi_{10}^2 - M_2 V_{22} \psi_{20}^2 \right)^2 + 4 V_{12} V_{21} M_1 M_2 \psi_{10}^2 \psi_{20}^2} \right);$$

both $\Omega_{k\pm}^2$ will be negative for every wave vector **k** verifying the condition

$$\mathbf{k}^2 < \frac{2}{M_1^2 + M_2^2} \left(M_1 V_{11} \psi_{10}^2 + M_2 V_{22} \psi_{20}^2 \right.$$
$$\left. - \sqrt{\left(M_1 V_{11} \psi_{10}^2 - M_2 V_{22} \psi_{20}^2 \right)^2 + 4 V_{12} V_{21} M_1 M_2 \psi_{10}^2 \psi_{20}^2} \right);$$

for any wave vector **k** verifying the condition

$$\frac{2}{M_1^2 + M_2^2} \left(M_1 V_{11} \psi_{10}^2 + M_2 V_{22} \psi_{20}^2 \right.$$
$$\left. + \sqrt{\left(M_1 V_{11} \psi_{10}^2 - M_2 V_{22} \psi_{20}^2 \right)^2 + 4 V_{12} V_{21} M_1 M_2 \psi_{10}^2 \psi_{20}^2} \right)$$
$$> \mathbf{k}^2 > \frac{2}{M_1^2 + M_2^2} \left(M_1 V_{11} \psi_{10}^2 + M_2 V_{22} \psi_{20}^2 \right.$$
$$\left. - \sqrt{\left(M_1 V_{11} \psi_{10}^2 - M_2 V_{22} \psi_{20}^2 \right)^2 + 4 V_{12} V_{21} M_1 M_2 \psi_{10}^2 \psi_{20}^2} \right),$$

we will have $\Omega_{k+}^2 < 0$ and $\Omega_{k-}^2 > 0$. These two examples show that both $\Omega_{k\pm}^2$ can be positive for some values of wave vector, and negative for some other values of wave vector **k**; also, one of the Ω_{k+}^2 and Ω_{k-}^2 can be positive/negative for all wave vectors **k** and the second one changes its sign with different values of k. Whence the following theorem.

Theorem 5.4 *If the conditions of each of Theorems 5.1, 5.2, and 5.3 are not met, then both Ω_{k-}^2 and Ω_{k+}^2 cannot be of the same sign for every wave-vector* **k**.

It is clear that for either the complex number Ω_{k+} or Ω_{k-}, $a_j = a_{j0} \exp\left[i \left(Kx + Qy - \Omega_k \tau\right)\right] + c.c.$ with $a_{j0} \neq 0$ will be unbounded, and the modulational instability occurs. We then conclude that Theorems 5.1 and 5.2 give the criteria of the modulational stability of the Stokes wave solutions $\psi_j = \psi_{j0} \exp\left[-i\left(V_{jj}\psi_{j0}^2 + V_{j3-j}\psi_{3-j0}^2 + \mu_j\right)\tau\right]$ of the coupled NLS equations (5.61), while the criteria of the modulational instability are given by Theorem 5.3.

Because the criteria on the modulational (in)stability we found in this section depends mainly on the coefficients of Eq. (5.61) which depend only on the parameters of the coupled nonlinear transmission line of Fig. 5.1b when $k_j^{X_0}$, $k_j^{Y_0}$, and ω_j are given, we conclude that for a given coupled nonlinear transmission line of Fig. 5.1b, a modulational instability may or may not occur, depending on the electrical properties of the line.

5.3.3 Computational Analysis

For the computational simulations, the following network parameters are used

$$C_0 = 370\,\text{pF},\ C_2 = 100\,\text{pF},\ L = 200\,\mu\text{H},\ V_0 = 2\,\text{V},\ b = 0.25\,\text{V}^{-1}, \qquad (5.66)$$

while the dispersive element C_S is taken variable. For these line parameters, we compute $v = 1/\sqrt{LC_0} = 3.6761 \times 10^6$. Table 5.1 gives the coefficients of the coupled NLS equations (5.61) for different values of C_S, when $k_B^{X_0} = -k_A^{X_0} = -0.4$ and $k_B^{Y_0} = -k_A^{Y_0} = -0.2$.

Using this table, we compute frequency square Ω_{k-}^2 and the modulational instability growth $\pm i\Omega_{k-}$ as function of squared modulation wave function $\mathbf{k}^2$. The frequency square Ω_{k-}^2 and the typical structure of the modulational instability growth rate $\Gamma = -i\Omega_{k-}(\mathbf{k})$ as a functions of squared modulation wave number $\mathbf{k}^2$ are depicted in Fig. 5.2 for the line parameters (5.66). The amplitudes of the Stokes wave solutions are taken in the proportion $|\psi_{20}|^2 = 2\,|\psi_{10}|^2 = 10^{-16}$. Figure 5.2a give the frequency square versus the squared modulation wave number $\mathbf{k}^2$. Here, the dotted black line, the solid black line, and the solid red line correspond to $C_S = 0$, $C_S = 50\ pF$, $C_S = 100\ pF$, respectively. Each square frequency shown in Fig. 5.2a admits both negative and positive values. The values of $\mathbf{k}^2$ for which the negative values are taken correspond to the modulational instability wave numbers. The value of $\mathbf{k}^2$ for which these curves

Table 5.1 Coefficients of the coupled NLS equations (5.61) for different values of the dispersive element C_s with $k_B^{X_0} = k_A^{X_0} = -0.4$ and parameters (5.66)

$C_s(in\,\text{pF})$	$M_1 \times 10^{-10}$	$M_2 \times 10^{-10}$	$V_{11} \times 10^5$	$V_{22} \times 10^5$	$V_{12} \times 10^5$	$V_{21} \times 10^5$
0	0.58378	−0.58378	3.0145	4.0226	12.688	−12.688
50	0.71753	−0.71753	1.3241	1.3052	6.6937	−6.6937
100	0.74904	−0.74904	0.84840	0.77895	4.546	−4.546

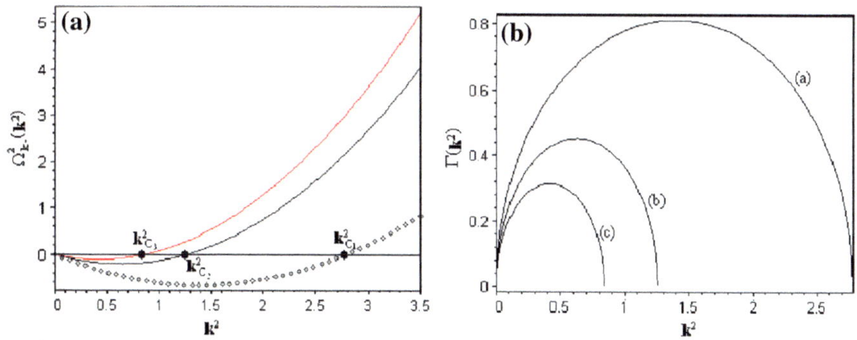

Fig. 5.2 The perturbation square frequency Ω^2_{k-} (a) and the typical structure of the modulational instability growth rate Γ (II) as a functions of squared modulation wave number $\mathbf{k}^2$ in the special case when $k_B^{X_0} = -k_A^{X_0} = -0.4$, $k_B^{Y_0} = -k_A^{Y_0} = -0.2$, and the line parameters $L = 200\,\mu\text{H}$, $V_0 = 2V$, $C_0 = 370\text{pF}$. In plot (a), the dotted black curve, the solid black curve, and the solid red curve correspond to $C_S = 0$, $C_S = 50$ pF, and $C_S = 100$ pF, respectively. Plot (a), curves (a), (b), and (c) correspond to $C_S = 0$, $C_S = 50$ pF, and $C_S = 100$ pF, respectively. The region of the modulational instability and the values of the modulational instability growth rate Γ decrease with the introduction of the linear capacitance C_S, reprint from Ref. [6], Copyright 2018, with permission from American Physical Society

cross zero are the critical wave number; they are $\mathbf{k}^2_{C_1} \approx 2.7854$, $\mathbf{k}^2_{C_2} \approx 1.2578$, and $\mathbf{k}^2_{C_3} \approx 0.83802$. Figure 5.2b gives the modulational instability growth versus $\mathbf{k}^2$. For Fig. 5.2b, curves (a), (b), and (c) of Fig. 5.2b correspond to $C_S = 0$, $C_S = 50\ pF$, and $C_S = 100\ pF$, respectively, and depicted for $0 \leq \mathbf{k}^2 \leq \mathbf{k}^2_{C_j}$, $j = 1, 2, 3$. Figure 5.2a, b show that the region of the modulational instability and the values of the modulational instability growth rate Γ decrease with the introduction of of the linear capacitance $C_S = 0$. We can then conclude that the effect of dispersive element C_S is to decrease the instability region and the instability growth rate.

In Fig. 5.3, we plot the electrical amplitudes $|\psi_j(x, y, \tau)|$ $(j = 1, 2)$ of the perturbed Stokes wave solutions of the coupled NLS equations (5.61) for different times. We have used the modulation wave vector $(K, Q) = (0.5, 0)$, which corresponds to the modulation wave number $\mathbf{k} = 0.5$. On each of figures a–f, the red horizontal line corresponds to the amplitude of the Stokes wave, the solid black curve denotes the electrical amplitude at the initial time $\tau = 0$, and the dotted black line denotes the electrical amplitude for $\tau > 0$. Cures (a), (c) and (e) are the amplitude of the component $\psi_1(x, y, \tau)$, while curves (b), (d) and (f) are the $\psi_2(x, y, \tau)$ component amplitudes. Figure 5.3a, b, c, d, e and f correspond respectively to the dispersive element $C_S = 0$, $C_S = 50\ pF$, and $C_S = 100$ pF. Curves 5.3b which correspond to $C_S = 0$ and $\tau = 63 \times 10^9$ show that the instability is intensified by $\tau = 63 \times 10^9$. Curves correspond to $C_S = 50$ pF and $\tau = 96 \times 10^9$ and show that the instability is observed by $\tau = 96 \times 10^9$. It is seen from curves (f) that correspond to $C_S = 100$ pF and $\tau = 131 \times 10^9$ that the instability is remarkable from $\tau = 131 \times 10^9$. For time $\tau = 63 \times 10^9$ or $\tau = 96 \times 10^9$, or $\tau = 131 \times 10^9$, at least one of the corresponding $|\psi_1(x, y, \tau)|$ and $|\psi_2(x, y, \tau|$ is far away from the Stokes wave amplitude $|\psi_{10}|$ or

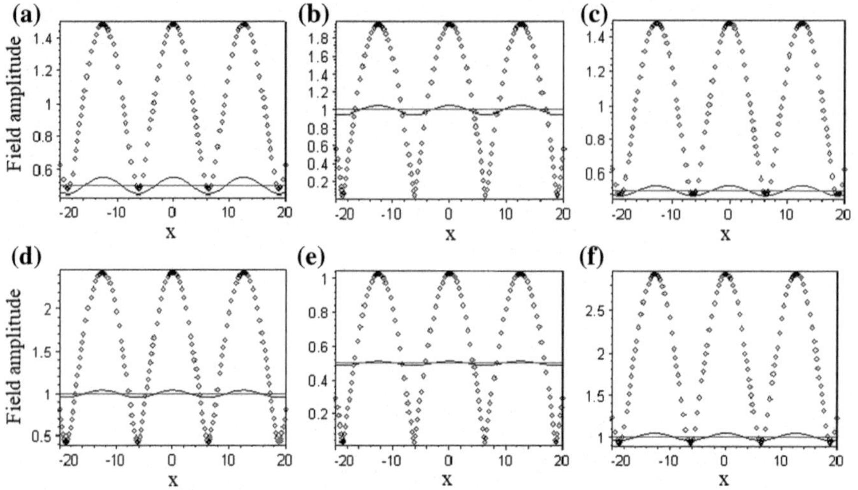

Fig. 5.3 Field amplitudes of perturbed Stokes wave solutions for different time and for $\varepsilon = 0.01$. The red curve is the Stokes wave amplitude $|\psi_{j0}|$, the solid black curve gives the amplitudes $|\psi_j(x, y, \tau)|$ of the perturbed solution at the initial time $\tau = 0$, and the dotted black curve corresponds to the amplitude of the perturbed solution at a given time $\tau > 0$ ($\tau = 63 \times 10^9$ for curves (a) and (b) corresponding to $C_S = 0$, $\tau = 96 \times 10^9$ for curves (c) and (d) that correspond to $C_S = 50$ pF, and $\tau = 131 \times 10^9$ for curves (e) and (f) for $C_S = 100$ pF). The onset of instability occurs at $\tau \approx 96 \times 10^9$, or $\tau \approx 131 \times 10^9$, or $\tau \approx 131 \times 10^9$ for $C_S = 0$ or $C_S = 50$ pF, or $C_S = 100$ pF, respectively, reprint from Ref. [6], Copyright2018, with permission from American Physical Society

$|\psi_{20}|$. We can then conclude that the introduction of the dispersive element C_S in the electrical system shown in Fig. 5.1b increases the time over which the onset of instability occurs. This fact is very important. Indeed, as it is well-known, the instability of a solution does not exclude its observation in experiments: if the time scale over which the onset of instability occurs far exceeds the duration of the experiments, then formally unstable solution is as relevant for the experiments as is the stable solution.

5.4 Analytical Solitons in Coupled Nonlinear Transmission Line of Fig. 5.1b

Base on the solitary wave solutions of a single NLS equation obtained from system (5.61), we investigate analytically in this section the transmission of solitary signals through the coupled NLTL of Fig. 5.1b. Setting $\psi_2 = 0$ in system (5.61) and replacing τ by t, we obtain the single NLS equation

$$i\frac{\partial \psi_1}{\partial t} = M_1 \nabla^2 \psi_1 + \left(\mu_1 + V_1 |\psi_j|^2\right) \psi_1, \tag{5.67}$$

where $\nabla^2 = \partial^2/\partial x^2 + \partial^2/\partial y^2$ and M_1, $V_1 = V_{11}$, and μ_1 are the parameters defined after system (5.61). We first transform Eq. (5.67) into a system of linearly coupled NLS equations by means of the transformation

$$\psi_1(x, y, t) = \widetilde{u}(x, t)\Phi_u(y) + \widetilde{v}(x, t)\Phi_v(y), \tag{5.68}$$

where $\Phi_u(y)$ and $\Phi_v(y)$ are two real valued functions satisfying the conditions [2]

$$\int \Phi_u(y)\Phi_v(y)dy \simeq 0, \quad \int \Phi_u^2(y)\, dy = \int \Phi_v^2(y)\, dy = 1. \tag{5.69}$$

Inserting Eq. (5.68) into Eq. (5.67) and using conditions (5.69), we obtain

$$i\frac{\partial \widetilde{u}}{\partial t} = M_1 \frac{\partial^2 \widetilde{u}}{\partial x^2} + \left(\mu_u + V_u |\widetilde{u}|^2\right)\widetilde{u} - \Gamma\widetilde{v}, \tag{5.70a}$$

$$i\frac{\partial \widetilde{v}}{\partial t} = M_1 \frac{\partial^2 \widetilde{v}}{\partial x^2} + \left(\mu_v + V_v |\widetilde{v}|^2\right)\widetilde{v} - \Gamma\widetilde{u}, \tag{5.70b}$$

where

$$\begin{aligned} \mu_j &= \int \left(\mu_1 \Phi_j^2 - M_1 \left(\Phi_j''\right)^2\right) dy, \quad V_j = V_1 \int \Phi_j^4 dy, \\ \Gamma &= \int \left(M_1 \Phi_u'' \Phi_v'' - \mu_1 \Phi_u \Phi_v\right) dy, \quad j = u, v. \end{aligned} \tag{5.71}$$

In what follows, we restrict our study to the case $M_1 > 0$, $\mu_1 < 0$, and $V_1 > 0$ [6]. Under these conditions, one can make $\Gamma < 0$ by changing, if necessary, the sign of either $\Phi_u(y)$ or $\Phi_v(y)$. Under the condition $V_1 > 0$, it is seen from Eq. (5.71) that $V_j > 0,\ j = u,\ v$. Other cases can be studied in a similar way. Now we make the following ansatz and notations

$$\begin{aligned} &\widetilde{u} = u\sqrt{-\frac{\Gamma}{V_u}} \exp\left(i\frac{\mu_u}{\Gamma}t'\right), \quad \widetilde{v} = v\sqrt{-\frac{V_u}{\Gamma}} \exp\left(i\frac{\mu_u}{\Gamma}t'\right), \\ &t' = -\Gamma t, \quad x' = \sqrt{-\frac{\Gamma}{M_1}}x, \quad \mu_0 = \frac{\mu_v - \mu_u}{\Gamma}, \quad a = \frac{V_v}{V_u}. \end{aligned} \tag{5.72}$$

Inserting Eq. (5.72) into Eqs. (5.70a)–(5.70b) and replacing x' and t' by x and t, respectively yield the following linearly couple NLS equations for u and v

$$i\frac{\partial u}{\partial t} = \frac{\partial^2 u}{\partial x^2} + |u|^2 u + v, \quad i\frac{\partial v}{\partial t} = \frac{\partial^2 v}{\partial x^2} + \left(\mu_0 + a|v|^2\right)v + u. \tag{5.73}$$

As one can see from Eqs. (5.71) and (5.72), the coefficient a in Eq. (5.73) is free from the line parameters M_1, V_{11}, and μ_1, while the coefficient μ_0 depends on the line parameters M_1 and μ_1. We note that Eq. (5.73) does not contain the explicitly the line parameter V_1. As shown Eqs. (5.71) and (5.72), this line parameter is contained implicitly in $u(x,t)$ and $v(x,t)$.

5.4.1 Propagation of Fundamental Solitons in the System

Here, we aim to find the analytical soliton solutions of the 1D linearly coupled NLS equations (5.73). Generally, the two linearly coupled NLS equations (5.73) are not integrable. However by imposing some restrictions on the coefficients of this system, one can construct some analytical solutions.

We start by seeking stationary solitary-pulse solutions of Eqs. (5.70a) and (5.70b). For this purpose, we set

$$u(x,t) = U(x)\exp(-i\omega t), \quad v(x,t) = V(x)\exp(-i\omega t), \tag{5.74}$$

where ω is the frequency of the propagation, and $U(x)$ and $V(x)$ are two real valued functions vanishing as $|x| \to +\infty$. Inserting Eq. (5.74) into Eqs. (5.70a) and (5.70b) yields

$$\begin{aligned} \frac{d^2U}{dx^2} + \left(U^2 - \omega\right)U + V = 0, \\ \frac{d^2V}{dx^2} + \left(\mu_0 - \omega + aV^2\right)V + U = 0. \end{aligned} \tag{5.75}$$

Considered as a dynamical system, Eqs. (5.75) possess the Hamiltonian

$$H = \left(\frac{dU}{dx}\right)^2 + \left(\frac{dV}{dx}\right)^2 + \left(-\omega + \frac{1}{2}U^2\right)U^2 + \left(\mu_0 - \omega + \frac{a}{2}V^2\right)V^2 + 2UV. \tag{5.76}$$

It is obvious that needed vanishing at $|x| \to +\infty$ solutions correspond to $H = 0$. We specify the class of soliton solutions we look for as even solutions, i.e., solutions that satisfy the conditions $U(-x) = U(x)$, $V(-x) = V(x)$, with a single maximum at $x = 0$ (for the moment, we consider single-humped, i.e., fundamental solitons). For definiteness' sake, we set $U(x=0) = U_0 > 0$. As the first derivative of the fields vanishes at the central point, from Eq. (5.76) we obtain the following quadric equation (with $V_0 = V(0)$) for the soliton amplitude U_0

$$\left(U_0^2 - 2\omega\right)U_0^2 + \left(2\left(\mu_0 - \omega\right) + aV_0^2\right)V_0^2 + 4U_0V_0 = 0. \tag{5.77}$$

Moreover, Eqs. (5.75) give the identities

$$\begin{aligned} \left(-2\omega + U_0^2\right)U_0^2 = 4\int_0^{+\infty} V(x)dU(x), \\ \left(2\left(\mu_0 - \omega\right) + aV_0^2\right)V_0^2 = 4\int_0^{+\infty} U(x)dV(x). \end{aligned} \tag{5.78}$$

Definition: *A soliton will be called either an "in-phase" (an "out–of-phase") soliton if* $V_0 > 0$ ($V_0 < 0$).

It is evident that the right-hand sides of Eqs. (5.78) are negative for the in-phase solitons and positive for the out-of-phase solitons. Thus, we conclude that the fundamental solitons must obey the two inequalities

$$U_0^2 < 2\omega \quad \text{if } V_0 > 0, \tag{5.79a}$$

$$aV_0^2 > 2(\omega - \mu_0) \quad \text{if } V_0 < 0. \tag{5.79b}$$

It is seen from Eq. (5.79a) that for the in-phase fundamental solitons, the frequency ω of the u-wave propagation is positive. Because $a = V_v/V_u > 0$, Eq. (5.79b) implies that for the out-of-phase solitons there is no restriction on the frequency ω of the wave propagation.

To obtain the domain where the soliton solutions are permitted, we use the dispersion law. Assuming the exponential decays $U(x) \sim \exp(-kx)$ and $V(x) \sim \exp(-kx)$ as $|x| \to +\infty$, and linearizing (5.75), we find two branches of the dispersion relation for the solitons:

$$k_1^2 = \left(-\omega + \frac{\mu_0}{2} - \sqrt{\frac{\mu_0^2}{4} + 1}\right), \quad k_2^2 = \left(-\omega + \frac{\mu_0}{2} + \sqrt{\frac{\mu_0^2}{4} + 1}\right). \tag{5.80}$$

It is seen from Eq. (5.80) that for k_1 to be positive, it is necessary and sufficient that $\omega < \min\left(\mu_0, \frac{\mu_0}{2} - \sqrt{\frac{\mu_0^2}{4} + 1}\right)$. It follows from the condition $\frac{\mu_0}{2} - \sqrt{\frac{\mu_0^2}{4} + 1} < 0$ that the frequency ω of the wave propagation for the k_1-branch is negative, and this excludes the in-phase solitons for this branch. For the k_2-branch, the wave propagation frequency ω must satisfy the condition $\omega < \frac{\mu_0}{2} + \sqrt{\frac{\mu_0^2}{4} + 1}$, and this allows the existence of both in-phase and out-of-phase solitons.

Finally, to have either the in-phase or the out-of-phase fundamental solitons, U_0 and V_0 must satisfy Eqs. (5.77) and (5.78) or Eqs. (5.77) and (5.79b), respectively, for given ω, with μ_0 satisfying either condition $k_1 > 0$ or condition $k_2 > 0$ and for a given $a > 0$.

By seeking an exact soliton solution in the form $(U(x), V(x))^T = (U_0, V_0)^T$ sech (βx), we find the following special in- and out-of-phase exact soliton solutions. The exact solutions for the in-phase solitons are given by

$$\begin{pmatrix} U(x) \\ V(x) \end{pmatrix} = \begin{pmatrix} \dfrac{U_{01}}{\cosh\left[\frac{U_{01}}{\sqrt{2}} x\right]} \\ \dfrac{U_{01}}{\sqrt{a}\cosh\left[\frac{U_{01}}{\sqrt{2}} x\right]} \end{pmatrix}, \quad U_{01} = \sqrt{2\left(\omega - \frac{1}{\sqrt{a}}\right)}, \tag{5.81}$$

which exist when a and μ_0 and a and ω are related as

$$\mu_0 = \frac{1}{\sqrt{a}} - \sqrt{a} = \mu_0(a), \quad \omega > \omega_{1\,\min} = \frac{1}{\sqrt{a}}. \tag{5.82}$$

The out-of-phase solitons are given by

$$\begin{pmatrix} U(x) \\ V(x) \end{pmatrix} = \begin{pmatrix} \frac{U_{02}}{\cosh\left[\frac{U_{02}}{\sqrt{2}}x\right]} \\ -\frac{U_{02}}{\sqrt{a}\cosh\left[\frac{U_{02}}{\sqrt{2}}x\right]} \end{pmatrix}, \quad U_{02} = \sqrt{2\left(\omega + \frac{1}{\sqrt{a}}\right)}, \tag{5.83}$$

and they exist when

$$\mu_0 = -\mu_0(a), \quad \omega > \omega_{2\,\min} = -1/\sqrt{a}, \tag{5.84}$$

where $\mu_0(a)$ is given in Eq. (5.82). Solving the equation $\mu_0 = 1/\sqrt{a} - \sqrt{a}$ for $\sqrt{a}$ and inserting the expression for $\sqrt{a}$ in U_{10} and U_{20}, we find that $U_{10}^2/2 = k_1^2$, while $U_{20}^2/2 = k_2^2$; this means that the in-phase solitons correspond to the k_1-branch of the dispersion law (5.80), while the out-of-phase solitons correspond to the k_2-branch of the dispersion law (5.80). By a simple substitution, one easily verifies that $(U_0, V_0) = \left(U_{10}, U_{10}/\sqrt{a}\right)$ and $(U_0, V_0) = \left(U_{20}, -U_{20}/\sqrt{a}\right)$ satisfy Eq. (5.77).

5.4.2 *Propagation of Higher Order Solitons in the System*

Now, we aim to construct some exact higher order soliton solutions of Eqs. (5.73), leading to the propagation of higher order soliton in the system showed in Fig. 5.1b. To accomplish this, we follow Hirota [5] and introduce the transformation

$$u(x,t) = \frac{F(x,t)}{\Psi(x,t)}, \quad v(x,t) = \frac{G(x,t)}{\Psi(x,t)}, \tag{5.85}$$

where F and G are two complex functions, and $\Psi(x,t)$ is a real function satisfying some additional condition. Inserting ansatz (5.85) into Eqs. (5.73) yields

$$\begin{aligned} i\left(\Psi\frac{\partial F}{\partial t} - F\frac{\partial \Psi}{\partial t}\right) - \left(\Psi\frac{\partial^2 F}{\partial x^2} + F\frac{\partial^2 \Psi}{\partial x^2} - 2\frac{\partial F}{\partial x}\frac{\partial \Psi}{\partial x}\right) = 0, \\ i\left(\Psi\frac{\partial G}{\partial t} - G\frac{\partial \Psi}{\partial t}\right) - \left(\Psi\frac{\partial^2 G}{\partial x^2} + G\frac{\partial^2 \Psi}{\partial x^2} - 2\frac{\partial G}{\partial x}\frac{\partial \Psi}{\partial x}\right) = 0, \end{aligned} \tag{5.86}$$

$$\begin{aligned} 2\left[\left(\frac{\partial \Psi}{\partial x}\right)^2 - \Psi\frac{\partial^2 \Psi}{\partial x^2}\right] &= -\Psi^2\frac{G}{F} - |U|^2, \\ 2\left[\left(\frac{\partial \Psi}{\partial x}\right)^2 - \Psi\frac{\partial^2 \Psi}{\partial x^2}\right] &= -\Psi^2\frac{F}{G} - \mu_0\Psi^2 - a|V|^2. \end{aligned} \tag{5.87}$$

Because the left-hand sides of Eqs. (5.87) are the same, the right-hand sides must also be the same. This requirement imposes the conditions

$$G = \pm\frac{F}{\sqrt{a}}, \quad \mu_0 = \pm\frac{1}{\sqrt{a}} \mp \sqrt{a}. \tag{5.88}$$

Analyzing solutions (5.81) and (5.83) and conditions (5.82) and (5.84), one finds that solutions (5.81) and (5.83) are also found under the integrability conditions (5.88). If we impose the additional condition $\frac{\Psi^2(x,t)}{\sqrt{a}} \simeq 0$ on the function $\Psi(x,t)$ and use Eq. (5.88), system (5.86) becomes

$$i\left(\Psi\frac{\partial F}{\partial t} - F\frac{\partial \Psi}{\partial t}\right) - \left(\Psi\frac{\partial^2 F}{\partial x^2} + F\frac{\partial^2 \Psi}{\partial x^2} - 2\frac{\partial F}{\partial x}\frac{\partial \Psi}{\partial x}\right) = 0, \tag{5.89}$$

while system (5.87) becomes

$$2\left(\frac{\partial \Psi}{\partial x}\right)^2 - 2\Psi\frac{\partial^2 \Psi}{\partial x^2} + |U|^2 = 0. \tag{5.90}$$

To generate the exact solutions, we expand the dependent variables F and Ψ in powers of real ϵ as $F(x,t) = \epsilon X(x,t)$, $\Psi(x,t) = 1 + Y(x,t)\epsilon^2$. Inserting this ansatz into Eq. (5.89) yields

$$i\frac{\partial X}{\partial t} - \frac{\partial^2 X}{\partial x^2} = 0, \quad i\left(Y\frac{\partial X}{\partial t} - X\frac{\partial Y}{\partial t}\right) - Y\frac{\partial^2 X}{\partial x^2} - X\frac{\partial^2 Y}{\partial x^2} + 2\frac{\partial X}{\partial x}\frac{\partial Y}{\partial x} = 0. \tag{5.91}$$

The first of Eqs. (5.91) gives $X(x,t) = \exp\left(ikx + ik^2t + k_{10}\right)$, where k and k_{10} are two complex constants. Inserting $X(x,t)$ into the second of Eqs. (5.91) and using Eq. (5.90), we obtain

$$Y(x,t) = -\frac{1}{2(k-k^*)^2}\exp\left[i\left(k-k^*\right)x + i\left(k^2 - k^{*2}\right)t + k_{10} + k_{10}^*\right],$$

where k^* is the complex conjugate of k. Thus under the integrability condition (5.88) and the additional condition $\frac{\Psi^2(x,t)}{\sqrt{a}} \simeq 0$, we have generated the following exact vectorial soliton solutions for Eqs. (5.73)

$$\begin{pmatrix} u(x,t) \\ v(x,t) \end{pmatrix} = \begin{pmatrix} \frac{2\epsilon(k-k^*)^2}{2(k-k^*)^2\exp\left[-i\left(kx+k^2t\right)-k_{10}\right]-\epsilon^2\exp\left[-i\left(k^*x+k^{*2}t\right)+k_{10}^*\right]} \\ \pm\frac{1}{\sqrt{a}}\frac{2\epsilon(k-k^*)^2}{2(k-k^*)^2\exp\left[-i\left(kx+k^2t\right)-k_{10}\right]-\epsilon^2\exp\left[-i\left(k^*x+k^{*2}t\right)+k_{10}^*\right]} \end{pmatrix}. \tag{5.92}$$

The nontrivial solutions (5.92) contain two complex parameters, k_{10} and k, with $k \notin \mathbb{R}$. Because $(k-k^*)^2 < 0$,

$$2\left(k-k^{*}\right)^{2}\exp\left[-i\left(kx+k^{2}t\right)-k_{10}\right]-\epsilon^{2}\exp\left[-i\left(k^{*}x+k^{*2}t\right)+k_{10}^{*}\right]\neq 0$$

Hence the only restriction on the parameters k and k_{10} is that $k\notin\mathbb{R}$. In the case when either $\epsilon\to 0$ or $\epsilon\to\infty$, solutions (5.92) amounts to $u(x,t)=0$ and $v(x,t)=0$.

By setting $k=k_r+ik_i$ and $k_{10}=k_{10r}+ik_{10i}$, Eq. (5.92) becomes

$$\begin{pmatrix}u(x,t)\\ v(x,t)\end{pmatrix}=\begin{pmatrix}r(z)\\ \pm r(z)/\sqrt{a}\end{pmatrix}\exp\left[i\left(Kz-\lambda t+\varphi_0\right)\right], \tag{5.93}$$

where

$$r(z)=\frac{\epsilon}{\exp(z)+\left(\frac{\epsilon}{2\sqrt{2}k_i}\right)^{2}\exp(-z)},$$

$$z=X-ct,\quad X=k_ix-k_{10r},\quad c=-2k_rk_i,\quad K=\frac{k_r}{k_i}, \tag{5.94}$$

$$\lambda=k_i^2-k_r^2+2k_rk_r,\quad \varphi_0=\frac{k_ik_{10i}+k_rk_{10r}}{k_i}.$$

In what follows, we will call $r(z)$ the "common factor amplitude". It follows from Eqs. (5.93) and (5.94) that $u(x,t)$ and $v(x,t)$ are two traveling waves with the same velocity $c=-2k_rk_i$, the same frequency λ, and the same wavenumber K. In Eq. (5.93) φ_0 is the initial phase. If k_r and k_i have the same sign, then the wave moves in the $-x$ direction, while it moves in the $+x$ direction if k_r and k_i have opposite signs. We thus have shown that system (5.73) admits non-periodic traveling wave solutions. If we now take $\epsilon=2\sqrt{2}k_2$, the first of Eq. (5.94) gives $r(z)=\sqrt{2}\,k_i\text{sech}(z)$, or $r(x,t)=\sqrt{2}\,k_i\text{sech}(k_ix-k_{10r}+2k_rk_it)$. This means that the traveling wave solutions (5.93)–(5.94) contain traveling wave solutions with moving soliton amplitude. As one can see from Eq. (5.94) the maximum amplitude of the common factor amplitude $r(z)$ depends on and k_i through the equality $r_{\max}=\sqrt{2}k_i$, its centre of symmetry depends on both ϵ and k_i through the equality $z_c=\ln\frac{\epsilon}{2\sqrt{2}k_i}$ (with $\epsilon k_i>0$: for simplicity, we consider only the case where $\epsilon>0$ and $k_i>0$). Another important quantity is the width of the common factor amplitude at one half the maximum amplitude $r_{\max}$, which we shall label L_{hw}. For a physical representation, Fig. 5.4 shows a common factor amplitude labelling $r_{\max}$ and the corresponding L_{hw}. To calculate L_{hw} from Eq. (5.94), we merely set $r(z)=\frac{r_{\max}}{2}$. Solving this equation for z gives z_1 and z_2, and we obtain $L_{hw}=|z_1-z_2|=\ln\left(\frac{2+\sqrt{3}}{2-\sqrt{3}}\right)\simeq 2.6339$, which is a constant for all ϵ and k_i.

Figure 5.5 shows the dependence of the common factor amplitude $r(z)$ on the parameters k and ϵ, respectively. Plots (a), (b), (c), and (d) are generated with $\epsilon=4$ and $k=0.81650$, $k=1$, $k=1.4142$, and $k=4.4721$, respectively. Plots of this figure show that the heights (i.e., the amplitude) and the width of the curve increase as k increases, and the centre of the curve (i.e., the centre of symmetry of the curve) moves in the $-x$ direction as k increases. This situation is confirmed by Fig. 5.6

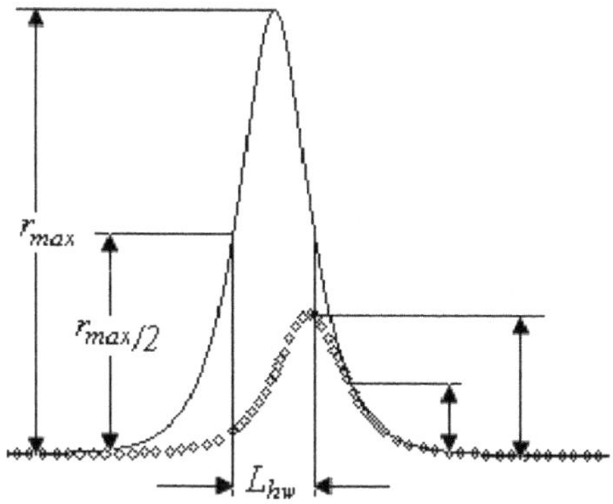

Fig. 5.4 The common factor amplitudes $r(z)$ with their corresponding maximal amplitude and width for $\epsilon = 4$ and for $k = 1.4142$ (solid line) and $k = 4.4721$ (dotted line)

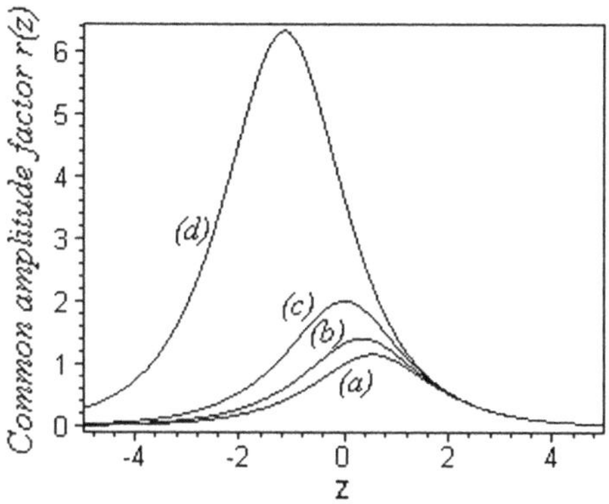

Fig. 5.5 Dependence of the common factor amplitude $r(z)$ on parameter k_i when $\epsilon = 4$. Curves **a**, **b**, **c**, and **d** correspond to $k = 0.816\,50$, $k = 1$, $k = 1.4142$, and $k = 4.4721$

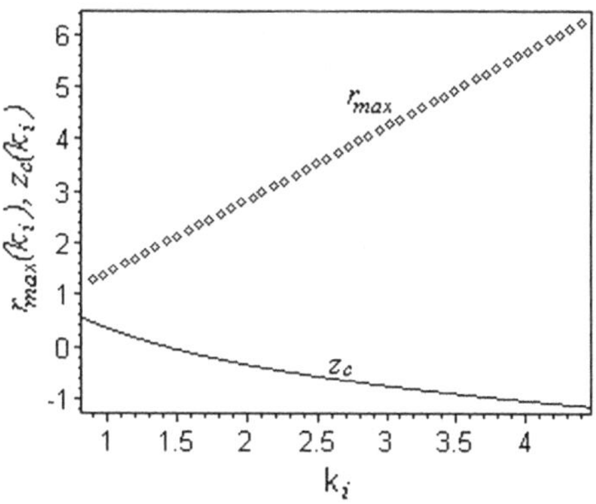

Fig. 5.6 Dependence of the centre z_c (solid line) and the amplitude $r_{\max}$ (dotted line) of the common amplitude $r(z)$ as function of the parameter k_i for $k_i \in [0.81650, 4.472\,1]$ and for $\epsilon = 4$

where we plot the centre $z_c = \ln \frac{\epsilon}{2\sqrt{2}k_i}$ (solid line) and the amplitude $r_{\max} = \sqrt{2}k_i$ (dotted line) of the common amplitude $r(z)$ as a function of the parameter k_i for $k_i \in [0.81650, 4.472\,1]$ and for $\epsilon = 4$.

As the expressions for z_c and $r_{\max}$ show, the centre of the common amplitude $r(z)$ depends both on ϵ and k, while its amplitude is constant with respect to ϵ. Thus if we need a common amplitude $r(z)$ with a given amplitude ($r_{\max}$) and a given centre (z_c), then we first vary k_i to obtain $r_{\max} : r_{\max} = \sqrt{2}k_{ir\,\max}$, which implies that $k_{ir\,\max} = r_{\max}/\sqrt{2}$ and, secondly, we insert k_i in the expression for z_c and vary ϵ in order to obtain the desired centre of symmetry.

If we take $\epsilon = 2k_i$ in the special case when k and k_{10} are pure imaginary numbers, (i.e., $k_r = k_{10r} = 0$), then Eqs. (5.93) and (5.94) give

$$\begin{pmatrix} u(x,t) \\ v(x,t) \end{pmatrix} = \begin{pmatrix} \dfrac{\sqrt{2}k_i}{\cosh[k_i x]} \\ \pm\dfrac{\sqrt{2}k_i}{\sqrt{a}\cosh[k_i x]} \end{pmatrix} \exp\left[i\left(-k_i^2 t + k_{10i}\right)\right]. \tag{5.95}$$

If we take either $k_i = U_{10}/\sqrt{2}$ or $k_i = U_{20}/\sqrt{2}$ (where U_{10} and U_{20} are given by Eqs. (5.81) and (5.83)), then Eq. (5.95) gives

$$\begin{pmatrix} u(x,t) \\ v(x,t) \end{pmatrix} = \begin{pmatrix} \dfrac{U_{10}}{\cosh\left[\frac{U_{10}}{\sqrt{2}}x\right]} \\ \dfrac{U_{10}}{\sqrt{a}\cosh\left[\frac{U_{10}}{\sqrt{2}}x\right]} \end{pmatrix} \exp\left(-i\frac{U_{10}^2}{2}t + ik_{10r}\right) \tag{5.96}$$

and

$$\begin{pmatrix} u(x,t) \\ v(x,t) \end{pmatrix} = \begin{pmatrix} \dfrac{U_{20}}{\cosh\left[\frac{U_{20}}{\sqrt{2}}x\right]} \\ -\dfrac{U_{20}}{\sqrt{a}\cosh\left[\frac{U_{20}}{\sqrt{2}}x\right]} \end{pmatrix} \exp\left[i\left(-\frac{U_{20}^2}{2}t + k_{10r}\right)\right], \tag{5.97}$$

and respectively. Moreover, solutions (5.96) correspond to $v(x,t) = u(x,t)/\sqrt{a}$, which implies, according to Eq. (5.88), that $\mu_0 = \frac{1}{\sqrt{a}} - \sqrt{a}$, while solutions (5.97) correspond to $v(x,t) = -u(x,t)/\sqrt{a}$, and this also gives $\mu_0 = \sqrt{a} - \frac{1}{\sqrt{a}}$. Thus we can conclude that the vector amplitudes of the traveling waves (5.96) and (5.97) coincide with the in- and the out-of-phase solitons given by Eqs. (5.81) and (5.83), respectively. In other words, the in-phase and the out-of-phase solitons (5.81) and (5.83) are the vector amplitudes of some vector traveling wave solutions of system (5.73).

References

1. Benjamin, T.B., Feir, J.F.: Ginzburg-Landau system of complex modulation equations for a distributed nonlinear-dispersive transmission line. J. Fluid Mech. **27**, 417 (1967)
2. Cornish, S.L., Claussen, N.R., Roberts, J.L., Cornell, E.A., Wieman, C.E.: Stable 85Rb Bose-Einstein condensates with widely tunable interactions. Phys. Rev. Lett. **85**, 1795 (2000)
3. Duan, W.-S.: Nonlinear waves propagating in the electrical transmission line. Europhys. Lett. **66**, 192 (2004)
4. Duan, W.-S., Hong, X.-R., Shi, Y.-R., Lu, K.-P., Sun, J.-A.: Weakly two-dimensional solitary waves on coupled nonlinear transmission lines. Chin. Phys. Lett. **19**, 1231 (2002)

5. Hirota, R.: Exact envelope-soliton solutions of a nonlinear wave equation. J. Math. Phys. **14**, 805–809 (1973)
6. Kengne, E., Chui, S.T., Liu, W.M.: Modulational instability criteria for coupled nonlinear transmission lines with dispersive elements. Phys. Rev. E **74**, 036614 (2006)
7. Manakov, S.V.: On the theory of two-dimensional stationary self-focusing of electromagnetic waves. Zh. Eksp. Theor. Fiz. **65**, 505 (1973); [Sov. Phys. JETP **38**, 248 (1974)]
8. Ody, M.S., Common, A.K., Sobhy, M.I.: Continuous symmetries of the discrete nonlinear telegraph equation. Eur. J. Appl. Math. **10**, 265 (1999)
9. Ostrovski, L.A., Papko, V.V., Stepanyants, Y.A.: Solitons and resonances in two dimensional lattices. Sov. Phys. JETP **51**, 417 (1980)
10. Taniuti, T., Yajima, N.: Perturbation method for a nonlinear wave modulation I. J. Math. Phys. **10**, 1369 (1969)

Chapter 6
Inhomogeneous Nonlinear Schrödinger Equations of Bose–Einstein Condensates with Two-Body Interactions

Abstract The present chapter deals with a time-dependent inhomogeneous cubic nonlinear Schrödinger equation with external real/complex potential which at ultralow temperatures, describes the time-dependent wavefunction of the Bose–Einstein condensates when only the two-body interatomic interactions are taken into account. Through the analytical investigation of this equation, we investigate (i) the generation of matter-wave solitons of BECs with time-dependent atomic scattering length in an expulsive parabolic potential, (ii) the dynamics of BECs with feeding/loss of atoms, (iii) the interaction of matter-wave solitons in BECs with time-dependent scattering length and varying potentials, and (iv) the nonautonomous matter waves in 1D BECs.

6.1 Introduction

It is well known that at low temperatures, the properties of an interacting Bose–Einstein condensate can be described by the time-dependent inhomogeneous nonlinear Schrödinger equations with external potential, known as nonlinear mean-field Gross–Pitaevskii equation, for both repulsive and attractive atomic interactions; this NLS equation describes the condensed wavefunction in the mean-field approximation. The effect of the interatomic interaction leads to a nonlinear term in the GP equation. The s-wave scattering length, $a_s(t)$, plays an important role in the description of an atom–atom interaction at ultralow temperatures. The magnitude and sign of the s-wave scattering length $a_s(t)$ can be tuned to any value, large or small, positive or negative by applying an external magnetic field. The presence of an attractive interaction ($a_s(t) < 0$) between the atoms has a profound effect on the stability of a BEC, since a large enough attractive interaction will cause the BEC to become unstable and collapse in some way. At low temperature and density, where interatomic distances are much greater than the distance scale of atom–atom interactions, the two-body interatomic interactions can be described by only the scattering length where the effects of the three-body interatomic interactions are negligible. If the atom density is considerably high, the three-body interatomic interactions can start to play an important role.

W.-M. Liu and E. Kengne, *Schrödinger Equations in Nonlinear Systems*,
https://doi.org/10.1007/978-981-13-6581-2_6

When both the two- and three-body interatomic interactions are taken into account, the time-dependent inhomogeneous nonlinear Schrödinger equation with external potential

$$i\hbar\frac{\partial\Psi(\widetilde{\mathbf{r}},t)}{\partial t}=\left[-\frac{\hbar^2\nabla^2}{2m}+\widetilde{V}(\widetilde{\mathbf{r}},t)+g(t)\left|\Psi(\widetilde{\mathbf{r}},t)\right|^2+k(t)\left|\Psi(\widetilde{\mathbf{r}},t)\right|^4\right]\Psi(\widetilde{\mathbf{r}},t), \tag{6.1}$$

will describe, at ultralow temperatures, the time-dependent wavefunction of the condensates $\Psi(\widetilde{\mathbf{r}},t)$ at position $\widetilde{\mathbf{r}}$ and time t [8, 10, 11, 20, 21]. In Eq. (6.1), $\widetilde{V}(\widetilde{\mathbf{r}},t)$ is the time-dependent external trapped potential and $g(t)=4\pi\hbar^2 a_s(t)/m$ and $k(t)$ are the strengths of time-dependent two-body and three-body interatomic interactions, respectively. $\hbar$ is Planck's constant, m is the mass of the single bosonic atom, and $a_s(t)$ is the time-dependent s-wave scattering length which can be tuned to any desired value by using the Feshbach resonance technique. $a_s(t)>0$ corresponds to the BEC with a repulsive two-body interaction between the atoms and $a_s(t)<0$ to an attractive two-body interaction. The normalization condition is $N^{-1}\int\left|\Psi(\widetilde{\mathbf{r}},t)\right|^2 d\widetilde{\mathbf{r}}=1$, where N is the total time-dependent number of atoms in the condensate. In this chapter, we mainly work with potential $\widetilde{V}(\widetilde{\mathbf{r}},t)=m\omega^2\left(y^2+z^2\right)/2+\widetilde{\alpha}(t)\left[m\omega_1^2x^2/2+V_1(x,t)\right]$, where $V_1(x,t)$ is a bounded complex potential or the optical lattice potential and $\widetilde{\alpha}(t)$ describes the time dependence of the potential.

In the simple but physically important case of the cigar-shaped trapping potential, the inhomogeneous NLS Eq. (6.1) can be integrated out, resulting in the quasi-one-dimensional dimensionless equation [8, 10, 11]

$$i\frac{\partial\psi}{\partial t}=\left[-\frac{\partial^2}{\partial x^2}+V(x,t)+g_0(t)\left|\psi\right|^2+\chi_0(t)\left|\psi\right|^4\right]\psi, \tag{6.2}$$

$$V(x,t)=\alpha(t)x^2+\lambda(t)x+i\gamma(t), \tag{6.3}$$

where the time t and the space coordinate x are measured in harmonic oscillator units, that is, in units of $2/\omega_\perp$ and $a_\perp$, respectively, where $a_\perp=\sqrt{\frac{\hbar}{m\omega_\perp}}$ and $a_0=\sqrt{\frac{\hbar}{m\omega_0}}$ are linear oscillator lengths in the transverse and cigar-axis directions, respectively. $\omega_\perp$ and ω_0 are corresponding axial oscillation frequencies. The quadratic term of Eq. (6.3) is the most physically relevant example of an external potential in the BEC case, giving the harmonic confinement of atoms by experimentally used magnetic traps. The strength of the magnetic trap $\alpha(t)$ may be positive (confining potential) or negative (repulsive potential). The linear term $\lambda(t)x$ of the potential may correspond to the gravitational field or some linear potentials. $\gamma(t)$ is a small parameter related to the feeding ($\gamma>0$) or loss ($\gamma<0$) of atoms in the condensate resulting from the contact with the thermal cloud and three-body recombination [14, 18, 39].

Throughout the present chapter, we focus our attention on BECs with only the two-body interatomic interactions; in other words, we are interested in this chapter to Eq. (6.2) with $\chi_0(t) = 0$. The normalized macroscopic wavefunction $\psi(x, t)$ is connected to the original order parameter $\Psi(\widetilde{\mathbf{r}}, t)$ as follows:

$$\Psi(\widetilde{\mathbf{r}}, t) = \frac{1}{\sqrt{2\pi a_B} a_\perp} \psi\left(\frac{x}{a_\perp}, \frac{\omega_\perp}{2} t\right) \exp\left[-i\omega_\perp t - \frac{y^2 + z^2}{2a_\perp}\right],$$

where a_B is the Bohr radius [31]. From the viewpoint of stability, 3D and 1D equations are very different. For a true 1D system, one does not expect the collapse of the BEC system when increasing the number of atoms [27, 32]. However, it happens that a realistic 1D limit is not a true 1D system, with the density of particles still increasing due to the strong restoring forces in the perpendicular directions. To avoid the collapse of the bright soliton [13], we must, throughout this chapter, restrict our investigation of BECs to a range of parameters, in which the system becomes effectively 1D [6].

6.2 Generation of Matter-Wave Solitons of BECs with Time-Dependent Atomic Scattering Length in an Expulsive Parabolic Potential

The present section deals with the analytical study of dynamics of matter-wave solitons in BECs whose wavefunction in the mean-field approximation is governed by the inhomogeneous cubic NLS equation with external expulsive parabolic potential

$$i\frac{\partial\psi}{\partial t} + \frac{\partial^2\psi}{\partial x^2} + g(t)\left|\psi\right|^2\psi + \frac{1}{4}\lambda^2 x^2\psi = 0, \tag{6.4}$$

where $0 < \lambda = 2\left|\omega_0\right|/\omega_\perp << 1$ and $g(t) = 2\left|a_s(t)\right|/a_B = 2g_0 \exp[\lambda t]$. In this section, we restrict our study of BECs to a safe range of parameters in which the energy of two-body interactions is much less than the kinetic energy in the transverse direction [6, 27]: $\epsilon^2 = a_\perp/\xi^2 \sim N\left|a_s\right|/a_0 << 1$, ξ being the healing length. The bright soliton in BECs has been created with the parameters of $N \approx \times 10^3$, $\omega_\perp \approx 2\pi \times 700\,\text{Hz}$ and $\omega_0 \approx 2\pi \times 7\,\text{Hz}$, and $a_{fin} = -4\ a_B$ for ^{7}Li [20], which provides a safe range of parameters. With the same experimental conditions in Ref. [20] and $a_s(t = 0) = -0.25\ a_B$, we can evaluate $\epsilon^2 = a_\perp/\xi^2 \sim N\left|a_s\right|/a_0 \approx 9.5 \times 10^{-3} << 1$. Then, the scattering length is increased in the form of $a(t) = g_0 \exp[\lambda t]$. After at least up to 50 dimensionless units of time, the absolute value of the atomic scattering length turns to $\left|a_s(t)\right| = 0.8\ a_B$, corresponding to $\epsilon^2 = a_\perp/\xi^2 \sim N\left|a_s\right|/a_0 \approx 3 \times 10^{-2} << 1$. Under the above conditions, the system is effectively 1D. So a safe range of parameters can be described as follows:

(i) with the same experimental conditions in Ref. [20];
(ii) ramp up the scattering length in the form of $a(t) = g_0 \exp[\lambda t]$ within 50 dimensionless units of time.

Also, we have to specify the terminology long-time dynamics. A unity of time, $\Delta t = 1$ in the dimensionless variables, corresponds to $2/\omega_\perp$ real seconds. This means, for example, that for a BEC in a trap with transversal size of the order of $a_\perp = 1.4$ μm, a unity of the dimensionless time corresponds to 4.5×10^{-4} s. The lifetime of a BEC in today's experiments is of the order of about 1 s, which is about 220 in our dimensionless units.

6.2.1 Study of the Modulational Instability

The modulational instability is an indispensable mechanism for understanding these relevant dynamic processes in the BEC systems, which include domain formation, generation and propagation of solitonic waves, and quantum phase transition. It indicates that due to the interplay between nonlinearity and the dispersive effects, a small perturbation on the envelope of a plane wave may induce an exponential growth of the wave amplitude, resulting in the carrier-wave breakup into a train of localized waves.

In order to examine the MI of BECs described by Eq. (6.4), we first perform a modified lens-type transformation of the form

$$\psi(x,t) = \frac{1}{\sqrt{\ell(t)}}\phi(X,T)\exp\left[if(t)x^2\right] \tag{6.5}$$

in which T, ℓ, and f are real functions of time t and $X = x/\ell(t)$. We demand that

$$\frac{dT}{dt} = \frac{1}{\ell^2}, \tag{6.6a}$$

$$\frac{1}{\ell}\frac{d\ell}{dt} - 4f = 0, \tag{6.6b}$$

$$\frac{df}{dt} + 4f^2 - \frac{1}{4}\lambda^2 = 0. \tag{6.6c}$$

The choice of Eq. (6.6a) is made to preserve the scaling. To investigate the MI of BECs through transformation (6.5), we impose to $T(t)$ to vary from zero to $+\infty$. Because λ is a positive real constant, Eq. (6.6c) possesses the special solutions $f(t) = \pm\lambda/4$. Inserting these solutions into Eqs. (6.6a) and (6.6b) yields $\ell(t) = \ell_0 \exp[\pm\lambda t]$ and $T(t) = \mp\frac{1}{2\lambda\ell_0^2}(\exp[\mp 2\lambda t] - 1)$, respectively. In order to that $T(+\infty) = +\infty$, we must have $T(t) = \frac{1}{2\lambda\ell_0^2}(\exp[2\lambda t] - 1)$, which corresponds to $f(t) = -\lambda/4$ and $\ell(t) = \ell_0 \exp[-\lambda t]$. Inserting transformation (6.5) into Eq. (6.4) and taking into account Eqs. (6.6a)–(6.6c) yield the standard NLS equation

$$i\frac{\partial\phi}{\partial T} + \frac{\partial^2\phi}{\partial X^2} + 2g_0\ell_0\left|\phi\right|^2\phi = 0. \tag{6.7}$$

Because $g_0\ell_0 > 0$, the plane-wave solution of Eq. (6.7) is unstable under the modulation. Therefore, the BECs described by Eq. (6.4) is unstable under modulation and admits envelope solitons (bright solitons).

6.2.2 Exact Bright Soliton Solutions

The analytical study of the dynamics of the bright soliton in BECs described by Eq. (6.4) is carried out through the analytical bright soliton solutions of the cubic NLS Eq. (6.4). For this purpose, we use the method of Darboux transformation [29], applied on the standard NLS Eq. (6.7). If we choose ℓ_0 from $2g_0\ell_0 = 1$, Eq. (6.7) will possess the plane-wave solution $\phi_c(X, T) = \widetilde{A}_c \exp[i\widetilde{\varphi}_c(X, T)]$, where $\widetilde{\varphi}_c(X, T) = \frac{\sqrt{2}k_c}{2}X + \frac{(2\widetilde{A}_c^2 - k_c^2)}{2}T$ and $\widetilde{A}_c$ is an arbitrary real constant. Inserting this plane-wave solution into transformation (6.5) leads to the following special solution of Eq. (6.4):

$$\psi_0(x, t) = A_c \exp\left[\frac{\lambda}{2}t + i\varphi_c(x, t)\right], \tag{6.8}$$

where

$$\varphi_c(x, t) = k_0 x \exp[\lambda t] - \frac{\lambda}{4}x^2 + \frac{k_0^2 - 2g_0 A_c^2}{2\lambda}(1 - \exp[2\lambda t]),$$

and $A_c = \sqrt{2g_0}\widetilde{A}_c$ and $k_0 = \sqrt{2}g_0 k_c$ are arbitrary real constants. Performing the Darboux transformation [29] and using solution $\phi_c(X, T)$ of Eq. (6.7) as the seed solution, we obtain an exact bright soliton solution of Eq. (6.7) as follows:

$$\phi(X, T) = \left[\widetilde{A}_c + A_s \frac{\gamma_0 \cosh\theta + \cos\varphi + i(\alpha_0 \sinh\theta + \beta\sin\varphi)}{\cosh\theta + \gamma_0\cos\varphi}\right]\exp[i\widetilde{\varphi}_c(X, T)], \tag{6.9}$$

where

$$\begin{aligned}
\theta &= \frac{M_R}{\sqrt{2}}X - \frac{(k_c + k_s)M_R - A_s M_I}{2}T, \\
\varphi &= \frac{M_I}{\sqrt{2}}X - \frac{(k_c + k_s)M_I + A_s M_R}{2}T, \\
\gamma_0 &= \frac{\widetilde{A}_c(M_R - A_S)}{\Lambda}, \quad \alpha_0 = \frac{\widetilde{A}_c(k_s - k_c + M_I)}{\Lambda}, \quad \beta = 1 - \frac{2\widetilde{A}_c^2}{\Lambda},
\end{aligned} \tag{6.10}$$

with $M = \sqrt{(-A_s + i(k_s - k_c))^2 - 4\widetilde{A}_c^2} = M_R + iM_I$ and $\Lambda = \widetilde{A}_c^2 + \frac{(M_R - A_s)^2}{4} + \frac{(k_s - k_c + M_I)^2}{4}$. Inserting Eq. (6.10) into transformation (6.5) and using the relationships $A_c = \sqrt{2g_0}\widetilde{A}_c$ and $k_0 = \sqrt{2}g_0 k_c$ lead to the following bright soliton solution of Eq. (6.4):

$$\psi(x,t) = \left[A_c + A_s \frac{\gamma_0 \cosh\theta + \cos\varphi + i\left(\alpha_0 \sinh\theta + \beta \sin\varphi\right)}{\cosh\theta + \gamma_0 \cos\varphi}\right] \exp\left[\frac{\lambda}{2}t + i\varphi_c(x,t)\right], \tag{6.11}$$

where

$$\begin{aligned}
\theta &= \frac{\left[(k_0 + k_s)\,\Delta_R - \sqrt{g_0}A_s\Delta_I\right]\left[1 - \exp\left[2\lambda t\right]\right]}{2\lambda} + \Delta_R x \exp\left[\lambda t\right], \\
\varphi &= \frac{\left[(k_0 + k_s)\,\Delta_I - \sqrt{g_0}A_s\Delta_R\right]\left[1 - \exp\left[2\lambda t\right]\right]}{2\lambda} + \Delta_I x \exp\left[\lambda t\right], \\
\alpha_0 &= \frac{\sqrt{g_0}A_c\left(k_0 - k_s + \Delta_I\right)}{\Lambda},\ \beta = 1 - \frac{2g_0A_c^2}{\Lambda}, \\
\gamma_0 &= \frac{A_c\sqrt{g_0}\left(\Delta_R - A_s\sqrt{g_0}\right)}{\Lambda},
\end{aligned} \tag{6.12}$$

with

$$\begin{aligned}
\Delta &= \sqrt{\left[-A_s\sqrt{g_0} + i(k_s - k_0)\right]^2 - 4g_0A_c^2} = \Delta_R + i\Delta_I, \\
\Lambda &= g_0A_c^2 + \frac{\Delta_R - A_s^2\sqrt{g_0}}{4} + \frac{(k_s - k_0 + \Delta_I)^2}{4},
\end{aligned} \tag{6.13}$$

k_s being an arbitrary real constant.

The solution (6.11) has the following properties. When A_s vanishes, it turns into the seed solution (6.8). On the other hand, when $A_c = k_0 = 0$, it turns into the bright soliton solution

$$\begin{aligned}
\psi_s(x,t) &= \frac{A_s}{\cosh\theta_s(x,t)} \exp\left[\frac{\lambda}{2}t + i\varphi_s(x,t)\right], \\
\theta_s(x,t) &= -\sqrt{g_0}A_s x \exp\left[\lambda t\right] + \frac{k_s A_s\sqrt{g_0}}{\lambda}\left(\exp\left[2\lambda t\right] - 1\right), \\
\varphi_s(x,t) &= \varphi_c + \frac{g_0A_s^2}{2\lambda}\left(1 - \exp\left[2\lambda t\right]\right).
\end{aligned} \tag{6.14}$$

Thus in general, the exact solution (6.11) represents a bright soliton embedded in the background $\psi_0(x,t)$.

6.2.3 *Dynamics of a Bright Soliton in BECs*

Now, we turn to the dynamics of a bright soliton in BECs described by Eq. (6.4). Considering the dynamics of the bright soliton in the background, the length $2L$ of the spatial background must be very large compared to the scale of the soliton. In the real experiment [36], the length of the background of BECs can reach at least $2L = 370$ μm. At the same time, in Fig. 6.1, the width of the bright soliton is about $2l = 2 \times$

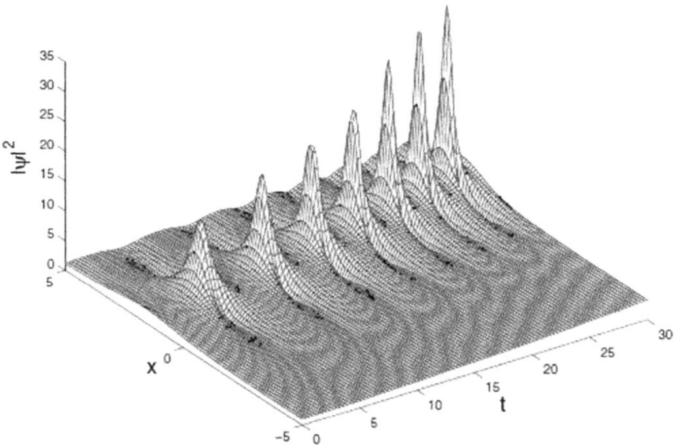

Fig. 6.1 The dynamics of the Feshbach resonance managed soliton in the expulsive parabolic potential given by Eq. (6.15). The parameters are given as follows: $\lambda = 2 \times 10^{-2}$, $g_0 = 0.25$, $A_c = 1$, $A_s = 2.4$, and $k_0 = 0.03$. Reprint from Ref. [27], Copyright2018, with permission from American Physical Society

1.4 μm [a unity of coordinate, $\Delta x = 1$ in the dimensionless variables, corresponds to $a_\perp = \sqrt{\hbar/m\omega_\perp} = 1.4\,\mu\text{m}$]. So we indeed to have $l << L$, a necessary condition for realizing our soliton in experiment. By utilizing the properties of Eq. (6.11), we demonstrate that the manipulation of the scattering length can be used to compress a bright soliton of BECs into an assumed peak matter density. It has been reported that an abrupt change of the scattering length can lead to the splitting of the soliton by generating the new solitons. The fragmentation of the soliton obviously decreases the numbers of atoms of the original soliton, which is undesirable for application [7]. However, in Eq. (6.11), the change of the scattering length preserves the soliton from splitting new ones. For simplicity, we assume $k_0 = k_s$ in Eq. (6.13) and consider only the case of $A_s^2 - 4A_c^2 > 0$, for in the case of $A_s^2 - 4A_c^2 < 0$, a small perturbation for Eq. (6.11) may lead to the modulation instability [42]. With the conditions above, Eq. (6.11) can be deduced into the following form:

$$\psi(x,t) = \left[-A_c + \delta_2 \frac{\delta_2 \cos\varphi - iA_s \sin\varphi}{A_s \cosh\theta - 2A_c \cos\varphi}\right] \exp\left[\frac{\lambda}{2}t + i\varphi_c(x,t)\right], \tag{6.15}$$

where

$$\begin{aligned} \theta &= \sqrt{g_0}\delta_2 \left[x \exp[\lambda t] - \tfrac{k_0}{\lambda}(\exp[2\lambda t] - 1)\right], \\ \varphi &= -\tfrac{g_0 A_s \delta_2}{2\lambda}(\exp[2\lambda t] - 1),\ \delta_2 = \sqrt{A_s^2 - 4A_c^2}. \end{aligned} \tag{6.16}$$

For a better understanding, we plot solution (6.15) in Fig. 6.1, which shows the dynamics of the Feshbach resonance managed bright soliton in the expulsive parabolic potential. As one can see from Fig. 6.1, with the increasing absolute value of the scattering length, the bright soliton has an increase in the peaking value and a compression in the width. As a result, we can get a bright soliton with the assumed peak matter density. It is interesting to observe that in the expulsive parabolic potential, the bright soliton is set into motion and propagates in the longitudinal direction, instead of oscillating in an attractive parabolic potential. The possibility of compressing the soliton of BECs into an assumed peak matter density could provide an experimental tool for investigating the range of validity of the 1D GP equation. Since the quasi-1D GP equation applies only for low densities, it would indeed be interesting to see how far one can compress a soliton in a real experiment by increasing the absolute value of the scattering length.

Inspired by two experiments [36], we can design the compression of a bright soliton in BECs with the following steps:

(i) Create a bright soliton in BECs with the parameters of $N \approx \times 10^3, \omega_\perp = 2\pi \times 700$ Hz and $\omega_0 = 2\pi \times 7$ Hz, $a_s = -0.25\, a_B$ for ^{7}Li. The main effect of this expulsive term is that the center of the BECs accelerates along the longitudinal direction.
(ii) Under the safe range of parameters discussed above, ramp up the absolute value of the scattering length according to $a(t) = g_0 \exp[\lambda t]$ due to Feshbach resonance, where $\lambda = 2\,|\omega_0|\,/\omega_\perp = 2 \times 10^{-2}$ is a very small value. A unity of time, $\Delta t = 1$ in the dimensionless variables, corresponds to $2/\omega_\perp = 4.5 \times 10^{-4}$ real seconds.
(iii) After at least up to 50 dimensionless units of time, the absolute value of the atomic scattering length turns to $|a_s(t)| = 0.8\, a_B$, which is less than $\left|a_{final}\right| = 4\, a_B$. This means that during the process of compressing, the bright soliton the stability of soliton and the validity of 1D approximation can be kept as displayed in Fig. 6.1. Therefore, the phenomena discussed in this section should be observable within the current experimental capability.

Furthermore, based on Eq. (6.15), we find that the peak matter density of bright soliton arrives at the maximum when $\sinh\theta = 0$ and arrives at the minimum when $\cosh\theta = \frac{A_s}{A_c\cos\varphi} - \frac{2A_c\cos\varphi}{A_s}$:

$$|\psi|^2_{\max} = \exp[\lambda t]\left(A_c^2 + \frac{\delta_2 A_s}{A_s - 2A_c\cos\varphi}\right), \tag{6.17a}$$

$$|\psi|^2_{\min} = \exp[\lambda t]\left(A_c^2 - \frac{A_c^2\delta_2^2\cos^2\varphi}{A_s^2 - 4A_c^2\cos^2\varphi}\right). \tag{6.17b}$$

Equations (6.17a) and (6.17b) mean that the bright soliton can only be squeezed into the assumed peak matter density between the minimum and maximum values. In order to investigate the stability of the bright soliton against the variation of the scattering length in the expulsive parabolic potential, we obtain

$$\lim_{L\to+\infty}\int_{-L}^{+L}\left[|\psi(x,t)|^2 - |\psi(\pm L,t)|\right]dx = \frac{2\delta_2}{\sqrt{g_0}}, \tag{6.18}$$

which is the exact number of the atoms in the bright soliton against the background described by Eq. (6.15). This indicates that during the process of the compression

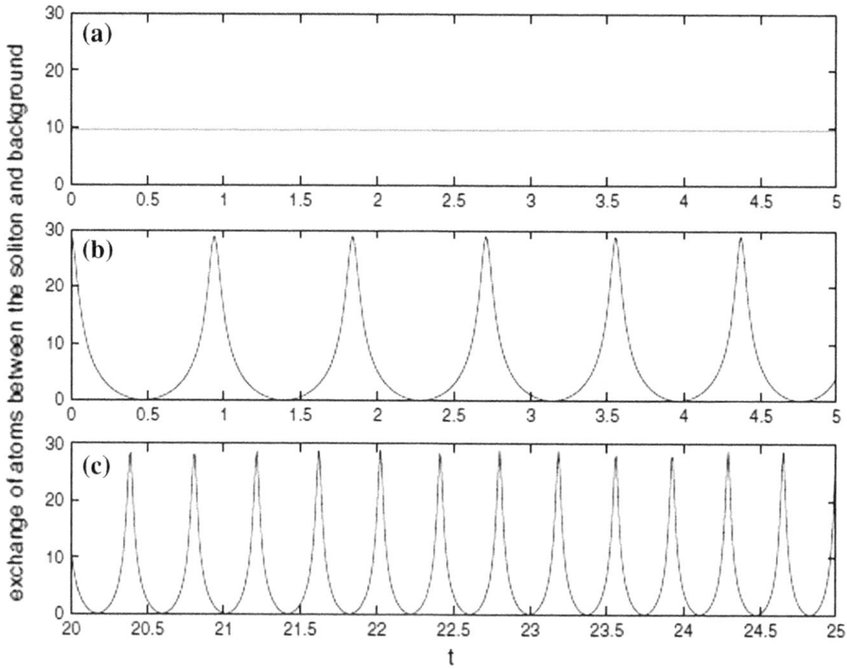

Fig. 6.2 Time-periodic atomic exchange between the bright soliton and the background given by Eq. (6.19). The range of time is **a**, **b** $t \in [0,\ 5]$ and **c** $t \in [20,\ 25]$. The parameters are given as follows: $\lambda = 0.02$, $g_0 = 1$, $A_s = 4.8$, **a** $A_c = 0$, and **b**, **c** $A_c = 2.3$. Reprint from Ref. [27], Copyright2018, with permission from American Physical Society

of the bright soliton, the number of atoms in the bright soliton remains invariant. In contrast, the quantity

$$\kappa = \lim_{L \to +\infty} \int_{-L}^{+L} |\psi(x,t) - \psi(\pm L, t)|^2 \, dx = \frac{2\delta_2}{\sqrt{g_0}} \left(1 + A_c \widetilde{M} \cos\varphi\right), \qquad (6.19)$$

where

$$\widetilde{M} = \frac{4 \arctan\left[\sqrt{A_s + 2A_c \cos\varphi} / \sqrt{A_s - 2A_c \cos\varphi}\right]}{\sqrt{(A_s + 2A_c \cos\varphi)\,(A_s - 2A_c \cos\varphi)}} \qquad (6.20)$$

counts the number of atoms in both the bright soliton and background under the condition of $\psi(\pm L, t) \neq 0$. Equation (6.19) displays that a time-periodic atomic exchange is formed between the bright soliton and the background. As shown in Fig. 6.2a, in the case of zero background, i.e., $A_c = 0$, there will not be the exchange of atoms. However, in the case of nonzero background as shown in Fig. 6.2b, c, the exchange of atoms between the bright soliton and the background becomes quicker when increasing the absolute value of the scattering length. The conclusion can be made that the number of atoms in the bright soliton in BECs keeps dynamic stability against the variation of the scattering length in the expulsive parabolic potential. The consideration provided above implies that we should construct the bright soliton in BECs on the background. A question arises about the possibility of the creation of such a state experimentally. We notice that Eq. (6.15) will take the particular form at the time

$$t_0 = \frac{1}{2\lambda} \ln \left[1 - \frac{\lambda (4n+1)}{g_0 A_s \delta_2} \right], \; n = -1, \; -2, \; \ldots$$

as follows:

$$\begin{aligned} \psi(x,t) &= -A_c \exp\left[\frac{\lambda t_0}{2}\right] \exp\left[i\varphi_c\right] - i\frac{\delta_2}{\cosh\theta} \exp\left[\frac{\lambda t_0}{2}\right] \exp\left[i\varphi_c\right] \\ &= -\psi_0(x,t_0) - \psi_{\text{soliton}}(x,t_0), \end{aligned} \quad (6.21)$$

which is the linear combination of the seed solution (6.8) and a bright soliton. So Eq. (6.21) means that Eq. (6.15) can be generated by coherently adding a bright soliton into the background of density described by Eq. (6.8).

6.3 Dynamics of BECs with Feeding/Loss of Atoms

This section deals with the dynamics of BECs with either the feeding or the loss of atoms whose wavefunction $\psi(x,t)$ is described by the cubic inhomogeneous NLS equation [18, 19]

$$i\frac{\partial \psi}{\partial t} + \frac{\partial^2 \psi}{\partial x^2} + 2g_0 \left|\psi\right|^2 \psi + \frac{1}{4}\lambda^2 x^2 \psi = \mp \frac{\lambda}{2} i\psi, \quad (6.22)$$

where $0 < \lambda = 2\left|\omega_0\right| / \omega_\perp << 1$ and g_0 is a real constant. The right-hand side (rhs) in Eq. (6.22) involves the nonconservative time-independent quantity $\mp\frac{\lambda}{2}$ (called the feeding parameter for $+\frac{\lambda}{2}$, and loss parameter for $-\frac{\lambda}{2}$), related to the feeding/loss of the condensate from the thermal cloud [14, 18]. The nonconservative time-independent quantity $\mp\frac{\lambda}{2}$ is present to take into account the decrease (increase) of the density due to the feeding (loss) process from the nonequilibrium thermal cloud. In the present section, the same range of parameters as in the previous section is used. In order to study the dynamics of BECs with feeding of atoms described by the NLS Eq. (6.22), we first find the soliton solution of this equation.

6.3.1 *Exact Soliton Solutions for BECs with Feeding of Atoms*

As in the previous section, we start by introducing the ansatz

$$\psi(x,t) = \frac{1}{\sqrt{\ell(t)}}\phi(X,T)\exp\left[\eta + if(t)x^2\right], \tag{6.23}$$

where $X = x/\ell(t)$, and $\ell(t), T(t), \eta(t)$, and $f(t)$ are real functions of time t satisfying the conditions

$$\frac{dT}{dt} = \frac{1}{\ell^2}, \tag{6.24a}$$

$$\frac{1}{\ell}\frac{d\ell}{dt} - 4f = 0, \tag{6.24b}$$

$$\frac{d\eta}{dt} - \frac{\lambda}{2} = 0, \tag{6.24c}$$

$$\frac{df}{dt} + 4f^2 - \frac{1}{4}\lambda^2 = 0. \tag{6.24d}$$

Because λ is a real constant, Eq. (6.24d) possesses the special solutions $f(t) = \pm\lambda/4$. Integrating system (6.24a)–(6.24c) with $f(t) = \pm\lambda/4$ yields $\ell(t) = \ell_0 \exp[\pm\lambda t]$, $T(t) = \pm\frac{1}{2\lambda\ell_0^2}(1 - \exp[\mp 2\lambda t])$, $\eta(t) = \mp\frac{\lambda}{2}t + \ln\sqrt{\eta_0}$, where $\ell_0 > 0$ and $\eta_0 > 0$ are real constants of integration. Inserting ansatz (6.23) into Eq. (6.22) and using the above solutions of system (6.24a)–(6.24d) lead to the standard cubic NLS equation

$$i\frac{\partial\phi}{\partial T} + \frac{\partial^2\phi}{\partial X^2} + p^2|\phi|^2\phi = 0, \tag{6.25}$$

if $f(t) = +\lambda/4$ and $f(t) = -\lambda/4$ are associated respectively with BEC with loss of atoms and with BEC with gain of atoms. In Eq. (6.25), $p^2 = \text{sign}(g_0)$, meaning that η_0 and ℓ_0 are chosen from condition $2g_0\eta_0\ell_0 = \text{sign}(g_0)$. Applying the Darboux transformation, the bright soliton ($p = 1$) solution of Eq. (6.25) is given by Eq. (6.9), while its dark soliton ($p^2 = -1$) solution is given by

$$\phi(X,T) = -\frac{\sqrt{2}}{2}(A_c + iA_s\tanh\zeta)\exp[i\Delta_2], \tag{6.26}$$

with

$$\zeta = -\frac{1}{2}A_sX + \left(\frac{A_sA_c}{2} + k_1A_s\right)T, \ \Delta_2 = k_1X - \left(\frac{(A_s^2 + A_c^2)}{2} + k_1^2\right)T, \tag{6.27}$$

where A_c, A_s, and k_1 are real constants. Returning to the original variables x and t through ansatz (6.23), we obtain the following soliton solutions of Eq. (6.22):

$$\psi_b(x,t) = \left[A_c + A_s\frac{\gamma_0\cosh\theta + \cos\varphi + i(\alpha_0\sinh\theta + \beta\sin\varphi)}{\cosh\theta + \gamma_0\cos\varphi}\right]\frac{\exp[\mp\lambda t + i\varphi_c]}{\ell_0\sqrt{2g_0}}, \tag{6.28}$$

if $g_0 > 0$, and

$$\psi_d(x,t) = -\frac{1}{2\ell_0\sqrt{-g_0}}\left(A_c + iA_s \tanh\zeta\right)\exp\left[\mp\lambda t + i\varphi_2\right], \tag{6.29}$$

if $g_0 < 0$, where

$$\begin{aligned}
\theta &= \frac{M_R}{\ell_0\sqrt{2}}x\exp\left[\mp\lambda t\right] - \frac{(k_c+k_s)\,M_R - A_s M_I}{4\lambda\ell_0^2}\left(\exp\left[\mp 2\lambda t\right] \pm 1\right),\\
\varphi &= \frac{M_I}{\ell_0\sqrt{2}}x\exp\left[\mp\lambda t\right] - \frac{(k_c+k_s)\,M_I + A_s M_R}{4\lambda\ell_0^2}\left(\exp\left[\mp 2\lambda t\right] \pm 1\right),\\
\varphi_c &= \frac{k_c}{\ell_0\sqrt{2}}x\exp\left[\mp\lambda t\right] \pm \frac{\lambda}{4}x^2 + \frac{2A_c^2 - k_c^2}{4\lambda\ell_0^2}\left(\exp\left[\mp 2\lambda t\right] \pm 1\right),
\end{aligned} \tag{6.30}$$

$$\begin{aligned}
\gamma_0 &= \frac{A_c\left(M_R - A_S\right)}{\Lambda},\ \alpha_0 = \frac{A_c\left(k_s - k_c + M_I\right)}{\Lambda},\ \beta = 1 - \frac{2A_c^2}{\Lambda},\\
\zeta &= -\frac{A_s}{2\ell_0}x\exp\left[\mp\lambda t\right] + \frac{A_s\left(A_c + 2k_c\right)}{4\lambda\ell_0^2}\left(\exp\left[\mp 2\lambda t\right] \pm 1\right),\\
\varphi_2 &= \frac{k_c}{\ell_0}x\exp\left[\mp\lambda t\right] \pm \frac{\lambda}{4}x^2 - \frac{A_s^2 + A_c^2 + 2k_c^2}{4\lambda\ell_0^2}\left(\exp\left[\mp 2\lambda t\right] \pm 1\right)
\end{aligned} \tag{6.31}$$

with $M = \sqrt{(-A_s + i\,(k_s - k_c))^2 - 4A_c^2} = M_R + iM_I$ and $\Lambda = A_c^2 + \frac{(M_R - A_s)^2}{4} + \frac{(k_s - k_c + M_I)^2}{4}$, A_c, A_s, k_c, and k_s being real constants. It is obvious that Eqs. (6.28) and (6.29) describe respectively a bright soliton and a dark soliton embedded in a background.

6.3.2 *Dynamics of a Bright Soliton in BECs with Feeding of Atoms*

Now, we study analytically, through the exact bright soliton solution (6.28), the dynamics of bright soliton in BECs with feeding of atoms described by Eq. (6.22). For the same reason introduced in the Sect. 6.2, we limit our study to the case when $k_c = k_s$ and $A_s^2 - 4A_c^2 > 0$. Under these conditions, Eq. (6.28) becomes

$$\psi(x,t) = \left[-A_c + \delta_2\frac{\delta_2\cos\varphi - iA_s\sin\varphi}{A_s\cosh\theta - 2A_c\cos\varphi}\right]\frac{\exp\left[\mp\lambda t + i\varphi_c\right]}{\ell_0\sqrt{2g_0}}, \tag{6.32}$$

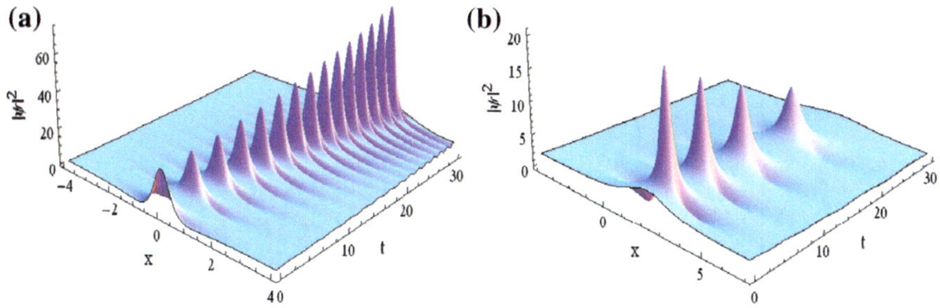

Fig. 6.3 The dynamics of the Feshbach resonance managed soliton in the expulsive parabolic potential given by Eq. (6.32). **a** Case of feeding of atoms in the condensate; **b** Case of loss of atoms in the condensate. The parameters are given as follows: $g_0 = 0.25$, $\lambda = 2 \times 10^{-2}$, $\ell_0 = 1$, $A_c = 1$, $A_s = 2.4$, and $k_c = 0.03$. Reprint from Ref. [19], Copyright2018, with permission from American Physical Society

where

$$\theta = \frac{\delta_2}{\ell_0\sqrt{2}} x \exp[\mp\lambda t] - \frac{k_c\delta_2}{2\lambda\ell_0^2}(\exp[\mp 2\lambda t] \pm 1),$$
$$\varphi = -\frac{A_s\delta_2}{4\lambda\ell_0^2}(\exp[\mp 2\lambda t] \pm 1), \tag{6.33}$$
$$\varphi_c = \frac{k_c}{\ell_0\sqrt{2}} x \exp[\mp\lambda t] - \frac{\lambda}{4}x^2 + \frac{2A_c^2 - k_c^2}{4\lambda\ell_0^2}(\exp[\mp 2\lambda t] \pm 1).$$

Analysis of Eq. (6.32) reveals that the center of the bright soliton is $\xi(t) = \frac{\sqrt{2}k_c}{\lambda\ell_0}\sinh[\lambda t]$ and satisfies the following equation $d^2\xi/dt^2 + 4k\xi(t) = 0$, where $k = -\lambda^2/4$ is the strength of the external harmonic trapping potential, as one can see from Eq. (6.22). This equation means that the center of mass of the macroscopic wave packet behaves like a classical particle, and allows one to manipulate the motion of bright solitons in the BEC systems by controlling the external harmonic trapping potential. Figure 6.3a, b shows the dynamics of the Feshbach resonance managed bright soliton with respectively the feeding and the loss of atoms in the expulsive parabolic potential for $A_s = 2.4$, $A_c = 1$, $k_c = 0.03$, $g_0 = 0.25$, $\lambda = 2 \times 10^{-2}$, and $\ell_0 = 1$. As one can see from Fig. 6.3a [from Fig. 6.3b], during its propagation, the bright soliton has an increase [an decrease] in the peaking value and a compression [enhancement] in the width. We can also see from Fig. 6.3a, b that bright solitons behave like a breather solitons propagating on a background, which corresponds to the atom exchanging between the bright soliton and the background. It is interesting to observe that in the expulsive parabolic potential, the bright soliton with either the feeding of atoms (Fig. 6.3a) or the loss of atoms (Fig. 6.3b) is set into motion and propagates in the longitudinal direction, instead of oscillating in an attractive parabolic potential.

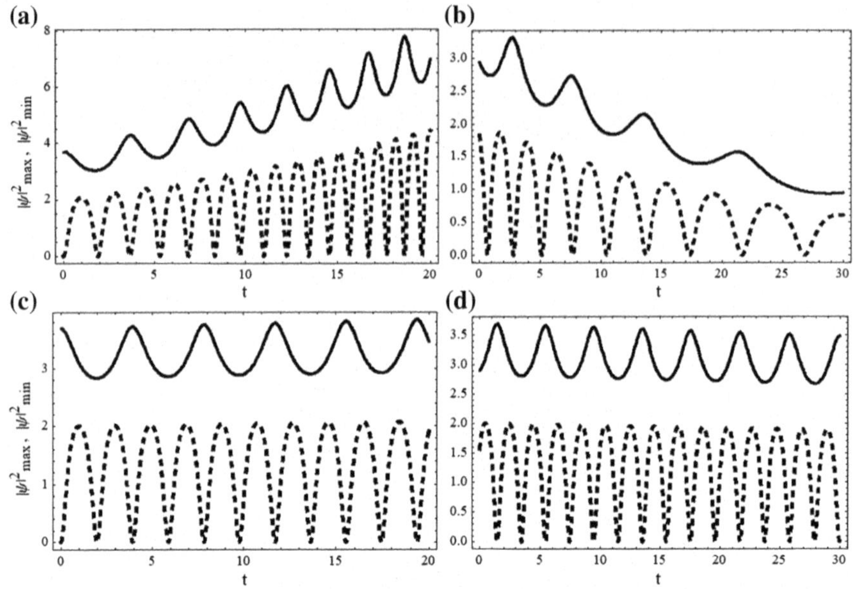

Fig. 6.4 The evolution plots of $|\psi|^2_{\max}$ (solid line) and $|\psi|^2_{\min}$ (dashed line) as function of time t. **a**, **c** Case of the feeding of atoms in the condensate; **b**, **d** Case of the loss of atoms in the condensate. Plots (**a**) and (**b**) are generated with $\lambda = 10^{-2}$, while plots (**c**) and (**d**) are obtained with $\lambda = 10^{-3}$. Other parameters are the same parameters as in Fig. 6.3

Based on Eq. (6.32), we find that the bright soliton can only be squeezed into the assumed peak matter density between the minimum and maximum values $|\psi|^2_{\min}$ and $|\psi|^2_{\max}$, obtained when $\cosh\theta = \frac{A_s}{A_c\cos\varphi} - \frac{2A_c\cos\varphi}{A_s}$ and $\sinh\theta = 0$,

$$|\psi|^2_{\min} = \frac{A_c^2}{2g_0\ell_0^2}\left(1 - \frac{\delta_2^2\cos^2\varphi}{A_s^2 - 4A_c^2\cos^2\varphi}\right)\exp\left[\mp 2\lambda t\right], \tag{6.34a}$$

$$|\psi|^2_{\max} = \frac{1}{2g_0\ell_0^2}\left(A_c^2 + \frac{\delta_2 A_s}{A_s - 2A_c\cos\varphi}\right)\exp\left[\mp 2\lambda t\right]. \tag{6.34b}$$

Plots of Fig. 6.4 present the evolution plots of the minimal and maximal intensities given by Eqs. (6.34a) and (6.34b), respectively. Figure 6.4a, b is associated with BECs with feeding and loss of atoms, respectively. From Fig. 6.4a obtained in the case of feeding of atoms, it is seen that with the time evolution, both the minimal and the maximal intensities increase. In contrast, in the case of loss of atoms in the condensate, as we can see from plots of Fig. 6.4b, both the minimal and the maximal intensities decrease with the time evolution. Meanwhile, in both cases of feeding and loss of atoms, the smaller λ, the longer the higher intensities can remain. Therefore, in order to keep a bright soliton in BECs with the feeding/loss of atoms for a longer time, we should reduce the gain/the loss of atoms and the potential intensity.

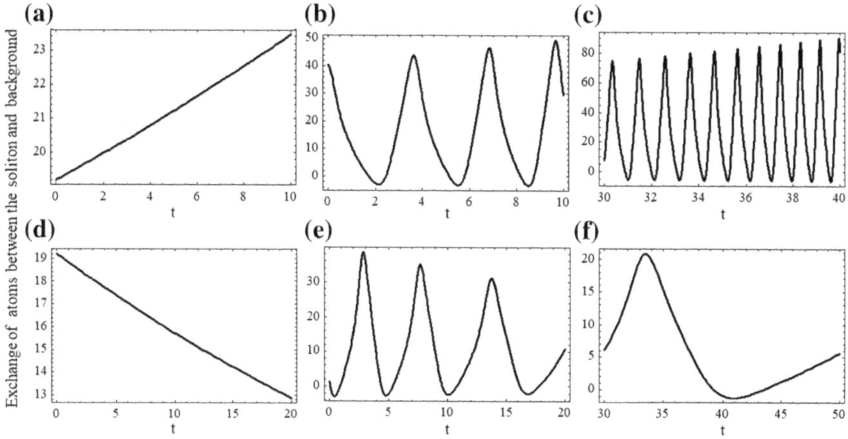

Fig. 6.5 Figure 6.6. The atomic exchange between the bright solitons and the background given by Eq. (6.35) for the same parameters as in Fig. 6.3. Plots (**a**), (**b**), and (**c**) correspond to BECs with gain of atoms, while plots (**d**), (**e**), and (**f**) are associated with BECs with loss of atoms

It follows from condition

$$\lim_{L\to+\infty}\int_{-L}^{+L}\left[|\psi(x,t)|^2-|\psi(\pm L,t)|\right]dx=\frac{\sqrt{2}\delta_2}{g_0\ell_0}\exp\left[\mp\lambda t\right]$$

which gives the exact number of the atoms in the bright soliton against the background described by Eq. (6.32) that during the process of the compression (enhancement) of the bright soliton, the number of atoms in the bright soliton increases (decreases) as the soliton propagates. In contrast, the quantity

$$\kappa=\lim_{L\to+\infty}\int_{-L}^{+L}|\psi(x,t)-\psi(\pm L,t)|^2\,dx=\frac{\sqrt{2}\delta_2}{g_0\ell_0}\exp\left[\mp\lambda t\right]\left(1+A_c\widetilde{M}\cos\varphi\right), \tag{6.35}$$

with $\widetilde{M}$ defined in Eq. (6.20) in which φ is given by Eq. (6.33), gives the number of atoms in both the bright soliton and background under the condition of $\psi(\pm L,t)\neq 0$. Equation (6.35) displays that a time-periodic atomic exchange is formed between the bright soliton and the background. As we can see from Fig. 6.5a, d, in the case of zero background, i.e., when $A_c=0$, the exchange of atoms increases for BECs with feeding of atoms (Fig. 6.5a), and decreases for BECs with loss of atoms (Fig. 6.5d). In the case of nonzero background as shown in Fig. 6.5b, c in the case of feeding of atoms [in Fig. 6.5e, f in the case of loss of atoms], the exchange of atoms between the bright soliton and the background becomes quicker [slower] as the bright soliton propagates. As we can see from Fig. 6.5b, c and from Fig. 6.5e, f obtains with $A_c\neq 0$, a slow–fast–slow process of atomic exchange is performed between the bright soliton and the background, but the whole trend of the atomic exchange between the bright soliton and the background decreases (increases) for

BECs with loss (gain) of atoms. The conclusion can be made that the number of atoms continuously injected (ejected) into (from) the BEC from (into) the reservoir depends on the linear gain or loss coefficient, and cannot be controlled by applying the external magnetic field via Feshbach resonance.

If in Eq. (6.32) we set $\cos\varphi = 0$, i.e., at time $t = t_n = \mp\frac{1}{2\lambda}\ln\left[-\frac{2\lambda\ell_0^2(1+2n)\pi \pm A_s\delta_2}{A_s\delta_2}\right]$, we obtain

$$\psi(x, t_n) = -A_c \frac{\exp\left[\mp\lambda t_n + i\varphi_c\right]}{\ell_0\sqrt{2g_0}} + i\frac{(-1)^{n-1}\,\delta_2}{\cosh\theta(x, t_n)}\frac{\exp\left[\mp\lambda t_n + i\varphi_c\right]}{\ell_0\sqrt{2g_0}}, \tag{6.36}$$

$n = 0, \pm 1, \pm 2, \ldots$, which is a superposition of a background and a bright soliton. This means that Eq. (6.36) can be generated by coherently adding a bright soliton into the background.

6.4 Interaction of Matter-Wave Solitons in Bose–Einstein Condensates with Time-Dependent Scattering Length and Varying Potentials

In this section, we consider a generalized cubic NLS equation, characterized by the time-dependent scattering length and varying potentials, which governs the dynamics of an effectively 1D BECs trapped in a cigar-shaped region, which is elongated in x and tightly confined in the transverse directions y and z. For such BECs, the motional degrees of freedom in the $y - z$ plane are frozen and transverse wavefunction can be well described by the normalized ground state of the transverse potential. Thus, the longitudinal wavefunction obeys the generalized NLS equation [26]

$$i\frac{\partial\psi}{\partial t} + \frac{\alpha}{2}\frac{\partial^2\psi}{\partial x^2} + \beta\,|\psi|^2\,\psi + \left[Mx^2 + Nx + iF\right]\psi = 0, \tag{6.37}$$

where $\alpha = \alpha(t)$, $\beta = \beta(t)$, $M = M(t)$, $N = N(t)$, and $F = F(t)$ are all real functions of time t.

In the context of nonlinear optics, α and β are the group velocity dispersion and the nonlinear parameter, respectively; while in the context of BEC, α is a constant. Here, we would like to point out that in our discussion α may be chosen as either a function of t or a constant. To investigate the dynamics of soliton in BECs, Eq. (6.37) is scaled as follows: time t and coordinate x are measured in units $1/\omega_\perp$ and $a_\perp$, where $a_\perp = \sqrt{\hbar/m\omega_\perp}$ and $a_0 = \sqrt{\hbar/m\omega_0(t)}$ are linear oscillator lengths in the transverse and cigar-axis directions, respectively. $\omega_\perp$ (without loss of generality, we set $\omega_\perp =$ const) is the radial oscillation frequency and ω_0 is the axial oscillation frequency. m is the atomic mass, and ψ is measured in units of $1/\left(\sqrt{2\pi a_B}a_\perp\right)$ with a_B the Bohr radius. $\beta = -2a_s(t)/a_B$ is the time-dependent scattering length. $M = \omega_0^2(t)/(2\omega_\perp^2)$, $Nx = N(t)x$ may correspond to the gravitational field or some linear potentials; thus, $Mx^2 + Nx$ represents a time-dependent parabolic and linear external potential [3]. $F = F(t)$ is a small parameter related to the feeding of condensate from the thermal cloud [14, 18]. It is obvious that Eq. (6.37) generalizes Eqs. (6.4) and (6.22).

6.4.1 Exact Soliton Solutions

In order to investigate analytically the dynamics of matter-wave solitons in BECs described by Eq. (6.37), we need to find exact soliton solutions of the generalized NLS Eq. (6.37).

Family of Bright One-Soliton Solution and Family of Dark One-Soliton Solution In order to find exact one-soliton solutions of Eq. (6.37), we use either the direct method introduced Li et al. [25] or the Darboux transformation method [18, 27].

One-Soliton Solutions by the Direct Method [25] According to the direct method proposed by Li et al. [25], we obtain the following family of bright one-soliton solution and a family of dark one-soliton solution for Eq. (6.37) as follows:

$$\psi_{11}(x,t) = \Omega\sqrt{\frac{\alpha}{\beta}}\left(A_c + A_s\frac{\rho\cosh\xi + \cos\eta + i\left[\theta\sinh\xi + \kappa\sin\eta\right]}{\cosh\xi + \rho\cos\eta}\right)\exp\left[i\Delta_1\right], \tag{6.38}$$

if $\alpha\beta > 0$, and

$$\psi_{21}(x,t) = \Omega\sqrt{-\frac{\alpha}{\beta}}\left(A_c + iA_s\tanh\zeta\right)\exp\left[i\Delta_2\right], \tag{6.39}$$

if $\alpha\beta < 0$, with

$$\begin{aligned}
\rho &= -\frac{2A_cA_s}{A_s^2+p_2^2},\ \kappa^2 = 1 - \frac{4A_c^2}{A_s^2+p_2^2},\ \theta = -\frac{2p_2A_c}{A_s^2+p_2^2},\\
\xi &= A_s\left[\kappa\Omega x - (p_2 + A_c\theta)v_1 - \kappa v_2\right] + \xi_0,\\
\eta &= p_2\Omega x - \frac{\left(p_2^2 - A_s^2\right)\kappa}{2}v_1 - p_2v_2 + \eta_0,\\
\Delta_1 &= k_2x^2 + k_1x + A_c^2v_1 - \frac{1}{2}v_3 + \delta_{10},\\
\zeta &= A_s\Omega x + A_sA_cv_1 - A_sv_2 + \zeta_0,\\
\Delta_2 &= k_2x^2 + k_1x - \left(A_s^2 + A_c^2\right)v_1 - \frac{1}{2}v_3 + \delta_{20},
\end{aligned} \tag{6.40a}$$

where Ω, k_1, k_2, v_1, v_2, and v_3 are time-dependent functions defined from the relationships

$$\frac{dv_1}{dt} = \alpha\Omega^2,\ \frac{dv_2}{dt} = \alpha k_1\Omega,\ \frac{dv_3}{dt} = \alpha k_1^2,\ \frac{1}{2\alpha\Omega}\frac{d\Omega}{dt} + k_2 = 0,$$
$$\Omega = \frac{\beta}{\alpha}\exp\left[-\int 2F(\tau)d\tau\right],\ \frac{dk_2}{dt} + 2\alpha k_2^2 - M = 0,\ \frac{dk_1}{dt} + 2\alpha k_1 k_2 - N = 0. \tag{6.40b}$$

Here, A_c, A_s, κ, p_2, δ_{10}, δ_{20}, ξ_0, and η_0 are real parameters.

One-Soliton Solutions by the Darboux Transformation Method [18] In the special when Eq. (6.37) describes the dynamics of matter-wave solitons in BECs, we can, under some restrictions on the coefficients α, β, M, N, and F, use the method of the Darboux transformation to derive other families of one-soliton solutions. Indeed, if $\alpha = 2$, $N = 0$, and β, M, and F satisfy the condition

$$\frac{1}{2}\frac{dF}{dt} + F^2 - F\frac{1}{\beta}\frac{d\beta}{dt} - \frac{1}{4}\frac{1}{\beta}\frac{d^2\beta}{dt^2} + \frac{1}{2}\frac{1}{\beta^2}\left(\frac{d\beta}{dt}\right)^2 - M = 0, \tag{6.41a}$$

Equation (6.37) will possess the special solution

$$\psi_{c1}(x,t) = \frac{1}{\ell_0\sqrt{\beta}}A_c\exp\left[-2\int_0^t F(\tau)d\tau + i\varphi_{c1}(x,t)\right],$$

if $\beta > 0$ and

$$\psi_{c2}(x,t) = \frac{1}{\ell_0\sqrt{-\beta}}A_c\exp\left[-2\int_0^t F(\tau)d\tau + i\varphi_{c2}(x,t)\right], \tag{6.42a}$$

if $\beta < 0$, with

$$\varphi_{c1}(x,t) = \frac{k_c}{\ell_0\sqrt{2}}x\exp\left[-4\int_0^t f(\tau)d\tau\right] + \left(\frac{1}{2}F(t) - \frac{1}{4\beta}\frac{d\beta}{dt}\right)x^2$$
$$+\frac{2A_c^2 - k_c^2}{2\ell_0^2}\int_0^t \exp\left[-8\int_0^y f(\tau)d\tau\right]dy,$$
$$\varphi_{c2}(x,t) = \frac{k_c}{\ell_0\sqrt{2}}x\exp\left[-4\int_0^t f(\tau)d\tau\right] + \left(\frac{1}{2}F(t) - \frac{1}{4\beta}\frac{d\beta}{dt}\right)x^2$$
$$-\frac{k_c^2 + 2A_c^2}{2\ell_0^2}\int_0^t \exp\left[-8\int_0^y f(\tau)d\tau\right]dy,$$

where A_c and k_c are two arbitrary real constants and ℓ_0 any positive number. Using the special solutions (6.41a) and (6.42a) as seed solutions and applying the method of Darboux transformation [27], we obtain the following families of one-soliton solutions of Eq. (6.37)

$$\psi_{12}(x,t) = \frac{1}{\ell_0\sqrt{\beta}}\left(A_c + A_s\frac{\rho\cosh\xi + \cos\eta + i\left[\theta\sinh\xi + \kappa\sin\eta\right]}{\cosh\xi + \rho\cos\eta}\right)\exp\left[-2\int_0^t F(\tau)d\tau + i\Delta_{c1}\right], \tag{6.43a}$$

if $\beta > 0$, and

$$\psi_{22}(x,t) = -\frac{1}{\ell_0\sqrt{-2\beta}}\left(A_c + iA_s \tanh\zeta\right)\exp\left[-2\int_0^t F(\tau)d\tau + i\Delta_{c2}\right], \tag{6.43b}$$

if $\beta < 0$, with

$$\begin{aligned}
\xi &= \frac{M_R}{\ell_0\sqrt{2}}x\exp\left[-4\int_0^t f(\tau)d\tau\right] - \frac{(k_c+k_s)M_R - A_sM_I}{2\ell_0^2}\int_0^t \exp\left[-8\int_0^y f(\tau)d\tau\right]dy,\\
\eta &= \frac{M_I}{\ell_0\sqrt{2}}x\exp\left[-4\int_0^t f(\tau)d\tau\right] - \frac{(k_c+k_s)M_I + A_sM_R}{2\ell_0^2}\int_0^t \exp\left[-8\int_0^y f(\tau)d\tau\right]dy,\\
\Delta_{c1} &= \frac{k_c}{\ell_0\sqrt{2}}x\exp\left[-4\int_0^t f(\tau)d\tau\right] + \left(\frac{1}{2}F(t) - \frac{1}{4\beta}\frac{d\beta}{dt}\right)x^2\\
&\quad + \frac{2A_c^2 - k_c^2}{2\ell_0^2}\int_0^t \exp\left[-8\int_0^y f(\tau)d\tau\right]dy, \qquad (6.43c)\\
\zeta &= -\frac{A_s}{2\ell_0}x\exp\left[-4\int_0^t f(\tau)d\tau\right] + \frac{A_s(A_c+2k_c)}{2\ell_0^2}\int_0^t \exp\left[-8\int_0^y f(\tau)d\tau\right]dy,\\
\Delta_{c2} &= \frac{k_c}{\ell_0}x\exp\left[-4\int_0^t f(\tau)d\tau\right] + \left(\frac{1}{2}F(t) - \frac{1}{4\beta}\frac{d\beta}{dt}\right)x^2\\
&\quad - \frac{A_c^2 + A_s^2 + 2k_c^2}{2\ell_0^2}\int_0^t \exp\left[-8\int_0^y f(\tau)d\tau\right]dy,
\end{aligned}$$

where $\rho = \frac{A_c(M_R - A_s)}{\Delta}$, $\theta = \frac{A_c(k_s - k_c + M_I)}{\Delta}$, $\kappa = 1 - \frac{2A_c^2}{\Delta}$, $M = \sqrt{(-A_s + ik_s - ik_c)^2 - 4A_c^2} = M_R + iM_I$, and $\Delta = A_c^2 + \frac{(M_R - A_s)^2 + (k_s - k_c + M_i)^2}{4}$.

Families of Two-Soliton Solutions

Families of two-soliton solutions can be obtained by using either the direct method or the Hirota method. Let us seek exact soliton solutions of Eq. (6.37) in the form

$$\psi(x,t) = \frac{g_1\cosh[\theta_1]\exp[\phi_2] + g_2\cosh[\theta_2]\exp[\phi_1]}{g_3\cosh[\mu] + g_4\cosh[\nu] + g_5\cos[\phi_1 - \phi_2]}, \tag{6.44}$$

where $\theta_1 = p_1x + q_1$, $\theta_2 = p_2x + q_2$, $\mu = \theta_1 + \theta_2$, $\nu = \theta_1 - \theta_2$, $\phi_1 = h_{12}x^2 + h_{11}x + h_{10}$, $\phi_2 = h_{22}x^2 + h_{21}x + h_{20}$, $p_i = p_i(t)$, $q_i = q_i(t)$, $h_{ij} = h_{ij}(t)$, $g_i = g_i(t)$ are all real functions of time t. Inserting Eq. (6.44) into Eq. (6.37), we obtain a series of ordinary differential equations for the determination of different functional parameters $p_i(t)$, $q_i(t)$, $h_{ij}(t)$, and $g_i(t)$. Solving this system of ODEs, we obtain the following family of bright two-soliton solutions ($\alpha\beta > 0$)

$$\psi_3(x,t) = \frac{2\lambda_1\lambda_2\Omega\sqrt{\frac{\alpha}{\beta}}\left\{\cosh[\theta_1]\exp[i\phi_2] + \sigma\cosh[\theta_2]\exp[i\phi_1]\right\}}{\lambda_1^2\cosh[\mu] + \lambda_2^2\cosh[\nu] + 4\sigma\cos[\phi_1 - \phi_2]}, \tag{6.45}$$

where $\lambda_1 = \sigma - 1$, $\lambda_2 = \sigma + 1$, $\theta_1 = \sigma(\Omega x - \upsilon_2) + \theta_{10}$, $\theta_2 = \Omega x - \upsilon_2 + \theta_{20}$, $\phi_1 = k_1x^2 + k_1x - (\upsilon_3 - \sigma^2\upsilon_1)/2 + \phi_{10}$, $\phi_2 = k_2x^2 + k_1x - (\upsilon_3 - \upsilon_1)/2 + \phi_{20}$, where

σ, θ_{10}, θ_{20}, ϕ_{10}, and ϕ_{20} are arbitrary real constant, Ω, k_1, k_2, υ_1, υ_2, and υ_1 are the same as in (6.40b).

Now, by using the Hirota method and performing some tedious deduction and computation, we obtain the following families of bright two soliton and dark two solitons of equations:

$$\psi_4(x,t) = \Omega\sqrt{\frac{\alpha}{2\beta}\frac{g_2}{f_2}}\exp\left[i\Delta_4\right], \tag{6.46a}$$

for $\alpha\beta > 0$, and

$$\psi_5(x,t) = \sigma\Omega\sqrt{-\frac{\alpha}{2\beta}\frac{G_2}{F_2}}\exp\left[i\Delta_5\right], \tag{6.46b}$$

for $\alpha\beta < 0$, with

$$\begin{aligned} g_2 &= e^{\xi_1} + e^{\xi_2} + e^{\xi_1+\xi_2+\xi_1^*+\theta_{12}+\theta_{13}+\theta_{23}} + e^{\xi_1+\xi_2+\xi_2^*+\theta_{12}+\theta_{14}+\theta_{24}}, \\ f_2 &= 1 + e^{\xi_1+\xi_1^*+\theta_{13}} + e^{\xi_1+\xi_2^*+\theta_{14}} + e^{\xi_2+\xi_1^*+\theta_{23}} + e^{\xi_2+\xi_2^*+\theta_{24}} \\ &\quad + e^{\xi_1+\xi_2+\xi_1^*+\xi_2^*+\theta_{12}+\theta_{13}+\theta_{14}+\theta_{23}+\theta_{24}+\theta_{34}}, \\ G_2 &= 1 + e^{\zeta_1+2i\vartheta_1} + e^{\zeta_2+2i\vartheta_2} + e^{\zeta_1+\zeta_2+2i(\vartheta_1+\vartheta_2)+A_{12}}, \\ F_2 &= 1 + e^{\zeta_1} + e^{\zeta_2} + e^{\zeta_1+\zeta_2+A_{12}}, \end{aligned} \tag{6.46c}$$

where

$$\begin{aligned} \xi_j &= p_j\left(\Omega x - \upsilon_2\right) + i\frac{p_j^2}{2}\upsilon_1 + \xi_j^{(0)},\ e^{\theta_{j,2+l}} = \frac{1}{2\left(p_j + p_l^*\right)^2}\ (j,\ l = 1,\ 2), \\ e^{\theta_{12}} &= 2\left(p_1 - p_2\right)^2,\ e^{\theta_{34}} = 2\left(p_1^* - p_2^*\right)^2,\ \Delta_4 = k_2x^2 + k_1x - \frac{1}{2}\upsilon_3, \\ \zeta_j &= p_j\left(\Omega x - \upsilon_2\right) \pm \frac{p_j}{2}\sqrt{2\sigma^2 - p_j^2}\upsilon_1 + \zeta_j^{(0)},\ p_j = \sqrt{2}\sigma\sin\left[\vartheta_j\right]\ (j = 1,\ 2), \\ e^{A_{12}} &= \sin^2\frac{\vartheta_1 - \vartheta_2}{2}\Big/\sin^2\frac{\vartheta_1 + \vartheta_2}{2},\ \Delta_5 = k_2x^2 + k_1x - \frac{1}{2}\upsilon_3 - \frac{\sigma^2}{2}\upsilon_1, \end{aligned}$$

$k_1, k_2, \Omega, \upsilon_1, \upsilon_2$, and υ_3 being given by Eq. (6.40b) and $\xi_j^{(0)}$ and $\zeta_j^{(0)}$ are real constants.

6.4.2 Propagation of One-Soliton in BECs

With the use of the above exact one-soliton solutions, especially the exact one-soliton solutions (6.38) and (6.39), we investigate here the propagation of one-soliton in BECs described by Eq. (6.37) (the case when the coefficients of Eq. (6.37) satisfies condition (6.41a) will be investigated in Sect. 6.5). In order to be close to the experimental condition and compare our theoretical prediction with experimental results in BECs, without loss of generality, we further set $\alpha = 1$ and $N = 0$ and discuss solutions (6.38) and (6.39) with some selected special control functions β, M, and F. With this advantage, the intensities of solitons can first increase to the peak value quickly and then decrease to the background value very slowly, so the lifetime of the solitons in BECs can be extended largely.

Propagation of a Bright One-Soliton in BECs

First of all, we note that when $A_s = 0$ or $A_c = 0$, solution (6.38) becomes respectively to the background

$$\psi_{11c}(x,t) = A_c\Omega\sqrt{\frac{\alpha}{\beta}}\exp\left[i\Delta_1\right] \tag{6.47a}$$

and the bright soliton

$$\psi_{11s}(x,t) = A_s\Omega\sqrt{\frac{\alpha}{\beta}}\frac{1}{\cosh\left[\xi\right]}\exp\left[i\Delta_1 \pm i\eta\right]. \tag{6.47b}$$

Thus, solution (6.38) represents a bright one-soliton embedded in the background, and when $A_s >> A_c$, the background is small within the existence of a bright one soliton.

To investigate the dynamics of the one-soliton embedded in the background (6.38), we set $\theta = p_2 = 0$ and $A_s^2 - 4A_c^2 > 0$. Under these conditions, Eq. (6.38) reduces to

$$\psi_{11}(x,t) = \Omega\sqrt{\frac{\alpha}{\beta}}\left(-A_c + \widetilde{\delta}_2\frac{\widetilde{\delta}_2\cos\left[\eta\right] \pm iA_s\sin\left[\eta\right]}{A_s\cosh\left[\xi\right] - 2A_c\cos\left[\eta\right]}\right)\exp\left[i\Delta_1\right] \tag{6.48}$$

where $\widetilde{\delta}_2 = \sqrt{A_s^2 - 4A_c^2}$, $\xi = \kappa A_s\left(\Omega x - \upsilon_2\right) + \xi_0$, and $\eta = \frac{A_s^2\kappa}{2}\upsilon_1 + \eta_0$. To be specific and translate our results into units relevant to the experiments, we investigate the dynamics of solitons in the ^{7}Li condensate. Under the realistic experiment parameters in Ref. [20], $\omega_\perp = 2\pi \times 710$ Hz, $\omega_0 = 2i\pi \times 70$ Hz, i.e., $M \approx 0.0097$. Then, from Eq. (6.40b), we obtain $k_2 = \sqrt{2M}/2\tanh\left[\sqrt{2M}t\right]$, $\Omega = c_1/\cosh\left[\sqrt{2M}t\right]$, $k_1 = c_2/\cosh\left[\sqrt{2M}t\right]$. If we further set $F = c_3 =$ constant, then $\beta = c_4/\cosh\left[\sqrt{2M}t\right]$ $\exp\left[2c_3t\right]$. As shown in Fig. 6.6a, one can see that the lifetime of the BEC is about $40 \times 2.5 \times 10^{-4}$ s $= 9$ ms, which is close to the experiment results: the lifetime of a BEC is about 8 ms. Considering the dynamics of the bright soliton in the background, the length $2L$ of the spatial background must be very large compared to the scale of the soliton. In the real experiment [36], the length of the background of the BEC can reach at least $2L = 370\,\mu$m. The compressed minimum width of the bright soliton is about $2l = 5.6\,\mu$m at $t = 0$ in Fig. 6.6a or $t = 100$ in Fig. 6.6b (a unit of coordinate, $\Delta x = 1$ in the dimensionless variables, corresponds to $a_\perp = \sqrt{\hbar/m\omega_\perp} = 1.4\,\mu$m). So, $l << L$, a necessary condition for realizing bright soliton in experiment. Here, by Eq. (6.50c), we can verify that the number of atoms in the bright soliton against the background is in the range of 2961 (at time $t = -15$) and 6589 (at time $t = 25$), which is a proper range of atoms where the soliton can be observed [20]. Now we turn to seek some parameter conditions to extend the lifetime of the bright solitons. To this end, we introduce the so-called control function Θ as

$$\Theta(\lambda_1, \lambda_2) = \frac{1}{\exp[\lambda_1 t] + \exp[-\lambda_2 t]}, \tag{6.49}$$

where λ_1 and λ_2 are positive real constants. If we select

$$\beta(t) = b_0 \Theta^{n_2}(\delta_1, \delta_2) \exp[-\tau(2n_1 - n_2)\delta_1 t],$$
$$\Omega(t) = \delta_0 \Theta^{n_1}(\delta_1, \delta_2),$$

where b_0, δ_0, n_1, n_2, and τ are real constants, $\delta_1 < \delta_2$ and $2n_1 - n_2 > 0$, then from Eq. (6.40b), we can deduce the expressions of $k_1 = K_1 \Theta^{n_1}(\delta_1, \delta_2)$ and other functions $\{M, k_2, F, v_1, v_2, v_3\}$. For simplification, we omit other function's expression in what follows. If we further select $A_c = 4$, $A_s = 200$, $\xi_0 = \eta_0 = 0$, $\delta_0 = 0.01$, $n_1 = 1$, $n_2 = 1.1$, $\delta_1 = b_0 = 0.001$, $\delta_2 = 0.1$, $K_1 = 0.4$, $\tau = 0.9$, and $\delta_2 = b_2 = -0.1$, the evolution of the bright soliton ψ_{11} is shown in Fig. 6.6b, in which (i) the six wavepackets from left to right denote the soliton evolution at time $t = 0, 100, 400, 1200, 2000, 2400$, respectively and (ii) the top red-thick line denotes the approximate evolution curve of the amplitude for the bright soliton ψ_{11} during the time $t \in [-10, 2800]$. As shown in Fig. 6.6b, the intensities of the bright soliton (6.48) increase rapidly to the peak, and then decay to the background very slowly. Thus, the lifetime of the bright soliton in BECs can be extended largely under the above parameter conditions. For example, the dimensionless time $t = 2000$ units, corresponding to a real time of 0.45 s, which arrives at the order of the lifetime of a BEC in today's experiments. Moreover, when t is varying from -10 to 2800, the number of atoms and the scattering length are all in the range of validity.

Figure 6.7a shows the evolution plots of control functions $\{\beta, F, M$ of $\psi_{11}\}$ used in Fig. 6.6b. It is easy to see that the values of the functions F and M are changed from large to small during a short time and then $F \to -4.5 \times 10^{-4}$ and $M \to 5 \times 10^{-7}$, while the scattering length β first increases to the maximum value fast and then decreases slowly to a fixed value. From ψ_{11}, we know that the evolution of the amplitudes of ψ_{11} is mainly controlled by $\Omega^2/\beta = \beta \exp\left[-4\int F(t)dt\right]$. Under the same parameters in Fig. 6.6b, the evolution of Ω^2/β is the line labeled $\tau = 0.9$ in Fig. 6.7b. From Fig. 6.7b, when $\tau = 1.1 > 1$, we can see that after the amplitude arrives the turning point, it still increases continuously. This is unreasonable for BEC experiments. While $\tau = 1$, once the amplitude arrives the turning point, the soliton will maintain its shape forever. In fact, when $\tau = 1$ and t arrives the condensate peak, $\beta \to 0.01 \exp[-0.002t]$ and $F \to -5 \times 10^{-4}$; thus, $\Omega^2/\beta \to 0.01$. Therefore, when $\beta \exp\left[-4\int F(t)dt\right] \to$ constant, the lifetime of the bright soliton in the BECs can be extended dramatically.

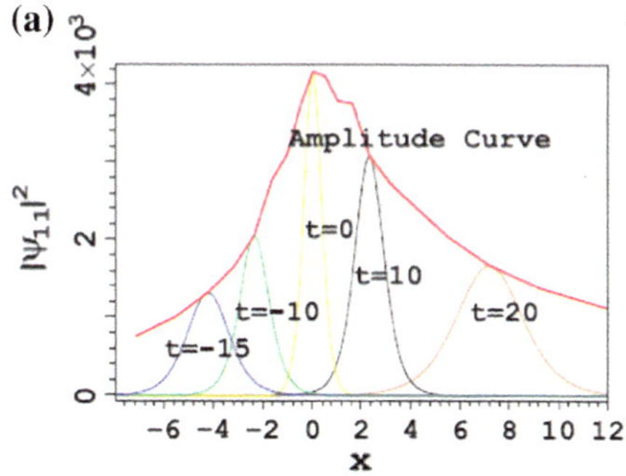

Fig. 6.6 The evolution plots of $|\psi_{11}(x,t)|^2$ with different parameters. **a** $A_c = 4$, $A_s = 200$, $\Omega = 0.01/\cosh[\varrho t]$, $\beta = 0.001/\cosh[\varrho t]\exp[-0.02t]$, $k_1 = 0.2/\cosh t$, $\varrho = 7\sqrt{2} \times 10^{-2}$; **b** $A_c = 4$, $A_c = 200$, $\Omega = 0.01\Theta(0.001, -0.1)$, $\beta = 0.001\Theta^{1.1}(0.001, -0.1)\exp[-0.00081t]$, $k_1 = 0.4$, Θ being given in Eq. (6.49). The top red-thick lines represent the approximate evolution curves of the amplitude for $|\psi_{11}|^2$ during the time period: (i) $t \in [-20, 25]$ in Fig. 6.6a; (ii) $t \in [-10, 2800]$, respectively. Reprint from Ref. [26], Copyright 2018, with permission from American Institute of Physics

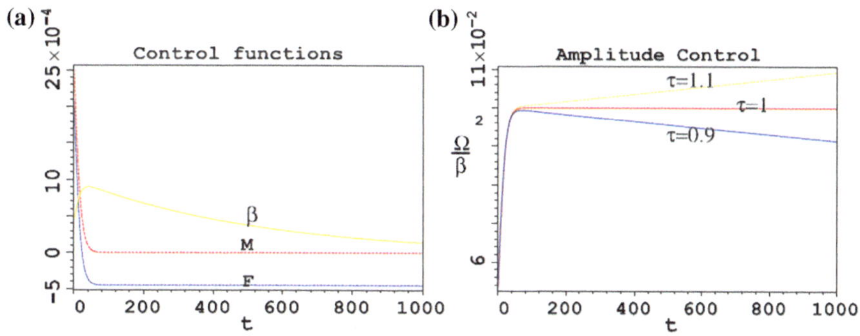

Fig. 6.7 The evolution plots of main control functions of ψ_{11}. **a** The evolution of control functions β, F, and M in **b**. The evolution trend of the amplitudes of ψ_{11} controlled by Ω^2/β under different τ. The case $\tau = 0.9$ is corresponding to the evolution trend of the amplitude in (**b**). Reprint from Ref. [26], Copyright 2018, with permission from American Institute of Physics

In most experiments, the atom scattering length β is almost constant during the lifetime of BECs. Then, from the amplitude control function $\Omega^2/\beta = \beta\exp\left[-2\int F(t)dt\right]$, the lifetime of BECs is mainly determined by the gain/loss term F. In Ref. [4], the crucial feature to the observed long lifetimes of solitons is the very low temperature, while F is a small parameter related to the feeding of the condensate from the thermal cloud. So it is reasonable to assume that the lower the temperature, the smaller F is, so the longer the lifetime of BECs is.

In addition, from the balance relation $\beta\exp\left[-4\int F(t)dt\right] \approx$ constant, we can conclude that the mean effect of the variation from β is equal to gain/loss atoms from BECs. Therefore, by adjusting the atom scattering length β through the FR technique, we can realize the effective control to gain/loss can realize the effective control to gain/loss atoms in BECs.

Furthermore, we find that when $\cosh\xi = A_s/(A_c\cos\eta) - 2A_c\cos\eta/A_s$ [$\sinh\xi = 0$], the intensity of (6.48) arrives at the minimum [maximum]

$$|\psi_{11}|_{\min} = \frac{\alpha\Omega^2}{\beta}A_c^2\left(1 - \frac{\widetilde{\delta}_2^2\cos^2\eta}{A_s^2 - 4A_c^2\cos^2\eta}\right), \tag{6.50a}$$

$$|\psi_{11}|_{\max} = \frac{\alpha\Omega^2}{\beta}\left(A_c^2 + \frac{A_s\widetilde{\delta}_2^2\cos^2\eta}{A_s - 2A_c\cos\eta}\right). \tag{6.50b}$$

Equations (6.50a) and (6.50b) reveal that bright solitons (6.48) can be only squeezed into the assumed peak matter density between the minimum and maximum values. From Eq. (6.50b) with the same parameters used in Fig. 6.7, we can deduce that, as time goes by, the intensities first increase until to the peak, and then decrease to the background very slowly. This further certifies the correctness of our solutions.

It is interesting to note that the exact number of the atoms in the bright soliton against the background described by Eq. (6.48) is

$$N(t) = \lim_{L\to+\infty}\int_{-L}^{L}\left(|\psi_{11}|^2 - |\psi_{11c}|^2\right)dx = 2\widetilde{\delta}_2\frac{\alpha}{\beta}\Omega. \tag{6.50c}$$

In contrast, the time-dependent function

$$\widetilde{\kappa}(t)\lim_{L\to+\infty}\int_{-L}^{L}\left(|\psi_{11} - \psi_{11c}|^2\right)dx = N(t)\left(1 + A_c\widetilde{M}\cos\eta\right), \tag{6.51}$$

where $\widetilde{M} = \frac{4\arctan\left[\sqrt{A_s+2A_c\cos\eta}/\sqrt{A_s-2A_c\cos\eta}\right]}{\sqrt{(A_s+2A_c\cos\eta)(A_s-2A_c\cos\eta)}}$, counts the number of atoms in both the bright soliton and background under the condition of $\psi_{11}(\pm L, t) \neq 0$. Equation (6.51) displays that a time-periodic atomic exchange is formed between the bright soliton and the background.

Propagation of a Dark One-Soliton in BECs

In order to investigate the dynamics of dark one-soliton of the BEC described by Eq. (6.37), we use, without loss of generality, the exact one-soliton solution (6.39). When either $A_s = 0$ or $A_c = 0$, the dark one-soliton (6.39) becomes

$$\psi_{21c}(x,t) = \Omega\sqrt{-\frac{\alpha}{\beta}}A_c\exp[i\Delta_2], \tag{6.52a}$$

$$\psi_{21s}(x,t) = \Omega\sqrt{-\frac{\alpha}{\beta}}iA_s\tanh\zeta\exp[i\Delta_2], \tag{6.52b}$$

respectively. This means that Eq. (6.39) describes a dark one-soliton embedded on a background.

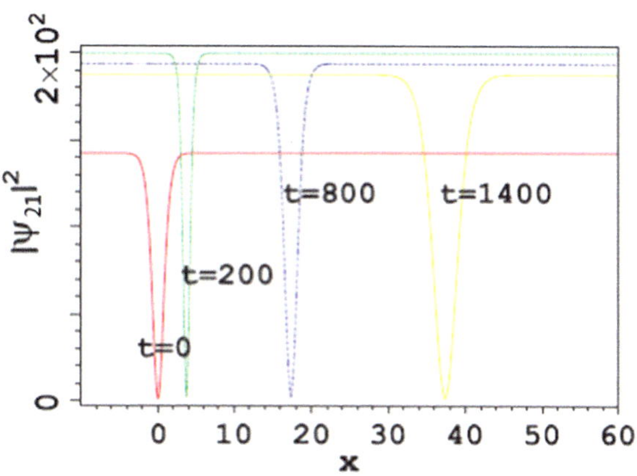

Fig. 6.8 The evolution plots of $|\psi_{21}|^2$, where $A_c = 1$, $A_s = 10$, $\Omega = 0.2\Theta(0.001, -0.1)$, $\beta = -0.02\Theta^{3/2}(0.001, -0.1)\exp[-0.00045t]$, and $k_1 = 1.1\Omega$, function Θ being given by Eq. (6.49). Reprint from Ref. [26], Copyright 2018, with permission from American Institute of Physics

From Eq. (6.52a), we can obtain the intensities of the background as follows:

$$|\psi_{21c}(x,t)|^2 = -\frac{\alpha}{\beta}A_c^2\Omega^2.$$

Proceeding as the case of the bright soliton, we obtain

$$N(t) = \lim_{L\to+\infty}\int_{-L}^{L}\left(|\psi_{21}(x,t)|^2 - |\psi_{21}(\pm L,t)|^2\right)dx = \frac{2\alpha\Omega A_s}{\beta},$$

which describes the region of decreased density and contains a negative "number of atoms" if $\Omega > 0$ [23]. As shown in Fig. 6.8, with the time going on, the intensity of the dark soliton ψ_{21} first increases rapidly until the condensate peak, and then decreases slowly to the background. Similar to the reason in the bright soliton ψ_{11}, $\beta\exp\left[-4\int F(t)dt\right] \to$ constant, the dark solitons in BECs can maintain their shape longer.

6.4.3 Interaction of Two Solitons in BECs

To study analytically the interaction of two solitons in BECs described by Eq. (6.37), we use its exact two-soliton solutions found above. Here, we mainly focus ourselves to the study of the collision properties of the two-soliton solutions.

Interaction of Bright Two Solitons

As shown in Fig. 6.9a, when the distance between two solitons $\psi_3(x,t)$ given by Eq. (6.45) is large, they will not collide with each other during the propagation. Figure 6.9b shows the collision between solitons with the collision periodic gradually increasing for a period of time. In this case, the effective external potential is weaker than the former case and the two solitons can collide with each other. Due to the

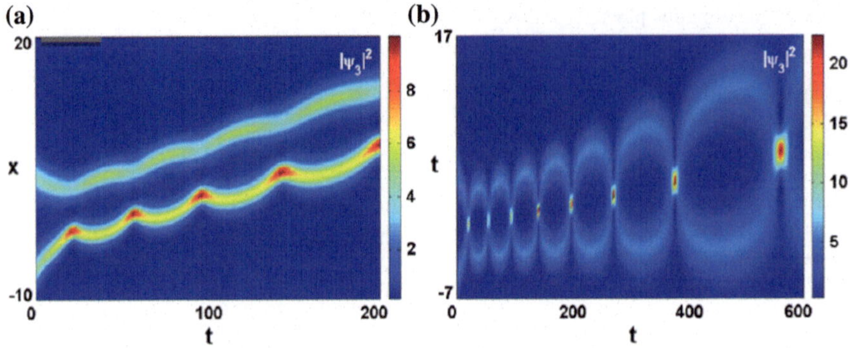

Fig. 6.9 The contour plots of $|\psi_3|^2$ with $\Omega = \Theta(0.002, -0.1)$, $\beta = -0.2\Theta^{3/2}(0.002, -0.1)$ $\exp[-0.0009t]$, $\sigma = 1.2$, and **a** $k_1 = 0.06$, $\theta_{10} = 1.4$, $\theta_{20} = 0.4$, $\phi_{10} = \phi_{20} = 0$; **b** $k_1 = 0.01\Omega$, $\theta_{10} = \theta_{20} = \phi_{10} = \phi_{20} = 0$. Reprint from Ref. [26], Copyright 2018, with permission from American Institute of Physics

faster velocity and weaker effective external potential, the spatial extent is larger than the former case. At the same time, regardless of collision or non-collision cases, the two-soliton $\psi_3(x, t)$ attract each other periodically along the $x-$direction with the same velocity v_2/Ω and their amplitudes increase to the peak value quickly, and then decrease to the background value very slowly.

Figure 6.10 shows the collision between two bright solitons $|\psi_4(x, t)|^2$ (see Eq. (6.46a)). In Fig. 6.10a, the two solitons attract each other periodically and do not collide with each other during the propagation, which is similar to Fig. 6.9a. Moreover, Fig. 6.10b, c shows the collision scenes. In Fig. 6.10b, the interaction between these two solitons is first attractive, and then becomes repulsive after collision. In this case, the spacing between them first decreases after they are initially generated on different positions, which is evidence for a short-range attractive interaction between solitons, and then gradually increases with time going on. Finally, from Fig. 6.10c, we can see that after the collision, the small soliton boosts up rapidly and the large one weakens down a little; therefore, the collision is inelastic. Here, we want to point out that, no matter whether the collision is an elastic or inelastic one, the common feature is that the amplitudes of two soliton in Fig. 6.10 increase to the peak value rapidly, and then decay to a background value very slowly.

Based on the above results, we conclude that the interaction between bright two solitons can present various forms, such as non-collision but quasi-periodic attraction, quasi-periodic collision, nonelastic collision, etc., but regardless of which cases, the lifetime of a bright two-soliton in BECs can be extended largely under the selected control functions so that these phenomena can all be observed in today's experiments.

Interaction of a Dark Two Solitons

To study the interactions of a dark two-soliton, we use that exact dark two-soliton solution (6.46b) for Eq. (6.37). Figure 6.11a shows the contour plots of $\psi_5(x, t)$. From this figure, it is seen that the depth of the dark two-soliton increases to a maximum

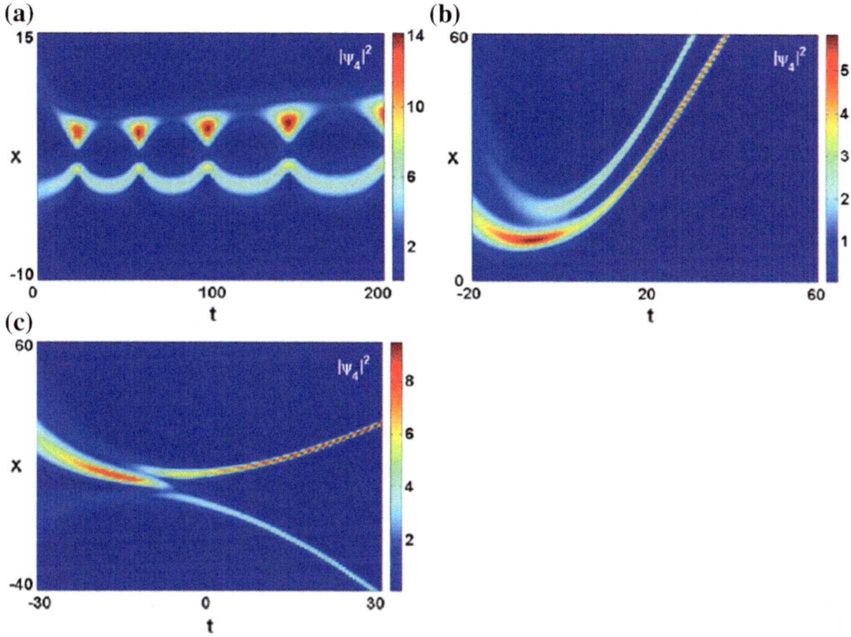

Fig. 6.10 The contour plots of $|\psi_2|^2$ with $\Omega = \Theta(0.002, -0.1)$, $\beta = 0.2\Theta^{3/2}(0.002, -0.1)$ $\exp[-0.0009t]$, and **a** $k_1 = 0.002\Omega$, $p_1 = \xi_1^{(0)} = 1.2$, $p_2 = \xi_2^{(0)} = 1$; **b** $k_1 = -0.002\Omega$, $p_1 = 1 + 2i$, $p_2 = 0.8 + 2.4i$, $\xi_1^{(0)} = 0$, $\xi_2^{(0)} = \pi$; (c): $k_1 = \Omega$, $p_1 = 1 - 2i$, $p_2 = 1.4 + 1.2i$, $\xi_1^{(0)} = 0$, $\xi_2^{(0)} = \pi$. Reprint from Ref. [26], Copyright 2018, with permission from American Institute of Physics

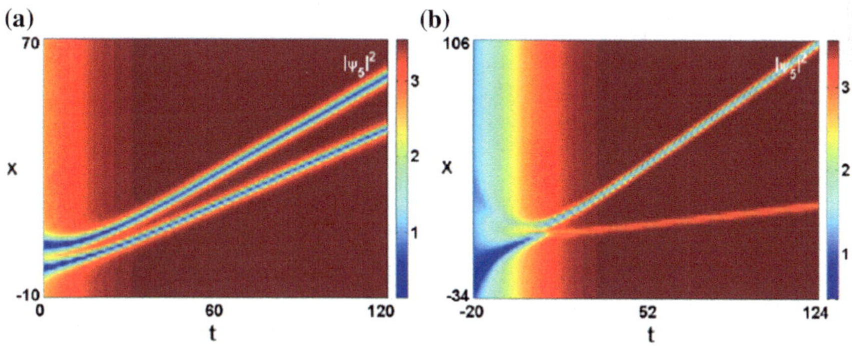

Fig. 6.11 The contour plots of $|\psi_5|^2$ with $\Omega = \Theta(0.002, -0.1)$, $\beta = -0.2\Theta^{3/2}(0.002, -0.1)$ $\exp[-0.0009t]$, $k_1 = \Omega$, $\sigma = -1.2$, and **a** $\theta_1 = \pi/4$, $\theta_2 = \pi/3$, $\xi_1^{(0)} = 1 + 2i$, $\xi_2^{(0)} = 0.1 + 0.2i$, **b** $\theta_1 = 2$, $\theta_2 = 4$, $\xi_1^{(0)} = 2 + i$, $\xi_2^{(0)} = 0.1 + 0.2i$. Reprint from Ref. [26], Copyright 2018, with permission from American Institute of Physics

value quickly, and then reaches the background value very slowly. At the same time, the distance between the two solitons becomes increasingly larger. From Fig. 6.11b, it is shown that the collision of the dark two-soliton is elastic and the depth of the dark two-soliton arrives at the maximum value at the moment of collision, and then the two dark solitons propagate along their former directions and decay to the background value very slowly. From Figs. 6.10 and 6.11, we conclude that regardless of the collision or no-collision cases, their dissipation is very slow, which can be used to keep a longer lifetime of solitons and to be observed in today's BECs experiments.

6.4.4 Special Case When $\alpha = 2$, $N = 0$

Let us go back to the special case when $\alpha = 2$, $N = 0$, and other parameters β, M, and F satisfy condition (6.41a). In other words, let us consider the cubic NLS equation

$$i\frac{\partial\psi}{\partial t} + \frac{\partial^2\psi}{\partial x^2} + \beta\,|\psi|^2\,\psi + \left[Mx^2 + iF\right]\psi = 0, \tag{6.53}$$

where β, M, and F are three time-dependent real function satisfying condition (6.41a). One can easily verify that sufficient conditions for condition (6.41a) to be satisfied are that β, M , and F be solutions of the ordinary differential system (ODS)

$$\frac{dF}{dt} - 2F^2 + 2M = 0, \tag{6.54a}$$

$$\frac{1}{\beta}\frac{d\beta}{dt} - 4F = 0. \tag{6.54b}$$

Under conditions (6.54a) and (6.54b), the NLS Eq. (6.53) is integrable. Indeed, let us make the ansatz (modified lens-type transformation)

$$\psi(x,t) = \frac{1}{\sqrt{\ell}}u(X,T)\exp\left[\eta - i\frac{F}{2}x^2\right], \tag{6.55}$$

where $X = x/\ell$, and $\ell > 0$, η, and T are solutions of the ODE

$$\frac{dT}{dt} - \frac{1}{\ell^2} = 0, \tag{6.56a}$$

$$\frac{1}{\ell}\frac{d\ell}{dt} + 2F = 0, \tag{6.56b}$$

$$\frac{d\eta}{dt} - \frac{1}{2\ell}\frac{d\ell}{dt} = 0. \tag{6.56c}$$

Solving system (6.56a)–(6.56c) yields

$$\begin{aligned}&\ell(t)=\ell_0\exp\left[-2\int_0^t F(\tau)d\tau\right],\quad \eta(t)=\ln\sqrt{\eta_0}-\int_0^t F(\tau)d\tau,\\ &T(t)=\frac{1}{\ell_0^2}\int_0^t\exp\left[4\int_0^y F(\tau)d\tau\right]dy+T_0,\quad X(x,t)=\frac{x}{\ell_0}\exp\left[2\int_0^t F(\tau)d\tau\right],\end{aligned}\tag{6.57}$$

where ℓ_0 and η_0 are arbitrary positive constants, and $T_0=T(0)$ is an arbitrary real constant. Inserting ansatz (6.55) into Eq. (6.37) and using the conditions (6.54a) and (6.54b), system (6.56a)–(6.56c), and Eq. (6.57) lead to the standard NLS equation

$$i\frac{dT}{dt}+\frac{\partial^2 u}{\partial X^2}+\mathrm{sign}\,[\beta_0]\,|u|^2\,u=0,\tag{6.58}$$

if ℓ_0 and η_0 are taken from condition $\beta_0\ell_0\eta_0=\mathrm{sign}\,[\beta_0]$ (here, $\beta_0=\beta(0)$). Equation (6.58) is an integrable equation. If $u(X,T)$ is a solution of Eq. (6.58), then by using transformation (ansatz) (6.55) and Eq. (6.57), we can write the corresponding solution of Eq. (6.53) as follows:

$$\psi(x,t)=\sqrt{\frac{\mathrm{sign}\,[\beta_0]}{\ell_0^2\beta_0}}u\left(\frac{x}{\ell_0}\exp\left[2\int_0^t F(\tau)d\tau\right],\frac{1}{\ell_0^2}\int_0^t\exp\left[4\int_0^y F(\tau)d\tau\right]dy+T_0\right)\exp\left[-i\frac{F}{2}x^2\right].\tag{6.59a}$$

The total number of atoms in the condensate will then be

$$N(t)=\frac{\mathrm{sign}\,[\beta_0]}{\ell_0\beta_0}\exp\left[-2\int_0^t F(\tau)d\tau\right]\int_{-\infty}^{+\infty}|u(X,T)|\,dX.$$

As before, we are here interested ourselves to the soliton solutions of Eq. (6.58) which describe solitons embedded on a continuous wave background. Applying the method of the Darboux transformation, we find the following soliton solutions of Eq. (6.58):

$$u(X,T)=\left(A_c+A_s\frac{\rho\cosh\xi+\cos\eta+i\,[\theta\sinh\xi+\kappa\sin\eta]}{\cosh\xi+\rho\cos\eta}\right)\exp\,[i\phi_b],\tag{6.60a}$$

if $\beta_0>0$, and

$$u(X,T)=-\frac{1}{\sqrt{2}}\,[A_c+iA_s\tanh\zeta]\exp\,[i\phi_d],\tag{6.60b}$$

if $\beta_0<0$, with

$$\begin{aligned}&\xi=\frac{M_R}{\sqrt{2}}X-\frac{(k_c+k_s)M_R-A_sM_I}{2}T,\ \eta=\frac{M_I}{\sqrt{2}}X-\frac{(k_c+k_s)M_I+A_sM_R}{2}T,\ \phi_b=\frac{k_c}{\sqrt{2}}X+\frac{2A_c^2-k_c^2}{2}T,\\ &\zeta=-\frac{A_s}{2}X+\frac{A_s(A_c+2k_c)}{2}T,\ \phi_d=k_cX-\frac{A_c^2+A_s^2+2k_c^2}{2}T,\end{aligned}\tag{6.60c}$$

where

$$\rho = \frac{A_c(M_R - A_s)}{\Delta},\ \theta = \frac{A_c(k_s - k_c + M_I)}{\Delta},\ \kappa = 1 - \frac{2A_c^2}{\Delta},$$
$$M = \sqrt{(-A_s + ik_s - ik_c)^2 - 4A_c^2} = M_R + iM_I,\ \Delta = A_c^2 + \frac{(M_R - A_s)^2}{4} + \frac{(k_s - k_c + M_i)^2}{4},$$

A_s, A_c, k_c, and k_s are real constants, and X and T are defined by Eq. (6.57).

As one can easily verify, solutions (6.60a) and (6.60b) describe respectively bright and dark solitons embedded on a cw background. Thus, by employing the expression (6.59a), we can obtain the bright and dark solitons which propagate on a cw background for Eq. (6.53) under the conditions (6.54a) and (6.54b). As in the previous section, we limit our studies in the special case when $k_s = k_c$ and $A_s^2 - 4A_c^2 > 0$. Under these conditions, solutions (6.60a) and (6.60b) lead to following bright and dark soliton solutions for Eq. (6.53):

$$\psi(x,t) = \left(-A_c + \delta_2 \frac{\delta_2 \cos[\eta(t)] - iA_s \sin[\eta(t)]}{A_s \cosh[\xi(x,t)] - 2A_c \cos[\eta(t)]}\right) \frac{\exp[i\Delta_b(x,t)]}{\ell_0\sqrt{|\beta_0|}}, \tag{6.61a}$$

$$\psi(x,t) = -\frac{1}{\ell_0\sqrt{2|\beta_0|}}[A_c + iA_s \tanh\zeta(x,t)]\exp[i\Delta_d(x,t)], \tag{6.61b}$$

respectively, with

$$\begin{aligned}
\xi(x,t) &= \frac{\delta_2}{\ell_0\sqrt{2}} x \exp\left[2\int_0^t F(\tau)d\tau\right] - \frac{k_c\delta_2}{\ell_0^2}\int_0^t \exp\left[4\int_0^y F(\tau)d\tau\right]dy + \widetilde{T}_0,\\
\zeta(x,t) &= -\frac{A_s}{2\ell_0} x \exp\left[2\int_0^t F(\tau)d\tau\right] + \frac{A_s(A_c + 2k_c)}{2\ell_0^2}\int_0^t \exp\left[4\int_0^y F(\tau)d\tau\right]dy + \widetilde{T}_0,\\
\eta(t) &= -\frac{A_s\delta_2}{2\ell_0^2}\int_0^t \exp\left[4\int_0^y F(\tau)d\tau\right]dy, \qquad (6.61c)\\
\Delta_b(x,t) &= -\frac{1}{2}F(t)x^2 + \frac{k_c}{\ell_0\sqrt{2}} x \exp\left[2\int_0^t F(\tau)d\tau\right] + \frac{2A_c^2 - k_c^2}{2\ell_0^2}\int_0^t \exp\left[4\int_0^y F(\tau)d\tau\right]dy,\\
\Delta_d(x,t) &= -\frac{1}{2}F(t)x^2 + \frac{k_c}{\ell_0} x \exp\left[2\int_0^t F(\tau)d\tau\right] - \frac{A_c^2 + A_s^2 + 2k_c^2}{2\ell_0^2}\int_0^t \exp\left[4\int_0^y F(\tau)d\tau\right]dy,
\end{aligned}$$

where $\delta_2 = \sqrt{A_s^2 - 4A_c^2}$.

It follows from Eq. (6.61c) that for $\widetilde{T}_0 = 0$, the center of bright and dark solitons (6.61a) and (6.61b) are respectively

$$\varkappa_b(t) = \frac{k_c\sqrt{2}}{\ell_0}\exp\left[-2\int_0^t F(\tau)d\tau\right]\int_0^t \exp\left[4\int_0^y F(\tau)d\tau\right]dy,$$
$$\varkappa_d(t) = \frac{(A_c + 2k_c)}{\ell_0}\exp\left[-2\int_0^t F(\tau)d\tau\right]\int_0^t \exp\left[4\int_0^y F(\tau)d\tau\right]dy,$$

and satisfy the equation $\frac{d^2\varkappa}{dt^2} - \left(\frac{1}{2}\frac{1}{\beta}\frac{d\beta}{dt}\right)^2 \varkappa = 0$. This means that the center of mass of the macroscopic wave packet behaves like a classical particle, and allows one to

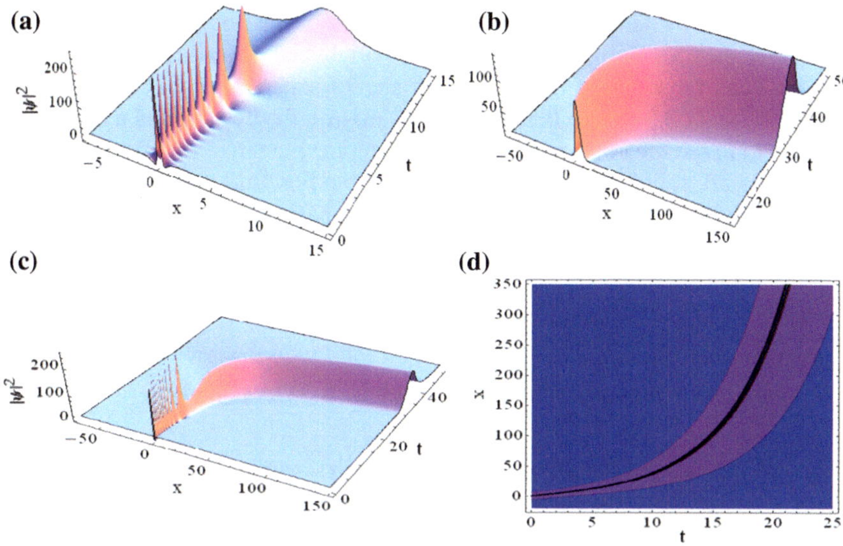

Fig. 6.12 **a**, **b**, **c** The dynamics of a bright soliton in a time-dependent harmonic trapping potential given by equation (6.61a) for different time intervals, **a** $t \in [0,\ 15]$, **b** $t \in [15,\ 50]$, and **c** $t \in [0,\ 50]$. **d** The contour plot of the dark soliton (6.61b). To generate these plots, we used the parameters $A_c = 1,\ A_s = 3,\ \beta_0 = 0.25,\ k_c = 0.03,\ \ell_0 = 0.5$, and $T_0 = 0$

manipulate the motion of bright and dark solitons in BEC systems by controlling the nonlinear parameter β. It is also seen from Eq. (6.61c) that Eqs. (6.61a) and (6.61b) can be used to describe the compression (broadening) of both bright and dark soliton when function $\exp\left[2\int_0^t F(\tau)d\tau\right]$ increases (decreases) with time.

Considering again the dynamics of matter-wave solitons in the ^{7}Li condensate, we have, under the realistic experiment parameters in Ref. [20], $\omega_\perp = 2\pi \times 710$ Hz, $\omega_0 = 2i\pi \times 70$ Hz, i.e., $M \approx 0.0097$. Solving the ODE (6.54a) with $M = 0.0097$ yields

$$F(t) = \frac{\sqrt{M}\left(C_2 \exp\left[-2\sqrt{M}t\right] - C_1 \exp\left[2\sqrt{M}t\right]\right)}{C_1 \exp\left[2\sqrt{M}t\right] + C_2 \exp\left[-2\sqrt{M}t\right]},$$

where C_1 and C_2 are arbitrary real constants so that $|C_1| + |C_2| > 0$ and $|F(t)| << 1$. In the special case of $C_1C_2 = 0$, we have $F = \pm\sqrt{M}$, which together with condition (6.54b) give $\beta(t) = \beta_0 \exp\left[\pm 4\sqrt{M}t\right]$. In this special case, the nonlinearity parameter is of form $\beta(t) = 2g_0 \exp[\lambda t]$ studied in Sect. 6.2, while the feeding/loss parameter is of the form $F = \lambda/2$ studied in Sect. 6.3.

Let us consider the special case when $C_1 = C_2 \neq 0$, leading to a time-dependent gain parameter $F(t) = -\sqrt{M}\tanh\left[2\sqrt{M}t\right]$. Using Eq. (6.54b) yields $\beta(t) = \frac{\beta_0}{\cosh^2\left[2\sqrt{M}t\right]}$. In the present special case, $\exp\left[2\int_0^t F(\tau)d\tau\right] = 1/\cosh\left[2\sqrt{M}t\right]$, which decreases with time $t \geq 0$. Therefore, Eqs. (6.61a) and (6.61b) can be used to describe the broadening of respectively the bright and the dark soliton. This phenomenon can be seen in Fig. 6.12 which show for parameters $A_c = 1$, $A_s = 3$, $\beta_0 = 0.25$, $k_c = 0.03$, $\ell_0 = 0.5$, and $T_0 = 0$, (a)–(c): the dynamics of a bright soliton (6.61a) in the time-dependent harmonic trapping potential, and (d): the contour plot of the dark soliton (6.61b). It is well seen from Fig. 6.12 that the width of both bright and dark soliton increases as the solitons propagate. Figure 6.12a–c show that with the time going on, the intensity of the bright soliton first oscillates around 200, and then decreases rapidly and is maintains around 100.

6.5 Nonautonomous Matter Waves in a Waveguide

In this section, we consider a model that simulates the coupling of a reservoir of Bose–Einstein-condensates atoms and the waveguide, as shown in Fig. 6.13. Condensate at a given chemical potential is injected into the waveguide from a reservoir at x_0. The reservoir emits a plane matter wave in both directions into the guide. Here, we go beyond previous studies by considering the case with both time- and space-modulated potential and inhomogeneous source, as well as time-modulated nonlinearity and gain or loss terms in BECs. The dynamics of such BECs are described by the NLS equation

$$i\hbar\frac{\partial\psi}{\partial t} = \left[-\frac{\hbar^2}{2m}\frac{\partial^2\psi}{\partial x^2} + V_{ext}(x,t) + g_{1D}(t)\,|\psi|^2 + i\Gamma(t)\right]\psi + S(t)\exp\left[i\varphi(x,t)\right], \tag{6.62}$$

where $\psi = \psi(x, t)$ is the macroscopic wave function of the condensate, m is the mass of atom, the nonlinearity $g_{1D}(t) = 2\hbar\omega_\perp a_s(t)$ is the effective $1D$ coupling strength with $\omega_\perp$ being the transverse confining frequency and $a_s(t)$ being the time-dependent s-wave scattering length modulated by a Feshbach resonance (FR) ($a_s(t) > 0$ for a

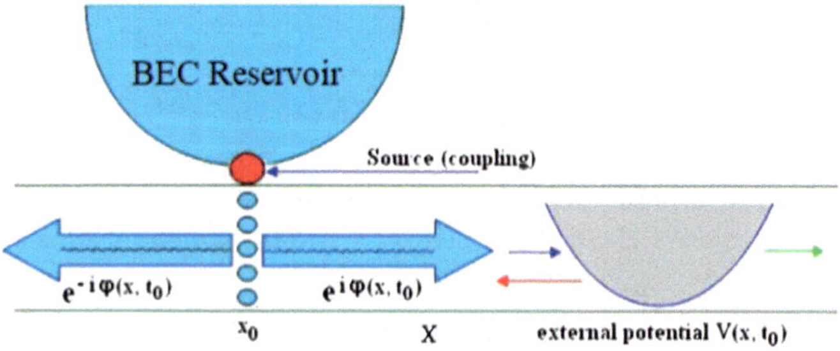

Fig. 6.13 Schematics of the coupling of a reservoir of Bose–Einstein-condensed atoms and the waveguide. Condensate at a given chemical potential is injected into the waveguide from a reservoir at x_0. The reservoir emits a plane matter wave in both directions into the guide. Shown in the right part is the external potential added along the axis direction of the waveguide

repulsive interaction and $a_s(t) < 0$ for an attractive interaction). $V_{ext}(x, t)$ denotes the time- and space-modulated external potential and can be chosen as the second-degree polynomial in space with all coefficients being functions of time. $\Gamma(t)$ stands for the time-dependent gain or loss distribution; the losses are caused by the collisions between thermal atoms, which is a three-body physical course. Finally, $S(t)\exp[i\varphi(x, t)]$, with the time-dependent amplitude $S(t)$ and phase $\varphi(x, t)$, simulates the coherent injection of matter waves from an external reservoir onto the scattering region [12].

Normalizing the length and time in Eq. (6.62) by $a_\perp = \sqrt{\hbar/(m\omega_\perp)}$ and $1/\omega_\perp$ leads to an effective GP equation with an inhomogeneous source term in the dimensionless form

$$i\frac{\partial\psi}{\partial t} = \left[-\frac{1}{2}\frac{\partial^2}{\partial x^2} + g(t)\,|\psi|^2 + v(x, t) + i\gamma(t)\right]\psi + s(t)\exp\left[i\varphi(x, t)\right], \quad (6.63)$$

where the external potential $v(x, t)$, the nonlinearity $g(t)$, the gain or loss distribution $\gamma(t)$ and the source $s(t)\ \exp[i\varphi(x, t)]$ are related to the external potential $V_{ext}(x, t)$, nonlinearity $g_{1D}(t)$, the gain or loss term $\Gamma(t)$, and the source term $S(t)$ $\exp[i\varphi(x, t)]$ in Eq. (6.62), respectively. Equation (6.63) is associated with $\delta\mathcal{L}/\delta\psi^\dagger$ in which the Lagrangian density can be written as $\mathcal{L} = i\left(\psi\psi_t^\dagger - \psi_t\psi^\dagger\right) + |\psi_x|^2 + 2\left[v(x, t) + i\gamma(t)\right]|\psi|^2 + g(t)\,|\psi|^4 + 2s(t)\exp[i\varphi(x, t)]\psi^\dagger$, where $\psi^\dagger$ denotes the complex conjugate of the wave function ψ.

To translate our results into units relevant to the experiments, our calculations are performed for ^{87}Rb atoms with $m = 1.44 \times 10^{-25}$ Kg, $\omega_\perp = 2\pi \times 10^3\ \text{s}^{-1}$ for repulsive interaction [30]. In this case, a unity of the dimensionless space corresponds to $a_\perp = \sqrt{\hbar/(m\omega_\perp)} \simeq 0.34\ \mu\text{m}$, and a unity of the dimensionless time corresponds to 1.6×10^{-4} s. For the attractive case, we focus on ^{7}Li atoms with $m = 1.17 \times 10^{-26}$ Kg, and a unity of the dimensionless space corresponds to $a_\perp = \sqrt{\hbar/(m\omega_\perp)} \simeq 1.19$ μm. Here, we want to point out that in the following discussion, both the space and time extent of the obtained solutions are all in the range of quasi-one-dimensional validity. Furthermore, we focus on various nonautonomous matter waves of Eq. (6.63), which generate new matter-wave structures and can be controlled by the modulation of the inhomogeneous source.

6.5.1 Search of Spatially Localized and Temporally Manipulated Nonautonomous Solutions

To derive the exact nonautonomous solutions of Eq. (6.63), we consider the general nonautonomous solutions in the form [44]

$$\psi(x, t) = \rho(t)\phi(\eta(x, t))\exp\left[i\varphi(x, t)\right], \quad (6.64)$$

where $\rho(t)$ is a time-dependent function that can be used to modulate the amplitude of nonautonomous matter waves, $\eta(x, t)$ stands for the nonautonomous matter-wave variables, and $\phi(\eta)$ is a real real function of a single variable η. We impose to $\rho(t)$ and $\eta(x, t)$ and to the distributed coefficients of Eq. (6.63) to satisfy the system

$$\frac{\partial^2\eta}{\partial x^2} = 0, \quad \frac{\partial\eta}{\partial t} + \frac{\partial\eta}{\partial x}\frac{\partial\varphi}{\partial x} = 0, \tag{6.65a}$$

$$2\frac{d\rho}{dt} + \rho\left(\frac{\partial^2\varphi}{\partial x^2} - 2\gamma\right) = 0, \quad s(t) - \delta_1\rho\left(\frac{\partial\eta}{\partial x}\right)^2 = 0, \tag{6.65b}$$

$$g(t)\rho^2 - \alpha_1\left(\frac{\partial\eta}{\partial x}\right)^2 = 0, \quad 2v(x,t) - 6\gamma_1\left(\frac{\partial\eta}{\partial x}\right)^2 + \left(\frac{\partial\varphi}{\partial x}\right)^2 + 2\frac{\partial\varphi}{\partial t} = 0, \tag{6.65c}$$

where γ_1 is a real constant, related with the chemical potential, $\alpha_1 \neq 0$ is a real constant having the same sign as $g(t)$ ($\alpha_1 < 0$ for the attractive interaction and $\alpha_1 > 0$ for the repulsive interaction), and the real constant δ_1 is related with the source amplitude $\delta_1 \neq 0$. Inserting ansatz (6.64) into Eq. (6.63) and taking into account system (6.65a)–(6.65c) lead to the following elliptic ordinary differential equation (EODE) for $\Phi(\eta)$:

$$\left(\frac{d\phi}{d\eta}\right)^2 = \alpha_1\phi^4 + 6\gamma_1\phi^2 + 4\delta_1\phi + \epsilon \equiv R(\phi), \tag{6.66}$$

where ϵ is an arbitrary real constant.

It follows from Eq. (6.65a) that

$$\eta(x,t) = \alpha(t)x + \chi(t), \tag{6.67}$$

$$\varphi(x,t) = -\frac{1}{2\alpha}\frac{d\alpha}{dt}x^2 - \frac{1}{\alpha}\frac{d\chi}{dt}x + \varphi_0(t). \tag{6.68}$$

Using Eqs. (6.67) and (6.68) to solve Eqs. (6.65b) and (6.65c) yields

$$\rho(t) = \rho_0^2\sqrt{\frac{g(t)}{\alpha_1}}\exp\left[2\int_0^t\gamma(\tau)d\tau\right], \quad |\alpha(t)| = \frac{\rho_0^2}{\alpha_1}g(t)\exp\left[2\int_0^t\gamma(\tau)d\tau\right], \tag{6.69}$$

$$v(x,t) = v_2(t)x^2 + v_1(t)x + v_0(t), \tag{6.70}$$

$$s(t) = \frac{\delta_1\rho_0^6}{\alpha_1^2}g^2(t)\sqrt{\frac{g(t)}{\alpha_1}}\exp\left[6\int_0^t\gamma(\tau)d\tau\right], \tag{6.71}$$

where ρ_0 is a constant of integration and can, without loss of generality, be chosen as $\rho_0 = 1$, the coefficients $v_j(t)$ ($j = 0, 1, 2$) of the external potential are given by

$$
\begin{aligned}
v_2(t) &= \frac{\alpha''}{2\alpha} - \left(\frac{\alpha'}{\alpha}\right)^2 = \frac{g\left(g'' + 4\gamma g' + 4\gamma^2 g + 2g\gamma'\right) - 2\left(g' + 2g\gamma\right)^2}{2g^2},\\
v_1(t) &= \frac{\chi''}{\alpha} - 2\frac{\chi'\alpha'}{\alpha^2},\\
v_0(t) &= -\frac{1}{2}\left(\frac{\chi'}{\alpha}\right)^2 + 3\gamma_1\alpha^2 - \varphi_0',
\end{aligned}
\tag{6.72}
$$

with $\varphi_0(t)$ and χ being two arbitrary differentiable functions of variable t. Without loss of generality, we assume $v_0(t) = 0$ so that φ_0 is defined as

$$
2\varphi_0' - 6\gamma_1\alpha^2 + \left(\frac{\chi'}{\alpha}\right)^2 = 0. \tag{6.73}
$$

Therefore, $\chi(t)$ will be an arbitrary differentiable functions of variable t. It follows from Eq. (6.73) that if the differentiable arbitrary function $\varphi_0(t)$ is taken from condition $6\gamma_1\alpha^2 - 2\varphi_0' = 0$, we will have $v_1(t) = 0$, which leads to the time-dependent parabolic potential $v(x,t) = v_2(t)x^2$. When $\varphi_0(t)$ is taken from condition $6\gamma_1\alpha^2 - 2\varphi_0' \neq 0$, the external potential of Eq. (6.63) will be $v(x,t) = v_2(t)x^2 + v_1(t)x$ for which the quadratic term is clearly the most physically relevant example of an external potential in the BEC case, giving the harmonic confinement of atoms by experimentally used magnetic traps, and the linear term, $v_1(t)x$, may correspond to the gravitational field or some linear potentials.

It follows from Eqs. (6.68) and (6.69) that the gain or loss term $\gamma(t)$ and $\alpha(t)$ (i.e., the inverse of the width of the nonautonomous matter waves) can be used to manipulate the amplitude $\rho(t)$ of the matter-wave solutions, and $\alpha(t)$ and $\chi(t)$ can adjust the phase offset and the frequency shift of the phase. Moreover, it follows from Eq. (6.70) that the external potential $v(x,t)$ is a second-degree polynomial in space with their coefficients being functions of time, which just satisfies the condition that the important potential in BECs is chosen as the time-dependent harmonic trapping potential. Equation (6.71) reveals that the gain/loss term $\gamma(t)$ can be used to modulate the source amplitude $s(t)$ and nonlinearity g(t). In addition, we find that the two functions $\alpha(t)$ and $\chi(t)$ do affect the amplitude $\rho(t)$, the external potential $v(x,t)$, the phase $\varphi(x,t)$, the nonlinearity coefficient $g(t)$, and the source amplitude $s(t)$.

The problem of finding the spatially localized and temporally manipulated nonautonomous solutions of the NLS Eq. (6.63) is thus reduced to that of finding solutions of the EODE (6.66). As is well known, solution ϕ of the EODE (6.66) can be expressed in terms of Weierstrass' elliptic function $\wp(\eta; g_2, g_3)$ according to (see for example Ref. [40], pp. 4–16)

$$
\phi(\eta) = \phi_0 + \frac{R'(\phi_0)}{4\left[\wp(\eta; g_2, g_3) - \frac{1}{24}R''(\phi_0)\right]}, \tag{6.74a}
$$

where the primes denote differentiation with respect to ϕ and ϕ_0 is a simple root of the polynomial $R(\phi)$. For simplicity, we assume $\epsilon = 0$ so that $\phi_0 = 0$ is a simple root of polynomial $R(\phi)$. The invariants g_2, g_3 of $\wp(\eta; g_2, g_3)$ and the discriminant Δ are related to the coefficients of $R(\phi)$ as follows:

$$g_2 = 3\gamma_1^2,\ g_3 = -\alpha_1\delta_1^2 - \gamma_1^3,\ \Delta = g_2^3 - 27g_3^2 = -27\left(\alpha_1\delta_1^2 + 2\gamma_1^3\right)\alpha_1\delta_1^2,$$

and are suitable to classify the behavior of $\phi(\eta)$ and to discriminate between periodic and solitary wave-like solutions of Eq. (6.66). The conditions

$$\Delta \neq 0 \text{ or } \Delta = 0,\ g_2 > 0, \text{ and } g_3 > 0$$

lead to periodic solutions

$$\phi(\eta) = \frac{\left(\alpha_1\phi_0^3 + 2e_3\phi_0 + 5\gamma_1\phi_0 + 2\delta_1\right)\operatorname{sn}^2\left[\sqrt{e_1 - e_3}\eta, \widetilde{m}\right] + 2\left(e_1 - e_3\right)\phi_0}{\left(-\alpha_1\phi_0^2 + 2e_3 - \gamma_1\right)\operatorname{sn}^2\left[\sqrt{e_1 - e_3}\eta, \widetilde{m}\right] + 2\left(e_1 - e_3\right)}, \tag{6.75}$$

where $\widetilde{m} = \frac{e_2 - e_3}{e_1 - e_3}$ and $e_1 \geq e_2 \geq e_3$ denote the roots of the equation

$$4s_1^3 - g_2 s_1 - g_3 = 0; \tag{6.76}$$

The conditions

$$\Delta = 0,\ g_2 \geq 0, \text{ and } g_3 \leq 0$$

are associated with solitary wave-like solutions

$$\phi(\eta) = \phi_0 + \frac{R'(\phi_0)}{4\left[e_1 - \frac{1}{24}R''(\phi_0) + \frac{3e_1}{\cosh^2\left[\sqrt{3e_1}\eta\right]}\right]}, \tag{6.77}$$

where $e_1 = \frac{1}{2}\sqrt[3]{-g_3}$ (if $g_3 \neq 0$); case $g_3 = g_2 = 0$ leads to an algebraic soliton solution. It is important to note that solutions (6.75) and (6.77) are valid only if ϕ_0 is a simple root of polynomial $R(\phi)$. Physical solutions $\phi(\eta)$ must be real and bounded. Conditions for real and bounded solutions $\phi(\eta)$ can be obtained by considering the phase diagrams of $R(\phi)$ [9].

It follows from Eqs. (6.67) and (6.72) that the center of the nonautonomous matter waves is $\xi(t) = -\chi(t)/\alpha(t)$, which satisfies the following equation:

$$\frac{d^2\xi}{dt^2} + 2\upsilon_2(t)\xi + \upsilon_1(t) = 0. \tag{6.78}$$

Equation (6.78) means that the center of mass of the macroscopic wave packet behaves like a classical particle, and allows one to manipulate the motion of nonautonomous matter waves in BEC systems by controlling the external trapping potential. Because $\upsilon_1(t)$ depends on an arbitrary functional parameter $\chi(t)$, we conclude that parameter $\chi(t)$ will guide the motion of the nonautonomous matter waves in BECs described by Eq. (6.63).

For simplicity, we focus ourselves to the solutions associated with the trivial simple root $\phi_0 = 0$ of polynomial $R(\rho)$. Solutions (6.75) and (6.77) become

$$\phi(\eta) = \frac{2\delta_1 \mathrm{sn}^2\left[\sqrt{e_1 - e_3}\eta, \widetilde{m}\right]}{(2e_3 - \gamma_1)\ \mathrm{sn}^2\left[\sqrt{e_1 - e_3}\eta, \widetilde{m}\right] + 2(e_1 - e_3)} \tag{6.79}$$

and

$$\phi(\eta) = -\frac{\delta_1}{\gamma_1} + \frac{\delta_1}{\gamma_1}\frac{3}{3 + 2\cosh^2\left[\sqrt{-\frac{3}{2}\gamma_1\eta}\right]}, \quad \gamma_1 < 0, \tag{6.80}$$

respectively. Here, $e_1 \geq e_2 \geq e_3$ are the roots of Eq. (6.76). In the situation when $\Delta = 0$ and $\gamma_1 > 0$, these roots are $e_1 = \gamma_1 > e_2 = -\frac{1}{2}\gamma_1 = e_3$, such that $\widetilde{m} = 0$, and solution (6.79) becomes

$$\phi(\eta) = -\frac{\delta_1}{\gamma_1} + \frac{\delta_1}{\gamma_1}\frac{3}{3 - 2\sin^2\left[\sqrt{\frac{3}{2}\gamma_1\eta}\right]}, \quad \gamma_1 > 0. \tag{6.81}$$

Thus, conditions $\Delta = 0$ and $\gamma_1 > 0$ lead to the trigonometric solution (6.81), while conditions $\Delta = 0$ and $\gamma_1 < 0$ lead to the soliton solution (6.80).

Choosing α_1, γ_1, and δ_1 from the condition $\alpha_1\delta_1^2 + \gamma_1^3 = 0$ and solving Eq. (6.76) lead to $e_1 = \frac{\sqrt{3}}{2}\gamma_1 > e_2 = 0 > e_3 = -\frac{\sqrt{3}}{2}\gamma_1$ if $\gamma_1 > 0$ and $e_1 = -\frac{\sqrt{3}}{2}\gamma_1 > e_2 = 0 > e_3 = \frac{\sqrt{3}}{2}\gamma_1$ if $\gamma_1 < 0$. Solution (6.79) becomes

$$\phi(\eta) = \frac{2\delta_1}{\left(1 + \sqrt{3}\right)\gamma_1}\frac{\mathrm{sn}^2\left[\sqrt{\sqrt{3}\gamma_1}\eta, \frac{1}{2}\right]}{\left(\frac{2\sqrt{3}}{1+\sqrt{3}} - \mathrm{sn}^2\left[\sqrt{\sqrt{3}\gamma_1}\eta, \frac{1}{2}\right]\right)}, \quad \text{if } \gamma_1 > 0, \tag{6.82}$$

$$\phi(\eta) = -\frac{2\delta_1}{\left(\sqrt{3} - 1\right)\gamma_1}\frac{\mathrm{sn}^2\left[\sqrt{-\sqrt{3}\gamma_1}\eta, \frac{1}{2}\right]}{\left(\frac{2\sqrt{3}}{\sqrt{3}-1} + \mathrm{sn}^2\left[\sqrt{-\sqrt{3}\gamma_1}\eta, \frac{1}{2}\right]\right)}, \quad \text{if } \gamma_1 < 0. \tag{6.83}$$

6.5.2 Dynamics of Nonautonomous Matter Waves in 1D BEC Systems with Inhomogeneous Source

In the following, we take some examples to demonstrate analytically the dynamics of nonautonomous matter waves in 1D BEC systems with gain/loss of atoms for different kinds of scattering length and harmonic trapping potential.

BECs in a Time-Independent Harmonic Potential

As the first example, we consider the time-independent harmonic potential which was used in the creation of bright BEC solitons [20]. In that experiment $\omega = 2\pi i \times 70$ Hz, $\omega_\perp = 2\pi \times 710$ Hz, so $v(x,t) = -2\kappa^2 x^2$ ($\kappa \simeq 0.05$). According to Eq. (6.70), $v_2(t) = -2\kappa^2$ and $v_1(t) = v_0(t) = 0$. From system (6.72), we assume that $\chi(t)$ is a constant and $\varphi_0' - 3\gamma_1\alpha^2 = 0$. It follows from the first equation of system (6.72) that either $g(t) = g_0 \exp\left[2\left(\pm\kappa - \gamma\right)t\right]$ for any real constant γ or $g(t) = g_0 =$ constant and $\gamma(t) = \pm\kappa/\sqrt{2}$. The situations when $g(t) = g_0 \exp\left[2\left(\pm\kappa - \gamma\right)t\right]$ with $\gamma = 0$ and when $g(t) = g_0$ and $\gamma(t) = \pm\kappa/\sqrt{2}$ are similar to those studied in Sects. 6.2 and 6.3, respectively.

Let us consider the situation when $g(t) = g_0 \exp\left[2\left(\pm\kappa - \gamma\right)t\right]$ with $\gamma \neq 0$. From system (6.69)–(6.71), we have $\rho(t) = \rho_0^2\sqrt{\frac{g_0}{\alpha_1}} \exp\left[\left(\pm\kappa + \gamma\right)t\right]$, $|\alpha(t)| = \frac{g_0\rho_0^2}{\alpha_1} \exp\left[\pm 2\kappa t\right]$, $s(t) = \frac{\delta_1 g_0^2 \rho_0^6}{\alpha_1^2}\sqrt{\frac{g_0}{\alpha_1}} \exp\left[\left(\pm 5\kappa + \gamma\right)t\right]$. Therefore, the wave amplitude $\rho(t)$ will increase (decrease) with time if the loss/gain parameter γ is taken from condition $\pm\kappa + \gamma > 0$ ($\pm\kappa + \gamma < 0$). When $\gamma = \mp\kappa$, the matter wave will maintain its amplitude during its propagation.

BECs with a Temporal Periodic Modulation of the s-wave Scattering Length

As the second example, we consider the BECs with a temporal periodic modulation of the s-wave scattering length [33], and the nonlinearity parameter takes the form $g(t) = 1 + m \sin\left[\omega t\right]$, with $0 < m < 1$. It then follows from Eq. (6.71) that, dependent on the choice of the gain/loss parameter $\gamma(t)$, the source $s(t)$ will be temporal modulated function. In the present example, we assume that $\gamma(t) = \kappa/2 - g'/g$ [18], where κ is a small parameter such that $|\gamma(t)| << 1$. Figure 6.14 shows the evolutions of nonautonomous matter waves for different values of parameter κ and $\chi(t) = 1 - m \sin\left[\omega t\right]$. It is seen from plots of Fig. 6.14 that the waves trajectories oscillate mainly due to the temporal periodic modulation of the s-wave scattering length and trapping potential.

6.6 NLS Equation with Spatiotemporally Modulated Nonlinearity

In the presents section, we consider a NLS equation with a spatiotemporal modulation of the nonlinearity and external potentials that describes the dynamics of the macroscopic wave function for the cigar-shaped BEC. A scheme for constructing exact soliton solutions to the NLS under consideration is presented. In this way, we construct stable higher order modes built as arrays of dark solitons nested in a finite-width background. We then show that the instability of those solitons which are not stable can be controlled by varying the nonlinearity's strength in time. The stabilization mechanism resembles the classical effect of the parametric stabilization, such as the inverted pendulum, which can be sustained by an oscillating pivot.

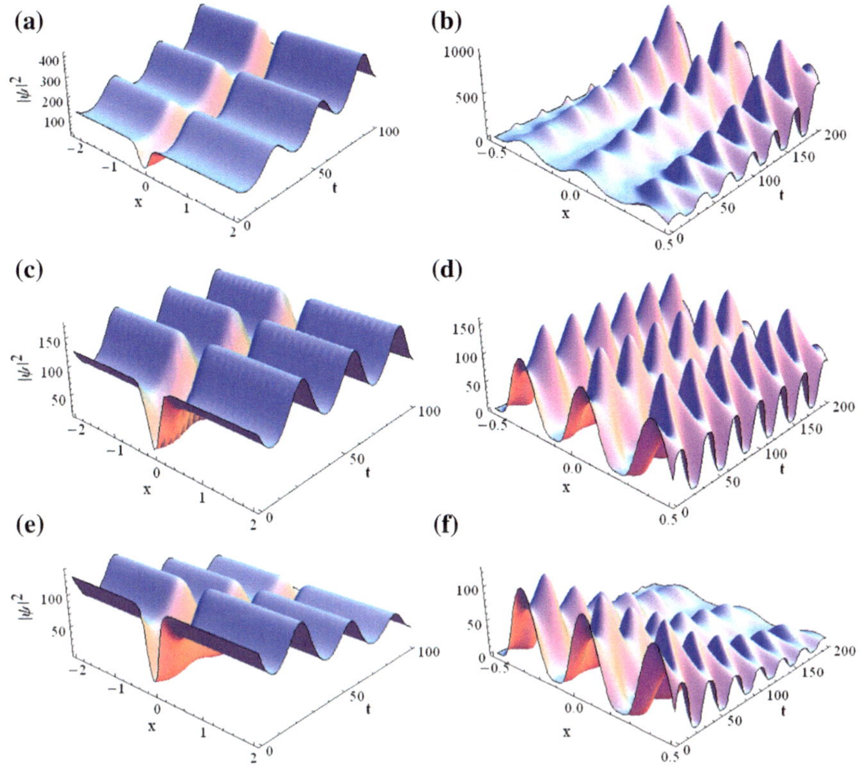

Fig. 6.14 The evolution of nonautonomous matter waves given by Eqs. (6.80) and (6.83) with $\gamma_1 = -1$, $\alpha_1 = 2$, and $\chi(t) = 1 - m \sin[\omega t]$ for different values of parameter κ. **a**, **b**: $\kappa = 0.005$, **c**, **d**: $\kappa = 0$, and **e**, **f**: $\kappa = -0.005$. Plots (**a**), (**c**), and (**e**) are obtained with $\delta_1 = 1$, while plots (**b**), (**d**), and (**f**) are generated with $\delta_1 = 2$. Other parameters are $m = 0.1$ and $\omega = 0.2$

6.6.1 Model and Stationary Solutions

The model to be studied in this section is the 1D normalized cubic NLS equation

$$i\frac{\partial \psi}{\partial \xi} = -\frac{\partial^2 \psi}{\partial x^2} + g(x, \xi)\,|\psi|^2\,\psi + v(x, \xi)\psi, \tag{6.84}$$

where $\psi(x, \xi)$ is the macroscopic wave function for the cigar-shaped BEC if ξ stands for time, $v(x, \xi)$ is the axial potential, and g is proportional to the interatomic s-wave scattering length. The spatiotemporal modulation of $g(x, \xi)$ can be induced by means of the Feshbach resonance technique [28]. On the other hand, Eq. (6.84) with $\xi = z$ governs the propagation of an optical beam in a planar waveguide along the z direction, the external potential being generated by a modulation of the local refractive index, while various forms of $g(x, z)$ may be realized via accordingly designed distributions of nonlinearity-enhancing dopants [16, 38].

Assuming that both g and v are functions of x only, we seek the stationary solutions of Eq. (6.84) in the form

$$\psi(x, \xi) = \rho(x)U(X(x)) \exp\left[-i\mu\xi\right], \tag{6.85}$$

where μ (chemical potential, in the case of BEC) is a real constant, and $X(x) = \int_{-\infty}^{x} \rho^{-2}(s)ds$, and $\rho(x)$ is a solution of equation $d^2\varphi/dx^2 + [\mu - v(x)]\,\varphi = 0$; we can then seek $\rho(x)$ in the form

$$\rho(x) = \sqrt{\alpha\varphi_1^2 + 2\beta\varphi_1\varphi_2 + \gamma\varphi_2^2}, \tag{6.86}$$

where $\varphi_1(x)$ and $\varphi_2(x)$ are two linearly independent solutions of equation $d^2\varphi/dx^2 + [\mu - v(x)]\,\varphi = 0$, α, β, and γ being real constants. Further, setting

$$g(x) = \rho^{-6}(x)F(U), \tag{6.87}$$

we conclude that $U(X)$ satisfies the ODE

$$EU = -\frac{d^2U}{dx^2} + F(U)U^3, \tag{6.88}$$

where $E \equiv (\alpha\gamma - \beta^2)W^2$, with constant Wronskian $W = \varphi_1(x)d\varphi_2/dx - \varphi_2(x)\,d\varphi_1/dx$. In what follows, we are interested in the case of $E > 0$, when Eq. (6.86) guarantees that $\rho(x)$ does not vanish; hence the nonlinearity coefficient (6.87) does not have singularities.

Exact soliton solutions of Eq. (6.84) are generated by exact solutions to Eq. (6.88), which may be obtained with many forms of $F(U)$. The widely used choice is that with $F(U)$ being a constant [5], whereas the sign of the nonlinearity coefficient $g(x)$ is either focusing (negative) or defocusing (positive) in the whole spatial domain. A more versatile choice is $F(U) = g_3 + g_5U^2$, which transforms Eq. (6.88) into a solvable cubic–quintic equation. The addition of the term g_5U^2 can generate a larger variety of exact soliton solutions to Eq. (6.84). Note that the respective cubic nonlinearity coefficient may even be made sign-changing, for $g_3g_5 < 0$. Yet more exact solitons can be generated by choosing other forms of $F(U)$. We will construct exact solutions taking

$$F(U) = U^{-3}[EU - \sin(EU)], \tag{6.89}$$

which casts Eq. (6.88) into the stationary sine-Gordon equation, $d^2U/dX^2 + \sin(EU) = 0$. This equation with$E > 0$ describes the motion of a pendulum, i.e., either oscillations, with U being a periodic function of X, or the rotation, with U

varying monotonically. The localized soliton requires U to be a periodical function; thus

$$U(X) = 2E^{-1} \arcsin\left[q\,\mathrm{sn}(\sqrt{q}X, q)\right] \quad (6.90)$$

where sn denotes the Jacobi's elliptic function with modulus q. The period of $U(X)$ in Eq. (6.90) is $4K(q)/\sqrt{E}$, where $K(q)$ is the complete elliptic integral of the first kind.

The cubic nonlinearity given by Eqs. (6.87) and (6.89) has the defocusing sign, i.e., $g(x) > 0$. Therefore, an external potential is needed to confine the system. Here, we consider the most physically relevant case of the harmonic potential, $v(x) = \omega_0^2 x^2$. In this case, solutions $\varphi_1(x)$ and $\varphi_2(x)$ may be expressed in terms of the M and W functions of Whittaker and Watson [41] and Gamma function Γ : $\varphi_1(x) = \omega_0^{-3/4} \mathrm{sign}(x)\,|x|^{-1/2} M$ and $\varphi_2(x) = \Gamma\left[(3\omega_0 - \mu)/(4\omega_0)\right] \omega_0^{-1/4}$ $(\pi\,|x|)^{-1/2} \left\{W + 2\sqrt{\pi}\Gamma\left[(3\omega_0 - \mu)/(4\omega_0)\right]^{-1} M\right\}$, when $\mu \neq \widetilde{\mu}^{(\ell)} \equiv (2\ell + 1)\,\omega_0$, $\ell = 0, 1, 2, \ldots$, with $\widetilde{\mu}^{(\ell)}$ being the ℓ th energy eigenvalue of the harmonic oscillator. Since $\varphi_1(0) = d\varphi_2(0)/dx = 0$ and $\varphi_2(0) = d\varphi_1(0)/dx = 1$, the Wronskian is $W = -1$. As $E > 0$, $\rho(x) = 0$ holds when $\alpha, \gamma > 0$.

We focus on the symmetric profiles of the nonlinearity modulation and the corresponding solitons by setting $\beta = 0$ in Eq. (6.86). To produce a typical example, we set $\omega_0 = 0.1$, $\alpha = 6^{1/3}$, and $\gamma = 1$, which makes $g(x) \sim 1$ around the center. Then, for given μ, $\rho(x)$ and $X(x)$ can be calculated. To meet the zero boundary condition at $x = \pm\infty$, modulus q must satisfy the constraint, $2nK(q) = \sqrt{E}R$, where n is a positive integer and $R = \int_{-\infty}^{+\infty} \rho^2(s)ds$. Since $K(q) > \pi/2$, it follows from here that $n < \sqrt{E}R/\pi$. This clearly means that there is only a finite number of the exact solutions for given μ. Integer n is the order of the soliton mode, which features $n - 1$ density nodes. Further analysis demonstrates that $\sqrt{E}R/\pi$ increases monotonically as μ increases,while other parameters are fixed, and n takes values up to $\ell + 1$ when $\widetilde{\mu}^{(\ell)} < \mu < \widetilde{\mu}^{(\ell+1)}$; when $\mu < \widetilde{\mu}^{(0)}$, there are no solutions (see Fig. 6.15). An explanation for this is that the defocusing nonlinearity pushes the energy levels up relative to the harmonic oscillator. Thus, the fundamental solitons ($n = 1$) exist when $\mu > \widetilde{\mu}^{(0)}$, first-order excited solitons ($n = 2$) exist when $\mu > \widetilde{\mu}^{(1)}$, etc. This conclusion coincides with the findings of Ref. [22] for the spatially homogeneous defocusing nonlinearity. A similar result holds for gap solitons in self-defocusing media: only first n families of the solitons exist when μ lies in the nth optical-lattice-induced bandgap [45].

In keeping with the preceding analysis, Fig. 6.15a shows that only one exact soliton exists when $\mu = 0.2$. The corresponding single-humped nonlinearity-modulation profile is localized near $x = 0$, and the exact soliton has a small dip, because of the self-repulsive sign of the nonlinearity. We checked the stability of the soliton against various perturbations by means of direct simulations of Eq. (6.84). In particular, if kicked with initial velocity 0.1, it remains stable, featuring periodic oscillations (Fig. 6.16a).

Using the numerical integration in imaginary time, one may show that this and other fundamental solitons in our model represent the ground state, which explains

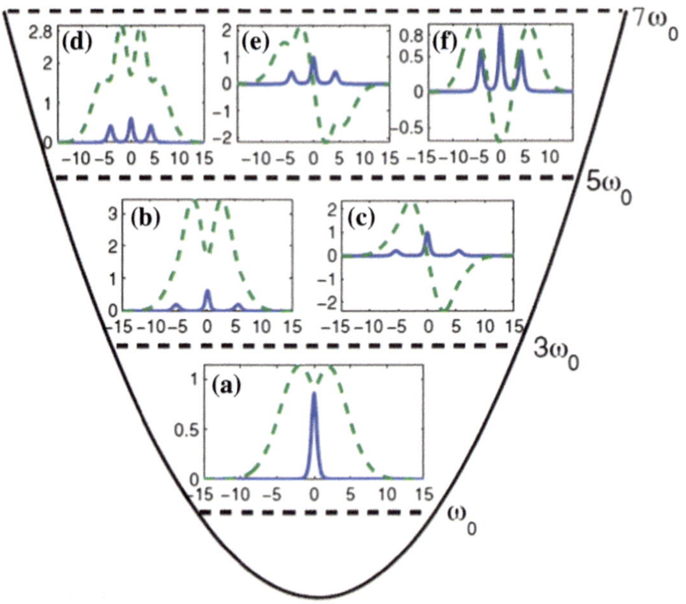

Fig. 6.15 The relation between the number of exact soliton solutions and the discrete energy levels of the harmonic oscillator potential (the underlying parabola). **a** Modulation profiles of the defocusing nonlinearity $g(x)$ (solid line), and the shape of the exact soliton (dashed line), when $\mu = 0.2$, $n = 1$, and $q = 0.7458$. The horizontal axis is coordinate x. Other insets display the same for **b** $\mu = 0.4$, $n = 1$, $q = 0.9984$, **c** $\mu = 0.4$, $n = 2$, $q = 0.8586$, **d** $\mu = 0.6$, $n = 1$, $q = 0.9998$, **e** $\mu = 0.6$, $n = 2$, $q = 0.9530$, and **f** $\mu = 0.6$, $n = 3$, $q = 0.5933$. The other parameters are $\alpha = 6^{1/3}$, $\beta = 0$, $\gamma = 1$, and $\omega_0 = 0.1$. Reprint from Ref. [38], Copyright 2018, with permission from American Physical Society

their robustness. As μ increases, more and more exact solitons appear, while the nonlinearity profile develops several peaks. In particular, Fig. 6.15b, c demonstrates that two exact solitons exist in the case of $\mu = 0.4$. Besides the stable fundamental soliton, the first-excited-state soliton, which looks like a dark soliton nested in a finite-width background, is stable too (see Fig. 6.16b). However, at $\mu = 0.6$, the exact first and second-excited-state solitons are not stable; see Fig. 6.17b for the former one. While it is not easy to analyze the stability of all the solitons corresponding to the excited states, at larger values of μ it is possible to generate arrays of nested dark solitons which maintain their stability—in particular, if they are kicked with small spatially homogeneous (Fig. 6.18b) or inhomogeneous (Fig. 6.18c) velocities. This finding suggests a new approach to constructing the Newton's cradle and exciting supersolitons in systems described by the scalar NLS equation, where the nested dark chains may be generated by the phase-imprinting method [35].

It is important to note that for a fixed chemical potential μ, there is one-to-one correspondence between the exact soliton and the nonlinearity profile $g(x)$. On the other hand, for fixed $g(x)$, one can find other soliton solutions numerically, by varying μ and using the relaxation method. We have checked that, if the exact soliton is stable,

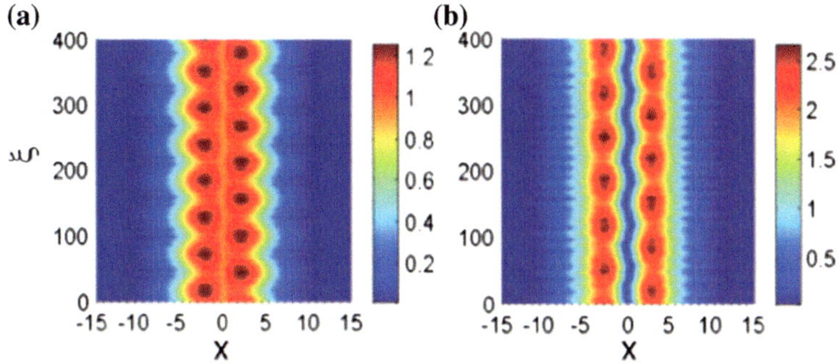

Fig. 6.16 **a** Oscillations of the soliton from Fig. 6.15a which was left-kicked with velocity 0.1, the corresponding initial condition for Eq. (6.84) being $\rho(x)U(X(x))\exp(-0.05ix)$. **b** Oscillations of the soliton from Fig. 6.15c which was left-kicked with initial velocity 0.05. Reprint from Ref. [38], Copyright 2018, with permission from American Physical Society

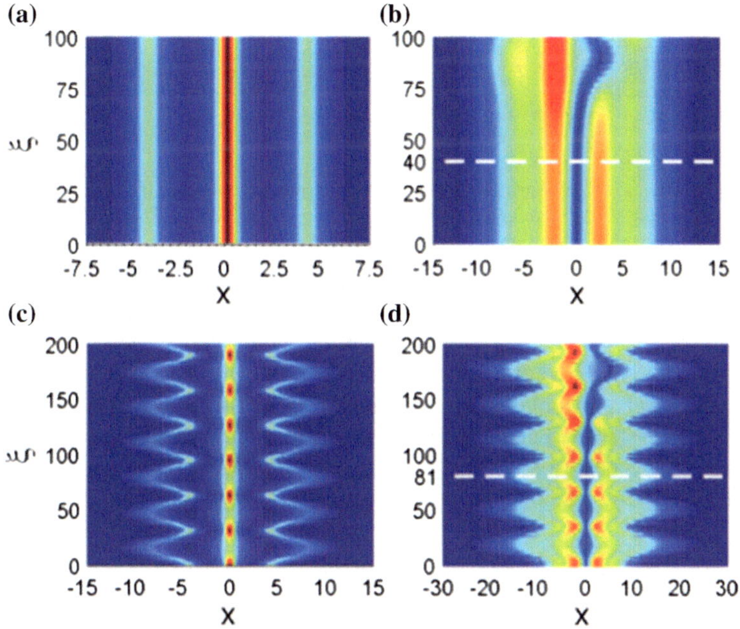

Fig. 6.17 **a** The profile of the spatially modulated nonlinearity and **b** the unstable evolution of the soliton from Fig. 6.15e. **c** The spatiotemporal nonlinearity modulation and **d** the unstable evolution of the soliton, for the initial condition the same as in panel (**b**), but with the nonlinearity taken as per Eq. (6.92) and half the trapping frequency, $\omega = \omega_0/2$. Reprint from Ref. [38], Copyright 2018, with permission from American Physical Society

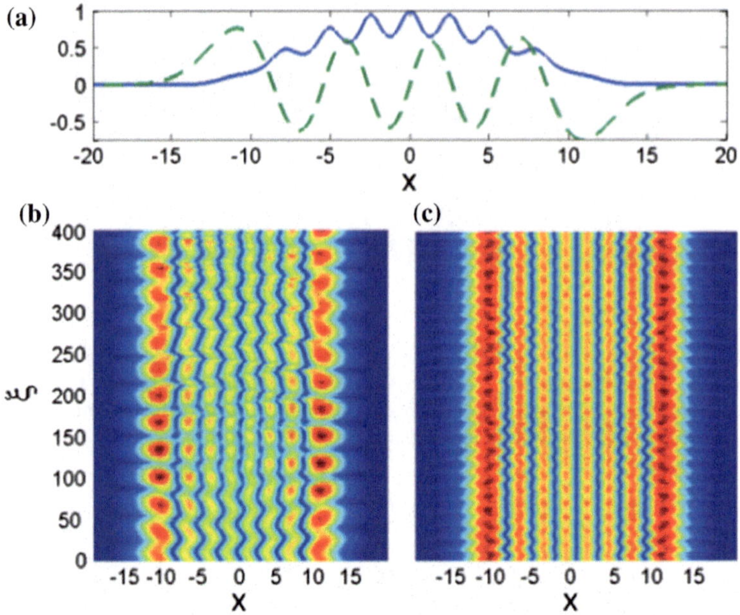

Fig. 6.18 **a** The modulation profile of the defocusing nonlinearity and the corresponding exact soliton (solid and dashed lines) for $\mu = 1.6$, $n = 8$, and $q = 0.4752$. **b**, **c** The quasistable evolution of the soliton. The initial condition in panel (**b**) for Eq. (6.84) is $\rho(x)U(X(x))\exp[-0.05ix]$, while that in panel (**c**) is $\rho(x)U(X(x))\exp[-0.05i\,|x|]$. Reprint from Ref. [38], Copyright 2018, with permission from American Physical Society

then its counterparts numerically found for the same $g(x)$ are stable too, provided that the chemical potential does not vary too much. Further, it was checked that these solitons remain stable when the nonlinearity profiles were taken somewhat different from the special form (6.87). Thus, higher order solitons are physically meaningful objects.

6.6.2 Stability Control of Nonautonomous Solitons

Now, we turn to the situation when v and g are functions of both x and ξ, reducing thus Eq. (6.84) to a nonautonomous NLS equation. Inspired by the finding in the previous section, we here seek a possibility to construct exact nonstationary solitons for Eq. (6.84) with the spatiotemporal modulation of the nonlinearity and harmonic external potential, transforming it into a NLS equation with time-independent coefficients,

$$i\frac{\partial \Psi}{\partial T} = -\frac{\partial^2 \Psi}{\partial Y^2} + g_2(Y)\,|\Psi|^2\,\Psi + \omega_0^2 Y^2 \Psi. \tag{6.91}$$

The reduction can be implemented if the nonlinearity coefficient and external potential in Eq. (6.84) are of the form

$$g(x,\xi) = g_1(\xi)g_2(x/W(\xi)),\ v(x,\xi) = \omega(\xi)x^2, \tag{6.92}$$

where $W(\xi)$ is the width of the nonlinearity modulation, which also determine the soliton's width. Thus, we introduce the self-similar transformation [42]

$$\psi(x,\xi) = \Psi(Y,T)W^{-1/2}\exp\left[i\frac{\partial W}{\partial \xi}\frac{x^2}{4W}\right], \tag{6.93}$$

with $T \equiv \int_0^\xi ds/W^2(s)$ and $Y \equiv xW^{-1}(\xi)$. It can be checked that Eq. (6.91) can be recovered, in this way, only for $g_1(\xi) = 1/W(\xi)$, with $W(\xi)$ satisfying equation $d^2W/d\xi^2 + 4\omega^2 W = 4\omega_0^2 W^{-3}$.

The exact stationary soliton solutions displayed in Fig. 6.15 can be borrowed to produce solutions to Eq. (6.91), replacing Y by x, g_2 by g, T by ξ, and Ψ by ψ. Moreover, given the functional form of $\omega(\xi)$, such as $a_0 + a_1\cos\xi$, the equation for $W(\xi)$ can be solved. Then, one can construct exact nonstationary soliton solutions of Eq. (6.84) as per Eq. (6.93).

It is worthy to note that the stability of the nonstationary soliton is related to that of the corresponding stationary solution and the functional form of $W(\xi)$. If the stationary soliton of Eq. (6.91) is stable, the respective nonstationary soliton of Eq. (6.84) is stable too. However, if the stationary soliton is unstable, with the instability setting in at, say, $T = T_0$, the instability of the nonstationary soliton may be delayed or even suppressed, choosing an appropriate form of $W(\xi)$. Namely, if $T = \int_0^\xi ds/W^2(s)$ always remains smaller than T_0, the instability is completely prevented, because the system does not have enough time (propagation distance) to reach the instability threshold. On the other hand, the onset of the instability will be delayed if equation $T = \int_0^\xi ds/W^2(s)$ yields $\xi > T_0$.

As a typical example, we take the exact soliton corresponding to Fig. 6.15e. Without the temporal nonlinearity modulation, its instability starts at $\xi \approx 40$ (Fig. 6.17b). If at $\xi = 0$, the trap frequency is abruptly reduced to a half, and after that the nonlinearity is varied as per Eq. (6.92) with $W(\xi) = \sqrt{1 + 3\sin^2(\omega_0\xi)}$ (Fig. 6.17c), we find that the soliton first experiences the exact self-similar evolution in accordance with Eq. (6.93), and then it starts to develop the instability at $\xi \approx 81$, according to equality $40 = \int_0^\xi W^{-2}(s)ds$ (Fig. 6.17d). Clearly, the instability onset is delayed. It can be delayed further if one decreases the trapping frequency more, varying the nonlinearity accordingly.

6.7 Wave Trains Generation in a Delayed Nonlinear Response of BECs with Spatiotemporally Nonlinearity

Most recently, using an appropriate additional phase imprint on the macroscopic wave function of BECs at the mean-field level as the zero temperature is approached, Kengne et al. [17] showed that the dynamical behavior of a BEC is accurately described by the following modified GP equation:

$$i\frac{\partial\psi}{\partial t}+\frac{\partial^2\psi}{\partial x^2}+V(x,t)\psi+g(x,t)\left|\psi\right|^2\psi+4i\beta(t)\frac{\partial\left(\psi^2\psi^*\right)}{\partial x}=0, \tag{6.94}$$

where ψ^* denotes the complex conjugate of ψ, the spatiotemporal functional parameter $g(x,t)$ of the cubic nonlinearity represents the two-body interatomic interaction coefficient, negative for repulsive interatomic interactions (or defocusing nonlinearities) and positive for attractive ones (focusing nonlinearities), and $V(x,t)$ is the external potential, and and $\beta(t)$ is a time-dependent real function which represents the relative magnitudes of the nonlinear dispersion term. The last term of Eq. (6.94), $4i\beta(t)\frac{\partial(\psi^2\psi^*)}{\partial x}$ represents the delayed nonlinear response of the system which offsets the modulation generally arising due to three-body interactions. This last term allows the investigation of the MI of BECs described by the GP Eq. (6.94) with the time-varying harmonic potential

$$V(x,t)=-\alpha(t)x^2 \tag{6.95}$$

for both focusing and defocusing nonlinearities; here, $\alpha(t)$ is a real function of time t. The potential (6.95) is relevant, in particular, to experimental setups in which the (magnetic) trap is strongly confined in the two directions, while it is much shallower in the third one. The strength of the magnetic trap $\alpha(t)$ may be negative (confining potential) or positive (repulsive potential); it is typically fixed in current experiments, but adiabatic changes in the strength of the trap are experimentally feasible, hence we examine the more general time-dependent case. The spatiotemporally nonlinearity $g(x,t)$ is assumed to be a first degree polynomial in x with time-dependent modulated coefficient parameter

$$g(x,t)=g_1(t)x+g_0(t), \tag{6.96}$$

where g_0 and g_1 are two real functions of time t. When $g_1(t)\equiv 0$, Eq. (6.96) gives a time-dependent s-wave scattering length, $g(x,t)=g_0(t)$, which has been used in many works. We therefore focus ourselves to $g_1(t)\neq 0$.

In this section, we investigate the phenomenon of the modulational instability of matter-wave condensates described by Eq. (6.94) with potential (6.95) when the spatiotemporally nonlinearity $g(x,t)$ is given by Eq. (6.96); the phenomenon MI is an indispensable mechanism for the understanding of some relevant dynamical processes in BEC systems such as generation and propagation of matter-wave solitons.

6.7.1 Modulational Instability of Matter-Wave Condensates in NLS Equation with a Delayed Nonlinear Response

In order to examine the MI of matter-wave condensates in NLS equation with a delayed nonlinear response, we first perform a modified lens-type transformation (generalized of lens transformation) of the form

$$\psi(x,t) = \frac{1}{\ell(t)}\phi(X,T)\exp\left[if(t)x^2 + \eta(t)\right], \tag{6.97}$$

where $T = T(t)$, $\ell(t)$, and $f(t)$ are three dependent real functions, and $X(x,t) = x/\ell(t)$ is a spatiotemporal real function. We demand that

$$\frac{dT}{dt} - \frac{1}{\ell^2(t)} = 0, \tag{6.98a}$$

$$\frac{df}{dt} + 4f^2 + \alpha = 0, \tag{6.98b}$$

$$\frac{1}{\ell}\frac{d\ell}{dt} - 4f = 0, \tag{6.98c}$$

$$\frac{d\eta}{dt} - 2f = 0, \tag{6.98d}$$

$$\beta(t) - \frac{g_1(t)}{8f(t)} = 0. \tag{6.98e}$$

The choice of Eq. (6.98a) is made to preserve the scaling. Solving Eqs. (6.98c), (6.98a), (6.98d) yields

$$\begin{aligned} &\ell(t) = \ell_0 \exp\left[4\int_0^t f(s)ds\right], \; T(t) = \tfrac{1}{\ell_0^2}\int_0^t \tfrac{ds}{\exp\left[8\int_0^s f(\tau)d\tau\right]} + T_0, \\ &\eta(t) = 2\int_0^t f(s)ds + \eta_0, \end{aligned} \tag{6.99}$$

where $\ell_0 = \ell(0)$, $T_0 = T(0)$, and $\eta_0 = \eta(0)$. Then, Eq. (6.94), in terms of rescaled variables X and T, is converted to the following derivative NLS equation with distributed (time-dependent) coefficients:

$$i\frac{\partial\phi}{\partial T} + \frac{\partial^2\phi}{\partial X^2} + \lambda_0 |\phi|^2 \phi + i\lambda\frac{\partial\left(\phi^2\phi^*\right)}{\partial X} = 0, \tag{6.100}$$

where

$$\lambda_0 = g_0(t)\exp\left[2\eta(t)\right], \; \lambda = 4\frac{\beta(t)}{\ell(t)}\exp\left[2\eta(t)\right], \tag{6.101}$$

with $\eta(t)$ and $\ell(t)$ given by Eq. (6.99). Generally, λ_0 and λ are real and depend on the time t. Thus, we retrieve the cubic derivative NLS equation with time-dependent coefficients.

Linear Stability Analysis and Criterion of the MI for the Cubic Derivative NLS

For MI investigations in the BECs with spatiotemporally nonlinearity described by Eq. (6.94), we look for the ansatz in the form

$$\phi = (\phi_0 + \varepsilon)\exp\left[-iQX - i\int_0^T \Omega(v)dv\right], \tag{6.102}$$

where $\Omega(T)$ is a real time-dependent function representing the nonlinear frequency shift, ϕ_0 is a real shift constant, Q is the wave number of the carrier, and $\varepsilon = \varepsilon(X, T)$ is a small perturbation of the wave amplitude ($|\varepsilon| << 1$). Substituting ansatz (6.102) into the NLS Eq. (6.100) and linearizing ε and its complex conjugate ε^*, we obtain, after assuming that

$$\Omega(T) = Q^2 - \phi_0^2\left[\lambda_0(t) + Q\lambda(t)\right], \tag{6.103}$$

the following linear equation for ε:

$$i\frac{\partial\varepsilon}{\partial T} + \frac{\partial^2\varepsilon}{\partial X^2} + 2i\left(\phi_0^2\lambda - Q\right)\frac{\partial\varepsilon}{\partial X} + i\phi_0^2\lambda\frac{\partial\varepsilon^*}{\partial X} + \phi_0^2\left(\lambda_0 + Q\lambda\right)\left(\varepsilon + \varepsilon^*\right) = 0. \tag{6.104}$$

Now, we seek the perturbation $\varepsilon(X, T)$ in the form

$$\varepsilon(X,T) = U_1\exp\left[i\left(KX - \int_0^T \omega(v)dv\right)\right] + U_2^*\exp\left[-i\left(KX - \int_0^T \omega^*(v)dv\right)\right], \tag{6.105}$$

where $KX - \int_0^T \omega(v)dv$ is the modulation phase in which K and ω are the wave number and the complex frequency of the modulation waves, respectively, and U_1 and U_2 are two complex numbers. The MI may set in provided the complex frequency ω has a non-null imaginary part. Inserting Eq. (6.105) into Eq. (6.104) and requesting that $|U_1| + |U_2| > 0$ yield the time-dependent dispersion relation

$$\left[\omega - 2K\left(\phi_0^2\lambda - Q\right)\right]^2 - K^2\left[K^2 + \phi_0^2\left(\phi_0^2\lambda^2 - 2Q\lambda - 2\lambda_0\right)\right] = 0. \tag{6.106}$$

For ω to have a non-null imaginary part, it is necessary and sufficient that

$$K^2 + \phi_0^2\left(\phi_0^2\lambda^2 - 2Q\lambda - 2\lambda_0\right) < 0. \tag{6.107}$$

Inequality (6.107) is the criterion of the MI for the cubic derivative NLS Eq. (6.100). Under the MI criterion (6.107), the local growth rate (gain) of modulational instability is given by the relation

$$\mathrm{Im}\left[\omega(t)\right] = |K|\sqrt{\phi_0^2\left[2\lambda_0(t) + 2Q\lambda(t) - \phi_0^2\lambda^2(t)\right] - K^2}. \tag{6.108}$$

MI for the Derivative Cubic NLS Equation with Constant Coefficients

A particularly simple and interesting case is when λ_0 and λ are constants. Then from system (6.98a)–(6.98e), it follows that $\beta(t)$ must be a constant and $g_0(t)$ must satisfy the nonlinear second-order ordinary differential equation

$$y\frac{d^2y}{dt^2} - 2\left(\frac{dy}{dt}\right)^2 - 4\alpha y^2 = 0. \tag{6.109}$$

In the case of constant λ_0 and λ, the problem of finding the time dependence of the parameters then reduces to solve Eq. (6.109). The simplest way to work with Eq. (6.109) is to solve it for $\alpha(t)$ when $g_0(t)$ is known. For example, any $g_0(t) = a_0 \exp\left[\widetilde{\lambda} t\right]$ corresponds to constant $\alpha(t) = -\widetilde{\lambda}^2/4$ [24]. As it is indicated in Ref. [37], one of the most interesting cases in the setting with the harmonic potential is the one with the inverse square dependence (of the trap amplitude, for a given x) on time of equation

$$\alpha(t) = A\left(t + t^*\right)^{-2}, \tag{6.110}$$

for an arbitrary nonzero real constant A. Inserting Eq. (6.110) into Eq. (6.109) yields

$$g_0(t) = \frac{4\lambda_0\beta}{\lambda\ell_0}\left(\frac{t+t^*}{t^*}\right)^m, \quad m = \frac{-1 \pm \sqrt{1-16A}}{2}. \tag{6.111}$$

For $g_0(t)$ to be a real function of time t, the parameter A of the strength of the magnetic trap $\alpha(t)$ must satisfy the condition $A < 1/16$. This condition on A allows us to investigate the MI of the cubic derivative NLS equation for both confining potential ($A < 0$) and repulsive potential ($A > 0$). In Eq. (6.110), t^* is an arbitrary constant which essentially determines the "width" of the trap at time $t = 0$. It is important to notice that $t^* < 0$ describes a BEC in a shrinking trap, while the case $t^* > 0$ corresponds to a broadening condensate. Inserting Eq. (6.110) into system (6.98b)–(6.98e), all the time-dependent parameters are explicitly defined; in particular, we have $T(t) = t^*\left(2m+1\right)^{-1}\ell_0^{-2}\left[\left(\frac{t+t^*}{t^*}\right)^{2m+1} - 1\right]$.To guarantee the variation of T from zero to infinity, we must take $m = \left(1 + \sqrt{1-16A}\right)/2$ in the case of broadening condensates, and $m = -\left(1 + \sqrt{1-16A}\right)/2$ for BECs in a shrinking trap. In the latter case, we focus our study on the case where t goes from zero to $-t^*$; this avoids a singularity at $t = -t^*$ and guarantees the variation of T from zero to $+\infty$.

In the case of constant λ_0 and λ, the growth rate (gain) of modulational instability is time independent but depends on the imprint parameter β as follows:

$$\mathrm{Im}\left[\omega(\beta)\right] = |K|\sqrt{\phi_0^2\left[2\lambda_0 + 2Q\lambda(\beta) - \lambda^2(\beta)\phi_0^2\right] - K^2}. \tag{6.112}$$

It is evident that the variation of the growth rate, related to the imprint parameter β, may significantly modify the instability domain and bring interesting effects. In fact, to different values of β are associated different instability diagrams, depending on whether $\lambda = \lambda(\beta)$ is positive or negative. Negative $\lambda(\beta)$ softens the instability, while positive one $[\lambda(\beta) > 0]$ relatively enhances the instability.

We can summarize the result for constant λ_0 and λ as follows. For the MI of constant amplitude plane-wave solutions $\phi = \phi_0 \exp\left[-iQX - i\left(Q^2 - \phi_0^2\lambda_0 - Q\phi_0^2\lambda\right)T\right]$, it is necessary and sufficient that the wave number K of the modulation waves satisfies the MI criterion (6.107). Moreover, for given ϕ_0, Q, and λ_0, the imprint parameter β should be chosen such that $\lambda^2(\beta)\phi_0^2 - 2Q\lambda(\beta) - 2\lambda_0 < 0$.

MI for the Derivative Cubic NLS Equation with Time-Dependent Coefficients

Let us now look at the case when at least one of λ_0 and λ is not constant. As in the previous case, if we consider the harmonic potential with the inverse square dependence on the time trap amplitude (6.110) with $A < 1/16$, we found as a particular solution of Riccati equation (6.98b)

$$f(t) = B/\left(t + t^*\right), \ B = \left(1 \pm \sqrt{1 - 16A}\right)/2. \tag{6.113}$$

With the help of the particular solution (6.113), the general solution of Eq. (6.98b) reads

$$f(t) = \frac{1 - 4B + CB\,(1 - 8B)\left(t + t^*\right)^{8B-1}}{C\,(1 - 8B)\,(t + t^*)^{8B} + 4\,(t + t^*)}, \ C = \frac{1 - 4B - 4f_0t^*}{f_0\,(1 - 8B)\,(t^*)^{8B} - B\,(1 - 8B)\,(t^*)^{8B-1}}, \tag{6.114}$$

with $f_0 = f(0)$. Using (6.110) and (6.113), we obtain the time dependence of the following parameters:

$$\ell(t) = \ell_0 \left(\frac{t + t^*}{t^*}\right)^{8B}, \tag{6.115}$$

$$\eta(t) = 2B \ln\left|\frac{t + t^*}{t^*}\right| + \eta_0, \tag{6.116}$$

$$\beta(t) = \frac{g_1(t)\,(t + t^*)}{8B}, \tag{6.117}$$

$$T(t) = \frac{1}{\ell_0\,(1 - 4B)}\left[\left(t + t^*\right)\left(\frac{t^*}{t + t^*}\right)^{4B} - t^*\right]. \tag{6.118}$$

It follows from system (6.115)–(6.118) that $\beta(t)$ is no longer a functional parameter, but depends of $g(t)$. In the case of positive t^* (broadening condensate), it is reasonable to take B such that $1 - 4B$ should be positive; a proper choice of ℓ_0 will then ensure a variation of $T(t)$ from zero to $+\infty$. For BECs in a shrinking trap ($t^* < 0$), a good choice of ℓ_0 and $B > 1/4$ and working with $0 \leq t < t^*$ will avoid singularity $t = t^*$ and ensure a variation of $T(t)$ from zero to $+\infty$.

In the case at least one of λ_0 and λ is not constant, the growth rate (gain) of modulational instability (6.108) is time dependent. In this case, the variation of the growth rate, related to the sign of t^*, i.e., on whether t^* describes a BEC in a shrinking trap ($t^* < 0$) or corresponds to a broadening condensate, may significantly affect the instability domain and bring new effects. Indeed, to different signs of t^* correspond different instability diagrams. Negative t^* relatively enhances the instability, while positive t^* softens the instability.

From our analysis, it is clear that the simplest and most interesting case in the setting with a time-dependent harmonic potential is the one with the inverse square dependence (of the trap amplitude, for a given x) on time of Eq. (6.110) with $A < 1/16$. In this case, the modified lens transformation suggests the equivalence with a cubic derivative NLS equation. In this special case of interest, the coefficients of the cubic derivative NLS equation are either constant or time dependent, suggesting that the frequencies of the modulation waves are either constant or time dependent.

6.7.2 Matter-Wave Solitons in a Delayed Nonlinear Response of BECs with Spatiotemporal Potential

Due to the suggested analogy with a cubic derivative NLS with a constant coefficient gain where the modulational stability analysis has been completely performed, the most interesting case of the spatiotemporal potential (6.95) is the case of $\alpha(t) = A\,(t + t^*)^{-2}$ with $A < 1/16$, which we now investigate analytically. We limit ourselves to constant λ_0 and λ in Eq. (6.100) and apply the F expansion method to find a large class of solitary-wave solutions. We are mainly interested in analytical solitary-wave solutions of constant coefficients derivative cubic NLS equation (6.100). In order to find a large class of solitary-wave solution of Eq. (6.100) with constant coefficients, we perturb the Stokes wave solution $\phi = \phi_0 \exp\left[-ik_0 X + i\left(\lambda_0\phi_0^2 + \lambda k_0 \phi_0^2 - k_0^2\right)T\right]$ of Eq. (6.100) to obtain its solitary-wave solutions. Writing a solution of Eq. (6.100) in polarform

$$\phi(X, T) = R(X, T)\exp[i\Phi(X, T)], \tag{6.119}$$

we obtain the following system for the definition of the real functions $R(X, T)$ and $\Phi(X, T)$:

$$\begin{cases} -R\frac{\partial \Phi}{\partial T} + \frac{\partial^2 R}{\partial X^2} - R\left(\frac{\partial \Phi}{\partial X}\right)^2 - \lambda R^3 \frac{\partial \Phi}{\partial X} + \lambda_0 R^3 = 0, \\ \frac{\partial R}{\partial T} + 2\frac{\partial \Phi}{\partial X}\frac{\partial R}{\partial X} + R\frac{\partial^2 \Phi}{\partial X^2} + 3\lambda R^2 \frac{\partial R}{\partial X} = 0. \end{cases} \tag{6.120}$$

Now, we use the above Stokes wave solution to perturb system (6.120) as follows:

$$\begin{cases} R(X,T) = \phi_0 + \rho(z = X - \upsilon T), \\ \Phi(X,T) = -k_0 z + \varphi(z) + \left(\lambda_0\phi_0^2 + \lambda k_0\phi_0^2 - k_0^2 - k_0\upsilon\right)T, \end{cases} \tag{6.121}$$

υ being an arbitrary constant (traveling wave velocity). Inserting (6.121) into (6.120) and integrating the resulting second equation yield

$$\frac{d\varphi}{dz} = \frac{C_0}{(\phi_0+\rho)^2} + \frac{\upsilon + 2k_0}{2} - \frac{3\lambda}{4}(\phi_0+\rho)^2, \tag{6.122}$$

where C_0 is a constant of integration. Inserting the expression for $d\varphi/dz$ into the first resulting equation yields

$$\left(\frac{d\zeta}{dz}\right)^2 = \widetilde{\alpha}\zeta^4 + 4\widetilde{\beta}\zeta^3 + 6\widetilde{\gamma}\zeta^2 + 4\widetilde{\delta}\zeta + \widetilde{\epsilon} = R(\zeta), \tag{6.123}$$

where

$$\begin{aligned} &\zeta(z) = (\phi_0+\rho)^2, \ \widetilde{\alpha} = -\lambda, \ \widetilde{\beta} = (\lambda\upsilon - 2\lambda_0)/4, \ \widetilde{\delta} = \text{const}, \\ &\widetilde{\gamma} = \frac{4\lambda_0\phi_0^2 + 2\lambda\left(2k_0\phi_0^2 - C_0\right) - 4\upsilon k_0 - \upsilon^2 - 4k_0^2}{6}, \ \widetilde{\epsilon} = -4C_0^2. \end{aligned} \tag{6.124}$$

It is important to notice that δ and C_0 are two arbitrary real constants (constants of integration).

As we know, the general solution to Eq. (6.123) is given in terms of Weierstrass' elliptic function $\wp(z; g_2, g_3)$ as [40]

$$\zeta(z) = \zeta_0 + \frac{\sqrt{R(\zeta_0)}\frac{d\wp}{dz} + \frac{R'(\zeta_0)}{2}\left[\wp(z; g_2, g_3) - \frac{1}{24}R''(\zeta_0)\right] + \frac{1}{24}R(\zeta_0)R'''(\zeta_0)}{2\left[\wp(z; g_2, g_3) - \frac{1}{24}R''(\zeta_0)\right]^2 - \frac{1}{48}R(\zeta_0)R^{(IV)}(\zeta_0)}, \tag{6.125}$$

where ζ_0 is an arbitrary real constant, not necessarily a zero of the polynomial R and $R' = dR/d\zeta$. The invariants g_2 and g_3 of the function $\wp(z; g_2, g_3)$ are related to the coefficients of $R(\zeta)$ as follows: [40]

$$g_2 = \widetilde{\alpha}\widetilde{\epsilon} - 4\widetilde{\beta}\widetilde{\delta} + 3\widetilde{\gamma}^2, \ g_3 = \widetilde{\alpha}\widetilde{\gamma}\widetilde{\epsilon} + 2\widetilde{\beta}\widetilde{\gamma}\widetilde{\delta} - \widetilde{\alpha}\widetilde{\delta}^2 - \widetilde{\gamma}^3 - \widetilde{\epsilon}\widetilde{\beta}^2. \tag{6.126}$$

The so-called discriminant

$$\Delta = g_2^3 - 27g_3^2 \tag{6.127}$$

of Weierstrass' elliptic function $\wp(z; g_2, g_3)$ is suitable to classify the behavior of the solution $\zeta(z)$ and to discriminate between periodic and solitary-wave-like solutions If $\Delta = 0$, $g_2 \geq 0$, and $g_3 \leq 0$, $\zeta(z)$ is a solitary-wave-like solution of Eq. (6.123) and is given by

$$\zeta(z) = \zeta_0 + \frac{R'(\zeta_0)}{4\left[e_1 - \frac{1}{24}R''(\zeta_0) + 3e_1\text{cosech}^2\left[\sqrt{3e_1}z\right]\right]}, \text{ if } g_3 < 0, \quad (6.128a)$$

$$\zeta(z) = \zeta_0 + \frac{R'(\zeta_0)z^2}{4\left[1 - \frac{R''(\zeta_0)}{24}z^2\right]}, \text{ if } g_3 = 0, \quad (6.128b)$$

where $e_1 = \sqrt{-g_3}/2$ and ζ_0 is any simple root of polynomial $R(\zeta)$. Because $\widetilde{\delta}$ and C_0 are two arbitrary constants of integration, we can, for simplicity, choose $\widetilde{\delta}$ and C_0 so that either $\widetilde{\delta} = \widetilde{\epsilon} = 0$ or $\widetilde{\delta} \neq 0, \widetilde{\epsilon} < 0$ and $g_2 = g_3 = 0$. In both cases,we have $g_2 \geq 0$, $g_3 0 \leq$, and $\Delta = 0$. Then Eq. (6.128a) defines solitary-wave-like solutions if and only if $\widetilde{\gamma} > 0$ and $2\widetilde{\beta}^2 - 3\widetilde{\alpha}\widetilde{\gamma} \geq 0$, while Eq. (6.128b) describes the intensity of the algebraic solitary waves if $R''(\zeta 0) < 0$. According to the definition of $\zeta(z)$ by system (6.124), the physical solution (6.128a) must be nonnegative. It is important to mention that a solution $\zeta(z)$ given by Eq. (6.128a) is said to be physical if it is real and bounded [34]. In what follows, we give examples of solitary-wave solutions when either $\widetilde{\delta} = \widetilde{\epsilon} = 0$ or $\widetilde{\delta} = 0, \widetilde{\epsilon} < 0$, and $g_2 = g_3 = 0$.

Examples of Solitary-Wave Solutions Under Conditions $\widetilde{\delta} = \widetilde{\epsilon} = 0$

According to Ref. [34], we have the following physical solitary-wave-like solutions when $\widetilde{\delta} = \widetilde{\epsilon} = 0$ of Eq. (6.123):

$$\zeta_{\pm}(z) = \frac{\widetilde{\gamma}\left[2\widetilde{\beta} \pm \sqrt{4\widetilde{\beta}^2 - 6\widetilde{\alpha}\widetilde{\gamma}}\right]\left[3\widetilde{\gamma}^2 + (\widetilde{\gamma}^2 - 1)\cosh^2\left(\sqrt{\frac{3}{2}\widetilde{\gamma}^3}z\right)\right]}{-3\alpha\widetilde{\gamma}^3 + \left[4\widetilde{\beta}^2 \pm 2\sqrt{4\widetilde{\beta}^2 - 6\widetilde{\alpha}\widetilde{\gamma}} - \widetilde{\alpha}\widetilde{\gamma}\,(5 + \widetilde{\gamma})^2\right]\cosh^2\left(\sqrt{\frac{3}{2}\widetilde{\gamma}^3}z\right)}. \quad (6.129)$$

Solutions (6.129) correspond to the two simple roots of the polynomial $R(\zeta)$. Let us now investigate the conditions of nonnegativity of solutions (6.129). Here, we distinguish two cases, namely, the case when $\widetilde{\gamma} = 1$ and the case when $0 < \widetilde{\gamma} \neq 1$. It is obvious that case $\widetilde{\gamma} = 1$ is associated only to the bright solitary-wave solution.

(A) If $\gamma = 1$, solutions (6.129) will be nonnegative if and only if
$\widetilde{\alpha}\left(2\widetilde{\beta}^2 \pm \sqrt{4\widetilde{\beta}^2 - 6\widetilde{\alpha}\widetilde{\gamma}} - 18\widetilde{\alpha}\right)^{-1} < 2/3$ and
$\left(2\widetilde{\beta} \pm \sqrt{4\widetilde{\beta}^2 - 6\widetilde{\alpha}\widetilde{\gamma}}\right)\left(2\widetilde{\beta}^2 \pm \sqrt{4\widetilde{\beta}^2 - 6\widetilde{\alpha}\widetilde{\gamma}} - 18\widetilde{\alpha}\right) > 0$.

An example of bright solitary-wave-like solution is obtained with the parameters $\lambda_0 = 1, \phi_0 = 1, \upsilon = 0.5, k_0 = 1$, and $\lambda = 8.25/4$. With this set of parameters, $\zeta_+(z)$ satisfies all the needed conditions (real, bounded, and nonnegative).

(B) If $0 < \widetilde{\gamma} \neq 1$, then solutions (6.129) will be nonnegative if and only if the following three conditions are simultaneously satisfied: (i) $3\widetilde{\gamma}^2\left(1 - \widetilde{\gamma}^2\right)^{-1} \leq 1$,
(ii) $3\widetilde{\alpha}\widetilde{\gamma}^3\left(4\widetilde{\beta}^2 \pm 2\sqrt{4\widetilde{\beta}^2 - 6\widetilde{\alpha}\widetilde{\gamma}} - \widetilde{\alpha}\widetilde{\gamma}\,(5 + \widetilde{\gamma})^2\right) < 1$, and
(iii) $\widetilde{\gamma}\left(\widetilde{\gamma}^2 - 1\right)\left(2\widetilde{\beta} \pm \sqrt{4\widetilde{\beta}^2 - 6\widetilde{\alpha}\widetilde{\gamma}}\right)\left(4\widetilde{\beta}^2 \pm 2\sqrt{4\widetilde{\beta}^2 - 6\widetilde{\alpha}\widetilde{\gamma}} - \widetilde{\alpha}\widetilde{\gamma}\,(5 + \widetilde{\gamma})^2\right) > 0$.
With the parameters $\lambda_0 = 1$, $\phi_0 = 1$, $\upsilon = 0.5$, $\lambda = 14.25/4$, and $k_0 = 1$, conditions

(i)–(iii) are simultaneously satisfied for $\zeta_+(z)$, and hence we have an example of the solitary-wave-like solution from Eq. (6.123) when $0 < \widetilde{\gamma} \neq 1$.

Examples of Solitary-Wave Solutions Under Conditions $\widetilde{\delta} \neq 0, \widetilde{\epsilon} < 0$, and $g_2 = g_3 = 0$

Now, we give examples of solitary-wave solutions when $\widetilde{\delta} \neq 0, \widetilde{\epsilon} < 0$, and $g_2 = g_3 = 0$. As we will see in what follows, one of the particularities of the present special case is that a bright and a dark solitary waves can exist for the same values of parameters $\widetilde{\alpha}$, $\widetilde{\beta}$, and $\widetilde{\gamma}$. According to Eq. (6.126), solving the system $g_1 = g_2 = 0$ for $\widetilde{\epsilon} < 0$ and $\widetilde{\delta} \neq 0$ yields

$$\widetilde{\epsilon} = \widetilde{\epsilon}(\widetilde{\alpha}, \widetilde{\beta}, \widetilde{\gamma}) = \frac{4\widetilde{\beta}^2\left(3\widetilde{\alpha}\widetilde{\gamma} - 2\widetilde{\beta}^2\right) - 3\widetilde{\alpha}^2\widetilde{\gamma}^2 \pm 8\widetilde{\beta}\left(\widetilde{\beta}^2 - \widetilde{\alpha}\widetilde{\gamma}\right)\sqrt{\widetilde{\beta}^2 - \widetilde{\alpha}\widetilde{\gamma}}}{\widetilde{\alpha}^3},$$

$$\widetilde{\delta} = \widetilde{\delta}(\widetilde{\alpha}, \widetilde{\beta}, \widetilde{\gamma}) = \frac{3\widetilde{\alpha}\widetilde{\beta}\widetilde{\gamma} - 2\widetilde{\beta}^3 \pm 2\left(\widetilde{\beta}^2 - \widetilde{\alpha}\widetilde{\gamma}\right)\sqrt{\widetilde{\beta}^2 - \widetilde{\alpha}\widetilde{\gamma}}}{\widetilde{\alpha}^2},$$

if $\widetilde{\beta}^2 - \widetilde{\alpha}\widetilde{\gamma} \geq 0$ and $4\widetilde{\beta}^2\left(3\widetilde{\alpha}\widetilde{\gamma} - 2\widetilde{\beta}^2\right) - 3\widetilde{\alpha}^2\widetilde{\gamma}^2 \pm 8\widetilde{\beta}\left(\widetilde{\beta}^2 - \widetilde{\alpha}\widetilde{\gamma}\right)\sqrt{\widetilde{\beta}^2 - \widetilde{\alpha}\widetilde{\gamma}} > 0$. Replacing $\widetilde{\delta}$ and $\widetilde{\epsilon}$ into Eq. (6.123) leads to the simple root $\zeta_0 = -\widetilde{\alpha}^{-1}\left[\widetilde{\beta} \pm 3\sqrt{\widetilde{\beta}^2 - \widetilde{\alpha}\widetilde{\gamma}}\right]$ of the polynomial $R(\zeta)$. Inserting the above simple root ζ_0 into solution (6.128b) leads to the algebraic solitary waves

$$\zeta(z) = \frac{-\widetilde{\alpha}\left[\widetilde{\beta} \pm 3\sqrt{\widetilde{\beta}^2 - \widetilde{\alpha}\widetilde{\gamma}}\right] + 4\left(\widetilde{\alpha}\widetilde{\gamma} - \widetilde{\beta}^2\right)\left[-\widetilde{\beta} \pm 3\sqrt{\widetilde{\beta}^2 - \widetilde{\alpha}\widetilde{\gamma}}\right]z^2}{\widetilde{\alpha}\left[\widetilde{\alpha} - 4\left(\widetilde{\beta}^2 - \widetilde{\alpha}\widetilde{\gamma}\right)z^2\right]} \tag{6.130}$$

satisfying the needed condition of the nonnegativity as soon as $\widetilde{\beta} > 0, \widetilde{\gamma} < 0, \widetilde{\alpha}\widetilde{\gamma} < \widetilde{\beta}^2 < 9\,\widetilde{\alpha}\widetilde{\gamma}/8$. In solution (6.130), signs "−" and "+" are associated with the dark and bright solitary waves, respectively. By taking $\alpha = -1, \beta = 1$, and $\gamma = -8.1/9$, all the conditions of the existence of solution (6.130) for both signs "−" and "+" are satisfied. In other words, parameters $\widetilde{\alpha} = -1, \widetilde{\beta} = 1$, and $\widetilde{\gamma} = -8.1/9$ together with $\widetilde{\epsilon}\,(-1, 1, -8.1/9) < 0$ and $\widetilde{\epsilon}(-1, 1, -8.1/9) < 0$ define a dark and a bright solitary wave. The above values for $\widetilde{\alpha}$, β, and $\widetilde{\gamma}$ are obtained from system (6.124) for $\lambda = \lambda_0 = 1, k_0 = 0, \upsilon = 6, \phi_0 = 2.7813$ for the dark solitary wave, and $\phi_0 = 2.8013$ for the bright solitary wave.

6.8 Dynamically Stabilized Matter-Wave Solitons in 1D BECs with Time-Dependent Atomic Scattering Length in Parabolic Potential

Investigating theoretically and numerically BECs with time-dependent atomic scattering length in parabolic potential, it has been showed in one hand that with the increasing (decreasing) absolute value of the scattering length, the bright soliton may have an increasing (decreasing) in the peaking value, and in the other hand, that one of the effects of loss (gain) on BECs solitons was to decrease (increase) the soliton peak during their propagation. In the this section, we consider 1D BECs with time-dependent scattering length in parabolic potential and demonstrate that matter-wave bright solitons can be dynamically stabilized in 1D free space by either injecting atoms in or ejecting atoms from the system. In the physically important case of the cigar-shaped BECs, the macroscopic wave function ψ of such BECs is governed by the 1D NLS equation

$$i\frac{\partial\psi}{\partial t}+\frac{1}{2}\frac{\partial^2\psi}{\partial x^2}+g\left|\psi\right|^2\psi-kx^2\psi=0, \tag{6.131}$$

where t and x are the temporal and spatial coordinates measured in units $1/\omega_\perp$ and $a_\perp=\sqrt{\hbar/m\omega_\perp}$, respectively; here, $a_\perp$ and $a_0=\sqrt{\hbar/m\omega_0}$ are linear oscillator lengths in the transverse and cigar-axis directions, respectively, while $\omega_\perp$ and ω_0 are corresponding harmonic oscillator frequencies, and m is the atomic mass. The normalized macroscopic wave function $\psi(x,t)$ is measured in units of $1/\left(\sqrt{2\pi a_B a_\perp}\right)$, the parameter $g=-\frac{2a_s}{a_B}$ of the cubic nonlinearity is negative (positive) for repulsive (attractive) interatomic interaction, the strength k of the magnetic trap may be positive (confining potential) or negative (repulsive potential) and is defined as $k=\pm\omega^2/\left(2\omega_\perp^2\right)$; a_s is the s-wave scattering length, and $a_B=\frac{a_\perp\int_{-\infty}^{+\infty}|\psi|^2dx}{2\int_{-\infty}^{+\infty}|\ |^2d\mathbf{r}}$ is the Bohr radius in which $\Psi(\mathbf{r},t)$ (with $\mathbf{r}=(y,z)$) is the original order parameter, connected to the normalized macroscopic wave function $\psi(x,t)$ as follows:

$$\Psi(\mathbf{r},t)=\frac{1}{\sqrt{2\pi a_B a_\perp}}\psi\left(\frac{x}{a_\perp},\omega_\perp t\right)\exp\left[-i\omega_\perp t-\frac{y^2+z^2}{2a_\perp^2}\right].$$

In Eq. (6.131), the parameter g of the cubic nonlinearity and the strength k of the magnetic trap can be time-dependent, and Eq. (6.131) can be used to describe the control and management of BEC by properly choosing these two functional parameters.

From the viewpoint of stability, 3D and 1D equations are very different. For a true 1D system, one does not expect the collapse of the system when increasing number of atoms. However, it happens that a realistic 1D limit is not a true 1D system, with the density of particles still increasing due to the strong restoring forces in the perpendicular directions. To either avoid the collapse or to improve the lifetime of the bright BEC soliton, it may be useful to consider a NLS equation with a complex potential; for such a BEC, the imaginary part of the potential has a clear physical interpretation as a localized gain or loss of atoms in the condensate resulting from the contact with the thermal cloud and three-body recombination. In other words, the absolute value of the nonzero imaginary part of a complex potential is the probability of the atoms feeding or loss from the thermal cloud into the condensates [note that the imaginary part of the complex potential must be in a practical situation, and the total number of atoms in the condensates is not conserved for nonzero imaginary part of the potential; generally, the condensate gains (losses) atoms for positive (negative) imaginary part of the potential]. It is reasonable to consider a symmetric complex potential with respect to the spatial variable x. This condition on the complex potential means that the gain or loss contributions are balanced in such a way that stationary states with real eigenvalues can exist. In this section, we answer the following question: "How can matter-wave bright solitons in 1D BECs with time-dependent atomic scattering length in parabolic potential be dynamically stabilized by either injecting atoms in or ejecting atoms from the system?"

6.8.1 BECs with Symmetric Complex Potential with Respect to the Spatial Variable x

Reduction of a Given BEC to a BEC with Symmetric Complex Potential with Respect to the Spatial Variable x

In this section, we formulate the condition under which a BEC described by the GP Eq. (6.131) can be reduced to a BECs with symmetric complex potential.

Theorem: If the nonlinearity parameter g and potential parameter k in Eq. (6.131) satisfy condition

$$2g\frac{d^2 g}{dt^2} - 3\left(\frac{dg}{dt}\right)^2 - 8kg^2 = 0, \tag{6.132a}$$

then the BEC with symmetric complex potential

$$i\frac{\partial \psi}{\partial t} + \frac{1}{2}\frac{\partial^2 \psi}{\partial x^2} + g\,|\psi|^2\,\psi = \left[kx^2 - i\frac{1}{4g}\frac{dg}{dt}\right]\psi, \tag{6.132b}$$

will admit solitons with stabilized peaks.

Remark: In Eq. (6.132b), the time-dependent dissipative parameter $\gamma(t) = -\frac{1}{4g}\frac{dg}{dt}$ relates the feeding ($\gamma > 0$) or loss ($\gamma < 0$) of atoms in the condensate.

Proof: Let condition (6.132a) be satisfied. In order to prove the above theorem, we introduce the modified lens-type transformation

$$\psi(x,t) = \frac{1}{\sqrt{\ell}}u(X,T)\exp\left[\widetilde{\eta} - i\frac{1}{4g}\frac{dg}{dt}x^2\right], \tag{6.133}$$

where $\eta = \widetilde{\eta}(t)$, $T = T(t)$ and $\ell = \ell(t) > 0$ are three real functions of time t, and $X = x/\ell(t)$ is a real function of x and t. Imposing to $\eta(t)$, $T(t)$, and $\ell(t)$ to satisfy the system

$$\frac{1}{\ell}\frac{d\ell}{dt}+\frac{1}{2g}\frac{dg}{dt}=0, \tag{6.134a}$$

$$\frac{dT}{dt}-\frac{1}{\ell^2}=0, \tag{6.134b}$$

$$\frac{d\widetilde{\eta}}{dt}-\frac{1}{2\ell}\frac{d\ell}{dt}=0, \tag{6.134c}$$

we obtain

$$\ell(t)=\ell_0\sqrt{\frac{g_0}{g(t)}},\; T(t)=\int_0^t\frac{g(\tau)}{\ell_0^2 g_0}d\tau+T_0,\; \widetilde{\eta}(t)=\frac{1}{4}\ln\frac{g_0}{g(t)}. \tag{6.135}$$

where $\ell_0=\ell(0)$, $T_0=T(0)$, and $g_0=g(0)$, with $g_0 g(t)>0$. Inserting ansatz (6.133) into Eq. (6.132b) and using either Eqs. (6.134a)–(6.134c) or Eq. (6.135) lead to standard integrable NLS equation

$$i\frac{\partial u}{\partial T}+\frac{1}{2}\frac{\partial^2 u}{\partial X^2}+p^2\left|u\right|^2 u=0, \tag{6.136}$$

where, without loss of generality, we set $g_0\ell_0=p^2=\text{sign}(g_0)$. Let $u(X,T)$ be any soliton solution to Eq. (6.135). Using ansatz (6.133) leads to the following soliton solution of Eq. (6.132b)

$$\psi(x,t)=\frac{1}{\sqrt{\ell_0}}u\left(\ell_0^{-1}\sqrt{\frac{g(t)}{g_0}}x,\int_0^t\frac{g(\tau)}{\ell_0^2 g_0}d\tau+T_0\right)\exp\left[-i\frac{1}{4g}\frac{dg}{dt}x^2\right]. \tag{6.137}$$

It follows from Eq. (6.137) that $|\psi|^2=\ell_0^{-1}\left|u\left(\ell_0^{-1}\sqrt{\frac{g(t)}{g_0}}x,\int_0^t\frac{g(\tau)}{\ell_0^2 g_0}d\tau+T_0\right)\right|^2$, meaning that the soliton given by Eq. (6.137) has a stabilized peaks during its propagation. Moreover, the total number of atoms in the condensate with symmetric complex potential will be $N(t)=\sqrt{\frac{g_0}{g(t)}}\int_{-\infty}^{+\infty}\left|u\left(X,T(t)\right)\right|^2 dX$. The theorem is proved.

Definition: Throughout this book, any BEC whose GP equation (6.131) satisfies the condition (6.132a) will be called **mother BEC**, while the BEC described by the GP Eq. (6.132b) will be called **daughter BEC** or BEC with symmetric complex potential with respect to the spatial variable x.

Applicability to Real Experiments

Here, we give examples of BECs leading to BECs with symmetric complex potential with respect to the spatial variable x. We focus ourselves to BECs with (a) a specific time-varying potential, (b) a time-independent harmonic potential, and (c) a temporal periodic modulation of the s-wave scattering length. Because many authors either use $2a(t)$ instead of $g(t)$ as the nonlinear parameter or use different letters to defined the strength k of the harmonic potential of the GP Eq. (6.131), we will, when using the results of some authors, introduce if necessary, some numbers or constant, that allow us to have the same GP equation as that used by these authors.

BECs with the Specific Time-Varying Potential $k(t) = a\,(t + t_0)^{-2}$ We start with BECs with the specific time-varying potential with strength $k(t) = a\,(t + t_0)^{-2} \neq 0$, which has been suggested as being of interest in quite different setups such as the study of explosion–implosion dualities for the quintic (critical) GP equation. Here, and t_0 is an arbitrary constant whose sign is related to the sign of $-\frac{1}{g}\frac{dg}{dt}$; $\frac{t_0}{g}\frac{dg}{dt} < 0$ and which essentially determines the width of the trap at time $t = 0$ [18]; notice that $t_0 < 0$ describes a BEC in a shrinking trap, while the case corresponds to a broadening condensate. This type of time-varying potential has been used by Theocharis et al. [18] when investigating the MI of BECs with a specific time-varying potential. Injecting the above potential strength $k(t)$ in Eq. (6.132a) leads to $g(t) = g_0\,(t + t_0)^{-1\pm\sqrt{1-8a}}$, where $g_0 \neq 0$ is any real constant $(1 - 8a \geq 0\,)$. With this nonlinearity functional parameter $g(t)$, the feeding or loss parameter of in Eq. (6.132b) becomes $\gamma(t) = \frac{-1\mp\sqrt{1-8a}}{4}\,(t + t_0)^{-1}$.

BECs with Time-Independent Harmonic Potential As the second example, let us consider the time-independent harmonic potential which was used by Khaykovich et al. [20] in the creation of bright BEC solitons. In that experiment, $\omega = 2\pi i \times 70$ Hz, $\omega_\perp = 2\pi \times 710$ Hz, so $k = -2\lambda^2$ $(\lambda \approx 0.05)$. Inserting $k = -2\lambda^2$ into Eq. (6.132a) and seeking $g(t)$ in the form $g(t) = 2g_0 \exp\left(\widetilde{\lambda t}\right)$, we obtain the special form of $g(t)$ and $k(t)$ $[g(t) = 2g_0 \exp[\pm 4\lambda t]$ and $k = -2\lambda^2]$ used by Liang et al. [27]. In this special case, Eq. (6.132b) becomes $i\frac{\partial\psi}{\partial t} + \frac{1}{2}\frac{\partial^2\psi}{\partial x^2} + g\,|\psi|^2\,\psi + 2\lambda^2 x^2\psi = \mp i\lambda\psi$, and describes a BEC with feeding $(\mp\lambda > 0)$ or loss $(\mp\lambda < 0)$ of atoms.

BECs with a Temporal Periodic Modulation of the s-wave Scattering Length As the last example, we consider the temporal periodic modulation of the s-wave scattering length. The oscillation of the scattering length with frequency $\omega \approx \omega_\perp$ is studied by Abdullaev et al. [1] in a different context. According to Ref. [15], we consider the nonlinearity parameter of the form $g(t) = 1 + m_0 \sin[\omega t]$ for $0 < m_0 < 1$. It is important to note that for this nonlinearity parameter, we have $g_0 = g(0) = 1$. This avoids the condensate from the collapse if the oscillating term $m_0 \sin[\omega t]$ of the nonlinearity parameter were absent [2]. For such a nonlinearity parameter, conditions (6.132a) are satisfied for a temporal periodic modulation of potential with strength $k(t) = -\frac{m_0\omega^2}{4}\frac{\sin[\omega t]}{1+m_0\sin[\omega t]} - \frac{3m_0^2\omega^2}{8}\frac{\cos^2[\omega t]}{(1+m_0\sin[\omega t])^2}$. In the present example, the time-dependent dissipative parameter of Eq. (6.132b) reads $\gamma(t) = -\frac{m_0\omega}{4}\frac{\cos[\omega t]}{1+m_0\sin[\omega t]}$.

6.8.2 *Dynamics of Matter-Wave Solitons in BECs with Symmetric Complex Potentials with Respect to the Spatial Variable x*

In this section, we investigate the dynamics of BECs with symmetric complex potentials with respect to the spatial variable x. We focus ourselves to BECs with the specific time-varying potential $k(t) = a\,(t + t_0)^{-2}$, BECs with time-independent harmonic potential, and BECs with a time-varying atomic scattering length in a time-varying expulsive parabolic potential.

We start with the exact soliton solutions of the standard NLS Eq. (6.136). Following Lei et al. in [46], the soliton solutions on the nonvanishing continuous wave (cw) background of the standard NLS Eq. (6.136) are

$$u(X,T) = \left[A_c + A_s \frac{\rho\cosh\xi + \cos\eta + i\,(\theta\sinh\xi + \kappa\sin\eta)}{\cosh\xi + \rho\cos\eta}\right]\exp\left[i\Delta_{2b}\right], \tag{6.138a}$$

if $p^2 > 0$, and

$$\left[A_c + iA_s\tanh\zeta\right]\exp\left[i\Delta_{2d}\right], \tag{6.138b}$$

if $p^2 < 0$, where

$$\rho = \frac{A_c\,(M_R - A_s)}{D},\ \theta = \frac{A_c\,(k_s - k_c + M_I)}{D},\ \kappa = 1 - \frac{2A_c^2}{D},$$

$$\xi = M_R X - \frac{(k_c + k_s)\,M_R - A_s M_I}{2}T + \xi_{20},\ \zeta = -A_s X + A_s\left(A_c + \widetilde{k}\right)T + \zeta_{20}$$

$$\eta = M_I X - \frac{(k_c + k_s)\,M_I + A_s M_R}{2}T + \eta_{20},$$

$$\Delta_{2b} = k_c X + \frac{2A_c^2 - k_c^2}{2}T,\ \Delta_{2d} = \widetilde{k}X - \frac{2\left(A_s^2 + A_c^2\right) + \widetilde{k}^2}{2}T + \delta_{2d0}$$

$$M = \sqrt{(A_s - ik_s + ik_c)^2 - 4A_c^2},\ D = \frac{4A_c^2 + (M_R - A_s)^2 + (k_s - k_c + M_I)^2}{4};$$

here, subscripts R and I stand for real and imaginary parts, respectively, and A_c, A_s, ξ_{20}, η_{20}, ζ_{20} , δ_{2d0}, and $\widetilde{k}$ are arbitrary real constants.

BEC with the Specific Time-Varying Potential $k(t) = a\,(t + t_0)^{-2}$

As the first application of result obtained in the previous section, we consider the BEC with the specific time-varying potential $k(t) = a\,(t + t_0)^{-2}$ with $a = 1/9$ and $t_0 = k_0/3$, k_0 being any positive real number (see Theocharis et al. in Ref. [37]). According to the results of the previous section, the functional parameters of Eq. (6.131) are

$$k(t) = \frac{1}{9\left(t + \frac{k_0}{3}\right)^2}\ \text{and}\ g(t) = g_0\frac{\left(\frac{k_0}{3}\right)^{\frac{2}{3}}}{\left(t + \frac{k_0}{3}\right)^{\frac{2}{3}}}. \tag{6.139}$$

According to Theocharis et al. in Ref. [37], the positive parameter k_0 essentially determines the "width" of the trap at time $t = 0$ and describes a BEC in a broadening condensate. For the nonlinearity parameter g and potential parameter k given by Eq. (6.139), condition (6.132a) is satisfied for any $g_0 \neq 0$. According to be above theorem, we can associate with the mother BEC described by Eq. (6.131) with parameters (6.139), the following daughter BEC with balanced gain and loss:

$$i\frac{\partial\psi}{\partial t}+\frac{1}{2}\frac{\partial^2\psi}{\partial x^2}+g\,|\psi|^2\,\psi=\left[kx^2+i\frac{1}{2\,(3t+k_0)}\right]\psi, \tag{6.140}$$

where g and k are given by Eq. (6.139). For Eq. (6.140), the complex potential $V(x,t)=kx^2+i\frac{1}{2(3t+k_0)}$ satisfies condition $V(-x,t)=V(x,t)$. In the practice, the absolute value of imaginary part of potential $V(x,t)$ must be small (as a probability), leading to condition $k_0>\frac{1}{2}$. Inserting Eq. (6.139) into Eq. (6.137) yields the following solution of Eq. (6.140):

$$\psi(x,t)=\ell_0^{-1}\sqrt{\frac{p^2}{g_0}}u\left(\frac{1}{\ell_0}\left(\frac{k_0}{3t+k_0}\right)^{\frac{1}{3}}x,\ \frac{k_0}{\ell_0^2}\left(\frac{3t+k_0}{k_0}\right)^{\frac{1}{3}}+\widetilde{T}_0\right)\exp\left[i\frac{1}{2\,(3t+k_0)}x^2\right], \tag{6.141}$$

where $\widetilde{T}_0=T_0-k_0\ell_0^{-2}$.

Now, let us investigate the dynamics of solitons in the both BECs described by Eqs. (6.131) and (6.140) with parameters (6.139). According to the direct method proposed by Li et al. in Refs. [37, 46], we obtain respectively the following families of bright soliton solution for $g_0>0$ and dark soliton solution for $g_0<0$ of the NLS equation (6.131) with parameters (6.139):

$$\psi(x,t)=\sqrt{g}\left[A_c+A_s\frac{\rho\cosh\xi+\cos\eta+i\,(\theta\sinh\xi+\kappa\sin\eta)}{\cosh\xi+\rho\cos\eta}\right]\exp\left[i\Delta_b\right], \tag{6.142a}$$

$$\psi(x,t)=\sqrt{-g}\,[A_c+i\,A_s\tanh\zeta]\exp\left[i\Delta_d\right] \tag{6.142b}$$

where

$$\rho=-\frac{2A_cA_s}{A_s^2+p_2^2},\ \kappa^2=1-\frac{4A_c^2}{A_s^2+p_2^2},\ \theta=-\frac{2p_2A_c}{A_s^2+p_2^2},$$

$$\xi=A_s\left[\kappa g_0\left(\frac{k_0}{3t+k_0}\right)^{\frac{2}{3}}x-(p_2+A_c\theta)v_1-\kappa v_2\right]+\xi_0,$$

$$\eta=p_2g_0\left(\frac{k_0}{3t+k_0}\right)^{\frac{2}{3}}x+\frac{\kappa\left(A_s^2-p_2^2\right)}{2}v_1-p_2v_2+\eta_0,$$

$$\Delta_b=k_2x^2+k_1x+A_c^2v_1-\frac{1}{2}v_3+\delta_{b0},$$

$$\zeta=A_sg_0\left(\frac{k_0}{3t+k_0}\right)^{\frac{2}{3}}x+A_sA_cv_1-A_sv_2+\zeta_0,$$

$$\Delta_d=k_2x^2+k_1x-\left(A_c^2+A_s^2\right)v_1-\frac{1}{2}v_3+\delta_{d0},$$

where A_c, A_s, p_2, κ, ξ_0, η_0, ζ_0, δ_{b0}, δ_{d0} and are arbitrary real constants, and

$$k_1 = k_{10}\left(\frac{k_0}{3t+k_0}\right)^{\frac{2}{3}},\ k_2 = \frac{1}{3t+k_0},\ \upsilon_1 = k_0 g_0^2\left[1-\left(\frac{k_0}{3t+k_0}\right)^{\frac{1}{3}}\right]+\upsilon_{10},$$

$$\upsilon_2 = k_0 g_0 k_{10}\left[1-\left(\frac{k_0}{3t+k_0}\right)^{\frac{1}{3}}\right]+\upsilon_{20},\ \upsilon_3 = k_0 k_{10}^2\left[1-\left(\frac{k_0}{3t+k_0}\right)^{\frac{1}{3}}\right]+\upsilon_{30},$$

k_{10}, υ_{10}, υ_{20}, k_{10}, and υ_{30} being arbitrary real constants.

To write the soliton solutions of Eq. (6.140) with parameters (6.139), we use the exact soliton solutions (6.138a) and (6.138b) of Eq. (6.136). Combining Eqs. (6.138a) and (6.138b) with Eq. (6.141) leads to the bright soliton solution for $g_0 > 0$ and dark soliton solution for $g_0 < 0$ of the GP Eq. (6.140) with parameters (6.139).

When $A_s = 0$ or $A_c = 0$, solutions (6.142a) and (6.138a) reduce respectively to the backgrounds $\psi_c(x,t) = \sqrt{g}A_c \exp[i\Delta_b]$ and $u_c(X,T) = A_c \exp[i\Delta_{2b}]$ and to the bright solitons $\psi(x,t) = \frac{A_s\sqrt{g}}{\cosh\xi}\exp[i(\Delta_b+\eta)]$ and $u(X,T) = \frac{A_s}{\cosh\xi}\exp[i(\Delta_{2b}+\eta)]$, while solutions (6.142b) and (6.138b) reduce respectively to the backgrounds $\psi_c(x,t) = \sqrt{-g}A_c \exp[i\Delta_d]$ and $u_c(X,T) = A_c \exp[i\Delta_{2d}]$ and to the dark solitons $\psi(x,t) = i\sqrt{-g}A_s \tanh\left[A_s g_0\left(\frac{k_0}{3t+k_0}\right)^{\frac{2}{3}} x - A_s\upsilon_2 + \zeta_0\right]\exp[i\Delta_d]$ and $u(X,T) = iA_s \tanh\left[-A_s X + A_s\widetilde{k}T + \zeta_{20}\right]\exp[i\Delta_{2d}]$. Therefore, each of solutions (6.142a) and (6.138a) describes a bright soliton embedded on a background, while each of solutions (6.142b) and (6.138b) describes a dark soliton embedded on a background, and when $A_c << A_s$, the background is small within the existence of the soliton.

In what follows, we restrict ourselves to the dynamics of bright solitons in the both BECs (that described by Eq. (6.131) and that described by Eq. (6.140). With the use of the properties of solutions (6.142a) and (6.138a), we wish demonstrate that the manipulation of the parameter k of the external harmonic trapping potential can be used to manipulate the motion of solitons in the above BECs systems. It has been reported that an abrupt change of the scattering length can lead to the splitting of the soliton by generating the new solitons. The fragmentation of the soliton obviously decreases the numbers of atoms of the original soliton, which is undesirable for application (See Liang in Refs. [7, 27]). However, in Eqs. (6.142a) and (6.141) for soliton (6.138a), the change of the parameter $g(t)$ given by Eq. (6.139) preserves the soliton from splitting new ones. For simplicity, we assume $p_2 = k_s - k_c = 0$ and consider only $A_s^2 - 4A_c^2 > 0$, for in the case $A_s^2 - 4A_c^2 < 0$ a small perturbation for Eqs. (6.142a) and (6.138a) may lead to the modulation instability (see Lei et al. in Ref. [46]). Under these conditions, solutions (6.142a) and (6.138a) become respectively

$$\psi(x,t) = \sqrt{g}\left[-A_c + \frac{(A_s^2-4A_c^2)\cos\eta \pm iA_s\sqrt{A_s^2-4A_c^2}\sin\eta}{A_s\cosh\xi - 2A_c\cos\eta}\right]\exp[i\Delta_b], \tag{6.143a}$$

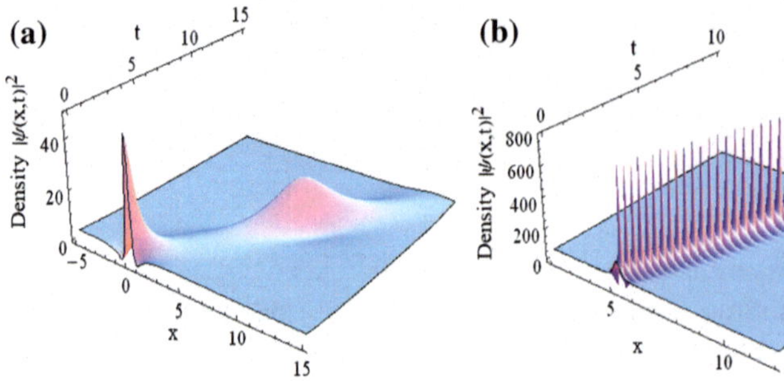

Fig. 6.19 (Color online) The evolution of bright solitons on nonvanishing backgrounds given by **a** Eq. (6.143a) with sign "−" for the NLS Eq. (6.131) and **b** Eq. (6.143b) for the NLS Eq. (6.140). The parameters are given as follows $k_0 = 10.1$, $g_0 = 0.25$, $A_c = 4.8$, $A_s = 10$, $k_s = k_c = 0.03$, $\upsilon_{10} = \upsilon_{20} = \upsilon_{30} = 0 = \xi_0 = \eta_0 = 0$, $\xi_{10} = \ell_0 = k_{10} = 1$, $\widetilde{T}_0 = -5$, $\eta_{01} = 0.1$

where $\xi = A_s\left[\kappa g_0\left(\frac{k_0}{3t+k_0}\right)^{\frac{2}{3}} x - \kappa \upsilon_2\right] + \xi_0$, $\kappa = \pm A_s^{-1}\sqrt{A_s^2 - 4A_c^2}$, $\eta = \frac{\kappa A_s^2}{2}\upsilon_1 + \eta_0$, and $\Delta_b = k_2 x^2 + k_1 x + A_c^2 \upsilon_1 - \frac{1}{2}\upsilon_3 + \delta_{b0}$, and

$$\psi(x,t) = \ell_0^{-1}\sqrt{\frac{1}{g_0}}\left[-A_c + \frac{\left(A_s^2 - 4A_c^2\right)\cos\eta - iA_s\sqrt{A_s^2 - 4A_c^2}\sin\eta}{A_s\cosh\xi - 2A_c\cos\eta}\right]\exp\left[i\varphi(x,t)\right], \tag{6.143b}$$

where $\varphi(x,t) = \frac{k_c}{\ell_0}\left(\frac{k_0}{3t+k_0}\right)^{\frac{1}{3}} x + \frac{(2A_c^2 - k_c^2)k_0}{2\ell_0^2}\left(\frac{3t+k_0}{k_0}\right)^{\frac{1}{3}} + \frac{(2A_c^2 - k_c^2)\widetilde{T}_0}{2} + \frac{x^2}{2(3t+k_0)}$, $\xi = \sqrt{A_s^2 - 4A_c^2}\left(\frac{1}{\ell_0}\left(\frac{k_0}{3t+k_0}\right)^{\frac{1}{3}} x - \frac{k_c k_0}{\ell_0^2}\left(\frac{3t+k_0}{k_0}\right)^{\frac{1}{3}} + \widetilde{T}_0\right) + \xi_{20}$, $\eta = -\frac{A_s\sqrt{A_s^2 - 4A_c^2}}{2}\left(\frac{k_0}{\ell_0^2}\left(\frac{3t+k_0}{k_0}\right)^{\frac{1}{3}} + \widetilde{T}_0\right) + \eta_{20}$. The center of solitons (6.143a) and (6.143b) are respectively $x_{sol1}(t) = \left(\frac{\upsilon_2}{g_0} - \frac{\xi_0}{\kappa g_0 A_s}\right)\left(\frac{3t+k_0}{k_0}\right)^{\frac{2}{3}}$ and $x_{sol2}(t) = \frac{k_c k_0}{\ell_0}\left(\frac{3t+k_0}{k_0}\right)^{\frac{2}{3}} - \frac{\ell_0\widetilde{T}_0\sqrt{A_s^2 - 4A_c^2} + \ell_0\xi_{20}}{\sqrt{A_s^2 - 4A_c^2}}\left(\frac{3t+k_0}{k_0}\right)^{\frac{1}{3}}$, each of which satisfies the condition

$$\frac{d^2x}{dt^2} + 2k(t)x = 0, \tag{6.144}$$

where $k(t)$ is the potential parameter given by Eq. (6.139). This last equation (6.144) means that the center of mass of the macroscopic wave packet behaves like a classical particle, and allows us to manipulate the motion of bright solitary waves in the above BEC systems by controlling the external harmonic trapping potential.

For a better understanding of the dynamics of bright solitons given by Eqs. (6.143a) and (6.143b), we plot these two equations in Fig. 6.19a, b, respectively, which show the dynamics of bright solitons in the confining parabolic potential. As one can see from Fig. 6.19a, with the decreasing absolute value of the scattering length, the bright soliton of the BEC described by Eq. (6.131) with parameters (6.139) has an decrease in the peaking value and a broadening in the width. Figure 6.19b reveals that with the decreasing absolute value of the scattering length, the bright soliton of the BEC described by Eq. (6.140) with parameters (6.139) has a preserving in the peaking value and a broadening in the width; thus, the lifetimes of the bright soliton in the BECs with balanced gain or loss can be extended largely at least to the order of the lifetime of a BECs in real experiments. We also observe from Fig. 6.19a, b that in the confining parabolic potential, the bright solitons of the two BECs are set into motion and propagates in the longitudinal direction, instead of oscillating in an expulsive parabolic potential. From Plots of Fig. 6.19, one can see the bright solitons behave like breather solitons propagating on backgrounds, which corresponds to the atom exchanging between the bright solitons and the background and this exchange keeps the bright solitons dynamically stable against the variation of the nonlinearity as demonstrated Liang et al. in Ref. [27].

It follows from Eqs. (6.143a) and (6.143b) that the peak matter density of each of the bright solitons arrives at the maximum and at the minimum when $\sinh\xi = 0$ and $\cosh\xi = \frac{A_s}{A_c\cos\eta} - \frac{2A_c\cos\eta}{A_s}$, respectively. Therefore, we obtain respectively for Eqs. (6.143a) and (6.143b)

$$|\psi|^2_{\max} = g(t)\left(A_c^2 + \frac{A_s\left(A_s^2 - 4A_c^2\right)}{A_s - 2A_c\cos\eta}\right) \text{ and } |\psi|^2_{\min} = g(t)\left(A_c^2 - \frac{A_c^2\left(A_s^2 - 4A_c^2\right)\cos^2\eta}{\left(A_s^2 - 4A_c^2\cos^2\eta\right)}\right) \tag{6.145}$$

where $\xi = A_s\left[\kappa g_0\left(\frac{k_0}{3t+k_0}\right)^{\frac{2}{3}}x - \kappa v_2\right] + \xi_0$ and $\eta = \frac{\kappa A_s^2}{2}v_1 + \eta_0$, and

$$|\psi|^2_{\max} = \frac{1}{\ell_0^2 g_0}\left(A_c^2 + \frac{A_s\left(A_s^2 - 4A_c^2\right)}{(A_s - 2A_c\cos\eta)}\right) \text{ and } |\psi|^2_{\min} = \frac{1}{\ell_0^2 g_0}\left(A_c^2 - \frac{A_c^2\left(A_s^2 - 4A_c^2\right)\cos^2\eta}{\left(A_s^2 - 4A_c^2\cos^2\eta\right)}\right), \tag{6.146}$$

where $\xi = \sqrt{A_s^2 - 4A_c^2}\left(\frac{1}{\ell_0}\left(\frac{k_0}{3t+k_0}\right)^{\frac{1}{3}}x - \frac{k_c k_0}{\ell_0^2}\left(\frac{3t+k_0}{k_0}\right)^{\frac{1}{3}} + \widetilde{T}_0\right) + \xi_{20}$ and $\eta = -\frac{A_s\sqrt{A_s^2-4A_c^2}}{2}\left(\frac{k_0}{\ell_0^2}\left(\frac{3t+k_0}{k_0}\right)^{\frac{1}{3}} + \widetilde{T}_0\right) + \eta_{20}$. We show in Fig. 6.20 the maximum and the minimum of the peak matter density given by Eqs. (6.145) and (6.146) with the same parameters as in Fig, 6.19. Plots of Fig. 6.20 show that each of bright solitons (6.143a) and (6.143b) can only be squeezed into the assumed peak matter density between the minimum and maximum values. Figure 6.20a confirms that the wave amplitude decreases as the soliton given by Eq. (6.143a) propagates, while plot Fig. 6.20b confirms that soliton (6.143b) of BEC with balanced gain or loss admits a stabilized peak during its propagation.

In order to verify the stability of the bright soliton against the variation of the scattering length in the confining parabolic potential, we evaluate the exact number of the atoms in the bright soliton against the background described by Eqs. (6.143a) and (6.143b) and respectively obtain

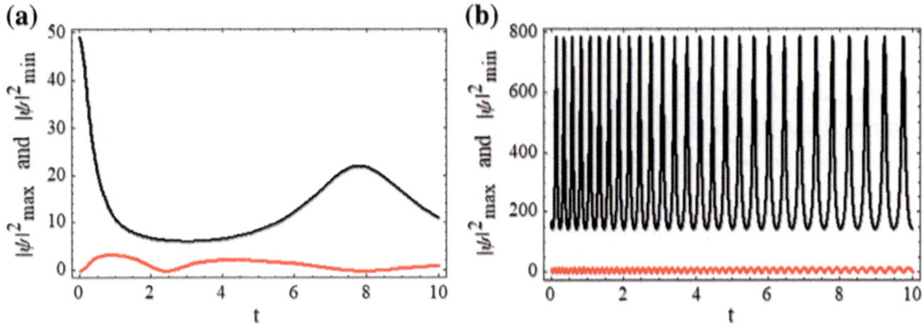

Fig. 6.20 (Color online) The evolution plots of maximum and the minimum of the peak matter density $|\psi|^2_{max}$ and $|\psi|^2_{min}$ given by Eqs. (6.145) and (6.146) for the same parameters as in Fig. 6.19. Plot (**a**) and plot (**b**) show maximum and the minimum of the peak matter density of the bright solitons given respectively by Eqs. (6.143a) and (6.143b). The solid lines show the $|\psi|^2_{min}$, while the dashed lines show the $|\psi|^2_{max}$

$$\lim_{L\to+\infty}\int_{-L}^{L}\left(|\psi(x,t)|^2-|\psi(\pm L,t)|^2\right)dx=\sqrt{A_s^2-4A_c^2}, \tag{6.147a}$$

$$\lim_{L\to+\infty}\int_{-L}^{L}\left(|\psi(x,t)|^2-|\psi(\pm L,t)|^2\right)dx=\frac{\sqrt{A_s^2-4A_c^2}}{\ell_0 g_0}\left(\frac{3t+k_0}{k_0}\right)^{\frac{1}{3}}, \tag{6.147b}$$

Equation (6.147a) indicates that during the process of the broadening of the bright soliton for the mother BEC, the number of atoms in the bright soliton keeps invariant. Equation (6.147b) reveals that during the process of the broadening of the bright soliton for the daughter BEC, the number of atoms in the bright soliton increases with time.

Under the condition of $\psi(\pm L,t)\neq 0$, we can evaluate the number of atoms in both the bright solitons and backgrounds:

$$\lim_{L\to+\infty}\int_{-L}^{L}|\psi(x,t)-\psi(\pm L,t)|^2\,dx=\sqrt{A_s^2-4A_c^2}\,(1+4A_cM\cos\eta_m) \tag{6.148a}$$

for the bright soliton (6.143a) of the mother BEC, where $\eta_m=\frac{\kappa A_s^2}{2}v_1+\eta_0$, and

$$\lim_{L\to+\infty}\int_{-L}^{L}|\psi(x,t)-\psi(\pm L,t)|^2\,dx=\frac{\sqrt{A_s^2-4A_c^2}\,(1+4A_cM\cos\eta_d)}{\ell_0 g_0}\left(\frac{3t+k_0}{k_0}\right)^{\frac{1}{3}} \tag{6.148b}$$

for the bright soliton (6.143b) of the daughter BEC, where $\eta_d=\eta=-\frac{A_s\sqrt{A_s^2-4A_c^2}}{2}\left(\frac{k_0}{\ell_0^2}\left(\frac{3t+k_0}{k_0}\right)^{\frac{1}{3}}+\widetilde{T}_0\right)+\eta_{20}$, M being given by

$$M=\frac{4\arctan\left[\sqrt{A_s+2A_c\cos\eta}/\sqrt{A_s-2A_c\cos\eta}\right]}{\sqrt{(A_s+2A_c\cos\eta)\,(A_s-2A_c\cos\eta)}},$$

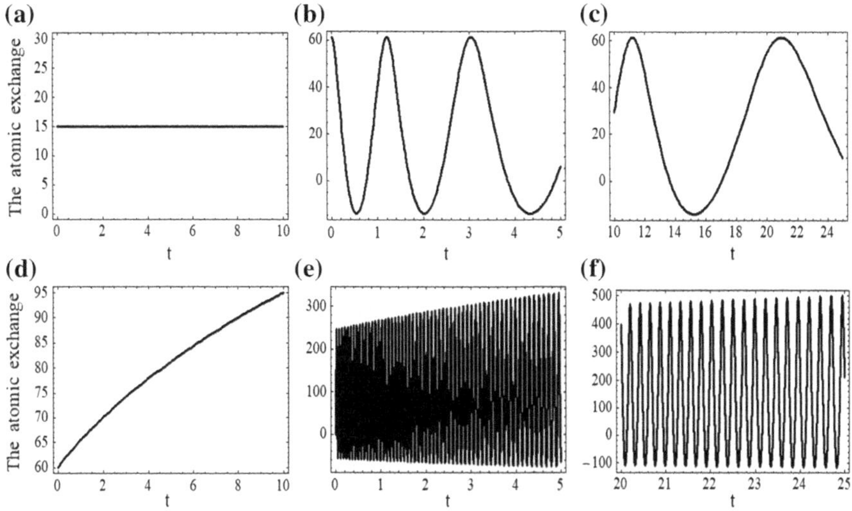

Fig. 6.21 (Color online) The atomic exchange between the bright solitons and the background given (**a**)–(**c**) by Eq. (6.148a) and (**d**)–(**f**) by Eq. (6.148b) for the same parameters as in Fig. 6.19. The range of time is (**a**), (**d**) [0, 10], (**b**), (**e**) [0, 5], (**c**) [10, 25], and (**f**) [20, 25]

with $\eta = \eta_m$ for Eq. (6.148a) and $\eta = \eta_d$ for Eq. (6.148b). Equation (6.148a) displays that a time-periodic atomic exchange is formed between the bright soliton of the mother BEC and the background. Equation (6.148b) reveals that a time-aperiodic atomic exchange is formed between the bright soliton of the daughter BEC and the background. Plots of Fig. 6.21 show the atomic exchange (top panels) between the bright soliton and the background given by Eq. (6.148a) and (bottom panels) between the bright soliton and the background given by Eq. (6.148b).

As shown in Fig. 6.21a, in the case of zero background, i.e., $A_c = 0$, there will not be the exchange of atoms between the bright soliton of the mother BEC and the zero background; Figure 6.21d shows that in the case of zero background the exchange of atoms of bright soliton of the daughter BEC increases. In the case of nonzero background as shown in Fig. 6.21b, c, and in Fig. 6.21e, f, the exchange of atoms between the bright solitons and the backgrounds becomes slower when decreasing the absolute value of the scattering length. It is important to note that in the case of the bright soliton of the daughter BEC, the whole trend of the atomic exchange between the bright soliton and the background increases. This means that the number of atoms continuously injected into the daughter Bose–Einstein condensate from the reservoir depends on the positive imaginary part of the complex potential (gain), and cannot be controlled by applying the external magnetic field via Feshbach resonance; this last fact is in concordance with the finding of Wu et al. [42].

With the choice of constants $\eta_0 = \upsilon_{10} = 0$, Eq. (6.143a) for the mother BEC at each time $t_n = \frac{k_0}{3}\left(\frac{\kappa k_0 g_0^2 A_s^2}{\kappa k_0 g_0^2 A_s^2-(2n+1)\pi}\right)^3 - \frac{k_0}{3}$, $n = 0, \pm 1, \pm 2, \ldots$ takes the form $\psi(x, t_n) = -\sqrt{g(t_n)}A_c \exp[\,i\Delta_b(x, t_n)\,] \pm i\,(-1)^n\sqrt{g(t_n)\left(A_s^2 - 4A_c^2\right)}\frac{\exp[i\Delta_b(x,t_n)]}{\cosh\xi(x,t_n)}$, which is a linear combination of the background and a bright soliton. This means that Eq. (6.143a) can be generated by coherently adding a bright soliton into the background of density described $\psi_c(x, t) = -\sqrt{g(t)}A_c \exp[i\Delta_b(x, t)]$. Equation (6.143b) for the daughter BEC at time $t_n' = \frac{k_0}{3}\left(\frac{\ell_0^2(2\eta_{20}-(1+2n)\pi)}{k_0 A_s\sqrt{A_s^2-4A_c^2}} - \frac{\ell_0^2}{k_0}\widetilde{T}_0\right)^3 - \frac{k_0}{3}$ takes the special form $\psi(x, t_n') = -\frac{A_c}{\ell_0\sqrt{g_0}}\exp\left[i\varphi(x, t_n')\right] + i\frac{(-1)^{n+1}\sqrt{A_s^2-4A_c^2}}{\ell_0\sqrt{g_0}}\frac{\exp\left[i\varphi(x,t_n')\right]}{\cosh\xi}$, which, as in the case of the mother BEC, is a linear combination of the background and a bright soliton. Therefore, Eq. (6.143b) for the daughter BEC can be generated by coherently adding a bright soliton into the background of density described $\psi_c(x, t) = -\frac{A_c}{\ell_0\sqrt{g_0}}\exp[i\varphi(x, t)]$.

Dynamics of a Bright Soliton in Bose–Einstein Condensates with Time-Independent Harmonic Potential

As the second example, we follow Liang et al. [27] and consider the BECs with parameters $g(t) = 2g_0 \exp[\pm 4\lambda t]$ and $k(t) = -2\lambda^2$, where g_0 is any positive real constant. As we have showed in section II, these two parameters satisfy Eq. (6.132a), leading to the BEC with symmetric complex potentials with respect to the spatial variable x described by the GP equation

$$i\frac{\partial\psi}{\partial t} + \frac{1}{2}\frac{\partial^2\psi}{\partial x^2} + 2g_0\exp[\pm 4\lambda t]\,|\psi|^2\,\psi + 2\lambda^2 x^2\psi \pm \lambda i\psi = 0. \tag{6.149}$$

Therefore, sign "+" and "−" in Eq. (6.149) correspond to BEC with loss of atoms and BEC with gain of atoms, respectively. Following Zhao et al. [46], soliton solutions of Eq. (6.149) read

$$\psi(x, t) = \sqrt{2g_0}\left[A_c + A_s\frac{\rho\cosh\xi + \cos\theta}{\cosh\xi + \rho\cos\theta} + i\,A_s\frac{d\sinh\xi + \kappa\sin\theta}{\cosh\xi + \rho\cos\theta}\right]\exp[i\Delta_1], \tag{6.150}$$

where $\rho = -\frac{2A_cA_s}{A_s^2+\rho_2^2}$, $\kappa^2 = 1 - \frac{4A_c^2}{A_s^2+\rho_2^2}$, $d = -\frac{2\rho_2 A_c}{A_s^2+\rho_2^2}$, $\xi = A_s\left[2\kappa g_0\exp[\pm 2\lambda t]\right.$ $x - (\rho_2 + A_c d)\,\upsilon_1 - \kappa\upsilon_2] + \xi_0$, $\theta = 2\rho_2 g_0\exp[\pm 2\lambda t]\,x - \left(\rho_2^2 - A_s^2\right)\frac{\kappa}{2}\upsilon_1 - \rho_2\upsilon_2$ $+ \theta_0$, $\Delta_1 = \mp\lambda x^2 + \exp[\pm 2\lambda t]\,x + A_c^2\upsilon_1 - \frac{\upsilon_3}{2} + \delta_{10}$, $\upsilon_1 = \pm\frac{g_0^2}{\lambda}\exp[\pm 4\lambda t] + \upsilon_{10}$, $\upsilon_2 = \frac{g_0}{\pm 2\lambda}\exp[\pm 4\lambda t] + \upsilon_{20}$, $\upsilon_3 = \pm\frac{1}{4\lambda}\exp[\pm 4\lambda t] + \upsilon_{30}$; here, A_c, A_s, ρ_2, ξ_0, θ_0, υ_{10}, υ_{20}, υ_{30}, and δ_{10} are arbitrary real constants. For simplicity, we assume $d = \rho_2 = 0$ and consider only the case of $A_s^2 > 4A_c^2$, for in the case of $A_s^2 < 4A_c^2$, a small perturbation for Eq. (6.150) may lead to the modulation instability. With these conditions, Eq. (6.150) can be deduced into the following form:

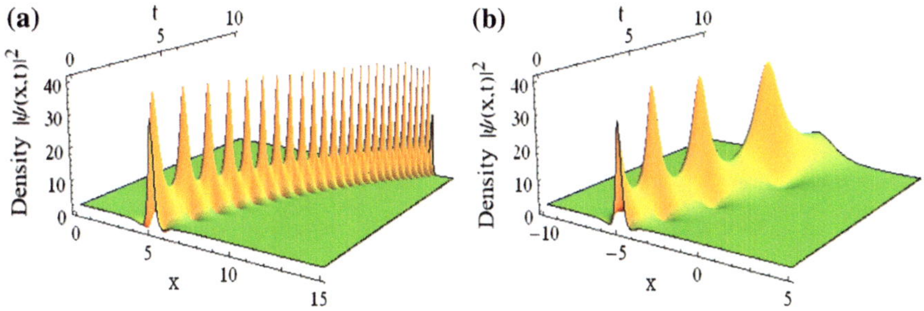

Fig. 6.22 (Color online) The evolution of bright solitons on cw background in a time-independent harmonic trapping potential given by equation (6.151) with the parameters $\lambda = 0.05$, $A_c = 5$, $A_s = 11$, $g_0 = 0.25$, and $\theta_0 = \xi_0 = 0$. **a** $g(t) = 2g_0 \exp[4\lambda t]$; **b** $g(t) = 2g_0 \exp[-4\lambda t]$

$$\psi(x,t) = \sqrt{2g_0}\left[-A_c + \delta \frac{\delta \cos[\theta] - i A_s \sin[\theta]}{A_s \cosh[\xi] - 2A_c \cos[\theta]}\right] \exp[i\Delta_1], \qquad (6.151)$$

where $\delta = \sqrt{A_s^2 - 4A_c^2}$, $\xi = \delta g_0 \left[2x \exp[\pm 2\lambda t] \mp \frac{1}{2\lambda} \exp[\pm 4\lambda t]\right] + \xi_0$, $\theta = \pm \frac{\delta A_s g_0^2}{2\lambda} \exp[\pm 4\lambda t] + \theta_0$, $\Delta_1 = \mp \lambda x^2 + x \exp[\pm 2\lambda t] \pm \frac{8A_c^2 g_0^2 - 1}{8\lambda} \exp[\pm 4\lambda t] + \delta_{10}$, where A_c, A_s, ξ_0, θ_0, and δ_{10} are arbitrary real constants. From Eq. (6.151), we find that the peak density of the bright soliton changes according to $P(t) = 2g_0 \left|-A_c + \delta \frac{\delta \cos[\theta] - i A_s \sin[\theta]}{A_s - 2A_c \cos[\theta]}\right|^2$, meaning that the peak density of the bright soliton (6.151) neither increases nor decreases with time t.

Figure 6.22a, b show the evolutions of bright solitons on cw backgrounds for $g(t) = 2g_0 \exp[4\lambda t]$ and $g(t) = 2g_0 \exp[-4\lambda t]$, respectively. Analysis of Eq. (6.151) reveals that the velocity for such solitons still satisfies Eq. (6.144). It is seen from Fig. 6.4 that the bright solitons behave like breather solitons propagating on a cw background, which corresponds to the atom exchanging between the bright solitons and the cw background, and this exchange keeps the bright solitons dynamically stable against the variation of the nonlinearity, as demonstrated by Liang et al. [27]. From Fig. 6.22, we can also see that with the increasing (decreasing) absolute value of the s-wave scattering length, the bright solitons have a constant in the peak value and a compression (broadening) in their width.

BEC Bright Solitons with Nonlinearity Parameter $g(t) = 1 + \tanh[\Omega t]$

We end our applications with controllable compression of BEC bright solitons. In order to compress the soliton in a controllable manner, we gradually increase the interaction and simultaneously turn off the longitudinal confinement potential. Following Xue [43], we consider the BECs with the nonlinearity parameter $g(t) = 1 + \tanh[\Omega t]$; this choice of parameter $g(t)$ leads to the BEC with the symmetric complex potential with respect to the spatial variable x described by the GP equation

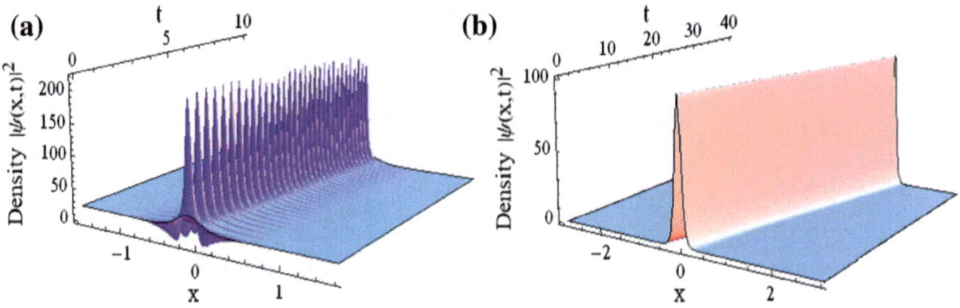

Fig. 6.23 (Color online) The dynamics of bright solitons with **stabilized peak** for BECs with a time-varying atomic scattering length in a time-varying expulsive parabolic potential with $\Omega = 0.1$, $A_s = 10, k_s = 0.03$, and $\eta_0 = \xi_0 = 0$. **a** The evolution of bright soliton on cw background given by Eq. (6.153) with the parameter $A_c = 4.8$. **b** The evolution of bright soliton on zero cw background ($A_c = 0$) given by Eq. (6.154)

$$i\frac{\partial\psi}{\partial t} + \frac{1}{2}\frac{\partial^2\psi}{\partial x^2} + g\,|\psi|^2\,\psi + \frac{1}{8}\left(3 - \tanh^2[\Omega t] - 2\tanh[\Omega t]\right)\Omega^2 x^2\psi = i\frac{\Omega}{4}\left(\tanh[\Omega t] - 1\right)\psi. \tag{6.152}$$

In Eq. (6.152), the positive parameter Ω represents the normalized longitudinal confinement frequency at the initial stage. Notice also that in Eq. (6.152), $\widetilde{k}(t) = \frac{1}{8}\left(3 - \tanh^2[\Omega t] - 2\tanh[\Omega t]\right) > 0$ and $g(t) = 1 + \tanh[\Omega t] > 0$, the focusing nature of the nonlinearity (which represents the attractive nature of the interatomic interactions) and the expulsive potential are considered. Also, the dissipative parameter $\gamma(t) = \frac{\Omega}{4}\left(\tanh[\Omega t] - 1\right) < 0$ so that Eq. (6.152) describes a BEC with loss of atoms.

Combining Eqs. (6.137) and (6.138a) with $\ell_0 = \mathrm{sign}(g_0)/g_0 = 1$ leads to the bright soliton solutions of Eq. (6.152). As in the above second example, we set, for simplicity, in Eq. (6.138a) $k_c = k_s$, and consider only A_c and A_s which satisfy condition $A_s^2 > 4A_c^2$. The solution for Eq. (6.152) then becomes

$$\psi(x,t) = \left(-A_c + \delta\frac{\delta\cos\eta - iA_s\sin\eta}{A_s\cosh\xi - 2A_c\cos\eta}\right)\exp\left[i\Delta_{2b} - i\frac{\Omega}{4}\left(1 - \tanh[\Omega t]\right)x^2\right], \tag{6.153}$$

where

$$\delta = \sqrt{A_s^2 - 4A_c^2},\; \xi = \delta\left(x\sqrt{1 + \tanh[\Omega t]} - \frac{k_s}{\Omega}\ln\frac{\exp[2\Omega t] + 1}{2} + \xi_0\right),$$

$$\eta = -\frac{A_s\delta}{2\Omega}\ln\frac{\exp[2\Omega t] + 1}{2} + \eta_0,$$

$$\Delta_{2b} = k_s x\sqrt{1 + \tanh[\Omega t]} + \frac{2A_c^2 - k_s^2}{2\ell_0^2\Omega}\ln\frac{\exp[2\Omega t] + 1}{2} + \Delta_0,$$

ξ_0, η_0, and Δ_0 being arbitrary real constants. It should be noted that, when the background wave vanishes (i.e., $A_c = 0$), the solution (6.153) is reduced to the bright soliton solution

$$\psi(x,t) = \frac{A_s}{\cosh\left[\frac{A_s}{\sqrt{2\ell_0}}\left(x\sqrt{\frac{2(1+\tanh[\Omega t])}{\ell_0}} - \frac{k_s}{\ell_0^2\Omega}\ln\frac{\exp[2\Omega t]+1}{2} + \xi_0\right)\right]} \times \exp\left[i\left(\Delta_{2b} - \eta - \frac{\Omega}{4}\left(1 - \tanh\left[\Omega t\right]\right)x^2\right)\right]. \quad (6.154)$$

It follows from Eq. (6.154) that (i) the bright soliton amplitude $\psi_m = A_s^2/\ell_0$ is constant amplitude (bright soliton with **stabilized peak**); (ii) the width $D(t)$ of the bright soliton changes according to $D(t) = \frac{1}{A_s\sqrt{2(1+\tanh[\Omega t])}} = \frac{1}{A_s\sqrt{2g(t)}}$. Because $D(t)$ is a decreasing function of time t, we conclude that Eq. (6.154) can be used to describe the compression of the bright soliton. From Eqs. (6.153) and (6.154), we find that the peak density of the bright soliton embedded in the cw background and that of the bright soliton in zero background change according to

$$P_{cw}(t) = \left|-A_c + \delta\frac{\delta\cos\eta - iA_s\sin\eta}{A_s - 2A_c\cos\eta}\right|^2 \text{ and } P_0(t) = A_s^2,$$

respectively; (iii) the center of bright solitons is $x = \frac{1}{\sqrt{1+\tanh[\Omega t]}}\left(\frac{k_s}{\Omega}\ln\frac{\exp[2\Omega t]+1}{2} - \xi_0\right)$ and satisfied Eq. (6.144), meaning that the center of mass of the macroscopic wave packet behaves like a classical particle, and allows one to manipulate the motion of bright solitons in BEC systems by controlling the external harmonic trapping potential.

For a better understanding, we respectively plot in Fig. 6.23a, b the dynamics of a bright solitons (6.153) and (6.154) in the harmonic trapping potential with $\Omega = 0.1$ (the parameter used in the experiment of Khaykovich et al. [20] and for simplicity we set $A_s = 10$). It is clear from plots Fig. 6.23a, b, with increasing absolute value of the scattering length and decreasing absolute value of the expulsive parabolic potential, the bright soliton has a constant in the peak matter density, and a compression in the width.

In summary, we have formulated in terms of the parameter g of the two-body interatomic interactions and the strength k of the magnetic trap the condition of the stabilization of BEC solitons by symmetric complex potentials with respect to the spatial variable x. As applications of our finding, we have discussed analytically the dynamics of matter-wave solitons for well-known BECs with symmetric complex potential whose imaginary part acts as either a gain or a loss parameter of the system. As predicted, we found that bright solitons in each considered BEC with balanced gain or loss has a stabilized peaks during their propagation.

References

1. Abdullaev, F.K., Bronski, J.C., Galimzyanov, R.M.: Dynamics of a trapped 2D Bose-Einstein condensate with periodically and randomly varying atomic scattering length. arXiv:cond-mat/0205464
2. Adhikari, S.K.: Numerical solution of the two-dimensional Gross-Pitaevskii equation for trapped interacting atoms. Phys. Lett. A **265**, 91 (2000)
3. Atre, R., Panigrahi, P.K., Agarwal, G.S.: Class of solitary wave solutions of the one-dimensional Gross-Pitaevskii equation. Phys. Rev. E **73**, 056611 (2006)
4. Becker, C., Stellmer, S., Soltan-Panahi, P., Dörscher, S., Baumert, M., Richter, E., Kronjäger, J., Bongs, K., Sengstock, K.: Oscillations and interactions of dark and dark-bright solitons in Bose-Einstein condensates. Nat. Phys. **4**, 496 (2008)
5. Belmonte-Beitia, J., Pérez-García, V.M., Vekslerchik, V., Torres, P.J.: Lie symmetries and solitons in nonlinear systems with spatially inhomogeneous nonlinearities. Phys. Rev. Lett. **98**, 064102 (2007)
6. Brazhnyi, V.A., Konotop, V.V.: Theory of nonlinear matter waves in optical lattices. Mod. Phys. Lett. B **18**, 627 (2004)
7. Carr, L.D., Castin, Y.: Dynamics of a matter-wave bright soliton in an expulsive potential. Phys. Rev. A **66**, 063602 (2002)
8. Dalfovo, F., Giorgini, S., Pitaevskii, L.P., Stringari, S.: Theory of Bose-Einstein condensation in trapped gases. Rev. Mod. Phys. **71**, 463 (1999)
9. Drazin, P.G., Johnson, R.S.: Solitons: An Introduction. Cambridge University Press, Cambridge (1988)
10. Fetter, A.L., Walecka, J.D.: Quantum Theory of Many-particle Systems. McGraw-Hill, New York (1971)
11. Gammal, A., Frederico, T., Tomio, L., Chomaz, P.: Atomic Bose-Einstein condensation with three-body interactions and collective excitations. J. Phys. B: At. Mol. Opt. Phys. **33**, 4053 (2000)
12. Hartung, M., Wellens, T., Müller, C.A., Richter, K., Schlagheck, P.: Coherent backscattering of Bose-Einstein condensates in two-dimensional disorder potentials. Phys. Rev. Lett. **101**, 020603 (2008)
13. Ilan, B., Fibich, G., Papanicolaou, G.: Self-focusing with fourth-order dispersion. SIAM J. Appl. Math. **62**, 1437 (2002)
14. Kagan, Y., Muryshev, A.E., Shlyapnikov, G.V.: Collapse and Bose-Einstein condensation in a trapped bose gas with negative scattering length. Phys. Rev. Lett. **81**, 933 (1998)
15. Kanamoto, R., Saito, H., Ueda, M.: Quantum phase transition in one-dimensional Bose-Einstein condensates with attractive interactions. Phys. Rev. A **67**, 013608 (2003)
16. Kartashov, Y.V., Malomed, B.A., Torner, L.: Solitons in nonlinear lattices. Rev. Mod. Phys. **83**, 247 (2011)
17. Kengne, E., Lakhssassi, A., Liu, W.M., Vaillancourt, R.: Phase engineering, modulational instability, and solitons of Gross–Pitaevskii-type equations in 1+1 dimensions. Phys. Rev. E **87**, 022914 (2013)
18. Kengne, E., Talla, P.K.: Dynamics of bright matter wave solitons in Bose-Einstein condensates in an expulsive parabolic and complex potential. J. Phys. B **39**, 3679 (2006)
19. Kengne, E., Liu, W.M.: Management of matter-wave solitons in Bose-Einstein condensates with time-dependent atomic scattering length in a time-dependent parabolic complex potential. Phys. Rev. E **98**, 012204 (2018)
20. Khaykovich, L., Schreck, F., Ferrari, G., Bourdel, T., Cubizolles, J., Carr, L.D., Castin, Y., Salomon, C.: Formation of a matter-wave bright soliton. Science **296**, 1290 (2002)
21. Kivshar, YuS, Agrawal, G.P.: Optical Solitons: From Fibers to Photonic Crystals. Academic, CA (2003)
22. Kivshar, YuS, Alexander, T.J., Turitsyn, S.K.: Nonlinear modes of a macroscopic quantum oscillator. Phys. Lett. A **278**, 225 (2001)

23. Konotop, V.V., Pitaevskii, L.: Landau dynamics of a grey soliton in a trapped condensate. Phys. Rev. Lett. **93**, 240403 (2004)
24. Kumar, V.R., Adha, R.R., Adati, M.W.: Phase engineering and solitons of Bose-Einstein condensates with two- and three-body interactions. J. Phys. Soc. Jpn. **79**, 074005 (2010)
25. Li, B., Zhang, X.-F., Li, Y.-Q., Chen, Y., Liu, W.M.: Solitons in Bose-Einstein condensates with time-dependent atomic scattering length in an expulsive parabolic and complex potentia. Phys. Rev. A **78**, 023608 (2008)
26. Li, B., Zhang, X.-F., Li, Y.-Q., Liu, W.M.: Propagation and interaction of matter-wave solitons in Bose–Einstein condensates with time-dependent scattering length and varying potentials. J. Phys. B: At. Mol. Opt. Phys. **44**, 175301 (2011)
27. Liang, Z.X., Zhang, Z.D., Liu, W.M.: Dynamics of a bright soliton in BECs with time-dependent atomic scattering length in an expulsive parabolic potential. Phys. Rev. Lett. **94**, 050402 (2005)
28. Malomed, B.A.: Soliton Management in Periodic Systems. Springer, New York (2006)
29. Matveev, V.B., Salli, M.A.: Darboux Transformations and Solitons. Springer Series in Nonlinear Dynamics. Springer, Berlin (1991)
30. Paul, T., Richter, K., Schlagheck, P.: Nonlinear resonant transport of Bose-Einstein condensates. Phys. Rev. Lett. **94**, 020404 (2005)
31. Pérez-García, V.M., Konotop, V.V., Brazhnyi, V.A.: Feshbach resonance induced shock waves in Bose-Einstein condensates. Phys. Rev. Lett. **92**, 220403 (2004)
32. Pérez-García, V.M., Michinel, H., Herrero, H.: Bose-Einstein solitons in highly asymmetric traps. Phys. Rev. A **57**, 3837 (1998)
33. Saito, H., Ueda, M.: Dynamically stabilized bright solitons in a two-dimensional Bose-Einstein condensate. Phys. Rev. Lett. **90**, 040403 (2003)
34. Schürmann, H.W., Serov, S.V.: Traveling wave solutions of a generalized modified Kadomtsev-Petviashvili equation. J. Math. Phys. **45**, 2181 (2004)
35. Stellmer, S., Becker, C., Soltan-Panahi, P., Richter, E.M., Dorscher, S., Baumert, M., Kronjager, J., Bongs, K., Sengstock, K.: Collisions of dark solitons in elongated Bose-Einstein condensates. Phys. Rev. Lett. **101**, 120406 (2008)
36. Strecker, K.E., Partridge, G.B., Truscott, A.G., Hulet, R.G.: Formation and propagation of matter-wave soliton trains. Nature (London) **417**, 150 (2002)
37. Theocharis, G., Rapti, Z., Kevrekidis, P.G., Frantzeskakis, D.J., Konotop, V.V.: Modulational instability of Gross-Pitaevskii-type equations in $1 + 1$ dimensions. Phys. Rev. A **67**, 063610 (2003)
38. Tian, Q., Wu, L., Zhang, J.F., Malomed, B.A., Mihalache, D., Liu, W.M.: Exact soliton solutions and their stability control in the nonlinear Schrödinger equation with spatiotemporally modulated nonlinearity. Phys. Rev. E **83**, 016602 (2011)
39. Wang, D.S., Zhang, X.-F., Zhang, P., Liu, W.M.: Matter-wave solitons of Bose-Einstein condensates in a time-dependent complicated potential. J. Phys. B **42**, 245303 (2009)
40. Weierstrass, K.: Mathematische Werke V. Johnson, New York (1915)
41. Whittaker, E.T., Watson, G.N.: A Course in Modern Analysis, 4th edn. Cambridge University Press, UK (1990)
42. Wu, L., Zhang, J.F., Li, L.: Modulational instability and bright solitary wave solution for Bose-Einstein condensates with time-dependent scattering length and harmonic potential. New J. Phys. **9**, 69 (2007)
43. Xue, J.-K.: Controllable compression of bright soliton matter waves. J. Phys. B: At. Mol. Opt. Phys. **38**, 3841 (2005)
44. Yan, Z.Y.: Envelope exact solutions for the generalized nonlinear Schrödinger equation with a source. J. Phys. A: Math. Gen. **39**, L401 (2006)
45. Zhang, Y., Wu, B.: Composition relation between gap solitons and bloch waves in nonlinear periodic systems. Phys. Rev. Lett. **102**, 093905 (2009)
46. Zhao, L.C., Yang, Z.-Y., Zhang, T., Chi, K.-J.: Dynamics of bright solitons in Bose-Einstein condensates with complicated potential. Chin. Phys. Lett. **26**, 120301 (2009)

Chapter 7
Dynamics of BECs with Two- and Three-Body Interactions

Abstract Based on the inhomogeneous cubic–quintic nonlinear Schrödinger equation with external real or complex potential, we investigate in this chapter the dynamics of Bose–Einstein condensates when both the two- and three-body interatomic interactions are taken into consideration. We start by performing different methods leading to the exact analytical periodic and solitary wave solutions of such equations. Via these exact solutions, we investigate quantitatively and qualitatively the dynamical properties including the coherent structures and the modulational instability of matter waves in BECs with both the two- and three-body interactions. The effects of the three-body interaction in weakly interacting BECs are also investigated.

7.1 Introduction

Three-body interactions are ubiquitous and play an important role in a wide variety of interesting physical phenomena, and yield new physics and many surprises not encountered in systems dominated by the two-body interactions. Most recently, many experimental and theoretical techniques have been proposed to observe and realize the three-body interactions in ultracold Bose gas [5, 10, 16, 42]. Inelastic three-body processes, including observations of Efimov quantum states and atom loss from recombination have been also reported in works [7, 12, 13]. Many proposals dealing with effects of effective three-body interactions in ultracold bosonic atoms in an optical lattice or a superlattice are reported in Refs. [30, 38]. It has been reported that the three-body interactions in dilute Bose gases may give rise to considerably modify the collective excitations at both zero and finite temperatures [1], the transition temperature, the condensate depletion and the stability of a BEC in a one- and two-dimensional trapping geometry [29, 34]. Akhmediev et al. [3] showed that the three-body interactions can extend considerably the region of stability for a condensate, and allow the number of condensed atoms to increase even when the strength of the three-body force is very small. Also, besides changes in the density profile of the condensate [39], the excitation spectrum of the collective oscillations can also be modified [1], as the compressibility of such a system is significantly affected by the three-body contribution to the ground state energy.

W.-M. Liu and E. Kengne, *Schrödinger Equations in Nonlinear Systems*,
https://doi.org/10.1007/978-981-13-6581-2_7

The three-body interactions are interesting also for the understanding of the fundamental limits for the functioning of devices using BEC [18, 34, 45]. Clearly, it is important to consider the stability of the BEC with the two- and three-body interactions for future applications of the ultracold gases. This chapter deals with dynamical properties of Bose–Einstein condensates in optical lattices when both two- and three-body interactions are taken into account.

At ultralow temperatures the time-dependent wavefunction $\Psi(\mathbf{r}, t)$ of the condensates trapped in a potential $V(\mathbf{r}, t)$ at position $\mathbf{r}$ and time t in the presence of both the two- and three-body interaction can be described by the following mean-field nonlinear GP equation:

$$i\hbar\frac{\partial\Psi(\mathbf{r},t)}{\partial t} = \left[-\frac{\hbar^2\nabla^2}{2m} + V(\mathbf{r},t) + g(\mathbf{r},t)\,|\Psi(\mathbf{r},t)|^2 + k(\mathbf{r},t)\,|\Psi(\mathbf{r},t)|^4\right]\Psi(\mathbf{r},t).$$

Here, $g(\mathbf{r},t) = 4\pi\hbar^2 a_s/m$ and $k(\mathbf{r},t)$ characterize the strengths of spatiotemporal two-body and three-body interatomic interactions, respectively; $\hbar$ is Planck's constant, m is the mass of the single bosonic atom and a_s is the (spatiotemporal) s-wave scattering length which can be tuned to any desired value by using the Feshbach resonance (FR) technique. The trapping potential we consider in this chapter is the cigar-shaped harmonic trapping potential with the elongated axis in the x-direction as used in experiment [6, 43], which takes the form $V = \left[\frac{m}{2}\omega^2 x^2 + V_1(x,t)\right] + \frac{m}{2}\omega_\perp^2(y^2+z^2)$ with $\omega << \omega_\perp$, where $V_1(x,t)$ is either a bounded complex potential or the optical lattice potential. In this case, since $\omega/\omega_\perp$ is small, one can expect that the variation of the profile of the order parameter is slow in the elongated direction. On the other hand, due to the strong confinement in the transverse (y and z) directions, the motion of the order parameter in the transverse directions is frozen, and can be described by the ground-state wavefunction of a two-dimensional quantum harmonic oscillator. Thus, by introducing the following transformation [23]

$$\Psi(\mathbf{r},t) = \frac{1}{\sqrt{2\pi a_B}a_\perp}\exp\left[-i\omega_\perp t - \frac{x^2+y^2}{2a_\perp^2}\right]\psi\left(\frac{x}{a_\perp}, \omega_\perp t\right),$$

one can obtain the model equation for the cigar-shaped BEC, which is the well-known 1D cubic–quintic NLS equation with an external trapping potential

$$i\frac{\partial\psi}{\partial t} = \left[-\frac{1}{2}\frac{\partial^2}{\partial x^2} + V(x,t) + g_0(x,t)\,|\psi|^2 + \chi_0(x,t)\,|\psi|^4\right]\psi.$$

Here, a_B is the Bohr radius [35], the time t and the space coordinate x are measured in harmonic-oscillator units, that is, in units of $1/\omega_\perp$ and $a_\perp$, respectively, where $a_\perp = \sqrt{\frac{\hbar}{m\omega_\perp}}$ and $a_0 = \sqrt{\frac{\hbar}{m\omega_0}}$ are linear oscillator lengths in the transverse and cigar-axis directions, respectively. $\omega_\perp$ and ω_0 are the corresponding axial oscillation frequencies.

7.2 Generalized Ginzburg–Landau Equation and Dynamic Behavior of BECs in Optical Lattices with Two- and Three-Body Interactions

The complex Ginzburg–Landau equations have drawn a great deal of attention in physics and applied mathematics communities as a class of universal models with a broad spectrum of applications. It arises in a variety of settings, including electro-convection in nematic liquid crystals [11], binary fluid convection [4], patterns near electrodes in gas discharges [31], oscillatory chemical reactions [25], nonlinear optical systems [33], mathematical biology [32], nonlinear transmission lines [17, 21], and spatiotemporal intermittency in spatially extended dissipative systems [9]. As reviewed in refs. [28], the CGL equation also applies to superfluidity, Bose–Einstein condensation, reaction–diffusion waves. Its general form is

$$\frac{\partial u}{\partial t} = (P_r + iP_i)\,\Delta u + (\lambda_r + i\lambda_i)\,u + (\gamma_r + i\gamma_i)\,|u|^2 u = 0, \tag{7.1}$$

where $u(\mathbf{x}, t)$ is a complex-valued function of time t and spatial variable $\mathbf{x}$ and P_r, P_i, λ_r, λ_i, γ_r, and γ_i are either real constants or real-valued functions. In the context of superconductivity, the unknown function $u(\mathbf{x}, t)$ of the CGL equation (7.1) is the order parameter, the constants P_i, λ_i, and γ_i are usually zero, and the bifurcation parameter $\lambda = \lambda_r$ is the GL parameter. The CGL equation (7.1) describes on a qualitative and often even on a quantitative level a vast variety of phenomena from nonlinear waves to second-order phase transitions, from superconductivity, superfluidity, and Bose–Einstein condensation to liquid crystals and strings in field theory (see for example Refs. [9, 25]).

The CGL equation (7.1) is a minimal, universal model that cannot be further simplified. However, there are many ways to generalize it in order to include qualitatively new features [4]. The main trends of generalization can be associated with the different terms of the CGL equation: generalization of the nonlinearity, generalization of the differential operator, generalization of the symmetry group, etc. The one-dimensional equation may be written in the general form

$$i\frac{\partial u}{\partial t} + (P_r + iP_i)\frac{\partial^2 u}{\partial x^2} + f_1\left(|u|^2\right)u + \frac{\partial}{\partial x}\left[f_2\left(|u|^2\right)u\right] + \left[\frac{\partial}{\partial x}f_3\left(|u|^2\right)\right]u = 0, \tag{7.2}$$

where f_n are generally complex functions of the real argument $|u|^2$ with $f_n = f_{nr} + if_{ni}$, $(n = 1, 2, 3)$, P_r and P_i being two real functions. In this section, we will focus attention on the following special case of Eq. (7.2), called generalized complex Ginzburg–Landau equation (GCGL) equation

$$i\frac{\partial u}{\partial t} + P\frac{\partial^2 u}{\partial x^2} + \Gamma u + Q|u|^2 u + Q_3|u|^4 u + Q_1 u^2\frac{\partial u^*}{\partial x} + Q_2\,|u|^2\frac{\partial u}{\partial x} = 0, \tag{7.3}$$

where P, Q, Γ, and Q_l are complex-valued functions of either x, or t, or both x and t, ($l = 1, 2, 3$), u^* being the complex conjugate of u. Equation (7.3) coincides with the generalized nonlinear Schrödinger equation if P, Γ, Q and Q_3 are real-valued functions and Q_1 and Q_2 purely imaginary functions [4, 31].

In the context of BEC, Eq. (7.3) models in the mean-field approximation the dynamics of the BEC. When $Q_1 \neq 0$, Eq. (7.3) may appear in BEC when one desires to investigate the impact of both two- and three-body interactions on the condensates. In this last situation, the strength of three-body interactions may be a complex-valued function. The imaginary part of the three-body interactions in this case represents the three-body loss. Because the three-body losses are dissipative terms they have to be neglected to conserve the energy and the total number of atoms from the viewpoint of integrability of the dynamical system. The derivative terms in Eq. (7.3) may represent the delayed nonlinear response of the system offsets the modulation instability arising due to three-body interactions [24]. When $|Q_1| + |Q_2| = 0$ and $QQ_3 \neq 0$, Eq. (7.3) models the condensate when both the two- and three-body interactions are taken into account [1].

In this section, we aim to explore via a "double-mapping method" exact solutions of Eq. (7.3). Using the similarity transformation, we first map this equation into an equation with constant coefficients. Then, the F-expansion method is applied to obtain exact Jacobi elliptic function solutions. The results are then applied on the BECs in optical lattices when both the two- and three-body interactions are taken into account.

7.2.1 Mapping Between GCGL Equation with Variable Coefficients and GCGL Equation with Constant Coefficients

We assume here that coefficients of the GCGL equation (7.3) are complex-valued functions of variables x and t, and seek nontrivial its solutions in the form [20]

$$u(x,t) = \psi(X,T)\,\rho(x,t)\exp\left[i\varphi(x,t)\right], \quad X = X(x,t), \quad T = T(t), \tag{7.4}$$

where $\rho(x,t)$, $\varphi(x,t)$, $X(x,t)$, and $T(t)$ are four real-valued functions and $\psi(X,T)$ is a complex-valued function of variables X and T. We assume that $\rho(x,t)$, $\varphi(x,t)$, $X(x,t)$, $T(t)$, and real and imaginary parts of coefficients of Eq. (7.3) are real smooth functions of their variables.

General Case When ψ is a Function of Two Variables X and T

In the general case when ψ is a complex function of two real variables X and T, we demand that

$$
\begin{aligned}
\rho(x,t)\frac{dT}{dt} - 1 &= 0,\\
P\rho(x,t)\left(\frac{\partial X}{\partial x}\right)^2 - p &= 0,\\
Q_3\rho^5(x,t) - q_3 &= 0,\\
Q_1\rho^3(x,t)\frac{\partial X}{\partial x} - q_1 &= 0,
\end{aligned}
\tag{7.5}
$$

$$
\begin{aligned}
Q_2\rho^3(x,t)\frac{\partial X}{\partial x} - q_2 &= 0,\\
\Gamma\rho + \left(i\frac{\partial\rho}{\partial t} - \rho\frac{\partial\varphi}{\partial t}\right) + P\left[\frac{\partial^2\rho}{\partial x^2} - \rho\left(\frac{\partial\varphi}{\partial x}\right)^2 + i\left(\rho\frac{\partial^2\varphi}{\partial x^2} + 2\frac{\partial\varphi}{\partial x}\frac{\partial\rho}{\partial x}\right)\right] - \gamma &= 0,\\
Q\rho^3 + Q_2\rho^2\left(\frac{\partial\rho}{\partial x} + i\rho\frac{\partial\varphi}{\partial x}\right) + Q_1\rho^2\left(\frac{\partial\rho}{\partial x} - i\rho\frac{\partial\varphi}{\partial x}\right) - q &= 0,\\
P\left(2\frac{\partial X}{\partial x}\frac{\partial\rho}{\partial x} + \rho\frac{\partial^2 X}{\partial x^2}\right) + i\rho\left(\frac{\partial X}{\partial t} + 2P\frac{\partial X}{\partial x}\frac{\partial\varphi}{\partial x}\right) &= 0,
\end{aligned}
$$

where $p \neq 0, \gamma, q, q_1, q_2, q_3$ are all complex constants. The first equation in system (7.5) is made to preserve the scaling. Inserting Eq. (7.4) into Eq. (7.3) and taking into account system (7.5) lead to the following GCGL equation with constant coefficients

$$
i\frac{\partial\psi}{\partial T} + p\frac{\partial^2\psi}{\partial X^2} + \gamma\psi + q\,|\psi|^2\,\psi + q_3\,|\psi|^4\,\psi + q_1\psi^2\frac{\partial\psi^*}{\partial X} + q_2\,|\psi|^2\,\frac{\partial\psi}{\partial X} = 0. \tag{7.6}
$$

For system (7.5) to have nontrivial solutions, it is necessary that

$$
\begin{aligned}
q_1 = 0 &\Leftrightarrow Q_1(x,t) = 0 \quad \text{for all } x \text{ and } t,\\
q_2 = 0 &\Leftrightarrow Q_2(x,t) = 0 \quad \text{for all } x \text{ and } t,\\
q_3 = 0 &\Leftrightarrow Q_3(x,t) = 0 \quad \text{for all } x \text{ and } t.
\end{aligned}
\tag{7.7}
$$

Conditions (7.7) are necessary for the similarity transformation (7.4).

1. If conditions (7.7) are satisfied and $q_1q_2q_3 \neq 0$, then system (7.5) gives

$$\begin{cases} \frac{dT}{dt} = \sqrt[5]{\frac{Q_3(x,t)}{q_3}}, \\ \rho(x,t) = \sqrt[5]{\frac{q_3}{Q_3(x,t)}}, \\ \frac{\partial X(x,t)}{\partial x} = \frac{q_1}{Q_1(x,t)} \sqrt[5]{\left(\frac{Q_3(x,t)}{q_3}\right)^3}, \\ \frac{\partial \varphi(x,t)}{\partial x} = \frac{i}{Q_1(x,t)-Q_2(x,t)} \left(q \sqrt[5]{\frac{Q_3(x,t)}{q_3}} - Q(x,t)\right), \\ \frac{\partial \varphi(x,t)}{\partial t} = \Gamma(x,t) + i \sqrt[5]{\frac{q_3}{Q_3(x,t)}} \frac{\partial \rho}{\partial t} - \gamma \sqrt[5]{\frac{q_3}{Q_3(x,t)}} - P(x,t) \sqrt[5]{\left(\frac{q_3}{Q_3(x,t)}\right)^2} \left[\left(\frac{\partial \varphi}{\partial x}\right)^2 - i \frac{\partial^2 \varphi}{\partial x^2}\right], \\ \frac{\partial X(x,t)}{\partial t} = i P \frac{\partial^2 X}{\partial x^2} - 2P \frac{\partial X}{\partial x} \frac{\partial \varphi}{\partial x}, \end{cases} \tag{7.8}$$

with the complementary conditions

$$\begin{cases} q_1 Q_2(x,t) = q_2 Q_1(x,t), \\ \frac{p}{P(x,t)} \sqrt[5]{\frac{Q_3}{q_3}} \geq 0, \\ q_1^2 Q_3(x,t) P(x,t) = p q_3 Q_1^2(x,t), \\ Q_1 - Q_2 = 0 \Leftrightarrow q \sqrt[5]{\left(\frac{Q_3}{q_3}\right)^3} - Q = 0 \text{ for all } x \text{ and } t. \end{cases} \tag{7.9}$$

Because each of $\rho(x,t)$, $\varphi(x,t)$, $X(x,t)$, and $T(t)$ is a real-valued function, the right-hand side of each equation of system (7.8) should be a real-valued function. The first equation of system (7.8) implies that $Q_3(x,t)$ should be a function of t only, which also implies that $\rho(x,t)$ is a function of t only, i.e., $Q_3 = Q_3(t)$ and $\rho = \rho(t)$.

2. If conditions (7.7) are satisfied and $q_1 = 0$ and $q_2 q_3 \neq 0$, then system (7.5) gives

$$\begin{cases} \frac{dT}{dt} = \sqrt[5]{\frac{Q_3(x,t)}{q_3}}, \\ \rho(x,t) = \sqrt[5]{\frac{q_3}{Q_3(x,t)}}, \\ \frac{\partial X}{\partial x} = \frac{q_2}{Q_2(x,t)} \sqrt[5]{\left(\frac{Q_3(x,t)}{q_3}\right)^3}, \\ \frac{\partial \varphi}{\partial x} = i \frac{1}{Q_2(x,t)} \left(Q(x,t) - q \sqrt[5]{\frac{Q_3(x,t)}{q_3}}\right), \\ \frac{\partial \varphi}{\partial t} = \Gamma(x,t) + \sqrt[5]{\frac{Q_3(x,t)}{q_3}} i \frac{\partial \rho}{\partial t} - \gamma \sqrt[5]{\frac{Q_3(x,t)}{q_3}} + P \left[-\left(\frac{\partial \varphi}{\partial x}\right)^2 + i \frac{\partial^2 \varphi}{\partial x^2}\right], \\ \frac{\partial X(x,t)}{\partial t} = i P \frac{\partial^2 X}{\partial x^2} - 2P \frac{\partial X}{\partial x} \frac{\partial \varphi}{\partial x}, \end{cases} \tag{7.10}$$

under the conditions

$$q_2^2 Q_3 P(x,t) = p q_3 Q_2^2(x,t) \quad \text{and} \quad \frac{p}{P(x,t)} \sqrt[5]{\frac{Q_3}{q_3}} \geq 0. \tag{7.11}$$

3. If conditions (7.7) are satisfied, $q_1 = q_2 = 0$ and $q_3 \neq 0$, then system (7.5) gives

$$\begin{cases} \frac{dT}{dt} = \sqrt[5]{\frac{Q_3(x,t)}{q_3}}, \\ \rho(x,t) = \sqrt[5]{\frac{q_3}{Q_3(x,t)}}, \\ \frac{\partial X}{\partial x} = \pm\sqrt{\frac{p}{P(x,t)}\sqrt[5]{\frac{Q_3(x,t)}{q_3}}}, \\ \frac{\partial X(x,t)}{\partial t} = iP\frac{\partial^2 X}{\partial x^2} - 2P\frac{\partial X}{\partial x}\frac{\partial \varphi}{\partial x}, \\ \frac{\partial \varphi}{\partial t} - iP\frac{\partial^2 \varphi}{\partial x^2} + P\left(\frac{\partial \varphi}{\partial x}\right)^2 = \Gamma(x,t) - \gamma\sqrt[5]{\frac{Q_3(x,t)}{q_3}} + i\sqrt[5]{\frac{Q_3(x,t)}{q_3}}\frac{\partial \rho}{\partial t}, \end{cases} \tag{7.12}$$

with the additional conditions

$$\left(\frac{q}{Q(x,t)}\right)^5 = \left(\frac{q_3}{Q_3(x,t)}\right)^3 \quad \text{and} \quad \frac{p}{P\,(x,t)}\sqrt[5]{\frac{Q_3(x,t)}{q_3}} \geq 0, \quad \text{for all } x \text{ and } t. \tag{7.13}$$

4. If conditions (7.7) are satisfied and $q_1 = q_2 = q_3 = 0$, then system (7.5) gives

$$\begin{cases} \frac{dT}{dt} = \sqrt[3]{\frac{Q(x,t)}{q}}, \\ \rho(x,t) = \sqrt[3]{\frac{q}{Q(x,t)}}, \\ \frac{\partial X}{\partial x} = \pm\sqrt{\frac{p}{P(x,t)}\sqrt[3]{\frac{Q(x,t)}{q}}}, \\ \frac{\partial X(x,t)}{\partial t} = iP\frac{\partial^2 X}{\partial x^2} - 2P\frac{\partial X}{\partial x}\frac{\partial \varphi}{\partial x}, \\ \frac{\partial \varphi}{\partial t} + P\left(\frac{\partial \varphi}{\partial x}\right)^2 - iP\frac{\partial^2 \varphi}{\partial x^2} = \Gamma(x,t) + i\sqrt[3]{\frac{Q(x,t)}{q}}\frac{\partial \rho}{\partial t} - \gamma\sqrt[3]{\frac{Q(x,t)}{q}}, \end{cases} \tag{7.14}$$

with the condition

$$\frac{p}{P(x,t)}\sqrt[3]{\frac{Q(x,t)}{q}} \geq 0 \quad \text{for all } x \text{ and } t.$$

We can see from (7.10), (7.12), and (7.14) that $\rho(x,t)$ and $Q_3(x,t)$ are functions of time t only.

Analogically, one can solve system (7.5) under any other condition obtained from $q_1q_2q_3 = 0$. As in the case $q_1q_2q_3 \neq 0$, complementary conditions are needed on Eqs. (7.10), (7.12), and (7.14) to have each of $X(x,t)$, $T(t)$, $\rho(x,t)$, and $\varphi(x,t)$ to be a real-valued function.

Case of Traveling Wave: $\psi\,(X,T) = y\,[z = k\,(X - \upsilon T)]$

In the case of traveling wave, we consider ansatz (7.4) in which ψ is a complex function of single real variable $z = k\,(X - \upsilon T)$, i.e.,

$$\psi\,(X,T) = y\,[z = k\,(X - \upsilon T)]\,, \tag{7.15}$$

where $\upsilon \neq 0$ is the propagating velocity, and $k \neq 0$ is any real parameter. Then, we demand that

$$
\begin{aligned}
P\rho\left(\frac{\partial X}{\partial x}\right)^2 - p = 0,\\
Q_1\rho^3\frac{\partial X}{\partial x} - q_1 = 0,\\
Q_2\rho^3\frac{\partial X}{\partial x} - q_2 = 0,\\
Q_3\rho^5 - q_3 = 0, \quad (7.16)
\end{aligned}
$$

$$
\Gamma\rho + \left(i\frac{\partial\rho}{\partial t} - \rho\frac{\partial\varphi}{\partial t}\right) + P\left[\frac{\partial^2\rho}{\partial x^2} - \rho\left(\frac{\partial\varphi}{\partial x}\right)^2 + i\left(\rho\frac{\partial^2\varphi}{\partial x^2} + 2\frac{\partial\varphi}{\partial x}\frac{\partial\rho}{\partial x}\right)\right] - \gamma = 0,
$$

$$
Q\rho^3 + Q_2\rho^2\left(\frac{\partial\rho}{\partial x} + i\rho\frac{\partial\varphi}{\partial x}\right) + Q_1\rho^2\left(\frac{\partial\rho}{\partial x} - i\rho\frac{\partial\varphi}{\partial x}\right) - q = 0,
$$

$$
P\left(2\frac{\partial X}{\partial x}\frac{\partial\rho}{\partial x} + \rho\frac{\partial^2 X}{\partial x^2}\right) + i\rho\left(\frac{\partial X}{\partial t} + 2P\frac{\partial X}{\partial x}\frac{\partial\varphi}{\partial x}\right) - iv\rho\frac{dT}{dt} + ikv = 0,
$$

where $p \neq 0, \gamma, q, q_1, q_2, q_3$ are all complex constants satisfying conditions (7.7). Inserting Eq. (7.4) with $\psi(X, T)$ given by Eq. (7.15) into Eq. (7.3) and taking into account the conditions (7.16) lead to the following stationary generalized complex GL equation for y

$$
k^2 p\frac{d^2y}{dz^2} + \gamma y + q\,|y|^2\,y + q_3\,|y|^4\,y + kq_1 y^2\,\frac{dy^*}{dz} + kq_2\,|y|^2\,\frac{dy}{dz} - ikv\frac{dy}{dz} = 0, \quad (7.17)
$$

where y^* stands for the complex conjugate of y.

7.2.2 F-Expansion Method for GCGL Equation with Constant Coefficients

The aim of this section is to map the use the GCGL equation with constant coefficients (7.6) into an elliptic ordinary differential equations (EODE) so that we may easily explore the various types of its solutions. We start by separating the variables as follows: [22]

$$
\psi(X, T) = \Theta(X)\exp(-i\omega T), \quad (7.18)
$$

where ω is a real constant to be determined and Θ is a complex-valued function of X. From the viewpoint of wave theory, ω is the signal wavenumber shift and $\Theta(X)$ is the envelope function, free from fast oscillations in T. Substituting ansatz (7.18) into Eq. (7.6) yields

$$
p\Theta'' + q\,|\Theta|^2\,\Theta + (\omega + \gamma)\,\Theta + q_3\,|\Theta|^4\,\Theta + q_1\Theta^2\Theta^{*\prime} + q_2\,|\Theta|^2\,\Theta' = 0. \quad (7.19)
$$

Seeking solutions of Eq. (7.19) in the form [22]

$$
\Theta(X) = r(X)\exp[i\theta(X)] \quad (7.20)
$$

leads to

$$\theta'(X) = -\frac{q_{2i} + q_{1i}}{4p_r} r^2(X) \tag{7.21}$$

and the EODE

$$\left(\frac{d\zeta}{dX}\right)^2 = \alpha\zeta^4 + 4\beta\zeta^3 + 6\widetilde{\gamma}\zeta^2 + 4\delta\zeta = f(\zeta), \tag{7.22}$$

where

$$\begin{array}{l} \zeta = r^2(X), \quad \alpha = -\frac{(q_{2i}+q_{1i})(3q_{3i}-5q_{1i})+16p_r q_{3r}}{12p_r^2}, \\ \beta = -\frac{q_r}{2p_r}, \quad \widetilde{\gamma} = -\frac{2(\omega+\gamma_r)}{3p_r}, \end{array} \tag{7.23}$$

δ being an arbitrary real constant (of integration). Equations (7.21)–(7.23) are valid only under the following conditions:

$$\begin{cases} p_r \neq 0, \\ p_i(\omega + \gamma_r) = p_r\gamma_i, \quad q_r p_i - p_r q_i = 0, \\ p_i(q_{2i} + q_{1i}) + p_r(q_{2r} + q_{1r}) = 0, \\ (q_{2r} + q_{1r})[p_i(q_{2i} - q_{1i}) + p_r(q_{2r} - q_{1r})] - p_i(p_i q_{3r} - p_r q_{3i}) = 0. \end{cases} \tag{7.24}$$

Because $\zeta(X) = r^2(X)$ (see Eq. (7.23)) we will focus ourselves only on nonnegative solutions of Eq. (7.22).

7.2.3 *F-Expansion Method for the Stationary GCGL Equation*

Let us now consider the stationary generalized complex GL Eq. (7.17). Looking for the solution of Eq. (7.17) in the polar form

$$y(z) = r(z)\exp(-i\omega_0 z) \tag{7.25}$$

and assuming that

$$q_1 = q_2 = p\omega_0 + \upsilon = 0 \tag{7.26}$$

lead to

$$k^2 pr'' + \gamma r + qr^3 + q_3 r^5 = 0. \tag{7.27}$$

Because υ and ω_0 are two real parameters, we conclude from Eq. (7.26) that p is also a real parameter. For nontrivial and nonconstant real solutions of Eq. (7.27) to exist, it is then necessary that γ, p, and q_3 should be real numbers. Integrating Eq. (7.27) yields the EODE (7.22):

$$\left(\frac{d\zeta}{dz}\right)^2 = \alpha\zeta^4 + 4\beta\zeta^3 + 6\widetilde{\gamma}\zeta^2 + 4\delta\zeta = f(\zeta),$$

where

$$\zeta = r^2(z), \ \alpha = -\frac{4q_3}{3k^2 p}, \ \beta = -\frac{q}{2k^2 p}, \widetilde{\gamma} = -\frac{2\gamma}{3k^2 p}. \tag{7.28}$$

According to [22], conditions (7.24) must be satisfied.

The EODE (7.22) with either (7.23) or (7.28) under either conditions (7.24) or, in the case of traveling wave, conditions (7.26) will be called the double-mapping method equation (D-MME) associated to Eq. (7.3).

7.2.4 Solutions of the Double-Mapping Method Equation for BECs with Both Two- and Three-Body Interactions

Here, we consider the NLS equation

$$i\frac{\partial u}{\partial t} + \frac{P(t)}{2}\frac{\partial^2 u}{\partial x^2} - g_c(t)|u|^2 u - g_q(t)|u|^4 u - V(t)x^2 u = 0 \tag{7.29}$$

which describes the time-dependent wave function of the condensates $u(x,t)$ at position x and time t in the presence of both the two- and three-body interatomic interactions, with the time-modulated dispersion $P(t)$, the time-dependent magnitude of the harmonic trap $V(t)$, and the strengths of time-dependent two-body and three-body interatomic interactions $g_c(t)$ and $g_q(t)$, respectively. Because (7.29) has neither the derivative terms nor the dissipative term, its corresponding similarity equation (7.6) has the form

$$i\frac{\partial\psi}{\partial T} + p\frac{\partial^2\psi}{\partial X^2} + \gamma\psi + q|\psi|^2\psi + q_3|\psi|^4\psi = 0 \tag{7.30}$$

and does not contain a dissipative term. Therefore all coefficients p, γ, q, and q_3 of Eq. (7.30) are real (hence, $p = p_r$, $q = q_r$, $q_3 = q_{3r}$, and $\gamma = \gamma_r$); moreover, $p\gamma q q_3 \neq 0$. Then the D-MME associated to Eq. (7.29) reads

$$\left(\frac{d\zeta}{dX}\right)^2 = \alpha\zeta^4 + 4\beta\zeta^3 + 6\widetilde{\gamma}\zeta^2 + 4\delta\zeta = f(\zeta), \tag{7.31}$$

with coefficients (7.28) in the case of traveling wave or,

$$\alpha = -\frac{4q_3}{3p}, \quad \beta = -\frac{q}{2p}, \quad \widetilde{\gamma} = -\frac{2(\omega+\gamma)}{3p}, \tag{7.32}$$

in general case. Here, δ is any real number. One easily verifies that conditions (7.24) are satisfied. Following the idea of Ref. [22], we may easily write down exact physical and mathematical solutions of EODE (7.31) with coefficients (7.32); we will distinguish two cases: case $\delta = 0$ and case $\delta \neq 0$.

Exact Physical and Mathematical Nonnegative Solutions When $\delta = 0$

(i) If

$$\widetilde{\gamma} > 0, \quad 2\beta^2 - 3\alpha\widetilde{\gamma} > 0, \quad \text{and} \quad (\alpha < 0 \quad \text{or} \quad \beta < 0), \tag{7.33}$$

then Eq. (7.31) admits the nonnegative bounded solitary wave solution

$$\zeta(X) = -\frac{6\widetilde{\gamma}}{2\beta - \sqrt{2\beta^2 - 3\alpha\widetilde{\gamma}}\cosh(6\widetilde{\gamma}X)}. \tag{7.34}$$

(ii) If

$$\widetilde{\gamma} > 0, \quad 2\beta^2 - 3\alpha\widetilde{\gamma} = 0, \quad \alpha > 0 \quad \text{and} \quad \beta < 0, \tag{7.35}$$

then Eq. (7.31) admits the nonnegative kink solitary wave solution

$$\zeta(X) = -\frac{\beta}{\alpha}\left[1 + \tanh\left(\frac{\beta}{\sqrt{\alpha}}X\right)\right]. \tag{7.36}$$

Exact Solitary Wave Solutions of the EODE When $\delta \neq 0$

In the case when $\delta \neq 0$, the EODE (7.31) with coefficients (7.32) will admit solitary wave solution as soon as (see [22])

$$\beta^2\left(16\beta^2 - 27\alpha\widetilde{\gamma}\right)^2 - 27\alpha^2\left(14\alpha\widetilde{\gamma} - 9\beta^2\right)\widetilde{\gamma}^2 \geq 0. \tag{7.37}$$

If condition (7.37) is satisfied and

$$\begin{array}{l} \delta(2e_1 - \widetilde{\gamma}) > 0 \quad \text{and either } (\widetilde{\gamma} + 10e_1)(2e_1 - \widetilde{\gamma}) > 0 \\ \text{or } (2e_1 - \widetilde{\gamma})(\widetilde{\gamma} + 10e_1) < 0 \quad \text{and} \quad e_1(2e_1 - \widetilde{\gamma}) > 0 \end{array} \tag{7.38}$$

with

$$\begin{cases} e_1 = -\frac{1}{2}\sqrt[3]{2\beta\widetilde{\gamma}\delta - \alpha\delta^2 - \widetilde{\gamma}^3}, \\ \delta \in \left\{\delta^-,\ \delta^+\right\},\ \delta^\pm = \frac{\left[2\beta(27\alpha\widetilde{\gamma} - 16\beta^2) \pm \sqrt{\beta^2(16\beta^2 - 27\alpha\widetilde{\gamma})^2 - 27\alpha^2(14\alpha\widetilde{\gamma} - 9\beta^2)\widetilde{\gamma}^2}\right]}{27\alpha^2}, \end{cases} \tag{7.39}$$

then EODE (7.31) with coefficients (7.32) will admit the solitary wave solution

$$\zeta(X) = \frac{2\delta\left[\cosh\left(2\sqrt{3e_1}X\right) - 1\right]}{\widetilde{\gamma} + 10e_1 + (2e_1 - \widetilde{\gamma})\cosh\left(2\sqrt{3e_1}X\right)}. \tag{7.40}$$

It should be noted that under condition (7.37), the EODE (7.31) with coefficients (7.32) admits, under complementary conditions, another nonnegative solitary wave solution of the form $\zeta(X) = \left[\widetilde{A} + \widetilde{B}\cosh(CX)\right] \Big/ \left[\widetilde{A'} + \widetilde{B}'\cosh(CX)\right]$.

Exact Elliptic Function Solutions of the EODE When $\delta \neq 0$

Now, we focus ourselves on exact nonnegative Jacobi elliptic function solutions of (7.31) with coefficients (7.32). Looking for nonnegative solutions of Eq. (7.31) equation in the form

$$\zeta(X) = \frac{1}{A + B\,\mathrm{sn}^2(\mu X, m)} \tag{7.41}$$

results in the following algebraic system for A, B, $\mu \neq 0$, and m:

$$\begin{aligned} m^2 &= \frac{\mu^2 \pm \sqrt{3\left(3\widetilde{\gamma}^2 - \mu^4 - 4\beta\delta\right)}}{2\mu^2}, \\ B &= \frac{\mu^2 \pm \sqrt{3\left(3\widetilde{\gamma}^2 - \mu^4 - 4\beta\delta\right)}}{2\delta}, \\ A &= -\frac{3\widetilde{\gamma} + 3\mu^2 \pm \sqrt{3\left(3\widetilde{\gamma}^2 - \mu^4 - 4\beta\delta\right)}}{6\delta}, \\ 0 &= 9\left(\widetilde{\gamma}^3 - 2\widetilde{\gamma}\beta\delta + \alpha\delta^2\right) \pm \left(3\widetilde{\gamma}^2 - 4\mu^4 - 4\beta\delta\right)\sqrt{3\left(3\widetilde{\gamma}^2 - \mu^4 - 4\beta\delta\right)}. \end{aligned} \tag{7.42}$$

The function $\mathrm{sn}(\xi, \eta)$ in expression (7.41) denotes the Jacobi elliptic sine function of argument ξ and with modulus $\eta \in [0, 1]$. It is clear that in the limit when $m \to 1$, solution (7.41) will be reduced to the following solitary wave solution

$$\zeta(X) = \frac{1}{A + B\tanh^2(\mu X)}. \tag{7.43}$$

We notice that the parameters α, β, $\widetilde{\gamma}$, and δ will be chosen such that

$$m \in [0, 1], \quad 3\widetilde{\gamma}^2 - \mu^4 - 4\beta\delta \geq 0, \quad A > 0, \quad \text{and} \quad A + B > 0. \tag{7.44}$$

We point out that to the solutions (7.34), (7.36), (7.40), (7.41), and (7.43) correspond the solutions

$$u_1(x,t) = \pm\rho(x,t)\sqrt{-\frac{6\widetilde{\gamma}}{2\beta - \sqrt{2\beta^2 - 3\alpha\widetilde{\gamma}}\cosh(6\widetilde{\gamma}X)}}\exp\left[i\varphi(x,t) + i\theta(X) - i\omega T\right], \tag{7.45a}$$

$$u_2(x,t) = \pm\rho(x,t)\sqrt{-\frac{\beta}{\alpha}\left[1 + \tanh\left(\frac{\beta}{\sqrt{\alpha}}X\right)\right]}\exp\left[i\varphi(x,t) + i\theta(X) - i\omega T\right], \tag{7.45b}$$

$$u_3(x,t) = \pm\rho(x,t)\sqrt{\frac{2\delta\left[\cosh\left(2\sqrt{3e_1}X\right) - 1\right]}{\widetilde{\gamma} + 10e_1 + (2e_1 - \widetilde{\gamma})\cosh\left(2\sqrt{3e_1}X\right)}}\exp\left[i\varphi(x,t) + i\theta(X) - i\omega T\right], \tag{7.45c}$$

$$u_4(x,t) = \pm \frac{\rho(x,t)}{\sqrt{A + B\,\mathrm{sn}^2(\mu X, m)}} \exp\left[i\varphi(x,t) + i\theta(X) - i\omega T\right], \tag{7.45d}$$

$$u_5(x,t) = \pm \frac{\rho(x,t)}{\sqrt{A + B\tanh^2(\mu X)}} \exp\left[i\varphi(x,t) + i\theta(X) - i\omega T\right], \tag{7.45e}$$

of GPG equation (7.29), respectively. In Eqs. (7.45a)–(7.45e), X should be replaced by $k(X - vT)$ in the EODE (7.25).

Definition: *The method that consists of finding exact analytic solutions of Eq.* (7.3) *by a combination of the above similarity and F-Expansion methods is called the double-mapping method (D-MM).*

7.2.5 Applications of the D-MM to the BECs with Two- and Three-Body Interactions

Here, we aim to present via similarity and F-Expansion methods exact analytical 1D solutions describing matter/optical waves in fully 1D. We focus ourselves on the cubic–quintic 1D NLS equation (7.29) for a complex-valued wave function $u(x,t)$ with time-modulated dispersion $P(t)$, cubic and quintic nonlinearities $g_c(t)$ and $g_q(t)$, respectively, and time-dependent harmonic potential $V(t)x^2$. Comparing Eq. (7.29) with Eq. (7.3) results in

$$\begin{aligned} &P(x,t) = \tfrac{P(t)}{2},\ \Gamma(x,t) = -V(t)x^2,\ Q(x,t) = -g_c(t), \\ &Q_3(x,t) = -g_q(t),\ \text{and}\ Q_1(x,t) = Q_2(x,t) = 0. \end{aligned} \tag{7.46}$$

According to condition (7.7), the similarity equation (7.6) coincides with Eq. (7.30) and we are in case 3 ($q_1 = q_2 = 0, q_3 \neq 0$). Conditions $q_1 = q_2 = 0$, $q_3 \neq 0$ are also conditions for traveling wave solution (see Eq. 7.26). Solving system (7.12) under conditions (7.13) with the use of (7.46) yields

$$\begin{aligned} g_q(t) &= q_3\left(\frac{g_c(t)}{q}\right)^{5/3}, \\ T(t) &= -\int \sqrt[3]{\frac{g_c(t)}{q}}\,dt + T_0, \\ \rho(x,t) &= \rho(t) = -\sqrt[3]{\frac{q}{g_c(t)}}, \\ X(x,t) &= \pm 2x \int_0^t \left[\frac{\partial \rho(\tau)}{\partial \tau}\sqrt[3]{\frac{g_c(\tau)}{q}}\sqrt{-\frac{2p}{P(\tau)}\sqrt[3]{\frac{g_c(\tau)}{q}}}\right] d\tau + X_0, \end{aligned} \tag{7.47}$$

$$\varphi(x,t) = \frac{1}{P(t)}\frac{\partial\rho(t)}{\partial t}\sqrt[3]{\frac{g_c(t)}{q}}x^2 + \gamma\int\sqrt[3]{\frac{g_c(t)}{q}}dt + c_0,$$

$$V(t) = -\frac{d}{dt}\left\{\frac{1}{P(t)}\frac{\partial\rho(t)}{\partial t}\sqrt[3]{\frac{g_c(t)}{q}}\right\} - \frac{2}{P(t)}\left(\frac{g_c(t)}{q}\right)^{2/3}\left(\frac{\partial\rho(t)}{\partial t}\right)^2,$$

$$0 \geq \frac{2p}{P(t)}\sqrt[3]{\frac{g_c(t)}{q}},$$

where T_0, c_0, and X_0 are three real constants (of integration). In traveling wave case, one easily solve system (7.16).

One can easily see from Eqs. (7.47) that in order to construct solutions $u(x,t)$ of Eq. (7.29), the diffraction P, the strengths of two-body $g_c(t)$ and three-body $g_q(t)$ interatomic interactions, and time-dependent magnitude of the harmonic trap $V(t)$ are not arbitrary, they are linked through the fourth and sixth relations of (7.47). Anyway, we have a large freedom to choose those functions to obtain physical meaningful solutions.

As pointed out by Gammal [14], the strength of the three-body interaction is usually very small when compared with the strength of the two-body interaction. Some authors take $g_q(t)$ as some percent of $g_c(t)$ in their works (for example Sabari et al. took in Ref. [37] $g_q(t) \approx 10\%\ g_c(t)$). Because $g_q(t)$ is expressed in terms of $g_c(t)$ (see the first equation of (7.47)) and depends on the similarity parameters q_3 and q, the three-body interactions can be controlled by adjusting q and q_3.

Example 1 Which Uses the Solution $u_1(x,t)$ Given by Eq. (7.45a)

As the first example, we take $p = -4$, $q = -8$, $q_3 = 1$, $\gamma = 4$, $\omega = 2$, $P(t) = 2 + \cos t$, and $g_c(t) = -\frac{3}{\sqrt{1+\tan^2\left(\frac{3\sin t+1}{3\sqrt{2}}\right)}}$. This set of data completely defines all functions appearing in Eqs. (7.47) and the last condition of these equations is satisfied. Conditions (7.33) are also verified. The plots in Fig. 7.1 show the time-evolution of the strengths of time-dependent two- and three-body interatomic interactions $g_c(t)$ (dashed line) and $g_q(t)$ (solid line), respectively. These two plots show how much smaller is the strength of the three-body interaction (in absolute value) compared with the strength of the two-body interaction. The plots of this figure show that our first example corresponds to BEC with attractive two-body interaction and repulsive three-body interaction.

In Fig. 7.2, we plot (a) the magnitude of the harmonic trap $V(t)$ and (b) $\rho(x,t) = \rho(t)$ as functions of time t. Plot (b) shows that $\rho(x,t)$ is periodically oscillating along the t-axis. It should be noted that the function $\rho(t)$ has an important effect on modulating the amplitude of the solitary wave solution $u_1(x,t)$ given by Eq. (7.45a) and, consequently, its density. The density $|u_1(x,t)|^2$ of the breathing solitary wave solution $u_1(x,t)$ as a function of the spatial variable x and time t is depicted in Fig. 7.3 where one can easily see its breathing behavior.

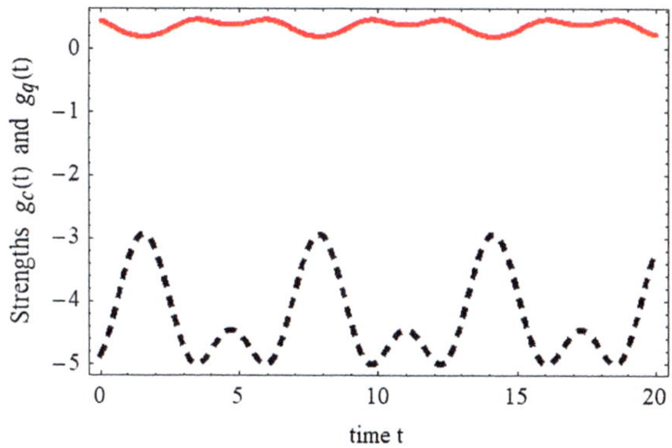

Fig. 7.1 Time-evolution of the strengths of the two-body interatomic interaction $g_c(t)$ (dashed line) and three-body interatomic interaction $g_q(t)$ (solid line) associated to solution (7.45a) with $p = -4, q = -8, q_3 = 1, \gamma = 4, \delta = 0$, and $\omega = 2$

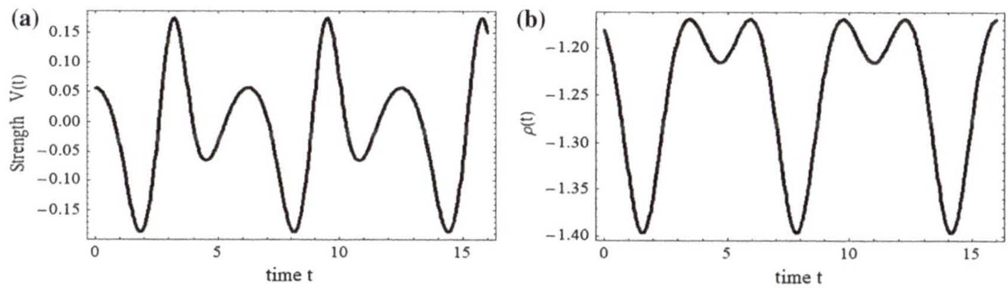

Fig. 7.2 Plot of **a** strength $V(t)$ of the potential and **b** function $\rho(x, t) = \rho(t)$ as functions of time t corresponding to solution (7.45a) with $p = -4, q = -8, q_3 = 1, \gamma = 4, \delta = 0$, and $\omega = 2$. Reprint from Ref. [20], Copyright 2018, with permission from American Institute of Physics

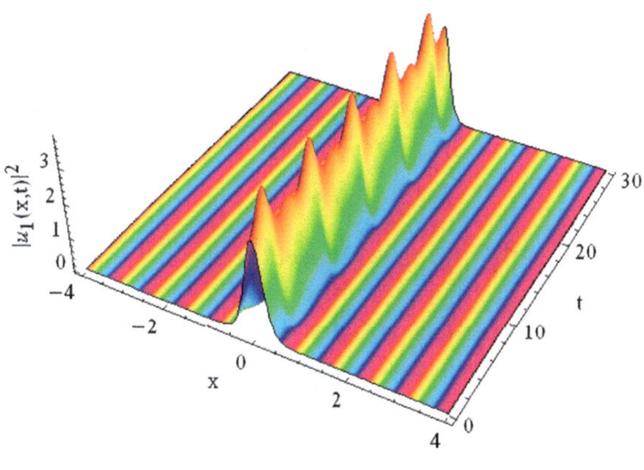

Fig. 7.3 Density $|u_1(x, t)|^2$ of the breathing solitary wave solution $u_1(x, t)$ given by Eq. (7.45a) as a function of the spatial variable x and time t with $p = -4, q = -8, q_3 = 1, \gamma = 4, \delta = 0$, and $\omega = 2$

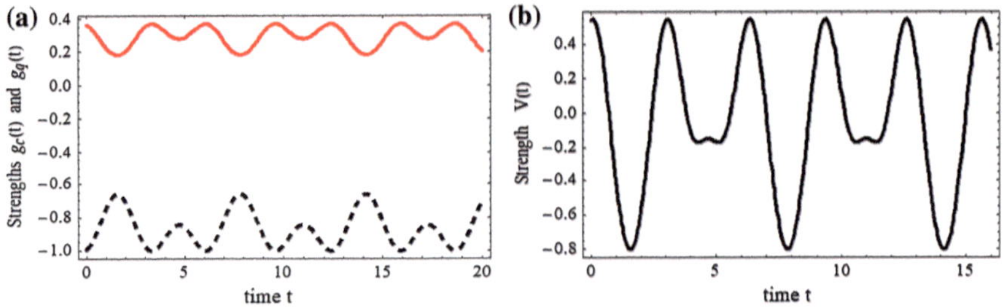

Fig. 7.4 Time-evolution of strengths of **a** two-body interaction $g_c(t)$ (dashed line) and three-body interaction $g_q(t)$ (solid line) and **b** potential $V(t)$ associated to solution (7.45b) with $p = \frac{1}{2}, q = 1$, $q_3 = -\frac{3}{8}$, $\gamma = 1$, $\delta = 0$, and $\omega = -\frac{3}{2}$

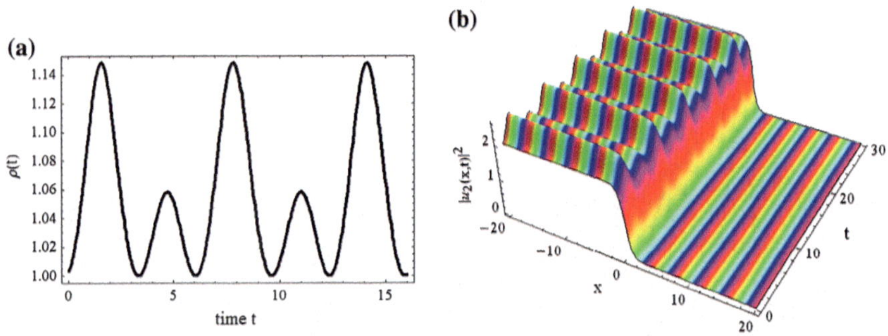

Fig. 7.5 Behavior of function $\rho(x, t) = \rho(t)$ as a function of time t (a) and density $|u_2(x, t)|^2$ of the kink solitary wave solution (7.45b) with $p = \frac{1}{2}, q = 1, q_3 = -\frac{3}{8}, \gamma = 1, \delta = 0$, and $\omega = -\frac{3}{2}$

Example 2 Which Uses the Solution $u_2(x, t)$ Given by Eq. (7.45b)

As the second example, we consider the kink solitary wave solution (7.45b). Here we focus on $p = \frac{1}{2}$, $q = 1$, $q_3 = -\frac{3}{8}$, $\gamma = 1$, $\omega = -\frac{3}{2}$, $P(t) = \frac{1}{2+\sin t+2\cos^2 t}$, and $g_c(t) = -\frac{3}{\sqrt{1+\tan^2\left(\frac{3\sin t+1}{3\sqrt{2}}\right)}}$. With these data, the conditions of validity of solution (7.45b) are all satisfied.

For this second example, we have showed in Fig. 7.4 the time-evolution of the strengths of (a) the two-body interatomic interaction $g_c(t)$ (dashed line) and the three-body interatomic interactions $g_q(t)$ (solid line) and (b) the strength of potential $V(t)$. It seems from this figure that the second example corresponds to a BEC with an attractive two-body interaction and a repulsive three-body interaction. Figure 7.5 shows (a) the behavior of the function ρ along the t-axis and (b) the density $|u_2(x, t)|^2$ of the kink solitary wave solution $u_2(x, t)$ as a function of the spatial variable x and time t. Figure 7.5b shows how much the function ρ affects the amplitude of the kink solitary wave and makes it modulating.

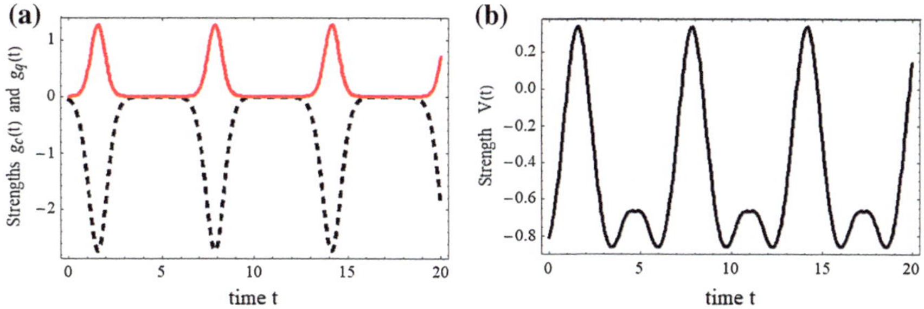

Fig. 7.6 Dynamics of two-body interaction $g_c(t)$ (dashed line) and three-body interaction $g_q(t)$ (solid line) associated to the solitary wave solution (7.45c) for the parameters $p = 1$, $q = 2$, $q_3 = -\frac{3}{4}$, $\gamma = 1$, $\omega = -1$, $\delta = \frac{16}{27}$. Reprint from Ref. [20], Copyright 2018, with permission from American Institute of Physics

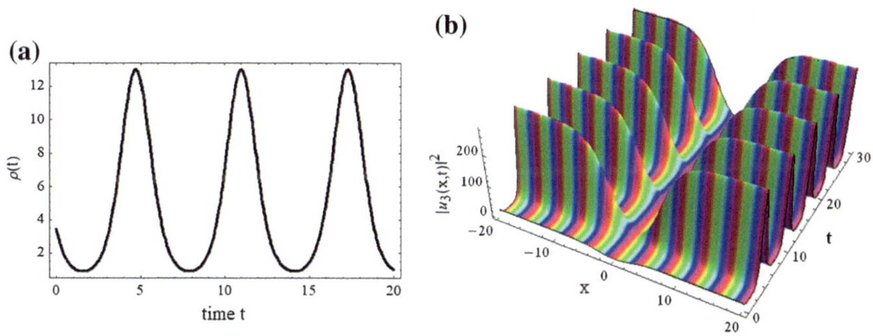

Fig. 7.7 **a** Strength $V(t)$ of potential and **b** behavior of density $|u_3(x, t)|^2$ given by solution (7.45c) as function of the space variable x and time t for $p = 1, q = 2, q_3 = -\frac{3}{4}, \gamma = 1, \omega = -1, \delta = \frac{16}{27}$

Example 3 Which Uses the Solution $u_3(x, t)$ Given by Eq. (7.45c)

The third example uses the soliton solution (7.45c) with $p = 1, q = 2, q_3 = -\frac{3}{4}, \gamma = 1$, $\omega = -1$, $\delta = \frac{16}{27}$, $P(t) = 2 + \sin t + 2\cos^2 t$, $\quad g_c(t) = -0.05 \exp(4 \sin t)$. For this example we plot in Fig. 7.6 the strengths (a) of two-body interatomic interactions $g_c(t)$ (dashed line) and three-body interatomic interaction $g_q(t)$ (solid line) and (b) magnitude of the harmonic trap $V(t)$ as functions of time t. As for the first two examples, the present one corresponds, as we may see from Fig. 7.6a, to a BEC with attractive two-body interaction and repulsive three-body interaction. Figure 7.7a corresponds to the plot of $\rho(t)$ as a function of time t, while Fig. 7.7b shows the behavior of the density $|u_3(x, t)|^2$ corresponding to the solitary wave solution (7.45c). In Fig. 7.7b, one easily sees how much the function $\rho(t)$ affects the amplitude of the solution $u_3(x, t)$. In fact, it has an important effect on modulating the amplitude of this solution. As in the first two examples, the function $\rho(t)$ periodically oscillates along the time axis and makes the corresponding solitary wave solution breathing.

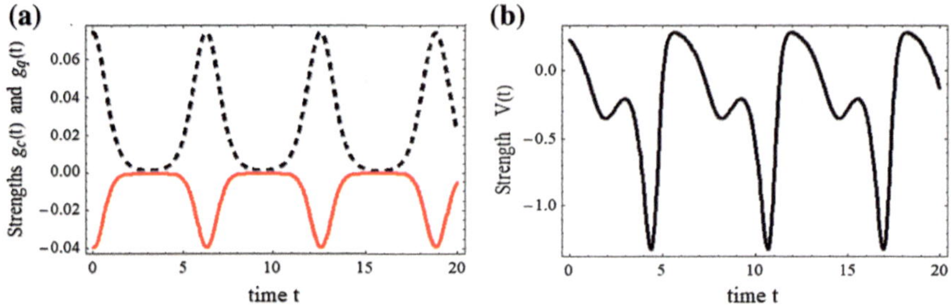

Fig. 7.8 Dynamic of the strengths of **a** two-body interaction $g_c(t)$ (dashed line) and three-body interaction $g_q(t)$ (solid line) and **b** potential $V(t)$ as functions of time t for a BEC with repulsive two-body interaction and attractive three-body interaction with parameters $p = 1, q = -1, q_3 = 3, \delta = 0, \gamma = -1, \omega = -\frac{1}{2}$

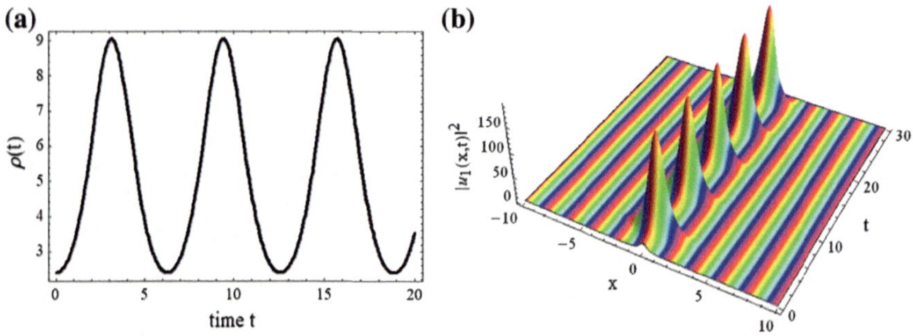

Fig. 7.9 **a** Dynamic of ρ as a function of time t and **b** behavior of density $|u_1(x,t)|^2$ of the solitary wave solution (7.45a) as function of the spatial variable x and time t with $p = 1, q = -1, q_3 = 3, \delta = 0, \gamma = -1, \omega = -\frac{1}{2}$

Example 4 Corresponding to a BEC with Repulsive Two-Body Interaction and Attractive Three-Body Interactions

With the parameters $p = 1$, $q = -1$, $q_3 = 3$, $\delta = 0$, $\gamma = -1$, $\omega = -\frac{1}{2}$, $P(t) = 2 + \sin t + 2\cos^2 t$, and $g_c(t) = 0.01 \exp(2\cos t)$,we associate to the solitary wave solution $u_1(x,t)$ given by Eq. (7.45a) a BEC with repulsive two-body interaction and attractive three-body interaction. For this example, the dynamics of the strengths (a) of the two-body interaction $g_c(t)$ (dashed line) and of the three-body interaction $g_q(t)$ (solid line) and (b) the time-dependent magnitude of the harmonic trap $V(t)$ are shown in Fig. 7.8. The plots in this figure show that the three strengths are periodic in time and oscillate along t-axis. In Fig. 7.9a, b, we depicted $\rho(t)$ as a function of time t and the density $|u_1(x,t)|^2$ associated to solution (7.45a), respectively.

This fourth example shows how we may manage the coefficients of the similarity equation to associate to the same solitary wave solution both BECs with attractive and repulsive two-body interactions and repulsive and attractive three-body interactions, respectively.

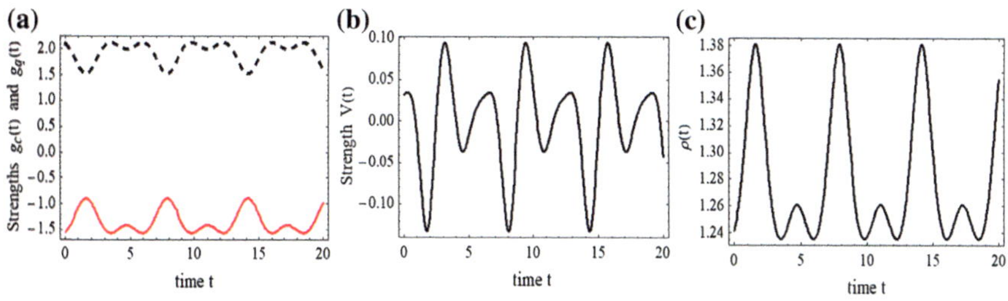

Fig. 7.10 Dynamic of strengths of $g_c(t)$ of the two-body interaction (dashed line) and $g_q(t)$ of the three-body interaction (solid line) (**a**), $V(t)$ of potential (**b**), and $\rho(t)$ as functions of time t for the parameters $p = 2,\ q = -4, q_3 = \frac{9}{2}, \gamma = 1, \omega = -4$

Example Which Uses the Jacobi Elliptic Function Solution (7.45d)

Our fifth example is associated with the Jacobi elliptic function solution $u_4(x, t)$ given by Eq. (7.45d). Here we use $p = 2,\ q = -4, q_3 = \frac{9}{2}, \gamma = 1, \omega = -4\ P(t) = 2 + \cos t$, and $g_c(t) = \frac{3}{\sqrt{1+\tan^2\left(\frac{3\sin t+1}{3\sqrt{2}}\right)}}$. Using these parameters, we found that $m = \sqrt{2}/2$, $A = \left(1 + \sqrt{7}\right)/2,\ B = -\sqrt{7}/2,\ k = \sqrt[4]{7}$. Figure 7.10a, b, c show respectively the dynamics of strengths $g_c(t)$ of two-body interatomic interaction (dashed line) and $g_q(t)$ of the three-body interatomic interaction (solid line), of potential $V(t)$, and $\rho(t)$ as functions of time t. These plots show that all the strengths are periodic in time t. We show in Fig. 7.11a, b the behavior of the density $|u_4|^2$ associated with the Jacobi elliptic function solution $u_4(x, t)$ defined by Eq. (7.45d) as a function of respectively the spatial variable x and time t and the propagation distance $X(x, t)$ and time t. As we can see from these Fig. 7.11a, b, the function $\rho(t)$ has an important effect on modulating the amplitude of this solution. Because $\rho(t)$ periodically oscillates along the $t-$axis, the corresponding solution $u_4(x, t)$ is the explicit breathing Jacobi elliptic function solution. Plots (a) and (b) of Fig. 7.11 show the effect of the propagation distance $X(x, t)$ on modulating the amplitude of the solution $u_4(x, t)$. They show that the amplitude of the solution is most modulated as a function of spatial variable than as a function of propagation distance $X(x, t)$. Thus, as function $\rho(t)$, the propagation distance $X(x, t)$ has an important effect on modulating the amplitude of the solution $u_4(x, t)$.

Example in the Limit of the Jacobi Elliptic Function Solution as Its Jacobi Modulus Tends to 1

In this last example, we use $p = 72, q = 60, q_3 = -7, \delta = -1, \gamma = 12, \omega = -120$, $P(t) = 2 + \cos t$, and $g_c(t) = -\frac{3}{\sqrt{1+\tan^2\left(\frac{3\sin t+1}{3\sqrt{2}}\right)}}$. Solving system (7.42) with these parameters yields $B = -1$, $A = 7/6$, $m = 1$, and $k = 1$. Figure 7.12a, b depict the density $|u_5(x, t)|^2$ of the solitary wave solution (7.45e) as a function of respectively (x, t) and (X, t). Comparing the plots of this figure, it is clear that the wave is

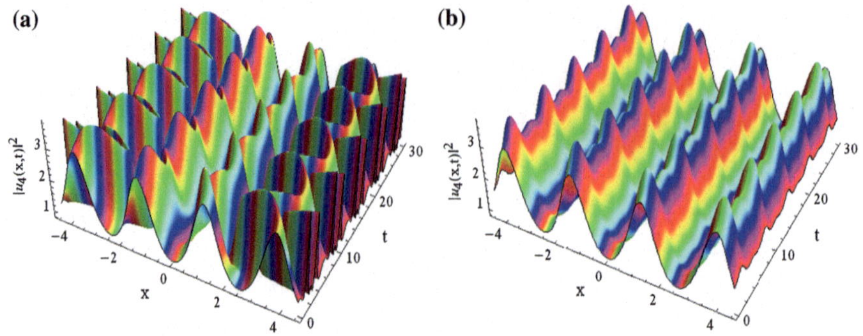

Fig. 7.11 **a** Density $|u_4(x,t)|^2$ of the Jacobi elliptic function solution (7.45d) as a function of the spatial variable **a** and as a function of the propagation distance $X(x,t)$ **b** and time t with parameters $p = 2, q = -4, q_3 = \frac{9}{2}, \gamma = 1, \omega = -4$

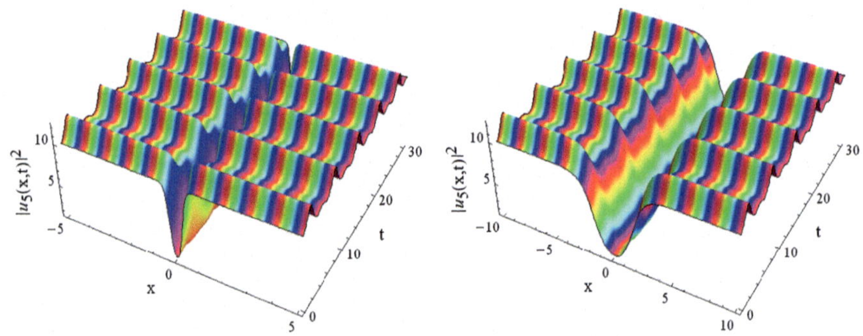

Fig. 7.12 Dynamic of density $|u_5(x,t)|^2$ of the solitary wave solution (7.45e) as a function of spatial variable x **a** and as a function of propagation distance X **b** and time t for $p = 72$; $q = 60$; $q_3 = -7$; $\delta = -1, \gamma = 12$; $\omega = -120$

most modulated as a function of the spatial variable x than it is as a function of the propagation distance $X = X(x,t)$.

7.3 Coherent Structures and MI in Bose–Einstein Condensates with Two- and Three-Body Interactions

This section deals with BECs for which the wavefunction is governed by the cubic–quintic NLS equation

$$i\frac{\partial \psi}{\partial t} + \frac{\partial^2 \psi}{\partial x^2} - V(x)\psi - g_1 |\psi|^2 \psi - g_2 |\psi|^4 \psi = 0, \tag{7.48}$$

with a time-independent trapping potential $V(x)$, formed by an optical lattice, of the form

$$V(x) = V_0 \,\mathrm{dn}\,(\gamma x, k), \tag{7.49}$$

where V_0 denotes the magnitude of the potential, or potential depth of the optical lattice, $\gamma = 2\pi/\lambda$ is the wavenumber of the laser light and λ is the wave length, corresponding to a lattice period $d = \lambda/2$, and dn (x, k) is the third Jacobian elliptic function with elliptic modulus $0 \leq k \leq 1$. In the limit $k \to 1 - 0$, $V(x)$ becomes an hyperbolic secant cosine potential barriers, while in the limit $k \to +0$, it becomes purely unity. The 1D lattice is created by a pair of counter-propagating laser beams with the wavelength $\lambda = 2\pi/\gamma$, and the lattice depth, V_0, is proportional to the intensity of the standing light wave. In Eq. (7.48), the nonlinearity parameters g_1 and g_2 represent respectively, the two- and three-body interatomic interaction coefficients, positive for repulsive interatomic interactions (or defocusing nonlinearities) and negative for attractive ones (focusing nonlinearities); they can be time- or spatiotemporal-dependent, and Eq. (7.48) can be used to describe the control and management of BEC by properly choosing the two time- or spatiotemporal-dependent parameters. In this section we consider a possible scenario of atomic systems with an arbitrary two-body interaction (g_1 can be of any sign) and with an attractive three-body interactions ($g_2 < 0$).

7.3.1 Coherent Structures

Many patterns that occur in experiments on traveling wave systems or numerical simulations of the single and coupled NLS equations exhibit local structures that have an essentially time-independent shape and propagate with a constant velocity v. For these so-called coherent structures, the spatial and temporal degrees of freedom are not independent: apart from a phase factor, they are stationary in the co-moving frame $z = x - vt$. Since the appropriate functions that describe the profiles of these coherent structures depend only on one variable z, these functions can be determined by ODEs. These are obtained by the substitution of an appropriate ansatz in the original NLS equations, which of course are partial differential equations. Here, we consider the uniformly propagating coherent structures with the ansatz

$$\psi(x, t) = R(x) \exp\left[i \int \Phi(x) dx - i\omega t \right], \tag{7.50}$$

where $R(x) \geq 0$ is the amplitude of the wave function, $\Phi(x)$ determines its phase, and ω is BECs chemical potential. When the (temporally periodic) coherent structure (7.50) is also spatially periodic, it is called a modulated amplitude wave MAW. By inserting Eq. (7.50) into Eq. (7.48) and separating real and imaginary terms, we obtain

$$R'' = R\Phi^2 + (V - \omega) R + g_1 R^3 + g_2 R^5, \tag{7.51a}$$

$$2R'\Phi + R\Phi' = 0 \Leftrightarrow \Phi = C_1/R^2, \tag{7.51b}$$

where C_1 is an arbitrary constant of integration. Case $C_1 = 0$ leads to $\Phi = 0$, which in what follows will be called the zero phase case.

Nonzero Phase Case ($C_1 \neq 0$)

Let $C_1 \neq 0$. Inserting the expression for Φ given by Eq. (7.51b) into Eq. (7.51a) and integrating the resulting equation yield

$$R'^2 = -\frac{C_1^2}{R^2} - \omega R^2 + \frac{g_1}{2} R^4 + \frac{g_2}{3} R^6 + \frac{C_2}{4} + 2\int V R dR, \tag{7.52}$$

where C_2 is a constant of integration. If we search $R(x)$ in the form

$$R(x) = \sqrt{aV(x) + b}, \tag{7.53}$$

then solving this last equation in V and inserting the result in Eq. (7.52) lead to

$$\zeta'^2 = -4C_1^2 + C_2\zeta - \frac{4(\omega a + b)}{a}\zeta^2 + \frac{2(ag_1 + 1)}{a}\zeta^3 + \frac{4g_2}{3}\zeta^4, \ \zeta = R^2. \tag{7.54}$$

Introducing the notations $A = -4C_1^2,\ B = C_2,\ C = -4(\omega a + b)/a,\ D = 2(ag_1 + 1)/a$, and $E = \frac{4g_2}{3}$, we rewrite Eq. (7.54) as

$$\zeta'^2 = A + B\zeta + C\zeta^2 + D\zeta^3 + E\zeta^4. \tag{7.55}$$

If coefficients A, B, C, D, and E of Eq. (7.55) satisfy conditions

$$\begin{gathered} 8BE^2 - 4CDE + D^3 = 0,\ E < 0,\ 8CE - 3D^2 > 0, \\ 9D^4 + 64C^2E^2 - 52D^2CE - 256AE^3 + 40BDE^2 \geq 0, \\ 10BDE^2 - 64AE^3 - CED^2 < 0, \end{gathered}$$

then it admits dn solutions

$$\zeta(x) = A_0 + A_1 \mathrm{dn}(\mu x, l), \tag{7.56}$$

where

$$A_0 = -\frac{D}{4E},\ A_1^2 = -\frac{\mu^2}{E},$$

$$l^2 = 2 + \frac{2\left(3D^2 - 8CE\right)}{8CE - 3D^2 \pm \sqrt{9D^4 + 64C^2E^2 - 52D^2CE - 256AE^3 + 40BDE^2}},$$

$$\mu^2 = \frac{8CE - 3D^2 + \sqrt{9D^4 + 64C^2E^2 - 52D^2CE - 256AE^3 + 40BDE^2}}{16E}.$$

Seeking particular solutions of Eq. (7.54) in form (7.56) and knowing that

$$\zeta(x) = R^2(x) = aV(x) + b = aV_0 \mathrm{dn}\,(\gamma x, k)$$

yield

$$A_0 = b,\ A_1 = aV_0,\ l = k,\ \mu = \gamma,$$
$$a = \pm\frac{\gamma}{2V_0}\sqrt{-\frac{3}{g_2}},\ b = -\frac{2\sqrt{-3g_2}V_0 \pm 3\gamma g_1}{\pm 8\gamma g_2},$$

under the restriction

$$h^{\pm} = h^{\pm}(g_1, g_2, V_0, k, \gamma) \geq 0, \tag{7.57}$$

where

$$h^{\pm} = g_2\left(k^2 - 1\right)\gamma^4 + \frac{\left(k^2 - 2\right)\left(2\sqrt{-g_2}V_0 \pm \gamma\sqrt{3}g_1\right)^2}{16}$$
$$-\frac{\left(2\sqrt{-g_2}V_0 \pm \gamma\sqrt{3}g_1\right)^4}{1024\gamma^4 g_2}.$$

The constants of integration C_1 and C_2 are chosen as

$$C_2 = \frac{3\gamma g_1 \pm 2\sqrt{-3g_2}V_0}{8\gamma g_2}\left(2\left(2 - k^2\right)\gamma^2 + \frac{1}{3}\frac{\left(2\sqrt{-3g_2}V_0 \pm \gamma 3g_1\right)^2}{16\gamma^2 g_2}\right),$$
$$C_1^2 = \frac{3\left(k^2 - 1\right)\gamma^4}{16g_2} + \frac{3\left(k^2 - 2\right)\left(2\sqrt{-g_2}V_0 \pm \gamma\sqrt{3}g_1\right)^2}{256g_2^2} - \frac{3\left(2\sqrt{-g_2}V_0 \pm \gamma\sqrt{3}g_1\right)^4}{4096\gamma^4 g_2^3},$$

while the frequency ω is

$$\omega = \frac{\left(k^2 - 2\right)\gamma^2}{4} + \frac{3\left(2V_0 \pm \gamma\sqrt{-\frac{3}{g_2}}g_1\right)^2 - 8V_0\left(2V_0 \pm \gamma\sqrt{-\frac{3}{g_2}}g_1\right)}{32\gamma^2}.$$

We then obtain the special dn solution

$$\zeta^{\pm}(x) = -\frac{3\gamma g_1 \pm 2\sqrt{-3g_2}V_0}{8\gamma g_2} \pm \frac{\gamma}{2}\sqrt{-\frac{3}{g_2}}\mathrm{dn}(\gamma x, k), \tag{7.58}$$

where sign "+" or "−" is chosen from the condition of the positivity of $\zeta(x)$. Therefore

$$R^{\pm}(x) = \sqrt{-\frac{3\gamma g_1 \pm 2\sqrt{-3g_2}V_0}{8\gamma g_2} \pm \frac{\gamma}{2}\sqrt{-\frac{3}{g_2}}\mathrm{dn}(\gamma x, k)}. \tag{7.59}$$

Passing to the limit $k \to 1$ in C_1, C_2, and ω under the condition

$$16g_2\gamma^4 + \left(2\sqrt{-g_2}V_0 \pm \gamma\sqrt{3}g_1\right)^2 \geq 0$$

leads

$$C_2 = \frac{\gamma 3g_1 \pm 2\sqrt{-3g_2}V_0}{8\gamma g_2}\left(2\gamma^2 + \frac{1}{3}\frac{(2\sqrt{-3g_2}V_0 \pm \gamma 3g_1)^2}{16\gamma^2 g_2}\right),$$

$$C_1^2 = -\frac{3\left(2\sqrt{-g_2}V_0 \pm \gamma\sqrt{3}g_1\right)^2}{256g_2^2}\frac{16\gamma^4 g_2 + \left(2\sqrt{-g_2}V_0 \pm \gamma\sqrt{3}g_1\right)^2}{16\gamma^4 g_2},$$

$$\omega = -\frac{\gamma^2}{4} + \frac{3\left(2V_0 \pm \gamma\sqrt{-\frac{3}{g_2}g_1}\right)^2 - 8V_0\left(2V_0 \pm \gamma\sqrt{-\frac{3}{g_2}g_1}\right)}{32\gamma^2}.$$

It follows then from Eq. (7.58) that at the limit $k \to 1$, Eq. (7.54) admits the solitary wave solution

$$\zeta_{\mathrm{lim}}(x) = -\frac{(3\gamma g_1 \pm 2\sqrt{-3g_2}V_0)}{8\gamma g_2} \pm \frac{\gamma}{2}\sqrt{-\frac{3}{g_2}}\frac{1}{\cosh(\gamma x)},$$

which is associated with the potential $\lim_{k\to 1} V(x) = \frac{V_0}{\cosh \gamma x}$. Here sign "+" or "−" is chosen from the condition of the positivity of $\zeta_{\mathrm{lim}}(x)$. Solution (7.59) then gives

$$R^{\pm}_{\mathrm{lim}}(x) = \sqrt{-\frac{(3\gamma g_1 \pm 2\sqrt{-3g_2}V_0)}{8\gamma g_2} \pm \frac{\gamma}{2}\sqrt{-\frac{3}{g_2}}\frac{1}{\cosh(\gamma x)}}. \tag{7.60}$$

The above dn solutions exist for a wide range of values of coefficients g_1 and g_2 of Eq. (7.48) and parameters V_0, k, and γ of potential (7.49). The ansatz (7.53) is the condition that restricts this range by imposing the relations (7.57) on them. The general solution can exist for both signs in expression (7.59) or (7.60). This means that the solutions must be analyzed separately in two case. For example, if $\gamma > 0$, g_1 and g_2 satisfy condition $3\gamma g_1 - 2\gamma^2\sqrt{-3g_2} > 0$, then by taking V_0 from condition

$$-\frac{3\gamma g_1}{2\sqrt{-3g_2}} \leq V_0 \leq \frac{3\gamma g_1}{2\sqrt{-3g_2}} - 2\gamma^2,$$

each of the square root in expressions $R(x)$ and $R_{\text{lim}}(x)$ will be well defined for both signs. Let us take, for example, $g_2 = -3$, $g_1 = 10$, $\gamma = 1$. Then $-5 \leq V_0 \leq 3$. In general, this interval for V_0 does not coincides with the region of existence of solutions $R^{\pm}$ or/and $R^{\pm}_{\text{lim}}$, because for the existence of these solutions, we need the positivity of $h^{\pm}$. Indeed, for $k = 1/2$, we can easily verify that both h^{+} and h^{-} are positive for $-5 \leq V_0 \leq -3.64$; when $-3.65 < V_0 \leq -0.12$, we have $h^{-} > 0$ and for $0.12 \leq V_0 \leq 3$, we have $h^{+} > 0$. Also, for $-0.12 < V_0 < 0.12$, both h^{-} and h^{+} are negative, meaning that for any V_0 taken from interval $]-0.12,\ 0.12[$, solution (7.59) does not exists neither for sign "+" nor for sign "−".

Remark 7.1 It is important to note that conditions (7.57) does not impose any restriction on the sign of g_1, so that the above solutions are for both attractive and repulsive BEC's. These solutions are not defined when $g_2 = 0$, that is, when consider only attractive or repulsive two-body interaction. Because function dn (x, k) lies on segment $\left[\sqrt{1-k^2},\ 1\right]$, we conclude that for the existence of $R^{+}(x)$, it is necessary and sufficient that

$$3\gamma g_1 + 2\sqrt{-3g_2}\left(V_0 + 2\gamma^2\sqrt{1-k^2}\right) \geq 0, \ \text{if}\ \gamma > 0, \ \text{and}$$
$$3\gamma g_1 + 2\sqrt{-3g_2}\left(V_0 + 2\gamma^2\right) \leq 0, \ \text{if}\ \gamma < 0.$$

For the same raison, for $R^{-}(x)$ to be defined, it necessary and sufficient that

$$3\gamma g_1 - 2\sqrt{-3g_2}\left(V_0 + 2\gamma^2\right) \geq 0, \ \text{for}\ \gamma > 0, \ \text{and}$$
$$2\sqrt{-3g_2}\left(V_0 + 2\gamma^2\sqrt{1-k^2}\right) - 3\gamma g_1 \geq 0, \ \text{for}\ \gamma < 0.$$

Thus, given g_1, V_0, k, and γ, one can always choose $g_2 < 0$ so that at least one of solutions $R^{+}(x)$ or $R^{-}(x)$ given by (7.59) should exist.

Zero Phase Case

Now, let us consider the zero phase case. In other words, let $C_1 = 0$. We then have $h^{\pm} = 0$, leading to

$$k^2 = \frac{1024\gamma^8 g_2^2 + \left(2\sqrt{-g_2}V_0 \pm \gamma\sqrt{3}g_1\right)^4 + 128\gamma^4 g_2\left(2\sqrt{-g_2}V_0 \pm \gamma\sqrt{3}g_1\right)^2}{64\gamma^4\left(g_2\left(2\sqrt{-g_2}V_0 \pm \gamma\sqrt{3}g_1\right)^2 + 16\gamma^4 g_2^2\right)}. \tag{7.61}$$

Using the condition $k \in [0,\ 1]$, we find

$$V_0 \in I_{V_0} = \left[\frac{-\gamma^2\sqrt{-\left(32\sqrt{3}+64\right)g_2} \mp \sqrt{3}g_1\gamma}{2\sqrt{-g_2}},\ \frac{-8\gamma^2\sqrt{-g_2} \mp \sqrt{3}g_1\gamma}{2\sqrt{-g_2}}\right]$$

$$\cup \left[\frac{-\gamma^2 \sqrt{\left(32\sqrt{3}-64\right) g_2} \mp \sqrt{3} g_1 \gamma}{2\sqrt{-g_2}}, \frac{\gamma^2 \sqrt{\left(32\sqrt{3}-64\right) g_2} \mp \sqrt{3} g_1 \gamma}{2\sqrt{-g_2}} \right]$$

$$\cup \left[\frac{8\gamma^2 \sqrt{-g_2} \mp \sqrt{3} g_1 \gamma}{2\sqrt{-g_2}}, \frac{\gamma^2 \sqrt{-\left(32\sqrt{3}+64\right) g_2} \mp \sqrt{3} g_1 \gamma}{2\sqrt{-g_2}} \right].$$

7.3.2 *Modulational Instability*

For the NLS equation (7.48) with external potential (7.49), we found in the previous section exact dn and soliton solutions having the form

$$\psi_0(x,t) = R(x) \exp\left[i \int \Phi(x) dx - i\omega t \right].$$

Fixing $x = x_0$ lead to the plane wave

$$\psi_0(x_0,t) = R(x_0) \exp\left[i \int_0^{x_0} \Phi(z) dz - i\omega t \right] = a_0 \exp\left(-i\omega t\right), \tag{7.62}$$

with zero phase and complex amplitude $a_0 = R(x_0) \exp\left[i \int_0^{x_0} \Phi(z) dz \right]$.

In what follows, we investigate the MI of the plane wave (7.62) in which $\psi_0(x,t)$ is any solution to NLS equation (7.48), (7.49) found in the previous section. To investigate the stability of any of solutions, we seek a perturbed solution of the form

$$\psi(x,t) = \left(R(x) + \epsilon g(t)\right) \exp\left[i \int \Phi(x) dx - i\omega t \right], \tag{7.63}$$

where $R(x) \exp\left[i \int \Phi(x) - i\omega t \right]$ is any of stationary solutions obtained in the previous section, $0 < \epsilon \ll 1$ is a small parameter, and $g(t)$ is perturbation (complex-valued) function. Here, x is a parameter with respect to the plane wave, and x is a variable with respect to the stationary solution $\psi_0(x,t) = R(x) \exp\left[i \int \Phi(x) - i\omega t \right]$ of the NLS equations (7.48), (7.49). Inserting Eq. (7.63) into Eq. (7.48) and linearizing in the small parameter ϵ, we obtain, after using the results of the previous section,

$$\frac{dg}{dt} = \left(\frac{a\omega + b}{a} - \frac{1 + 2ag_1}{a} R^2 - \frac{C_1^2}{R^2} - 2iC_1 \frac{R'}{R^3} - 3g_2 R^4 \right) g - \left(g_1 R^2 + 2g_2 R^4\right) g^*, \tag{7.64}$$

where g^* is the complex conjugate of g. Seeking the perturbation function $g(t)$ of Eq. (7.64) in the form

$$g(t) = u_1 \exp(i\Omega t) + u_2^* \exp\left(-i\Omega^* t\right) \tag{7.65}$$

and demanding that $|u_1| + |u_2| > 0$ yield

$$\Omega = \Omega_\pm = \mp B + i\left(-\frac{a\omega + b}{a} + \frac{C_1^2}{R^2} + \frac{1 + 2ag_1}{a}R^2 + 3g_2R^4 \pm A\right),$$

where

$$\pm A = \left[\frac{1}{2}\left(\left(g_1R^2 + 2g_2R^4\right)^2 - 4C_1^2\frac{R'^2}{R^6}\right) + \frac{1}{2}\left(\left(\left(g_1R^2 + 2g_2R^4\right)^2 - 4C_1^2\frac{R'^2}{R^6}\right)^2 \right.\right.$$
$$\left.\left. +16C_1^2\frac{R'^2}{R^6}\left(\frac{a\omega + b}{a} - \frac{C_1^2}{R^2} - \frac{1 + 2ag_1}{a}R^2 - 3g_2R^4\right)^2\right)^{1/2}\right]^{1/2},$$

$$\mp B = \left[-\frac{1}{2}\left(\left(g_1R^2 + 2g_2R^4\right)^2 - 4C_1^2\frac{R'^2}{R^6}\right) + \frac{1}{2}\left(\left(\left(g_1R^2 + 2g_2R^4\right)^2 - 4C_1^2\frac{R'^2}{R^6}\right)^2 \right.\right.$$
$$\left.\left. +16C_1^2\frac{R'^2}{R^6}\left(\frac{a\omega + b}{a} - \frac{C_1^2}{R^2} - \frac{1 + 2ag_1}{a}R^2 - 3g_2R^4\right)^2\right)^{1/2}\right]^{1/2}.$$

Whence

$$\mathrm{Re}\,(i\Omega_\pm) = \mathrm{Re}\left(-i\Omega_\pm^*\right) = \frac{a\omega + b}{a} - \frac{C_1^2}{R^2} - \frac{1 + 2ag_1}{a}R^2 - 3g_2R^4 \mp A = F(x). \tag{7.66}$$

We conclude that if

$$^\pm F = F(x) = \frac{a\omega + b}{a} - \frac{C_1^2}{R^2} - \frac{1 + 2ag_1}{a}R^2 - 3g_2R^4 \mp A \leq 0,$$

then the perturbation function $g(t)$ will be bounded as $t \to +\infty$. In this case, we say that the plane wave (7.62) is stable under modulation. In contrast, if

$$^\pm F = \frac{a\omega + b}{a} - \frac{C_1^2}{R^2} - \frac{1 + 2ag_1}{a}R^2 - 3g_2R^4 \mp A > 0,$$

then the perturbation function $g(t)$ will be unbounded as $t \to +\infty$, and we say that the plane wave (7.62) is modulational unstable. Functions $^\pm F(x)$ is called perturbation growth rate.

Definition 7.1 We say that solution $\psi_0(x,t)$ of the NLS equations (7.48), (7.49) is **uniformly modulational stable (uniformly modulational unstable)**, if the plane waves (7.62) are modulational stable (modulational unstable) for every x.

According to this definition, we conclude that solution $\psi_0(x,t)$ will be uniformly stable under modulation (uniformly unstable under modulational) if and only, if $\sup^{\pm} F(x) \leq 0$ ($\inf^{\pm} F(x) > 0$).

For general solution of form (7.50), giving a general condition that determines the sign of F (modulational instability criterion) is beyond the scope of current method. However, some cases of zero phase solutions ($C_1 = 0$) are amenable to analysis. In what follows, we investigate the MI for zero phase solutions.

Zero Phase Solutions

In this subsection we derive the modulational instability criterion of plane waves in the case of zero phase solutions. In this case, we take $C_1 = 0$, and from above results, we have

$$^{\pm}F = (3 \pm 2)\, g_2 R^4 - \frac{1 + (2 \pm 1)\, a g_1}{a} R^2 + \frac{a\omega + b}{a}.$$

Hence $^{\pm}F$ is a fourth-order polynomial in R and according to results of Sect. 2, coefficient $(3 \pm 2)\, g_2$ of higher degree is negative. Hence a necessary condition uniformly modulational instability of zero phase solution of the GP Equation (7.48), (7.66) is that either

$$\frac{a\omega + b}{a} > 0 \text{ or } \frac{1 + (2 \pm 1)\, a g_1}{a} < 0.$$

If either

$$4\,(3 \pm 2)\, g_2 \left(a^2\omega + ab\right) - \left(1 + (2 \pm 1)\, a g_1\right)^2 \geq 0,$$

or

$$\frac{1 + (2 \pm 1)\, a g_1}{a} > 0 \text{ and } \frac{(a\omega + b)}{a} < 0,$$

then $^{\pm}F \leq 0$ and plane waves (7.62) are modulational stable for every x. Therefore zero phase solution $\psi_0(x,t)$ is uniformly modulational stable. Let

$$4\,(3 \pm 2)\, g_2 \left(a^2\omega + ab\right) - \left(1 + (2 \pm 1)\, a g_1\right)^2 < 0$$

and either

$$\frac{1 + (2 \pm 1)\, a g_1}{a} < 0 \text{ or } \frac{(a\omega + b)}{a} > 0$$

and let us introduce the notations $=\ _{\pm}r^{\pm}$, $m^{\pm}$, $M^{\pm}\ _{\pm}F_{\min}$ and $_{\pm}F_{\max}$ as follows

$$
\begin{aligned}
{}_{\pm}r^{+} &= \frac{1}{2\,(3\pm 2)\,g_2}\left(\frac{1+(2\pm 1)\,ag_1}{a}\right.\\
&\quad\left.+\sqrt{\frac{(1+(2\pm 1)\,ag_1)^2}{a^2}-\frac{4\,(3\pm 2)\,g_2\,(a\omega+b)}{a}}\right);\\
{}_{\pm}r^{-} &= \frac{1}{2\,(3\pm 2)\,g_2}\left(\frac{1+(2\pm 1)\,ag_1}{a}\right.\\
&\quad\left.-\sqrt{\frac{(1+(2\pm 1)\,ag_1)^2}{a^2}-\frac{4\,(3\pm 2)\,g_2\,(a\omega+b)}{a}}\right);\\
m^{\pm} &= \inf_{\mathbb{R}}\left\{-\frac{3\gamma g_1\pm 2\sqrt{-3g_2}V_0}{8\gamma g_2}\pm\frac{\gamma}{2}\sqrt{-\frac{3}{g_2}dn(\gamma x,k)}\right\};\\
M^{\pm} &= \sup_{\mathbb{R}}\left\{-\frac{3\gamma g_1\pm 2\sqrt{-3g_2}V_0}{8\gamma g_2}\pm\frac{\gamma}{2}\sqrt{-\frac{3}{g_2}dn(\gamma x,k)}\right\};\\
{}_{\pm}F_{\min} &= \inf_{\mathbb{R}}\left\{(3\pm 2)\,g_2R^4-\frac{1+(2\pm 1)\,ag_1}{a}R^2+\frac{a\omega+b}{a}\right\};\\
{}_{\pm}F_{\max} &= \sup_{\mathbb{R}}\left\{(3\pm 2)\,g_2R^4-\frac{1+(2\pm 1)\,ag_1}{a}R^2+\frac{a\omega+b}{a}\right\}\\
{}^{+}R &= \left\{r={}_{\pm}r^{\pm}:r>0\right\}.
\end{aligned}
$$

If ${}^{+}R\cap\left[m^{\pm},\ M^{\pm}\right]=\varnothing$ for both signs "−" and "+" then, ${}_{\pm}F\leq 0$ and zero phase solution $\psi_0(x,t)$ is uniformly modulational stable. In contrast, if one of the sign "−" and "+" we have $\left[{}_{\pm}F_{\min},\ {}_{\pm}F_{\max}\right]\subset\left[m^{\pm},\ M^{\pm}\right]$, then ${}_{\pm}F>$ and zero phase solution $\psi_0(x,t)$ is uniformly modulational unstable. We call condition

$$
4\,(3\pm 2)\,g_2\left(a^2\omega+ab\right)-(1+(2\pm 1)\,ag_1)^2\geq 0 \tag{7.67}
$$

or

$$
\frac{1+(2\pm 1)\,ag_1}{a}>0 \text{ and } \frac{(a\omega+b)}{a}<0 \tag{3.6}
$$

or

$$
{}^{+}R\cap\left[m^{\pm},\ M^{\pm}\right]=\varnothing \tag{7.68}
$$

the "uniformly modulational stability criterion" of zero phase solution $\psi_0(x,t)$, and condition

$$
\left[{}_{\pm}F_{\min},\ {}_{\pm}F_{\max}\right]\subset\left[m^{\pm},\ M^{\pm}\right] \tag{7.69}
$$

the condition of uniformly modulational instability. In Eqs. (7.67)–(7.69) the upper signs $\pm$ refer to $R^{\pm}$, and lower signs $\pm$ refer to ${}_{\pm}F$. We notice that to one sign for

$R^{\pm}$ correspond two signs of ${}_{\pm}F$. In other words, the signs of R and the signs of F are independent. In Eqs. (7.67) and (3.6), we have, after using Eq. (7.61),

$$a = \pm\frac{\gamma\sqrt{3}}{2V_0\sqrt{-g_2}},\quad b = -\frac{2\sqrt{-3g_2}V_0 \pm 3\gamma g_1}{\pm 8\gamma g_2},$$

and

$$\omega = \frac{1024\gamma^8 g_2^2 + \left(2\sqrt{-g_2}V_0 \pm \gamma\sqrt{3}g_1\right)^4 + 128\gamma^4 g_2\left(2\sqrt{-g_2}V_0 \pm \gamma\sqrt{3}g_1\right)^2}{256\gamma^2\left(g_2\left(2\sqrt{-g_2}V_0 \pm \gamma\sqrt{3}g_1\right)^2 + 16\gamma^4 g_2^2\right)}$$
$$-\frac{\gamma^2}{2} + \frac{3\left(2V_0 \pm \gamma\sqrt{-\frac{3}{g_2}}g_1\right)^2 - 8V_0\left(2V_0 \pm \gamma\sqrt{-\frac{3}{g_2}}g_1\right)}{32\gamma^2}.$$

We have then obtained the following criterion of uniformly modulational stability of zero phase solutions of the NLS equation (7.48) with the external potential (7.49).

Theorem 7.1 (modulational stability of zero phase solutions). *In order that a zero phase solution of the NLS equation (7.48) with the external potential (7.49) should be uniformly stable under modulation [uniformly unstable under modulation], it is necessary and sufficient that at least one of conditions (7.67) and (3.6) and (7.68) occurs [condition (7.69) occurs].*

In the following, we carry out some computational simulations for zero phase soliton solutions of both repulsive and attractive BESc.

Zero phase: Case of soliton solutions Soliton solutions of NLS equation (7.48), (7.49) are obtained early in the limit $k \to 1$. When $k \to 1$, we have either

$$V_0 \to \mp\frac{\gamma\sqrt{3}g_1}{2\sqrt{-g_2}} \text{ or } V_0 \to \pm\frac{8\gamma^2\sqrt{-g_2} - \gamma\sqrt{3}g_1}{2\sqrt{-g_2}}.$$

The case $V_0 \to \mp\frac{\gamma\sqrt{3}g_1}{2\sqrt{-g_2}}$ corresponds to decreasing to zero at infinity solutions. In this case, conditions (7.67), (3.6), (7.68) and (7.69) become respectively

$$-\gamma^2 g_2 \geq 0,\ \text{for sign “ − ”, and } -4g_1^2 - 5\gamma^2 g_2 \geq 0,\ \text{for sign “ + ”},$$
$$g_1 > 0,$$
$${}^{+}R \cap \left[0,\ \frac{\sqrt{3}\,|\gamma|}{2\sqrt{-g_2}}\right] = \varnothing,\ \left[{}_{\pm}F_{\min},\ {}_{\pm}F_{\max}\right] \subset \left[m^{\pm},\ M^{\pm}\right].$$

We then have the following conclusion:

Proposition 7.1 *Let the potential strength* $V_0 \in \left\{\mp\frac{\gamma\sqrt{3}g_1}{2\sqrt{-g_2}}\right\}$. *Then zero phase soliton solution* $\psi_0(x,t)$ *of the repulsive Bose–Einstein condensates* ($g_1 > 0$) *in potential* $V(x) = \frac{V_0}{\cosh x}$ *is uniformly modulational stable. For the uniformly modulational stability (instability) of the mentioned solution for the attractive BECs* ($g_1 < 0$) *in the above potential, it is necessary and sufficient that either* $4g_1^2 + 5\gamma^2 g_2 \leq 0$ *or* ${}^{+}R \cap \left[0,\ \frac{\sqrt{3}|\gamma|}{\sqrt{-g_2}}\right] = \varnothing$ ($4g_1^2 + 5\gamma^2 g_2 >$ *and* $\left[\pm F_{\min},\ \pm F_{\max}\right] \subset \left[m^{\pm},\ M^{\pm}\right]$ *for at least one sign "±").*

We notice that in the case when $V_0 \to \mp\frac{\gamma\sqrt{3}g_1}{2\sqrt{-g_2}}$, only one soliton solution exists: solution corresponding to

$$R^{+}_{\lim}(x) = \sqrt{\frac{\gamma}{2}\sqrt{-\frac{3}{g_2}}\frac{1}{\cosh(\gamma x)}},\ \text{if}\ \gamma > 0\ \text{and}$$

$$R^{-}_{\lim}(x) = \sqrt{-\frac{\gamma}{2}\sqrt{-\frac{3}{g_2}}\frac{1}{\cosh(\gamma x)}},\ \text{if}\ \gamma < 0.$$

By taking $\gamma = 1$, $g_2 = -3$, $g_1 = 1$, condition (3.5) as condition (3.6) occurs In this case the stable evolution is depicted in Fig. 7.13a with $V_0 = -1/2$ and $\epsilon = 10^{-4}$. If we take $\gamma = 1$, $g_2 = -3$, $g_1 = -2$, then neither condition (7.67) nor (3.6) is satisfied, and the unstable evolution is depicted in Fig. 7.13b with $V_0 = 1$ and $\epsilon = 0.01$. From Fig. 7.13b, we observe three peaks, one centered at the $x^0 = 0$, another at $x_0 \approx -2.75$, and the last one at $x_1 \approx 2.75$. The two last peaks are a consequence of the fact that, F^+ as a function of variable x admits positive local maxima at x_0 and x_1. Considering the case of the uniformly modulational stable soliton solution for BECs with attractive two-body interactions ($g_1 < 0$), we take $g_1 = -1$, which leads to $4g_1^2 + 5\gamma^2 g_2 = -11 \leq 0$ if $\gamma = 1$ and $g_2 = -3$. According to the above

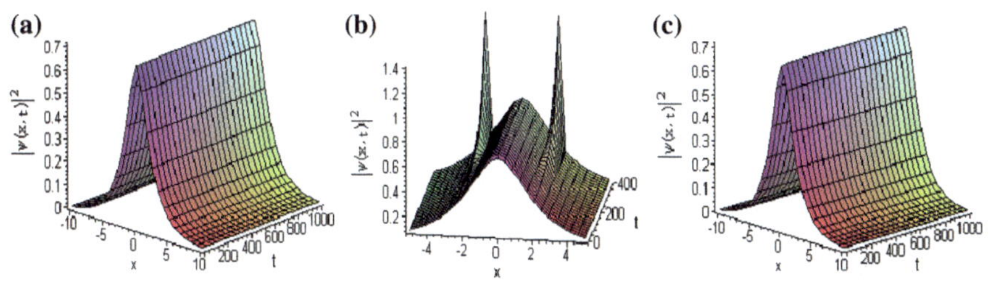

Fig. 7.13 **a**: Uniformly stable evolution of soliton solution associated with the soliton solution Eq. (7.60) for $V_0 = -1/2$ and $g_1 = 1$ (case of repulsive BECs); **b**: Uniformly stable evolution of soliton associated with the soliton solution given by Eq. (7.60) for $V_0 = 1$, $\epsilon = 0.001 =$ and $g_1 = -1$ (case of attractive BECs); **c**: Uniformly stable evolution of soliton associated with the soliton solution given by Eq. (7.60) with $V_0 = 1/2$, $\epsilon = 0.001$, and $g_1 = -1$ (case of attractive BECs)

proposition, the corresponding zero phase soliton solution is uniformly stable under modulation. Figure 7.13c shows the evolution of such a zero phase soliton solution with $V_0 = \pm 1/2$.

We now consider the case when $V_0 \to \pm \frac{8\gamma^2\sqrt{-g_2} - \gamma\sqrt{3}g_1}{2\sqrt{-g_2}}$, we obtain

$$R^{\pm}_{\text{lim}}(x) = \sqrt{\frac{\gamma\sqrt{3}}{\sqrt{-g_2}}\left(1 \pm \frac{1}{2\cosh(\gamma x)}\right)},$$

and this means that $\gamma > 0$. Moreover,

$$^{-}F = g_2 R^4 - \frac{8\gamma\sqrt{-g_2}}{\sqrt{3}} R^2 - \frac{41}{4}\gamma^2 + \frac{2\gamma\sqrt{3}g_1}{\sqrt{-g_2}},$$

$$^{+}F = 5g_2 R^4 - \left(\frac{8\gamma\sqrt{-g_2}}{\sqrt{3}} + 2g_1\right) R^2 - \frac{41}{4}\gamma^2 + \frac{2\gamma\sqrt{3}g_1}{\sqrt{-g_2}}.$$

Hence $^{-}F \le 0$ for $g_1 < 0$ and $^{+}F(R = 0) < 0$. Then the solution that corresponds to $R^{+}_{\text{lim}}(x)$ cannot be uniformly unstable under modulation. For the attractive BECs. Moreover, $^{+}F\left(\inf_{\mathbb{R}} R^{-}_{\text{lim}}(x)\right) < 0$, if $g_1 < 0$. Hence, zero phase solution that corresponds to $R^{-}_{\text{lim}}(x)$ cannot be uniformly unstable under modulation. Therefore, zero phase soliton solution for the attractive BECs cannot be uniformly unstable under modulation, if $V_0 \in \left\{\pm \frac{8\gamma^2\sqrt{-g_2} - \gamma\sqrt{3}g_1}{2\sqrt{-g_2}}\right\}$.

For $\gamma = 1$, $g_2 = -3$, then conditions (7.67) are satisfied if and only, $g_1 \in \left[-\frac{1}{2}\sqrt{1995} - 19, \frac{59}{24}\right]$, and conditions (3.6) are verified if and only, $-4 < g_1 < \frac{41}{8}$. From here we conclude that both the bright soliton ($g_1 < 0$) and the dark soliton ($g_1 > 0$) can be uniformly stable under modulation. These two solitons can also be unstable under modulation. Figure 7.14a–d gives the evolution of uniformly stable dark soliton (with $g_1 = 4$) and stable bright soliton (with $g_1 = -1$), respectively, when $\gamma = 1$, and $g_2 = -3$. Plots (a) and (c) correspond to $R^{+}(x)$, while plots (b) and (d) are associated with $R^{-}(x)$.

The unstable evolution of dark soliton (with $g_1 = 9$) and bright soliton (with $g_1 = -42$) when $\gamma = 1$, and $g_2 = -3$ are depicted in Fig. 7.15a–d, respectively. With these values of parameters, Eq. (7.60) defines two-soliton solutions, R^{+}_{lim} and R^{-}_{lim}. For these two groups of parameters, $R^{\pm}_{\text{lim}}(x;\, \gamma = 1,\, g_1 = 4,\, g_2 = -3,\, V_0 = \pm 2) = R^{\pm}_{\text{lim}}(x;\, \gamma = 1,\, g_1 = -1,\, g_2 = -3,\, V_0 = \pm 9/2,)$. Figure 7.15 shows an instability that occurs at about $t \approx 10^{-3}$.

By taking $g_1 = 40$, $g_2 = -3$, and $\gamma = 1$, condition (7.69) is satisfied for $R^{+}_{\text{lim}}(x)$, and the corresponding zero phase solution is uniformly unstable under modulation. Figure 7.16 illustrates this behavior. This gives an example of a zero phase soliton solution of repulsive BECs that is uniformly unstable under modulation. It is seen from Fig. 7.16 that the instability occurs very early near the initial time.

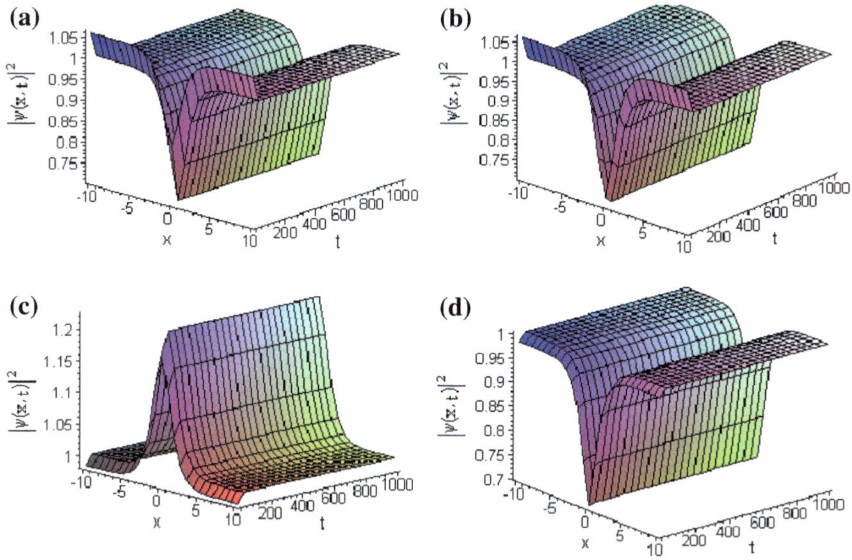

Fig. 7.14 Dynamics of type $R^+_{\rm lim}$ uniformly stable soliton solution (**a**) and dynamics of type $R^-_{\rm lim}$ uniformly stable soliton solution (**b**) associated with Eq. (7.60) with $V_0 = \pm 2$, $g_1 = 4$, $g_2 = -3$, $\gamma = 1$, and $\epsilon = 0.01$. Dynamics of type $R^+_{\rm lim}$ uniformly stable soliton solution (**c**) and dynamics of type $R^-_{\rm lim}$ uniformly stable soliton solution (**d**) given by Eq. (7.60) with $V_0 = \pm 9/2$, $g_1 = -1$, $g_2 = -3$, $\gamma = 1$, and $\epsilon = 0.001$

7.4 Effects of Three-Body Interaction in Weakly Interacting BECs in Harmonic Potential

In this section, we consider the 1D NLS equation with external parabolic potential

$$i\frac{\partial u}{\partial t} + \frac{1}{2}\frac{\partial^2 u}{\partial x^2} + g\,|u|^2\,u - kx^2 u = 0, \tag{7.70}$$

which governs the dynamics of BECs with time-dependent scattering length in a harmonic potential. We assume that Eq. (7.70) describes the evolution of the wave function of a quasi-one-dimensional cigar-shaped BEC, so that the nonlinearity parameter g and/or potential parameter k can be time-dependent. Therefore Eq. (7.70) can be used to describe the control and management of BEC by properly choosing these two time-dependent parameters g and k. In what follows, we suitably modify Eq. (7.70) by including both the nonlinear dispersion and the three-body interatomic interactions in the system to investigate the dynamics of BECs with time-dependent scattering length in harmonic trapping potential. In particular, we derive an explicit criterion for the MI for such a system and find its instability diagrams. We then present the explicit, analytical solutions to describe the dynamics of matter-wave solitons for the BECs under consideration. Based on the obtained exact solutions, we discuss

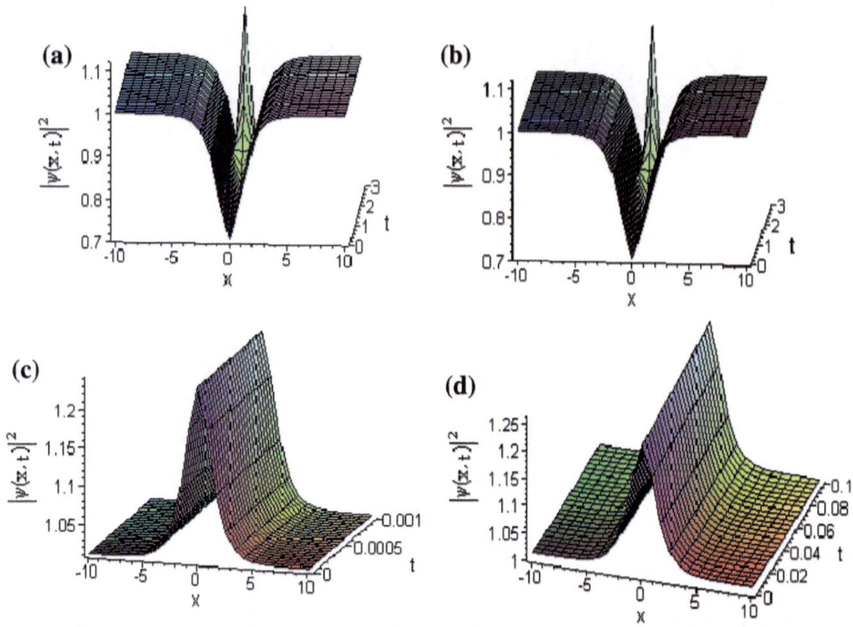

Fig. 7.15 Unstable evolution of type $R^{+}_{\lim}(x)$ solutions (**a**) and of type $R^{-}_{\lim}(x)$ (**b**) for $V_0 = \mp 1/2$, $g_1 = 9$, $g_2 = -3$, $\gamma = 1$, and $\epsilon = 0.0001$; Unstable evolution of type $R^{+}_{\lim}(x)$ solutions over $t \in [0,\ 0.001]$ (**c**) and over $t \in [0,\ 0.1]$ (**d**) for $V_0 = \mp 1/2$, $g_1 = -42$, $g_2 = -3$, $\gamma = 1$

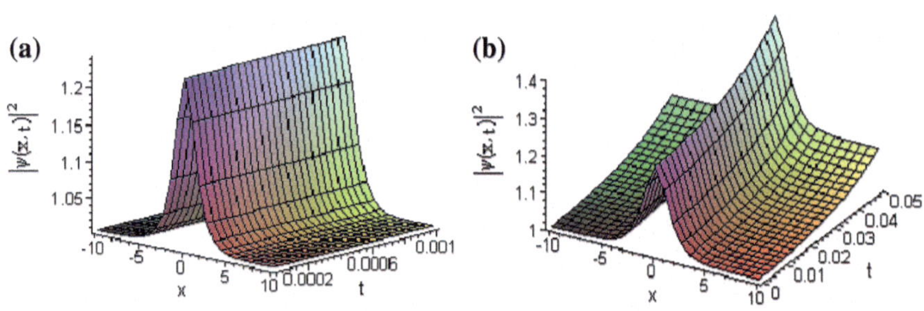

Fig. 7.16 Uniformly unstable evolution of type $R^{+}_{\lim}(x)$ solution with $V_0 = -6$, $g_1 = 40$, $g_2 = -3$, $\gamma = 1$ and $\epsilon = 0.0001$ over $t \in [0,\ 0.001]$ (**a**) and over $0 \leq t \leq 0.05$ (**b**)

analytically the effects of both the nonlinear dispersion and the three-body interactions on the dynamics of the BECs.

7.4.1 *Modified NLS Equation and Linear Stability Analysis*

We start by introducing a suitably ansatz to transform Eq. (7.70) into a cubic–quintic derivative NLS equation. Using the obtained modified equation, we investigate the linear stability of the condensates.

Derivative Cubic–Quintic NLS Equation

In order to introduce the new model describing the impact of three-body interactions on the condensates, we follow the idea introduced in Refs. [19, 36] and consider an additional phase imprint on the order parameter u to generate a new order parameter ψ as

$$u(x,t) = \psi(x,t)\exp\left[-2i\theta(x,t)\right], \tag{7.71a}$$

$$\frac{\partial\theta}{\partial x} = -\frac{2c_0}{3}|u|^2, \tag{7.71b}$$

$$\frac{\partial\theta}{\partial t} = i\frac{c_0}{6}\left[u\frac{\partial u^*}{\partial x} - u^*\frac{\partial u}{\partial x}\right] + \frac{c_1}{2}|u|^4, \tag{7.71c}$$

$\theta(x,t)$ being the phase imprint on the old order parameter u, and c_0 and c_1 being two real functional parameters. The spatiotemporal phase imprint $\theta(x,t)$ which arises from the first expression of Eq. (7.71b) can be realized experimentally by giving an instantaneous exposure of condensates to electromagnetic fields while the time-dependent phase imprint originates from the second expression of Eq. (7.71c) and one has to determine how this can be experimentally implemented [24].

Inserting ansatz (7.71a)–(7.71c) into Eq. (7.70) yields

$$i\frac{\partial\psi}{\partial t} + \frac{1}{2}\frac{\partial^2\psi}{\partial x^2} + g|\psi|^2\psi - kx^2q + c_1|\psi|^4\psi + ic_0\psi\frac{\partial|\psi^2|}{\partial x} = 0. \tag{7.72}$$

In Eq. (7.72), the quintic term $c_1|q|^4q$ accounts for both attractive ($c_1 > 0$) and repulsive ($c_1 < 0$) three-body interactions, while the cubic derivative term $ic_0q\frac{\partial|q^2|}{\partial x}$ represents the delayed nonlinear response of the system (nonlinear dispersion term of the system). The imprint parameter c_0 represents the relative magnitudes of the nonlinear dispersion term. When $c_0 = c_1 = 0$ (in this case we take $\theta(x,t) = 0$), Eq. (7.72) reduces to the cubic NLS equation with external harmonic potential involving two-body interatomic interactions alone and hence we call Eq. (7.72) as "modified NLS equation", or "derivative cubic–quintic NLS (DCQ-NLS) equation". Under some proper experimental condition, it is advised to take c_1 as a function of $g(t)$, $c_1 = N\%$ of g, so that the three-body interactions can be controlled by the tuning of s-wave scattering length [14].

Linear Stability Analysis

Here, we aim to use Eq. (7.72) to investigate the linear stability of condensates described by the DCQ-NLS equation (7.72). For the investigation of linear stability of BECs described Eq. (7.72), we first perform a modified lens-type transformation of the form

$$\psi(x,t) = \frac{1}{\sqrt{\ell(t)}}\phi(X,T)\exp\left[-if(t)x^2\right], \tag{7.73}$$

where $\ell(t) > 0$, T, and $f(t)$ are real functions of time t, and we take $X = \widetilde{\alpha}(t)x$, $\widetilde{\alpha}(t)$ being a positive real function t. To preserve the scaling and demanding that

$$\frac{df}{dt} - 2f^2 - k = 0, \tag{7.74a}$$

$$\frac{d\alpha}{dt} - 2f\alpha = 0, \tag{7.74b}$$

$$\frac{d\ell}{dt} + 2\ell f = 0, \tag{7.74c}$$

$$\frac{dT}{dt} - \frac{1}{\ell^2} = 0. \tag{7.74d}$$

Inserting ansatz (7.73) into the DCQ-NLS equation (7.72) and using conditions (7.74a)–(7.74d) lead to the DCQ-NLS equation

$$i\frac{\partial\phi}{\partial T} + \alpha_0(t)\frac{\partial^2\phi}{\partial X^2} + \beta_0(t)\left|\phi\right|^2\phi + \gamma_0(t)\left|\phi\right|^4\phi + i\delta_0(t)\frac{\partial\left|\phi\right|^2}{\partial X}\phi = 0, \tag{7.75}$$

with distributed (time-dependent) coefficients

$$\alpha_0(t) = \frac{1}{2}\ell^2\widetilde{\alpha}^2,\ \beta_0(t) = g\ell,\ \gamma_0(t) = c_1,\ \delta_0(t) = c_0\ell\widetilde{\alpha}. \tag{7.76}$$

For simplicity, we limit ourselves to the situation when both c_0 and c_1 are two real constants; moreover, we assume that parameters $g(t)$ and $k(t)$ of Eq. (7.70) satisfy the relation

$$g\frac{d^2g}{dt^2} - 2\left(\frac{dg}{dt}\right)^2 - 2kg^2 = 0, \tag{7.77}$$

so that system (7.74a)–(7.74d) will possess the special solution

$$\ell = \frac{c_{02}}{g},\ T = \frac{1}{c_{02}^2}\int_0^t g^2(\tau)d\tau,\ f = \frac{1}{2g}\frac{dg}{dt},\ \text{and}\ \widetilde{\alpha} = \frac{\sqrt{2c_{01}}}{c_{02}}g, \tag{7.78}$$

if c_{01}, c_{02}, and $c_{03}$are three real constants satisfying the relationship $c_{03}^2 = 2c_0^2c_{01}$. It follows from the positivity of $\ell(t)$ that c_{02} and $g(t)$ must have the same sign. In the special case, when the strengths g and satisfy Eqs. (7.75), (7.77) becomes a derivative cubic–quintic NLS equation with constant coefficients

$$i\frac{\partial\phi}{\partial T} + c_{01}\phi_{XX} + c_{02}\left|\phi\right|^2\phi + c_1\left|\phi\right|^4\phi + ic_{03}\frac{\partial\left|\phi\right|^2}{\partial X}\phi = 0, \tag{7.79}$$

and different functional parameters appearing in transformation (7.73) and satisfying system (7.74a)–(7.74d) are given by Eq. (7.78). The DCQ-NLS equation of type (7.79) generally appears in asymptotic derivation and its derivative coefficient c_{03} represents the relative magnitude of the nonlinear dispersion. The DCQ-NLS equation (7.79) demonstrates a large variety of dynamical behaviors, including bright and dark solitons and shock waves. It should be noted that the special cases of strengths $g(t)$ and $k(t)$ such that $g(t) = g_0 \exp[\lambda t]$ and $k(t) = -\lambda^2/2$, and $g(t) = g_0 (\tanh(\mu_0 t) \pm 1)$ and $k(t) = \pm\mu_0^2 (\tanh(\mu_0 t) \mp 1)$ satisfy Eq. (7.77). In our simulations, we may also use the particular case when $g(t) = g_0 \cosh(\lambda t)$ and $k(t) = \lambda^2/2 - 1$. In this particular case Eq. (7.78) becomes

$$\begin{cases} \ell(t) = \frac{c_{02}}{g_0 \cosh(\lambda t)}, \\ T(t) = \frac{g_0^2}{2c_{02}^2}\left(t + \frac{1}{2\lambda}\sinh(2t\lambda)\right), \\ f(t) = \frac{\lambda}{2}\tanh(t\lambda), \\ \widetilde{\alpha}(t) = \sqrt{2c_{01}}\frac{g_0}{c_{02}}\cosh(\lambda t). \end{cases} \tag{7.80}$$

For the set of functional parameters given by Eq. (7.80), it is obvious that $T(0) = 0$ and $T(t) \to +\infty$ as $t \to +\infty$. It is important to point out that Eq. (7.77) coincides with the so-called integrable condition obtained by Lei Wu et al. [26] when seeking for analytical solutions of Eq. (7.72) in the case of absence of the two last terms. We thus call (7.77) integrable condition and any functions $g(t)$ and $k(t)$ that satisfy Eq. (7.77) are said to verify integrable condition.

In what follows, we investigate the question of the instability of the Stokes wave solution with constant complex amplitude ϕ_0 and real wavenumber k_0

$$\phi(X,T) = \phi_0 \exp[ik_0 X - i\omega_0 T], \omega_0 = k_0^2 c_{01} - c_{02}|\phi_0|^2 - c_1|\phi_0|^4 \tag{7.81}$$

of the DCQ-NLS equation (7.79). The sinusoidal wave (7.81) is nonlinear and the principle of superposition is useless. To analyze the modulational instability of the carrier, we perturb the amplitude ϕ_0 as

$$\phi(X,T) = [\phi_0 + a]\exp[ik_0 X - i\omega_0 T], \tag{7.82}$$

where $a = a(X,T)$ is a small complex perturbation on the wave amplitude, $|a| << 1$. Substituting Eq. (7.82) into the (DCQ-NLS) equation (7.79) and keeping only linear terms in a and its complex conjugate a^* yield

$$\begin{aligned} i\frac{\partial a}{\partial T} + c_{01}\frac{\partial^2 a}{\partial X^2} + ic_{03}\phi_0^2\frac{\partial a^*}{\partial X} + i\left(2c_{01}k_0 + c_{03}|\phi_0|^2\right)\frac{\partial a}{\partial X} \\ + |\phi_0|^2\left(c_{02} + 2c_1|\phi_0|^2\right)a + \phi_0^2\left(c_{02} + 2c_1|\phi_0|^2\right)a^* = 0. \end{aligned} \tag{7.83}$$

Seeking the solution $a(X,T)$ of Eq. (7.83) in the form

$$a(X,T) = b_1 \exp[i(KX + \Omega T)] + b_2^* \exp\left[-i(KX + \Omega^* T)\right], \tag{7.84}$$

with the wavenumber K and the angular complex frequency Ω, we obtain the following condition for the nontrivial solution:

$$\left(\Omega + \left(2k_0 c_{01} + c_{03} |\phi_0|^2\right) K\right)^2 - \left[c_{01}^2 K^2 - \left(2c_{01} |\phi_0|^2 \left(c_{02} + 2c_1 |\phi_0|^2\right) - c_{03}^2 |\phi_0|^4\right)\right] K^2 = 0.$$

The MI may set in provided that the frequency of modulation is complex with a non-nil imaginary part, leading to

$$c_{02} + \left(2c_1 - c_0^2\right) |\phi_0|^2 > 0, \tag{7.85}$$

$$K^2 - \frac{2 |\phi_0|^2 \left(c_{02} + 2c_1 |\phi_0|^2\right) - 2c_0^2 |\phi_0|^4}{c_{01}} < 0. \tag{7.86}$$

Under conditions (7.85)–(7.86), $|a(X, T)|$ increases exponentially when T tends to infinity and the system remains unstable under modulation. Conditions (7.85) and (7.86) are respectively the necessary and the sufficient conditions for Stokes wave (7.81) to remain unstable under modulation. It is clear that the necessary condition (7.85) is violated for any BEC with both repulsive two- and three-body interactions, that is, when c_{02} and c_1 are negative (see Eq. (7.78) and condition of the positivity of $\ell(t)$). From the above dispersion relation we then obtain the following growth rate (gain) of modulational instability

$$|\mathrm{Im}\,(\Omega)| = |K| \sqrt{c_{01}} \sqrt{2 |\phi_0|^2 \left(c_{02} + 2c_1 |\phi_0|^2\right) - 2c_0^2 |\phi_0|^4 - c_{01} K^2}. \tag{7.87}$$

It is clear that the sign of c_0 does not affect the instability growth rate (7.87). It follows from Eq. (7.87) that the physical wavenumber $K \in [-K_0,\ K_0]$, where $K_0 = \sqrt{c_{01}^{-1} \left[2 |\phi_0|^2 \left(c_{02} + 2c_1 |\phi_0|^2\right) - 2c_0^2 |\phi_0|^4\right]}$.

The instability growth rate defined by Eq. (7.87) is shown in Fig. 7.17 as a function of wavenumber K for different values of the strength c_1 of the three-body interaction (top plots) and different values of the relative magnitude of the nonlinear dispersion c_0 (bottom plots). To generate plots of Fig. 7.17, we use $|\phi_0| = 1$, $c_{01} = 1/2$, and $c_{02} = +1$ and $c_{02} = -1$ for attractive and repulsive two-body interatomic interactions. All dot-dashed lines in Fig. 7.17 are obtained with $c_0 = c_1 = 0$ and correspond to the instability growth rate in the absence of three-body interactions and nonlinear dispersion.

As we can see from plots of Fig. 7.17, the introduction of the three-body interactions and nonlinear dispersion in the system enormously affect the instability growth rate and, consequently, the MI domain. It is seen from Fig. 7.17 that for BECs with repulsive two-body interatomic interactions, attractive three-body interatomic interactions ($c_1 > 0$) relatively enhances the instability [see plots (a) and (b)], while repulsive three-body interatomic interactions ($c_1 < 0$) softens the instability [see plot (c)]. In fact, the gain increases when the strength c_1 of attractive three-body interatomic interactions increases, and decreases when the strength c_1 of repulsive three-

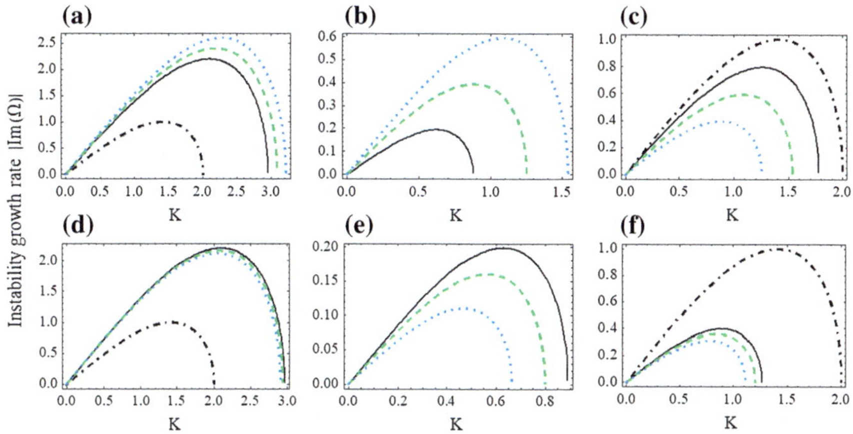

Fig. 7.17 Instability growth rate $|\text{Im}\,(\Omega)|$ according to Eq. (7.87) for attractive two-body and three-body interactions [plots (**a**), (**c**), (**d**), and (**f**)] and repulsive two-body and three-body interactions [plots (**b**) and (**e**)]. Top plots are obtained for $c_0 = 0.08$ with three values of the strength of the three-body interatomic interactions: (**a**), (**b**) $c_1 = 0.6$ (solid line), 0.7 (dashed line), 0.8 (dotted line) and (**c**) $c_1 = -0.1$ (solid line), -0.2 (dashed line), -0.3 (dotted line), while the bottom plots are obtained for three values of the strength of the relative magnitude of the nonlinear dispersion $c_0 = 0.04$ (solid line), 0.2 (dashed line), 0.3 (dotted line) with (**d**) and (**e**) $c_1 = 0.6$ and (**f**) $c_1 = -0.3$. Dot-dashed line of plots (**a**), (**c**), (**d**) and (**f**) are obtained with $c_0 = c_1 = 0$ and give the instability growth rate of the MI in the absence of nonlinear dispersion and three-body interactions in the system. Other parameters are given in the text

body interatomic interactions increases. On the other hand, the nonlinear dispersion c_0 always softens the instability; this last behavior can be seen from plots (d)–(f). Indeed, as one can see from plots (d)–(f), the gain decreases when the strength c_0 of the nonlinear dispersion increases. We can then conclude that the attractive three-body interatomic interactions ($c_1 > 0$) and the nonlinear dispersion have opposite effects: While the attractive three-body interatomic interactions ($c_1 > 0$) contributes to the MI of the system, the nonlinear dispersion contributes to its stabilization. On the other hand, the repulsive three-body interatomic interactions ($c_1 < 0$) and the nonlinear dispersion modulationally stabilize the system.

7.4.2 Solitary Wave Solutions Under the Integrable Condition

Assuming that the strengths $g(t)$ and $k(t)$ of respectively the two-body interactions and the harmonic trap satisfy the integrable condition (7.77), we present in what follows analytical solitary wave solutions of the GP equation (7.70). It should be noted that if we know a solution $\phi(X, T)$ of the DCQ-NLS equation of type (7.79), the corresponding solution of Eq. (7.70) is written as $\psi(x,t) = \sqrt{\frac{g(t)}{c_{02}}}\phi\left(\frac{c_{03}}{c_0 c_{02}}g(t)x, \frac{1}{c_{02}^2}\int_0^t g^2\right.$

$(\tau)d\tau)\exp\left[-i\frac{x^2}{2g(t)}\frac{dg}{dt}-2i\theta(x,t)\right]$, with $c_{03}^2=2c_0^2c_{01}$. We will limit ourselves to Weierstrass' solutions of Eq. (7.79), that is, solutions that are expressed compactly in terms of Weierstrass's elliptic function $\wp$ and include (real, bounded) periodic and solitary wavelike solutions [2].

Seeking the traveling wave solution of Eq. (7.79) in the form

$$\phi(X,T)=\rho(\xi=\mu X+\upsilon T)\exp\left[iF(\xi,T)\right] \tag{7.88}$$

where

$$F(\xi,T)=\frac{k_0\left[\xi-(\upsilon+\mu c_{01}k_0)\,T\right]}{\mu}+\frac{1}{\mu}\int\left[\left(\frac{\mu A_0}{\zeta}-\frac{\upsilon}{2c_{01}\mu}-\frac{2c_{03}}{4c_{01}}\zeta\right)\right]d\xi$$

and μ, υ, k_0, and A_0 are four arbitrary real constants, we obtain the EODE

$$\left(\frac{d\zeta}{d\xi}\right)^2=\alpha\zeta^4+4\beta\zeta^3+6\gamma\zeta^2+4\delta\zeta+\epsilon=R(\zeta), \tag{7.89}$$

where

$$\begin{aligned}&\zeta=\rho^2(\xi),\ \alpha=\tfrac{2c_0^2-4c_1}{3c_{01}\mu^2},\ \beta=-\tfrac{c_{02}}{2c_{01}\mu^2},\\ &\gamma=-\tfrac{(\upsilon+2c_{01}k_0\mu)^2+4A_0c_{01}c_{03}\mu^3}{6c_{01}^2\mu^4},\ \delta=B_0,\ \epsilon=-4A_0^2,\end{aligned} \tag{7.90}$$

B_0 being an arbitrary real constant. It is obvious that Eq. (7.89) possesses the special soliton solution

$$\zeta(\xi)=-\frac{4\beta}{\alpha}\frac{1}{\cosh\left(\frac{4\beta}{\sqrt{-\alpha}}\xi\right)},\ \alpha<0,\ \beta>0 \tag{7.91}$$

as soon as $A_0=B_0=0$ and $\upsilon+2c_{01}k_0\mu=0$. Because $c_{01}>0$ and c_{02} and $g(t)$ have the same sign, condition $\beta>0$ leads to $g(t)<0$; on other hand, a the necessary condition for $\alpha<0$ is that $c_1>0$. Thus, the bright solitary wave solution (7.91) can exist only for BECs with repulsive two-body and attractive three-body interatomic interactions.

A large class of solitary wave solutions of Eq. (7.89) with coefficients (7.90) in terms of Weierstrass' elliptic function can be obtained as follows. If we denote by g_2 and g_3 the invariant of Weierstrass' elliptic function $\wp(\xi;g_2,g_3)$ [2]

$$g_2=\alpha\epsilon-4\beta\delta+3\gamma^2,\ g_3=\alpha\gamma\epsilon+2\beta\gamma\delta-\alpha\delta^2-\gamma^3-\epsilon\beta^2, \tag{7.92}$$

then the general solution of Eq. (7.89) reads

$$\zeta(\xi)=\zeta+\left.\frac{\sqrt{R}\frac{d\wp(\xi;g_2,g_3)}{d\xi}+\frac{1}{2}R'\left[\wp(\xi;g_2,g_3)-\frac{1}{24}R''+\frac{1}{24}RR'''\right]}{2\left[\wp(\xi;g_2,g_3)-\frac{1}{24}R''\right]^2-\frac{1}{48}RR^{(iv)}}\right|_{\zeta=\zeta_0}, \tag{7.93}$$

where ζ_0 is an arbitrary real constant. In the special case when ζ_0 is a simple root of polynomial $R(\zeta)$, Eq. (7.93) becomes

$$\zeta(\xi) = \zeta_0 + \frac{R''(\zeta_0)}{4\left[\wp(\xi; g_2, g_3) - \frac{1}{24}R''(\zeta_0)\right]}. \tag{7.94}$$

Te discriminant $\Delta = g_2^3 - 27g_3^2$ of the Weierstrass' elliptic function $\wp(\xi; g_2, g_3)$ is suitable to classify the behavior of $\zeta(\xi)$, conditions $\Delta \neq 0$ or $\Delta = 0$, $g_2 > 0$, $g_3 > 0$ leading to periodic solutions, and conditions

$$\Delta = 0, \;\; g_2 \geq 0, \;\; g_3 \leq 0 \tag{7.95}$$

leading to solitary wavelike solutions. Because we are interesting in the solitary wave solutions of Eq. (7.70), we limit ourselves to the case when the discriminant satisfies conditions (7.95). For simplicity, we assume that ζ_0 is a simple root of polynomial $R(\zeta)$ so that the corresponding solitary wave solution is given by Eq. (7.94). For the solitary wavelike solutions of Eq. (7.89), we use the Weierstrass' elliptic function ($g_2 > 0$ and $g_3 < 0$)

$$\wp(\xi; g_2, g_3) = e_1 + \frac{3e_1}{\sinh^2\left(\sqrt{3e_1}\xi\right)}, \;\; e_1 = \frac{1}{2}\sqrt[3]{-g_3},$$

which reduces solution (7.94) to the form

$$\zeta(\xi) = \zeta_0 + \frac{R'(\zeta_0)}{4\left[e_1 - \frac{R''(\zeta_0)}{24} + 3e_1\cosh^2\left(\sqrt{3e_1}\xi\right)\right]}. \tag{7.96}$$

In Eq. (7.89) with coefficients (7.90), $\delta = B_0$ and $\epsilon = -4A_0^2$ are two arbitrary constant with $\epsilon \geq 0$ so that a large set of its solitary wave solutions (7.96) can be explicitly written by choosing either $\delta = \epsilon = 0$ or $\delta\epsilon = 0$ and $|\delta| + |\epsilon| > 0$. When $\delta = \epsilon = 0$, we have $g_2 = 3\gamma^2 > 0$ and $g_3 = -\gamma^3 = \frac{(v+2c_{01}k_0\mu)^6}{6^2c_{01}^6\mu^{12}} > 0$ and conditions (7.95) leading to solitary wave solutions are violated. Let $\delta\epsilon = 0$ and $|\delta| + |\epsilon| > 0$ and let us evaluate Eq. (7.96) explicitly. For simplicity, we limit ourselves to the special case when $\zeta_0 = 0$ is a simple root of polynomial $R(\zeta)$, that is, $\epsilon = 0$ and $\delta \neq 0$. In this case, we obtain

$$\zeta(\xi) = \begin{cases} -\dfrac{(v+2c_{01}k_0\mu)^2}{4c_{01}c_{02}\mu^2\left[1+\cosh^2\left(\frac{|v+2c_{01}k_0\mu|}{2\sqrt{2}c_{01}\mu^2}\xi\right)\right]}, & \text{if } c_1 = c_0^2/2, \\ \dfrac{2\delta}{2e_1 + \frac{(v+2c_{01}k_0\mu)^2}{6c_{01}^2\mu^4} + 6e_1\cosh^2\left(\sqrt{3e_1}\xi\right)}, & \text{if } c_1 \neq c_0^2/2, \end{cases} \tag{7.97}$$

where

$$\delta = \begin{cases} -\frac{(\upsilon+2c_{01}k_0\mu)^4}{32c_{02}c_{01}^3\mu^6}, \text{ if } c_1 = c_0^2/2, \\ \delta_* = \frac{c_{02}\left(9\widetilde{\upsilon}^2\alpha+8c_{02}^2\right)\pm\sqrt{c_{02}^2\left(9\widetilde{\upsilon}^2\alpha+8c_{02}^2\right)^2+27\widetilde{\upsilon}^4\alpha^2\left(\widetilde{\upsilon}^2\alpha+c_{02}^2\right)}}{54c_{01}^3\mu^6\alpha^2}, \text{ if } c_1 \neq c_0^2/2 \end{cases} \tag{7.98}$$

with $\widetilde{\upsilon} = \upsilon + 2c_{01}k_0\mu$. Taking into account the condition $c_{03}^2 = 2c_0^2c_{01}$, the positivity of $\zeta(\xi)$ impose on solution (7.97) with δ defined by Eq. (7.98) the condition $\delta > 0$, which can explicitly be written as

$$\begin{cases} g(t) < 0 \text{ if } c_1 = c_0^2/2, \\ c_{02}^2\left(9\widetilde{\upsilon}^2\alpha + 8c_{02}^2\right)^2 + 27\widetilde{\upsilon}^4\alpha^2\left(\widetilde{\upsilon}^2\alpha + c_{02}^2\right) \geq 0 \text{ and } \delta_* > 0, \text{ if } c_1 \neq c_0^2/2, \end{cases} \tag{7.99}$$

with $\widetilde{\upsilon} = \upsilon + 2c_{01}k_0\mu$. The last conditions in system (7.97) are the positivity of $g_2 = \frac{\upsilon^4}{12c_{01}^4\mu^8} + \frac{2c_{02}}{c_{01}\mu^2}\delta$ and the negativity of $g_3 = \frac{c_{02}\upsilon^2}{6c_{01}^3\mu^6}\delta + \frac{\upsilon^6}{6^3c_{01}^6\mu^{12}} - \alpha\delta^2$. These two conditions are satisfied as soon as parameter μ and υ are chosen from conditions

$$\begin{cases} \upsilon^4 + 24c_{02}c_{01}^3\mu^6\delta > 0, \\ 6^2c_{02}^2\widetilde{\upsilon}c_{01}^3\mu^6\delta + \widetilde{\upsilon}^6 - 6^3c_{01}^6\mu^{12}\alpha\delta^2\Big|_{\widetilde{\upsilon}=\upsilon+2c_{01}k_0\mu} < 0. \end{cases} \tag{7.100}$$

Thus, for any δ defined by Eq. (7.98), for any real parameters μ and υ satisfying conditions (7.99) and (7.100), Eq. (7.70) admits the solitary wave solution

$$\begin{aligned} u(x,t) = \pm&\sqrt{\frac{g(t)}{c_{02}}\zeta\left(\frac{\mu\sqrt{2c_{01}}}{c_{02}}g(t)x + \frac{\upsilon}{c_{02}^2}\int_0^t g^2(\tau)d\tau\right)} \\ &\times \exp\left[-2i\theta(x,t) - i\frac{1}{2g(t)}\frac{dg}{dt}x^2\right] \\ &\exp\left[ik_0\left(\frac{\sqrt{2c_{01}}}{c_{02}}g(t)x - \frac{c_{01}k_0}{c_{02}^2}\int_0^t g^2(\tau)d\tau\right)\right] \\ &\times \exp\left[\frac{i}{\mu}\int\left[\left(\frac{\mu A_0}{\zeta(\xi)} - \frac{\upsilon}{2c_{01}\mu} - \frac{2c_{03}}{4c_{01}}\zeta(\xi)\right)\right]d\xi\right], \end{aligned} \tag{7.101}$$

where $\zeta(\xi) = \zeta\left(\frac{\mu\sqrt{2c_{01}}}{c_{02}}g(t)x + \frac{\upsilon}{c_{02}^2}\int_0^t g^2(\tau)d\tau\right)$ is defined by Eq. (7.97).

7.4.3 Effects of Three-Body Interactions and Nonlinear Dispersion on the Dynamics of BECs

Using the above analytical solitary wave solutions, we aim here to investigate the effects of three-body interactions and nonlinear dispersion on the dynamics of BECs. First of all, it is important to notice that the traveling wave solution (7.88) is just a special case of solution

$$\phi(X,T) = (\phi_0 + a(X,T))\exp\left[i\,(k_0X - \omega_0 T)\right], \tag{7.102}$$

obtained by setting $\phi_0 = 0$ and $\omega_0 = k_0^2 c_{01}$.As we have showed in previous section, this Eq. (7.102) is the perturb solution (7.82) of the Stokes wave solution (7.81). Any function (7.102) that satisfies Eq. (7.79) is known as a solution on a continuous wave background. On the other hand, when ϕ_0 vanishes, which means there is no cw background, it turns into solution $\phi(X,T) = a(X,T)\exp\left[i\,(k_0X - \omega_0 T)\right]$. Therefore in general, solution (7.102) describes a nonlinear wave embedded on a cw background. We can thus say that all solutions found in the previous section are solitary waves embedded in a vanishing cw background. It follows from the results of Sect. 2 that the trivial Stokes wave solution $\phi(X,T) = \phi_0 \exp\left[i\,(k_0X - \omega_0 T)\right]|_{\phi_0=0}$ of Eq. (7.79) is stable under modulation [see conditions (7.85) and (7.86)].

Without loss of generality, we will limit our analysis on the bright solitary wave solution (7.91) which was obtained under conditions $g(t) < 0$ and $c_1 > 0$, that is, we are in presence of BECs with repulsive two-body and attractive three-body interatomic interactions; moreover, the strength c_1 of the three-body interatomic interactions and the relative magnitudes c_0 of the nonlinear dispersion must satisfy condition $2c_1 - c_0^2 > 0$. Let us remember that for BECs with repulsive two-body interatomic interactions and attractive three-body interatomic interactions, the nonlinear dispersion always softens the instability while the three-body interactions enhances the instability (see Sect. 2). By using transformations (7.71a)–(7.71c), (7.73), and (7.88) with parameters (7.77), the density function $|u(x,t)|^2$ of condensates associated with function (7.91) reads

$$|u(x,t)|^2 = \frac{3g(t)}{c_0^2 - 2c_1} \frac{1}{\cosh\left(2\sqrt{\frac{3}{2c_1 - c_0^2}}\,g(t)\left(x - \frac{\sqrt{2c_{01}}k_0}{c_{02}g(t)}\int_0^t g^2(\tau)d\tau\right)\right)}. \tag{7.103}$$

The total number N of atoms in the condensate is then $N = \int_{-\infty}^{+\infty} |u(x,t)|^2\,dx = \pi\sqrt{\frac{3}{2c_1 - c_0^2}}$ which is a constant and depends on only the strength c_1 of the three-body interactions and the relative magnitudes c_0 of the nonlinear dispersion. It is seen from the expression of N that the attractive three-body interatomic interactions ($c_1 > 0$) and the nonlinear dispersion have opposite effects on the number of atoms in the condensate: the condensate loses atoms when c_1 increases and gain atoms when c_0 increases. Therefore, for condensate described by Eq. (7.70), parameter c_1 of the attractive three-body interatomic interactions and parameter c_0 of the nonlinear dispersion play the role of loss and feeding parameter, respectively. The density function (7.103) is just a bright soliton with the following properties: (i) the amplitude of bright solitary wave is proportional to $g(t)$, while its width is inversely proportional to $g(t)$, and this justifies the fact that the total number of atoms in the condensate is a constant. We can then say that function (7.103) can be used to describe the compression of bright solitary waves when the absolute value of the

s-wave scattering length increases with time. (ii) the center of bright solitary wave is $\tilde{\xi}(t) = \frac{\sqrt{2c_{01}}k_0}{c_{02}g(t)} \int_0^t g^2(\tau)d\tau$. By using Eq. (7.77), it is evident that $\tilde{\xi}(t)$ satisfies the second-order ODE $\frac{d^2\tilde{\xi}(t)}{dt^2} + 2k(t)\tilde{\xi}(t) = 0$. This means that the center of mass of the macroscopic wave packet behaves like a classical particle, and allows one to manipulate the motion of bright solitary waves in BECs systems by controlling the external harmonic trapping potential via its strength $k(t)$. In what follows, we use different kinds of strengths $k(t)$ and $g(t)$ of the harmonic trapping potential and two-body interatomic interactions to demonstrate the dynamics of bright solitary waves in one-dimensional BEC systems; the impact of the three-body interatomic interactions and nonlinear dispersion on the dynamics of BECs is analyzed by giving different values to the strength c_1 and c_0, respectively. For all our simulations, we will use $\sqrt{2c_{01}}k_0 + c_{02} = 0$.

Effects of Three-Body Interatomic Interactions on the Dynamics of BECs

We start with the study of the impact of the three-body interactions on the dynamics of BECs. Here, we fix $c_0 = 0.8$ and vary $c_1 > c_0^2/2$: $c_1 = 0.4, 0.5, 0.6$ (see Ref. 2 in [27]).

For the first example we focus on the case of the time-independent harmonic potential which was used in the creation of bright BEC solitons [6]. Here, authors used $\omega = 2\pi \times 70$ Hz, $\omega_\perp = 2\pi \times 710$ Hz, $k = -2\lambda^2$ ($\lambda \simeq 0.05$). Solving Eq. (7.77) yields the particular solution $g(t) = g_0 \exp(\pm 2\lambda t)$, g_0 being an arbitrary real constant. For our example, we use $g(t) = -\exp(\pm 2\lambda t)$. In Fig. 7.18, we show the dynamics of a bright one-solitary wave in the harmonic trapping potential with $c_0^2 = 0.64$, $\mu = 1$, $\lambda = \pm 0.05$. Top and bottom plots in Fig. 7.18 show that with the increasing (top plots) or decreasing (bottom plots) of the absolute value of the s-wave scattering length respectively has an increase or decrease in the peak value of the solitary wave and, as we have mentioned above, has a compression or a broadening in its width. Also, it is seen that for both the increasing and decreasing of the absolute value of the s-wave scattering, the solitary wave has a decrease in the peak value when the strength c_1 of the three-body interactions increases. Therefore, parameter c_1 of the three-body interatomic interactions can be used to control the dynamics of the bright soliton, especially when a long time is needed (for example, lower or enhance the wave amplitude in the case of top or bottom plots, respectively).

For the second example, we consider the case when both $g(t)$ and $k(t)$ are periodic functions of time t. We follow Gao et al. [15] and consider $g(t) = g_0 \exp(2\lambda \sin t)$ and $k(t) = \lambda \sin t - 2\lambda^2 \cos^2 t$, where g_0 and λ are two real constant. One easily verify that g and k satisfy Eq. (7.77). The dynamics and evolution plots of the corresponding solitary wave are is shown in Fig. 7.19 with $g_0 = -1$, $\lambda = 1/15$, and for three different values of c_1.

As we can see from the plots of Fig. 7.19, the density of the solitary wave has strong variations on its propagation, as well as on its trajectory: the variations are getting more and more significant as the wave propagates; the density $|u(x,t)|^2$ oscillates along both the t-axis and $x-$axis. As one can easily seen from the bottom plots, the solitary wave has an oscillating trajectory; this oscillating in its trajectory is due to the

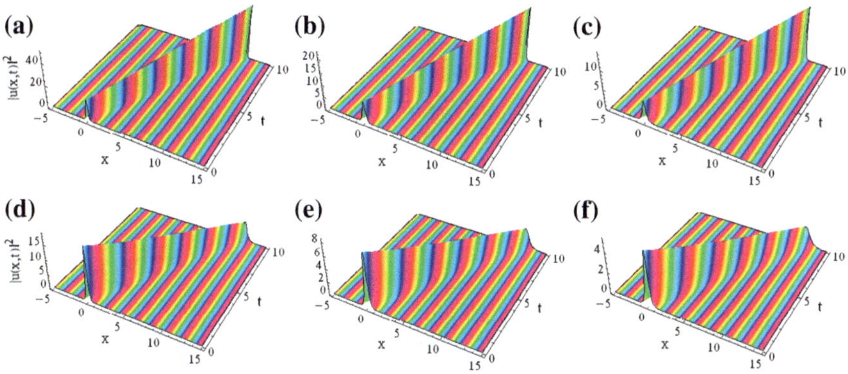

Fig. 7.18 Density $|u(x,t)|^2$ given by Eq. (7.103) as a function of spatial variable x and time t with $g(t) = -\exp(2\lambda t)$, $k = -2\lambda^2$, $c_0 = 0.8$, and for three different values of parameter c_1. **a**, **d**: $c_1 = 0.4$, **b**, **e**: $c_1 = 0.5$, and **c**, **f**: $c_1 = 0.6$. Top and bottom plots are obtained with $\lambda = 0.05$ and $\lambda = -0.05$, respectively. Other parameters are given in the text

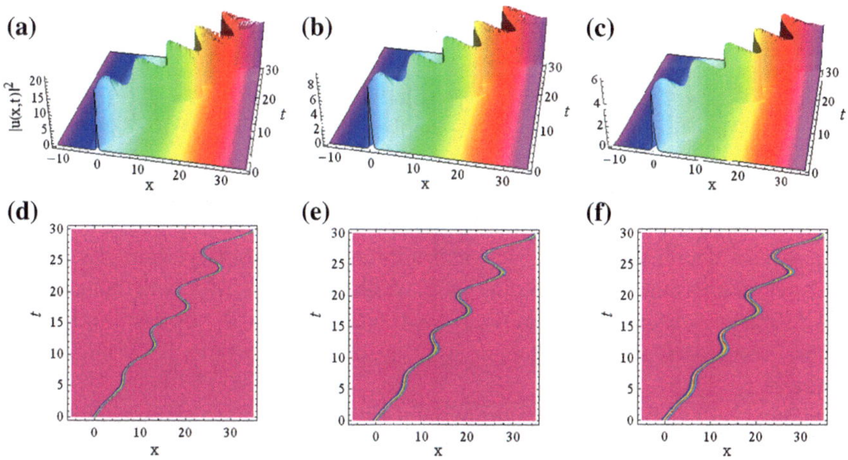

Fig. 7.19 The dynamics (top plots) and evolution plots (bottom plots) of a bright solitary wave (7.103) as a function of spatial variable x and time t with $g(t) = -\exp(2\lambda \sin t)$, $k(t) = \lambda \sin t - 2\lambda^2 \cos^2 t$, $c_0 = 0.8$, and for three different values of parameter c_1. **a**, **d**: $c_1 = 0.4$, **b**, **e**: $c_1 = 0.5$, and **c**, **f**: $c_1 = 0.6$. Other parameters are given in the text

temporal periodic modulation of the s-wave scattering length and harmonic trapping potential. Plots of Fig. 7.19 (top) show a decrease in the density of condensates is observed when the strength of attractive three-body interactions increases. Bottom plots show that as the wave propagates, it trajectory oscillates and moves along the positive $x-$axis; moreover the width of the matter-wave soliton broadens when the strength c_1 of the three-body interactions increases.

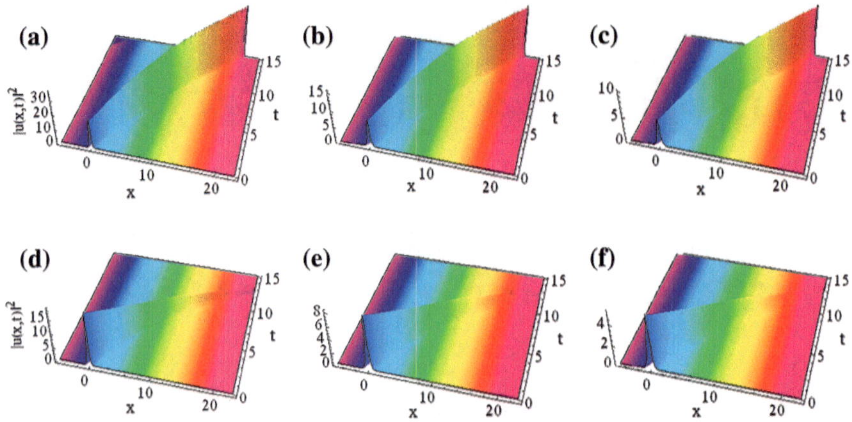

Fig. 7.20 Density $|u(x,t)|^2$ given by Eq. (7.103) as function of spatial variable x and time t with $g(t) = g_0\,(\tanh(2\lambda t) + 1)$, $k(t) = 4\lambda^2\,(\tanh(2\lambda t) - 1)$ for three different values of strength c_1 of the three-body interactions: **a**, **d**: $c_1 = 0.4$, **b**, **e**: $c_1 = 0.5$, **c**, **f**: $c_1 = 0.6$. Top and bottom plots are obtained with $\lambda = 0.05$ and $\lambda = -0.05$, respectively. Other parameters are given in the text

Now, we turn to the case $g(t) = g_0\,(\tanh(2\lambda t) + 1)$ and $k(t) = 4\lambda^2\,(\tanh(2\lambda t) - 1)$ [44]. The dynamics of the corresponding density $|u(x,t)|^2$ is plotted in Fig. 7.20 with $g_0 = -1$ and $\lambda = \pm 0.05$, for three different values of parameter c_1.

As in the first example, plots of Fig. 7.20 reveal that the peak value of the solitary wave increase or decreases when the absolute value of the s-wave scattering length increases or decreases, respectively and, moreover, the solitary wave has respectively a compression or a broadening in its width. For BECs with both an increasing and decreasing absolute value of the s-wave scattering length, the amplitude of the solitary wave decreases when the strength c_1 of attractive three-body interactions increases.

Figures 7.18, 7.19, and 7.20, portraying the density of the condensates for different kinds of strength $g(t)$ of the two-body interactions and the strength $k(t)$ of the harmonic potential, show that in the presence of the attractive three-body interactions in addition to the repulsive two-body interactions, the density of condensates decreases when the strength of attractive two-body interactions increases; in other words, the attractive three-body interactions contribute for the ejection of atoms from the condensates. Therefore, an instability sets in the condensates leading to the ejection of the atoms resulting in the decrease of matter-wave density. Thus, to ensure the stability of the condensates, the strength of the attractive three-body interactions should be minimum ($c_1 \to c_0^2/2$). Moreover, the attractive three-body interactions contribute to the broadening in the width of the matter-wave solitons.

Effects of the Nonlinear Dispersion on the Dynamics of BECs

Now, we focus on the study of the nonlinear dispersion on the dynamics of BECs. Here, we fix $c_1 = 0.16$ and vary c_0 with $c_0^2 < 2c_1$: $|c_0| = 0.2,\ 0.4,\ 0.53$ [24, 27]. We firstly consider the time-dependent s-wave scattering length with strength

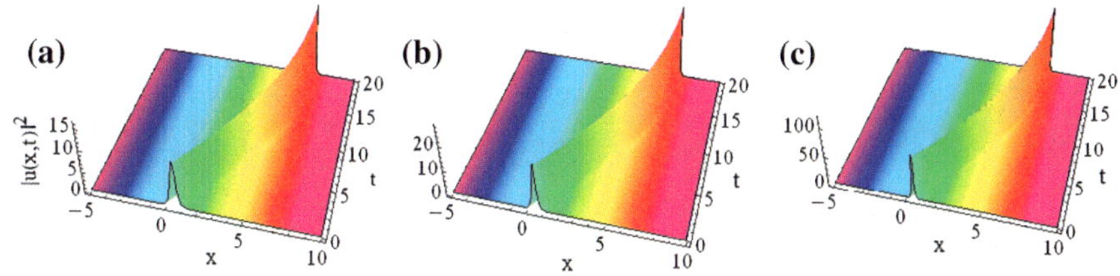

Fig. 7.21 The dynamics of the bright solitary wave (7.103) of BECs with time-dependent s-wave scattering length in a time-independent harmonic trapping potential with $g(t) = -|g_0|\cosh(\lambda t)$, $k(t) = \lambda^2/2 - 1$ for three different values of strength c_0 of the nonlinear dispersion parameter **a**: $c_0 = 0.2$, **b**, **e**: $c_0 = 0.4$, **c**, **f**: $c_0 = 0.53$. Other parameters are given in the text

$g(t) = -|g_0|\cosh(\lambda t)$ and a time-independent harmonic potential with strength $k(t) = \lambda^2/2 - 1$. All functional parameters of transformation (7.73) are then given by Eq. (7.80). The dynamics of the corresponding bright one-solitary wave is shown in Fig. 7.21 with $g_0 = 1$, $\lambda = 0.05$, for three different values of nonlinear dispersion parameter c_0. For our second example, we consider the temporal periodic modulation of the s-wave scattering length and trapping potential of the form $g(t) = g_0(1 + m\cos(\lambda t))$ and $k(t) = \frac{\lambda^2 m\left(m\cos^2(\lambda t) - \cos(\lambda t) - 2m\right)}{2(1+m\cos(\lambda t))^2}$, for any real λ and $0 < m < 1$ [26]. Plots of Fig. 7.22 display the dynamics (top plots) and the evolution (bottom plots) of the corresponding bright solitary wave with $g_0 = -1$, $m = 0.8$, $\lambda = 2$ for three values of strength c_0 of the nonlinear dispersion.

From the dynamic of the bright solitary wave shown in Figs. 7.20 and 7.21 (top), it is seen that the increasing in the nonlinear dispersion contributes to an increase in the matter-wave density (that is, the density of the matter wave increases as the strength c_0 of the nonlinear dispersion increases). Bottom plots of Fig. 7.21 displaying the evolution plots of the bright solitary wave shown that the solitary wave has an oscillating trajectory; the oscillation of the wave trajectory is due to the temporal periodic modulation of the s-wave scattering length and trapping potential. From these evolution plots, it is seen that the solitary wave has a compression in its width as the nonlinear dispersion increases. We then conclude that the nonlinear dispersion contributes (i) to the compression of the matter-wave solitons and (ii) to the accumulation of atoms in the condensates. The fact that the condensates with repulsive two-body interactions and attractive three-body interactions can hold enormous number of atoms means that one can extend the region of stability of the condensates in one-dimensional regime by manipulating the nonlinear dispersion.

In our investigation of the effect of three-body interactions and nonlinear dispersion on the dynamics of BECs, we have seen that (i) The nonlinear dispersion contributes to the compression of the matter-wave solitons and the solitary wave has a broadening in its width for the BECs with repulsive two-body interactions having a decreasing absolute value of the s-wave scattering length. We then conclude that on can overcome the broadening in the width of matter-wave solitons by increasing the strength of the nonlinear dispersion. (ii) The attractive three-body interactions

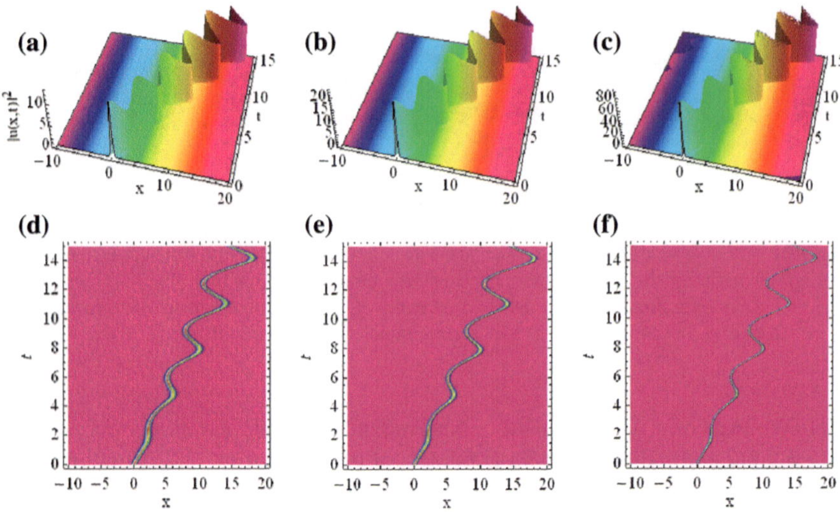

Fig. 7.22 The dynamics (top plots) and evolution plots (bottom plots) of a bright solitary wave (7.103) in a temporal periodic modulation of the s-wave scattering length and trapping potential for $g(t) = g_0\,(1 + m\cos(\lambda t))$ and $k(t) = \frac{\lambda^2 m(m\cos^2(\lambda t) - \cos(\lambda t) - 2m)}{2(1+m\cos(\lambda t))^2}$ with $g_0 = -1$, $m = 0.2$, $\lambda = 2$, for three different values of nonlinear dispersion parameter c_0: **a**, **d**: $c_0 = 0.2$, **b**, **e**: $c_0 = 0.4$, **c**, **f**: $c_0 = 0.53$. Other parameters are given in the text

and the nonlinear dispersion have opposite effect on both the number of atoms in the condensates and on the width of the matter-wave solitons: The attractive three-body interactions contribute to the ejection of atoms from the condensate while the nonlinear dispersion contributes to the accumulation of atoms in the condensates; on the other hand, the nonlinear dispersion contributes to the compression of the matter-wave solitons, while the attractive three-body interactions contribute to the broadening of the matter-wave solitons. We conclude that with a best management of the strengths c_1 and c_0, the number of atoms in the condensate and the width of the matter-wave solitons of the BECs can be remained unchange after the introduction of both the three-body interactions and the nonlinear dispersion in the system.

7.5 Dynamics of Soliton Structures in 1D Distributed NLS Equation

The present section deals with BECs for which the wavefunction is governed by the 1D distributed cubic–quintic NLS equation

$$i\frac{\partial\psi}{\partial t} + \frac{1}{2}\frac{\partial^2\psi}{\partial x^2} + g\,|\psi|^2\,\psi + \chi_0\,|\psi|^4\,\psi - \left[k(t)x^2 + i\gamma_0(t)\right]\psi = 0, \tag{7.104}$$

where the time-dependent strengths g and χ_0 of respectively the two- and three-body interactions are negative for repulsive interactions (or defocusing nonlinearities) and positive for attractive ones (focusing nonlinearities); we assume that the strength χ_0 of the three-body interaction is very small when compared with the strength g of the two-body interaction [14]. $k(t)$ is the strength of the magnetic trap, can be positive (confining potential) or negative (repulsive potential), and $\gamma_0(t)$ is a small parameter related to the feeding ($\gamma_0 > 0$) or loss ($\gamma_0 < 0$) of atoms in the condensate resulting from the contact with the thermal cloud and three-body recombination. In our studies, we assume that the nonlinearity parameters g and χ_0 can be time- or spatiotemporal-dependent, so that Eq. (7.104) can be used to describe the control and management of BEC by properly choosing these two functional parameters.

7.5.1 Analytical Solitons of BECs in Dissipative Cubic–Quintic Distributed NLS Equation

To derive analytical soliton solutions of Eq. (7.104), we introduce the ansatz

$$\psi(x,t) = \frac{1}{\ell(t)} a\,(T)\,u\,(X) \exp\left[\eta(t) + i\left(f(t)x^2 + \frac{K}{\ell(t)}x - \varphi\,(T)\right)\right], \quad (7.105)$$

where $X = x/\ell(t) - KT + K_0$ with K and K_0 being two real constants, $a(T) \neq 0$ and $\varphi\,(T)$ are functions of T, $u\,(X)$ is a function of X, and $\ell(t)$, $T(t)$, $\eta(t)$, $f(t)$ are functions of time t satisfying system

$$\frac{df}{dt} + 2f^2 + k = 0, \quad (7.106a)$$

$$\frac{1}{\ell}\frac{d\ell}{dt} - 2f = 0, \quad (7.106b)$$

$$\frac{dT}{dt} - \frac{1}{\ell^2} = 0, \quad (7.106c)$$

$$\frac{d\eta}{dt} - \gamma_0 = 0, \quad (7.106d)$$

$$\frac{1}{a}\frac{da}{dt} - f = 0. \quad (7.106e)$$

Inserting the ansatz (7.105) into Eq. (7.104) and taking into account constraints (7.106a)–(7.106e) yield the EODE

$$\left(\frac{d\rho}{dX}\right)^2 = \alpha\rho^4 + 4\beta\rho^3 + 6\gamma\rho^2 + 4\delta\rho \equiv R(\rho), \text{ with } \rho = u^2(X) \quad (7.107a)$$

where

$$\alpha = -\frac{8a^4\chi_0}{3\ell^2}\exp[4\eta], \ \beta = -a^2 g\exp[2\eta], \ \gamma = \frac{2}{3}\left(K^2 - 2\frac{d\varphi}{dT}\right), \tag{7.107b}$$

$\delta = \delta(t)$ being an arbitrary real function of time t. In what follows, we assume that $\alpha \neq 0$, leading to BECs with both two- and three-body interactions.

A large set of nonnegative analytical soliton solutions of Eq. (7.107a) can be found by choosing either $\gamma = 0$ or $\delta(t) = 0$. If we choose $\gamma = 0$ (i.e., if $\varphi(t)$ is taken from condition $\frac{d\varphi}{dT} - \frac{K^2}{2} = 0$), then the conditions leading to soliton solutions read $\alpha > 0$ (meaning that $\chi_0 < 0$) and $\delta = -\left(\frac{8}{\alpha}\right)^2\left(\frac{\beta}{3}\right)^3 \neq 0$. Therefore, $\rho_0 = 0$ is a simple root of polynomial $R(\rho)$. To this simple root corresponds the following nonnegative bright solitary wavelike solution of Eq. (7.107a) [2]

$$\rho(X) = -\frac{\ell^2 g}{a^2\chi_0}\frac{\exp[-2\eta]}{\left[1 + 3\cosh^2\left[\frac{g\ell}{\sqrt{-\chi_0}}X\right]\right]}, \quad \chi_0 < 0, \quad g > 0. \tag{7.108a}$$

In the situation when $\delta(t) = 0$ and $\gamma \neq 0$, polynomial $R(\rho)$ admits two simple roots if $\ell^2 g^2 + 4\chi_0\gamma > 0$. According to Refs. [2, 41], exact nonnegative bright solitary wavelike solution of (7.107a) under condition $\delta(t) = 0$ can be given by

$$\rho(X) = \frac{3\gamma\rho_0\cosh^2\left[\sqrt{\frac{3}{2}\gamma}X\right]}{6\gamma + 2\beta\rho_0 + 3\gamma\cosh^2\left[\sqrt{\frac{3}{2}\gamma}X\right]}, \ \text{if } \gamma > 0 \text{ and } \ell^2 g^2 + 4\chi_0\gamma > 0, \tag{7.108b}$$

where $\rho_0 = -\frac{6\ell\left(\ell g \pm \sqrt{\ell^2 g^2 + 4\chi_0\gamma}\right)\exp[-2\eta]}{8a^2\chi_0}$. For $\rho(X)$ to be nonnegative, it is necessary and sufficient that $9\gamma + 2\beta\rho_0 > 0$ and $\rho_0 > 0$. Condition $\rho_0 > 0$ is satisfied if and only, if either $\chi_0 < 0$, $g > 0$ and $\rho_0 = -\frac{3\ell\left(\ell g \pm \sqrt{\ell^2 g^2 + 4\chi_0\gamma}\right)\exp[-2\eta]}{4a^2\chi_0}$ or $\chi_0 > 0$ and $\rho_0 = -\frac{3\ell\left(\ell g - \sqrt{\ell^2 g^2 + 4\chi_0\gamma}\right)\exp[-2\eta]}{4a^2\chi_0}$. Therefore parameter K and the functional parameter $\varphi(T)$ must be chosen from

$$K^2 - 2\frac{d\varphi}{dT} > 0 \text{ and } 4\chi_0\left(K^2 - 2\frac{d\varphi}{dT}\right) + \ell g\left(\ell g \pm \sqrt{\ell^2 g^2 + 4\chi_0\gamma}\right) < 0,$$

if $\chi_0 < 0$ and $g > 0$, and

$$K^2 - 2\frac{d\varphi}{dT} > 0 \text{ and } 4\chi_0\left(K^2 - 2\frac{d\varphi}{dT}\right) + \ell g\left(\ell g - \sqrt{\ell^2 g^2 + 4\chi_0\gamma}\right) > 0,$$

if $\chi_0 > 0$.

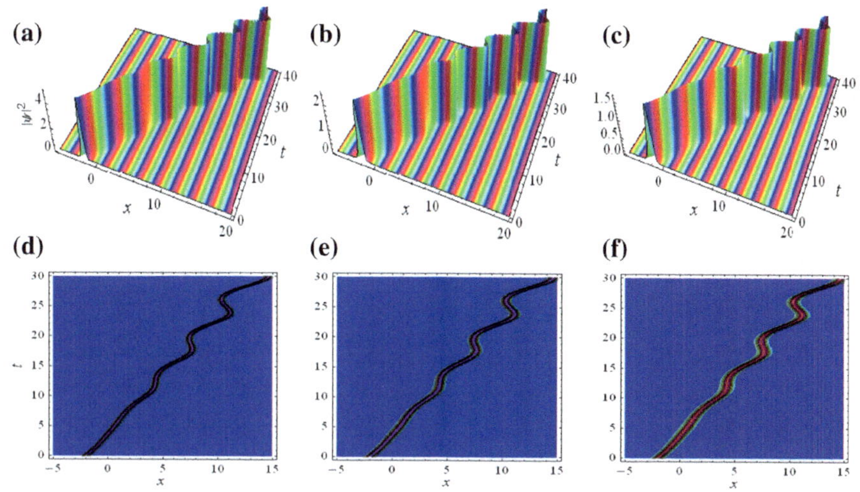

Fig. 7.23 (Plots of the BEC density $|\psi(x,t)|^2$ associated with solution (7.109) for $g(t) = 1 + m\sin[\omega t]$, $k(t) = -\frac{m\omega^2}{2}\frac{[m+\sin[\omega t]+m\cos^2[\omega t]]}{(1+m\sin[\omega t])^2}$, and $f(t) = -\frac{m\omega}{2}\frac{\cos[\omega t]}{1+m\sin[\omega t]}$, $(1 < m < 1)$, leading to $\ell(t) = \frac{\ell_0}{1+m\sin[\omega t]}$ and $T(t) = \ell_0^{-2}\left(\frac{2+m^2}{2}t + \frac{4m}{\omega}\sin^2\frac{\omega t}{2} - \frac{m^2}{4\omega}\sin[2\omega t]\right)$ with different strengths of the three-body interatomic interactions. **a**, **d**: $\chi_0(t) = -5\%g(t)$; **b**, **e**: $\chi_0(t) = -10\%g(t)$; **c**, **f**: $\chi_0(t) = -15\%g(t)$. We used the parameters $\ell_0 = 1$, $m = 0.1$, and $\omega = 1$

7.5.2 Dynamics of Solitons in 1D Cubic–Quintic Distributed NLS Equation

Now, we use the above exact soliton solutions to investigate the dynamics of solitons in BECs described by the 1D cubic–quintic distributed NLS equation (7.104). In our simulations, we may use the following three strength of the magnetic trap: (i) the constant strength of the magnetic trap $k = -2\lambda^2$ ($\lambda \simeq 0.05$) [40], leading to the particular solutions $f(t) = \pm\lambda$ of the Riccati equation (7.106a). In this case, we will use $g(t) = \exp[\pm\lambda t]$; (ii) the temporal periodic modulation of strength of the magnetic trap $k(t) = -\frac{m\omega^2}{2}\frac{[m+\sin[\omega t]+m\cos^2[\omega t]]}{(1+m\sin[\omega t])^2}$ with $0 < m < 1$ [8], leading to the special solution $f(t) = -\frac{m\omega}{2}\frac{\cos[\omega t]}{1+m\sin[\omega t]}$ of the Riccati equation (7.106a). Here, we will use $g(t) = 1 + m\sin[\omega t]$; and (iii) the interesting case of time-dependent potential which corresponds to the strength of the magnetic trap $k(t) = (t+t_0)^{-2}/8$ [40], where t_0 is any real constant whose sign is related to the sign of $f(t)$; $t_0 f(t) > 0$ and which essentially determines the width of the trap at time $t = t_0$ according to $k(t)$. Case $t_0 < 0$ describes a BEC in a shrinking trap while case $t_0 > 0$ corresponds to a broadening condensate. Inserting these expressions for $k(t)$ into the Riccati equation (7.106a) leads to the following particular solutions: $f(t) = \pm\lambda$ for constant k, $f(t) = -\frac{m\omega}{2}\frac{\cos[\omega t]}{1+m\sin[\omega t]}$ for the temporal periodic $k(t)$, and $f(t) = (t+t_0)^{-1}/4$ for the last case of k.

For simplicity, we consider only the exact bright solitary wavelike solution associated with solution (7.108a) of the EODE (7.107a) for the strength of the magnetic trap $k(t) = -\frac{m\omega^2}{2}\frac{[m+\sin[\omega t]+m\cos^2[\omega t]]}{(1+m\sin[\omega t])^2}$ with $0 < m < 1$ and the strength of the two-body interaction $g(t) = 1 + m\sin[\omega t]$. Thus, using Eqs. (7.107a) and (7.105) and remembering that $\rho(X) = u^2(X)$ yield

$$\psi(x,t) = \sqrt{-\frac{g}{\chi_0}}\frac{1}{\sqrt{1+3\cosh^2\left[\frac{g}{\sqrt{-\chi_0}}(x-\ell(t)(KT-K_0))\right]}}\exp[i\theta], \quad (7.109)$$

where $\theta = f(t)x^2 + \frac{K}{\ell(t)}x - \varphi(t)$, $f(t)$, $\ell(t)$, and $T(t)$ are functions defined by Eqs. (7.106a), (7.106b), and (7.106c), respectively, and $\varphi(t)$ is any real solution of the differential equation $2\ell^2\frac{d\varphi}{dt} - K^2 = 0$. The bright solitary wavelike solution (7.109) is defined under conditions $\chi_0 < 0$ and $g(t) > 0$. Because the strength of the three-body interaction is usually very small when compared with strength of the two-body interaction as pointed out by Gammal [14], we consider in this study that $\chi_0(t) \approx -\chi\%$ of $g(t)$ so that $|\chi_0(t)/g(t)| << 1$. Thus, $\chi_0(t) = -\frac{\chi}{100}g(t)$, $0 < \chi < 100$. It is seen from solution (7.109) that when the strength of repulsive three-body interactions is increased, one observes a decrease in the density of condensates as shown in the top panel of Fig. 7.23. One then understands that the strength of the repulsive three-body interactions can be used to control the total number of BEC atoms. The plots of bottom panel of Fig. 7.23 confirms, as one can see from Eq. (7.109), that the bright soliton will have a broadening in its width when χ increases. Therefore, the strength $\chi_0(t)$ of the repulsive three-body interactions can be used to manage the soliton motion in the BECs described by the GP equation (7.104). It is important to notice that during its propagation, the bright soliton obtained from the exact analytical solution (7.109) has a constant in the peak value, meaning that, despite the presence of the feeding/loss, the total number of BEC atoms remains unchanged. This means that the expulsive three-body interaction can be used to ensure the stability of the condensates over the time.

References

1. Abdullaev, FKh, Gammal, A., Tomio, L., Frederico, T.: Stability of trapped Bose-Einstein condensates. Phys. Rev. A **63**, 043604 (2001)
2. Abramowitz, M., Stegun I.A.: Handbook of Mathematical Functions, 9th edn., pp. 651–652. Dover Publications, New York (1972)
3. Akhmediev, N., Das, M.P., Vagov, A.: Bose-Einstein condensation of atoms with attractive interaction. Int. J. Mod. Phys. B **13**, 625 (1999)
4. Aranson, I.S., Kramer, L.: The world of the complex Ginzburg-Landau equation. Rev. Mod. Phys. **74**, 99 (2002)
5. Boudjemaa, A.: Dipolar Bose gas with three-body interactions at finite temperature. arXiv:1705.09272 [cond-mat.quant-gas]

6. Burger, S., Bongs, K., Dettmer, S., Ertmer, W., Sengstock, K., Sanpera, A., Shlyapnikov, G.V., Lewenstein, M.: Dark solitons in Bose-Einstein condensates. Phys. Rev. Lett. **83**, 5198 (1999)
7. Burt, E.A., Ghrist, R.W., Myatt, C.J., Holland, M.J., Cornell, E.A., Wieman, C.E.: Coherence, correlations, and collisions: what one learns about Bose-Einstein condensates from their decay. Phys. Rev. Lett. **79**, 337 (1997)
8. Chong, G.S., Hai, W.H., Xie, Q.T.: Breathing bright solitons in a Bose-Einstein condensate. Chin. Phys. Lett. **20**, 2098 (2003)
9. Cross, M.C., Hohenberg, P.C.: Pattern formation outside of equilibrium. Rev. Mod. Phys. **65**, 851 (1993)
10. Daley, A.J., Simon, J.: Effective three-body interactions via photon-assisted tunneling in an optical lattice. Phys. Rev. A **89**, 053619 (2014)
11. Dennin, M., Ahlers, G., Cannell, D.S.: Chaotic localized states near the onset of electroconvection. Phys. Rev. Lett. **77**, 2475 (1996)
12. Efimov, V.: Low-energy properties of three resonantly interacting particles. Sov. J. Nucl. Phys. **12**, 589 (1971)
13. Efimov, V.: Energy levels arising from resonant two-body forces in a three-body system. Phys. Lett. B **33**, 563 (1970)
14. Gammal, A., Frederico, T., Tomio, L., Chomaz, P.: Atomic Bose-Einstein condensation with three-body interactions and collective excitations. J. Phys. B: At. Mol. Opt. Phys. **33**, 4053 (2000)
15. Gao, Y., Lou, S.-Y.: Analytical solitary wave solutions to a (3 + 1)-dimensional Gross–Pitaevskii equation with variable coefficients. Commun. Theor. Phys. (Beijing, China) **52**, 1031 (2009)
16. Hammer, H.-W., Nogga, A., Schwenk, A.: Colloquium: three-body forces: from cold atoms to nuclei. Rev. Mod. Phys. **85**, 197 (2013)
17. Kengne, E.: Ginzburg-Landau system of complex modulation equations for a distributed nonlinear-dispersive transmission line. J. Phys. A: Math. Gen. **37**, 6053 (2004)
18. Kengne, E., Lakhssassi, A.: Impact of three-body interactions and nonlinear dispersion on the dynamics of Bose-Einstein condensates with time-dependent scattering length in harmonic potential. Chin. J. Phys. **53**(1), 020402 (2015)
19. Kengne, E., Lakhssassi, A., Liu, W.M., Vaillancourt, R.: Phase engineering, modulational instability, and solitons of Gross-Pitaevskii-type equations in 1+1 dimensions. Phys. Rev. E **87**, 022914 (2013)
20. Kengne, E., Lakhssassi, A., Vaillancourt, R., Liu, W.M.: Exact solutions for generalized variable-coefficients Ginzburg-Landau equation: application to Bose-Einstein condensates with multi-body interatomic interactions. J. Math. Phys. **53**, 123703 (2012)
21. Kengne, E., Vaillancourt, R.: 2D Ginzburg-Landau system of C complex modulation for coupled nonlinear T transmission lines. J. Infrared, Millim. Terahertz Waves **30**(7), 679 (2009)
22. Kengne, E., Vaillancourt, R.: Modulational stability of solitary states in a lossy nonlinear electrical line. Can. J. Phys. **87**, 1191 (2009)
23. Kevrekidis, P.G., Frantzeskakis, D.J.: Pattern forming dynamical instabilities of Bose-Einstein condensates. Mod. Phys. Lett. B **18**, 173 (2004)
24. Kumar, V.R., Adha, R.R., Adati, M.W.: Phase engineering and solitons of Bose-Einstein condensates with two- and three-body interactions. J. Phys. Soc. Jpn. **79**(7), 074005 (2010)
25. Kuramoto, Y.: Chemical Oscillations, Waves and Turbulence. Springer, Berlin (1984)
26. Lei, W., Zhang, J.F., Li, L.: Modulational instability and bright solitary wave solution for Bose-Einstein condensates with time-dependent scattering length and harmonic potential. New J. Phys. **9**, 69 (2007)
27. Liang, Z.X., Zhang, Z.D., Liu, W.M.: Dynamics of a bright soliton in BECs with time-dependent atomic scattering length in an expulsive parabolic potential. Phys. Rev. Lett. **94**, 050402 (2005)
28. Malomed, B.A.: In: Encyclopedia of Nonlinear Science. Scott, A., (ed.), vol. 157. Routledge, New York (2005)
29. Mashayekhi, M.S., Bernier, J.-S., Borzov, D., Song, J.-L., Zhou, F.: Two-dimensional Bose gases near resonance: competing two-body and three-body effects. Phys. Rev. Lett. **110**, 145301 (2013)

30. Mazza, L., Rizzi, M., Lewenstein, M., Cirac, J.I.: Emerging bosons with three-body interactions from spin-1 atoms in optical lattices. Phys. Rev. A **82**, 043629 (2010)
31. Mülle, K.G.: Structures at the electrodes of gas discharges. Phys. Rev. A **37**, 4836 (1988)
32. Murray, J.D.: Mathematical Biology. Biomathematics, vol. 19. Springer, Berlin (1989)
33. Newell, A.C., Moloney, J.V.: Nonlinear Optics. Addison-Wesley, Redwood City (1992)
34. Peng, P., Li, G.-Q.: Effects of three-body interaction on collective excitation and stability of Bose-Einstein condensate. Chin. Phys. B **18**, 3221 (2009)
35. Pérez-García, V.M., Konotop, V.V., Brazhnyi, V.A.: Feshbach resonance induced shock waves in Bose-Einstein condensates. Phys. Rev. Lett. **92**, 220403 (2004)
36. Roy, U., Atre, R., Sudheesh, C., Kumar, C.N., Panigrahi, P.K.: Complex solitons in Bose-Einstein condensates with two- and three-body interactions. J. Phys. B **43**, 025003 (2010)
37. Sabari, S., Vasantha Jayakantha Raja, R., Porsezian, K., Muruganandam, P.: Stability of trapless Bose–Einstein condensates with two- and three-body interactions. J. Phys. B: At. Mol. Opt. Phys. **43** 125302 (2010)
38. Singh, M., Dhar, A., Mishra, T., Pai, R.V., Das, B.P.: Quantum phases of attractive bosons on a Bose-Hubbard ladder with three-body constraint. Phys. Rev. A **85**, 051604 (2012)
39. Tewari, S.P., Silotia, P., Saxena, A., Gupta, L.K.: Effect of three-body interaction in strongly interacting Bose-Einstein condensate in isotropic and highly anisotropic magnetic trap. Phys. Lett. A **359**, 658 (2006)
40. Theocharis, G., Rapti, Z., Kevrekidis, P.G., Frantzeskakis, D.J., Konotop, V.V.: Modulational instability of Gross-Pitaevskii-type equations in 1+1 dimensions. Phys. Rev. A **67**, 063610 (2003)
41. Whittaker, E.T., Watson, G.N.: A Course of Modern Analysis, pp. 452–455. Cambridge University Press, Cambridge (1927)
42. Will, S., Best, T., Schneider, U., Hackermüller, L., Lü hmann, D.S., Bloch, I.: Collapse and revival oscillations of the matter-wave coherence of ultracold atoms in optical lattices. Nature (London) **465**, 197 (2010)
43. Wu, L., Zhang, J.F., Li, L.: Modulational instability and bright solitary wave solution for Bose-Einstein condensates with time-dependent scattering length and harmonic potential. New J. Phys. **9**, 69 (2007)
44. Xue, J.-K.: Controllable compression of bright soliton matter waves. J. Phys. B: At. Mol. Opt. Phys. **38**, 3841 (2005)
45. Zhang, W.P., Wright, E.M., Pu, H., Meystre, P.: Fundamental limit for integrated atom optics with Bose-Einstein condensates. Phys. Rev. A **68**, 023605 (2003)

Chapter 8
Multicomponent NLS Type Equations and BECs

Abstract To investigate the dynamical behaviors of multi-species Bose–Einstein condensates in an optical lattice, we consider, in this chapter, the multicomponent nonlinear Schrödinger type equations. We successively study the phase diagram for superfluid and insulator phases of two-species BECs in a one-dimensional optical lattice, the vector solitons and their dynamics in two-species BECs, the dynamics and collisions of the symmetric vector solitons, the dynamics of breathing solitons and quasibreathing solitons in multi-species BECs, as well as the dynamics of solitons in heteronuclear two-species BECs.

This chapter deals with the use of multicomponent NLS type equations in the investigation of dynamics of multi-species Bose–Einstein condensates in an optical lattice.

8.1 Phase Diagram for Superfluid and Insulator Phases of Two-Species BECs in a 1D Optical Lattice

We start this chapter with the study of the phase diagram for superfluid and insulator phases of two-species BECs in a one-dimensional optical lattice and the property of persistent current as well. In this 1D geometry, we assume that the confinement along the radial direction is so tight that the trap frequency ω_0 along the radial direction is much greater than the mean-field interaction energy. At low temperatures, the dynamics of the atoms in the radial direction is essentially "frozen," with all the atoms occupying the ground state of the harmonic trap with the wave function that [52]

$$\phi_0(y, z) = \sqrt{\frac{1}{\pi l_0^2}} \exp\left[-\frac{y^2 + z^2}{2l_0^2}\right], \tag{8.1}$$

for which the extension of the wave function in the radial direction is given by the length scale $l_0 = \sqrt{\hbar/m\omega_0}$ of a harmonic oscillator, m being the mass of the atoms. In the mean-field regime, l_0 is much greater than the radius of the interatomic potential R_e. The scattering of atoms in this effective 1D system is thus still a process of three-dimension. The effective coupling constant in this 1D system is [31]

W.-M. Liu and E. Kengne, *Schrödinger Equations in Nonlinear Systems*,
https://doi.org/10.1007/978-981-13-6581-2_8

$$g_{1D} = \frac{2\hbar^2}{m} \frac{a_s}{l_0(l_0 - Ca_s)}, \tag{8.2}$$

where a_s is the s-wave scattering length and C is a numerical constant of the order unit. The term Ca_s in Eq. (8.2) is negligible for $l_0 >> R_e$. In this limit, the expression for g_{1D} is the same as that obtained by averaging over the radial wave function given in Eq. (8.1),

$$g_{1D} = g_{3D} \int_{\mathbf{R}} \int_{\mathbf{R}} \phi_0^4(y, z) dy dz = \frac{2\hbar^2 a}{m l_0^2}. \tag{8.3}$$

This last expression will be used throughout this section.

We consider the two-species BECs in a 1D periodic potential with the macroscopic wave functions ψ_1 and ψ_2. The energy functional is seen to be

$$E(\psi_1, \psi_2) = \int dx \left\{ \sum_{j=1}^{2} \left[\frac{\hbar^2}{2m_j} \left| \frac{\partial \psi_j}{\partial x} \right|^2 + V_j(x) \left| \psi_j \right|^2 + \frac{\hbar^2 a_j}{m_j l_j^2} \left| \psi_j \right|^4 \right] + \frac{2\hbar^2 a_{12}}{\sqrt{m_1 m_2 l_1 l_2}} \left| \psi_1 \right|^2 \left| \psi_2 \right|^2 \right\}, \tag{8.4}$$

where m_j and $l_j = \sqrt{\hbar / m_j \omega_0}$ are, respectively, the mass and the harmonic oscillator lengths in the radial direction of the j-th species ($j = 1, 2$), a_1, a_2, and ma$_{12}$ denote the s-wave scattering lengths between same species and interspecies collisions, and $V_1(x)$ and $V_2(x)$ are the periodic potentials

$$V_j(x) = V_{0,j} \mathrm{sn}^2(k_L x, k), \tag{8.5}$$

with $V_{0,j}$ denoting the magnitude of potentials, where $k_L = 2\pi/\lambda$ is the wave vector of the laser light and λ is the wavelength, corresponding to a lattice period $d = \lambda/2$. $\mathrm{sn}(k_L x, k)$ is the Jacobian elliptic sine function with modulus k ($0 \leq k \leq 1$). In the limit $k \to 0$, the Jacobian elliptic sine reduces to the sinusoidal function, and thus $V(x)$ possesses a standard form of the standing light wave. For values of $k < 0.9$, the potential is virtually indistinguishable from a standing light wave. Finally, in the limit $k \to 1$, $V(x)$ becomes an array of well-separated hyperbolic secant potential barriers or wells.

The governing equations of the trapped BECs are obtained in terms of the variational procedure

$$i\hbar \frac{\partial \psi_j}{\partial t} = \frac{\delta E(\psi_1, \psi_2)}{\delta \psi_j^*}, \tag{8.6}$$

which leads to the coupled NLS equations

$$\begin{aligned} i\hbar \tfrac{\partial \psi_1}{\partial t} &= \left[-\tfrac{\hbar^2}{2m_1} \tfrac{\partial^2}{\partial x^2} + \tfrac{2\hbar^2 a_1}{m_1 l_1} |\psi_1|^2 + \tfrac{2\hbar^2 a_{12}}{\sqrt{m_1 m_2 l_1 l_2}} |\psi_2|^2 + V_1(x) \right] \psi_1, \\ i\hbar \tfrac{\partial \psi_2}{\partial t} &= \left[-\tfrac{\hbar^2}{2m_2} \tfrac{\partial^2}{\partial x^2} + \tfrac{2\hbar^2 a_2}{m_2 l_2} |\psi_2|^2 + \tfrac{2\hbar^2 a_{12}}{\sqrt{m_1 m_2 l_1 l_2}} |\psi_1|^2 + V_2(x) \right] \psi_2. \end{aligned} \tag{8.7}$$

8.1.1 Exact Solution of the Coupled NLS Equations

For the case of weakly coupled condensates in an optical lattice, the wave function ψ can be decomposed as a sum of wave functions localized in each well of the periodic potential straight-binding approximation with the assumption relying on the fact that the height of the interwell barrier is much higher than the chemical potential. We, however, do not restrict ourselves to the low-energy case and look for the global condensate wave functions of excitations as $\psi_j(x,t) = \phi_j(x)\exp\left[-i\mu_j t/\hbar\right]$, where μ_j $(j = 1,\ 2)$ are the chemical potentials. Thus, the spatial wave functions $\phi_j(x)$ satisfy the stationary coupled NLS equations that

$$\begin{aligned} \mu_1\phi_1 &= \left[-\frac{\hbar^2}{2m_1}\frac{d^2}{dx^2} + \frac{2\hbar^2 a_1}{m_1 l_1}|\psi_1|^2 + \frac{2\hbar^2 a_{12}}{\sqrt{m_1 m_2 l_1 l_2}}|\psi_2|^2 + V_1(x)\right]\psi_1, \\ \mu_2\phi_2 &= \left[-\frac{\hbar^2}{2m_2}\frac{d^2}{dx^2} + \frac{2\hbar^2 a_2}{m_2 l_2}|\psi_2|^2 + \frac{2\hbar^2 a_{12}}{\sqrt{m_1 m_2 l_1 l_2}}|\psi_1|^2 + V_2(x)\right]\psi_2. \end{aligned} \tag{8.8}$$

If we write the wave functions $\phi_j(x)$ in its polar form $\phi_j(x) = r_j(x)\exp\left[i\varphi_j(x)\right]$, then system (8.8) can be separated into real and imaginary parts. Integrating once the imaginary part of the resulting system (8.8) yields

$$\frac{d\varphi_j}{dx} - \frac{\alpha_j}{r_j^2} = 0, \quad j = 1,\ 2, \tag{8.9}$$

where α_1 and α_2 are two constants of integration. Inserting Eq. (8.9) into the real part and integrating the resulting equation yield

$$\begin{aligned} \left(r_1\frac{dr_1}{dx}\right)^2 &= \frac{2\alpha_1}{l_1^2}r_1^6 - \frac{2m_1\mu_1}{\hbar^2}r_1^4 + \beta_1 r_1^2 - \alpha_1^2 + \frac{4a_{12}\sqrt{m_1}}{\sqrt{m_2}l_1 l_2}r_1^2\int r_2^2 d(r_1^2) + \frac{2m_1}{\hbar^2}r_1^2\int V_1(x)d(r_1^2), \\ \left(r_2\frac{dr_2}{dx}\right)^2 &= \frac{2\alpha_2}{l_2^2}r_2^6 - \frac{2m_2\mu_2}{\hbar^2}r_2^4 + \beta_2 r_2^2 - \alpha_2^2 + \frac{4a_{12}\sqrt{m_2}}{\sqrt{m1}l_2 l_1}r_2^2\int r_1^2 d(r_2^2) + \frac{2m_2}{\hbar^2}r_2^2\int V_2(x)d(r_2^2), \end{aligned} \tag{8.10}$$

where β_1 and β_2 are other constants of integration.

Based on the form of the external potentials $V_j(x)$ given by Eq. (8.5), we seek "periodic" solutions of system (8.10) in the form

$$r_j^2(x) = A_j \mathrm{sn}^2(k_L x, k) + B_j, \quad j = 1,\ 2, \tag{8.11}$$

where B_j $(j = 1,\ 2)$ determine the mean amplitudes and act as the dc offsets for the numbers of the condensed atoms [4], and the parameters A_j $(j = 1,\ 2)$ are to be determined.

Inserting Eq. (8.11) into Eq. (8.10) and using properties of the Jacobian elliptic functions lead to the following system for α_j, β_j, μ_j, and A_j

$$\begin{aligned} A_1 &= \frac{\frac{\sqrt{m_1}l_1 l_2 a_{12}}{\sqrt{m_2}}\left(m_2 V_{0,2} - \hbar^2 k_L^2 k^2\right) - a_2 l_1^2\left(m_1 V_{0,1} - \hbar^2 k_L^2 k^2\right)}{2\hbar^2\left(a_1 a_2 - a_{12}^2\right)}, \\ A_2 &= \frac{\frac{\sqrt{m_2}l_2 l_1 a_{12}}{\sqrt{m_1}}\left(m_1 V_{0,1} - \hbar^2 k_L^2 k^2\right) - a_1 l_2^2\left(m_2 V_{0,2} - \hbar^2 k_L^2 k^2\right)}{2\hbar^2\left(a_2 a_1 - a_{12}^2\right)}, \end{aligned} \tag{8.12a}$$

$$\begin{aligned} \alpha_1^2 &= B_1 k_L^2 \left[\frac{k^2}{A_1} B_1^2 + \left(1 + k^2\right) B_1 + A_1 \right], \\ \alpha_2^2 &= B_2 k_L^2 \left[\frac{k^2}{A_2} B_2^2 + \left(1 + k^2\right) B_2 + A_2 \right], \end{aligned} \tag{8.12b}$$

$$\begin{aligned} \mu_1 &= \frac{\hbar^2 k_L^2}{2m_1} \left(1 + k^2 + \frac{6a_1}{l_1^2 k_L^2} B_1 + \frac{4a_{12}\sqrt{m_1}}{l_1 l_2 k_L^2 \sqrt{m_2}} B_2 + \frac{2a_{12}\sqrt{m_1}}{l_1 l_2 k_L^2 \sqrt{m_2}} \frac{A_2}{A_1} B_1 + \frac{m_1 V_{0,1}}{\hbar^2 k_L^2} \frac{B_1}{A_1} \right), \\ \mu_2 &= \frac{\hbar^2 k_L^2}{2m_2} \left(1 + k^2 + \frac{6a_2}{l_2^2 k_L^2} B_2 + \frac{4a_{12}\sqrt{m_2}}{l_1 l_2 k_L^2 \sqrt{m_1}} B_1 + \frac{2a_{12}\sqrt{m_2}}{l_1 l_2 k_L^2 \sqrt{m_1}} \frac{A_1}{A_2} B_2 + \frac{m_2 V_{0,2}}{\hbar^2 k_L^2} \frac{B_2}{A_2} \right). \end{aligned} \tag{8.12c}$$

In the limit $k \to 0$, Eq. (8.11) leads to $r_j^2(x) \to A_j \sin^2(k_L x) + B_j$. Therefore, the corresponding solution reads

$$\psi_j(x,t) = \sqrt{A_j^0 \sin^2(k_L x) + B_j} \exp\left[i\left(\varphi_j^0 - \mu_j^0 t/\hbar\right)\right], \tag{8.13}$$

where A_j^0, B_j, and μ_j^0 are obtained from Eqs. (8.12a)–(8.12c) by setting $k = 0$, and the phases φ_j^0 is any solution of the ODE (8.9):

$$\begin{aligned} A_1^0 &= \frac{l_1\sqrt{m_1}\left(\sqrt{m_2}\, l_2 a_{12} V_{0,2} - a_2 l_1 \sqrt{m_1} V_{0,1}\right)}{2\hbar^2\left(a_1 a_2 - a_{12}^2\right)}, \\ A_2^0 &= \frac{l_2\sqrt{m_2}\left(\sqrt{m_1}\, l_1 a_{12} V_{0,1} - a_1 l_2 \sqrt{m_2} V_{0,2}\right)}{2\hbar^2\left(a_2 a_1 - a_{12}^2\right)}, \end{aligned} \tag{8.14}$$

$$\tan\left[\varphi_j^0(x)\right] \mp \sqrt{1 + \frac{A_1^0}{B_1}} \tan\left[k_L x\right] = 0, \tag{8.15}$$

$$\begin{aligned} \mu_1 &= \frac{\hbar^2 k_L^2}{2m_1} \left(1 + \frac{6a_1}{l_1^2 k_L^2} B_1 + \frac{4a_{12}\sqrt{m_1}}{l_1 l_2 k_L^2 \sqrt{m_2}} B_2 + \frac{2a_{12}\sqrt{m_1}}{l_1 l_2 k_L^2 \sqrt{m_2}} \frac{A_2^0}{A_1^0} B_1 + \frac{m_1 V_{0,1}}{\hbar^2 k_L^2} \frac{B_1}{A_1^0} \right), \\ \mu_2 &= \frac{\hbar^2 k_L^2}{2m_2} \left(1 + \frac{6a_2}{l_2^2 k_L^2} B_2 + \frac{4a_{12}\sqrt{m_2}}{l_1 l_2 k_L^2 \sqrt{m_1}} B_1 + \frac{2a_{12}\sqrt{m_2}}{l_1 l_2 k_L^2 \sqrt{m_1}} \frac{A_1^0}{A_2^0} B_2 + \frac{m_2 V_{0,2}}{\hbar^2 k_L^2} \frac{B_2}{A_2^0} \right). \end{aligned} \tag{8.16}$$

The constants A_j and B_j are related by restrictions such that $B_1 \geq -A_1^0$ for $A_1^0 < 0$ and $B_1 \geq 0$ for $A_1^0 > 0$; $B_2 \geq -A_2^0$ for $A_2^0 < 0$ and $B_2 \geq 0$ for $A_2^0 > 0$.

8.1.2 The Phase Diagram

The average particle number densities n_i for the two species are obtained as

$$n_j = \frac{1}{L} \int_0^L \left|\psi_j(x,t)\right|^2 dx = \frac{1}{h\pi} \int_0^{h\pi} \left[A_j^0 \sin^2 x' + B_j\right] dx', \tag{8.17}$$

where $x' = k_L x$ and $L = hd$ denotes the length of the optical lattice with $h = 1, 2, 3, \ldots$. This leads to

$$B_j = n_j - \frac{A_j^0}{2}. \tag{8.18}$$

Going back to Eq. (8.13), it is clear that the wave function (8.13) can exist only for nonnegative $A_j^0 \sin^2(k_L x) + n_j - \frac{A_j^0}{2}$, that is, when

$$n_1 \geq \frac{|A_1^0|}{2} = \frac{|a_{12}V_{0,2} - a_2 V_{0,1}|}{4\hbar^2\omega_0 \left|a_1 a_2 - a_{12}^2\right|}, \quad n_2 \geq \frac{|A_2^0|}{2} = \frac{|a_{12}V_{0,1} - a_1 V_{0,2}|}{4\hbar^2\omega_0 \left|a_1 a_2 - a_{12}^2\right|}. \tag{8.19}$$

The condensate atom currents can be evaluated from the usual definition $\ell = (\hbar/m)\mathrm{Im}[\psi^* (\partial\psi/\partial x)]$ [7], with the exact wave functions Eq. (8.13) which are seen to be traveling matter waves. The result is

$$\ell_j = \pm\frac{\hbar k_L}{m_j}\sqrt{B_j\left(B_j + A_j^0\right)} = \pm\frac{\hbar k_L}{m_j}\sqrt{n_j^2 - \frac{\left(A_j^0\right)^2}{4}}, \tag{8.20}$$

which are independent of space time variables and therefore persistent currents. We may demand that the wave functions Eq. (8.13) satisfy the periodic boundary condition $\psi_j(x,t) = \psi_j(x+L,t)$, which is naturally fulfilled as the total length L is an integer times the lattice constant d ($L = hd$, $h = 1, 2, 3, \ldots$). These periodic solutions in 1D space with spatial period L are equivalent to the solutions in a ring of circumstance L. The persistent currents then can be viewed as in the optical lattice ring.

It is found from Eq. (8.20) that the persistent currents are valid only for conditions in which the number density of atoms is greater than critical values, $n_j \geq \left|A_j^0\right|/2$, i.e., when the macroscopic wave functions of the condensates exist. These persistent currents are similar to the 1D Fröhlich superconductivity induced by the traveling lattice wave [15] and can be controlled by adjusting the barriers height of the periodic potentials and parameters of the bosonic atoms. The currents increase with the decrease of $\left|A_j^0\right|$ and approach the asymptotic maximum value $\ell_{j,\max} = \pm\hbar n_j k_L/m_j$, when $\left|A_j^0\right|$ become vanishingly small. With the recent progress made on confinement of atoms in the light-induced [30] as well as the magnetic-field-induced [11] atom waveguides [22], the persistent currents may be observed experimentally in the future.

The energy spectrum for the two species is obtained as

$$\mu_1^0 = E_{R,1} + \frac{V_{0,1}}{2} + 2\hbar\omega_0\left(a_1 n_1 + a_{12}n_2\right), \quad \mu_2^0 = E_{R,2} + \frac{V_{0,2}}{2} + 2\hbar\omega_0\left(a_2 n_2 + a_{12}n_1\right), \tag{8.21}$$

where $E_{R,j} = \hbar^2 k_L^2/2m_j$ are the recoil energy of an atom absorbing one of the lattice phonons [6].

In this section, we consider the two species both with repulsive interactions, namely, $a_1 > 0$, $a_2 > 0$, and $a_{12} > 0$. Then the macroscopic wave functions of the condensates can exist only when

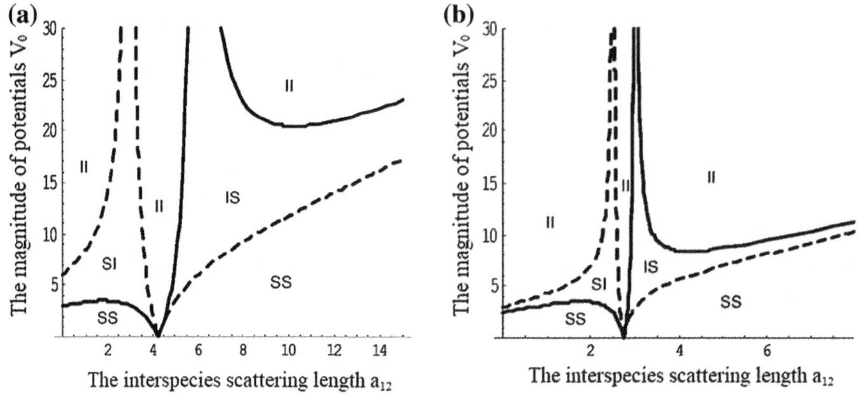

Fig. 8.1 Phase diagrams of two-species BECs in a 1D optical lattice. The magnitude of potentials V_0 is in units of $4nh\omega_0$. The interspecies scattering length is in units of nm. Dashed curves: phase boundary of species 1; solid curves: phase boundary of species 2. The s-wave scattering lengths between the same species are as follows: **a** $a_1 = 6$ nm (^{87}Rb), $a_2 = 3$ nm (^{23}Na). **b** $a_1 = 3$ nm (hyperfine state $|f = 1, m_f = 1\rangle$ of ^{23}Na), $a_2 = 2.5$ nm hyperfine state $|f = 1, m_f = 0\rangle$ of ^{23}Na). Reprint from Ref. [52], Copyright 2018, with permission from American Physical Society

$$\begin{aligned} \mu_1^0 &\geq E_{R,1} + \frac{V_{0,1}}{2} + \frac{a_1\left|a_{12}V_{0,2} - a_2V_{0,1}\right|}{2\left|a_1a_2 - a_{12}^2\right|} + \frac{a_{12}\left|a_{12}V_{0,1} - a_1V_{0,2}\right|}{2\left|a_1a_2 - a_{12}^2\right|}, \\ \mu_2^0 &\geq E_{R,2} + \frac{V_{0,2}}{2} + \frac{a_2\left|a_{12}V_{0,1} - a_1V_{0,2}\right|}{2\left|a_1a_2 - a_{12}^2\right|} + \frac{a_{12}\left|a_{12}V_{0,2} - a_2V_{0,1}\right|}{2\left|a_1a_2 - a_{12}^2\right|}. \end{aligned} \tag{8.22}$$

The macroscopic wave functions of the condensates $\psi_j(x, t)$ are complex functions defined as the expectation value of the boson field operators: $\psi_j(x, t) \equiv \langle\widehat{\Psi}_j(x, t)\rangle$, which have the meaning of order parameters and characterize the off-diagonal long-range behavior of the one-particle density matrix $\rho_j(x', x, t) = \langle\widehat{\Psi}_j^\dagger(x', t)\widehat{\Psi}_j(x, t)\rangle$ [9]. So the condensates can be kept in the superfluid phase only when the macroscopic wave functions exist. Otherwise, the phase coherence and the currents vanish and therefore the condensates are in the insulator phase. Then we obtain four cases of phases for two-species BECs in a 1D optical lattice as follows.

Case 1. The Two Species are Both in the Superfluid Phase

Our first case is associated with a coupled BECs for which the two species are both in the superfluid phase, namely

$$n_1 \geq \frac{\left|a_{12}V_{0,2} - a_2V_{0,1}\right|}{4\hbar\omega_0\left|a_1a_2 - a_{12}^2\right|}, \; n_2 \geq \frac{\left|a_{12}V_{0,1} - a_1V_{0,2}\right|}{4\hbar\omega_0\left|a_1a_2 - a_{12}^2\right|}. \tag{8.23}$$

which is labeled as SS in the phase diagram, Fig. 8.1.

Case 2. Species 1 is in the Superfluid Phase While Species 2 is in the Insulator Phase

This second case corresponds to the situation when

$$n_1 \geq \frac{|a_{12}V_{0,2} - a_2 V_{0,1}|}{4\hbar\omega_0 |a_1 a_2 - a_{12}^2|}, \quad n_2 \leq \frac{|a_{12}V_{0,1} - a_1 V_{0,2}|}{4\hbar\omega_0 |a_1 a_2 - a_{12}^2|}, \tag{8.24}$$

which is labeled as SI in Fig. 8.1.

Case 3. Species 1 is in the Insulator Phase and Species 2 is in the Superfluid Phase

Case 3 corresponds to the situation when the total number of atoms n_1 and n_2 satisfy the relationships

$$n_1 \leq \frac{|a_{12}V_{0,2} - a_2 V_{0,1}|}{4\hbar\omega_0 |a_1 a_2 - a_{12}^2|}, \quad n_2 \geq \frac{|a_{12}V_{0,1} - a_1 V_{0,2}|}{4\hbar\omega_0 |a_1 a_2 - a_{12}^2|}, \tag{8.25}$$

and is labeled as IS in Fig. 8.1.

Case 4. The Two Species Are Both in the Insulator Phase

The last case, i.e., Case 4 occurs when the two species are both in the Insulator Phase, that is, when

$$n_1 \leq \frac{|a_{12}V_{0,2} - a_2 V_{0,1}|}{4\hbar\omega_0 |a_1 a_2 - a_{12}^2|}, \quad n_2 \leq \frac{|a_{12}V_{0,1} - a_1 V_{0,2}|}{4\hbar\omega_0 |a_1 a_2 - a_{12}^2|}, \tag{8.26}$$

which is labeled as II in Fig. 8.1.

The quantum phases of the condensates can be determined by all parameters of BECs and optical lattice as shown above. In Fig. 8.1, we show the phase diagram with the various same-species s-wave scattering lengths and equal particle number density ($n_1 = n_2 = n$) and the magnitudes of potentials ($V_{0,1} = V_{0,2} = V_0$) for simplicity. Thus the conditions for the two species in the superfluid phase are given by

$$\frac{V_0}{4n\hbar\omega_0} \leq \frac{|a_1 a_2 - a_{12}^2|}{|a_{12} - a_2|}, \tag{8.27}$$

$$\frac{V_0}{4n\hbar\omega_0} \leq \frac{|a_1 a_2 - a_{12}^2|}{|a_{12} - a_1|}, \tag{8.28}$$

respectively.

Component separation in two-species BECs has been predicted by means of the mean field theory and observed in experiments when the relation of the scattering lengths that $a_{12} > \sqrt{a_1 a_2}$ is fulfilled. From the above conditions Eqs. (8.27) and (8.28) and the phase diagram Fig. 8.1, we find that the larger values of a_{12} favor the superfluid phase in the two species mixture and the component separation according to the experimental observation [40]. Particularly, when the interspecies scattering length approaches the value of the same species such that $a_{12} = a_2$ or a_1, the conditions Eqs. (8.27) and (8.28) result in the superfluid phase independent of the potential magnitude V_0. One should not be surprised by this result since we consider the case in which the chemical potential is always higher than the potential magnitude V_0 seen from Eq. (8.22).

8.2 Solitons in Two-Species BECs with Tunable Interactions and Harmonic Potential

In this section, we aim to study vector solitons and their dynamics in two-species BECs with tunable interaction parameters and harmonic trapping potential. As it is known, two interacting dilute Bose condensates can be well described by the zero-temperature mean field theory, in which the collisions between the condensate atoms and the thermal cloud are neglected. Considering a two-species BECs each of mass m trapped in a quasi-one-dimensional harmonic potential, the evolution of this system is govern by a double-component dimensionless NLS equations with external potentials

$$\begin{aligned} i\frac{\partial\psi_1}{\partial t} &= \left[-\frac{\partial^2}{\partial x^2} + V_1 + g_{11}\left|\psi_1\right|^2 + g_{12}\left|\psi_2\right|^2\right]\psi_1, \\ i\frac{\partial\psi_2}{\partial t} &= \left[-\frac{\partial^2}{\partial x^2} + V_2 + g_{22}\left|\psi_2\right|^2 + g_{21}\left|\psi_1\right|^2\right]\psi_2, \end{aligned} \tag{8.29}$$

where ψ_j denotes the macroscopic wave functions of the j-th component, with the normalization conditions $\int_{-\infty}^{+\infty}|\psi_1|^2\,dx = 1$ and $\int_{-\infty}^{+\infty}|\psi_2|^2\,dx = N_2/N_1$. Here we do not allow for the components to transform into each other and $N_1 = N_2$, so the number of atoms is conserved for both components. The external harmonic time-dependent potential associated with system (8.29) is $V_j = \lambda_j^2(t)x^2 = \omega_j^2(t)x^2/\omega_{j\perp}^2$, where ω_j and $\omega_{j\perp}$ are the angular frequencies in the axial and radial directions. Temporal and spatial coordinates t and x are measured in units $2/\omega_\perp$ and $\ell_\perp = \sqrt{\hbar/m\omega_\perp}$, where $\ell_\perp$ is linear oscillator length in the transverse direction. The interactions between atoms are described by a self-interaction $g_{jj} = 4a_{jj}N_j/\ell_\perp$ and the interaction between different components $g_{12} = g_{21} = 4a_{jk}N_j/\ell_\perp$, where a_{jj} are the scattering lengths of component j and a_{12} is that between components 1 and 2. In the present book, we assume $m_1 = m_2, \omega_1 = \omega_2$, and define variables $\sigma_1 = g_{12}^2 - g_{11}g_{22}$, $\sigma_2 = g_{12} - g_{11}$, and $\sigma_3 = g_{12} - g_{22}$ for the facilitation of expression.

8.2.1 Exact Soliton Solutions

Because we are interested in the exact soliton solutions of the coupled NLS equations (8.29), we first consider its integrability. The symmetric case, that is, the case when all the interactions take exactly the same form, is the simplest one. Then we can directly extend study on single-component BECs to the two-component case and obtain a similar integrability conditions for the existence of the exact symmetric solitons. However, in typical BEC experiments, the strength of the interactions is always unequal to each other; but fortunately, we can employ the extend mapping deformation method, which does not depend on the integrability, to obtain the exact asymmetric vector-soliton solutions and determine the regions where these solutions are existing.

Symmetric Case

In the situation when the interactions take equal values, Eq. (8.29) can be simply written as

$$i\frac{\partial \psi_j}{\partial t} = \left[-\frac{\partial^2}{\partial x^2} + \lambda^2(t)x^2 + g(t)\left(\left|\psi_j\right|^2 + \left|\psi_{3-j}\right|^2\right)\right]\psi_j,\ j = 1,\ 2. \quad (8.30)$$

Seeking the solutions of system (8.30) in the form

$$\psi_j = \frac{1}{\sqrt{g(t)\ell(t)}}\phi_j(X, T)\exp\left[if(t)x^2\right], \quad (8.31)$$

where $X = x/\ell(t)$ and $\ell(t)$, $T(t)$, and $f(t)$ are three time-dependent functions satisfying the system

$$\frac{dT}{dt} - \frac{1}{\ell^2} = 0,\ \frac{df}{dt} + 4f^2 + \lambda^2 = 0,\ \frac{1}{\ell}\frac{d\ell}{dt} - 4f = 0, \quad (8.32)$$

yields

$$i\frac{\partial \phi_j}{\partial T} = \left[-\frac{\partial^2 \phi_j}{\partial X^2} + \sigma\left(|\phi_1|^2 + |\phi_2|^2\right)\right]\phi_j,\quad \sigma = \pm 1, \quad (8.33)$$

if $g(t)$ and $\lambda(t)$ satisfy the condition

$$-\frac{1}{2g}\frac{d^2 g}{dt^2} + \frac{1}{g^2}\left(\frac{dg}{dt}\right)^2 + 2\lambda^2 = 0. \quad (8.34)$$

In Eq. (8.33), $\sigma = \pm 1$ corresponds to the cases of positive and negative scattering lengths. The coupled equations (8.30) are nothing but the well-known integrable model proposed by Manakov [26] or the defocusing–defocusing NLS equations. Since the exact solutions for these systems have been well studied, the exact vector-soliton solutions for Eq. (8.30) can be easily obtained by combining the well-known solutions for the Manakov system or the defocusing–defocusing NLS equations through Eq. (8.31).

The virtue of the transformation is that the variable ϕ_j satisfies the integrable Manakov system or the defocusing NLS equations while the varying of the scattering length and external potential appear in the remains of Eq. (8.31) and variables X and T. It is interesting to observe that the amplitude and width of the bright–bright vector solitons vary with function $g(t)$ according to Eq. (8.31), which offers us also a possibility to change the soliton's parameters in a controllable manner by choosing a proper external potential to satisfy the integrability condition.

Asymmetric Case

Now, we consider the asymmetric case, which is the more general situation when $g_{11}(t) \neq g_{12}(t) \neq g_{22}(t)$. To carry out the analysis, we start from the constant amplitude solutions of Eq. (8.29) without external potential,

$$\psi_j = A_j \exp\left[i\Omega_j t\right], \quad \Omega_j = g_{jj}\left|A_j\right|^2 + g_{j,3-j}\left|A_{3-j}\right|^2. \tag{8.35}$$

The instability of this coupled constant amplitude solutions with small perturbations is discussed in [34]. When a harmonic trapping potential is added along the longitudinal direction, the exact vector-soliton solutions only exist under specific assumptions. Does the introduction of the external potential and arbitrary interaction parameters make any qualitative change in the behavior of the integrability conditions? The answer is yes and we find that the integrability conditions, in this case, have more constrains on the soliton's parameters. In the following, we take advantage of the extend mapping deformation method, which does not depend on the integrability, to obtain the exact vector-soliton solutions for the general equations (8.29).

Expressing the order parameters in terms of their modulus and phases, i.e., $\psi_j = \sqrt{n_j}\exp\left[i\Delta_j\right]$ and then separating real and imaginary parts, we obtain a set of coupled nonlinear equations for n_j and Δ_j

$$\begin{aligned} \sqrt{\frac{\partial^2 n_j}{\partial x^2}} - \sqrt{n_j}\frac{\partial \Delta_j}{\partial t} - \sqrt{n_j}\left(\frac{\partial \Delta_j}{\partial x}\right)^2 - g_{12}n_{3-j}\sqrt{n_j} - g_{jj}\left(\sqrt{n_j}\right)^3 - \lambda^2 x^2\sqrt{n_j} = 0, \\ \sqrt{\frac{\partial n_j}{\partial t}} + \sqrt{n_j}\frac{\partial^2 \Delta_j}{\partial x^2} + 2\sqrt{\frac{\partial n_j}{\partial x}\frac{\partial \Delta_j}{\partial x}} = 0, \end{aligned} \tag{8.36}$$

where $j = 1, 2$. By setting $\sqrt{n_j} = A_0(t) + A_j(t)\phi(\xi)$, $\xi = p_1(t)x + p_2(t)$, and $\Delta_j = k_j(t) = \Gamma_j(t)x + \Lambda_j(t)x^2$ and using the auxiliary equation $(d\phi/dt)^2 = c_0 + c_2\phi^2 + c_4\phi^4$, we derive a set of overdetermined partial differential equations with respect to $p_1(t)$, $p_2(t)$, $k_j(t)$, $\Gamma_j(t)$, $\Lambda_j(t)$, $A_0(t)$, $A_1(t)$, and $A_2(t)$. Finally, solving these equations, we can obtain the following constrains for the existence of the exact vector-soliton solutions:

$$\begin{aligned} g_{11}A_1^2(t) + g_{12}A_2^2(t) = g_{22}A_2^2(t) + g_{12}A_1^2(t), \\ -\frac{d^2 p_1}{dt^2}p_1(t) + 2\left(\frac{dp_1}{dt}\right)^2 + 4p_1^2(t)\lambda^2 = 0, \\ p_1^2(t) - g_{11}p_1(t)\frac{\sigma_2}{\sigma_3}p_1(t) = 0, \end{aligned} \tag{8.37}$$

where λ is the trapping frequency. Equation (8.37) implies that the exact vector-soliton solutions exist with many constrains. For example, the first equation of Eq. (8.37) imposes a constrain that the amplitudes of the vector solitons must satisfy $A_1^2/A_2^2 = (g_{12} - g_{22})\,/\,(g_{12} - g_{11})$; in particular, the others impose a relation between the scattering length and trapping potential which reads

$$\frac{g_{12} - g_{22}}{g_{12}^2 - g_{11}g_{22}} = \frac{a_1 \sin(2\lambda t) - a_2 \cos(2\lambda t)}{2\lambda}. \tag{8.38}$$

In the case of all the interaction parameters $g_{jk} < 0$, Eq. (8.29) has bright–bright vector solitons [51]

$$\psi_{1B} = \frac{C_1\sqrt{-a(t)}}{\cosh[\xi(x,t)]} \exp[i\upsilon_{1B}(x,t)], \ \psi_{2B} = \frac{C_2\sqrt{-\sigma a(t)}}{\cosh[\xi(x,t)]} \exp[i\upsilon_{2B}(x,t)], \tag{8.39}$$

where

$$\begin{aligned} \upsilon_{jB} &= -\frac{1}{4a}\frac{da}{dt}x^2 + C_2 ax + \left(C_1^2 - C_2^2\right)\int a^2(t)dt + C_{j+2}, \\ \xi(x,t) &= C_1 ax - 2C_1C_2 \int a^2(t)dt + C_5, \\ a(t) &= \frac{\sigma_1}{\sigma_3} < 0, \ \sigma = \frac{\sigma_2}{\sigma_3}. \end{aligned} \tag{8.40}$$

When $a(t) > 0$ and $\sigma(t) > 0$, Eq. (8.29) has dark–dark vector solitons

$$\begin{aligned} \psi_{1D} &= D_1\sqrt{a(t)} \tanh[\xi(x,t)] \exp[i\upsilon_{1D}(x,t)], \\ \psi_{2D} &= D_2\sqrt{\sigma a(t)} \tanh[\xi(x,t)] \exp[i\upsilon_{2D}(x,t)], \end{aligned} \tag{8.41}$$

where $\xi(x,t)$, $a(t)$, and $\sigma(t)$ are the same as Eq. (8.40), but the phase functions are given by

$$\upsilon_{jD} = -\frac{1}{4a}\frac{da}{dt}x^2 + D_3 ax - \left(2D_1^2 + D_2^2\right)\int a^2(t)dt + D_{j+3}. \tag{8.42}$$

In Eqs. (8.39) and (8.41), the coefficients C_1, C_2, C_3, C_4, C_5, D_1, D_2, D_3, D_4, and D_5 are arbitrary constants. Based on the above results, we can conclude the following conditions for the existence of the exact vector-soliton solutions:

$$\begin{aligned} &\Delta_1 > 0, \Delta_2 < 0: \text{ bright–bright BB}, \\ &\Delta_1 < 0, \Delta_2 > 0: \text{ dark–dark DD}, \end{aligned} \tag{8.43}$$

where $\Delta_1 \equiv -\sigma_3/\sigma_1$ and $\Delta_2 \equiv \sigma_2/\sigma_1$. As long as Eqs. (8.37) and (8.43) are satisfied, we can obtain the exact vector soliton solutions, either bright–bright or dark–dark, for an arbitrary periodic time dependence of the scattering length since we can choose an appropriate λ to satisfy the above equations. It is relevant to mention that all interaction parameters g_{jj} and g_{jk} can be functions of time t in our cases.

Now, we conclude the regions for the existence of these vector solitons. In the case of bright–bright vector solitons, we consider a simple case where one of the

Table 8.1 Table for the existence of dark–dark solitons versus intercomponent interaction g_{12}. $+$ $(-)$ indicates the repulsive attractive atom–atom interaction in single component. The single-component interaction parameters g_{11} and g_{22} are assumed to be unvaried and $g_{12} > g_{11}$

Parameters	g_{11}	g_{12}	g_{12}
Case 1	$+$	$+$	$g_{12} > g_{11}, -\sqrt{g_{11}g_{22}} < g_{12} < g_{22}$
Case 2	$-$	$-$	$g_{12} > \sqrt{g_{11}g_{22}}$
Case 3	$+$	$-$	$g_{12} > g_{11}$

intracomponent interaction parameters g_{11} is time-dependent while others are time-independent. According to the integrability condition, the bright–bright vector solitons can exist in two regions: (i) $g_{12} < g_{11} < 0$ (here we assume that $0 > g_{22} > g_{12}$) and (ii) $g_{11} < g_{12} < 0$ (here we assume that $0 > g_{12} > g_{22}$). In real experiments, we consider a two-component condensates with $N_1 = N_2 = 5000$, $a_{12} = -2$ nm, $a_{22} = -1$ nm, $\omega_\perp = 2\pi \times 100$ Hz, and $\omega_j = 2\pi \times 2$ Hz. Then the nonlinearities can be taken as $g_{12} = -40$, $g_{22} = -20$, and $-40 < g_{11} < 0$.

Table 8.1 shows the regions where the dark–dark vector solitons can exist when the intercomponent interaction parameter g_{jk} is time dependent and intracomponent ones g_{jj} are time independent. Case 1 has been extensively studied in [13, 32]. In cases 2 and 3, it is interesting to find that dark soliton can be formed in a condensate with self-attractive atom–atom interaction. The common sense is that the dark solitons only exist in a condensate with self-repulsive atom–atom interaction; however, in the two-component BECs, the repulsive interaction coming from the second component or the intercomponent interaction induces an effective repulsive interaction in the self-attractive one. This leads to the formation of dark solitons in the component with self-attractive interaction. It is clear that the vector solitons in case 2 are different from the Manakov solitons, where vector solitons are formed by all the nonlinear coefficients cooperation. In our case, the mutual repulsive interaction between the two components solely supports the existence of such dark–dark vector solitons. At last, we would like to point out that the formation of this type of vector solitons has a threshold value in g_{12}, which indicates that the intercomponent repulsive interaction must be strong to induce an effective repulsive interaction in the self-attractive ones.

Furthermore, when a linear gain term, which contributes to the growth of each component during the simultaneous evaporative cooling is added to the coupled GP equations, our methods will also allow us to obtain the exact vector-soliton solutions, but with a more complicated integrability condition. These solutions may help us to better understand the nonlinear dynamics of the condensates and the dynamical pattern formation during growth of a dual-component BECs, which has been experimentally realized in [33] and theoretically investigated in [36]. This needs a further investigation and is beyond the scope of the present section.

8.2.2 *Dynamics and Collisions of the Symmetric Vector Solitons*

One extension of application of the above transformation leading to the integrability condition is to study the dynamics of a single soliton and the collisions between two orthogonal solitons. In real experiments, the s-wave scattering length can be changed with linear, exponential, and periodic time dependence of the magnetic field via Feshbach resonance. As mentioned above, once a function of the scattering length is given, we can choose a proper external potential according to Eq. (8.34) to satisfy the integrability condition. In the following, we focus on the most natural periodic time dependence of the scattering length.

Dynamics of a Single Soliton

For a better understanding of the influence of Feshbach resonance on soliton's parameters, we study the dynamics of a single solitons with Feshbach management. As discussed in works [1, 38], a periodic scattering length is obtained by means of an ac magnetic field, which allows one to create a self-confined two-dimensional BEC without the magnetic trap. We suppose that the intercomponent interaction parameter g_{12} is modulated periodically as

$$\overline{g}_{12}(t) = \pm g_{12}\left(1 + m\sin\left[\omega t\right]\right),\ \ 0 < m < 1, \tag{8.44}$$

where m and ω are the amplitude and frequency of the modulation. The symmetric vector solitons can be realized in a two-component system constituting from different hyperfine spin states of the same atom, such as ^{87}Rb with two hyperfine spin states which can be chosen as $\left|F = 1, m_f = -1\right\rangle$ and $\left|F = 2, m_f = 1\right\rangle$. In this case, the trapping potential has the same shape for these two states, which is in accordance with our previous assumption. We choose the atom number $N_1 = N_2 = 10^4$, and $a_{11} = a_{22} = a_{12} = -5.36\left(1 + m\sin\left[\omega t\right]\right)$ nm. According to Eq. (8.34), the axial frequency of the harmonic potential should be

$$\omega_1^2(t) = -\frac{m\omega^2\omega_\perp^2}{4}\frac{\sin\left[\omega t\right] + m + m\cos^2\left[\omega t\right]}{\left(1 + m\sin\left[\omega t\right]\right)^2}. \tag{8.45}$$

We now investigate how the amplitude of the ac drive can be used to control the parameters of the vector solitons. When the amplitude is small, $m = 0.1$, this leads to a small periodically modulation of the scattering length and the trapping frequency. As shown in Fig. 8.2, we observe that the amplitude and width of the vector solitons vary periodically with time. When the amplitude takes a bigger value, the vector solitons will have a larger oscillation in amplitudes and widths. A natural extension are the cases where the magnetic field varies exponentially and linearly with time; in these cases, the amplitude and width of the vector solitons are also proportional and inversely proportional to the scattering length, respectively.

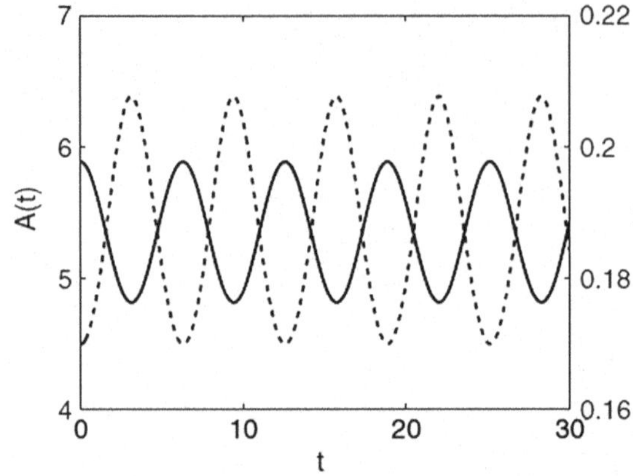

Fig. 8.2 The variation of amplitude $A(t)$ and width $W(t)$ of the vector solitons when the scattering length varies periodically with time t. The parameters are given as follows: $\ell(0) = g(0) = 1$, $\omega = 1$, $m = 0.1$, $a_{11} = a_{22} = a_{21} = -5.36\,(1 + m\cos[\omega t])$ nm, and the initial amplitude $A(0)$ and width $W(0)$ are taken equal to 1. The amplitude and width are measured in units $4aN/\ell_\perp$. Time t and coordinate x are the temporal and spatial coordinates measured in units $a/\omega_\perp$ and $\ell_\perp = \sqrt{\hbar/m\omega_\perp}$. Reprint from Ref. [51], Copyright 2018, with permission from American Physical Society

Collisions Between Solitons

Finally, we study the collision between solitons. Interactions between solitons are fascinating since in many aspects solitons interact like particles: they pass through one another and accomplish elastic collision. As shown in Eq. (8.33), the nonlinearity appears solely through the invariant combination $\left(|\psi_1|^2 + |\psi_2|^2\right)$. The isotopic invariance ensures that once we obtain a soliton solution for one component, the other one can be obtain in an obvious way

$$\begin{pmatrix} \psi_1 \\ \psi_2 \end{pmatrix} = \begin{pmatrix} \cos\theta \\ \sin\theta \end{pmatrix} \begin{pmatrix} \exp[-i\mu t] \\ \exp[-i\mu t] \end{pmatrix} \phi(x) \tag{8.46}$$

with $0 \le \theta \le \pi/2$. Furthermore, the general vector solitons with unequal chemical potentials are also possible. We consider a special case with equal chemical potential between two orthogonal solitons which reads $\theta = 0$ and $\theta \le \pi/2$ and set the initial velocities as, respectively. In this case, the initial conditions at $t = 0$ are given by

$$\begin{pmatrix} \psi_1^{(0)} \\ \psi_2^{(0)} \end{pmatrix} = \begin{pmatrix} \phi(x - x_0/2) \\ \phi(x - x_0/2) \end{pmatrix}, \tag{8.47}$$

where x_0 is the initial separation. Figure 8.3 shows the interaction between two orthogonal solitons when the scattering length varies periodically with time. Our results show that the parameters of the two colliding solitons do not change after collision, which remarkably indicates no energy exchange between the two solitons. Shown in Fig. 8.4 is the collision scenario when the amplitude of the modulation has

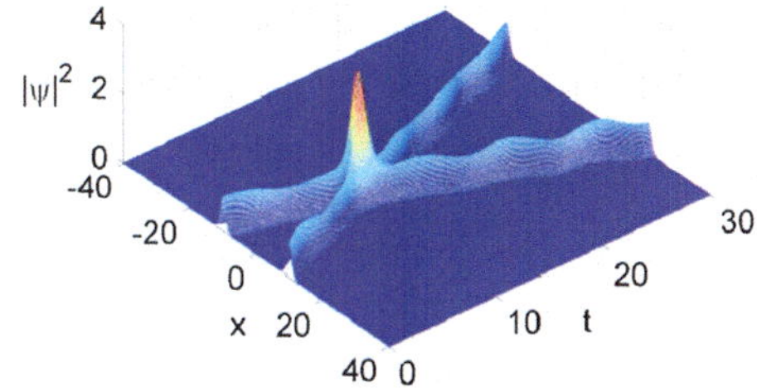

Fig. 8.3 The interaction between two orthogonal solitons when the scattering length varies periodically with time t. The parameters are given as follows: $x_0 = 20$, $v = 1$, $m = 0.1$, $a_{11} = a_{22} = a_{21} = -(1 + m\cos[\omega t])$ nm, $\omega = 1$, and the other parameters are the same as in Fig. 8.2. Reprint from Ref. [51], Copyright 2018, with permission from American Physical Society

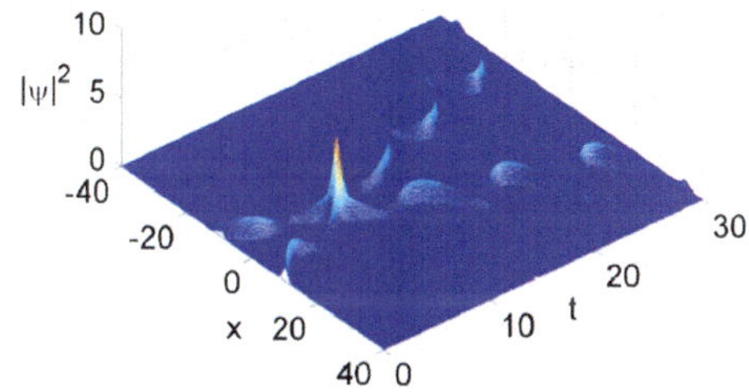

Fig. 8.4 Elastic collision between two orthogonal solitons with a larger oscillation of the soliton's parameters when the modulation is increased from $m = 0.1$ to $m = 0.5$. The other parameters are the same as in Fig. 8.3. Reprint from Ref. [51], Copyright 2018, with permission from American Physical Society

a large value as $m = 0.5$. It demonstrates that the intrinsic oscillations and elastic collision do not depend on the amplitude of the modulation. The only difference between the two cases is that the parameters of the two solitons, such as amplitude, width, and velocity, have a larger oscillation in the latter. Another interesting issue is whether the initial velocity and separation will destroy the intrinsic elastic collision. Our results show that the outcome does not depend on the initial velocity and separation except for a large enough value of x_0. The elastic collisions between solitons originate from the integrable Manakov system which is obtained with the condition $g\left(|\varphi_1|^2 + |\varphi_2|^2\right) << 1$. These collision scenarios are similar to to the cases in nonlinear optics where vector solitons collide with each other elastically, except that their polarization may change after collision if the vector solitons have the same or orthogonal polarization, such change will not occur. While, if the condition $g\left(|\varphi_1|^2 + |\varphi_2|^2\right) << 1$ is not satisfied, a more accurate coupled non polynomial Schrödinger equations, which are non integrable, can be applied to investigate the inelastic collision between solitons. The various inelastic collision scenarios in BEC system and fiber optics were studied in works [24, 25, 39], which will help us to better understand the physics behind the equations.

8.3 Two-Component Bose–Einstein Condensates with Time- and Space-Modulated Nonlinearities

We consider in this section the two-component BECs each of mass m_a trapped in an external potential whose dynamics is governed by the dimensionless two components 1D NLS equations with external potential

$$\begin{aligned} i\frac{\partial\psi_1}{\partial t} &= \left[-\frac{1}{2}\frac{\partial^2}{\partial x^2} + \frac{\gamma^2}{2}x^2 + b_{11}|\psi_1|^2 + b_{12}|\psi_2|^2\right]\psi_1, \\ i\frac{\partial\psi_2}{\partial t} &= \left[-\frac{1}{2}\frac{\partial^2}{\partial x^2} + \frac{\gamma^2}{2}x^2 + b_{22}|\psi_2|^2 + b_{12}|\psi_1|^2\right]\psi_2, \end{aligned} \tag{8.48}$$

where $b_{11} = 2a_{11}$, $b_{22} = 2a_{22}$, $b_{12} = 2a_{12}$, and the units for length and time are $\sqrt{\hbar/m\omega_\perp}$ $1/\omega_\perp$, respectively. Here, we have assume that the intracomponent coupling parameter is characterized by the scattering lengths a_{11} and a_{22} between atoms of the same species, while the intercomponent coupling parameter between atoms of different species is determined by the scattering length a_{12} where an atom in the ψ_1 component scatters from another atom in the ψ_2 component. In what follows, we consider the scattering lengths a_{11}, a_{22}, and a_{12} to be spatiotemporally inhomogeneous, which is experimentally feasible due to the flexible and precise control of the scattering lengths achievable in quasi-1D BECs with magnetically tuning the Feshbach resonances. Our aim in this section is to construct the explicitly exact solutions of model (8.48) under the special interaction parameters b_{11}, b_{12}, and b_{22}, and investigate their unique dynamics.

8.3.1 Analytic Spatially Localized Solutions

We aim here to construct exact spatially localized solutions of Eq. (8.48) for which $\lim_{|x|\to+\infty} \psi_j(x,t) = 0$. For this purpose, we consider the similarity transformation [48]

$$\psi_1(x,t) = \beta_1(x,t)U(X(x,t))\exp\left[i\alpha_1(x,t)\right], \tag{8.49a}$$

$$\psi_2(x,t) = \beta_2(x,t)V(X(x,t))\exp\left[i\alpha_2(x,t)\right], \tag{8.49b}$$

to reduce Eq. (8.48) to the two stationary NLS equations

$$\frac{d^2U}{dX^2} + g_{11}U^3 + g_{12}UV^2 = 0, \tag{8.50a}$$

$$\frac{d^2V}{dX^2} + g_{22}V^3 + g_{12}VU^2 = 0, \tag{8.50b}$$

where g_{11}, g_{12}, and g_{22} are constants, and $\alpha_1, \alpha_2, \beta_1, \beta_2, X$ are functions of x and t to be determined. For brevity, we define variables $\sigma_1 = g_{12} - g_{11}$, $\sigma_2 = g_{12} - g_{22}$, and

$\sigma_{12} = g_{12}^2 - g_{11}g_{22}$. Inserting Eqs. (8.49a), (8.49b) into Eq. (8.48) and demanding $U(X)$, $V(X)$ to satisfy Eqs. (8.50a), (8.50b), we have a set of PDEs. By solving this set of PDEs, we conclude that when

$$b_{11} = g_{11}\theta(t,x),\ b_{12} = g_{12}\theta(t,x),\ b_{22} = g_{22}\theta(t,x), \tag{8.51}$$

with $\theta(t,x) = -\lambda^2 \exp\left[-3\lambda^2 x^2 - 6\lambda\delta x - 2\delta^2\right] / \left[2\zeta_3^2(t)\right]$, one has

$$\begin{aligned}&\alpha_1 = \zeta_1(t) - \tfrac{1}{2\lambda}\tfrac{d\lambda}{dt}x^2 - \tfrac{1}{\lambda}\tfrac{d\delta}{dt}x,\ \alpha_2 = \zeta_2(t) - \tfrac{1}{2\lambda}\tfrac{d\lambda}{dt}x^2 - \tfrac{1}{\lambda}\tfrac{d\delta}{dt}x,\\ &\beta_1 = \beta_2 = \zeta_3(t)\exp\left[\tfrac{\lambda x(\lambda x+2\delta)}{2}\right]\ X = \sqrt{\tfrac{\pi}{2}}\mathrm{erf}(\lambda x + \delta),\end{aligned} \tag{8.52}$$

where $\mathrm{erf}(s) = \frac{2}{\sqrt{\pi}}\int_0^s \exp\left[-\tau^2\right]d\tau$ is called an error function, $\zeta_1(t)$, $\zeta l(t)$, and $\zeta_3(t)$ satisfy

$$\begin{aligned}\zeta_1(t) &= \int\left[\lambda^4 - \left(\frac{d\delta}{dt}\right)^2 + \lambda^4\delta^2\right] / \left(2\lambda^2\right)dt + C_1,\\ \zeta_2(t) &= \int\left[\lambda^4 - \left(\frac{d\delta}{dt}\right)^2 + \lambda^4\delta^2\right] / \left(2\lambda^2\right)dt + C_2,\ \zeta_3(t) = \sqrt{\lambda}\exp\left[\frac{\delta^2}{2}\right],\end{aligned}$$

with C_1 and C_2 arbitrary constants, and λ, δ satisfy

$$\lambda^6 - \gamma^2\lambda^2 - 2\left(\frac{d\lambda}{dt}\right)^2 + \lambda\frac{d^2\lambda}{dt^2} = 0, \tag{8.53a}$$

$$2\lambda\frac{d^2\delta}{dt^2} + 2\lambda^5\delta - 4\frac{d\delta}{dt}\frac{d\lambda}{dt} = 0. \tag{8.53b}$$

When setting $\sigma_1/\sigma_{12} > 0$ and $\sigma_2/\sigma_{12} > 0$, Eqs. (8.50a), (8.50b) have two families of exact solutions as

$$U^{(1)}(X) = \sqrt{\sigma_2/\sigma_{12}}\nu_1\mathrm{cn}\left(\nu_1 X, \sqrt{2}/2\right), \tag{8.54a}$$

$$V^{(1)}(X) = \sqrt{\sigma_1/\sigma_{12}}\nu_1\mathrm{cn}\left(\nu_1 X, \sqrt{2}/2\right), \tag{8.54b}$$

and

$$U^{(2)}(X) = \frac{\sqrt{2}}{2}\sqrt{\sigma_2/\sigma_{12}}\nu_2\mathrm{sd}\left(\nu_2 X, \sqrt{2}/2\right), \tag{8.55a}$$

$$V^{(2)}(X) = \frac{\sqrt{2}}{2}\sqrt{\sigma_1/\sigma_{12}}\nu_2\mathrm{sd}\left(\nu_2 X, \sqrt{2}/2\right), \tag{8.55b}$$

where ν_1 and ν_2 are arbitrary constants, sd = sn/dn with sn, cn, and dn being Jacobi elliptic functions. When imposing the bounded condition $\lim_{|x|\to+\infty} \psi_j(x,t) = 0$, we have $\nu_1 = 2(2n+1)K(\sqrt{2}/2)/\sqrt{\pi}$ for Eqs. (8.54a), (8.54b) and $\nu_2 = 4mK(\sqrt{2}/2)/\sqrt{\pi}$ for Eqs. (8.55a), (8.55b), where n and m are integer numbers and $K(\sqrt{2}/2) = \int_0^{\pi/2}\left[1-\left(\sqrt{2}/2\right)^2 \sin^2\tau\right]^{-1/2} d\tau$ is an elliptic integral of the first kind.

After some algebra, we find Eq. (8.53b) has a solution of the following form

$$\delta(t) = c_1 \exp\left[i\int \lambda^2 dt\right] + c_2 \exp\left[-i\int \lambda^2 dt\right] \tag{8.56}$$

with c_1 and c_2 arbitrary constants.

Next we set $\lambda = 1/\xi$ to rewrite Eq. (8.53a) in the form of the Ermakov–Pinney equation [35] as

$$\frac{d^2\xi}{dt^2} + \gamma^2\xi = 1/\xi^3. \tag{8.57}$$

According to the result in Ref. [18], to obtain the explicit solutions of Eq. (8.57) we impose to γ to satisfy the equation

$$\gamma^2 = \gamma_0^2 + \epsilon\cos(\gamma_1 t), \tag{8.58}$$

with $\epsilon \in (-1, 1)$ and $\gamma_0, \gamma_1 \in \mathbf{R}$. Therefore, the general solution to Eq. (8.53a) is

$$\lambda = \left(A\xi_1^2 + B\xi_2^2 + 2C\xi_1\xi_2\right)^{-1/2}, \tag{8.59}$$

where A, B, C are constants satisfying $AB - C^2 = 1/W^2$, and the Wronskian $W = \xi_1\frac{d\xi_2}{dt} - \xi_2\frac{d\xi_1}{dt}$ with (ξ_1, ξ_2) being two linearly independent solutions of a homogeneous ordinary differential equation

$$\frac{d^2\xi}{dt^2} + \gamma^2\xi = 0. \tag{8.60}$$

Combining Eqs. (8.49a), (8.49b) and (8.52) with (8.54a), (8.54b) and (8.55a), (8.55b), we arrive at two families of explicitly exact solutions for the coupled quasi-1D NLS equations (8.48) as

$$\begin{aligned} \psi_1^{(j)}(x,t) &= \sqrt{\lambda}\exp\left[\frac{\delta^2}{2} + \frac{\lambda x(\lambda x + 2\delta)}{2}\right] U^{(j)}(X)\exp\left[i\alpha_1(x,t)\right], \\ \psi_2^{(j)}(x,t) &= \sqrt{\lambda}\exp\left[\frac{\delta^2}{2} + \frac{\lambda x(\lambda x + 2\delta)}{2}\right] V^{(j)}(X)\exp\left[i\alpha_2(x,t)\right], \end{aligned} \tag{8.61}$$

where $U^{(j)}(X)$ and $V^{(j)}(X)$ are given by Eqs. (8.54a), (8.54b) and (8.55a), (8.55b), index $j = 1, 2$, α_1, α_2 satisfy Eq. (8.52), and X, δ, λ satisfy Eqs. (8.52), (8.56), and

(8.59), respectively. It is easy to check that $\lim_{|x|\to+\infty} \psi_j(x,t) = 0$ for $j = 1, 2$, so these two families of exact solutions are localized nonlinear wave solutions.

8.3.2 Dynamics of the Exact Localized Nonlinear Matter Waves

Now, we discuss the dynamics and important physical applications of the exact localized nonlinear wave solutions (8.61) of the coupled quasi-1D NLS equations (8.48). We also propose how to control dynamics of localized nonlinear waves in the two-component BECs by the harmonic potentials and the spatiotemporal inhomogeneous s-wave scattering lengths in future experiments.

Similar to the single BECs cases [3], considering different choices of the parameters, γ_0, and γ_1 in Eq. (8.58), we can also single out several different types of soliton behaviors, such as breathing solitons, resonant solitons, quasi-periodic solitons, and moving solitons. Here we discuss the existence regions of the exact localized nonlinear matter waves by assuming the two constraint conditions $\sigma_1/\sigma_{12} > 0$ and $\sigma_2/\sigma_{12} > 0$ in Eqs. (8.49a), (8.49b) and (8.55a), (8.55b). Without loss of generality, we always assume $g_{11} > g_{22}$. So we have four cases of parameters g_{11}, g_{22}, and g_{12} as given in the following:

Case 1. $g_{11} > g_22 > 0$ and $g_{12} > g_{11}$.
Case 2. $g_{11} > g_{22} > 0$ and $\sqrt{g_{11}g_{22}} > g_{12} > -\sqrt{g_{11}g_{22}}$.
Case 3. $g_{22} < g_{11} < 0$ and $g_{12} > \sqrt{g_{11}g_{22}} > 0$.
Case 4. $g_{11} > 0$, $g_{22} < 0$ and $g_{12} > g_{11}$.

These correspond to four cases of the intracomponent interaction parameters $b11$, $b22$, and intercomponent interaction parameters $b12$ [see Eq. (8.51)] as follows:

Case a. $b_{11} < b_{22} < 0$ and $b_{12} < b_{11}$.
Case b. $b_{11} < b_{22} < 0$ and $-\sqrt{b_{11}b_{22}} < b_{12} < \sqrt{b_{11}b_{22}}$.
Case c. $b_{22} > b11 > 0$ and $b_{12} < -\sqrt{b11b22}$.
Case d. $b_{11} < 0$, $b_{22} > 0$, and $b_{12} < b_{11}$.

This demonstrates the regions where the exact localized nonlinear matter waves of the two-component BECs can exist when the intracomponent interaction parameters b_{11}, b_{22}, and intercomponent interaction parameters b_{12} are spatiotemporally inhomogeneous. Cases a and b show that the two-component BEC with two self-attractive atom–atom interactions supports the exact localized nonlinear matter waves, which is similar to the single-component one [3]. However, for cases c and d, it is surprising to see that the exact novel localized nonlinear matter waves can be formed in the two-component BECs with self-repulsive atom–atom interactions. This is because in the two-component BECs, the attractive interactions from the intercomponent interaction b_{12} (see case c), the second component (see case d), or both, induce an effective attractive interaction in the self-repulsive one.

Now we only consider case c, which denotes two self-repulsive atom–atom interactions and attractive intercomponent interactions. The other cases can be studied in the same way. Let us first provide the experimental parameters for producing the quasi-1D two-component condensates composed of $N_1 = N_2 = 5 \times 10^4$ ^{87}Rb atoms [29], confined in a cigar-shaped trap, with the ratio of the confining frequencies, $\gamma = \omega_x/\omega_\perp$ of order $O(10^{-2})$. Typically, we choose axial frequency $\omega_x = 70\pi$ Hz and radial frequency $\omega_\perp = 800\pi$ Hz [41]. Then when taking the scattering lengths a_{jk} to be of the order of nanometer, it turns out that the normalized trap strength γ is typically of order $O(10^{-2})$. The behavior of the scattering lengths near a Feshbach resonant magnetic field B_0 is typically of the form as $a_s(B) = a[1 + \Delta/(B_0 - B)]$, with a being the asymptotic value of the scattering length far from the resonance, B_0 being the resonant value of the magnetic field, and being the width of the resonance. In our case, the scattering lengths are space and time-dependent, i.e., $a_{jk} = a_{jk}(x, t)$, so the magnetic field B that we will use should vary following the space and time. In real experiments, such magnetic field may be generated by a microfabricated ferromagnetic structure integrated on an atom chip [47]. In the following, within the safe region we will always take the parameters g_{ij} in Eq. (8.51) to be $g_{11} = -1$, $g_{22} = -3$, $g_{12} = 2$.

Breathing Solitons

In order to investigate the dynamics of the explicitly exact solutions (8.61), we take special parameters γ_0, γ_1, and ϵ in Eq. (8.58). When the ratio of the confining frequencies γ of the harmonic potential is time independent, that is, parameters $\gamma_1 = \epsilon = 0$, by solving Eq. (8.60) we have $\lambda = 1/\xi$ with $\xi = [A - (A - B)\cos^2(\gamma_0 t) + \sqrt{AB\gamma_0^2 - 1}\sin(2\gamma_0 t)/\gamma_0]^{1/2}$. Here ξ is the width of the explicitly exact solutions (8.61) and $\sqrt{\lambda}$ is its amplitude. We further suppose $c_1 = c_2 = 0$, i.e., $\delta = 0$. When the cigar-shaped trap with axial frequency $\omega_x = 70\pi$ Hz and radial frequency $\omega_\perp = 800\pi$ Hz is considered, the ratio of the confining frequencies $\gamma = \omega_x/\omega_\perp = 7/80$. So from Eq. (8.58) the parameter $\gamma_0 = \gamma = 7/80$. In order to determine parameters A and B in Eq. (8.59), we further consider the initial condition of Eq. (8.57) as $\xi(0) = \sqrt{3}$ and $\frac{d\xi(0)}{dt} = 1/80$.

We now investigate how the time- and space-modulated nonlinearities b_{ij} control the dynamics of the localized nonlinear matter waves. In the case of parameters $\gamma_1 = \epsilon = \delta = 0$, the nonlinearities in Eq. (8.48) become $b_{11} = g_{11}\theta(t, x)$, $b_{12} = g_{12}\theta(t, x)$, $b_{22} = g_{22}\theta(t, x)$ with $\theta(t, x) = -\frac{\lambda}{2}\exp\left[-3\lambda^2 x^2\right]$, which are space localized and time-periodic Gaussian nonlinearity, as shown in Fig. 8.5. In all figures of this Section, the units of space length and time are 5.34 μm and 0.4 ms, respectively. In real BECs experiments, the Gaussian nonlinearity can be generated by controlling the Feschbach resonances optically or magnetically using a Gaussian beam.

In Fig. 8.6, we show the evolution of condensate density of the wave functions $\psi_1^{(1)}$ and $\psi_1^{(2)}$ in (8.61) with the above parameters. The density profiles of the other component $\psi_2^{(1)}$ and $\psi_2^{(2)}$ can also be plotted in the same way, which is similar to that of $\psi_1^{(1)}$ and $\psi_1^{(2)}$ with different amplitudes. Figure 8.6a and b demonstrate the density profiles of the wave function $\psi_1^{(1)}$ for $n = 0$, 1, respectively, and Fig. 8.6c and d

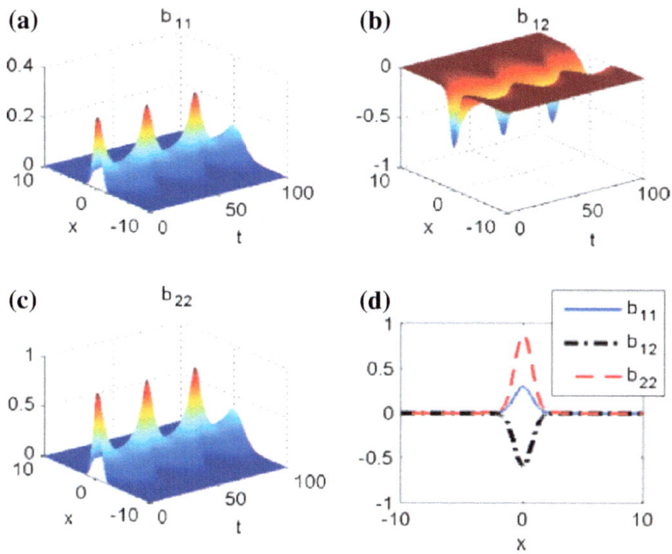

Fig. 8.5 **a–c** Demonstrate the examples of spatiotemporal-dependent nonlinearities b_{11}, b_{12}, and b_{22} given by Eq. (8.51) with parameters $g_{11} = -1, g_{22} = -3, g_{12} = 2, \gamma_1 = \epsilon = 0$, and $\gamma_0 = 7/80$. **d** Demonstrates the nonlinearities $b11$, $b12$, and b_{22} for $t = 0$. Here and after, the units of space length and time are 5.34 μm and 0.4 ms, respectively. Reprint from Ref. [48], Copyright 2018, with permission from American Physical Society

demonstrate the density profiles of the wave function $\psi_1^{(2)}$ for $m = 1, 2$, respectively. Figure 8.6e demonstrates the width $\xi = 1/\lambda$ and amplitude $\sqrt{\lambda}$ of the wave functions. It is observed that the localized nonlinear matter waves are space localized and time periodic, which are usually called breathing solitons. Therefore, we have constructed some exact multisoliton breathing solutions of the quasi-1D two-component BECs. Here n and m are the order of the breathing solitons. It is also observed that the amplitude and width of the localized nonlinear matter waves vary periodically with respect to time.

Quasibreathing Solitons

When the ratio of the confining frequencies γ of the harmonic potential is time-dependent, we can select proper parameters γ_0, γ_1, and ϵ in Eq. (8.58) to formulate a type of quasibreathing solitons in the two-component BECs. We especially choose parameters $\gamma_0 = 7/80$, $\gamma_1 = \sqrt{2}$, and $\epsilon = 1/2$. In this case, the two solutions of Eq. (8.60) are Mathieu functions $\xi_1 = C_M(49/3200, -1/2, \sqrt{2}t/2)$, $\xi_2 = S_M(49/3200, -1/2, \sqrt{2}t/2)$, where C_M and S_M refer to the cosine and sine Mathieu functions, respectively, which are in the stability region of Eq. (8.60). With these in mind, we can construct the function λ from Eq. (8.59). In this section, we still assume parameter $\delta = 0$. To determine the unknown parameters A and B in λ, we consider the initial condition of Eq. (8.57) to be $\xi(0) = \sqrt{3}$ and $\frac{d\xi(0)}{dt} = \sqrt{6}/6$.

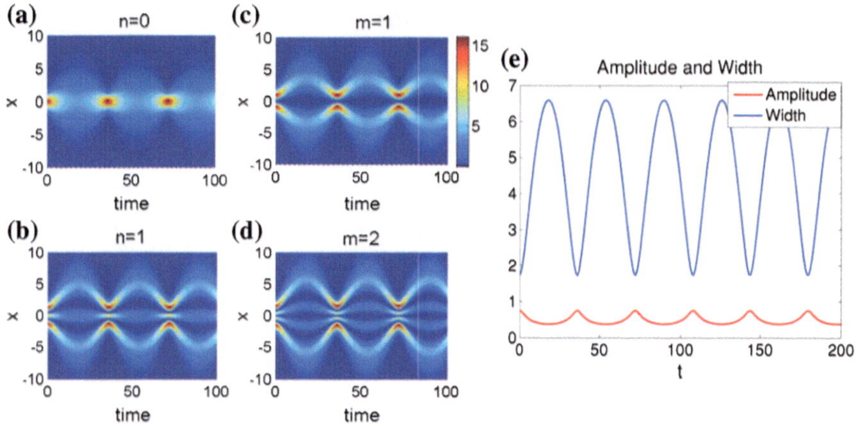

Fig. 8.6 Dynamics of breathing solitons in two-component BECs in quadratic potential with spatiotemporal-dependent nonlinearities shown in Fig. 8.5. **a** and **b** Demonstrate the evolution of condensate density $\left|\psi_1^{(1)}\right|^2$ for order $n = 0$ and 1, respectively. **c** and **d** Demonstrate examples of condensate density $\left|\psi_1^{(2)}\right|^2$ for order $m = 1$ and 2, respectively. **e** demonstrates the width $\xi(t) = 1/\lambda(t)$ (upper line) and amplitude $\sqrt{\lambda(t)}$ (lower line) of wave functions. The parameters are $\gamma_1 = \epsilon = c_1 = c_2 = 0$ and $\gamma_0 = 7/80$. The initial data for Eq. (8.57) are $\xi(0) = \sqrt{3}$ and $\frac{d\xi(0)}{dt} = 1/80$. Reprint from Ref. [48], Copyright 2018, with permission from American Physical Society

In this case, the nonlinearities in Eq. (8.48) are still space- and time-dependent and the external potential is time-dependent. Next, we study how these nonlinearities control the dynamics of the localized nonlinear matter waves. In Fig. 8.7, we show the development of density profiles for the wave functions $\psi_1^{(1)}$ and $\psi_2^{(2)}$ (8.61). The density profiles of the wave functions $\psi_1^{(2)}$ and $\psi_2^{(1)}$ can also be plotted in the same way, which is very similar to that of $\psi_1^{(1)}$ and $\psi_2^{(2)}$ except the amplitudes. Figure 8.7a and b demonstrate the density profiles of the wave functions $\psi_1^{(1)}$ for order $n = 0$, and Fig. 8.7c and d demonstrate the density profiles of the wave functions $\psi_2^{(2)}$ for order $m = 1$. It is observed that the localized nonlinear matter waves are space localized and time quasi-periodic, which are quasibreathing solitons with orders n or m. We have constructed some exact quasibreathing soliton solutions of the quasi-1D two-component BECs. It is also observed that the amplitude and width of the quasibreathing solitons vary quasi-periodically with respect to time.

Resonant Solitons

If we let the parameters $\delta = 0$, $\gamma_0 = 7/80$, $\gamma_1 = 4$, and $\epsilon = 1/2$, we are in the instability region of Eq. (8.57). However, the solutions of Eq. (8.60) are also Mathieu functions $\xi_1 = C_M(49/25600, -1/16, 2t)$, $\xi_2 = S_M(49/25600, -1/16, 2t)$. After some algebra, we can also get the analytical expression of function λ. In order to

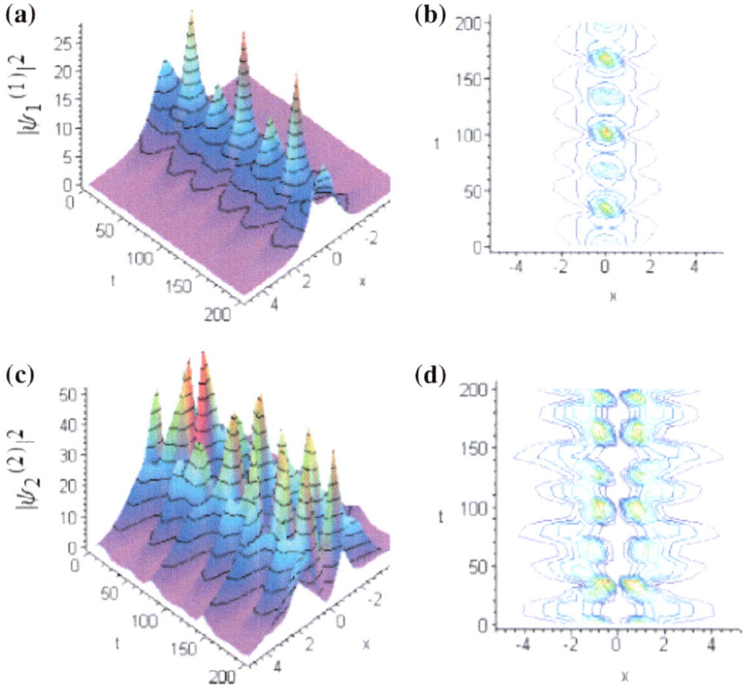

Fig. 8.7 Examples of quasibreathing solitons in two-component BECs in quadratic potential with spatiotemporal-dependent nonlinearities in Eq. (8.51). **a** and **b** Demonstrate the evolution of condensate density $\left|\psi_1^{(1)}\right|^2$ for order $n = 0$. **c** and **d** Demonstrate the evolution of condensate density $\left|\psi_2^{(2)}\right|^2$ for order $m = 1$. The parameters are $c_1 = c_2 = 0$, $\gamma_0 = 7/80$, $\gamma 1 = \sqrt{2}$, and $\epsilon = 1/2$. The initial data for Eq. (8.57) are $\xi(0) = \sqrt{3}$ and $\frac{d\xi(0)}{dt} = \sqrt{6}/6$. Reprint from Ref. [48], Copyright 2018, with permission from American Physical Society

determine the unknown parameters A and B in λ, we consider the initial condition of Eq. (8.57) as $\xi(0) = 1$ and $\frac{d\xi(0)}{dt} = 1$.

Figure 8.8a, b, c, and d describe the evolution of the condensate density profiles for the wave functions $\psi_1^{(1)}$ with order $n = 0$ and $\psi_2^{(2)}$ with order $m = 1$, respectively. It is shown that the density wave packets are space localized and time resonant, which have the behaviors of resonant solitons. At the beginning, resonant solitons are localized nonlinear matter waves with low amplitude and large width. After some time, their amplitudes become high but widths become small; as time goes on their amplitudes decrease and the widths increase gradually. The nonlinear matter waves demonstrate transitory resonant soliton behaviors. These resonant soliton behaviors come from the coaction of the time-dependent harmonic potential and the spatiotemporal inhomogeneous interactions.

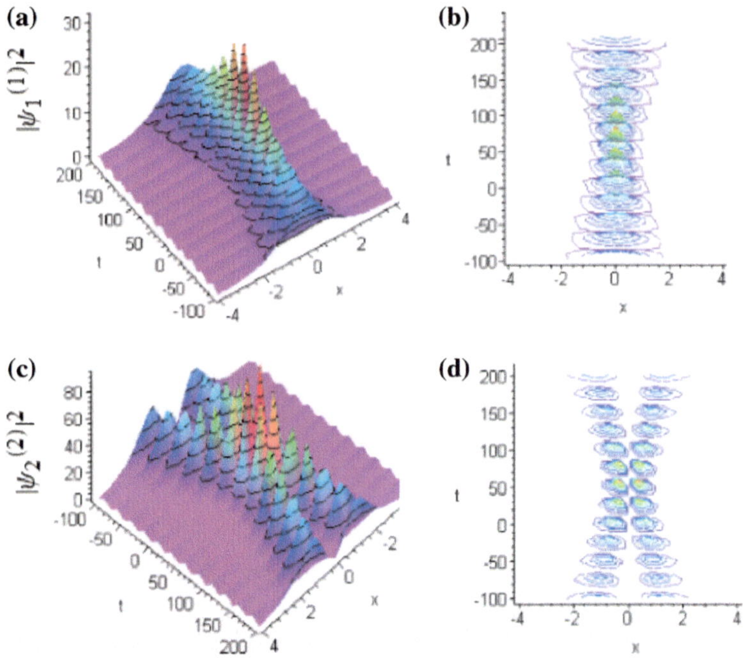

Fig. 8.8 Examples of resonant solitons in two-component BEC in quadratic potential with spatiotemporal-dependent nonlinearities in Eq. (8.51). **a** and **b** Describe the evolution of condensate density $\left|\psi_1^{(1)}\right|^2$ for order $n = 0$. **c** and **d** Describe the evolution of condensate density $\left|\psi_2^{(2)}\right|^2$ for order $m = 1$. The parameters here are $c_1 = c_2 = 0$, $\gamma_0 = 7/80$, $\gamma_1 = 4$, and $\epsilon = 1/2$. The initial data for Eq. (8.57) are $\xi(0) = 1$ and $\frac{d\xi(0)}{dt} = 1$. Reprint from Ref. [48], Copyright 2018, with permission from American Physical Society

Moving Solitons

Finally, we consider the case of $\delta = 0$, which will produce more nonlinear matter waves. The time-dependent function δ can be determined by Eq. (8.56) with Eq. (8.59). In this case, the nonlinearities in Eq. (8.48) become more complicated, the amplitude of the nonlinear matter wave seems more complex, i.e., $\sqrt{\lambda}\exp\left[\delta^2/2\right]$, and the center of the solitons can move following the time because of variable $X = \frac{\sqrt{\pi}}{2}\mathrm{erf}(\lambda x + \delta)$. So for $\delta \neq 0$, our localized nonlinear matter waves are localized moving solitons. By choosing different parameters γ_0, γ_1, and ϵ one can again derive moving breathing soliton solutions, moving quasibreathing soliton solutions, and moving resonant soliton solutions of the quasi-1D two-component BECs (8.48). For brevity, we assume the ratio of the confining frequencies γ of the harmonic potential is time independent to show the dynamics of the moving breathing solitons. To do so, we still choose $\gamma_0 = 7/80$ and $\gamma 1 = \epsilon = 0$.

Figure 8.9a and b describe the evolution of the density profiles for the wave function $\psi_1^{(1)}$ with orders $n = 0, 1$, respectively, and Fig. 8.9c and d describe the evolution of the density profiles for the wave function $\psi_1^{(2)}$ with orders $m = 1, 2$, respectively.

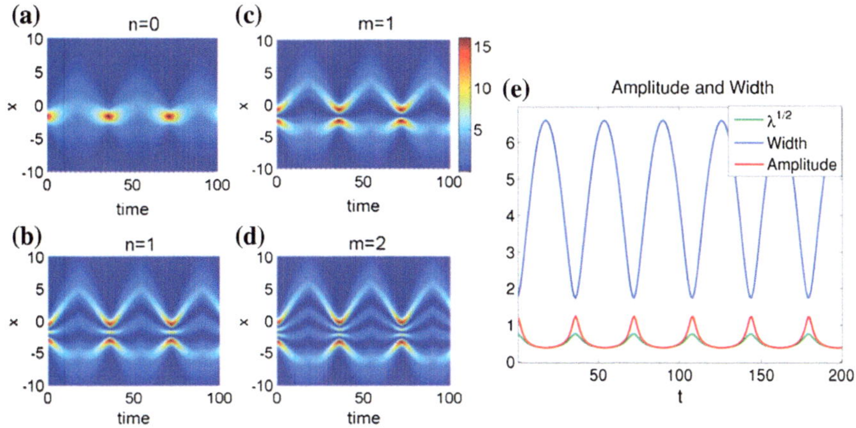

Fig. 8.9 Dynamics of moving breathing solitons in two-component BECs in a quadratic potential with spatiotemporal-dependent nonlinearities in Eq. (8.51). **a** and **b** Describe the evolution of condensate density $\left|\psi_1^{(1)}\right|^2$ for order $n = 0$ and 1, respectively. **c** and **d** Describe the examples of condensate density $\left|\psi_1^{(2)}\right|^2$ for order $m = 1$ and 2, respectively. **e** describes the width $\xi(t) = 1/\lambda(t)$ (upper line), amplitude $\sqrt{\lambda(t)}\exp\left[\delta^2/2\right]$ (middle line), and function $\sqrt{\lambda(t)}$ (lower line). The parameters are the same as those in Fig. 8.6 except for $c_1 = c_2 = 1/2$. Reprint from Ref. [48], Copyright 2018, with permission from American Physical Society

Figure 8.9e demonstrates the shapes of the function $\sqrt{\lambda}$, the width $\xi = 1/\lambda$, and the amplitude $\sqrt{\lambda}\exp\left[\delta^2/2\right]$ of the moving breathing solitons. It is found that the localized nonlinear matter waves are space localized and time periodically moving. The amplitude and width of the localized nonlinear matter waves vary periodically with respect to time. It is also observed that the amplitudes of the moving breathing solitons are higher than the breathing solitons in Fig. 8.6.

8.3.3 Numerical Simulations

We now turn to the numerical study the two-component NLS equations with external potentials (8.48). First, the dynamical stability of the exact localized nonlinear wave solutions (8.61) in response to perturbation by initial stochastic noise is done by direct numerical simulations.

Figure 8.10a and b show the numerical results for the evolution of a breathing soliton solution (8.61) for $j = 1$ and $n = 0$ with initial Gaussian noise. The other parameters are $g_{11} = 1, g_{22} = -3, g_{12} = 5, \gamma_0 = 7/80, \gamma_1 = \epsilon = 0$, and $c_1 = c_2 = 0$, i.e., $\delta = 0$. Figure 8.11a and b show the numerical results for the evolution of a moving breathing soliton solution (8.61) for $j = 1$ and $n = 0$ also with initial Gaussian noise. The other parameters are the same as Fig. 8.10 but $c_1 = c_2 = 1/2$, i.e., $\delta \neq 0$. Here the Gaussian noise is included by adding to the first component a Gaussian-distributed

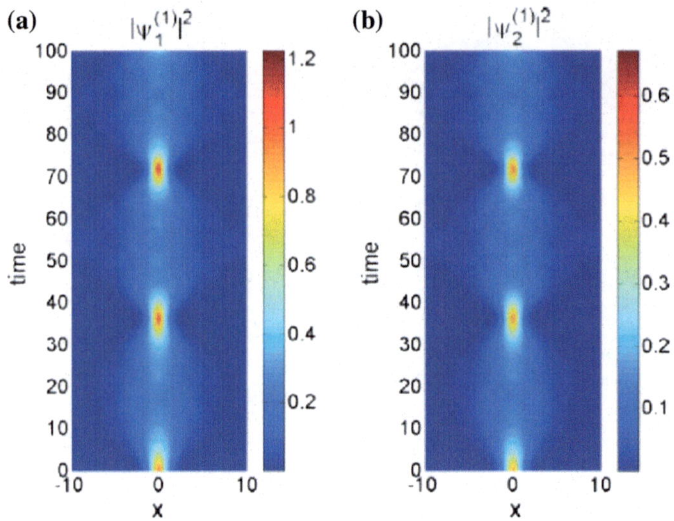

Fig. 8.10 The evolution of a breathing soliton solution (8.61) for $j = 1$ and $n = 0$ with an initial Gaussian noise of level 5%. The other parameters are $g_{11} = 1$, $g_{22} = -3$, $g_{12} = 5$, $\gamma_0 = 7/80$, $\gamma_1 = \epsilon = 0$, and $c_1 = c_2 = 0$, i.e., $\delta = 0$. Reprint from Ref. [48], Copyright 2018, with permission from American Physical Society

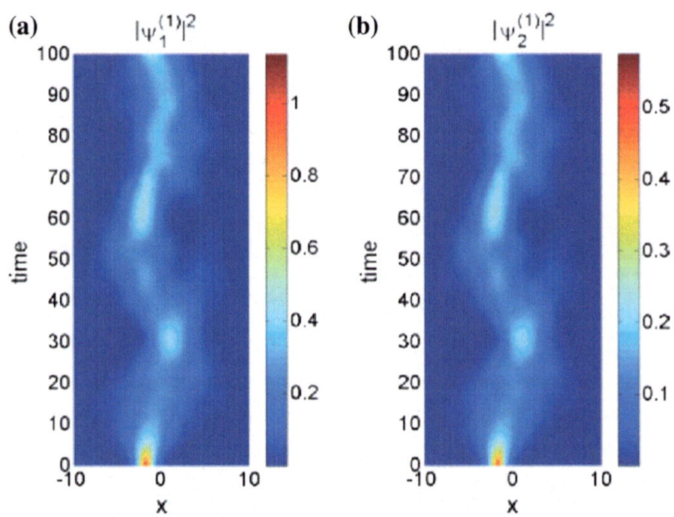

Fig. 8.11 The evolution of a moving breathing soliton solution (8.61) for $j = 1$ and $n = 0$ with an initial Gaussian noise of level 5%. The other parameters are $g_{11} = 1, g_{22} = -3, g_{12} = 5, \gamma_0 = 7/80$, $\gamma_1 = \epsilon = 0$, and $c_1 = c_2 = 1/2$, i.e., $\delta = 0$. It is shown that the moving breathing solitons are unstable. Reprint from Ref. [48], Copyright 2018, with permission from American Physical Society

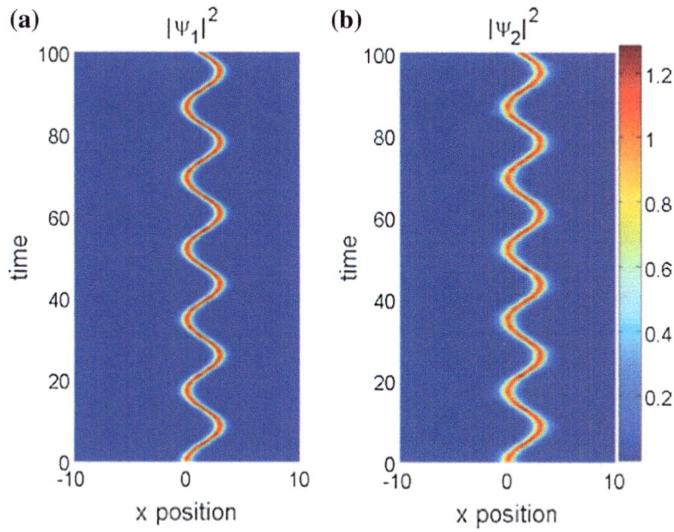

Fig. 8.12 Evolutions of stable vector bright solitons in two-component BECs in quadratic potential with space-periodic nonlinearities $b11 = -5\rho$, $b12 = -2\rho$, $b22 = -4.5\rho$, with $\rho = 0.1 + 0.01 \sin x$. Reprint from Ref. [48], Copyright 2018, with permission from American Physical Society

random number with mean $1/2$ and unit variance, and the second component with mean $2/3$ and unit variance multiplied level 5%. It is observed that the exact solution (8.61) for $j = 1$ and $n = 0$ is dynamically stable for $\delta = 0$, i.e., one order breathing soliton is stable, and the exact solution (8.61) for $j = 1$ and $n = 0$ is dynamically unstable for $\delta \neq 0$, i.e., one order moving breathing soliton is unstable.

We have tested the dynamical stability of the exact localized nonlinear wave solutions (8.61) for other parameters by adding certain initial stochastic noises with various intensities, such as Gaussian-distributed noise and uniform distribution noise. Our numerical calculations show that only for order $n = 0$ (ground state) is the breathing soliton solution (8.61) with $\delta = 0$ dynamically stable, while the moving soliton solution (8.61) with $\delta \neq 0$ is dynamically unstable for all n.

Second, we illustrate the effect of the general time- and space-modulated nonlinearities on the nonlinear evolution of the condensates. We have earlier shown that the exact localized nonlinear wave solutions (8.61) exist only for special interaction parameters in Eq. (8.51). But for other cases of interaction parameters b_{jk}, the two-component NLS equations (8.48) are nonintegrable. Thus we have to appeal to the numerical simulation. Here, we still consider the quasi-one-dimensional geometry characterized by the aspect ratio of the confining frequencies $\gamma = \omega_x/\omega_\perp = 7/80$.

In what follows, we investigate the effect of the interaction parameters $b11 = -5\rho$, $b_{12} = -2\rho$, $b_{22} = -4.5\rho$ with three cases of special functions ρ. In the first case, we study the situation in which the scattering lengths are space periodically modulated nonlinearities with $\rho = 0.1 + 0.01 \sin x$. As shown in Fig. 8.12, we have performed numerical simulations in the case of the so-called vector bright soliton solutions

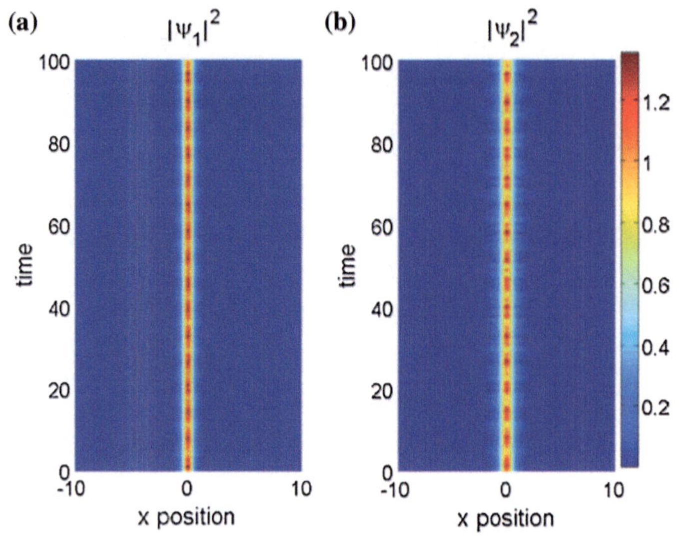

Fig. 8.13 Evolutions of stable breathing states in two-component BECs in quadratic potential with time-periodic and space-Gaussian nonlinearities $b11 = -5\rho, b12 = -2\rho, b22 = -4.5\rho$, with $\rho = 0.1 + 0.01\sin(t)\exp\left[-x^2\right]$. Reprint from Ref. [48], Copyright 2018, with permission from American Physical Society

with initial soliton amplitude $A = 3$ for the first component and $A = 1$ for the second component. It is observed that the space-periodic nonlinearities can support stable snake-shaped vector bright solitons. In the second case, we demonstrate the effect of time-periodic and space-Gaussian nonlinearities with $\rho = 0.1 + 0.01\sin t \exp\left[-x^2\right]$. It is shown that these nonlinearities can support stable breathing states of the two-component BECs, see Fig. 8.13. The third case demonstrates the effect of space weakly linear-modulated nonlinearities with $\rho = 0.1 + 0.001x$. As shown in Fig. 8.14, it is seen that these nonlinearities can also support stable snake-shaped vector bright solitons of the two-component BECs.

8.4 Matter-Wave Solitons in Heteronuclear Two-Species BECs

With the help of a combination of the homogeneous balance principle and the F-expansion technique, we investigate here the dynamics of matter-wave soliton pairs of two-species heteronuclear atomic BECs with tunable interactions and harmonic potentials.

Let us consider a two-species interacting atomic BEC trapped in external potentials during the simultaneous evaporative cooling process. In the zero-temperature approximation, the systems can be well described by the following two-component

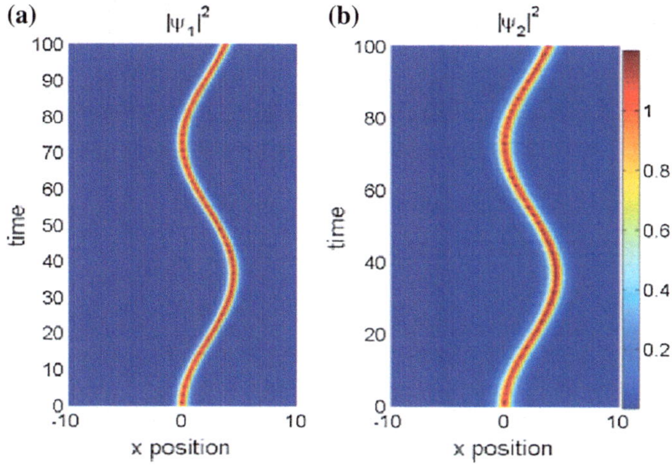

Fig. 8.14 Evolutions of stable vector bright solitons in two-component BECs in quadratic potential with space weakly linear nonlinearities $b_{11} = -5\rho,\ b_{12} = -2\rho,\ b_{22} = -4.5\rho$, with $\rho = 0.1 + 0.001x$. Reprint from Ref. [48], Copyright 2018, with permission from American Physical Society

NLS equations with external potential

$$i\hbar\frac{\partial\Psi_1}{\partial t} = \left[-\frac{\hbar^2}{2m_1}\nabla^2 + U_{11}|\Psi_1|^2 + U_{12}|\Psi_2|^2 + V_{ext}^{(1)} + i\hbar\Gamma_1\right]\Psi_1,$$
$$i\hbar\frac{\partial\Psi_2}{\partial t} = \left[-\frac{\hbar^2}{2m_2}\nabla^2 + U_{22}|\Psi_2|^2 + U_{21}|\Psi_1|^2 + V_{ext}^{(2)} + i\hbar\Gamma_2\right]\Psi_2,$$

where the condensate wave functions are normalized by the particle numbers $N_j = \int d\mathbf{r}\,|\Psi_j|^2$ $(j = 1, 2)$ and $U_{jj} = 4\pi\hbar^2 a_{jj}/m_j$ and $U_{12} = U_{21} = 2\pi\hbar^2 a_{12}/m_R$ represent intraspecies and interspecies interactions strengths, respectively, with a_{jk} $(k = 1, 2)$ being the corresponding scattering lengths and $m_R = m_1 m_2/(m_1 + m_2)$. The external trapped potentials are assumed to be $V_{ext}^{(j)} = m_i[\omega_{jx}^2 x^2 + \omega_{j\perp}^2(y^2 + z^2)]/2$. Here we assume that the condensates are trapped in elongated and regulatable harmonic oscillating external potentials; namely, $\omega_{i,\perp} >> \omega_{i,x}$, which means that their transverse motion is frozen to the ground state of the transverse harmonic trapping potentials. Γ_j are gain (or loss) terms which give rise to the growth of the two condensates and which can be used to describe the evaporative cooling process.

Hence the system is quasi-one-dimensional. Integrating out the transverse coordinates, the resulting equations for the axial wave. Functions $\psi_j(x, t)$ in dimensionless form can be written as

$$i\frac{\partial\psi_1}{\partial t} = \left[-\frac{1}{2}\frac{\partial^2}{\partial x^2} + g_{11}|\psi_1|^2 + g_{12}|\psi_2|^2 + \frac{\lambda_1^2}{2}x^2 + i\gamma_1\right]\psi_1, \quad (8.62a)$$

$$i\frac{\partial\psi_2}{\partial t} = \left[-\frac{1}{2}\frac{\partial^2}{\partial x^2} + g_{22}|\psi_2|^2 + g_{21}|\psi_1|^2 + \frac{\lambda_2^2}{2}x^2 + i\gamma_2\right]\psi_2, \quad (8.62b)$$

where x and t are the spatial and temporal coordinates measured in units $a_0 = \sqrt{\hbar/(m_1\omega_{1\perp})}$ and $1/\omega_{1\perp}$, respectively. The other parameters in Eqs. (8.62a) and (8.62b) are defined as $g_{11} = 2a_{11}$, $g_{22} = 2a_{22}m_1\eta/m_2$, $g_{12} = g_{21} = 2m_1a_{12}/[(1+\eta)m]$, $\eta = \omega_{2\perp}/\omega_{1\perp}$, $\lambda_j = \omega_{jx}/\omega_{1\perp}$, $\gamma_j = \Gamma_j/\omega_{1\perp}$, and $\kappa = m_1/m_2$. In addition, $a_{11}, a_{22}, a_{12}, \lambda_j$, and γ_j are controllable time-dependent parameters in experiments.

8.4.1 Exact Soliton-Pair Solutions

To investigate the dynamics of matter-wave solitons in the BECs described by system (8.62a) and (8.62b), we start by finding exact soliton-pair solutions of this system. As a starting point, we write $\psi_j(x,t)$ in their polar form:

$$\psi_j(x,t) = A_j(x,t)\exp\left[iB_j(x,t)\right], \quad j = 1,\ 2, \tag{8.63}$$

where $A_j(x,t)$ and $B_j(x,t)$ are two real functions of variables x and t, $j = 1,\ 2$. Inserting ansatz (8.63) into the NLS system (8.62a) and (8.62b) yield the following system for A_j and B_j

$$\frac{\partial A_1}{\partial t} + \frac{\partial A_1}{\partial x}\frac{\partial B_1}{\partial x} + \frac{1}{2}A_1\frac{\partial^2 B_1}{\partial x^2} - \gamma_1 A_1 = 0, \tag{8.64a}$$

$$-A_1\frac{\partial B_1}{\partial t} + \frac{1}{2}\frac{\partial^2 A_1}{\partial x^2} - \frac{1}{2}A_1\left(\frac{\partial B_1}{\partial x}\right)^2 - g_{11}A_1^3 - g_{12}A_2^2A_1 - \frac{1}{2}\lambda_1^2x^2A_1 = 0, \tag{8.64b}$$

$$\frac{\partial A_2}{\partial t} + \kappa\frac{\partial A_2}{\partial x}\frac{\partial B_2}{\partial x} + \frac{\kappa}{2}A_2\frac{\partial^2 B_2}{\partial x^2} - \gamma_2 A_2 = 0, \tag{8.64c}$$

$$-A_2\frac{\partial B_2}{\partial t} + \frac{\kappa}{2}\frac{\partial^2 A_2}{\partial x^2} - \frac{\kappa}{2}A_2\left(\frac{\partial B_2}{\partial x}\right)^2 - g_{22}A_2^3 - g_{21}A_1^2A_2 - \frac{1}{2}\lambda_2^2x^2A_2 = 0. \tag{8.64d}$$

Based on the homogeneous balance principle [53] and the F-expansion technique [30], we set $A_j(x,t) = f_j(t)F(\theta) + g_j(t)/F(\theta)$ with $\theta = k(t)x + \omega(t)$. Here $k(t)$ and $\omega(t)$ are the inverse spatial width and velocity of the wave, respectively. The functions $fi(t)$ and $gi(t)$ are parameters to be determined below. In the expressions for A_j, $F(\theta)$ denotes a Jacobi elliptic function, which satisfies the following EODE

$$\left(\frac{dF}{d\theta}\right)^2 = c_0 + c_2F^2(\theta) + c_4F^4(\theta), \tag{8.65}$$

where c_0, c_2, and c_4 are real constants related to the elliptic modulus m of the Jacobi elliptic functions (JEFs; see Table 8.1). In addition, it is assumed that $B_j(x,t) = a_j(t)x^2 + b_j(t)x + e_j(t)$, where a_j, b_j, and e_j are time-dependent chirp functions, wave numbers, and phase shifts, respectively, which will be determined self-consistently. Then we substitute the expressions of Ai and Bi into Eqs. (8.64a)–(8.64d) and collect the coefficients of the polynomial $x^pF^q(\theta)$, where p and q are nonnegative integers. One can determine those parameters included in the expressions of $A_j(x,t)$ and $B_j(x,t)$ by taking the coefficients of $F(\theta)$ as zero:

$$\begin{aligned}
k(t) &= D_1 \exp\left[-2\int a(t)dt\right],\\
\omega(t) &= -D_2 D_1 \exp\left[-4\int a(t)dt\right] + C_4,\\
b_1(t) &= D_2 \exp\left[-2\int a(t)dt\right], \quad b_2(t) = \frac{b_1(t)}{\kappa},\\
f_j(t) &= D_{3j} \exp\left[\int \left(\gamma_j(t) - a(t)\right) dt\right],\\
g_j(t) &= D_{5j} \exp\left[\int \left(\gamma_j(t) - a(t)\right) dt\right],\\
e_j(t) &= \int h_j(t) \exp\left[-2\int a(t)dt\right] dt,
\end{aligned} \tag{8.66}$$

where $a(t) = a_1(t) = \kappa a_2(t)$ are the chirp functions. D_1 and D_{3j} $(j = 1, 2)$ are real constants and represent the initial inverse spatial width and the initial amplitudes of the waves, respectively. D_2 and D_4 are arbitrary real constants. The value of D_{5j} will be discussed for the case of soliton solutions (see below). In addition, the expressions of $h_j(t)$ in the phase shifts are given by

$$\begin{aligned}
h_1(t) &= D_{11} g_{12}(t) \exp\left[2\int \gamma_2(t)\right] dt - 3D_5 1 D_{31} g_{11}(t) \exp\left[2\int \gamma_1(t)dt\right]\\
&\quad + \frac{1}{2}\left(c_2 D_1^2 - D_2^2\right) \exp\left[-2\int a(t)dt\right],\\
h_2(t) &= D_{22} g_{12}(t) \exp\left[2\int \gamma_1(t)dt\right] - 3D_{32} D_{52} g_{22}(t) \exp\left[2\int \gamma_2(t)dt\right]\\
&\quad + \frac{\kappa}{2}\left(c_2 D_1^2 - \frac{D_2^2}{\kappa 2}\right) \exp\left[-2\int a(t)dt\right].
\end{aligned} \tag{8.67}$$

Here one defines $D_{11} = -D_{32}(2D_{52}D_{31} + D_{32}D_{51})/D_{31}$ and $D_{22} = -D_{31}(2D_{32}D_{51} + D_{52}D_{31})/D_{32}$.

Furthermore, it is worth noting that the chirp function satisfies the following standard Ricatti-type equation

$$\frac{da(t)}{dt} + 2a^2(t) + \frac{\lambda^2(t)}{2} = 0, \tag{8.68}$$

with $\lambda(t) = \lambda_1(t) = \kappa\lambda_2(t)$. It can thus be seen that the external potentials for the two components are significantly different in heteronuclear systems, which is caused by the mass of two components being different. In homonuclear systems, due to the mass ratio of the two components being unity (i.e., $\kappa = 1$), then the potentials for the two components are identical.

At the same time, the existence of exact solutions requires the following set of constraint conditions

$$D_{3j}^2 \exp\left\{2\int [\gamma_j(t) - a(t)]dt\right\} = c_4\Delta_j k^2(t), \tag{8.69a}$$

$$D_{5j}^2 \exp\left\{2\int [\gamma_j(t) - a(t)]dt\right\} = \epsilon^2 c_0\Delta_j k^2(t), \tag{8.69b}$$

where $\epsilon = 0, \pm 1$ and

$$\Delta_1 = \frac{g_{22}(t) - \kappa g_{12}(t)}{g_{22}(t)g_{11}(t) - g_{12}(t)g_{21}(t)}, \tag{8.70a}$$

$$\Delta_2 = \frac{\kappa g_{11}(t) - g_{12}(t)}{g_{22}(t)g_{11}(t) - g_{12}(t)g_21(t)}. \tag{8.70b}$$

The similar conditions are usually called integrable conditions, which are quite complicated in the general case. From Table 8.1, Eqs. (8.62a) and (8.62b) have various solutions. When $0 < m < 1$, the JEFs are periodic traveling wave solutions. When $m \to 0$, the periodic traveling wave solutions evolve into the periodic trigonometric functions. When $m \to 1$, the periodic traveling wave solutions become the time-dependent soliton solutions of the two-component NLS equations.

In the following, we mainly focus on the soliton solutions by taking the elliptic modulus m as 1. According to the conditions of Eqs. (8.69a) and (8.69b), c_0, c_4, and Δ_j take the same sign. After some careful analysis, we find that only the first three lines in Table 8.2 are able to give out soliton-pair solutions. In the first line of Table 8.2, because $c_0 = 0$ and $F(\theta)$ has the tanh form, one takes $\epsilon = 0$ to avoid a possible divergence caused by $1/F(\theta)$ in the second term of $A_j(x, t)$. As a consequence, D_{5j} are zero, thus g_j are zero, and a dark–dark soliton-pair solution can be obtained. In the second and third lines, because c_0 is zero, the values of D_{5j} are bound to zero according to the condition of Eq. (8.69b) and a bright–bright soliton-pair solution is obtained. Thus, for these two classes of soliton-pair solutions, only Eq. (8.69a) survives, which is the so-called integrable condition. Finally, two types of soliton-pair solutions can be written as

$$\begin{aligned} \psi_1(x,t) &= f_1(t)F(\theta)\exp\left\{i[a(t)x^2 + b_1(t)x + e_1(t)]\right\}, \\ \psi_2(x,t) &= f_2(t)F(\theta)\exp\left\{i[a(t)x^2 + b_2(t)x + e_2(t)]\right\}. \end{aligned} \tag{8.71}$$

For the two types of soliton-pair solutions (8.71), Eq. (8.69a) proposes explicit constraint conditions on the interaction parameters g_{jk} and κ, as shown in Table 8.3. Note that, for the heteronuclear system, the phase shifts for the two components are different, which is in contrast to the case of homonuclear systems. We thus obtain the symbiotic solitons in heteronuclear two-component systems with the time-dependent interactions and external potentials. Considering the possible

Table 8.2 All possible Jacobi elliptic functions (JEFs, fifth column) satisfy Eq. (8.65), in which the values of c_0, c_2 and c_4 are determined uniquely for a given JEF form. Heremdenotes the elliptic modulus of the JEFs. The last two columns represent the limiting cases of the JEFs. When $m \to 0$, the JEFs reduce to the trigonometric functions and when $m \to 1$, the JEFs become hyperbolic functional forms or unity (the last two cases)

	c_0	c_2	c_4	$F(\theta)$	$m = 0$	$m = 1$
1	1	$-(1+m)$	m	sn	sin	tanh
2	$1-m$	$2m-1$	$-m$	cn	cos	sech
3	$m-1$	$2-m$	-1	dn	1	sech
4	m	$-(1+m)$	1	ns	cosec	coth
5	$-m$	$2m-1$	$1-m$	nc	sec	cosh
6	-1	$2-m$	$m-1$	nd	1	cosh
7	1	$2-m$	$1-m$	sc	tan	sinh
8	$1-m$	$2-m$	1	cs	cot	csch
9	1	$-(1+m)$	m	cd	cos	1
10	m	$-(1+m)$	1	dc	sec	1

Table 8.3 Explicit constraint conditions on the "intra" and "inter" interactions and the parameter κ for the bright–bright and dark–dark soliton-pair solutions and their phase shifts

Soliton pairs	Existence	Phase shifts
Bright–bright	$g_{22} < \kappa g_{12},\ \kappa g_{11} < g_{12},\ g_{22}g_{11} > g_{12}^2$ or $g_{22} > \kappa g_{12},\ \kappa g_{11} > g_{12},\ g_{22}g_{11} < g_{12}^2$	$e_1 = \frac{1}{2}\left(D_1^2 - D_2^2\right)\int e^{\int -4a(t)dt}dt$ $e_2 = \frac{\kappa}{2}\left(D_1^2 - \frac{D_2^2}{\kappa^2}\right)\int e^{\int -4a(t)dt}dt$
Dark–dark	$g_{22} < \kappa g_{12},\ \kappa g_{11} < g_{12},\ g_{22}g_{11} < g_{12}^2$ or $g_{22} > \kappa g_{12},\ \kappa g_{11} > g_{12},\ g_{22}g_{11} > g_{12}^2$	$e_1 = -\left(D_1^2 - \frac{D_2^2}{2}\right)\int e^{\int -4a(t)dt}dt$ $e_2 = -\kappa\left(D_1^2 - \frac{D_2^2}{2\kappa^2}\right)\int e^{\int -4a(t)dt}dt$

preparation in experiments, we can first load the two condensates into the harmonic cigar trap (during the simultaneous evaporative cooling of a mixture of two gases, with this case corresponding to $\gamma = 0$). Then we use techniques like phase imprinting to prepare the initial soliton-pair and change the trap frequencies, together with the scattering lengths changed by the Feshbach resonance, according to Eq. (8.68) and the integrable conditions of Eq. (8.69a). The bright–bright or dark–dark soliton with time-varying amplitudes and speeds may appear in the systems. Hence their dynamical behavior can be controlled by tuning experimentally the interaction and harmonic external potential parameters under the constraint conditions, which is explicitly discussed later.

8.4.2 About the Integrable Condition

The integrable condition Eq. (8.69a) provides many possibilities to tune the interaction and external potential parameters. According to the condition, a relationship between two components can be written as

$$\frac{\Delta_1}{\Delta_2} = \frac{D_{31}^2}{D_{32}^2} \exp\left\{2 \int [\gamma_1(t) - \gamma_2(t)]\, dt\right\}, \tag{8.72}$$

from which one can conclude that Δ_1 and Δ_2 have the same sign. In the following we explicitly discuss two classes of interaction parameter cases.

First, we consider that all interaction parameters are tuned homogeneously; namely, $g_{11}(t) = \delta_{11} g_0(t)$, $g_{22}(t) = \delta_{22} g_0(t)$, and $g_{12}(t) = \delta_{12} g_0(t)$, where δ_{11}, δ_{22}, and δ_{12} are time-independent interaction parameters dependent on the explicit systems and $g_0(t)$ is a modulated interaction parameter. In this case, Eqs. (8.70a) and (8.70b) can be simplified to

$$\Delta_1 = \frac{1}{g_0(t)} \frac{\delta_{22} - \kappa\delta_{12}}{\delta_{22}\delta_{11} - \delta_{12}^2}, \quad \Delta_2 = \frac{1}{g_0(t)} \frac{\kappa\delta_{11} - \delta_{12}}{\delta_{22}\delta_{11} - \delta_{12}^2}. \tag{8.73}$$

In addition, the initial amplitudes of soliton pairs can be identified as $D_{31}^2 = c_4 \Delta_1 g_0(t) D_1^2$ and $D_{32}^2 = c_4 \Delta_2 g_0(t) D_1^2$. After these parameters are given, Eqs. (8.68) and (8.69a) can be recombined into

$$-\frac{1}{g_0}\frac{d^2 g_0}{dt^2} + \frac{2}{g_0}\frac{dg_0}{dt} - 2\frac{d\gamma}{dt} + 4\gamma^2 g_0^2 + 4\gamma \frac{1}{g_0}\frac{dg_0}{dt} + \lambda^2 = 0. \tag{8.74}$$

This is a condition under which the soliton-pair solutions can exist when the interaction and external potential parameters are time-dependent.

As the second case, we let $g_{11}(t) = \delta_{11}$ and $g_{22}(t) = \delta_{22}$ be real constants and $g_{12}(t) = g_{21}(t) = g(t)$ are time-dependent. These correspond to the inhomogeneous case. Then one has

$$\Delta_1 = \frac{\kappa}{\kappa\delta_{11} + g(t)}, \quad \Delta_2 = \frac{1}{\kappa\delta_{11} + g(t)}, \tag{8.75}$$

and $\delta_{22} = \kappa^2 \delta_{11}$. Moreover, we take the constant amplitudes of soliton pairs $D_{31}^2 = c_4 \kappa D_1^2$ and $D_{32}^2 = c_4 D_1^2$. As a result, Eqs. (8.68) and (8.69a) can be merged into

$$\begin{aligned} \lambda^2(t) = {} & 2\frac{d\gamma}{dt} - \frac{1}{[\kappa\delta_{11} + g(t)]^2}\left(\frac{dg}{dt}\right)^2 + \frac{1}{\kappa\delta_{11} + g(t)}\frac{d^2 g}{dt^2} \\ & - \left[2\gamma + \frac{1}{\kappa\delta_{11} + g(t)}\frac{dg}{dt}\right]^2. \end{aligned} \tag{8.76}$$

Likewise, this is a condition under which the soliton-pair solutions exist. In the above cases, $\gamma_1 = \gamma_2 = \gamma$ can be obtained from Eq. (8.72). In what follows, we discuss explicit collision dynamics of the soliton pairs under these conditions.

8.4.3 Collision Dynamics of Soliton-Pairs

Interactions between two solitons can exhibit a very intriguing collision dynamics behavior, like particles which preserve their identify after passing through each other. For same-atom systems the collision is, in essence, partially coherent due to a direct overlapping of the atoms but, for different atoms, due to the incoherent nature, the collision is incoherent, which is the case we study here. In this situation, the total intensity of two solitons is $I = |\psi_1|^2 + |\psi_2|^2$. Explicitly, we investigate the elastic collisions of soliton pairs with opposite velocities by synchronizing control over the interactions and external potentials mentioned above. For comparison, we consider two experimentally accessible systems to study the influence on the soliton-pair collision dynamics of related parameters that are tuned experimentally.

Case of Homogeneously Tuned Interactions

As the first case, we consider that all interaction parameters are modulated periodically by the Feshbach resonance technique according to

$$g_0(t) = 1 + A \sin[\Omega t],$$

where A and Ω are the amplitude and frequency of the modulation and here we take $A = 0.7$ and $\Omega = 0.25$. Using this modulation we investigate the collision dynamics of two types of soliton pairs for the special system of an ^{87}Rb–^{41}K mixture with the dimensionless parameters $\delta_{11} = 69$ and $\delta_{22} = 99$ during simultaneous evaporative cooling. Here and hereafter, all parameters are dimensionless. According to Table 8.2 and the definition of the initial amplitudes, the existence of a bright–bright soliton pair requires $\Delta_1 < 0$ and $\Delta_2 < 0$. Hence we take $\delta_{12} = -84$. The chosen parameters are related to those used in work [12, 34], where $g_0(t) = 1$ was assumed and the bright–bright soliton pair with constant velocities, amplitudes, and widths formed as a consequence of modulation instability (MI). In this book the collision dynamics of the bright–bright soliton pair is shown in Fig. 8.15a. It is found that the solitons pass through each other after merging at about $t = 30$, as also shown in Fig. 8.15b for a corresponding density plot. In this process, the solitons oscillate and deform due to the synchronous modulation of the homogeneous interaction parameters and the external potentials, which are shown in Fig. 8.15e.

Next we study the dark–dark soliton and consider the same physical system and parameters δ_{11}, δ_{22} as for the bright–bright soliton pair. In this situation, $\Delta_1 > 0$ and $\Delta_2 > 0$ are necessary for the existence of the dark–dark soliton pair. So we set $\delta_{12} = 200$ to meet the requirements. Shown in Fig. 8.15c and d is the collision scene between the dark–dark soliton pair, which is obviously elastic.

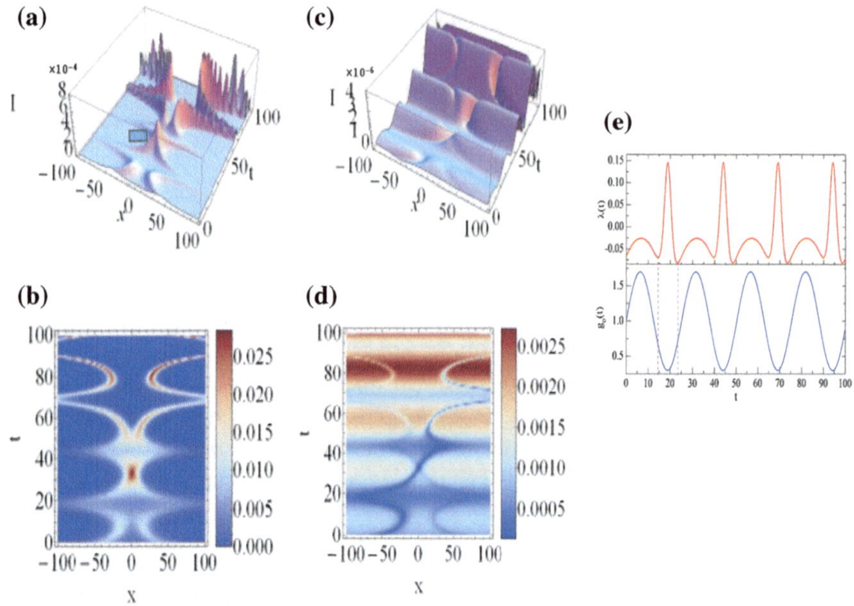

Fig. 8.15 Elastic collision dynamics of bright–bright and dark–dark soliton pairs with opposite velocities, and synchronous evolution of external potential with modulation of homogeneous interaction via Feshbach resonance in ^{87}Rb–^{41}K mixture with growth of condensates. The parameters used are given by $A = 0.7, \Omega = 1/4, \gamma = 1/150$, and $D_4 = 4$. **a** Collision dynamics of bright–bright soliton pair with $D_1 = 1/150$, $D_2 = 1/20$, $c_2 = 1$, and $c_4 = -1$. **b** Corresponding density plot for plot **a**. **c** Collision dynamics of dark–dark soliton pair with $D_1 = 1/200$, $D_2 = 1/15$, $c_2 = -2$, and $c_4 = 1$. **d** Corresponding density plot of plot **c**. **e** Synchronous controls of the external harmonic potential (upper panel) with the homogenous interaction (lower panel). Reprint from Ref. [12], Copyright 2018, with permission from American Physical Society

In the above case, the MI also occurs for the existence of the two types of solitons. From Fig. 8.15, it is easy to see that each soliton pair is sharply asymmetric due to the mass ratio of two components and because $\delta_{11} \neq \delta_{22} \neq \delta_{12}$, and their amplitudes grow periodically with time along their propagation directions. In addition, the growth of the two condensates in each period makes the depth of the potential well shallow in order to keep the basic excitation of the two condensates unchanged, which leads to two asymmetric dips of the strength of the external potential around $t = 15$ and $t = 25$. From the upper panel of Fig. 8.15e, the potential well $[\lambda(t) > 0]$ and barrier $[\lambda(t) < 0]$ alternate with the periodic modulation of the homogeneous interaction parameter. Furthermore, Eq. (8.74) shows that the strength only depends on the time-dependent interaction and the gain term, but has nothing to do with the constant interaction parameters in the two-component BECs and their mass ratio. Thus the evolution behavior of the strengths for different soliton pairs is identical under the modulation of the same homogeneous interaction and gain term.

Finally, we study the collision dynamics of such soliton pairs without evaporative cooling (in this case, $\gamma = 0$) by controlling the homogeneous interaction parameter

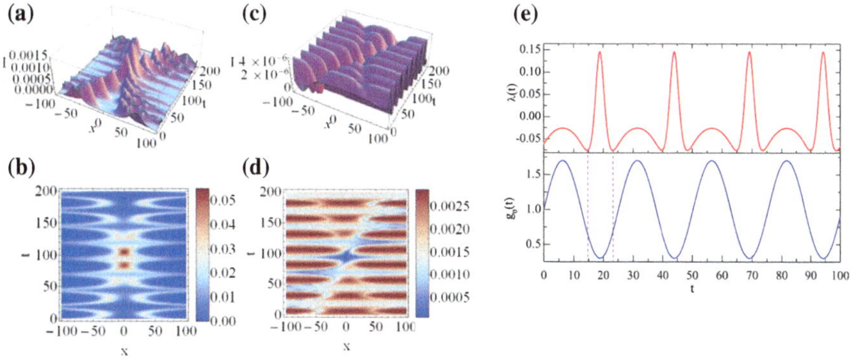

Fig. 8.16 Collision dynamics between soliton pairs and evolution of external potential after evaporative cooling. The parameters used are $D_4 = 4$, and $\gamma = 0$ for the collision dynamics of the bright–bright soliton pair [(**a**) and (**b**)] and the dark–dark soliton pair [(**c**) and (**d**)]. **e** Synchronous control of the external potential (upper panel) with homogenous interaction (lower panel). The other parameters used are the same as for Fig. 8.15. Reprint from Ref. [12], Copyright 2018, with permission from American Physical Society

as mentioned above and as shown in Fig. 8.16. Comparison of the condensates during the simultaneous evaporation shows that the peak density of each soliton in adjacent periods is the same in the absence of condensate growth, and the strength of the harmonic external potential is perfectly symmetric at each period, as shown at the top of Fig. 8.16e. Moreover, it is found that the collision times of soliton pairs without a gain term are delayed in the same initial conditions.

Inhomogeneous Case

Next we consider the case with fixed intraspecies interactions, but with a time-dependent interspecies interaction modulated as

$$g(t) = \xi + A \sin[\Omega t]. \tag{8.77}$$

Here we take the modulation amplitude $A = 3$ and Ω to be the same as above. According to these values and the existence conditions [i.e., $\kappa\delta_{11} < -g(t)$ for the bright–bright solitons pair and $\kappa\delta_{11} > -g(t)$ for dark–dark soliton pair], ξ is different for the former and for the latter in explicit systems. The modulation type is easy to realize under the present experimental condition for the heteronuclear system of an ^{87}Rb–^{41}K mixture [42].

In this system, when $\delta_{11} = 99$ for a ^{87}Rb atom, it requires $\delta_{22} = \kappa^2\delta_{11} \sim 446$ for ^{41}K. ξ is taken to be -215 for the bright–bright soliton pair and -197 for the dark–dark soliton pair. Under synchronous modulation of the interspecies interactions and external potentials, the collision dynamics of the bright–bright and dark–dark soliton pairs are shown in Fig. 8.17a–d with the gain terms and in Fig. 8.18a–d with no gain terms. Figures 8.17e and 8.18e show the synchronous behavior of the time-dependent external potentials for the bright–bright soliton-pair dynamics (left) and the dark–

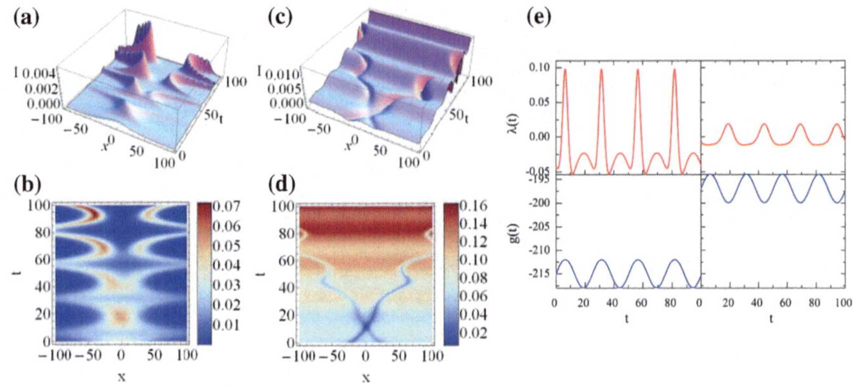

Fig. 8.17 Collision dynamics between soliton pairs under interspecies interactions via Feshbach resonance management and synchronous evolution of time-dependent external potentials in ^{87}Rb–^{41}K mixture with growth of condensates. The parameters used are $\gamma = 1/200$, $\Omega = 1/4$, $A = 3$, $D_1 = 1/6000$, and $D_2 = 1/400$. **a** Collision dynamics of bright–bright soliton pair with $D_4 = 1$ and $\xi = -215$. **b** Corresponding density plot for plot **a**. **c** Collision dynamics of dark–dark soliton pair with $D_4 = 8$ and $\xi = -197$. **d** Corresponding density plot for plot **c**. **e** Synchronous controls of external harmonic potential and interaction for bright–bright soliton (left panel) and dark–dark soliton pairs (right panel). Reprint from Ref. [12], Copyright 2018, with permission from American Physical Society

dark soliton-pair dynamics (right). With the same initial conditions, the collision times are also delayed in Fig. 8.18a–d compared with that in Fig. 8.17a–d. After the collisions, the soliton pairs in Figs. 8.17 and 8.18 show strong oscillations along their propagation directions, which are similar to those in Figs. 8.15 and 8.16. In contrast to Figs. 8.15 and 8.16, the amplitudes of the two solitons in each soliton pair are more sharply asymmetric in Figs. 8.17 and 8.18 due to different modulation schemes. As mentioned above, the soliton-pair dynamics discussed above could be realized in the present experiments. Two key steps are necessary. First, at the initial time the soliton pair should be created experimentally by possible techniques like phase imprinting, then one synchronously modulates the interspecies interaction and the strength of the harmonic external potential, as shown in our figures.

It is also interesting to study an isotopic system consisting of ^{85}Rb and ^{87}Rb with $\delta_{11} = 103.7$ and $\delta_{22} = 99$, which is related to those in Ref. [33]. This system has been used to investigate the phase separation in experiment and to numerically simulate the growth of dual-species BECs during simultaneous evaporative cooling [36], in which the results show that the growth of dual-species BECs during simultaneous evaporative cooling makes the condensates far from the true ground state of the system under the case of unchanged interaction and trap parameters.

In this system, ξ is taken as -106 for the bright–bright soliton pair and -95 for the dark–dark soliton pair. We explore the collision dynamics of two types of soliton pairs during the growth of two condensates are displayed in Fig. 8.19a, d, from which their amplitudes also increase periodically with the modulation of the

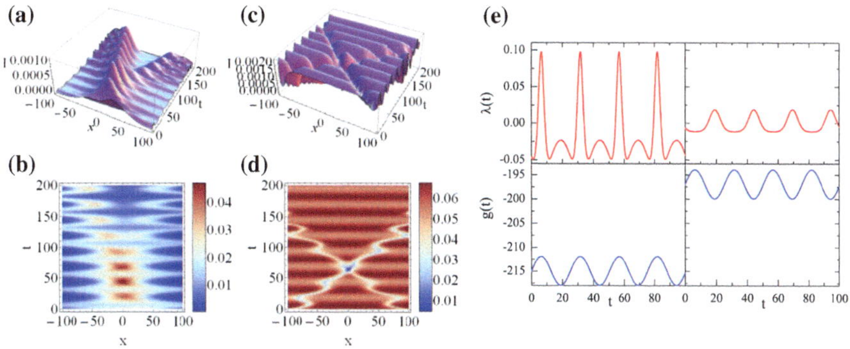

Fig. 8.18 Collision dynamics between soliton pairs and synchronous change with time of external potentials in absence of gain term. The parameters are as follows: Panels **a** and **b** show the collision dynamics of the bright–bright pair with $D_4 = 1$ and $\xi = -215$. Panels **c** and **d** show the collision dynamics of the dark–dark pair with $D_4 = 8$ and $\xi = -197$. **e** Synchronous controls of the external harmonic potential and the interaction for the bright–bright soliton (left panel) and the dark–dark soliton pairs (right panel). The other parameters used are the same as for Fig. 8.17. Reprint from Ref. [12], Copyright 2018, with permission from American Physical Society

interspecies interactions and harmonic external potentials, which to some extent explains the experimental phenomena in Ref. [36]. After evaporative cooling ($\gamma = 0$), the condensates are no longer growing. The collisional dynamics of soliton pairs are sequentially studied by modulating the interspecies interactions as shown in Fig. 8.20a–d, from which the collision times of the soliton pairs without gain are also delayed compared with those during evaporation cooling. In addition, it can be seen that the difference between the amplitudes of the two solitons in each pair only depends on the mass ratio of the two species. Because the mass ratio of ^{85}Rb and ^{87}Rb is much smaller compared with that of the heteronuclear system ^{87}Rb–^{41}K, the asymmetry between the two components is not as clear as for the two types of soliton pairs shown in Figs. 8.19a, d, 8.20a and d compared with those in Figs. 8.17a, d, 8.18a and d. In Figs. 8.19e and 8.20e, the left upper and lower panels of these plots show the synchronous controls of the external potential and the interspecies interaction parameters for the bright–bright soliton dynamics shown in Figs. 8.19a, b, 8.20a and b, respectively. The controls given in the right-hand side correspond to the cases of the dark–dark soliton shown in Figs. 8.19c, d, 8.20c and 8.20d.

From the condition Eq. (8.75), one can see that the strength of the external potential is related to the mass ratio, the interspecies interaction, and the gain term. Hence the different modulations of the interactions for the bright–bright and dark–dark soliton pair directly result in the distinct evolutions of the external potential, as shown in Figs. 8.17e, 8.18e, 8.19e, and 8.20e. Moreover, in comparison with Figs. 8.17e and 8.19e, in Figs. 8.18e and 8.20e, respectively, the external potentials are symmetric over a period of time.

In order to compare the above isotopic system with ^{85}Rb and ^{87}Rb, we consider a homonuclear system with the mixture of two different hyperfine states of ^{87}Rb and

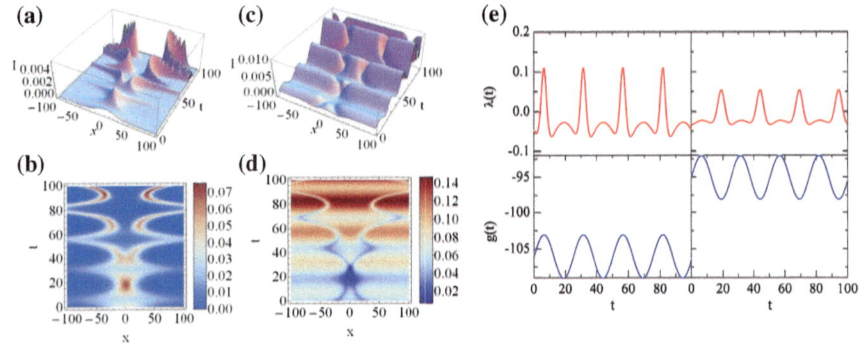

Fig. 8.19 Collision dynamics between soliton pairs under interspecies interactions via Feshbach management and synchronous evolution of time-dependent external potentials in isotopic mixture with growth of condensates. The parameters used are $\gamma = 1/200$, $\Omega = 1/4$, $A = 3$, $D_1 = 1/8000$, and $D_2 = 1/800$. **a** Collision dynamics of bright–bright soliton pair with $D_4 = 2$ and $\xi = -106$. **b** Corresponding density plot for plot **a**. **c** Collision dynamics of dark–dark soliton pair with $D_4 = 4$ and $\xi = -95$. **d** Corresponding density plot for plot **c**. **e** Synchronous control of external harmonic potential and interaction for bright–bright soliton pair (left panel) and dark–dark soliton pair (right panel). Reprint from Ref. [12], Copyright 2018, with permission from American Physical Society

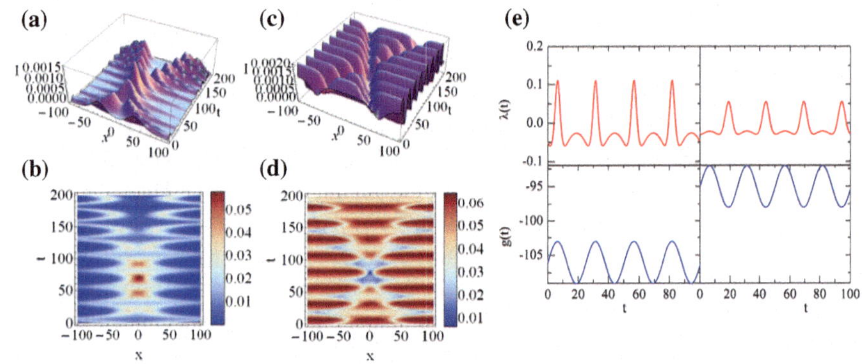

Fig. 8.20 Collision dynamics between soliton pairs and synchronous change with time of external potentials in absence of the gain term. The parameters are as follows: Panels **a** and **b** show collision dynamics of bright–bright pair with $D_4 = 2$ and $\xi = -106$. Panels **c** and **d** show collision dynamics of dark–dark pair with $D_4 = 4$ and $\xi = -95$. **e** Synchronous controls of external harmonic potential and interaction for bright–bright soliton (left panel) and dark–dark soliton pairs (right panel). The other parameters used are the same as for Fig. 8.19. Reprint from Ref. [12], Copyright 2018, with permission from American Physical Society

set $\delta_{22} = 99$. Because $\kappa = 1$, then $\delta_{11} = 99$. These quantities are basically close to the experimental observations. In addition, ξ is the same as that in ^{85}Rb and ^{87}Rb. In the homonuclear system, the dynamical evolution of two types of soliton pairs is shown in Fig. 8.21a–d for the case of $\gamma = 1/200$ and in Fig. 8.22a–d for $\gamma = 0$. It is obvious that each soliton pair is perfectly symmetric in panels (a)–(d) of Figs.

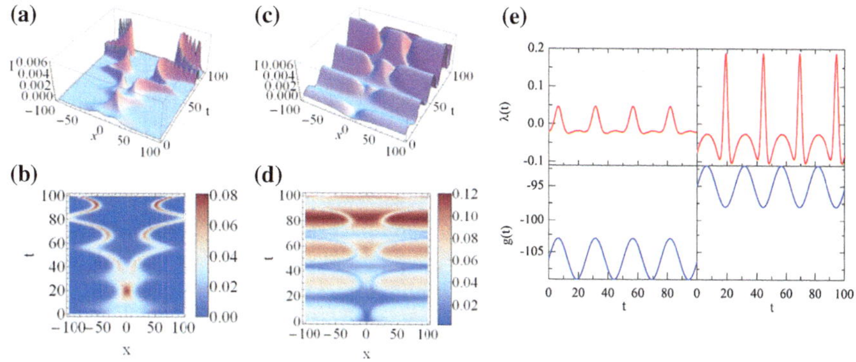

Fig. 8.21 Collision dynamics between soliton pairs and synchronous change with time of external potentials in homonuclear system with gain term. The parameters are as follows: $\gamma = 1/200$, $D_1 = 1/100$, $D_2 = 1/10$, $A = 3$, and $\Omega = 1/4$. **a** Collision dynamics of bright–bright soliton pair with $D_4 = 4$ and $\xi = 106$. **b** Corresponding density plot for plot **a**. **c** Collision dynamics of dark–dark soliton pair with $D_4 = 1$ and $\xi = -95$. **d** Corresponding density plot for plot **c**. **e** Strengths of external potentials in upper panel and interspecies interactions in lower panel corresponding to bright–bright and dark–dark soliton ones. Reprint from Ref. [12], Copyright 2018, with permission from American Physical Society

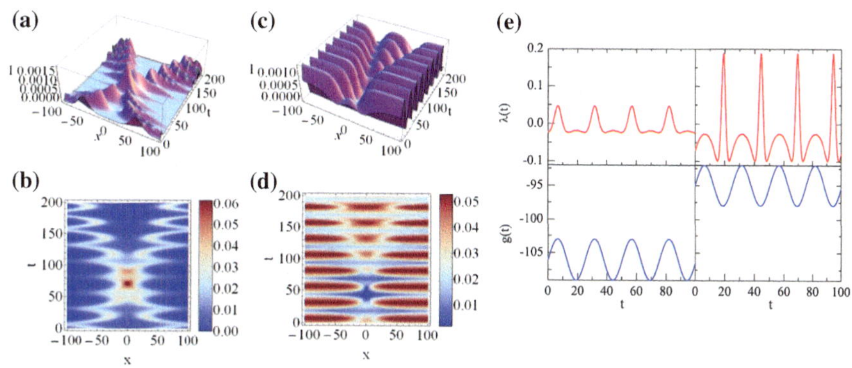

Fig. 8.22 Collision dynamics between soliton pairs and evolution of relevant external potentials after stopping growth of homonuclear condensates. The parameter $\gamma = 0$ for the collision dynamics of bright–bright soliton pair [(**a**) and (**b**)] and dark–dark soliton pair in [(**c**) and (**d**)]. The other parameters are the same as those for Fig. 8.21. **e** Potentials and interspecies interactions for bright–bright and dark–dark soliton pairs, respectively. Reprint from Ref. [12], Copyright 2018, with permission from American Physical Society

8.21 and 8.22. Their modulated interactions in the lower panels of Figs. 8.21e and 8.22e are the same as those in the lower panels of Figs. 8.19e and 8.20e. However, the changes of the external potential in the upper panels of Figs. 8.21e and 8.22e are very different from those in the upper panels of Figs. 8.19e and 8.20e due to the equal masses for the two components.

8.5 Effects of the Three-Body Interactions on the Localized Spatially Matter Waves in Atomic-Molecular BECs with Space-Modulated Nonlinearity

The intrinsic nonlinearity is the most remarkable characteristic of the BECs systems. Here, we investigate the nonlinear matter waves in the two-dimensional atomic-molecular BECs with trapping potential and space-modulated nonlinearity, which can be described by the two-component NLS equations with space-modulated nonlinearities. We focus ourselves on three kinds of localized nonlinear wave solutions for both the attractive spatially inhomogeneous interactions and the repulsive ones by using the similarity transformation.

8.5.1 Description of the Model and Useful Transformations

In real experiment, the coherent free-bound stimulated Raman transition can cause atomic BECs of ^{87}Rb to produce a molecular BECs of $^{87}Rb_2$. If the molecular spontaneous emission and the light shift effect can be ignored, according to the mean field theory, the two-species atomic-molecular BECs with three-body interaction term can be written as [16, 43]

$$
\begin{aligned}
i\hbar\frac{\partial\Psi_a}{\partial t} &= \left[-\frac{\hbar^2}{2M_a}\nabla^2 + g_a\left|\Psi_a\right|^2 + g_{am}\left|\Psi_m\right|^2 + \gamma_a\left|\Psi_a\right|^4 + V_{ext}^{(a)}\right]\Psi_a + \sqrt{2}\chi\Psi_a^*\Psi_m,\\
i\hbar\frac{\partial\Psi_m}{\partial t} &= \left[-\frac{\hbar^2}{2M_m}\nabla^2 + g_m\left|\Psi_m\right|^2 + g_{am}\left|\Psi_a\right|^2 + \gamma_m\left|\Psi_m\right|^4 + V_{ext}^{(m)}\right]\Psi_m + \frac{\chi}{\sqrt{2}}\Psi_a^2 + \varepsilon\Psi_m,
\end{aligned} \tag{8.78}
$$

where $\nabla^2 = \partial^2/\partial x^2 + \partial^2/\partial y^2$, Ψ_j $(j = a, m)$ denotes the macroscopic wave function of atomic condensate and molecular condensate, respectively, $g_a = \frac{4\pi\hbar^2}{M_a}a_a$ and $g_{am} = \frac{4\pi\hbar^2 a_{am}}{M_a M_m/(M_a+M_m)}$, and γ_a and γ_m represent, respectively, the cubic and quintic nonlinearity the strengths of interaction, $V_{ext}^{(j)} = \frac{M_j\omega^2(x^2+y^2)}{2}$ $(j = a, m)$ are the trapping potentials, M_a (M_m) is the mass of atomic (molecule), χ is the parametric coupling coefficient which describes the conversions of atoms into molecules due to stimulated Raman transitions. The parameter ε characterizes Raman detuning for a two photon resonance [19, 27]. Integrating along the transverse coordinates, Eq. (8.78) for the wave functions Ψ_j $(j = a, m)$ in dimensionless form can be written as the two-component NLS equations

$$
\begin{aligned}
i\frac{\partial\Psi_a}{\partial t} &= \left[-\frac{1}{2}\nabla^2 + g_a\left|\Psi_a\right|^2 + g_{am}\left|\Psi_m\right|^2 + \gamma_a\left|\Psi_a\right|^4 + V_{ext}^{(a)}\right]\Psi_a + \sqrt{2}\chi\Psi_a^*\Psi_m,\\
i\frac{\partial\Psi_m}{\partial t} &= \left[-\frac{1}{2}\nabla^2 + g_m\left|\Psi_m\right|^2 + g_{am}\left|\Psi_a\right|^2 + \gamma_m\left|\Psi_m\right|^4 + V_{ext}^{(m)}\right]\Psi_m + \frac{\chi}{\sqrt{2}}\Psi_a^2 + \varepsilon\Psi_m.
\end{aligned} \tag{8.79}
$$

Here, the unit of length, time and energy correspond respectively to $\sqrt{\hbar/(m\omega)}$ ($\approx$1.07 um), $1/\omega$ ($\approx 1.6\times 10^{-3}$ s), and $\hbar\omega$. Throughout this section, we use the parameters of atomic-molecular BECs of ^{87}Rb system with $M_m = 2M_a = 2m$, ($m = 144.42 \times 10^{-27}$ Kg), $g_m = 2g_a$ ($a_a = 101.8\ a_B$), where a_B is the Bohr radius.

Let us consider the spatially localized stationary solution

$$(\Psi_a, \Psi_m) = (\phi_a(x, y) \exp[-i\mu_a t], \phi_m(x, y) \exp[-i\mu_m t]) \tag{8.80}$$

of system (8.79), where $\phi_j(x, y)$ ($j = a, m$) are two real functions satisfying condition $\lim_{|x|,|y| \to +\infty} \phi_j(x, y) = 0$, and μ_a and μ_m are chemical potentials. Ansatz (8.80) maps system (8.79) into the following stationary system

$$\begin{aligned} &\Delta_2\phi_a - 2g_a\phi_a^3 - 2g_{am}\phi_m^2\phi_a - 2\gamma_a\phi_a^5 + \left[2\mu_a - \omega^2\left(x^2 + y^2\right)\right]\phi_a + 2\sqrt{2}\chi\phi_a\phi_m = 0, \\ &-\Delta_2\phi_m + 4g_m\phi_m^3 + 4g_{am}\phi_m\phi_a^2 + 4\gamma_m\phi_m^5 + \left[\omega^2\left(x^2 + y^2\right) - 8\mu_m + 4\varepsilon\right]\phi_m - 2\sqrt{2}\chi\phi_a^2 = 0, \end{aligned} \tag{8.81}$$

where $\Delta_2 = \frac{\partial^2}{\partial x^2} + \frac{\partial^2}{\partial y^2}$. In order to solve system (8.81), we take the similarity transformation

$$(\phi_a, \phi_m) = (\beta_a(x, y)U(X(x, y)), \beta_m V X(x, y)) \tag{8.82}$$

to transform system (8.81) to the ODE

$$\begin{cases} \frac{d^2U}{dX^2} + b_{11}U^3 + b_{12}UV^2 + b_{13}UV + b_{14}U^5 = 0, \\ \frac{d^2V}{dX^2} + b_{22}V^3 + b_{21}U^2V + b_{23}U^2 + b_{24}V^5 = 0, \end{cases} \tag{8.83}$$

where b_{ij}, $i = 1, 2$, $j = 1, 2, 3$ are constants. Substituting (8.82) into (8.81) and demanding that $U(X)$ and $V(X)$ satisfy system (8.83) lead to a set of PDEs. After solving this set of PDEs, we have

$$\begin{aligned} (g_a, g_m, g_{am}) &= -\frac{\omega}{2F^2}\left(b_{11}, \frac{b_{22}}{2}, b_{12}\right) \times \exp\left[-\omega\left(x^2 + 4xy + y^2\right)\right], \\ (\gamma_a, \gamma_m) &= -\frac{\omega}{2F^4}(b_{14}, b_{24}) \times \exp\left[-\omega\left(x^2 + 6xy + y^2\right)\right], \\ F &= c_1 \text{Kummer}M\left(\frac{\omega - \mu_a}{2\omega}, \frac{3}{2}, \frac{\omega(y - x)^2}{2}\right)(y - x) \\ &\quad + c_2 \text{Kummer}U\left(\frac{\omega - \mu_a}{2\omega}, \frac{3}{2}, \frac{\omega(y - x)^2}{2}\right)(y - x), \\ X &= \frac{1}{2}\sqrt{\pi}\,\text{erf}\left(\frac{\sqrt{2\omega}(x + y)}{2}\right), \quad \varepsilon = \mu_m - \frac{1}{2}\mu_a, \end{aligned} \tag{8.84}$$

where $\text{erf}(s) = \int_0^s \exp\left[-\tau^2\right] d\tau$ is the error function, $\text{Kummer}M\left(\frac{\omega - \mu_a}{2\omega}, \frac{3}{2}, \frac{\omega(y-x)^2}{2}\right)$ $(y - x)$ and $\text{Kummer}U\left(\frac{\omega - \mu_a}{2\omega}, \frac{3}{2}, \frac{\omega(y-x)^2}{2}\right)(y - x)$ are solutions of the ODE

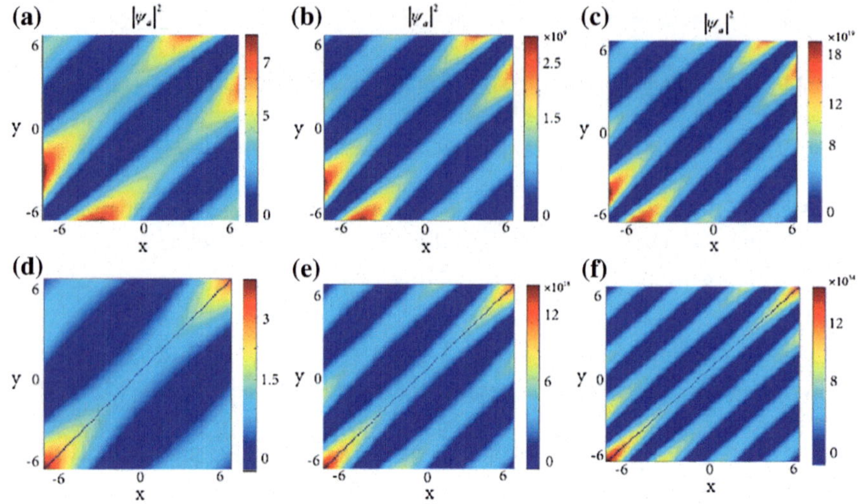

Fig. 8.23 The density distributions $|\Psi_a|^2$ of the atomic-molecular BEC with three-body interaction term as the function of ε and μ_a with $\omega = 0.02$, $b_{11} = 3$, $b_{22} = 12$, $b_{12} = 1$. The energy packets are striped distribution. (a–c) show that the number of the energy stripes increases with chemical potential μ_a when ε is fixed. Plots **a**, **b**, and **c** are generated with $\mu_a = 0.1, 0.2, 0.3$, respectively for $\varepsilon = 0.01$. **d–f** illustrate that some zero points appear on the middle density stripe when the number of the density stripes is odd. **d**, **e**, and **f** are obtained with $\mu_a = 0.1, 0.2, 0.6$, respectively with $\varepsilon = 0.02$ [46]

$$\frac{d^2F}{dY^2} - \omega Y \frac{dF}{dY} + \mu_a F = 0, \tag{8.85}$$

where $Y = y - x$. Specially, when $\frac{\omega - \mu_a}{\omega} = 3$, the Kummer$M$ function can be simplified as exponential function $\exp\left[\frac{\omega(y-x)^2}{2}\right]$. In this case, the interactions become as

$$(g_a, g_m, g_{am}) = -\frac{\omega}{2c_1^2\,(y-x)^2}\left(b_{11}, \frac{b_{22}}{2}, b_{12}\right) \times \exp\left[-\omega\left(x^2 + 4xy + y^2\right)\right],$$

which is experimentally feasible due to the flexible and precise control of the scattering lengths achievable in BECs with magnetically tuning the Feshbach resonances [45, 46].

8.5.2 Exact Solutions of the Two-Component Cubic–Quintic NLS Equations

With the help of the above transformation, we present here exact solutions of the two-component cubic–quintic NLS equations (8.79). Here, we focus ourselves on two families of solutions, the family of rational solutions and the family of Jacobi elliptic solutions.

Family of Rational Solution

When the coupling $\chi = 0$, Eqs. (8.82) and (8.83) give the rational formal solution of (8.79)

$$\begin{aligned}\Psi_a &= \frac{\sqrt[6]{\left(g_{11}g_{22} - 2g_{12}^2\right)(g_{12} - g_{22})F}}{\sqrt{3\left(g_{11}g_{22} - 2g_{12}^2\right)^2\left(c - \frac{\sqrt{\pi}}{2}\mathrm{erf}^2\left(\frac{\sqrt{2\omega}}{2}(x+y)\right) + 4g_{14}(g_{12} - g_{22})^2\right)}} \\ &\quad \times \exp[\omega xy]\exp[-i\mu_a t], \\ \Psi_m &= \frac{\sqrt[6]{(2g_{12} - g_{11})\left(g_{11}g_{22} - 2g_{12}^2\right)(g_{12} - g_{22})}}{\sqrt{3(g_{12} - g_{22})\left(\left(g_{11}g_{22} - 2g_{12}^2\right)^2\left(c - \frac{\sqrt{\pi}}{2}\mathrm{erf}^2\left(\frac{\sqrt{2\omega}}{2}(x+y)\right) + 4g_{14}(g_{12} - g_{22})^2\right)\right)}} \\ &\quad \times \exp[\omega xy]\exp[-i\mu_m t],\end{aligned} \tag{8.86}$$

where c is arbitrary constant and F is given in (8.84).

In order to investigate the topological properties of the exact spatially localized stationary solution (8.86), we plot their density distributions. In Fig. 8.23, it can be observed that the energy packets are striped distribution, and the number of the energy stripes increases with the chemical potential μ_a when ε is fixed. It can also be seen that some zero points appear on the middle density stripe along line $y = x$ when the number of the density stripes is odd.

Families of Jacobi Elliptic Function Solution

When the three-body effect is very weak and the coupling $\chi = 0$, we obtain the following exact solutions of (8.83)

$$(U, V) = \left(\sqrt{\theta_1}, \sqrt{\theta_2}\right) c_0 \mathrm{cn}\left[c_0 X, \frac{\sqrt{2}}{2}\right], \tag{8.87a}$$

$$(U, V) = \left(\sqrt{\frac{\theta_1}{2}}, \sqrt{\frac{\theta_2}{2}}\right) d_0 \mathrm{sd}\left[d_0 X, \frac{\sqrt{2}}{2}\right], \tag{8.87b}$$

where c_0, d_0 are arbitrary constants, $\theta_1 = \frac{b_{22} - b_{12}}{b_{11}b_{22} - b_{12}b_{21}}$, $\theta_2 = \frac{b_{11} - b_{21}}{b_{11}b_{22} - b_{12}b_{21}}$ and cn and sd $=$ sn/dn are Jacobi elliptic functions. Imposing the vanishing condition $\lim_{|x|,|y| \to +\infty} \phi_j(x, y) = 0$ leads to $c_0 = \frac{2(2N-1)K\left(\frac{\sqrt{2}}{2}\right)}{\sqrt{\pi}}$, $d_0 = \frac{4NK\left(\frac{\sqrt{2}}{2}\right)}{\sqrt{\pi}}$, where N is a natural number and $K\left(\frac{\sqrt{2}}{2}\right) = \int_0^{\pi/2} 1/\sqrt{1 - \frac{1}{2}\sin^2\xi}\, d\xi$.

From Eqs (8.82), (8.84), (8.87a) and (8.87b), we obtain the Jacobi elliptic function solutions for the atomic-molecular BEC described by system (8.79)

$$(\Psi_a, \Psi_m) = \frac{2nK\left(\frac{\sqrt{2}}{2}\right) F \exp[\omega xy]}{\sqrt{\pi}} \left(\sqrt{\theta_1}\exp[-i\mu_a t], \sqrt{\theta_2}\exp[-i\mu_m t]\right)$$
$$\times \mathrm{cn}\left[nK\left(\frac{\sqrt{2}}{2}\right)\mathrm{erf}\left(\frac{\sqrt{2\omega}}{2}(x+y)\right), \frac{\sqrt{2}}{2}\right], \quad n = 2N-1, \tag{8.88a}$$

$$(\Psi_a, \Psi_m) = \frac{2nK\left(\frac{\sqrt{2}}{2}\right) F \exp[\omega xy]}{\sqrt{\pi}} \left(\sqrt{\theta_1}\exp[-i\mu_a t], \sqrt{\theta_2}\exp[-i\mu_m t]\right)$$
$$\times \mathrm{sd}\left[nK\left(\frac{\sqrt{2}}{2}\right)\mathrm{erf}\left(\frac{\sqrt{2\omega}}{2}(x+y)\right), \frac{\sqrt{2}}{2}\right]\exp[\omega xy], \quad n = 2N. \tag{8.88b}$$

Here we discuss the existence regions of the spatially localized stationary solution (8.88a) and (8.88b) by assuming the two constraint conditions $\theta_1 > 0$ and $\theta_2 > 0$. We have the eight cases of parameters b_{11}, b_{12} and b_{22}. According to the real experiment, we consider the following two cases:

(1) $b_{22} < b_{11} < 0$ and $\sqrt{2}b_{12} > \sqrt{b_{11}b_{22}}$,
(2) $b_{22} > b_{11} > 0$ and $-\sqrt{b_{11}b_{22}} < 2b12 < b11$.

These correspond to two cases of the intercomponent interaction parameters g_a, g_m and g_{am} are

(a) $g_m > g_a > 0$ and $\sqrt{2}g_{am} < \sqrt{g_a g_m}$,
(b) $g_m < g_a < 0$ and $\sqrt{g_a g_m} > \sqrt{2}g_{am} > g_a$.

These are the regions that the exact spatially localized stationary solutions (8.88a) and (8.88b) exist. Now we only consider case (b), which denotes two self-attractive atom–atom interactions, two self-attractive molecular–molecular interactions, and attractive and repulsive atomic-molecular interactions. The other cases can be analyzed in the same way.

In the following, we will see that the integer n and the number of the zero points of function F which equals to that of the Kummer$U$ and Kummer$M$ functions determine the topological properties of the atom and molecular packets, so we call n and l as the principal quantum number and the secondary quantum number, respectively. In order to investigate the topological properties of the exact spatially localized stationary solution (8.88a) and (8.88b), we plot their density distributions by manipulating the principal quantum number n when the secondary quantum number l is fixed. In Fig. 8.24, we analyze the atomic BEC when the secondary quantum number l is fixed and the principal quantum number n is modulated. It is easy to see that the number of density packets for each quantum states is equal to $2n$. And the number of density packets on each quantum states increases two by two when the principal quantum number n increases. The properties of the molecular BEC are similar to that of the atomic BEC. In Fig. 8.25, we analyze the interactions of the atomic BEC and the molecular BEC when the secondary quantum number l is fixed. It is shown that the interaction is stronger when $N = 1$ and becomes weaker with the increasing of N.

When the principal quantum number n is fixed, we can adjust the secondary quantum number l to observe the properties of the atomic-molecular BEC. Figure

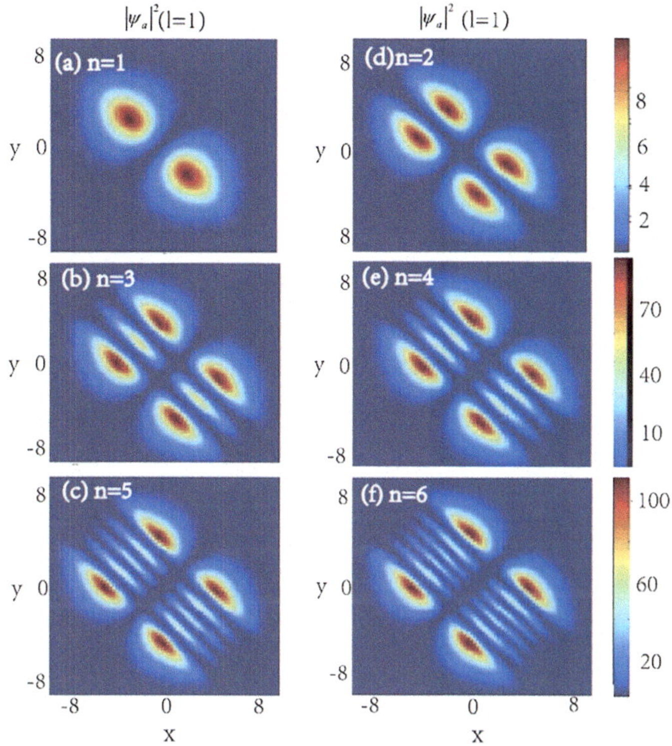

Fig. 8.24 The density distributions $|\Psi_a|^2$ of the atomic BEC as the function of the principal quantum number n when the secondary quantum number is fixed. The wave function Ψ_a take the form in Eqs. (8.88a) and (8.88b) with $b_{11} = 3$, $b_{22} = 12$, $b12 = 1$ and $\omega = 0.2$. The number of density packets for each quantum states is equal to $2n$. **a–c** show the density distributions of the odd parity wave functions (8.88a) for $n = 1, 3, 5$, respectively. **d–f** illustrate the density distributions of the even parity wave functions (8.88b) for $n = 2, 4, 6$, respectively. The solution displayed in figure **a** is linear stable. The unit of the length is 1.07 μm [46]

8.26 demonstrates the density distributions of atomic-molecular BEC for different secondary quantum number l. It is easy to find that the number of energy packets increases when l increases, and the number of the nodes for each quantum state equals to the secondary quantum number l. And some zero points appear on the middle density packets along the line $y = x$ when the number of the secondary quantum number l is even. Figure 8.27 demonstrates the interaction of the atomic BEC and molecular BEC when the principal quantum number n is fixed. It is shown that the number of the atomic-molecular pair is the function of the secondary quantum number l and some zero points appear on the middle atomic-molecular pair along the line $y = x$ when the number of the secondary quantum number l is even.

Now we analyze the effect of Raman detuning ε for the atomic-molecular BEC. From Fig. 8.28, we can see that when $\varepsilon < \mu_a$ and ε is fixed, the number of the density packets increases one by one with the increasing of the chemical potential μ_a. When

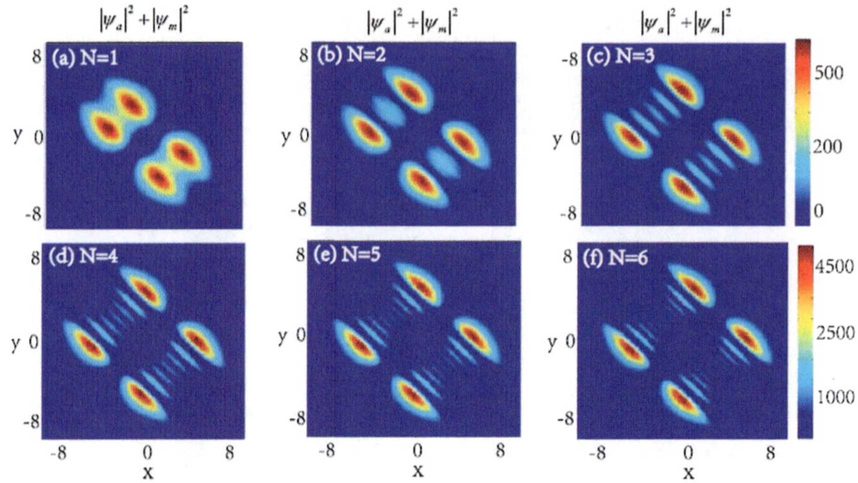

Fig. 8.25 The density distributions $|\Psi_a|^2 + |\Psi_m|^2$ of the atomic-molecular pair as the function of N when the secondary quantum number $l = 1$. Here $b_{11} = 3$, $b_{22} = 12$, $b_{12} = 1$, and $\omega = 0.2$. **a** shows the interaction of the atomic-molecular pair for $N = 1$. **b** shows the interaction of the middle atomic-molecular pair for $N = 2$, and it also displays that the interactions are weaker than the interaction in **a**. **a–f** illustrate that the interaction becomes weaker with the increasing of N. The unit of the length is 1.07 μm [46]

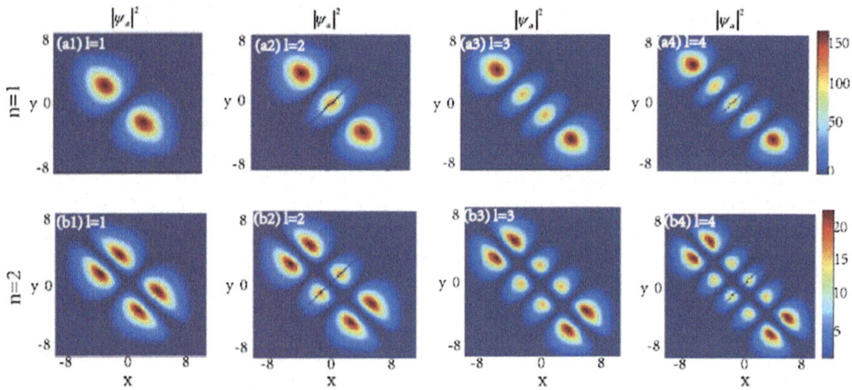

Fig. 8.26 The density distributions $|\Psi_a|^2$ of the atomic-molecular BEC as the function of the secondary quantum number l. Here $\omega = 0.2$, $b_{11} = 3$, $b_{22} = 12$, $b_{12} = 1$. The number of the nodes for each quantum state equals to the secondary quantum number l and some zero points appear on the middle one when the number of the density packets is odd. **a1–a4** show the density distributions of the odd parity wave functions (8.88a) for $n = 1$ and $l = 1, 2, 3, 4$, respectively. **b1–b4** show the density distributions of the even parity wave functions (8.88b) for $n = 2$ and $l = 1, 2, 3, 4$, respectively. The solutions displayed in the first and third figures on the upper row are linear stable. The unit of the length is 1.07 μm [46]

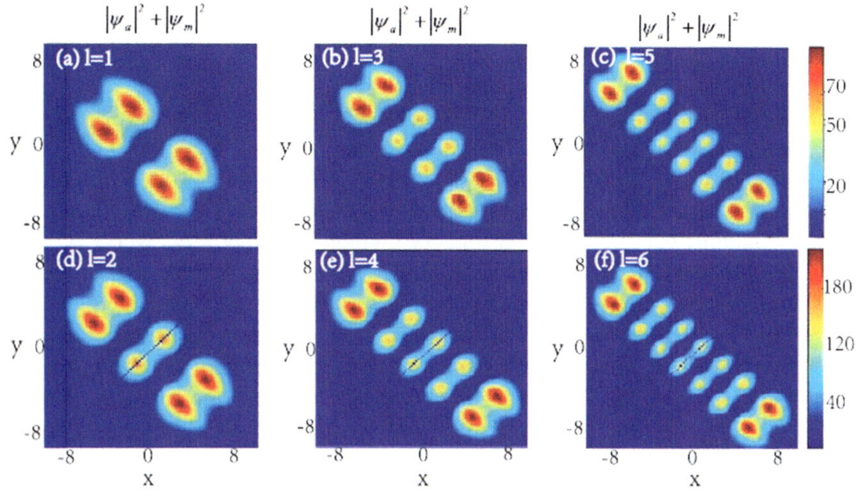

Fig. 8.27 The density distributions $|\Psi_a|^2 + |\Psi_m|^2$ of the atomic-molecular pair as the function of the secondary quantum number l with $\omega = 0.2$, $b_{11} = 3$, $b_{22} = 12$, $b_{12} = 1$. The number of the atomic-molecular pairs equals to $l + 1$. **a–c** show the density profiles of the atomic-molecular pair for l is odd. **d–f** show the density profiles of the atomic-molecular pair for l is even, and it also displays that some zero points appear on the middle one. The unit of the length is 1.07 μm [46]

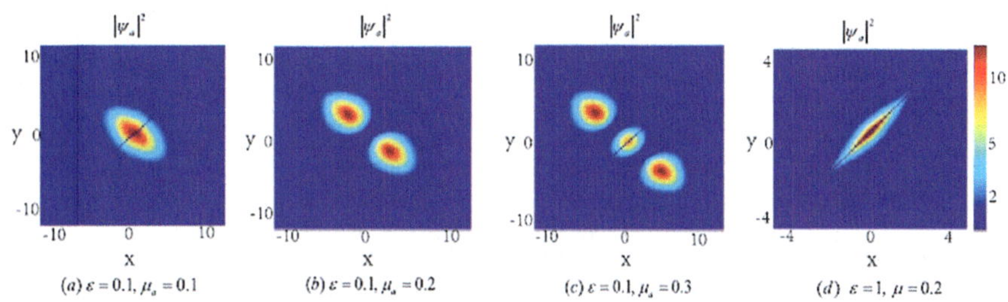

Fig. 8.28 The effect of Raman detuning ε for the atomic-molecular BEC. **a–c** show that the number of the density packets increases with the chemical potential μ_a when $\varepsilon < \mu_a$. **a, d** reveal that the number of the density packets don't depends on the chemical potential μ_a and there is only one density packet for each quantum state when $\varepsilon \geq \mu_a$, it also show the value of $\varepsilon - \mu a$ effects the shape of the density packet. The solution displayed in figure **b** is linear stable. The unit of the length is 1.07 μm [46]

$\varepsilon \geq \mu_a$, there is only one density packets for each quantum states. The absolute of $\varepsilon - \mu_a$ affect the shape of the energy packets: when the absolute of $\varepsilon - \mu_a$ is small, the shape of the density packet is like circle, and when the absolute of $\varepsilon - \mu_a$ is larger, the shape of the density packet becomes narrow and long.

Rational Formal Solution

When the three-body effect is very weak and the coupling $\chi \neq 0$, Eqs. (8.82) and (8.83) also gives the rational formal solution of system (8.79)

$$\begin{aligned}(\Psi_a, \Psi_m) &= -\frac{12\sqrt{2}b_{13}F\exp[\omega xy]}{b_{13}^2\left(2c-\sqrt{\pi}\,\mathrm{erf}^2\left(\frac{\sqrt{2\omega}}{2}(x+y)\right)\right)+27b_{11}}\left(\exp\left[-i\mu_a t\right], 2\exp\left[-2i\mu_a t\right]\right),\\ \chi &= \frac{\omega\sqrt{2}b_{13}}{8F}\exp\left[-\omega\left(x^2+3xy+y^2\right)\right],\end{aligned} \tag{8.89}$$

where $b_{13} = b_{23}$, $b_{11} > 0$, c is arbitrary constant and F is given in (8.84).

In order to investigate the topological properties of the exact spatially localized stationary solution (8.89), we plot their density distributions by adjusting the secondary quantum number l. The secondary quantum number l is always zero for $\varepsilon < \mu_a$, and can be taken different values for $\varepsilon \geq \mu_a$. In Fig. 8.9, it can be observed that the energy packets are striped distribution, and the number of the energy stripes increases with the chemical potential μ_a when the secondary quantum number $l = 0$ and ε is fixed. When the secondary quantum number $l \neq 0$, there is only one energy stripe and the energy stripe becomes more narrower with the increasing of the secondary quantum number l. It can also be seen that some zero points appear on the middle density stripe along line $y = x$ when the number of the density stripes is odd. Figures 8.23 and 8.29 show that the rational solution (8.86) and (8.89) have similar topological properties, which implies that three-body interactions does not hinder the formation of the localized nonlinear matter wave solutions.

8.5.3 Linear Stability Analysis

In what follows, we analyze the linear stability of the solutions (8.88a), (8.88b) and (8.89) by using the linear stability analysis. A perturbed solution is constructed as [2]

$$\begin{aligned}\Psi_a &= \left\{\phi_a(x,y) + u_1(x,y)\exp\left[i\lambda t\right] + w_1^*(x,t)\exp\left[-i\lambda t\right]\right\}\exp\left[-i\mu_a t\right],\\ \Psi_m &= \left\{\phi_m(x,y) + u_2(x,y)\exp\left[i\lambda t\right] + w_2^*(x,t)\exp\left[-i\lambda t\right]\right\}\exp\left[-i\mu_m t\right],\end{aligned} \tag{8.90}$$

where $|u_1| \ll 1$, $|u_2| \ll 1$, $|w_1| \ll 1$, $|w_2| \ll 1$ are small perturbation. Substituting this perturbed solution into system (8.79) and neglecting the higher order terms in u_1, u_2, w_1, and w_2, we obtain the eigenvalue problem

$$\begin{pmatrix} L_1 & \sqrt{2}\chi\phi_m - g_a\phi_a^2 & \sqrt{2}\chi\phi_a - g_{am}\phi_a\phi_m & -g_{am}\phi_a\phi_m \\ g_a\phi_a^2 - \sqrt{2}\chi\phi_m & -L_1 & g_{am}\phi_a\phi_m & g_{am}\phi_a\phi_m - \sqrt{2}\chi\phi_m \\ \sqrt{2}\chi\phi_a - g_{am}\phi_a\phi_m & -g_{am}\phi_a\phi_m & L_2 & -g_m\phi_m^2 \\ g_{am}\phi_a\phi_m & g_{am}\phi_a\phi_m - \sqrt{2}\chi\phi_m & g_m\phi_m^2 & -L_2 \end{pmatrix} \times \begin{pmatrix} u_1 \\ w_1 \\ u_2 \\ w_2 \end{pmatrix} = \lambda \begin{pmatrix} u_1 \\ w_1 \\ u_2 \\ w_2 \end{pmatrix}, \tag{8.91}$$

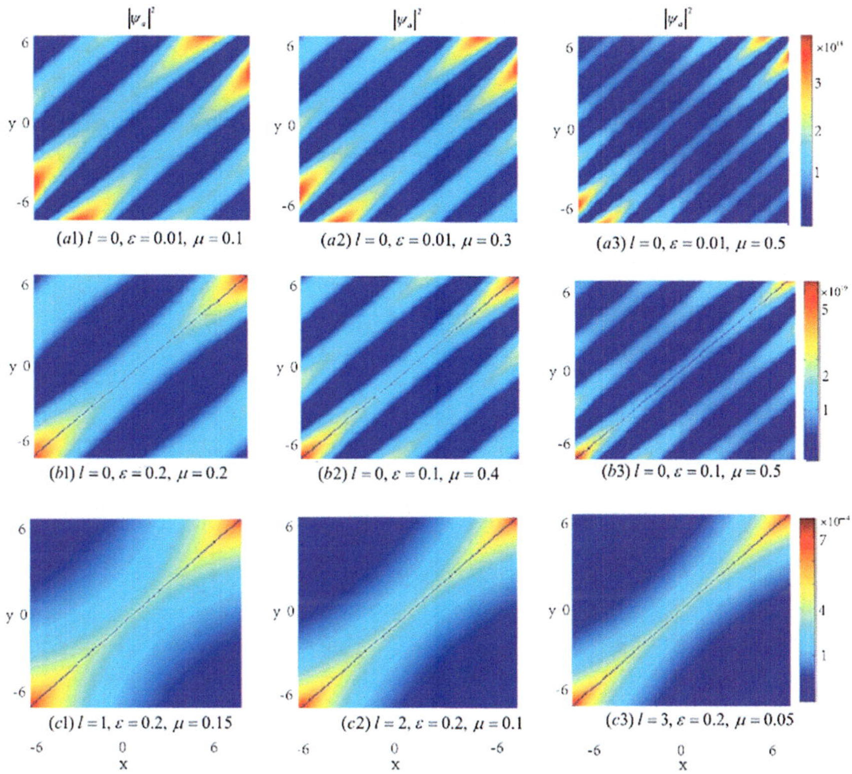

Fig. 8.29 The density distributions $|\Psi_a|^2$ of the atomic-molecular BEC as the function of ε and μ_a with $\omega = 0.02$, $b_{11} = 3$, $b_{22} = 12$, $b_{12} = 1$. The energy packets are striped distribution. **a1**–**b3** show that the number of the energy stripes increases with chemical potential μ_a when the secondary quantum number $l = 0$. **c1**–**c3** show that there is only one density stripe when the secondary quantum number $l \neq 0$. **b1**, **c3** illustrate that some zero points appear on the middle density stripe when the number of the density stripes is odd. The solutions displayed in figures **a1**, **b1** are linear stable [46]

where

$$
\begin{aligned}
L_1 &= \frac{1}{2}\Delta_2 - 2g_a\phi_a^2 - g_{am}\phi_m^2 - \frac{\omega^2}{2}\left(x^2 + y^2\right) + \mu_a, \\
L_1 &= \frac{1}{4}\Delta_2 - 2g_m\phi_m^2 - g_{am}\phi_a^2 - \frac{\omega^2}{4}\left(x^2 + y^2\right) + 2\mu_a - \varepsilon.
\end{aligned}
$$

Numerical experiments show that the eigenvalue λ of the eigenvalue problem (8.91) is real for $n = 1$. This suggests that the localized nonlinear matter wave solution (8.88a) is linearly stable for $n = 1$ and solution (8.88b) is unstable. For the solution (8.89), it can be shown that the linear stability rests on the chemical potential μ_a and the Raman detuning ε (see Fig. 8.30).

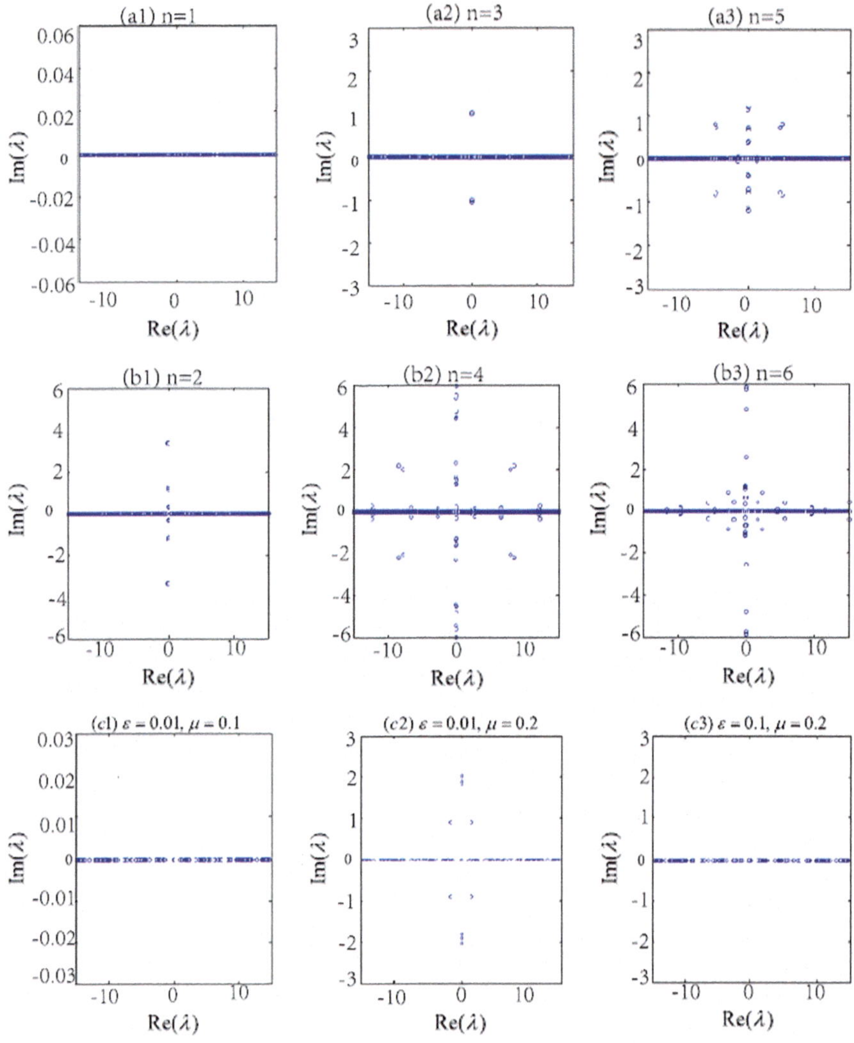

Fig. 8.30 Linear stability. Eigenvalue for different principal quantum numbers n with parameters $b_{11} = 3$, $b_{22} = 12$, $b_{12} = 1$. **a1**–**a3** show that the exact solution (8.88a) is linearly stable only for $n = 1$; **b1**–**b3** show that the exact solution (8.88b) is linearly unstable for all n; **c1**–**c3** illustrate that the solution (8.89) are linearly stable in the two group parameters $\varepsilon = 0.01$, $\mu_m = 0.1$ and $\varepsilon = 0.1$, $\mu_m = 0.2$ [46]

8.6 Crystallized and Amorphous Vortices in Rotating Atomic-Molecular BECs

Amorphous vortices are the result of the considerable deviation induced by the interaction of atomic-molecular vortices. The phase diagram indicates that atom–molecule interaction can control the atomic-molecular vortices to suffer a dramatic dissociation transition from an overlapped atomic-molecular vortices with interlaced molecular vortices to the carbon-dioxide-type atomic-molecular vortices, then to the atomic vortices with interstitial molecular vortices, and finally to the completely separated atomic-molecular vortices. In this section, we investigate the crystallized and amorphous vortices in rotating atomic-molecular BECs [23, 27, 44].

8.6.1 Characterization of Atomic-Molecular BECs

For the characterization of atomic-molecular BECs, we ignore throughout this section the molecular spontaneous emission and the light shift effect [16]. Here, we use the same notations as in the previous section. According to the mean field theory, the two-component NLS equations for characterizing atomic-molecular BECs can be written as [16]

$$\begin{aligned}
i\hbar\frac{\partial\Psi_a}{\partial t} &= \left[-\frac{\hbar^2}{2M_a}\nabla^2 + \frac{M_a\omega^2\left(x^2+y^2\right)}{2}\right]\Psi_a - \Omega\widehat{L}_z\Psi_a \\
&\quad + \left[g_a\left|\Psi_a\right|^2 + g_{am}\left|\Psi_m\right|^2\right]\Psi_a + \sqrt{2}\chi\Psi_a^*\Psi_m, \\
i\hbar\frac{\partial\Psi_m}{\partial t} &= \left[-\frac{\hbar^2}{2M_m}\nabla^2 + \frac{M_m\omega^2\left(x^2+y^2\right)}{2}\right]\Psi_m - \Omega\widehat{L}_z\Psi_m \\
&\quad + \left[g_{am}\left|\Psi_a\right|^2 + g_m\left|\Psi_m\right|^2\right]\Psi_m + \frac{\chi}{\sqrt{2}}\Psi_a^2 + \varepsilon\Psi_m,
\end{aligned} \tag{8.92}$$

where Ψ_j $(j = a,\, m)$ denotes the macroscopic wave function of atomic condensate and molecular condensate respectively, the cubic and quintic nonlinearity the strengths of interaction g_a and g_m and the coupling parameter g_{am} are defined as in the previous section; M_a and M_m are the mass of, respectively, the condensate atom and the molecule, ω is the trapped frequency, Ω is the rotation frequency $\widehat{L}_z\left[\widehat{L}_z = -i\hbar\left(x\frac{\partial}{\partial y} - y\frac{\partial}{\partial x}\right)\right]$ is the z component of the orbital angular momentum. The parameter τ describes the conversions of atoms into molecules due to stimulated Raman transitions. ε is a parameter to characterize Raman detuning for a two photon resonance [16, 19].

In real experiment, it is observed that the coherent free-bound stimulated Raman transition can cause atomic BEC of ^{87}Rb to generate a molecular BEC of ^{87}Rb [50]. In numerical simulations, we use the parameters of atomic-molecular BECs of ^{87}Rb system with $M_m = 2M_a = 2m$, ($m = 144.42 \times 10^{-27}$ Kg), $g_m = 2g_a$ ($a_a = 101.8\ a_B$), where a_B is the Bohr radius, $\chi = 2 \times 10^{-3}$, and the trapped frequency $\omega = 100 \times 2\pi$. Note that if the change in energy in converting two atoms into one molecule ($\Delta U = 2U_{Ta} - U_{Tm}$) [50], not including internal energy, approaches zero, we can obtain the value $2g_a = g_m$. The unit of length, time, and energy correspond to $\sqrt{\hbar/(m\omega)}$ ($\approx$1.07 µm), $1/\omega$ ($\approx 1.6 \times 10^{-3}$ s), and $\hbar\omega$, respectively.

8.6.2 Crystallized and Amorphous Vortices in Rotating Atomic-Molecular BECs

With rotation frequency $\Omega = 0.8\omega$, we study the influence of atom–molecule interaction on the formation of vortices. Figure 8.31 displays the densities and phases obtained under the equilibrium state with different atom–molecule interactions. The first and the second columns are the densities of the atomic BEC and the molecular

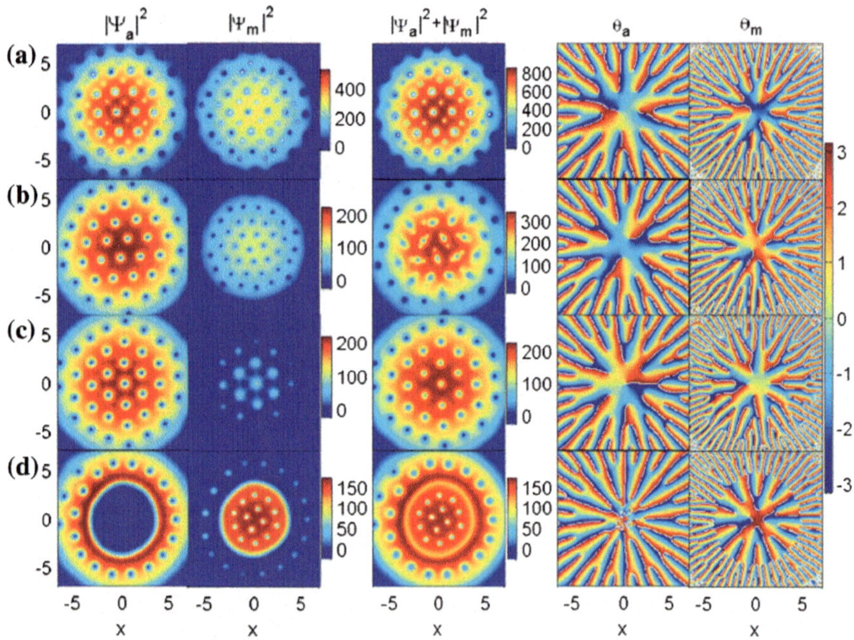

Fig. 8.31 The densities and phases of atomic-molecular BECs of ^{87}Rb when the system reaches the equilibrium state. The rotation frequency is $\Omega = 0.8\omega$. The strength of atom–atom interaction is g_a with the scattering length $a_a = 101.8\ a_B$. The strength of molecule–molecule interaction g_m is twice as much as that of atom–atom interaction. The strength of atom–molecule interaction and Raman detuning is set as **a** $g_{am} = -0.87\ g_a$, $\varepsilon = 14\hbar\omega$, **b** $g_{am} = 0$, $\varepsilon = 14\hbar\omega$, **c** $g_{am} = 0.87$, $\varepsilon = 14\hbar\omega$ and **d** $g_{am} = 2\ g_a$, $\varepsilon = 7\hbar\omega$. Note that the fourth and fifth columns are the phases of atomic and molecular BECs, respectively. The unit of length is 1.07 µm [23]

BEC, respectively. The third column is the total density. The fourth and the fifth column are the corresponding phases of atomic and molecular BECs, respectively. The vortices can be identified in the phase image of BECs.

The composite of atomic and molecular vortices locates at the trap center to lower the system's energy. For the case of attractive atom–molecule interaction ($g_{am} = -0.87 g_a$), vortices form a square lattice [see Fig. 8.31a]. Interestingly, we can observe that the size of the molecular vortices can be divided into two types: one is big, and the other is small. Each atomic vortex has approximately the same size. Figure 8.32a further plots the vortices, where the atomic vortices overlap with a molecular one. The overlapping of atomic and molecular vortices causes the size of molecular vortices to become big. Thus, we obtain different size of molecular vortices in the same experiment.

It is easy to understand the size enlargement of molecular vortices which are overlapping with atomic ones. The size of vortex reflects the healing length [$\xi = \hbar/\sqrt{2mg\overline{n}}$, where $\overline{n}$ is the uniform density in a nonrotating cloud] of the BEC because within this distance, the order parameter 'heals' from zero up to its bulk value. The attractive interspecies interaction implies that the densities of the two BECs would have a similar trend to decrease and increase. It also causes some molecular vortices to overlap with atomic ones. In addition, the density of atomic BEC forms local nonzero minima at the region of the left molecular vortex [see Fig. 8.32a]. Here, the size of atomic vortices is obvious bigger than that of molecular vortices. Therefore, the local density of molecular vortices follows that of atomic vortices and the size becomes big when molecular vortices overlap with atomic vortices.

When $g_{am} = 0$, atomic vortex lattices are triangular and the molecular vortices are amorphous state [see Figs. 8.31b and 8.32d]. Meanwhile, the total density (the third column) indicates that the molecular vortices and the atomic vortices form some structure like the carbon dioxide, which is also observed by a different method [49]. Figure 8.32b shows an enlarged configuration of the carbon dioxide vortices. Here, the size of atomic vortices is much larger than that of the molecular one. With repulsive interaction ($g_{am} = 0.87 g_a$), vortex lattices are approximately hexagonal with a little deviation [see Fig. 8.31c]. Increasing atom–molecule interaction up to $g_{am} = 2 g_a$ [see Fig. 8.31d], atomic-molecular BECs separate into two parts, molecular BEC locating at the center and atomic BEC rounding it. The results are understandable, since the mass of a molecule is twice as that of atom, molecular BEC tends to locate at the center. This is much different from that of the normal two-component BECs, where the same mass and intraspecies interactions are considered [21].

Figure 8.32c–f further illuminate the position of vortices. Note that we do not point out the vortices where the densities of BECs are very low. We approximately view the vortex lattice as triangular, square, etc, although some vortices may deviate from the regular lattice slightly. The distance of adjacent lattice sites of atomic vortices is $\sqrt{2}$ times of that of molecular vortices [see Fig. 8.32c]. We have plotted a green circle to differentiate these vortices as two parts. The atomic vortices construct an approximately quadrangle lattice, especially near the center region, the atomic vortices overlap with a molecular one locating among four adjacent molecular

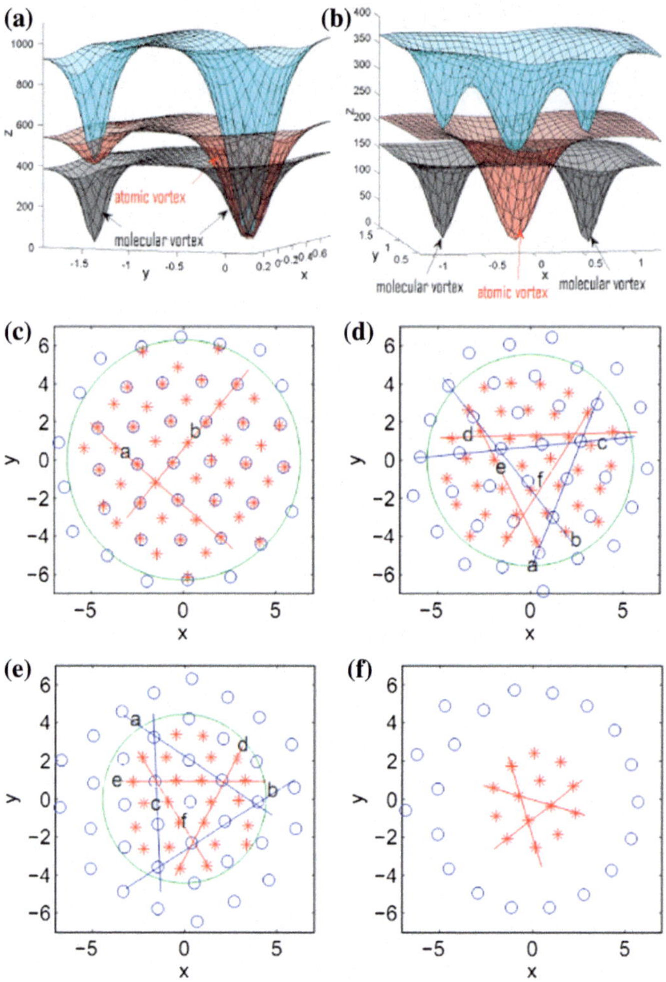

Fig. 8.32 Vortex configurations and vortex position. **a** The scheme of composite vortices in Fig. 8.31a. The size of the right molecular vortex which overlaps with an atomic vortex is bigger than the left one. **b** The scheme of carbon-dioxide-type vortex structure in Fig. 8.31b. The size of the atomic vortex is bigger than that of the molecular vortex. The red, black, and blue indicate the densities of atomic BEC, molecular BEC and the sum, respectively. **c**, **d**, **e** and **f** show the position of vortices in Fig. 8.31a–d, respectively. The circle (∘) and asterisk (∗) are the position of vortices formed by atomic BEC and molecular BEC, respectively. In **c**, the red lines indicate that vortices can array in the square lattice. In **d**, the blue lines show atomic vortices form the triangular lattice. While, the deviation of molecular vortices from the red lines indicates they form the amorphous state. In **e**, atomic vortices form the triangular lattice and molecular vortices form the honeycomb lattices. Similarly, molecular vortices display the triangular lattice in **f**. The unit of length is 1.07 μm [23]

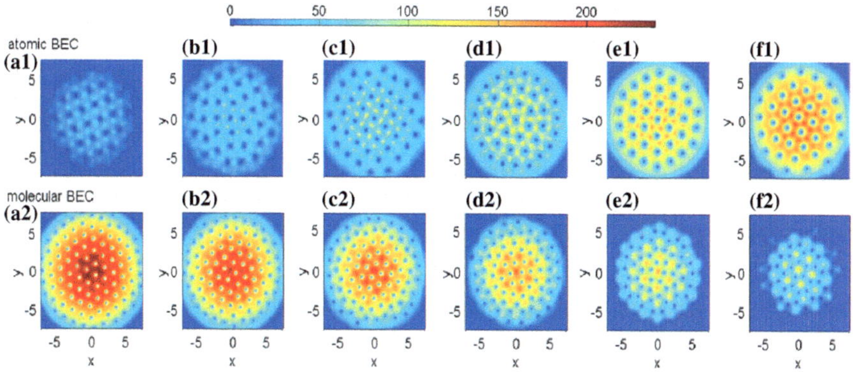

Fig. 8.33 The effect of Raman detuning on the lattice of atomic vortices with interstitial molecular vortices at the equilibrium state. Here the strength of interactions are $g_m = 2g_a$ and $g_{am} = 0.87ga$. The rotation frequency is $\Omega = 0.8\omega$. The upper plots **a1**–**f1** show the densities of atomic BEC and the lower plots **a2**–**f2** indicate the corresponding densities of molecular BEC. The value of Raman detuning varies. **a** $\varepsilon = -2\hbar\omega$, **b** $\varepsilon = 0\hbar\omega$, **c** $\varepsilon = 2\hbar\omega$, **d** $\varepsilon = 4\hbar\omega$, **e** $\varepsilon = 7\hbar\omega$, and **f** $\varepsilon = 10\hbar\omega$. **a1**, **b1** and **c1** are the square lattice. **a2**, **b2**, **e1** and **f1** are the triangular lattice. Other plots show the amorphous state. The unit of length is 1.07 μm [23]

vortices. Thus, vortex position indicates that vortices density of atomic BEC is half of that of molecular vortices. The atomic vortices expand over to the outskirts of the lattice where no overlapped molecular vortices appear [see Fig. 8.32c].

Figure 8.32d indicates that the carbon dioxide structure is not fixed in the same orientation. Similarly to Fig. 8.32c, the carbon dioxide structure only exists at the center. However, the deviation of molecular vortices from the red lines d, e, and f is so large that we have to view the molecular vortices as an amorphous state. Vortex position in Fig. 8.32e shows that atomic vortices form the triangle lattice. All molecular vortices are distributed among atomic vortices, forming the hexagonal lattices without overlapping. Certainly, atomic vortices and molecular vortices are separated in Fig. 8.32f according to the immiscibility of atomic-molecular BECs with strong gam. We can conclude that the strength of atom–molecule interaction can adjust the composite degrees of vortices, and cause the overlapping composite, carbon-dioxide-type composite, interstitial composite, and separation.

Furthermore, we find that the lattice configuration of vortices is very complex when atomic vortices and interstitial molecular vortices coexist. In Fig. 8.31c, atomic vortices form the triangular lattice and interstitial molecular vortices display the honeycomb lattice. We further plot the densities of atomic BEC and molecular BEC at various cases in Fig. 3. When the number of atoms is much more than that of molecules, vortices in atomic BEC tend to form the triangular lattice, and vice versa. The lattice configurations are triangular in Fig. 8.33a2, b2, e1 and f1. Atomic vortices display square lattice in Fig. 8.33a1–c1. In all other subplots, the lattices are irregular and can be viewed as the amorphous state. For example, the number of adjacent molecular vortices which form bubbles [8] around some atomic vortices is not six but five in Fig. 8.33d2–f2. In fact, the regular structures imply that both long-range order and short-range order should remain. Thus, the observed random configuration is really amorphous.

8.6.3 The Phase Diagram of Rotating Atomic-Molecular BECs

To explore the phase diagram of atomic-molecular vortices, we first show the modulation effect of Raman detuning for the number of vortices in rotating atomic-molecular BECs. Figure 8.34a–d show the relationship between vortices number and Raman detuning. Generally speaking, the number of molecular vortices decreases monotonously as Raman detuning increases. As we can see for $g_{am} = -0.87ga$, 0, 0, $0.87g_a$ and $2g_a$, the slope of the number of molecular vortices is -2.7, -3.4, -4.2 and -7.8, respectively. Increasing of the strength of atom–molecule interaction, the faster the number of molecular vortices decreases as Raman detuning increasing. The number of atomic vortices approaches to 40 as the Raman detuning increasing. Thus, the ratio of atomic vortices and molecular vortices is not fixed as the Raman detuning changes in atomic-molecular BECs. Furthermore, we calculate the number of composite vortices, i.e., the atom-vortex number C'_a and the molecule-vortex number C'_m in the green circles in Fig. 8.32c–e, and define the parameter $P_m = 100C'_a/C'_m$. P_a in Fig. 8.34a, b, and c are almost around the value of 50, i.e, C'_a: $C'_m \approx 1:2$. Thus, the composite vortices keep the ratio 1:2 approximately. The deviation of P_a from dash black line with the value of 50 mainly comes from vortices at the boundary. Certainly, vortex number in pure atomic BEC is independent of both atom–molecule interaction and Raman detuning.

Phase diagram of rotating atomic-molecular BECs of ^{87}Rb when the system reaches the equilibrium state. AMBEC (I) denotes the miscible mixture of atomic-molecular BECs, and AMBEC (II) is immiscible atomic-molecular BEC. Furthermore, based on the phase diagram of atomic-molecular BECs, we further plot the phase diagram of atomic-molecular vortices when the atomic-molecular BECs of ^{87}Rb reaches the equilibrium state. Then, the region of AMBEC (I) is divided into three parts: (1), (2), and (3). The overlapped atomic-molecular vortices, carbon-dioxide-type atomic-molecular vortices and atomic vortices with the interstitial molecular vortices occur in region (1), region (2) and region (3), respectively. In the green region, atomic and molecular vortices match fully with the rough ratio 1:2. The parameters are $\Omega = 0.8\omega$, $g_m = 2g_a$ ($a_a = 101.8a_B$), and $\chi = 2 \times 10^{-3}$. The units of detuning parameter and gam are $\hbar\omega$ and g_a, respectively. Now, we further indicate the modulation effect of Raman detuning for particle numbers of atomic-molecular BEC of ^{87}Rb [see Fig. 8.34e–h]. The particle numbers in equilibrium state depend on the system itself. For attractive atom–molecule interaction $g_{am} = -0.87g_a$, both of atom number and molecule number decrease when the Raman detuning increases. Meanwhile, atom number always is greater than molecule number [see Fig. 8.34e].

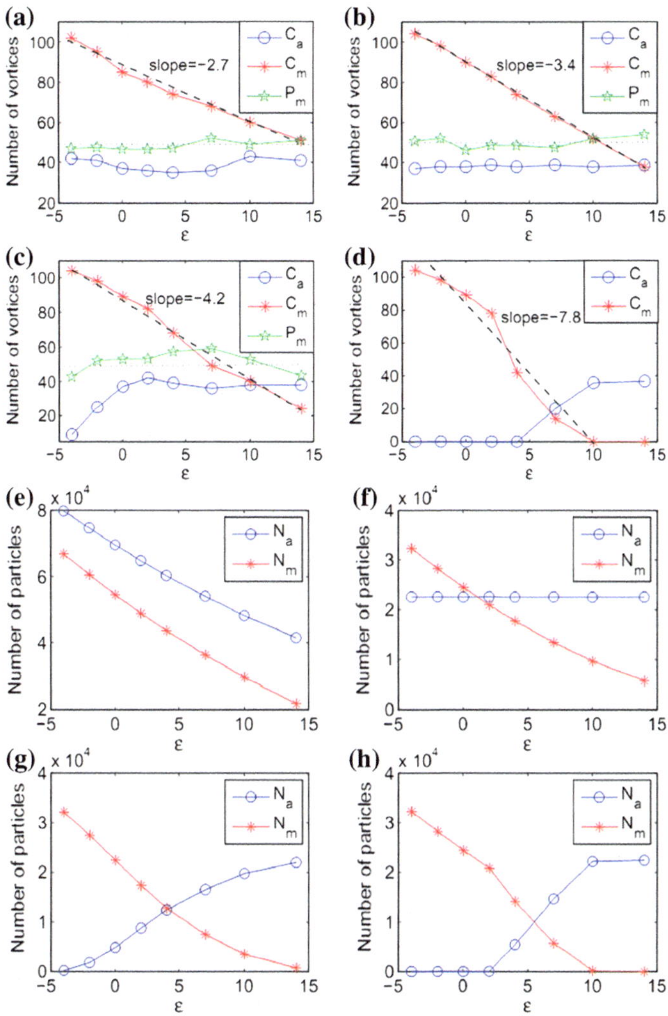

Fig. 8.34 The number of vortices and particles. **a**–**d** show the number of atomic vortices Ca and molecular vortices Cm in atomic-molecular BECs of ^{87}Rb with the detuning parameter **e** when the system reaches the equilibrium state. **a** $g_{am} = -0.87ga$, **b** $g_{am} = 0$, **c** $g_{am} = 0.87g_a$ and **d** $g_{am} = 2g_a$. **e**–**h** indicate the corresponding particle number of atomic-molecular BECs of ^{87}Rb, respectively. The rotation frequency is $\Omega = 0.8\omega$, the strength of molecule–molecule interactions are $g_m = 2g_a$ with the atom–atom scattering length $a_a = 101.8a_B$, and the parameter χ is fixed to be 2×10^{-3}. The unit of detuning parameter is $\hbar\omega$ [23]

For the limit case of $g_{am} = 0$, atom number keeps unchanged and only molecule number decreases as the Raman detuning increases [see Fig. 8.34f]. When the interaction is repulsive ($g_{am} > 0$), molecule number keeps decreasing but atom number increases [see Fig. 8.34g and h]. When the repulsive interaction is up to $g_{am} = 2g_a$, the single molecular BEC or the single atomic BEC can be obtained by adjusting Raman detuning from $-4\hbar\omega$ to $14\hbar\omega$. The particle numbers can characterize the possible regions for the existence of atomic-molecular vortices.

Figure 8.35 plots the phase diagram of atomic-molecular BECs. The stable atomic-molecular BECs system exists only when atom–molecule interaction is larger than $-\sqrt{g_a g_m}$. When the Raman detuning is large enough, single atomic BEC occurs. Oppositely, if the Raman detuning is low enough, production changes into pure molecular BEC. Between these two regions, it is atomic-molecular BECs, where AMBEC (I) denotes the miscible mixture and AMBEC (II) stands for the phase separated mixture. Therefore, to explore atomic-molecular vortices, we mainly focus on AMBEC (I) region.

According to above analysis about atomic-molecular vortices and the corresponding atomic-molecular BECs, we calculate lots of other results and finally give a vortex phase diagram to summarize the vortex structures in Fig. 8.35. In (1) region $[-\sqrt{g_a g_m} < g_{am} < 0.1 g_a]$, atomic-molecular vortices form the square lattice where the overlapped atomic-molecular vortices and the molecular vortices interlacedly exist. The carbon-dioxide-type atomic-molecular vortices occur in (2) region $[-0.1 g_a \le g_{am} < 0.5 g_a]$. Atomic vortices with interstitial molecular vortices emerge in region (3). In the AMBEC (II) region, atomic vortices and molecular vortices are separated. Certainly, in the region of atomic BEC (molecular BEC), atomic vortices (molecular vortices) favor to form the triangular lattice. The green region indicates that the created atomic and molecular vortices fully match with each other by roughly the ratio 1:2. Above the green region, more atomic vortices occur. Below the green region, more molecular vortices appear. Table 8.4 shows a summary of the details of various vortices in rotating atomic-molecular BECs.

Interestingly, the vortex phase diagram indicates some exotic transitions. (i) Imbalance transition: The increase of Raman detuning causes more atomic BEC. Thus, the pure molecular vortices change into carbon-dioxide-type atomic-molecular vortices (atom vortices with interstitial molecular vortices, and separated atomic-molecular vortices), and finally into single atomic vortices in region of $0 < g_{am} < 0.5 g_a$ ($0.5 g_a \le g_{am} \le \sqrt{g_a g_m}$ and $g_{am} > \sqrt{g_a g_m}$, respectively). In the region of $-\sqrt{g_a g_m} < g_{am} < -0.1 g_a$ ($-0.1 g_a \le g_{am} \le 0$), the interlaced-overlapped atomic-molecular vortices (the carbon-dioxide-type atomic-molecular vortices) become into atomic vortices under a very high detuning parameter. (ii) Dissociation transition: By changing the atom–molecule interaction from attractive to repulsive, the composite atomic-molecular vortices change from overlapped to carbon-dioxide-type and finally into the independent separated ones.

8.6.4 Discussion

In this section, we focus on the strength of atom–molecule interaction and the Raman detuning term. The form of Hamiltonian here is like that in the Ref. [49]. In fact, a real experiment would include lots of other factors such as the light shift effect [41],

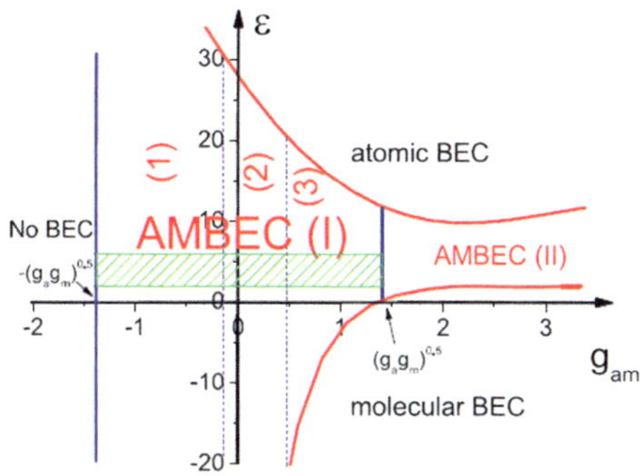

Fig. 8.35 Phase diagram of rotating atomic-molecular BECs of ^{87}Rb when the system reaches the equilibrium state. AMBEC (I) denotes the miscible mixture of atomic-molecular BECs, and AMBEC (II) is immiscible atomic-molecular BEC. Furthermore, based on the phase diagram of atomic-molecular BECs, we further plot the phase diagram of atomic-molecular vortices when the atomic-molecular BECs of ^{87}Rb reaches the equilibrium state. Then, the region of AMBEC (I) is divided into three parts: (*1*), (*2*), and (*3*). The overlapped atomic-molecular vortices, carbon-dioxide-type atomic-molecular vortices and atomic vortices with the interstitial molecular vortices occur in region (*1*), region (*2*) and region (*3*), respectively. In the green region, atomic and molecular vortices match fully with the rough ratio 1 : 2. The parameters are $\Omega = 0.8\omega$, $g_m = 2g_a$ ($a_a = 101.8a_B$), and $\chi = 2 \times 10^{-3}$. The units of detuning parameter and gam are $\hbar\omega$ and g_a, respectively [23]

Table 8.4 A summary of the properties of vortices in the rotating atomic-molecular BECs of ^{87}Rb when the system reaches the equilibrium state. The atomic-molecular vortices are composite in the matching region [inside the green circle in Fig. 8.32c–e]

Region (in Fig. 8.35)	Vortex state (in the matching region)	Lattice of atomic vortex	Lattice of molecular vortex	Vortex lattice out of the matching region
(1)	Overlapped atomic-molecular vortices with interstitial molecular vortices	Square	Square	Triangular
(2)	Carbon-dioxide-type atomic-molecular vortices	Triangular	Amorphous	Triangular
(3)	Atomic vortices with interstitial molecular vortices	Square/amorphous/triangular	Triangular/amorphous/ honeycomb	Triangular
AMBEC (II)	Separated atomic vortices and molecular vortices	Triangular	Triangular	No
Atomic BEC	Pure atomic vortices	Triangular	No	No
Molecular BEC	Pure molecular vortices	No	Triangular	No

decay due to spontaneous emission. In Ref. [41], Gupta and Dastidar have considered a more complicated model when they study the dynamics of atomic and molecular BECs of ^{87}Rb in a spherically symmetric trap coupled by stimulated Raman photo association process. In fact, the light shift effect almost has the same function as the Raman detuning term. Thus, it can be contributed to the Raman detuning term. This is the reason why we do not consider the light shift term in Hamiltonian like that in Ref. [41], but follows the form in Ref. [49].

In real experiment, it is believed that the single molecular BEC would occur when the Raman detuning goes to zero. However, the measure of the remaining fraction of atom does not reach the minimum when Raman detuning is zero. With the adiabatic consideration, the dynamical study also agrees with this point [41]. In fact, they show the evolutionary process of creating a molecular BEC from a single atomic BEC. Thus, particle number of molecular BEC varies with time but not fixed. The resonance coupling would cause the atomic BEC to convert into a molecular one as much as possible, but the molecular BEC also will convert into the atomic one. Therefore, the results in Ref. [41] only shows a temporary conversion of atoms into molecules. In fact, when we use single atomic BEC as the initial condition and set $\frac{\gamma_j}{k_B T} = 0$, the temporary conversion of atomic BEC into molecular BEC can be observed with current damped projected NLS equations.

It is obvious that the Raman detuning term in the Hamiltonian behaves just like the chemical potential to control the system's energy. The external potential for atomic BEC is fixed to be $V_a(r)$ and molecular BEC experiences the trap potential $V_m(r) + \varepsilon$. Here, our method initially derives from the finite-temperature consideration: the system is divided into the coherent region with the energies of the state below E_R and the noncoherent region with the energies of the state above E_R [37]. So, our method will behavior just likes to catch the particles with a shallow trap and exchange particles with an external thermal reservoir. But ultimately we remove the external thermal reservoir to get system to the ground state. Raman detuning changes the depth of shallow trap to $\mu_m - \varepsilon$. The molecular BEC will be converted by atoms until the system reaches the equilibrium state. Therefore, a maximum of creating molecular BEC does not occur at the equilibrium state when Raman detuning varies. Instead, molecule number decreases monotonously when Raman detuning increases.

Why do atomic-molecular vortices display so rich lattice configurations? In fact, atomic vortices and molecular vortices tend to be attractive in region (1) and (2). Otherwise, the overlapped atomic-molecular vortices and the carbon-dioxide-type ones cannot occur. The attractive force makes atomic vortices and molecular vortices behave similarly. Thus, both atomic and molecular vortex lattices in region (1) are square. In region (2), atomic vortices display the triangular lattice. Molecular vortices seem to follow the triangular lattice but the interaction among vortices causes the considerable deviation. Obviously, the CO_2-type structures do not follow the fixed direction, i.e., long-range order vanishes but there is still short-range order. Thus, we have to view molecular vortices as the amorphous state. In region (3), atomic vortices and molecular vortices cannot form the carbon dioxide structure. Because the size of molecular vortices is smaller than that of atomic vortices, it tends to locate at the interval of the lattice of atomic vortices. When the number of one component is much more than that of the other, the vortices of this component dominate over the vortices of the other component. The former is easy to form the regular vortex lattice. The latter has to follow the interaction of the former and forms the vortex lattice. The amorphous state originates from the competition between atomic vortices and molecular vortices, especially when the number of atom and molecule has the considerable proportion [see Fig. 8.33d1 and d2]. In that case, short-range order is only partly kept and ultimately long-range order is destroyed. Certainly, this also causes the distribution of vortices in one component is relatively regular and that in the other component is amorphous.

The structural phase transitions of vortex lattices are explored through tuning the atom–molecule coupling coefficient and the rotational frequency of the system [41]. Certainly, the Archimedean lattice of vortices in Ref. [41] is one of the interstitial-composite-structures. Here, we show the crystallized and amorphous vortices by the combined control of Raman detuning and atom–molecule interaction. In fact, when we increase the value of χ, the CO_2-type structure of vortices are easy to be created. Even the interstitial-composite structure we now obtain in Fig. 8.33 would transfer into the CO_2-type structure if χ is big enough. We have also considered the effect of rotation frequency. With the attractive interaction of atom–molecule ($g_{am} = -0.87g_a$), Fig. 8.36 shows various rotation frequencies to produce the vortices. Figure 8.36a indicates that no vortex would occur with $\Omega = 0$. For $\Omega = 0.2\omega$, only one molecular vortex is induced. In atomic BEC, the phase indicates no vortex is created although there is a local minimum of density near the center. For $\Omega = 0.4\omega$, the phase indicates that there is an atomic vortex. In fact, we find the atomic vortex is overlapped with a molecular vortex. Undoubtedly, more and more vortices emerge when rotation frequency increases. When the rotation frequency is up to $\Omega = 0.8\omega$, we can obtain a regular square vortex lattice. Meanwhile, each atomic vortex is overlapped with a corresponding molecular vortex. Obviously, vortices and vortex lattice may not be induced with a slow rotation. This is the reason why we favor to investigate the vortices with a fast rotation in Figs. 8.31, 8.32, 8.33 and 8.34.

We now show that ultracold Bose gases of ^{87}Rb atoms are a candidate for observing the predicted atomic-molecular vortices. By initially using a large atomic BEC of ^{87}Rb (the atom number is up to 3.6×10^5 in Wynar's experiment), the Raman photoassociation of atoms can produce the corresponding molecular BEC with partial of the atoms. By loading a pancake-like optical trap $V_{j=a,m}(x, y, z) = \frac{M_j}{2}\left[\omega^2(x^2+y^2)+\omega_z^2 z^2\right]$, with trapping frequencies $\omega_z >> \omega^{1-3}$, the 2D atomic-molecular BECs may be prepared. It is convenient to use the laser to rotate the atomic-molecular BECs and induce the atomic-molecular vortices. Meanwhile, the whole system should be further quenched to a lower temperature to approach the ground state by the evaporative cooling techniques. The resulting atomic-molecular vortices may be visualized by using the scanning probe imaging techniques. All the techniques are therefore within the reach of current experiments.

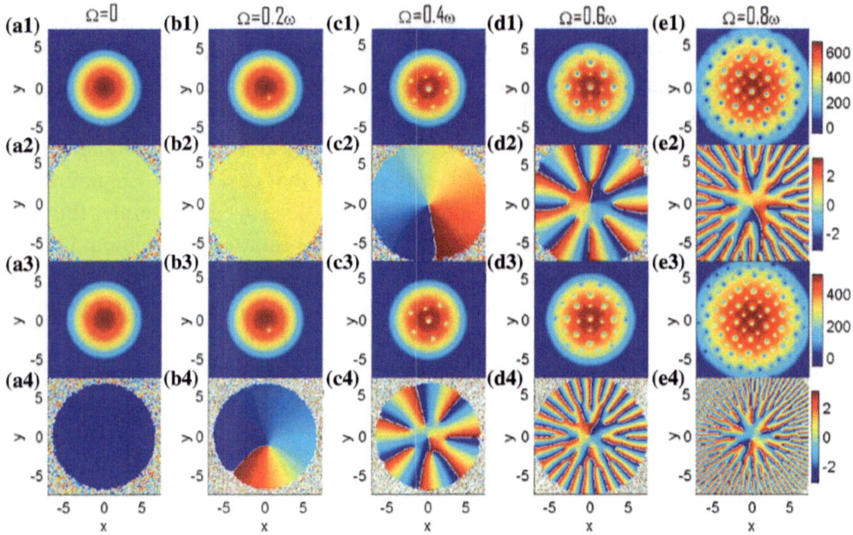

Fig. 8.36 The densities and phases of the atomic-molecular BECs of ^{87}Rb under various rotating frequencies when the system reaches the equilibrium state. The rotating frequencies are indicated at the title of the subplots. **a1**–**e1** are the densities of atomic BEC, **a2**–**e2** are the corresponding phases of atomic BEC, **a3**–**e3** are the densities of molecular BEC, and **a4**–**e4** are the corresponding phases of molecular BEC, respectively. The critical rotating frequencies for inducing molecular vortex and atomic vortex are about 0.1ω and 0.3ω, respectively. The strength of atom–molecule interaction is $g_{am} = -0.87g_a$ with the atom–atom scattering length $a_a = 101.8a_B$, molecule–molecule interactions is $g_m = 2g_a$, the parameter χ is fixed to be 2×10^{-3} and Raman detuning is $\varepsilon = 0\hbar\omega$. The unit of length is 1.07 μm [23]

8.7 Half-Vortex Sheets of the Rotating Two-Component Bose–Einstein Condensates in Spin-Dependent Optical Lattices

In this section, we consider the rotating two-component BECs in 1D spin-dependent optical lattices and study their half-vortex sheets.

8.7.1 Description of the Model

We consider two-component BECs with two hyperfine spin states. In the weak interaction limit, the two-component condensates in a frame rotating at an angular frequency Ω around the z axis can be described by the two-component NLS equations

$$i\hbar\frac{\partial\Psi_j(\mathbf{r},t)}{\partial t}=\left[-\frac{\hbar^2}{2m}\nabla^2+V_H+\sum_{k=1,2}g_{jk}\left|\Psi_k\right|^2+V_{OLj}-\Omega\widehat{L}_z\right]\Psi_j(\mathbf{r},t), \tag{8.93}$$

where $\Psi_j(\mathbf{r},t)$ is the macroscopic wave function of the j-th component ($j=1,\,2$). $g_{jk}=4\pi\hbar^2 a_{jk}/m$ represents the strength of interatomic interactions characterized by the intracomponent and intercomponent s-wave scattering lengths a_{jk} and the mass m of an atom. $\widehat{L}_z=-i\hbar^2\left(x\partial/\partial y-y\partial/\partial x\right)$ is the z component of the angular momentum operator. The external potential consists of two parts, the harmonic trapping potential $V_H=\frac{1}{2}m\left[\omega_\perp^2\left(x^2+y^2\right)+\omega_z^2 z^2\right]$ and the spin-dependent optical lattice potential V_{OLj}, where $V_{OL1}=I_0\sin^2[kn]$, $V_{OL2}=I_0\cos^2[kx]$. Here k is the wave vector of the laser light used for the optical lattice potentials and I_0 is the potential depth of the lattices. The wave functions are normalized as $\sum_j\int\left|\Psi_j\right|^2 d\mathbf{r}=N$, where N is the total number of condensate atoms.

For simplicity, we assume that the harmonic trapping frequencies satisfy $\omega_z>>\omega_\perp$. Then, the condensates are pressed into a pancake. This allows us to reduce Eq. (8.93) to a two-dimensional form as

$$i\hbar\frac{\partial\psi_j(x,y,t)}{\partial t}=\left[-\frac{\hbar^2}{2m}\Delta_2+\widetilde{V}_H+\sum_{k=1,2}\eta g_{jk}\left|\psi_k\right|^2+V_{OLj}-\Omega\widehat{L}_z\right]\psi_j(x,y,t), \tag{8.94}$$

where $\eta=1/\sqrt{\hbar/m\omega_\perp}$ is a reductive parameter. The two-dimensional wave functions are normalized as $\sum_j\int\left|\psi_j\right|^2 dxdy=N$. The harmonic trapping potential is reduced to its 2D form $\widetilde{V}_H=\frac{1}{2}m\omega_\perp^2\left(x^2+y^2\right)$, where the tilde will be omitted in the following discussions for simplicity.

For numerical calculations, it is convenient to make the following parameter transformations

$$\begin{aligned}x=a_h\widetilde{x},\ y=a_h\widetilde{y},\ t=\widetilde{t}/\omega_\perp,\ k=\widetilde{k}/a_h,\ I_0=\widetilde{I}_0\hbar\omega_\perp,\\ \Omega=\omega_\perp\widetilde{\Omega},\ \psi_j=\sqrt{N}\widetilde{\psi}_j/a_h,\ \beta_{jk}=4\pi N\eta a_{jk},\end{aligned} \tag{8.95}$$

where $a_h=\sqrt{\hbar/m\omega_\perp}$ is the characteristic length of the harmonic trap. Then the two-component NLS equation Eq. (8.94) is reduced to a dimensionless form as

$$\begin{aligned}i\hbar\frac{\partial\widetilde{\psi}_1}{\partial t}&=\left[-\frac{1}{2}\widetilde{\Delta}_2+\frac{1}{2}\left(\widetilde{x}^2+\widetilde{y}^2\right)+\sum_{k=1,2}\beta_{1k}\left|\psi_k\right|^2+\widetilde{I}_0\sin^2\left[\widetilde{k}\widetilde{x}\right]\right.\\&\quad\left.+i\widetilde{\Omega}\left(\widetilde{x}\partial/\partial\widetilde{y}-\widetilde{y}\partial/\partial\widetilde{x}\right)\right]\widetilde{\psi}_1,\\ i\hbar\frac{\partial\widetilde{\psi}_2}{\partial t}&=\left[-\frac{1}{2}\widetilde{\Delta}_2+\frac{1}{2}\left(\widetilde{x}^2+\widetilde{y}^2\right)+\sum_{k=1,2}\beta_{2k}\left|\psi_k\right|^2+\widetilde{I}_0\sin^2\left[\widetilde{k}\widetilde{x}\right]\right.\\&\quad\left.+i\widetilde{\Omega}\left(\widetilde{x}\partial/\partial\widetilde{y}-\widetilde{y}\partial/\partial\widetilde{x}\right)\right]\widetilde{\psi}_2,\end{aligned} \tag{8.96}$$

where β_{jk} represents the 2D dimensionless effective interaction parameters, determined by the s-wave scattering lengths, the total number of condensate atoms, and the axial trapping frequency. The two-dimensional dimensionless wave functions are normalized as $\sum_j \int |\psi_j|^2 \, d\widetilde{x} d\widetilde{y} = 1$.

In order to describe the system more clearly, the intuitive pictures of the external potentials are presented in Fig. 8.37a–f. Experimentally, the spin-dependent optical lattice potentials V_{OL1} and V_{OL2} can be realized by employing two counterpropagating blue-detuned laser beams with the same frequency but perpendicular linear polarization vectors [13, 17, 53]. A schematic of the spin-dependent optical lattices is presented in Fig. 8.37g.

In our simulations, stemming from the consideration of experiments, we choose a two-level ^{87}Rb BEC system with $|F = 1, m_f = -1\rangle \equiv |1\rangle$ and $|F = 2, mf = 1\rangle \equiv |2\rangle$ [28]. The ^{87}Rb atoms are assigned to the two states equally and the total number of them is $N = 105$. The radial and axial trapping frequencies are $\omega_\perp = 2\pi \times 15$ Hz and $\omega_z = 2\pi \times 150$ Hz, respectively. We use the scattering lengths [28]: $a_{11} = 100.40a_0$, $a_{22} = 95.00a_0$, and $a_{12} = a_{21} = 97.66a_0$ (a_0 is the Bohr radius), The intensity of the laser light used for the optical lattice potentials is chosen as $I_0 = k_B \times 27.5$ nK (k_B is the Boltzmann's constant) [5], which is powerful enough such that the two states are phase separated. With these experimental parameters, according to the transformation relation Eq. (8.95), we can calculate the corresponding computational parameters for our numerical simulations as: $\beta_{11} = 3025$, $\beta_{22} = 2862$, $\beta_{12} = \beta_{21} = 2942$, and $\widetilde{I}_0 = 38.2$.

8.7.2 *Half-Vortex Sheets*

Here, we wish to present a stable vortex configuration, straight half-vortex sheets. By using the imaginary time propagation method [10], we solve Eq. (8.96) numerically and obtain the ground state of the two-component condensates. Without loss of generality, we choose the rotating angular frequency $\Omega = 0.6\omega_\perp$ and the period of the optical lattice potential $T = \pi\xi$, where $\xi = (\hbar/m\omega_\perp)^{-1}$ is the spatial scale. The density profiles are presented in Fig. 8.38. All the vortices are denoted by crosses ($\times$), whose positions are determined by the singularities of the phase. As the presence of the spin-dependent optical lattices, the two-component condensates undergo phase separation in the form of stripes arranged alternately. It is energetically favorable for the vortices to site in the low-density regions, so we find that all the vortices of one component are aligned in lines in the trough of the optical lattice potential and filled with the other component. This results in alternately arranged straight vortex sheets in the two components. Obviously, all the positions of the vortices in one component are vortex-free regions in the other component, so all the vortices are half quantized. We refer to this vortex configuration as straight half-vortex sheets. It should be indicated that the rotational frequency Ω for the realization of the straight half-vortex sheets is not limited to $0.6\omega_\perp$, but it can take the value from 0 to $\omega_\perp$. Even

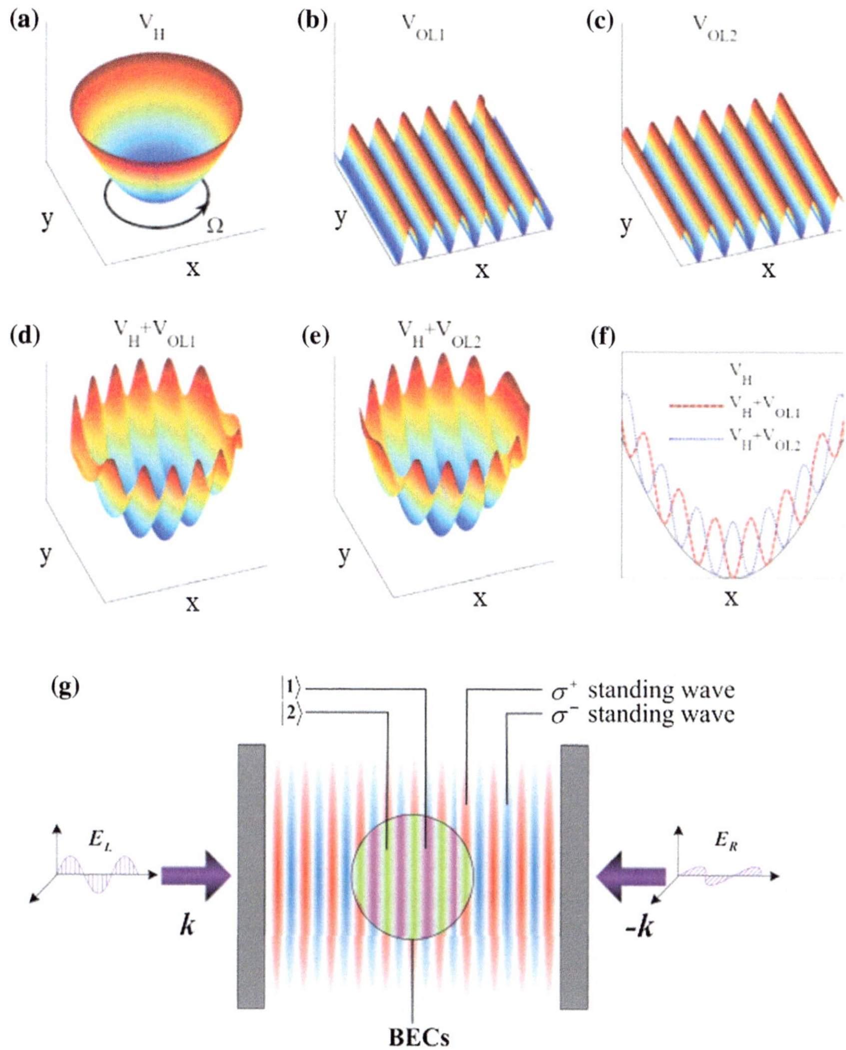

Fig. 8.37 **a** The harmonic trapping potential V_H is rotated about the z axis at a frequency Ω. **b** The spin-dependent optical lattice potential $V_{OL1} = I_0 \sin^2[kx]$, which is experienced by the $|1\rangle$ state. **c** The spin-dependent optical lattice potential $V_{OL2} = I_0 \cos^2[kx]$, which is experienced by the $|2\rangle$ state. **d** The composite potential $V_H + V_{OL1}$. **e** The composite potential $VH + VOL2$. **f** Cross sections of V_H (solid line), $V_H + V_{OL1}$ (dashed line), and $V_H + V_{OL2}$ (dotted line) along the x axis. **g** Schematic of the spin-dependent optical lattices. Two polarized standing wave laser fields σ^+ (red) and σ^- (blue) are formed by two counterpropagating blue-detuned laser beams with the same frequency but perpendicular linear polarization vectors. This gives rise to the optical lattice potentials V_{OL1} and V_{OL2}, which are experienced by the $|1\rangle$ state (pink) and the $|2\rangle$ state (green), respectively

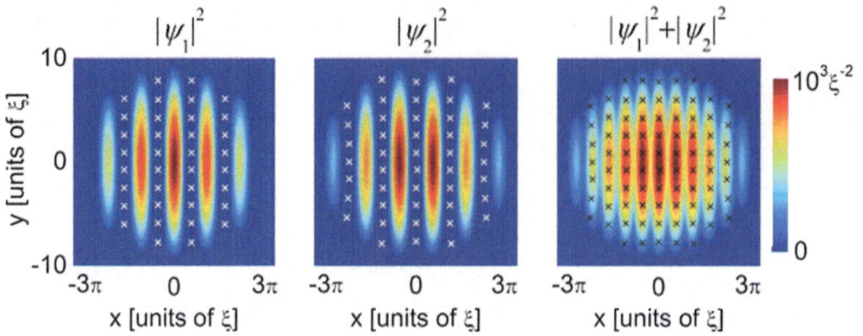

Fig. 8.38 The ground state density profiles of $|\psi_1|^2$, $|\psi_2|^2$, and $|\psi_1|^2 + |\psi_2|^2$ for the rotating angular frequency $\Omega = 0.6\omega_\perp$ and the period of the optical lattice potential $T = \pi\xi$ with $\xi = \sqrt{\hbar/m\omega_\perp}$. The locations of the vortices are marked by crosses ($\times$)

though the half-vortex sheets do not influence the total density distribution of the condensates, they are crucial for the formation of the spatially periodic spin textures on the domain walls.

In the absence of the spin-dependent optical lattices, similar straight vortex sheets configuration can also be observed in the pure harmonic trap [20]. However, in that case, the intercomponent and intracomponent interaction strengths are required to be nearly equal, so the parameter region of this phase is extremely narrow. At the same time, the straight vortex sheets are unstable as there exist many different metastable bent vortex sheet configurations with almost the same energy as the straight one. This is because the energy of the vortex sheets is mainly determined by two factors, the intervortex spacing within a vortex sheet and the inter-sheet spacing. The shape of the vortex sheets has slight influence on the energy of the system. Thus, it is easy for the straight vortex sheets to bend under external disturbances. In contrast, when the spin-dependent optical lattices are present, the shape of the vortex sheets becomes an important factor in determining the energy of the system, and any bending in the straight vortex sheets will cost much energy. So the straight half-vortex sheets configuration obtained in our system can be maintained stably. In our imaginary time propagations, for any trial initial configurations, the straight vortex sheets configuration is always uniquely obtained after sufficient convergence of the energy. In addition, this configuration does not face the intracomponent interaction strength limitations. Although the intercomponent and intracomponent interaction strengths, which are chosen in Fig. 8.38, happen to be approximately equal, we find that for a large difference between intercomponent and intracomponent interaction strengths, stable straight vortex sheets configuration can also be obtained.

Next, we study the unique superfluid velocity behavior of this structure. The most well-known character of a vortex sheet is the discontinuity of the tangential component of the velocity across the sheet. The regular arrangement of the straight vortex sheets allows us to observe this phenomenon clearly. In Fig. 8.39a, the tangential components (the components along the y axis) υ_{1y} and υ_{2y} of the superfluid velocities $\mathbf{v}_1$ and $\mathbf{v}_2$ of the two states are presented. We can see that both υ_{1y} and υ_{2y}

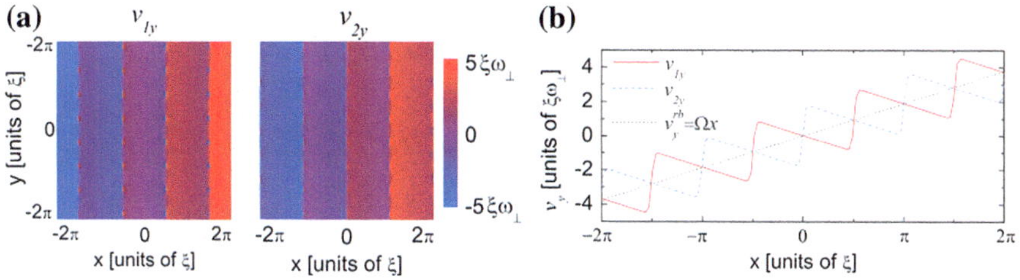

Fig. 8.39 **a** The tangential components υ_{1y} (left) and υ_{2y} (right) of the superfluid velocities $\mathbf{v}_1$ and $\mathbf{v}_2$ of the two states for $\Omega = 0.6\omega_\perp$ and $T = \pi\xi$. **b** Section views of υ_{1y} (solid line) and υ_{2y} (dashed line) along the x axis. The y component $\upsilon_y^{rb} = \Omega x$ of the rigid body rotation velocity $\mathbf{v}^{rb}$ is shown by the dotted line for comparison

discontinuously jump across every sheet. In order to describe the tangential velocities in detail, the section views of υ_{1y} and υ_{2y} along the x axis are shown in Fig. 8.39b. The y component $\upsilon_y^{rb} = \Omega x$ of the rigid body rotation velocity $\mathbf{v}^{rb} = \Omega \times \mathbf{r}$ is presented for comparison. We find that both the tangential velocities υ_{1y} and υ_{2y} have a sawtooth-like change following the rigid body value υ_y^{rb}. The value of υ_{jy} jumps $2\Omega b$ across the sheet in each component and then decreases Ωb linearly in an inter-sheet spacing b. Here, the inter-sheet spacing b is defined as the length between two neighboring sheets in the same component.

The numerical results obtained above can be understood analytically. Considering that the y component of the rigid body rotation velocity is $\upsilon_y^{rb} = \Omega x$, which is independent of y, we suppose that the tangential components of the superfluid velocity on both sides of the sheet υ_{jy}^- and υ_{jy}^+ are also independent of y. According to Onsager–Feynman quantization condition [14]

$$\int_C \mathbf{v}_s \cdot d\mathbf{l} = \frac{2\pi\hbar}{m} N_\upsilon, \tag{8.97}$$

if we choose the two sides of the sheet as the integration path, we can calculate that the tangential velocity jump across a sheet is

$$\Delta\upsilon_{jy} = \upsilon_{jy}^+ - \upsilon_{jy}^- = \frac{2\pi\hbar}{m}\frac{1}{d_\upsilon}, \tag{8.98}$$

where d_υ is the intervortex spacing within a vortex sheet. This implies that the tangential velocity jump across the sheet is only determined by the intervortex spacing within the sheet. According to the Feynman relation, the mean vortex density of each component can be estimated as

$$n_1 = n_2 = \frac{m\Omega}{\hbar\pi}, \tag{8.99}$$

so we can obtain that the intervortex spacing within a vortex sheet is

$$d_v = \frac{\pi \hbar}{m\Omega T}. \tag{8.100}$$

Substituting Eq. (8.100) into Eq. (8.98) yields

$$\Delta v_{jy} = 2\Omega T. \tag{8.101}$$

Equation (8.101) suggests that the velocity jump can be accurately controlled by adjusting the rotating angular frequency Ω and the period of the optical lattice potential T. As the inter-sheet spacing b is just equal to the period of the optical lattice potential T, the tangential velocity jump across the sheet can also be expressed as

$$\Delta v_{jy} = 2\Omega b. \tag{8.102}$$

Meanwhile, in order to follow the rigid body value v_y^{rb}, v_{jy} must decrease Ωb in an inter-sheet spacing. These analytical results agree well with the numerical simulations above.

Further numerical simulations with different choices of computational parameters in Eq. (8.96) show that the straight half-vortex sheets can also be obtained in any other two-component BEC system consisting of two hyperfine spin states. For example, a two-level ^{23}Na BEC system with $|F = 1, m_f = 1\rangle$ and $|F = 1, m_f = 0\rangle$. In addition, for the case of a two-component BEC system consisting of different atomic species, as the atomic masses of the two components are unequal, the vortex densities of the two components will differ from one another. We expect that more complicated straight half-vortex sheet structures, with the sheets of different components carrying a different number of vortices, would be obtained.

References

1. Abdullaev, F.Kh., Caputo, J.G., Kraenkel, R.A., Malomed, B.A.: Controlling collapse in Bose-Einstein condensates by temporal modulation of the scattering length. Phys. Rev. A **67**, 013605 (2003)
2. Afanasjev, V.V., Malomed, B.A., Chu, P.L.: Stability of bound states of pulses in the Ginzburg-Landau equations. Phys. Rev. E **56**, 6020 (1997)
3. Belmonte-Beitia, J., Perez-Garcia, V.M., Vekslerchik, V., Konotop, V.V.: Localized nonlinear waves in systems with time- and space-modulated nonlinearities. Phys. Rev. Lett. **100**, 164102 (2008)
4. Bronski, J.C., Carr, L.D., Deconinck, B., Kutz, J.N.: Bose-Einstein condensates in standing waves: the cubic nonlinear Schrö dinger equation with a periodic potential. Phys. Rev. Lett. **86**, 1402 (2001)

5. Burger, S., Cataliotti, F.S., Fort, C., Minardi, F., Inguscio, M., Chiofalo, M.L., Tosi, M.P.: Superfluid and dissipative dynamics of a Bose-Einstein condensate in a periodic optical potential. Phys. Rev. Lett. **86**, 4447 (2001)
6. Cataliotti, F.S., Berger, S., Fort, C., Maddaloni, P., Minardi, F., Trombrttoni, A., Smerzi, A., Inguscio, M.: Josephson junction arrays with Bose-Einstein condensates. Science **293**, 843 (2001)
7. Choi, D.I., Niu, Q.: Bose-Einstein condensates in an optical lattice. Phys. Rev. Lett. **82**, 2022 (1999)
8. Cooper, N.R., Rezayi, E.H., Simon, S.H.: Vortex lattices in rotating atomic Bose gases with dipolar interactions. Phys. Rev. Lett. **95**, 200402 (2005)
9. Dalfovo, F., Giorgini, S., Pitaevskii, L.P., Stringari, S.: Theory of Bose-Einstein condensation in trapped gases. Rev. Mod. Phys. **71**, 463 (1999)
10. Dalfovo, F., Stringari, S.: Bosons in anisotropic traps: ground state and vortices. Phys. Rev. A **53**, 2477 (1996)
11. Denschlag, J., Cassettari, D., Schmiedmayer, J.: Guiding neutral atoms with a wire. Phys. Rev. Lett. **82**, 2014 (1999)
12. Ding, C.-Y., Zhang, X.-F., Zhao, D., Luo, H.-G., Liu, W.M.: Matter-wave solitons in heteronuclear atomic Bose-Einstein condensates with synchronously controllable interactions and potentials. Phys. Rev. A **84**, 053631 (2011)
13. Dutton, Z., Clark, C.W.: Effective one-component description of two-component Bose-Einstein condensate dynamics. Phys. Rev. A **71**, 063618 (2005)
14. Fetter, A.L., Svidzinsky, A.A.: Vortices in a trapped dilute Bose-Einstein condensate. J. Phys.: Condens. Matter **13**, R135 (2001)
15. Fröhlich, H.: On the theory of superconductivity: the one-dimensional case. Proc. R. Soc. Lond. Ser. A **223**, 296 (1954)
16. Gupta, M., Dastidar, K.R.: Control of the dynamics of coupled atomic-molecular Bose-Einstein condensates: modified Gross-Pitaevskii approach. Phys. Rev. A **80**, 043618 (2009)
17. Han, W., Zhang, S., Jin, J., Liu, W.M.: Half-vortex sheets and domain-wall trains of rotating two-component Bose-Einstein condensates in spin-dependent optical lattices. Phys. Rev. A **85**, 043626 (2012)
18. Hawkins, R.M., Lidsey, J.E.: Ermakov-Pinney equation in scalar field cosmologies. Phys. Rev. D **66**, 023523 (2002)
19. Heinzen, D.J., Wyna, R., Drummond, P.D., Kheruntsyan, K.V.: Superchemistry: dynamics of coupled atomic and molecular Bose-Einstein condensates. Phys. Rev. Lett. **84**, 5029 (2000)
20. Kasamatsu, K., Tsubota, M.: Vortex sheet in rotating two-component Bose-Einstein condensates. Phys. Rev. A **79**, 023606 (2009)
21. Kasamatsu, K., Tsubota, M., Ueda, M.: Vortex phase diagram in rotating twocomponent Bose-Einstein condensates. Phys. Rev. Lett. **91**, 150406 (2003)
22. Kasevich, M.A.: Coherence with atoms. Science **298**, 1363 (2002)
23. Liu, C.-F., Fan, H., Gou, S.-C., Liu, W.M.: Crystallized and amorphous vortices in rotating atomic-molecular Bose-Einstein condensates. Sci. Rep. **4**, 29566 (2014)
24. Malomed, B.A.: Inelastic collisions of polarized solitons in a birefringent optical fiber. J. Opt. Soc. Am. B **9**, 2075 (1992)
25. Malomed, B.A., Wabnitz, S.: Soliton annihilation and fusion from resonant inelastic collisions in birefringent optical fibers. Opt. Lett. **16**, 1388 (1991)
26. Manakov, S.V.: On the theory of two-dimensional stationary self-focusing of electromagnetic waves. Sov. Phys. JETP **38**, 248 (1974)
27. McKenzie, C., Hecker Denschlag, J., Häffner, H., Browaeys, A., de Araujo, L.E., Fatemi, F.K., Jones, K.M., Simsarian, J.E., Cho, D., Simoni, A., Tiesinga, E., Julienne, P.S., Helmerson, K., Lett, P.D., Rolston, S.L., Phillips, W.D.: Photoassociation of sodium in a Bose-Einstein condensate. Phys. Rev. Lett. **88**, 120403 (2002)
28. Mertes, K.M., Merrill, J.W., Carretero-Gonzalez, R., Frantzeskakis, D.J., Kevrekidis, P.G., Hall, D.S.: Nonequilibrium dynamics and superfluid ring excitations in binary Bose-Einstein condensates. Phys. Rev. Lett. **99**, 190402 (2007)

29. Myatt, C.J., Burt, E.A., Ghrist, R.W., Cornell, E.A., Wieman, C.E.: Production of two overlapping Bose-Einstein condensates by sympathetic cooling. Phys. Rev. Lett. **78**, 586 (1997)
30. Ol'shanii, M.A., Ovchinnikov, Y.B., Letokhov, V.S.: Laser guiding of atoms in a hollow optical fiber. Opt. Commun. **98**, 77 (1993)
31. Ol'shanii, M.A.: Atomic scattering in the presence of an external confinement and a gas of impenetrable bosons. Phys. Rev. Lett. **81**, 938 (1998)
32. Papp, S.B., Wieman, C.E.: Observation of heteronuclear feshbach molecules from a ^{85}Rb - ^{87}Rb gas. Phys. Rev. Lett. **97**, 180404 (2006)
33. Papp, S.B., Pino, J.M., Wieman, C.E.: Tunable miscibility in a dual-species Bose-Einstein condensate. Phys. Rev. Lett. **101**, 040402 (2008)
34. Pérez-García, V.M., Beitia, J.B.: Symbiotic solitons in heteronuclear multicomponent Bose-Einstein condensates. Phys. Rev. A **72**, 033620 (2005)
35. Pinney, E.: The nonlinear differential equation $y'' + p(x)y + cy^{-3} = 0$. Proc. Am. Math. Soc. **1**, 681 (1950)
36. Ronen, S., Bohn, J.L., Halmo, L.E., Edwards, M.: Dynamical pattern formation during growth of a dual-species Bose-Einstein condensate. Phys. Rev. A **78**, 053613 (2008)
37. Rooney, S.J., Bradley, A.S., Blakie, P.B.: Decay of a quantum vortex: test of nonequilibrium theories for warm Bose-Einstein condensates. Phys. Rev. A **81**, 023630 (2010)
38. Saito, H., Ueda, M.: Dynamically stabilized bright solitons in a two-dimensional Bose-Einstein condensate. Phys. Rev. Lett. **90**, 040403 (2003)
39. Salasnich, L., Malomed, B.A.: Vector solitons in nearly one-dimensional Bose-Einstein condensates. Phys. Rev. A **74**, 053610 (2006)
40. Stamper-Kurn, D.M., Andrews, M.R., Chikkatur, A.P., Inouye, S., Miesner, H.-J., Stenger, J., Ketterle, W.: Optical confinement of a Bose-Einstein condensate. Phys. Rev. Lett. **80**, 2027 (1998)
41. Strecker, K.E., Merrill, J.W., Carretero-Gonzalez, R., Frantzeskakis, D.J., Kevrekidis, P.G., Hall, D.S.: Formation and propagation of matter-wave soliton trains. Nature (London) **417**, 150 (2002)
42. Thalhammer, G., Barontini, G., De Sarlo, L., Catani, J., Minardi, F., Inguscio, M.: Double species Bose-Einstein condensate with tunable interspecies interactions. Phys. Rev. Lett. **100**, 210402 (2008)
43. Tikhonenkov, I., Vardi, A.: Atom-molecule dephasing in an $SU(1, 1)$ interferometer based on the stimulated dissociation of a molecular Bose-Einstein condensate. Phys. Rev. A **80**, 051604(R) (2009)
44. Timmermans, E., Tommasini, P., Côté, R., Hussein, M., Kerman, A.: Raried liquid properties of hybrid atomicatomic-molecular Bose-Einstein condensates. Phys. Rev. Lett. **83**, 2691 (1999)
45. Yamazaki, R., Taie, S., Sugawa, S., Takahashi, Y.: Submicron spatial modulation of an interatomic interaction in a Bose-Einstein condensate. Phys. Rev. Lett. **105**, 050405 (2010)
46. Yao, Y.-Q., Li, J., Han, W., Wang, D.-S., Liu, W.M.: Localized spatially nonlinear matter waves in atomic-molecular Bose-Einstein condensates with spacemodulated nonlinearity. Sci. Rep. **6**, 29566 (2016)
47. Vengalattore, M., Conroy, R.S., Rooijakkers, W., Prentiss, M.: Ferromagnets for integrated atom optics. J. Appl. Phys. **95**, 4404 (2004)
48. Wang, D.-S., Hu, X.-H., Liu, W.M.: Localized nonlinear matter waves in two-component Bose-Einstein condensates with time- and space-modulated nonlinearities. Phys. Rev. A **82**, 023612 (2010)
49. Woo, S.J., Park, Q.H., Bigelow, N.P.: Phases of atom-molecule vortex matter. Phys. Rev. Lett. **100**, 120403 (2008)
50. Wynar, R., Freeland, R.S., Han, D.J., Ryu, C., Heinzen, D.J.: Molecules in a Bose-Einstein condensate. Science **287**, 1016 (2000)
51. Zhang, X.-F., Hu, X.-H., Liu, X.-X., Liu, W.M.: Vector solitons in two-component Bose-Einstein condensates with tunable interactions and harmonic potential. Phys. Rev. A **79**, 033630 (2009)

52. Zheng, G.-P., Liang, J.-Q., Liu, W.M.: Phase diagram of two-species Bose-Einstein condensates in an optical lattice. Phys. Rev. A **71**, 053608 (2005)
53. Zhou, Y.B., Wang, M.L., Miao, T.D.: The periodic wave solutions and solitary wave solutions for a class of nonlinear partial differential equations. Phys. Lett. A **323**, 77 (2004)

Chapter 9
Two-Component Nonlinear Schrödinger Equations for Spin-1 BECs

Abstract A single-component BEC is typically well-described by a scalar order parameter, the BEC wave function, whose dynamics are governed by a one-component nonlinear Schrödinger equation with an external potential. Atomic BECs with internal spin degrees of freedom (spinor condensates) offer a new form of coherent matter with complex internal quantum structures. For these multicomponent BEC systems, the formalism is extended to a vector order parameter which possesses $2F + 1$ components for spin-F condensates and is invariant under rotation in spin space. Based on two-component nonlinear Schrödinger equations, we investigate in this chapter the dynamics of spin-1 BECs.

9.1 Introduction

The experimental realization of spinor condensates [6, 29] opened the possibility to observe various phenomena which are not seen in a single-component BEC. Spinor condensates are distinguished by an additional internal degree of freedom represented by the spin of the atoms. In the case of spin-1 BECs, the dynamics is described by three spin degrees of freedom with $m_F = 1, 0, -1$ of the $F = 1$ atomic hyperfine state. Theoretically, the spin-1 BECs were first studied by Ho [10] through generalizing the three-coupled NLS equations with external potential under the restriction of gage and spin-rotation symmetry. Moreover, within the mean-field theory, Ho [10] and Ohmiand Machida [21] predicted a rich set of novel phenomena such as spin textures and topological excitations. In recent years, much work [16, 30, 32, 33] has focused on the solitons solutions, dynamics, and novel quantum states of spin-1BECs.

Most recently, a synthetic spin–orbit (SO) coupling for a BEC has been engineered in a landmark experiment [14]. In this experiment, two Raman laser beams were used to couple a two-component BEC consisting of (predominantly) two hyperfine states of ^{87}Rb. The momentum transfer between laser beams and atoms contributes to the rich possibility of creating synthetic electric and magnetic fields. Recent

W.-M. Liu and E. Kengne, *Schrödinger Equations in Nonlinear Systems*,
https://doi.org/10.1007/978-981-13-6581-2_9

investigations have explored the possibility of identifying tunable spin–orbit-coupled (SOC) BECs [7, 26] with various trapping potentials and stable regimes of condensates have been observed. The identification of striped phase in the investigation of spin–orbit-coupled BECs [31] has only reignited the enthusiasm in this domain of interest as the striped phase is believed to be an indication of the phase transition taking place in a spin–orbit-coupled BEC. Also, the influence of spin–orbit coupling in a 1D BEC, particularly the interplay of the SOC, Raman coupling, and nonlinearity-induced precession of the soliton's spin has been studied recently [34]. The existence of "Striped phase", which essentially consists of a linear combination of plane waves and this has contributed to the idea of using ultracold atoms for the implementation of a quantum simulator [20].

9.2 Dual Equal Rashba–Dresselhaus SOC in the Atomic-Molecular BECs

By combining the spin-dependent photo-association [12] and Raman coupling [22], we present a way of creating a dual equal Rashba–Dresselhaus SOC in the atomic-molecular BECs [17].

9.2.1 Creation of Dual Equal Rashba–Dresselhaus Spin–Orbit Coupling in Atomic-Molecular BECs

We consider a Bose gas of ultracold ^{87}Rb atoms, commonly used in both experiments on the SOC atomic BEC and the atomic-molecular BECs [17], to demonstrate the scheme for creating SOC atomic-molecular BECs (see Fig. 9.1a). We take a BEC of up to 3×10^5 atoms and select two atomic internal spin states with $m_f = -1$ and $m_f = 0$ of the ^{87}Rb 5 $S_{1/2}$ ground state $f = 1$ manifold, to be described as the (pseudo) spin-up ($|\uparrow\rangle$) and spin-down ($|\downarrow\rangle$) states (see Fig. 9.1b). Here, f and m_f are the quantum numbers of the atomic total spin and its projection along a quantization axis. A pair of counterpropagating Raman lasers (ω_{L3} and ω_{L4}) couple these spin states with the matrix element (Raman Rabi frequency) Ω_A inducing the SOC of the atomic BECs (see Fig. 9.1b). The third spin state with $m_f = 1$ is detuned away from the Raman resonance due to the quadratic Zeeman shift and hence can be neglected. On the other hand, the counterpropagating Raman lasers (ω_{L5}, ω_{L6}, and ω_{L7}) are used to induce the SOC for the spin-1 ($F = 1$) molecular BECs (see Fig. 9.1c). An extra laser ω_{L7} is needed to couple the molecular states with $m_F = 0$ and $m_F = 1$ detuned from the resonance between the states $m_F = -1$ and $m_F = 0$ due to a quadratic Zeeman effect. To prevent population of the excited state, the laser fields are sufficiently detuned from the single photon resonance.

The molecular BECs of $^{87}Rb_2$ with hyperfine state $|F = 1, m_F = 0, \pm 1\rangle$ are produced by the spin-dependent photo-association of atoms (see Fig. 9.1d–f). A pair of atoms with various combinations of internal states are coupled to an excited

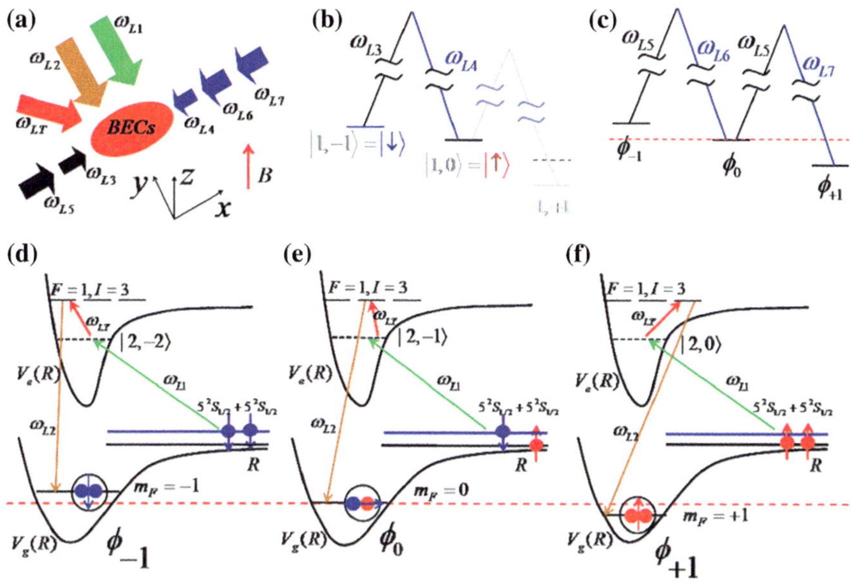

Fig. 9.1 Experimental setup for creating the spin–orbit-coupled atomic-molecular BECs of ^{87}Rb. **a** The experimental geometry. The formation of spin molecular BECs: Laser field $L1$ finishes the spin-dependent photo-association [17], laser field LT performs the $\Delta F = -1$, $\Delta m_F = +1$ is the transiting process of the excited molecules, and laser field $L2$ induces the excited molecules to emit a photon and forms the stable spin-1 molecular BECs. The creation of dual SOC: The laser fields $L3$ and $L4$ ($L5$, $L6$, and $L7$) are used to create the SOC of atomic (molecular) BECs along the x axis. B denotes a weak magnetic field along the z-axis, which is used to induce the Zeeman shift. **b** Level diagram of Raman coupling within the atomic BECs. Two lasers ($L3$ and $L4$) are used to couple states of $|\uparrow\rangle$ (i.e., $|1,-1\rangle$) and $|\downarrow\rangle$ (i.e.,$|1,0\rangle$). **c** Level diagram of Raman coupling within the $F = 1$ molecular BECs. The three lasers ($L5$, $L6$, and $L7$) couple states of ϕ_{-1}, ϕ_0, and ϕ_{+1}. **d** The procedure of creating $|F = 1, mF = -1\rangle$ state molecular BEC. A close pair of atoms from the condensate in a state $|\downarrow\rangle$ absorb a photon from laser field $L1$ and then transit to a bound excited molecular state $|F = 2, mF = -2\rangle$, i.e., the spin-dependent photo-association process $|1,-1\rangle \otimes |1,-1\rangle \to |2,-2\rangle$ [17]. Via the laser field LT, the excited molecule transits to $|F = 1, mF = -1\rangle$ state. Finally, laser field $L2$ induces the excited molecular state to emit a photon into molecular state ϕ_{-1} which keeps $F = 1$ and $m_F = -1$. Panels **e** and **f** are the procedure of creating $|F = 1, m_F = 0\rangle$ and $|F = 1, m_F = +1\rangle$ state molecular BECs, respectively. Reprint from Ref. [17], Copyright 2018, with permission from American Physical Society

molecule state $|\phi_{em}\rangle$ via a laser beam ω_{L1}. Here, the photo-association through the total spin-2 scattering channel gives molecular F numbers of 2. The three spin-dependent photo-association processes can be written as follows: $|1,0\rangle \otimes |1,0\rangle \to |2,0\rangle |1,-1\rangle \otimes |1,0\rangle \to |2,-1\rangle$, and $|1,-1\rangle \otimes |1,-1\rangle \to |2,-2\rangle$. Under the transition $F = -1$ and $m_F = +1$, the excited molecule states further change into $|\phi'_{em}\rangle$ with $F = 1$ and $I = 3$, where F, m_F, and I denote the total spin, its projection, and the nuclear spin quantum numbers of a molecule. Finally, $|\phi'_{em}\rangle$ is coupled to a ground molecule state $|\phi_m\rangle$ via another laser beam ω_{L2} (see Fig. 9.1d–f). As a result, the two-component atomic BEC can be coherently converted into a three-component (spin-1) molecular BEC with three internal spin states $|F = 1, m_F = 0, \pm 1\rangle$.

Our schematic has shown a possible method to create dual SOC atom–molecule BECs. In real experiments, the realization of dual SOC atom–molecule BECs would be affected by many factors, such as spontaneous Raman scattering by molecules, the inelastic collision channels between atoms and molecules, heating due to spontaneous scattering, some undesired light shifts, and inhomogeneous detuning of the laser fields. For simplicity, we ignore the above factors. The dual SOC atomic-molecular BECs can be described by the two-component NLS equations as follows:

$$
\begin{aligned}
i\hbar\frac{\partial\psi_a}{\partial t} &= \left[-\frac{\hbar^2}{2M_A}\nabla_d^2 + V_A + \sum_{q=\uparrow,\downarrow} g_{a,q}\left|\psi_q\right|^2 + \sum_{m=0,\pm1} g_{a,m}\left|\phi_m\right|^2\right]\psi_a \\
&\quad + \sum_{q=\uparrow,\downarrow}\left[\lambda\left(\widehat{\sigma}_z\right)_{aq} p_x + \hbar\Omega_A\left(\widehat{\sigma}_x\right)_{aq}\right]\psi_q + K_a,
\end{aligned} \tag{9.1a}
$$

$$
\begin{aligned}
i\hbar\frac{\partial\psi_m}{\partial t} &= \left[-\frac{\hbar^2}{2M_M}\nabla_d^2 + V_M + \sum_{q=\uparrow,\downarrow} g_{m,q}\left|\psi_q\right|^2 + g_n\left|\phi\right|^2\right]\psi_m \\
&\quad + \sum_{n=0,\pm1}\left[\left(2\hbar\Omega_A\widehat{F}_x\right)_{mn} + \lambda\left(\widehat{F}_z\right)_{mn} p_x\right]\phi_n \\
&\quad + g_s\sum_{\alpha=x,y,z}\sum_{n,k,l=0\pm1}\left(\widehat{F}_\alpha\right)_{mn}\left(\widehat{F}_\alpha\right)_{kl}\phi_n\phi_k^*\phi_l + K_m,
\end{aligned} \tag{9.1b}
$$

where $K_\uparrow = \sqrt{2}\chi\psi_\uparrow^*\phi_{+1} + \chi\psi_\downarrow^*\phi_0$, $K_\downarrow = \sqrt{2}\chi\psi_\downarrow^*\phi_{-1} + \chi\psi_\uparrow^*\phi_0$, $K_{=1} = \frac{\chi}{\sqrt{2}}\psi_\uparrow^2 + \left(\varepsilon + M_A\lambda^2\right)\phi_{+1}$, $K_0 = \chi\psi_\uparrow\psi_\downarrow + \varepsilon\phi_0$, and $K_{-1} = \frac{\chi}{\sqrt{2}}\psi_\downarrow^2 + \left(\varepsilon + M_A\lambda^2\right)\phi_{-1}$. Here, $a = \uparrow, \downarrow$ and $m = 0, \pm1$ label, respectively, the atomic and molecular spins; $d = 1$, 2, 3 refers to the dimension of the system; M_A (M_M) is the mass of the atom (molecule); $V_A = M_A\omega^2 r^2/2$ ($V_M = M_M\omega^2 r^2/2$) is the harmonic trap for atomic (molecular) BECs; $\widehat{F}_{\alpha=x,y,z}$ ($\widehat{\sigma}_{\alpha=x,y,z}$) are the spin-1 (Pauli) matrices; $g_{a,q} = \frac{4\pi\hbar^2 a_{aq}}{M_A}$ and $g_{a,m} = \frac{2\pi\hbar^2 a_{am}}{M_A M_M/(M_A+M_M)}$ are, respectively, the strengths of atom–atom and atom–molecule interactions; and g_n and g_s are the strengths of the spin-independent and spin-dependent interactions between the molecules. Here, also ψ_a and ϕ_m denote the macroscopic wave functions of the atomic BEC in the internal state $|a\rangle$ ($a = \uparrow, \downarrow$) and those of the molecular BEC in the internal state $|m_F = m\rangle$ ($m = 0, \pm1$). The parameters λ, Ω_A, χ, ε, and ω are the strengths of dual SOC, Raman coupling, atom–molecule conversion, Raman detuning, and trap frequency, respectively.

9.2.2 *Numerical Method for Simulations*

The damped PNLS equations [25] are used to obtain the ground state of atomic-molecular BECs. By neglecting the noise term according to the corresponding stochastic PNLS equations [25], the damped PNLS equations for the atomic-molecular BECs is described as

$$d\psi_a = \mathcal{P}\left\{-\frac{i}{\hbar}\widehat{H}_a\psi_a dt + \frac{\gamma_a}{k_B T}\left(\mu_a - \widehat{H}_a\right)\psi_a dt\right\}, \tag{9.2a}$$

$$d\phi_m = \mathcal{P}\left\{-\frac{i}{\hbar}\widehat{H}_m\phi_m dt + \frac{\gamma_m}{k_B T}\left(\mu_m - \widehat{H}_m\right)\phi_m dt\right\}, \tag{9.2b}$$

where $\widehat{H}_a\psi_a = i\hbar\partial\psi_a/\partial t \ \ (a = \uparrow, \downarrow)$, $\widehat{H}_m\phi_m = i\hbar\partial\phi_m/\partial t \ \ (m = 0, \pm 1)$, T is the final temperature, k_B is the Boltzmann constant, and μ_a and μ_m are the chemical potential of atom and molecule, respectively. γ_a (γ_m) is the growth rate for the component (m-th component). To obtain the ground state, we set the parameter $\frac{\gamma_a}{k_B T} = \frac{\gamma_m}{k_B T} = 0.05$ directly. The projection operator $\mathcal{P}$ is used to restrict the dynamics of atomic-molecular BEC in the coherent region. The initial state of each component is generated by sampling the grand-canonical ensemble for a free ideal Bose gas with the chemical potential $\mu_{m,0} = 2\mu_{a,0}$. The final chemical potential of the noncondensate band is altered to the values $\mu_m = 2\mu_a$.

In this section, we mainly focus on the two-dimensional BECs. We use the parameters of atomic-molecular BECs of ^{87}Rb with $M_M = 2M_A$ ($M_A = 144.42 \times 10^{-27}$ Kg), the trapping frequency $\omega = 200 \times 2\pi$ Hz, and the strength of the atom–atom interaction $g_{\uparrow,\uparrow} = g_A$ with the scattering length $a_A = 101.8a_B$, where a_B is the Bohr radius. In addition, when the change in energy of converting two atoms into one molecule ($\Delta U = 2U_{Ta} - U_{Tm}$), without including internal energy, approaches zero, we can obtain the value $2g_A = g_M$, and we set $g_n = g_M = 2g_A$. For simplicity, we set the atom–molecule interaction $g_{\uparrow,-1} = g_{\uparrow,0} = g_{\uparrow,+1} = g_{\downarrow,-1} = g_{\downarrow,0} = g_{\downarrow,+1} = g_{AM}$. The unit of length, time, and energy correspond to $\sqrt{\hbar/(M_A\omega)}$ (≈ 0.76 μm), ω^{-1} ($\approx 0.96 \times 10^{-3}$ s), and $\hbar\omega$, respectively. To excite vortices in this system, we add the rotation mechanism [37]. To obtain the rotating SOC BECs in real experiments, one method is to rotate the lasers creating SOC for an isotropic trap or to rotate both the lasers creating SOC and an anisotropic trap [24]. This leads to an effective time-independent Hamiltonian, $H_{eff} = H - \Omega_{\text{rotation}} L_z$, which describes the system in a rotating frame of reference [$L_z = -i\hbar(x\partial/\partial y - y\partial/\partial x)$ being the orbital angular momentum operator and rotation is the frequency of rotation.

9.2.3 A Single Carbon-Dioxide-Like Skyrmion in the Dual SOC Atomic-Molecular BECs

Spin texture excitations, such as skyrmion and half-skyrmion, are particle-like topological entities in a continuous field and they play an important role in many condensed matter systems [13]. Skyrmion has a nontrivial topological charge of absolute value 1 and half-skyrmion proposes the nontrivial topological charge of absolute value 1/2. Usually, spin texture excitations relate to some vortices structure [15, 18]. Especially, spin texture excitations would display variable configurations but the topological charge is fixed. The dual SOC atomic-molecular BECs provide a new platform for exploring the spin texture excitations coming from atomic BECs and molecular BECs simultaneously. Now, we present a way to obtain a single carbon-dioxide-like skyrmion, which consists of one skyrmion with the topological charge -1 in the atomic BECs and two half-skyrmions with the topological charge $-1/2$ in the molecular BECs. On both sides of the skyrmion are the two half-skyrmions and they couple as a unity resembling the carbon dioxide molecule.

In Fig. 9.2, we set the initial chemical potential $\mu_{m,0} = 2\mu_{a,0} = 8\hbar\omega$ in the numerical simulations. The final chemical potential of the noncondensate band is increased to $\mu_m = 2\mu_a = 28\hbar\omega$. The densities and the corresponding phases of the dual SOC atomic-molecular BECs are obtained for the equilibrium state (see Fig. 9.2a and b). One can see that both components of atomic BECs contain a vortex excitation. The spin-up (spin-down) component fills in the vortex of the spin-down (spin-up) component. Thus, this structure is similar to a vortices molecule [11]. At the same time, two vortices are induced in each molecular component, respectively. The densities of all the components are symmetric with respect to the y-axis. In addition, only the density of the $m_F = 0$ component is symmetric with respect to the x-axis. In this way, these vortices form a cluster.

The emergence of the vortices cluster results from a combination of the dual SOC, Raman coupling, atom–molecule conversion, and rotation effect. The dual equal Rashba–Dresselhaus SOC causes a translational displacement along the y-axis on the rotating BECs. The translational displacement for different spin state is opposite. Thus, the spin-up and $m_F = +1$ components tend to move to the $y < 0$ region, and the spin-down and $m_F = -1$ components go to the $y > 0$ region. Note that the $m_F = 0$ component is not affected by the SOC. Rotation induces vortices. In the atomic BECs, the spin-up and the spin-down components have to fill with each other in the region of the vortices. A similar case occurs in the molecular BECs with the vortices of the $m_F = 0$ component locating at the center. The atom–molecule conversion forces the atomic vortex of spin-up (spin-down) component to couple with two molecular ones of $m_F = +1$ ($mF = -1$) components, just like that in the normal atomic-molecular BECs [35]. This leads to the formation of the above vortices cluster.

Since our system includes the pseudospin-1/2 atomic BECs and the spin-1 molecular BECs, there will be spin textures coming from atomic BECs and molecular BECs, respectively. Thus, this mixed system will show the interaction of the spin textures induced by different hyperfine spin states. In pseudospin-1/2 BECs and spin-1 BECs, the spin textures [11] are defined by $\mathbf{S}_\alpha^{\text{atom}} = \sum_{m,n=\uparrow\downarrow} \psi_m^* (\widehat{\sigma}_\alpha)_{m,n} \psi_n / |\psi|^2$ and $\mathbf{S}_\alpha^{\text{molecule}} = \sum_{m,n=0,\pm1} \phi_m^* \left(\widehat{F}_\alpha\right)_{m,n} \phi_m / |\phi|^2$, $(\alpha = x, y, z)$, where $\widehat{\sigma}_\alpha$ and $\widehat{F}_\alpha$ are the Pauli and spin-1 matrices, respectively. The spin textures in the atomic BECs and molecular BECs are shown in Fig. 9.2c and d, respectively. In Fig. 9.2c, the arrows form a circle locating near the position $(0, 0)$. The arrows tend to point down in the region $y < 0$ and to point up in the region $y > 0$. Similarly, in Fig. 9.2d, the arrows form two half circles locating near the positions (0, –0.7) and (0, 0.7), respectively. Performing a transformation $(\mathbf{S}'_x, \mathbf{S}'_y, \mathbf{S}'_z) = (\mathbf{S}_z, \mathbf{S}_y, -\mathbf{S}_x)$, one can identify clearly a skyrmion (see Fig. 9.2e) and two half-skyrmions (see Fig. 9.2f). In the Supplemental Material [28], it has been proved that the transformation does

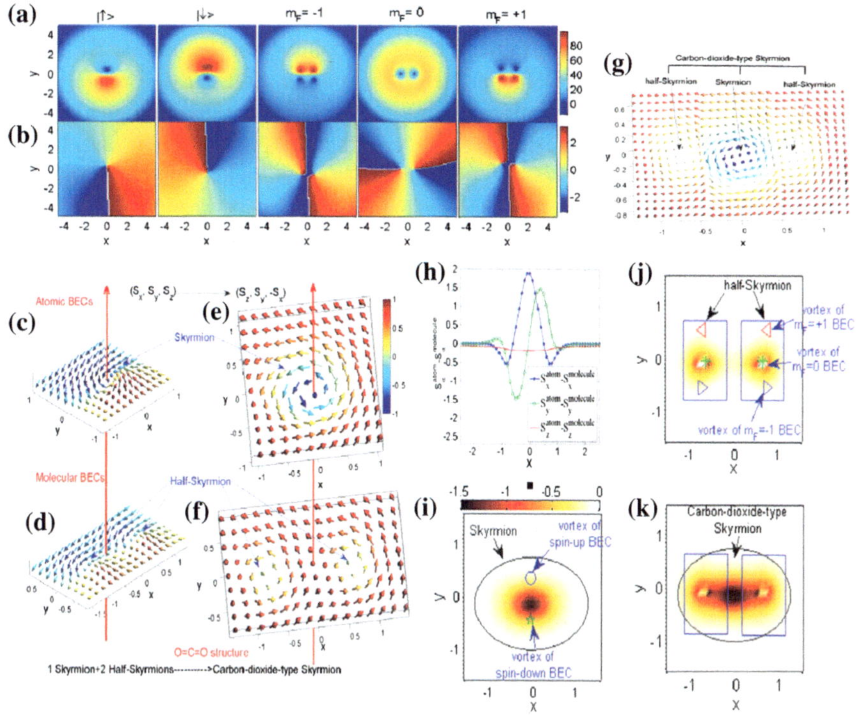

Fig. 9.2 Creating a carbon-dioxide-like skyrmion in the dual SOC atomic-molecular BECs. **a** The densities of each component when the system reaches the equilibrium state. The particles number $(N_{\uparrow}, N_{\downarrow}, N_{-1}, N_0, N_{+1})$ is $(2.56 \times 10^3, 2.98 \times 10^3, 1.72 \times 10^3, 3.22 \times 10^3, 1.59 \times 10^3)$. **b** The corresponding phases of each component in **a**. **c** and **d** The spin texture of the atomic BECs and molecular BECs, respectively. The color of each arrow indicates the magnitude of S_z. **e** and **f** The corresponding spin texture under the transformation: $(\mathbf{S}'x, \mathbf{S}'y, \mathbf{S}'z) = (\mathbf{S}'z, \mathbf{S}'y, -\mathbf{S}'x)$. (g) A combined scheme of the carbon-dioxide-like skyrmion. A skyrmion (center) couples with two half-skyrmions, which have to be arranged at the two sides of the skyrmion. **h** The difference of spin vectors between the atomic BECs and the molecular BECs, where $y = 0$. **i** and **j** The topological charge density of the atomic BECs and molecular BECs, respectively. The ellipse and rectangles mark out the regions of a skyrmion and two half-skyrmions, respectively. The symbol of "circle", "star", "triangleright", "asterisk", and "triangleleft" are the position of vortices formed by the $\uparrow$, $\downarrow$, $m_F = -1$, $m_F = 0$, and $m_F = +1$ components, respectively. **k** The sum of the topological charge densities. It denotes the creation of carbon-dioxide-like skyrmion that the ellipse overlaps with the two rectangles. The unit of length, strength of χ, Ω_A, and dual SOC are 0.76 μm, $\hbar\omega$, ω, and 0.96×10^{-3} m/s, respectively. Reprint from Ref. [17], Copyright 2018, with permission from American Physical Society

not affect the value of the topological charge Q [$Q = \frac{1}{4\pi} \int \int q(x, y)dxdy$, where $q(x, y) = \mathbf{s} \cdot \left(\frac{\partial \mathbf{s}}{\partial x} \times \frac{\partial \mathbf{s}}{\partial y}\right)$, $\mathbf{s} = \mathbf{S}/|\mathbf{S}|$]. Additionally, we provide approximate solutions for the carbon-dioxide-like skyrmion (see the Supplemental Material [28]). Figure 9.2g is a combined schematic of a carbon-dioxide-like skyrmion, where a skyrmion of atomic BECs couples with two half-skyrmions of molecular BECs.

Figure 9.2h shows the differences of spin vectors between the atomic BECs and the molecular BECs ($y = 0$). It indicates that the spin vectors of atomic BECs and molecular BECs would tend to be parallel to each other, except those at the region of the carbon-dioxide-like skyrmion. Figure 9.2i and j show the topological charge density $q(x, y)$. The positions of vortices are pointed out to highlight the relationship between skyrmion and vortices. We find that the topological charge approaches -1 and $-1/2$ for the two types of spin texture, respectively. The two zero points in Fig. 9.2j further prove that we obtain two half-skyrmions. Figure 9.2k shows the sum of topological charge density [i.e., $q_{\mathrm{sum}}(x, y) = q_{\mathrm{atom}}(x, y) + q_{\mathrm{molecule}}(x, y)$], which further clearly indicates the carbon-dioxide-like structure.

9.2.4 The Chain of Carbon-Dioxide-Like Skyrmion in the Dual SOC Atomic-Molecular BECs

To fully show why SOC changes the spin textures, we now further demonstrate the effect of the dual SOC on the rotating atomic-molecular BECs in Fig. 9.3. We change the initial chemical potential and final chemical potential to be $\mu_m = 2\mu_a = 3.5\mu_{m,0} = 7\mu_{a,0} = 42\hbar\omega$. Figure 9.3a, c, and e display densities of the atomic-molecular BECs obtained for the equilibrium state at $\Omega_{\mathrm{rotation}} = 0.5\,\omega$. Figure 9.3 b, d, and f demonstrate the corresponding phases. Without the dual SOC ($\lambda = 0$), two atomic and three molecular components of the BECs do not display the phase separation (see Fig. 9.3a). Yet there are a lot of vortices in each component, and they can be identified by the phase image (see Fig. 9.3b). However, adding the dual SOC ($\lambda = 1$), the two components of atomic BECs separate (see Fig. 9.3c). The component described by $\psi_\uparrow$ mainly locates below the x-axis, whereas $\psi_\downarrow$ mainly locates above the x-axis. Similarly, the phase separation also occurs in the molecular BECs. Because the spin-up (spin-down) atoms can form the ϕ_{+1} (ϕ_{-1}) component, the corresponding molecular BECs of ϕ_{+1} (ϕ_{-1}) also tends to locate below (above) $y = 0$ axis. The ϕ_0 component is spread around the x-direction and squeezed in the y-direction, and the area is filled with vortices. When the strength of the dual SOC is up to $\lambda = 2$, the phase separation becomes more obvious (see Fig. 9.3e). In fact, the phase separation does not take place at all if the rotation is absent. Therefore, the combination of dual SOC and rotation induces the phase separation in the atomic-molecular BECs.

To detect the underlying relationship between the topological excitations, Fig. 9.3g–i shows the position of vortices, where the results correspond to Fig. 9.3a, c, and e, respectively. Note that we do not indicate the vortices where the densities of BECs are very low. In that case, the contribution of energy and momentum to the system almost can be neglected. Without the dual SOC (see Fig. 9.3g) five types of vortices form certain clusters (a black rectangle shows an example). The repulsion among vortices of each component forces them to spread around the whole BECs, but they do not form a regular lattice. At the same time, the repulsion among vortex clusters also leads to the deformation of the clusters shown in Fig. 9.3g. In Fig. 9.3h ($\lambda = 1$), a black rectangle shows some regular structure of vortices occurring

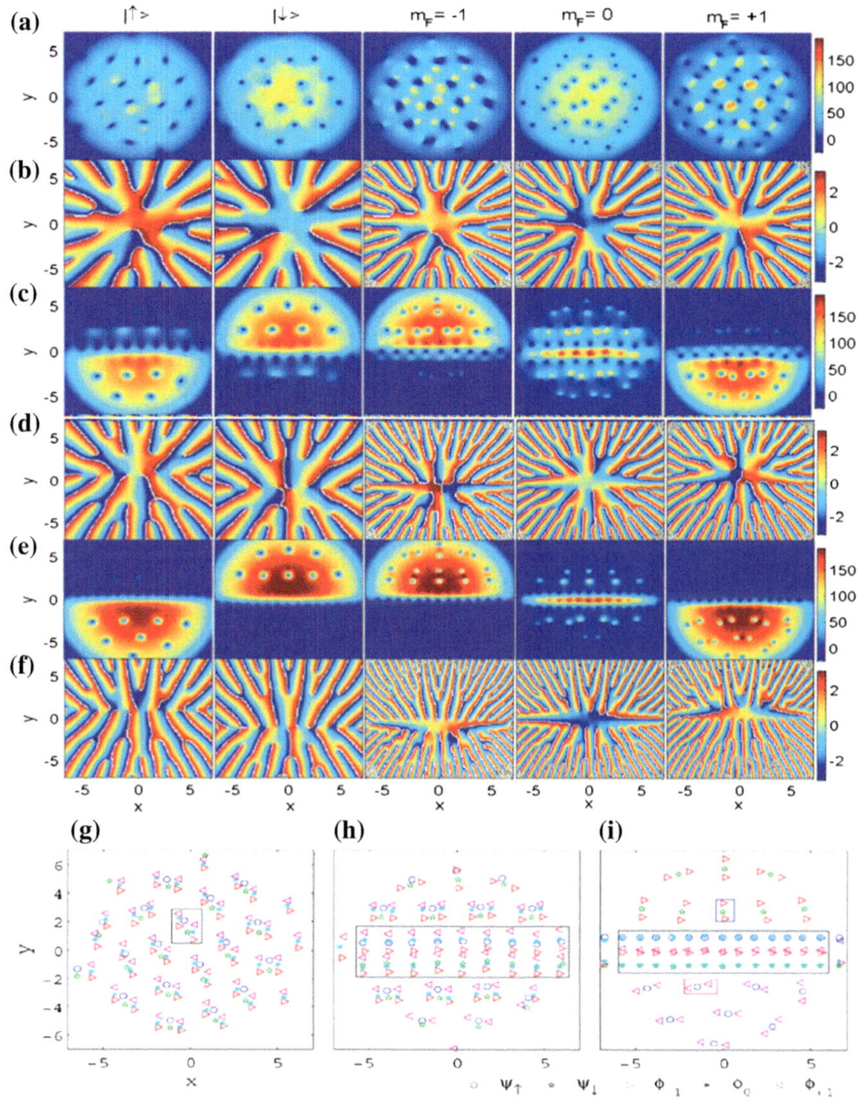

Fig. 9.3 (Continued)

◀ **Fig. 9.3** (Continued) The effect of dual SOC on the vortices in rotating atomic-molecular BECs. The rotation frequency is $\Omega_{\text{rotation}} = 0.5\omega$, where $\omega = 200 \times 2\pi$ Hz. We set the parameters $g_{\uparrow,\uparrow} = g_{\uparrow,\downarrow} = g_A$ with the scattering length $a_A = 101.8a_B$, $g_{\downarrow,\downarrow} = 0.95g_A$, $g_{AM} = 0.5g_A$, $g_n = 2g_A$, $g_s = 0.03g_A$, $\chi = 0.08$, $\varepsilon = 0$, $A = 0.8$, and $\mu_m = 2\mu_a = 3.5\mu_{m,0} = 7\mu_{a,0} = 42\hbar\omega$. **a** Densities of each components under the equilibrium state, the strength of dual SOC $\lambda = 0$. Note that the components are pointed out by the title: $|\uparrow, |\downarrow, m_F = -1, m_F = 0$, and $m_F = +1$, respectively. **b** The corresponding phases of each components in **a**. **c** and **d** The corresponding densities and phases with $\lambda = 1$. **e** and **f** The corresponding densities and phases with $\lambda = 2$. **g** and **i** The position of vortices in **a**, **c**, and **e**, respectively. In **g**, the rectangle points out a vortex cluster. In **h**, the rectangle points out the region where the vortices form a lattice. In **i**, the red (blue) rectangle marks out a carbon dioxide vortex, and the black rectangle points out the vortices lattice. The particles number $(N_\uparrow, N_\downarrow, N_{-1}, N_0, N_{+1})$ for the three cases are $(5.82 \times 10^3, 8.01 \times 10^3, 4.96 \times 10^3, 8.23 \times 10^3, 4.93 \times 10^3)$, $(7.86 \times 10^3, 8.80 \times 10^3, 7.95 \times 10^3, 3.93 \times 10^3, 7.85 \times 10^3)$, and $(11.42 \times 10^3, 12.29 \times 10^3, 10.03 \times 10^3, 2.00 \times 10^3, 10.04 \times 10^3)$, respectively. The unit of length, strength of χ, Ω_A, and dual SOC are 0.76 μm, $\hbar\omega$, ω, and 0.96×10^{-3} m/s, respectively. Reprint from Ref. [17], Copyright 2018, with permission from American Physical Society

around the region along the x-axis. In fact, they are essentially vortex clusters which degenerate into a line. The number of vortices in each column is the same as that of the cluster in Fig. 9.2a. The dual SOC leads the vortices to occur along the x-axis, but the repulsion among vortices squeezes them to distribute in a line. Thus, several lines of vortices form a 2D lattice. In the periphery of the lattice, there are the remaining vortex clusters. Outside the BECs, the vortices fail to form the full vortex clusters, because the phase separation causes some components to disappear locally. When the strength of the dual SOC increases up to $\lambda = 2$, the structure of the vortices lattice [a black rectangle region in Fig. 9.3i] differs from that in Fig. 9.3h. Here, each vortex of ϕ_0 component tends to overlap with an atomic vortex. The vortex number of ϕ_{+1} (ϕ_{-1}) component decreases to 1 in each column. The blue rectangle and the red rectangle in Fig. 9.3i indicate that vortices form some carbon dioxide structures far from the center of the vortex lattice (black rectangle). Above the black rectangle, vortices in components of $\psi_\downarrow$ and ϕ_{-1} form the carbon dioxide structure, whereas the atomic vortex is situated in the middle, and two molecular vortices are located at the two sides symmetrically. Similarly, the carbon dioxide structure of vortices in components of $\psi_\uparrow$ and ϕ_{+1} occurs below the black rectangle. These results indicate that the strong dual SOC deepens the phase separation. At the same time, it forces the vortices cluster to transform into the lattice around the x-axis and even into the carbon dioxide structures in the remaining regions.

We now discuss the effect of the dual SOC on the spin textures. Figure 9.4a and d shows the spin textures in atomic BECs and molecular BECs without dual SOC ($\lambda = 0$), respectively. The topological charge density in Fig. 9.5a [Fig. 9.5d] shows that they are skyrmions (half-skyrmions) with $Q = -1$ ($Q = -1/2$). The position of vortices proves that the skyrmions (half-skyrmions) originate from the vortices dipole of atomic BECs (the three vortices structure of molecular BECs). The sum of topological charge density in Fig. 9.5g indicates that the skyrmion of atomic BECs and the half-skyrmion of molecular BECs form the carbon dioxide structure, which is marked by the blue ellipses.

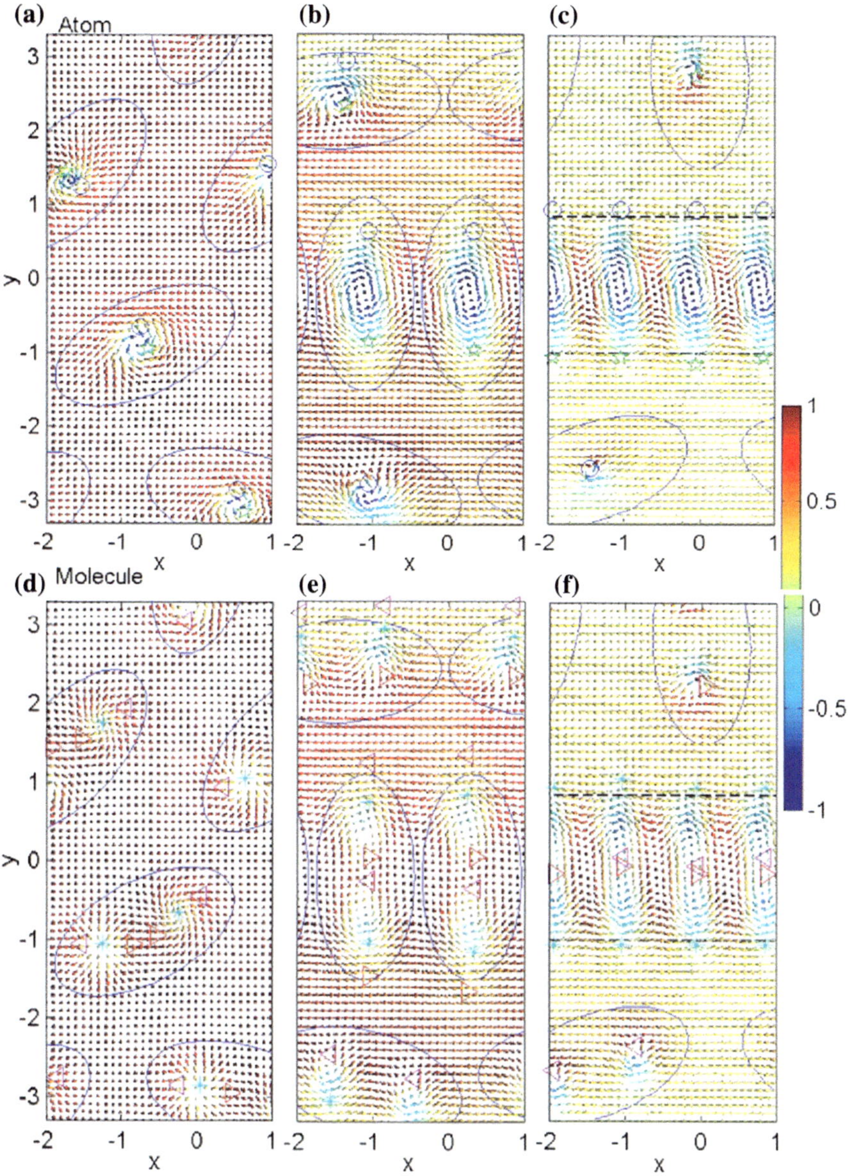

Fig. 9.4 The effect of SO coupling on the spin texture in the atomic-molecular BECs. **a–c** The corresponding spin textures of atomic BECs in Fig. 9.43a, c, and e, respectively. **d–f** The corresponding spin textures of molecular BECs in Fig. 9.43a, c, and e, respectively. Note that the spin textures are under the transformation: $(\mathbf{S}'x, \mathbf{S}'y, \mathbf{S}'z) = (\mathbf{S}'z, \mathbf{S}'y, -\mathbf{S}'x)$. The color of each arrow indicates the magnitude of $\mathbf{S}'$. The blue ellipses point out the regions of the carbon-dioxide-type skyrmion that skyrmion of atomic BECs couples two half-skyrmion of molecular BECs. In addition, we mark the position of vortices in order to illuminate the relationship between spin texture and position of vortices clearly. The meanings of the marks are the same as that in Fig. 9.3g–i. The unit of length, strength of χ, Ω_A, and SO coupling are 0.76 μm, $\hbar\omega$, ω, and 0.96×10^{-3} m/s, respectively. Reprint from Ref. [17], Copyright 2018, with permission from American Physical Society

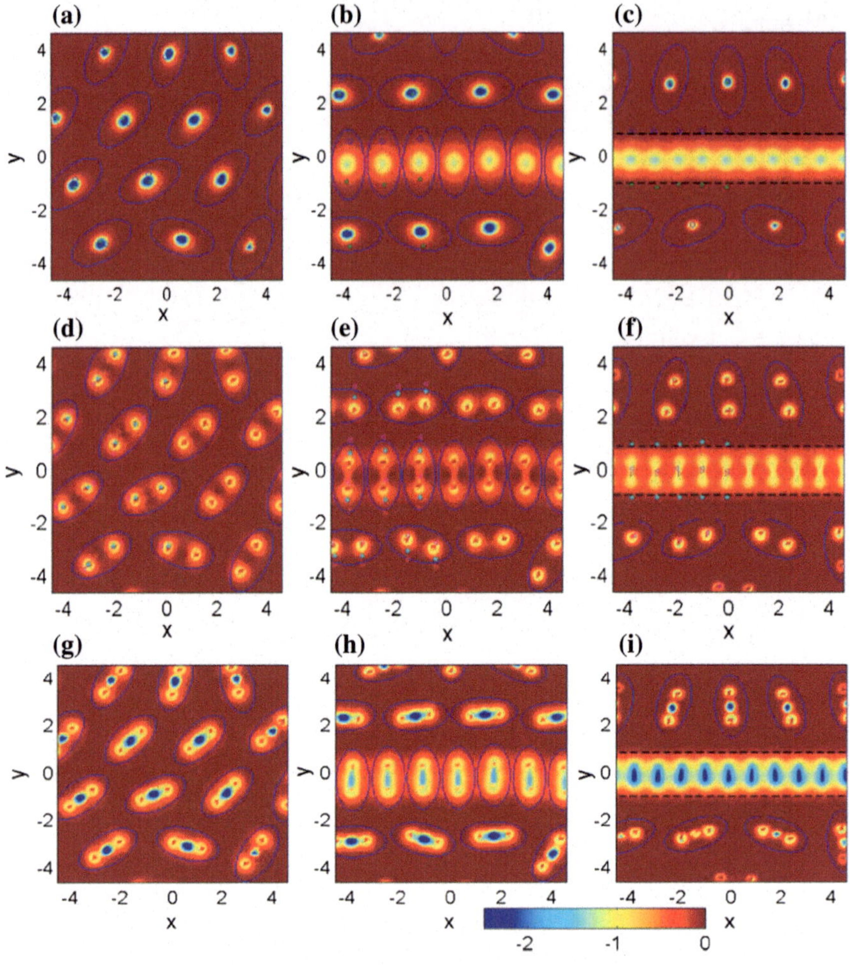

Fig. 9.5 The topological charge density of spin textures in SO-coupled atomic-molecular BECs. **a–c** The corresponding topological charge density in Fig. 9.4a–c, respectively. **d–f** The corresponding topological charge density in Fig. 9.4d–f, respectively. In addition, we mark the position of vortices in $x < 0$ region in order to illuminate the relationship between spin texture and position of vortices clearly. The meanings of the marks are the same as that in Fig. 9.3g–i. **g–i** The total topological charge density of **a** and **d**, **b** and **e**, and **c** and **f**, respectively. We use the blue ellipses to point out the regions of the carbon-dioxide-type skyrmion. The unit of length and SOC are 0.76 μm and 0.96×10^{-3} m/s, respectively. Reprint from Ref. [17], Copyright 2018, with permission from American Physical Society

The dual SOC can redistribute the carbon-dioxide-like skyrmions in the atomic-molecular BECs. For $\lambda = 1$, we also obtain some skyrmions and half-skyrmions in the atomic BECs and molecular BECs, respectively. Due to the phase separation caused by the dual SOC and rotation, some skyrmions are forced to distribute along the x-axis in Fig. 9.4b. Similarly, some half-skyrmions distribute parallel along the x-axis in Fig. 9.4e. The topological charge density in Fig. 9.5b, e, and h shows the carbon-dioxide-like skyrmions distributed into a line along the x-axis. In other words, the dual SOC induces a chain of carbon-dioxide-like skyrmions.

The strong dual SOC can even redesign the spin excitations of the textures. When the strength of the dual SOC λ reaches 2, the skyrmion can occur in the atomic BECs (see Figs. 9.4c and 9.5c). The skyrmion along the x-axis is induced by the vortices dipole, but the skyrmion away from the x-axis is induced by a single vortex. In the molecular BECs, the half-skyrmion is induced by a single vortex of ϕ_{-1} component in the region of $y > 1.5$ (see Figs. 9.4f and 9.5f). The sum of the topological charge density in Fig. 9.5i further proves that the carbon-dioxide-like skyrmion indeed occurs in the region of $y > 1.5$ and $y < -1.5$. Along the x-axis (between the black lines), the spin texture transforms into a structure with a fractional topological charge. This structure is not constructed by the normal half-skyrmion because the singular points in the topological charge densities disappear. Therefore, the extremely strong dual SOC can destroy the half-skyrmion excitations in the original chain of carbon-dioxide-like skyrmions.

Why do the carbon-dioxide-like skyrmions form only one chain? Initially, the atomic-molecular BECs are miscible. Under combined interaction of the rotation and dual SOC, the phase separation occurs. Around the domain wall, the energy of interaction is lower than that in other parts. Thus, the vortices and the corresponding skyrmions tend to distribute along the domain wall to form the ordered structure. Meanwhile, there is only one domain wall formed when the BECs separate. Therefore, there are no vortice lattices or a chain of carbon-dioxide-like skyrmions in other parts of the BECs.

9.2.5 Chern Number Indicates That Spin–Orbit Coupling Induces Topological Excitations in High-$|k|$ Region

Chern number is one of the most canonical topological invariant characterizations. The topological properties can be understood in terms of the Chern number [1] in k space, which is defined as

$$C_\delta = \frac{1}{2\pi i}\int\int A_\delta d^2k, \quad A_\delta = i\left(\langle \partial u_\delta/\partial k_x|\partial u_\delta/\partial k_y\rangle - \langle \partial u_\delta/\partial k_y|\partial u_\delta/\partial k_x\rangle\right), \tag{9.3}$$

where A_δ is the Berry curvature and $u_\delta = \frac{\phi((k,\delta)}{|\phi((k,\delta)|}$ with the $k-$space occupied state $\phi(k,\delta)$. Because BECs are described by the macroscopic wave functions of the lowest energy, δ only denotes the corresponding ground states. Here, we consider the five components of the spin-dependent atomic-molecular BECs. For example, the k space occupied state of the ϕ_m $(m = -1, 0, 1)$ component is $\phi(k,\delta)$, where $\phi(k,\delta) = FFT(\phi_m)$. We use the $\phi(k,\delta)$ of each component to obtain the corresponding Berry curvature and Chern number. Figure 9.6a–c shows the values of Chern number as $|k|$ increases, where $|k| = \sqrt{k_x^2 + k_y^2}$. These figures correspond to the three cases in Fig. 9.4. The ladder-like changes of the Chern number mean the variation of the number

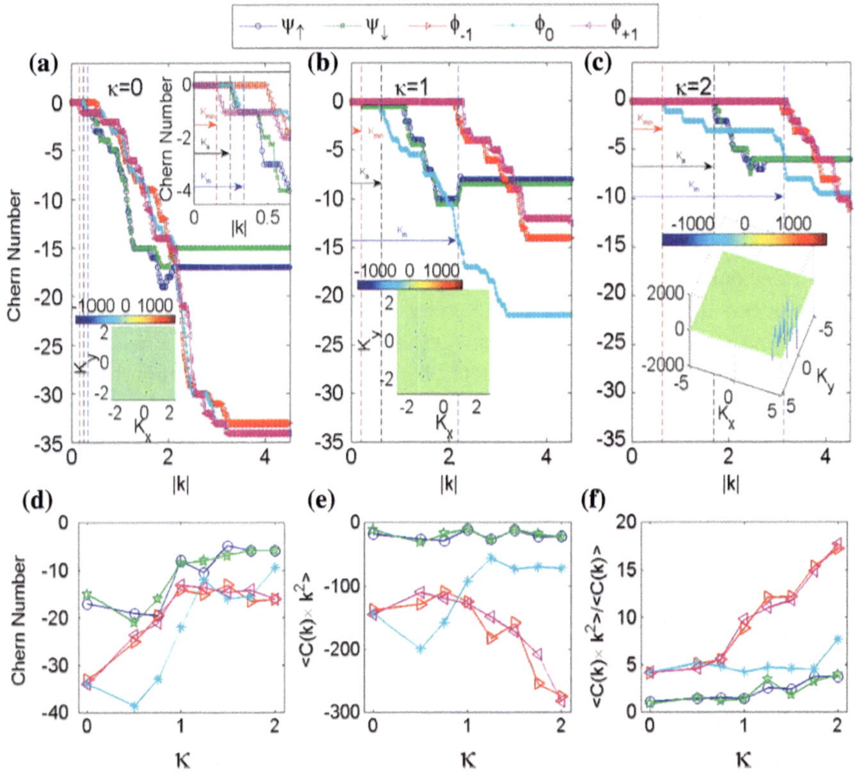

Fig. 9.6 Chern number verifying the high-$|k|$ topological excitations in the dual SOC atomic-molecular BECs. **a–c** Chern number in k space of atomic-molecular BECs in Fig. 9.3a, c, and e, respectively. The top inset in Fig. 9.6a is a local enlargement. The red dotted line and the red arrow indicate the minimum value K_{min} for the first hop as $|k|$ increases. The black dotted line and black arrow show the first hop of average position K_a in atomic BECs of $\psi_\uparrow$ and $\psi_\downarrow$. Similarly, the blue dotted line and blue arrow show the first hop of average value K_m in molecular BECs of ϕ_{+1} and ϕ_{-1}. The bottom inset in Fig. 9.6a is $A_\uparrow/i$ under $\lambda = 0$, where $A_\uparrow$ is the Berry curvature of $\psi_\uparrow$. Similarly, the inset in Fig. 9.6b is $A_\uparrow/i$ under $\lambda = 1$. The inset in Fig. 9.6c is A_{-1}/i under $\lambda = 2$. **d** Chern number as function of strength of SOC λ. **e** $\langle C(k) \times k^2 \rangle$ as function of strength of SOC λ. **f** $\langle C(k) \times k^2 \rangle / \langle C(k) \rangle$ as function of strength of SOC λ. Reprint from Ref. [17], Copyright 2018, with permission from American Physical Society

of topological excitations in k space as the integral radius increases. The platform denotes that there is no change as $|k|$ varies. Finally, the Chern number tends to be some invariants up to 10 or more as $|k|$ is large enough. The topological properties of the system in Fig. 9.6a–c differ considerably from each other because the final Chern numbers which imply the number of topological excitations in k space differ. Additionally, the bottom inset in Fig. 9.6a is $A_\uparrow/i$ in k space, where $A_\uparrow$ is the Berry curvature of $\psi_\uparrow$. Obviously, most of its value is 0, except some points whose values are up to thousands. These points almost are discretely and equally distributed around the position (0, 0). Thus, the topological excitations have different momentums in k space. This phenomenon is easy to be understood because the topological excitations tend to occur with the momentum as low as possible. Meanwhile, the topological excitations also have a nonzero size and have the interaction with each other in k space, which lead to the discrete distribution. When SOC increases to $\lambda = 1$, the points almost separate into two parts. One part is in $k_x > 0.5$ and the other is in $k_x < -0.5$, while there are no points around the region of k_y axis. The number of points in the region of $k_x < -0.5$ is much more than that in the region of $k_x > 0.5$ [see the inset in Fig. 9.6b]. This indicates that the topological excitations with k_x around 0 are restrained by SOC. And they have to get a higher momentum of k_x. Thus, the topological excitations have to get together along the x-direction. This is a main reason for the creation of topological spin textures's chain in Figs. 9.4 and 9.5. Therefore, SOC changes the properties of topological excitations according to the momentum. When SOC is up to $\lambda = 2$, almost all the points are in the region of $k_x > 1.5$ [see the inset in Fig. 9.6c]. In this case, the topological spin textures form a more serried chain.

We can define three parameters K_a, K_m, and K_{min} to characterize the properties of topological excitations represented by Chern number. K_a is the average position of the first hop of Chern number in atomic BECs with $\psi_\uparrow$ and $\psi_\downarrow$. Similarly, K_m is that in molecular BECs with ϕ_{+1} and ϕ_{-1}, and K_{min} is the minimum value of the first hop of the Chern number in the whole atomic-molecular BECs. Comparing these three cases, we find that the values of K_a (K_m, K_{min}) are the largest in Fig. 9.6c, the middle in Fig. 9.6b, and the least in Fig. 9.6a (see the top inset). These phenomena further show that SOC forces the Chern number to be induced in the high-$|k|$ region and restrain it in the low-$|k|$ region.

Figure 9.6d shows the value of Chern number as a function of SOC. SOC changes the topological properties of the whole system because the values of Chern number depend on it. SOC would finally restrict the Chern number when SOC is strong enough, but SOC cannot restrict its value to be zero completely. We also examine the value of $\langle C(k) \times k^2 \rangle$ in Fig. 9.6e and $\langle C(k) \times k^2/C(k) \rangle$ in Fig. 9.6f. $\langle C(k) \times k^2 \rangle$ implies the total properties of energy of all topological excitations in the system, while $\langle C(k) \times k^2 \rangle / \langle C(k) \rangle$ implies the average properties. Figure 9.6f shows that the value of $\langle C(k) \times k^2 \rangle / \langle C(k) \rangle$ tends to become larger as SOC increases. Thus, the topological excitations which are characterized by the Chern number should be induced in the high-$|k|$ region. Therefore, SOC plays a role in changing energy of the topological excitations. With the application of SOC, we can have the separated carbon-dioxide-like skyrmions constructed as the chain, whose energy is obviously much higher than the former.

9.2.6 Phase Diagram of the Topological Spin Textures in the Spin–Orbit-Coupled Atomic-Molecular BECs

Figure 9.7 plots the phase diagrams of the SOC atomic-molecular BECs with various products in our numerical simulations. In Fig. 9.7a, there is no skyrmion excitations when the rotation frequency is very low (see the red region). The white region shows the occurrence of only a single carbon-dioxide-like skyrmion. As long as the rotation

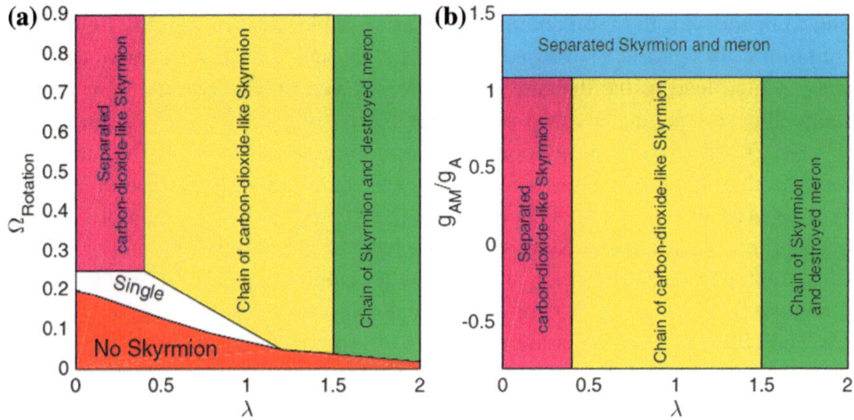

Fig. 9.7 Phase diagram of topological spin textures in the dual SOC atomic-molecular BECs. **a** Ration frequency-SOC ($\Omega_{\text{rotation}} - \lambda$) phase diagram for the products. We set the parameters $\omega = 200 \times 2\pi$ Hz, $\chi = 0.03$, $\Omega_A = 0.8$, $g_{\uparrow,\uparrow} = g_{\uparrow,\downarrow} = g_A$ with the scattering length $a_A = 101.8a_B$, $g_{\downarrow,\downarrow} = 0.95g_A$, $g_n = 2g_A$, $g_s = 0.03g_A$, $g_{AM} = 0.5g_A$, $\varepsilon = 0$, and $\mu_m = 2\mu_a = 3.5\mu_{m,0} = 7\mu_{a,0} = 28\hbar\omega$. **b** Atom–molecule interaction-SOC $[(g_{AM}/g_A) - \lambda]$ phase diagram for the products. We set the parameters $\omega = 200 \times 2\pi\, Hz$, $\chi = 0.03$, $\Omega_A = 0.8$, $g_{\uparrow,\uparrow} = g_{\uparrow,\downarrow} = g_A$ with the scattering length $a_A = 101.8a_B$, $g_{\downarrow,\downarrow} = 0.95g_A$, $g_n = 2gA$, $g_s = 0.03g_A$, $\varepsilon = 0$, and $\mu_m = 2\mu_a = 3.5\mu_{m,0} = 7\mu_{a,0} = 28\hbar\omega$. The unit of length, strength of χ, Ω_A, and SOC are $0.76\,\mu_m$, $\hbar\omega$, ω, and 0.96×10^{-3} m/s, respectively. Reprint from Ref. [17], Copyright 2018, with permission from American Physical Society

frequency is not too low, the skyrmion excitations appear. The induced carbon-dioxide-like skyrmions are separated when SOC is very weak (see the magenta region). At the yellow region, we can obtain the chain formed by carbon-dioxide-like skyrmions across the BECs. If SOC is strong enough (see the green region), then the half-skyrmions attached to the center skyrmion chain would be destroyed.

We further examine the effect of the interaction g_{am} on the products with a fixed rotation frequency $\Omega_{\text{rotation}} = 0.5\omega$ in Fig. 9.7b. Similarly, as its strength increases, SOC can make the products be separated carbon-dioxide-like skyrmion (see the magenta region), the carbon-dioxide-like skyrmion chain (see the yellow region), and the destroyed half-skyrmions (see the green region), respectively. However, if g_{am}/g_a is larger than 1.1, then the phase separation destroys the combinations among the topological excitations. So, we only obtain the separated skyrmions and half-skyrmions.

9.3 Supersolid with Nontrivial Topological Spin Textures in Spin–Orbit-Coupled BECs with a Spin-Dependent Periodic Potential

We study the hard-core Bose gas interacting via a contact (zero-range) potential. The atoms experience a spin-dependent periodic potential and are subjected to 2D SOC of the Rashba–Dresselhaus type $V_{so} = -i(\kappa_x\sigma_x\partial/\partial x + \kappa_y\sigma_y\partial/\partial y)$. Here, $\sigma_{x,y}$ are the Pauli matrices and $\kappa_{x,y}$ represents the corresponding SO-coupling strengths. We demonstrate that a supersolid phase characterized by the coexistence of periodic density modulation and superfluidity can be stabilized by strong SOC [8]. The supersolid phase in the present system involving a spin-dependent periodic potential is accompanied by the spontaneous generation of a lattice composed of meron pairs and antimeron pairs, and hence featuring a topologically nontrivial spin configuration. With decreasing the SO-coupling strength, this supersolid phase gives way to a state consisting of alternating spin domains separated by chiral Bloch walls. Depending on the sign of SOC, the chirality of the Bloch walls can be either right handed or left handed. We also discuss the influence of asymmetric interatomic interaction and SO-coupling anisotropy ($\kappa_x \neq \kappa_y$) on the properties of the supersolid phase.

9.3.1 SO-Coupling-Induced Supersolid

We consider SO-coupled two-component BECs in a spin-dependent periodic potential. The SOC is of the Rashba–Dresselhaus type, which can be realized by using a periodic pulsed magnetic field [17, 26]. The spin-dependent periodic potential is usually produced by means of the counterpropagating cross-polarized laser beams [19]. The Hamiltonian reads in the Gross–Pitaevskii mean-field approximation as

$$\mathcal{H} = \int d\mathbf{r}\Psi^\dagger\left(-\frac{\hbar^2}{2m}\nabla^2 + \nu_{so}\right)\Psi + \int d\mathbf{r}\sum_{\alpha=\uparrow,\downarrow} V_\alpha \Psi_\alpha^*\Psi_\alpha$$
$$+\frac{1}{2}\int d\mathbf{r}\sum_{\alpha,\beta=\uparrow,\downarrow} g_{\alpha\beta}\Psi_\alpha^*(\mathbf{r})\Psi_\beta^*(\mathbf{r})\Psi_\beta(\mathbf{r})\Psi_\alpha(\mathbf{r}), \tag{9.4}$$

where the complex-valued order parameter $\Psi = \left[\Psi_\uparrow(\mathbf{r}),\ \Psi_\downarrow(\mathbf{r})\right]^T$ is normalized to the total particle number N as $\int d\mathbf{r}\Psi^\dagger\Psi = \mathbf{N}$. The strength of the atom–atom interaction $g_{\alpha\beta} = 4\pi^2 a_{\alpha\beta}/m$ is characterized by the s-wave scattering length $a_{\alpha\beta}$. The SO-coupling term can be written as $V_{so} = -i(\kappa_x\sigma_x\partial/\partial x + \kappa_y\sigma_y\partial/\partial y)$, where $\sigma_{x,y}$ are the Pauli matrices and $\kappa_{x,y}$ denote the SO-coupling strengths. In the isotropic case when $\kappa x = \kappa y$, the SOC belongs to the Rashba type. The spin-up and spin-down atoms are subjected to the spin-dependent periodic potentials $V_\uparrow = V_0\sin^2[\pi x/a]$ and $V_\downarrow = V_0\cos^2[\pi x/a]$, respectively.

The many-body ground state can be obtained by numerically minimizing the Hamiltonian functional given by Eq. (9.4). In our calculation, we additionally introduce a weak harmonic trap $V_H = m[\omega_\perp^1(x^2+y^2) + \omega_z^2 z^2]/2$ with $\omega_\perp = \pi^2\hbar/ma^2$ to simulate realistic configurations of cold atom experiments. When $\lambda = \omega_z/\omega_\perp >> 1$, the condensates can be regarded as quasi-2D, and the effective interaction parameter in a 2D dimensionless form is $\widetilde{g}_{\alpha\beta} = 2\sqrt{2\pi}Na_{\alpha\beta}/a_{h_z}$, where $a_{h_z} = \sqrt{\hbar/m\omega_z}$. Considering that the differences in $a_{\uparrow\uparrow}$, $a_{\downarrow\downarrow}$, and $a_{\uparrow\downarrow}$ are within 1% in typical experiments involving the magnetic sublevels of alkali atoms, first we focus on the case of SU(2) symmetric interactions with $\widetilde{g} = \widetilde{g}_{\uparrow\uparrow} = \widetilde{g}_{\downarrow\downarrow} = \widetilde{g}_{\uparrow\downarrow}$.

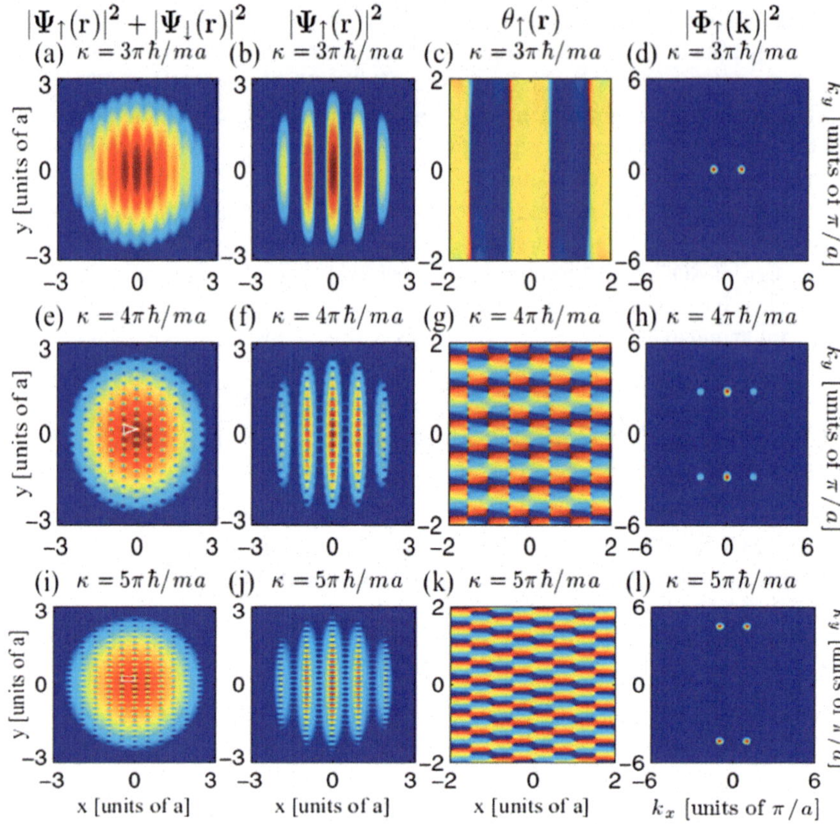

Fig. 9.8 Spin–orbit-coupling-induced transition from superfluid to supersolid. Various ground-state properties of Rashba SO-coupled Bose gases loaded in a 1D spin-dependent periodic potential are shown, with the SO-coupling strengths $\kappa = 3\pi\hbar/ma$ (top), $4\pi\hbar/ma$ (middle), and $5\pi\hbar/ma$ (bottom). With increasing SO-coupling strength, the system undergoes a phase transition from the superfluid phase (top) to a supersolid phase (middle and bottom) characterized by density modulation along the y-direction (first and second columns). The supersolid phase features a triangular lattice (middle) and a rectangular lattice (bottom) at intermediate and high SO-coupling strengths, respectively. The supersolid is accompanied by a spontaneous generation of vortex and antivortex chains in the spin-up and spin-down domains, respectively, as can be seen from the condensate phase modulation (third column). The occurrence of the supersolid phase is signaled by the separation of the momentum distribution peaks (fourth column), and can be readily observed via time-of-flight measurements. Other parameters used in these plots are $V_0 = 20\pi^2\hbar^2/ma^2$ and $\widetilde{g} = 6000$. Reprint from Ref. [8], Copyright 2018, with permission from American Physical Society

For a fixed value of atom–atom interaction, we observe a transition from the superfluid phase to a supersolid phase with increasing Rashba SO-coupling strength κ, as one can see in Fig. 9.8. Specifically, when the SOC is weak, the ground state of the system consists of alternating spin domains, where stripes filled with spin-up and spin-down atoms are segregated (see Fig. 9.8a and b). While the translational symmetry along the x-direction is explicitly broken by the spin-dependent periodic potential, the system preserves its translational symmetry along the y-direction. If the

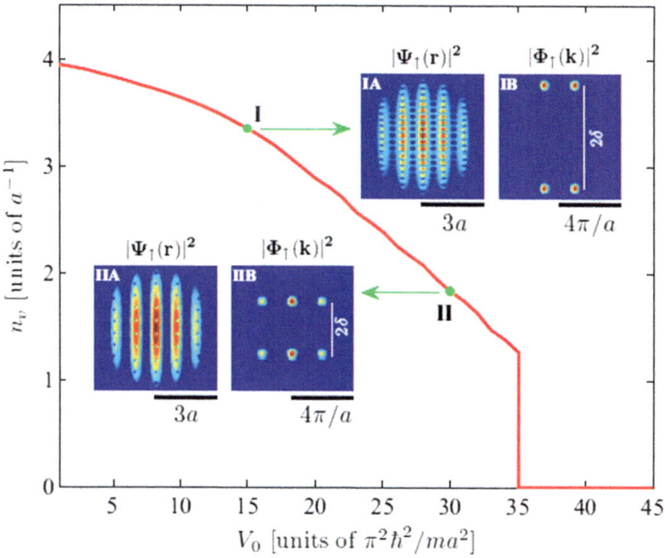

Fig. 9.9 Line density of vortices as a function of periodic potential depth. With increasing the depth V_0 of the periodic potential, the line density n_v of the vortices decreases gradually, and drops to zero suddenly at the transition point $V_0 = 35\pi^2\hbar^2/ma^2$ between the supersolid and superfluid phases. The insets depict the change of vortex density and atomic momentum distributions. The line density nv of the vortices is proportional to the separation δ of the momentum distribution peaks, and can be expressed as $n_v = \delta/\pi$. The Rashba SO-coupling strength is fixed at $\kappa = 4\pi\hbar/ma$, and the dimensionless interaction parameter is taken as $\widetilde{g} = 6000$. Reprint from Ref. [8], Copyright 2018, with permission from American Physical Society

strength of the SO coupling is increased beyond a critical value, the translational symmetry along the y-direction is spontaneously broken. As a result, a new phase with periodic density modulation along the y-direction is stabilized (see Fig. 9.8e, f, and i, j), and hence can be considered as a supersolid state. The emergence of such a density modulation can be understood as a stripe phase along the y-direction induced by SO coupling. However, this supersolid phase also exhibits exotic spin textures, which will be discussed below. The momentum distribution of the supersolid phase features a qualitative difference from the superfluid phase. In the superfluid phase, the atoms are condensed at a set of discrete points on the edge of Brillouin zones with finite momenta ($k_x \in K, k_y = 0$), where $K = \{\pm\pi/a, \pm 3\pi/a, \pm 5\pi/a, ...\}$ (see Fig. 9.8d). In the supersolid phase, the momentum distribution peaks are separated from $k_y = 0$ to $k_y = \pm\delta$. The separation distance $\delta \in (0, m\kappa/\hbar)$ depends on the SO-coupling strength (see Fig. 9.8h and l) and the periodic potential depth (see Fig. 9.9). This qualitative difference can be detected using conventional time-of-flight imaging technique.

In addition to the density modulation along the y-direction, the supersolid phase is also characterized by a vortex lattice structure consisting of vortex and antivortex chains in the spin-up and spin-down domains, respectively (see Fig. 9.8e–g and i–k). Depending on the competition between the SOC strength and periodic potential

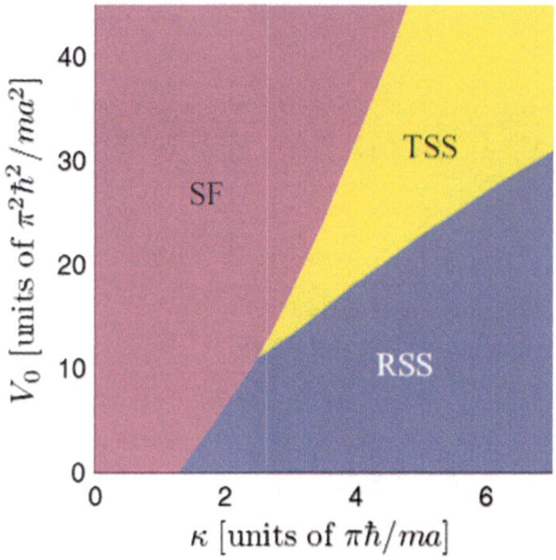

Fig. 9.10 The ground-state phase diagram spanned by the Rashba SO-coupling strength κ and the periodic potential depth V_0. Three phases can be identified on this phase diagram, including the superfluid (SF) phase, the triangular supersolid (TSS) phase, and the rectangular supersolid (RSS) phase. The dimensionless interaction parameter is taken as $\widetilde{g} = 6000$. Reprint from Ref. [8], Copyright 2018, with permission from American Physical Society

depth, two different arrangements of vortices can be stabilized. In one case, the vortices of the neighboring chains are staggered, forming a triangular lattice [Fig. 9.8e]. In the other case, the vortices of the neighboring chains are parallel, forming a rectangular lattice [Fig. 9.8i]. As shown in Fig. 9.8(h, i), these two different vortex lattices correspond to qualitatively different momentum distributions, and hence can be distinguished by experiments. In Fig. 9.10, we present the ground-state phase diagram spanned by the SO-coupling strength κ and the periodic potential depth V_0, with the effective interaction parameter being $\widetilde{g} = 6000$.

We stress that a vortex lattice is not directly associated with the supersolid phase, as it is absent in the supersolid droplet crystals. In the present system, the generation of vortices is a direct consequence of the interplay between the SOC, spin-dependent periodic potential, and interatomic interactions. This is very different from the usual manner of creating supersolid vortices by rotation or artificial magnetic fields. The alternating arrangement of vortex and antivortex chains can be viewed as alternating plane waves propagating in opposite directions along the y-axis, as shown in Fig. 9.8g and k. According to the Onsager–Feynman quantization condition [23] $\int_C \mathbf{v}_s \cdot d\mathbf{l} = 2\pi\hbar N_v/m$, we can express the line density of the vortices as $n_v = k_y/\pi$, where $k_y = \delta$ is the wave number of the plane waves. Numerical simulations show that for a given SO-coupling strength κ the line density of vortices decreases from $m\kappa/\pi\hbar$ to 0 with increasing the periodic potential depth V_0, as one can see in Fig. 9.9.

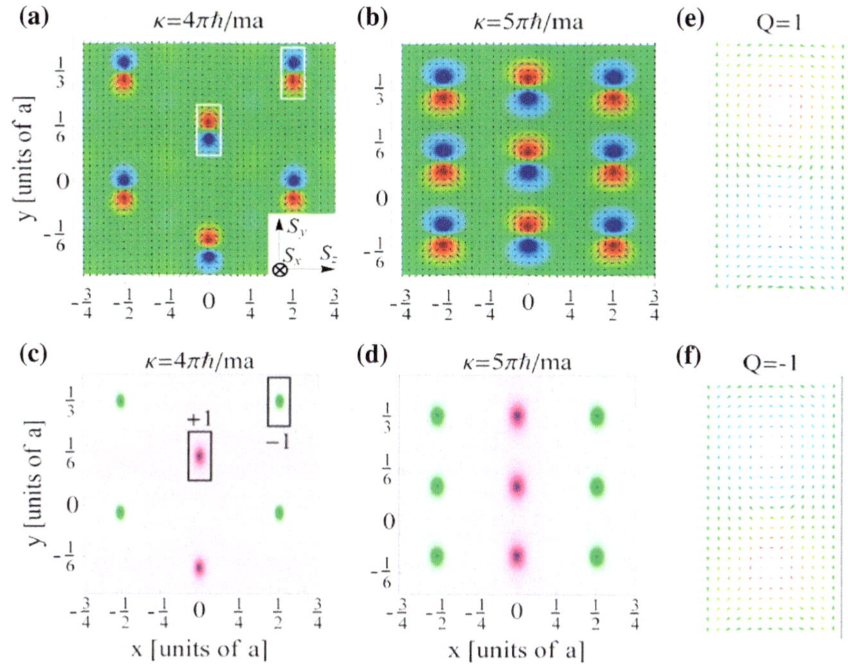

Fig. 9.11 Topological spin textures. **a**, **b** Spin configurations of the triangular and rectangular supersolid phases showing the spontaneous emergence of meron pairs and antimeron pairs. In these plots, the arrows represent the directions of spin vector (S_z, S_y) and the colors ranging from blue to red describe the values of S_x from -1 to 1. **c**, **d** Density distributions $q(\mathbf{r})$ of topological charge for the spin textures shown in **a** and **b**. The pink and green bubbles indicate that the meron pairs and antimeron pairs carry positive and negative topological charges, respectively. **e**, **f** Schematic spin configurations of a meron pair **e** and an antimeron pair **f**. Both the meron pairs and antimeron pairs have the same "circular-hyperbolic" structure, but with opposite spin orientations. Parameters used in these plots are identical to those used in Fig. 9.48e–h and i–l. Reprint from Ref. [8], Copyright 2018, with permission from American Physical Society

9.3.2 *Topological Spin Textures*

The two-component Bose gas can be considered as a magnetic system. Thus, one might naturally think that the supersolid transition would be associated with some magnetic ordering. We next demonstrate that the supersolid indeed features topologically nontrivial spin textures. To see this, we define a spin density vector $S = \mathbf{\Psi}^{\dagger}\boldsymbol{\sigma}\mathbf{\Psi}/|\mathbf{\Psi}|^2$ in the pseudospin representation, where œ is the Pauli matrix vector. Vectorial plots of **S** (under a pseudospin rotation $\sigma_x \rightarrow \sigma_z$ and $\sigma_z \rightarrow -\sigma_x$) are shown in Fig. 9.11a and b for the triangular and rectangular lattices, respectively, where the parameters are the same as in Fig. 9.8e and i. In Fig. 9.11a and b, the spin texture represents a spontaneous magnetic ordering in the form of crystals of meron pairs and antimeron pairs. The meron pairs reside in the spin-up domains, while the antimeron pairs reside in the spin-down ones. We note that a meron is a topological configuration in which the spin points up or down at the meron core and rotates away from the core. Both a meron pair and an antimeron pair have a "circular-hyperbolic" structure, and the only difference is that they have exactly opposite spin orientations (see Fig. 9.11e and f).

The topological nature of the spin textures can be characterized by the topological charge Q (Chern number), which is defined as a spatial integral of the topological charge density $q(\mathbf{r}) = (1/8\pi)\epsilon^{ij}\mathbf{S} \cdot \partial_i \mathbf{S} \times \partial_j \mathbf{S}$. In Fig. 9.11c and d, we present the topological charge density distribution for the triangular and rectangular lattices, respectively. Notice that both meron pairs and antimeron pairs are topologically nontrivial. A meron pair carries a topological charge 1, while an antimeron pair carries a topological charge -1. As a comparison, the topological charge density is zero everywhere in the topologically trivial superfluid phase. Topological spin texture lattices, such as meron-pair and skyrmion lattices, are usually stabilized by bulk rotation [11, 27]. Recently, it has been also suggested that skyrmion lattices can be realized by the combined effects of SO coupling and harmonic trap, provided that the trapping potential energy $\omega_\perp$ is higher than the characteristic interaction energy $\widetilde{g}\hbar\omega_\perp$ (i.e., $\widetilde{g} < 1$) [8]. Our results demonstrate that meron-pair lattices can also be stabilized by the combined effects of SO-coupling and spin-dependent periodic potential, within a large regime of interatomic interaction strength. For example, the meron-pair lattices in Fig. 9.11a and b are obtained at the large effective interaction strength with $\widetilde{g} = 6000 \gg 1$. This interaction strength can be naturally realized in experiments under realistic conditions without resort to the Feshbach resonance. This observation hence provides a way to create and manipulate topological spin textures in SO-coupled systems.

9.3.3 Chiral Domain Walls

After discussing the novel properties of the supersolid state in the previous sections, here we investigate the superfluid phase appearing at weak SOC. In this phase, the translational symmetry along the y-axis (orthogonal to the direction of the 1D periodic potential) is preserved, such that the system does not support density modulation along this axis. However, the presence of SOC breaks the spin-rotational symmetry in the $S_x - S_y$ plane, and leads to spontaneous chiral domain walls.

In order to give a clear description of this phenomenon, we first consider the effect of SOC on the relative phase $\theta_\uparrow - \theta_\downarrow$ of the two-component condensates, where $\theta_\uparrow$ and $\theta_\downarrow$ represent the phases of the spin-up and spin-down wave functions, respectively. In the absence of SOC, the Hamiltonian of Eq. (9.4) does not depend on the relative phase between the two spin components. In the presence of SOC, for the superfluid phase illustrated in Fig. 9.8a–d, the phase of the components does not alter except periodic jumps in the x-direction (see Fig. 9.8c), thus we have $\nabla\theta_\uparrow = \nabla\theta_\downarrow = 0$. Due to the translational symmetry, the gradient of the density along the y-direction can be approximately considered as $\partial_y|\Psi_\uparrow|^2 \simeq \partial_y|\Psi_\downarrow|^2 \simeq 0$, so the SO-coupling term in Eq. (9.4) can be represented as

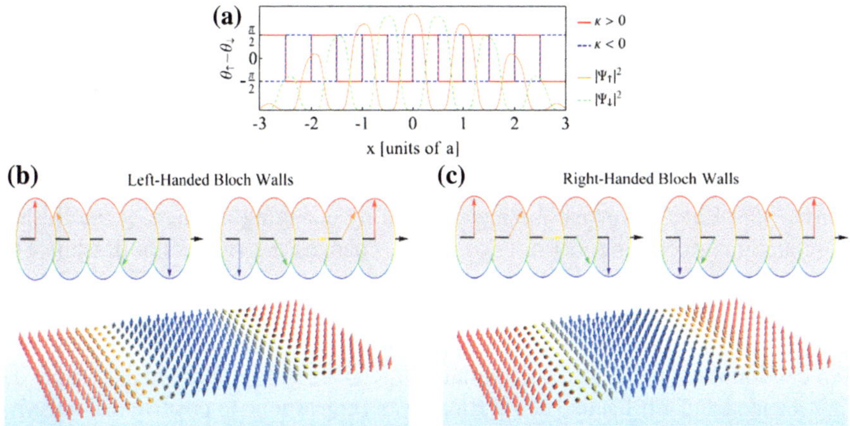

Fig. 9.12 Domain wall chirality. **a** Section view of the relative phase $\theta_\uparrow - \theta_\downarrow$ along the x-axis. The presence of Rashba SOC locks the relative phase at $\pm\pi/2$, and the periodic change of the densities along the x-direction induces a periodic modulation of the relative phase between $\pi/2$ and $-\pi/2$. By tuning the sign of the Rashba SOC from positive ($\kappa > 0$) to negative ($\kappa < 0$), the relative phase will be changed from $\pm\pi/2$ (red thick line) to $\mp\pi/2$ (blue thick dashed line). **b** An illustration of left-handed chiral Bloch walls stabilized by negative Rashba SO coupling with $\kappa < 0$. By crossing a right-handed (left-handed) chiral Bloch wall, the spin vector flips like a right-rotating (left-rotating) spiral. **c** An illustration of right-handed chiral Bloch walls stabilized by positive Rashba SO coupling with $\kappa > 0$. Reprint from Ref. [8], Copyright 2018, with permission from American Physical Society

$$\int \Psi^\dagger \nu_{so} \Psi d\mathbf{r} = 2\kappa \int |\Psi_\downarrow|\, \partial_x |\Psi_\uparrow| \sin\left[\theta_\uparrow - \theta_\downarrow\right] d\mathbf{r}. \tag{9.5}$$

One can easily see that in the presence of SOC, the Hamiltonian depends on the relative phase $\theta_\uparrow - \theta_\downarrow$. By minimizing the energy functional, the relative phase of the ground-state wave functions has to be locked at $\pm\pi/2$, where the sign $\pm$ is determined by the sign of $\partial_x|\Psi_\uparrow|$. As a result, the periodic density modulation along the x direction leads to a relative phase alternating between $\pi/2$ and $-\pi/2$ (see Fig. 9.12a).

The relative phase plays an important role in determining the type of the domain walls, which separate the spin-up and spin-down domains [8]. From the definition of the spin density vector $\mathbf{S}$, one finds that S_z is uniquely determined by the relative density, while the direction of the spin projection on the $S_x - S_y$ plane is determined by the relative phase and can be represented by an azimuthal angle $\alpha = \arctan\left[S_y/S_x\right] = \theta_\downarrow - \theta_\uparrow$. In the absence of SOC, the two-component condensates have an arbitrary relative phase, such that the spin projection on the $S_x - S_y$ plane within the domain wall can take arbitrary directions. In the presence of SOC, the relative phase is locked at $\pm\pi/2$, thus the spin-rotational symmetry in the $S_x - S_y$ plane is broken. As $S_x = 2|\Psi_\uparrow||\Psi_\downarrow| \cos\left[(\theta_\uparrow - \theta_\downarrow)/(|\Psi_\uparrow|^2 + |\Psi_\downarrow|^2)\right]$, obviously we have $S_x = 0$. This implies that the spins on the domain wall are confined within the $S_y - S_z$ plane and form a Bloch wall, crossing which the spin vector rotates like a spiral.

One important feature of a domain wall is its chirality, which distinguishes the right-handed rotation from the left-handed rotation as moving between domains. Domain wall chirality has been recently investigated in ultrathin ferromagnetic films [8]. As a new controllable degree of freedom, domain wall chirality opens up new opportunities for spintronics device designs, and has potential application in information processing and storage. In the present system, we find that the chirality of the Bloch walls can be manipulated by changing the sign of the Rashba SO-coupling. According to Eq. (9.5), if one changes the sign of the Rashba SO-coupling constant, the relative phase will jump between $\pm\pi/2$ (see Fig. 9.12a). As $S_y = -2|\Psi_\uparrow||\Psi_\downarrow| \sin\left[(\theta \uparrow -\theta \downarrow)/(|\Psi_\uparrow|^2 + |\Psi_\downarrow|^2)\right]$, changing the sign of the relative phase will change the sign of S_y, and hence the chirality of the Bloch walls. Typical examples of the spin configurations are given in Fig. 9.12b and c, where the right-handed and left-handed chiral Bloch walls correspond to positive and negative Rashba SO-coupling constants, respectively. In a realistic experiment, the sign of the Rashba SOC can be varied by tuning the phase of the r_f field [8].

9.3.4 *Effects of Asymmetric Interaction and Anisotropic SOC*

In the discussion above, we have focused on the case of SU(2) symmetric interaction with $g_{\uparrow\uparrow} = g_{\downarrow\downarrow} = g_{\uparrow\downarrow}$. It is important to consider also the non-SU(2) symmetric interaction with $g_{\uparrow\uparrow} = g_{\downarrow\downarrow} = g_{\uparrow\downarrow}$. We find that if a supersolid phase can be stabilized with a proper combination of SOC strength κ and periodic potential depth V_0 with a SU(2) symmetric interaction, an asymmetric interaction with $g_{\uparrow\uparrow} > g_{\uparrow\downarrow}$ always favors the supersolid phase, as shown in Fig. 9.13a. The supersolid phase is also stable for $g_{\uparrow\uparrow} < g_{\uparrow\downarrow}$ provided that the difference in $g_{\uparrow\downarrow}$ and $g_{\uparrow\uparrow}$ is sufficiently small, $g_{\uparrow\downarrow} - g_{\uparrow\uparrow}g_{\uparrow\downarrow}$. Such a situation corresponds to Fig. 9.13b. As one further increases the asymmetry such that $g_{\uparrow\uparrow}g_{\uparrow\downarrow}$, the supersolid phase becomes unfavorable and is replaced by the superfluid phase [Fig. 9.13c and d].

Additionally, we have also considered the anisotropy effects of the SOC. By decreasing κ_x, we find that the supersolid phase, if it exists in the Rashba case, remains stable for a certain range of $\kappa_x < \kappa_y$, as shown in Fig. 9.13e. By further increasing the anisotropy, the system undergoes a phase transition and becomes a superfluid, as shown in Fig. 9.13f. This superfluid phase can be regarded as a plane-wave state characterized by a phase modulation in the y-direction, as shown in Fig. 9.13g. The momentum distribution for this case is illustrated in Fig. 9.13h. In particular, when $\kappa_x = 0$, the SOC becomes unidirectional and reduces to that of the National Institute of Standards and Technology scheme [14], and the supersolid phase with nontrivial topological spin texture is no longer formed.

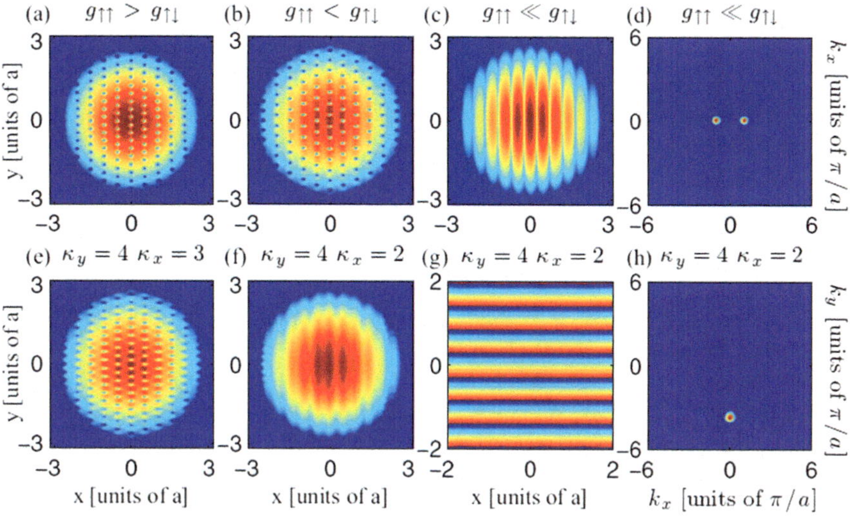

Fig. 9.13 Effects of asymmetric interaction and anisotropic spin–orbit coupling. **a–c** Density distributions of Bose condensates with Rashba SO coupling and asymmetric interaction. While the intracomponent interaction is fixed with $\widetilde{g}_{\uparrow\uparrow} = \widetilde{g}_{\downarrow\downarrow} = 6000$, the intercomponent interactions are varied to be **a** $\widetilde{g}_{\uparrow\downarrow} = 3000$, **b** $\widetilde{g}_{\uparrow\downarrow} = 7000$, and **c** $\widetilde{g}_{\uparrow\downarrow} = 18,000$. The Rashba SO-coupling strength is taken as $\kappa = 4\pi\hbar/ma$ in these plots. Notice that a sufficiently strong intercomponent interaction can drive the system from the supersolid phase to the superfluid phase. **e**, **f** Density distributions of Bose condensates in the presence of anisotropic SO interaction with $\kappa_x \neq \kappa_y$. Parameters used in these two plots are **e** $\kappa_x = 3\pi\hbar/ma$, $\kappa_y = 4\pi\hbar/ma$, and **f** $\kappa_x = 2\pi\hbar/ma$, $\kappa_y = 4\pi\hbar/ma$. Here, the interatomic interactions are considered to be SU(2) symmetric with $\widetilde{g} = 6000$. By increasing the SO-coupling anisotropy beyond a certain value, the system will undergo a phase transition and become a superfluid which can be regarded as a plane-wave state characterized by phase modulation shown in **g**. We emphasize that the superfluid phases driven by asymmetric interaction (**c**) and by anisotropic SO coupling (**f**) acquire different properties, as can be easily distinguished from the momentum distribution depicted in **d** and **h**, respectively. In this figure, the periodic potential depth is taken as $V_0 = 20\pi^2\hbar^2/ma^2$. Reprint from Ref. [8], Copyright 2018, with permission from American Physical Society

9.3.5 Discussion

The system considered in this section can be realized experimentally in ^{87}Rb condensates using two magnetic states $|F = 1, m_F = 1\rangle$ and $F = 1, m_F = -1\rangle$ of the $F = 1$ ground-state manifold. The Rashba SOC and spin-dependent periodic potential can be implemented by a combination of magnetic pulses [2], and a pair of cross-linear polarized counterpropagating laser beams [19]. The detailed proposal is given in Appendix A of Ref. [8]. Considering a typical experimental situation in which a total of $N = 1.7 \times 10^5$ atoms with the s-wave scattering length $a_{\alpha\beta} \approx 100a_B$ (a_B is the Bohr radius) are confined in a harmonic trap with the frequencies $\omega_\perp \approx 2\pi \times 40$ Hz and $\omega_z \approx 2\pi \times 200$ Hz, we obtain the effective interaction parameter $\widetilde{g} \approx 6000$. By using a CO_2 laser operated at a wavelength of 10.6 µm, one can produce a lattice constant a coinciding with $\pi\sqrt{\hbar/m\omega_\perp}$. These are consistent with the parameters used in our calculations.

9.4 Feynman Relation of Bose–Einstein Condensates with SOC

We investigate the Feynman relation of Bose–Einstein condensates with SOC, which relates the energy of excitations, the static structure factor in condensed phase, and the dispersion of free bosons. We show that the Feynman relation of these BECs is not satisfied in the whole momentum space.

With the presence of two-dimensional isotropic Rashba SOC, the ground state of free bosons is a ring with $|k| = \lambda$ in momentum space, which has infinite degeneracy, where λ is the strength of the SOC. Taking into account s-wave scattering processes between bosons, the full Hamiltonian of the system possesses $O(2)$ symmetry in momentum space. Mean-field analysis shows that the interactions between bosons break the $O(2)$ symmetry of the ground state and lift the infinite degeneracy. In this approximation, it is found that bosons prefer to condense at single momentum on the ring if the interactions between particles satisfy $g_{\uparrow\uparrow}g_{\downarrow\downarrow} > g_{\uparrow\downarrow}^2$, where $g_{\uparrow\uparrow}$ and $g_{\downarrow\downarrow}$ are the strengths of interactions between the same components of the bosons, and $g_{\uparrow\downarrow}$ is that between different components of the bosons. In this case, the condensate has plane-wave (PW) order. When $g_{\uparrow\uparrow}g_{\downarrow\downarrow} < g_{\uparrow\downarrow}^2$, bosons prefer to condense at two points with opposite momenta on the ring. This gives the striped phase (SP). The properties of the excitations over the condensates include the dispersion relations, superfluid critical velocity, mapping of the system to the chiral magnetism, effective scattering vertex using renormalization-group analysis, and effect of excitations on the stability of condensates [9].

However, the relation between the energy of the fluctuations and the corresponding correlation functions has not been investigated. In bosonic condensed systems without SOC, there is a simple Feynman relation between the energy of the excitations $\omega_{\mathbf{k}}$ and the static structure factor $S(\mathbf{k})$ at the same momentum [9]

$$\omega_{\mathbf{k}} = \frac{\varepsilon_0(\mathbf{k})}{S(\mathbf{k})}, \tag{9.6}$$

where $\varepsilon_0(\mathbf{k})$ is the dispersion for free bosons. In what follows, we investigate whether the Feynman relation still holds when SOC is introduced. We show that the excitations over condensates consist of two bands. And due to breaking of the $O(2)$ symmetry of the ground state, the dispersion is highly anisotropic and more divergent in the infrared limit than that without SOC. The dispersion also breaks time-reversal symmetry because the bosons are condensed onto a single momentum with a finite value. SOC enhances the coherence of bosons and the effect is more significant when its strength is larger. The Feynman relation is not applicable in the whole momentum space.

9.4.1 Description of the Model (Nonlinear σ Model)

The model of a two-dimensional homogeneous boson system with the presence of SOC is written in Lagrangian form as

$$Z = \int \left[d\Psi^*, d\Psi\right] \exp\left[-L\right],$$
$$L = \int_0^\beta d\tau d\mathbf{r} \Psi^*(\mathbf{r}, \tau) \left(\frac{\partial}{\partial \tau} - \frac{\hbar^2}{2m} \nabla^2 + 2i\lambda\nabla \cdot \sigma - \mu \right) \Psi(\mathbf{r}, \tau) \tag{9.7}$$
$$+ \frac{1}{2} g \int_0^\beta d\tau d^2\mathbf{r} n^2(\mathbf{r}, \tau),$$

where $\Psi(\mathbf{r}, \tau) = \left(\psi_\uparrow, \psi_\downarrow\right) \cdot \Psi(\mathbf{r}, \tau)$ are boson fields . They are complex numbers in a path integral scheme due to the bosonic properties of the particles. $\Psi^*(\mathbf{r}, \tau)$ are complex conjugates of fields. $\Psi(\mathbf{r}, \tau)\cdot$ $n(\mathbf{r}, \tau) = \Psi^*(\mathbf{r}, \tau)\Psi(\mathbf{r}, \tau)$ is the density of bosons. $\mathbf{r}$ and τ are the space and imaginary time coordinates. $\beta = \frac{1}{k_B T}$, where k_B is the Boltzmann constant and T is the temperature of the system. Here, we consider the zero-temperature case. The spin indexes $\uparrow$, $\downarrow$ are pseudospins, which usually refer to degenerate bands. λ is the strength of SOC. It couples the psedudospins with the movement directions of bosons. μ is the chemical potential of the system. m is the mass of bosons. $g = 4\pi a\hbar^2/m$ is the positive interacting coupling constant fixed by the s-wave scattering length a. We consider here the symmetric point for particle–particle interactions, that is, the interaction strength between different spins are the same as that between the same spins. In typical experimental situations, the strengths of spin-dependent interactions are always much lower than those for spin-independent terms. We expect that the result derived with symmetric interactions should also be applicable to cases where the interactions are not symmetric but PW order is still preferred. We set $\hbar = 2m = 1$ for simplicity.

In boson systems without SOC, the method of the nonlinear σ model has been proved successful [37]. With SOC, we write the bosonic field $\Psi(\mathbf{r}, \tau)$ into its amplitude $\nu(\mathbf{r}, \tau) = \sqrt{n(\mathbf{r}, \tau)}$ and phase $\mathbf{z}(\mathbf{r}, \tau)$,

$$\Psi(\mathbf{r}, \tau) = \nu(\mathbf{r}, \tau)\mathbf{z}(\mathbf{r}, \tau), \tag{9.8}$$

where $\mathbf{z}(\mathbf{r}, \tau)$ are complex two-component vectors under the constraint $|\mathbf{z}|^2 = 1$. In terms of ν and $\mathbf{z}$, the Lagrangian in Eq. (9.6) can be written as

$$L = \int d\tau d\mathbf{r} \left[\nu^2 \mathbf{z}^* \frac{\partial \mathbf{z}}{\partial \tau} + (\nabla\nu)^2 + \nu^2 \nabla\mathbf{z}^* \nabla\mathbf{z} \right.$$
$$+ 2i\lambda z_\uparrow^* z_\downarrow \nu\nabla_- \nu + 2i\lambda z_\downarrow^* z_\uparrow \nu\nabla_+ \nu \tag{9.9}$$
$$\left. + 2i\nu^2 \mathbf{z}^* \nabla \cdot \sigma\mathbf{z} - \mu\nu^2 + \frac{1}{2} g\nu^4 \right],$$

where $\nabla_\pm = \nabla_x \pm i\nabla_y$. It is a nonlinear σ model modified by SOC. The nonlinear effects caused by the particle–particle interactions are, for the most part, contained in the constraint $|z| = 1$.

9.4.2 Dispersion for Excitations Over Plane-Wave Condensates

The mean-field approximation cannot distinguish whether the condensates have PW or SP order when $g_{\uparrow\uparrow} = g_{\downarrow\downarrow} = g_{\uparrow\downarrow}$. However, after taking into account Gaussian fluctuations, it is showed that [36] the condensates with PW order have a lower ground-state energy than those with SP order, so that the condensates prefer PW to SP order.

Based on the result in Ref. [36], we here wish to study the properties of the excitations over condensates with PW order. We choose a coordinate system such that the momentum of the condensates is located on the k_x axis with coordinates $k_x = -\lambda$ and $k_y = 0$. The corresponding wave function of the condensate is $\Psi_0(\mathbf{r}, \tau) = \nu_0 \frac{1}{\sqrt{2}} \exp[-i\lambda x] (1, 1)^T$, where $\varphi_{\mathbf{k}} = \arg(k_x + ik_y) = \pi$ and ν_0 is the square root of the density of the condensates n_0, $\nu_0 = \sqrt{n_0}$. Then, the boson fields can be written as

$$\Psi(\mathbf{r}, \tau) = [\nu_0 + \delta\nu(\mathbf{r}, \tau)] \exp[-i\lambda x + i\theta] \begin{pmatrix} \cos\left[\frac{\pi}{4} + \phi\right] \exp[-i\xi] \\ \sin\left[\frac{\pi}{4} + \phi\right] \exp[i\xi] \end{pmatrix}. \quad (9.10)$$

Here, $\delta\nu(\mathbf{r}, \tau)$ are density fluctuations over the condensates and $\theta(\mathbf{r}, \tau)$ are the corresponding phase fluctuations, while $\phi(r, \tau)$ are spin fluctuations and $\xi(\mathbf{r}, \tau)$ are the corresponding phase fluctuations.

We substitute the above expression into the Lagrangian, Eq. (9.9), and expand it in orders of fluctuations. In the zeroth order, the Lagrangian for the condensates is $L_0 = \beta V\left[-(\mu + \lambda^2)n_0 + \frac{1}{2} g n_0^2\right]$, which takes a minimum value at $n_0 = \mu + \lambda^2 g$. This means that it must satisfy $\mu + \lambda^2 > 0$ for bosons to condense. Interestingly, for negative μ, the bosons do not condense in the case without SOC, but they do with the inclusion of large enough SOC.

The Lagrangian for fluctuations of quadratic order after taking Fourier transformation is

$$L' = \int d\mathbf{k} d\omega V(-\mathbf{k}, -\omega)^T M(\mathbf{k}, \omega) V(\mathbf{k}, \omega),$$

with

$$V(\mathbf{k}, \omega) \equiv (\theta(\mathbf{k}, \omega), \xi(\mathbf{k}, \omega), \phi(\mathbf{k}, \omega), \delta\nu(\mathbf{k}, \omega))^T,$$

and

$$M(\mathbf{k},\omega)=\begin{pmatrix} n_0k^2 & -in_0\lambda k_y & 0 & -\omega\sqrt{n_0} \\ in_0\lambda k_y & n_0\left(\frac{k^2}{4}+\lambda^2\right) & -n_0\left(\frac{\omega}{2}-2i\lambda k_x\right) & 0 \\ 0 & n_0\left(\frac{\omega}{2}-2i\lambda k_x\right) & n_0\left(k^2+4\lambda^2\right) & 2i\sqrt{n_0}\lambda k_y \\ \omega\sqrt{n_0} & 0 & -2i\sqrt{n_0}\lambda k_y & k^2+m_\nu \end{pmatrix}, \tag{9.11}$$

where $k=|\mathbf{k}|$. Above, $\delta\nu$ is gapped with mass $m_\nu=2gn_0$. In the mean-field approximation $m_\nu=2(\mu+\lambda^2)$, and it does not depend on g. Matrix elements with ω reflect the Heisenberg uncertainty relation between amplitude and phase fluctuations. The $\delta\nu,\theta$ section is composed of density-wave (DW) excitations. It becomes soft for a momentum around $k=(0,0)$, The ϕ,ξ section is composed of spin-wave (SW) excitations. It goes soft around $\mathbf{k}=(2\lambda,0)$. Differently from the case without SOC, the two sections $\delta\nu,\theta$ and ϕ,ξ are bound together by SOC.

We must emphasize that the fluctuations are over condensates with momentum $\mathbf{k}=(-\lambda,0)$. This means that the momenta of the above fluctuations have been shifted by $-\lambda$ in the k_x direction. The realistic momenta for the soft modes are around $(\pm\lambda,0)$. In what follows, we use the relative momentum unless otherwise specified.

9.4.3 Overall Band Structure

The dispersion relations of the fluctuations are determined by the poles of the corresponding propagators. These are just the solutions of the equation det $M(\mathbf{k},\omega_\mathbf{k})=0$, where the analytic continuum $i\omega\to\omega_\mathbf{k}+i\delta$ has been taken. It has

$$\omega_\mathbf{k}^4+b\omega_\mathbf{k}^3+c\omega_\mathbf{k}^2+d\omega_\mathbf{k}+e=0. \tag{9.12}$$

Here, for simplicity, the detailed expression for the coefficient is omitted. The four roots of this quartic equation are the same as the solutions of the following two quadratic equations:

$$\omega_\mathbf{k}^2+f_{1+}\omega_\mathbf{k}+f_{2+}=0,\ \omega_\mathbf{k}^2+f_{1-}\omega_\mathbf{k}+f_{2-}=0, \tag{9.13}$$

$$f_{1\pm}=\frac{b\pm\sqrt{8y+b^2-4c}}{2},\ f_{2\pm}=y\pm\frac{by-d}{\sqrt{8y+b^2-4c}}, \tag{9.14}$$

where y is any real solution of the cubic equation,

$$8y^3-4cy^2+(2bd-8e)+e\left(4c-b^2\right)-d^2=0. \tag{9.15}$$

By careful analysis, we find that if $\omega_\mathbf{k}$ is a solution of the first Eq. (9.13), then $-\omega_{-\mathbf{k}}$ is also a solution of the second Eq. (9.13). This particle–hole symmetry is also shown in BECs without SOC. For BECs without SOC, both the path integral method and the

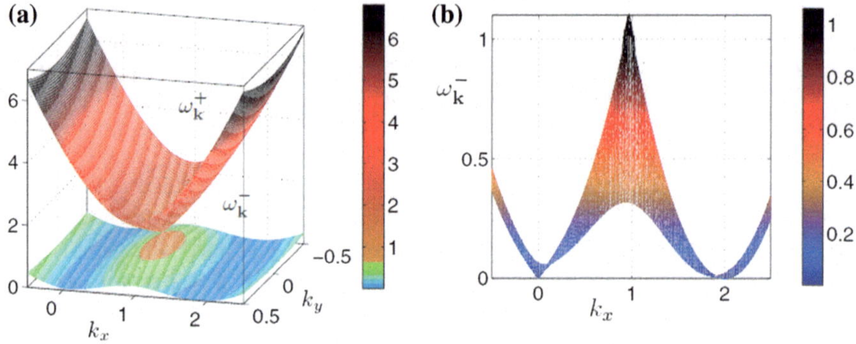

Fig. 9.14 **a** Dispersion of the upper band $\omega_{\mathbf{k}}^{+}$ and lower band $\omega_{\mathbf{k}}^{-}$. $\omega_{\mathbf{k}}^{+}$ is gapped for all momenta. $\omega_{\mathbf{k}}^{-}$ has some kind of flatness in the k_y direction. There are two soft modes. One is the density wave located around $\mathbf{k} = (0, 0)$, and the other is the spin wave located around $\mathbf{k} = (2, 0)$. **b** Dispersion of ω_{k}^{-} mapped onto the $k_x - \omega_{\mathbf{k}}$ plane. The positions for the soft modes are clearly shown. The momenta have been scaled in units of λ, and $\omega_{\mathbf{k}}^{\pm}$ in units of $\hbar^2\lambda^2/2m$. Reprint from Ref. [9], Copyright 2018, with permission from American Physical Society

BdG equation method give hole solutions. It is purely a consequence of mathematical disposal, and the "holes" do not exist physically. So solutions with a negative value are neglected. Here, with SOC, for the same reason, we neglect solutions with a negative value. Therefore, we only need to solve the second Eq. (9.13), for it gives a nonnegative $\omega(\mathbf{k})$ in the whole momentum space. This is different from the result using the hydrodynamical method [36] but agrees with the Bogoliubov method [4].

The overall band structure for excitations is shown in Fig. 9.14, where the momenta are in units of λ. The upper band ω_k^{+} is gapped for all momenta. The lower band ω_k^{-} is somewhat flat in the k_y direction. Because the coefficient e in Eq. (9.12) is 0 only at momenta $(0, 0)$ and $(2\lambda, 0)$, ω_k^{-} are gapless only at these two points. This means that the condensation lifts the infinite degeneracy of the ring ground state to twofold. The additional breaking of the $O(2)$ symmetry of the ring-shaped ground state makes the dispersion of the Goldstone modes more singular in the infrared limit compared to that without SOC. This is the reason for the flatness of the band, and the flattest direction is the k_y direction.

Because m_ν does not depend on the value of the particle–particle interaction strength g, we find that the dispersion does not rely on the value of g. However, for large enough g there will be three-body losses and the situation is different, and our calculation is limited to weak enough interactions g.

The mapping of the lower band $\omega_{\mathbf{k}}^{-}$ onto the $k_x - \omega_{\mathbf{k}}^{-}$ plane shows asymmetry about $k_x = \lambda$, as shown clearly in the lower panel in Fig. 9.14. We must emphasize that the momenta have been shifted by $-\lambda$ in the k_x direction, so the dispersion at the realistic momentum is asymmetric about $k_x = 0$. This is a reflection of the breaking of the time-reversal symmetry due to the fact that the bosons are condensed onto a single point of the ring in momentum space. By contrast, for BECs without SOC, $\omega_{\mathbf{k}} = \omega_{-\mathbf{k}}$, and the time-reversal symmetry is maintained.

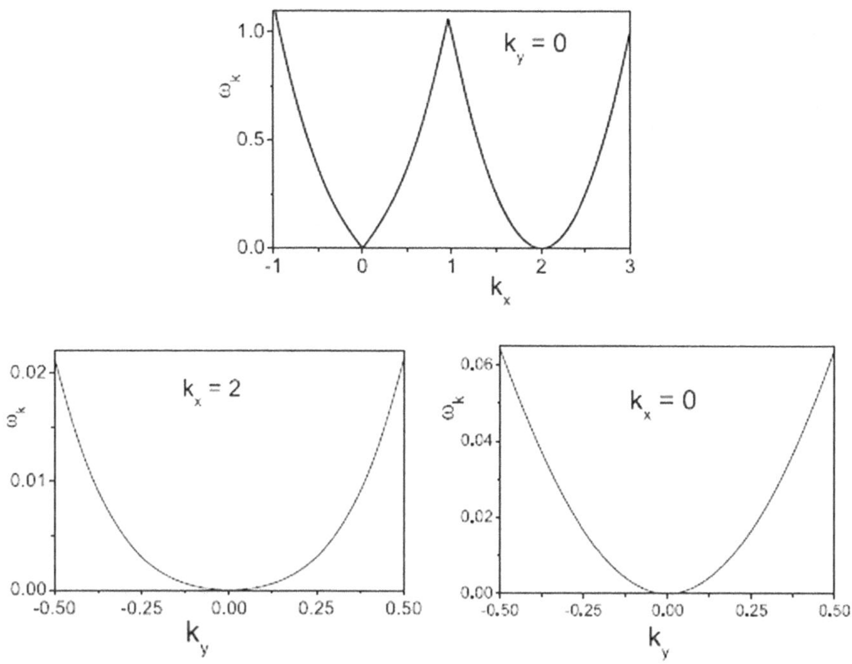

Fig. 9.15 Top: Dispersion of the lower band $\omega_{\mathbf{k}}^-$ along the k_x direction at $k_y = 0$. Bottom: Dispersion of the lower band $\omega_{\mathbf{k}}^-$ along the ky direction: (left) $k_x = 2$; (right) $k_x = 0$. We see that around $\mathbf{k} = (2, 0)$, the dispersion is quadratic and anisotropic in two directions. And around $\mathbf{k} = (0, 0)$, it is linear in the k_x direction and quadratic in the k_y direction. The momenta have been scaled in units of λ, and $\omega_{\mathbf{k}}^-$ in units of $\hbar^2\lambda^2/2m$. Reprint from Ref. [9], Copyright 2018, with permission from American Physical Society

9.4.4 *Dispersion for Soft Modes*

The excitations become soft around $\mathbf{k} = (0, 0)$ and $\mathbf{k} = (2\lambda, 0)$, which are DW and SW, respectively. Now, we consider the dispersion of these soft modes in the k_x and k_y directions. We choose $k_y = 0$ in the k_x direction and $k_x = 0, 2\lambda$ in the k_y direction. The result is shown in Fig. 9.15.

In the k_x direction with $k_y = 0$, the $\theta, \delta\nu$ and ϕ, ξ sections are decoupled. We only need to solve two quadratic equations to obtain the dispersion. For $k = (k_x, 0)$, the ϕ, ξ section is gapped, while the $\theta, \delta\nu$ section has the dispersion $E_{\mathbf{k}} = |k_x| \sqrt{k_x^2 + m_\nu}$. At small k_x, it is linear in k_x. For $\mathbf{k} = (2\lambda + k_x, 0)$, the $\theta, \delta\nu$ section is gapped, while the ϕ, ξ section is gapless, with dispersion $E_{\mathbf{k}} \sim k_x^2$. In the k_y direction with $k_x = 2\lambda$ and $k_x = 0$, from numerical calculations we find that the dispersion is quadratic in k_y for both DW and SW excitations. This can also be obtained analytically by calculating the expression of coefficient e in Eq. (9.12) which equals the multiplication of four roots.

In summary, the dispersion of DWs around $\mathbf{k} = (0, 0)$ is linear in k_x at fixed $k_y = 0$, and quadratic in k_y at fixed $k_x = 0$. Note that the dispersion for free bosons, $\varepsilon_0(\mathbf{k}) =$

$\left(\sqrt{(k_x-\lambda)^2+k_y^2}-\lambda\right)^2$, near the condensation point $\mathbf{k}=(0,0)$ has $\varepsilon_0(k_x,0)\sim k_x^2$ and $\varepsilon_0(0,k_y)\sim\frac{k_y^4}{4\lambda^2}$. We see that the dispersion for excited states over condensates is just approximately the square of the one in the free boson case in these two directions. This is the same as that without SOC. SWs around $\mathbf{k}=(2\lambda,0)$ have an anisotropic quadratic dispersion in the k_x and k_y directions. For a two-component boson system without SOC, the SW section is decoupled from the DW section, which makes the SWs behave like free bosons. However, SOC induces scattering processes with momentum exchange $\mathbf{q}\simeq(\pm2\lambda,0)$, which couple the DW and SW sections together. As a result, the SW excitations are no longer free bosons. Because of the highly infrared divergence of the SW modes, there is the natural question of the effect of these SW excitations on the stability of the condensates. The result needs detailed calculation of the depletion.

9.4.5 Static Structure Factor and Feynman Relation

The static structure factor $S(\mathbf{k})$ is the Fourier transform of the pair distribution function

$$\frac{N}{V}S(\mathbf{k})=\int d^d\mathbf{r}\langle\delta n(\mathbf{r})\delta n(\mathbf{0})\rangle\exp\left[i\mathbf{k}\cdot\mathbf{r}\right], \tag{9.16}$$

where V is the volume of the system, N is the total number of bosons, and $\delta n(\mathbf{r})$ is the density fluctuations over the mean density of bosons $\overline{n}=N/V$. It measures the strength of scattering pairs of atoms out of the condensates to form the quantum depletion. There is

$$\begin{aligned}\frac{N}{V}S(\mathbf{k}) &= \langle\delta n(\mathbf{r})\delta n(-\mathbf{k})\rangle\simeq 4n_0\delta\nu(\mathbf{k})\delta\nu(-\mathbf{k})\\ &=\frac{2n_0}{\beta}\sum_{n=-\infty}^{+\infty}\exp\left[i\omega_n\delta^+\right]\frac{k^2(k^2+4\lambda^2)^2-4\lambda^2k_y^2(k^2+4\lambda^2)-k^2(i\omega_n+4\lambda k_x)^2}{(i\omega_n-\omega_{\mathbf{k}}^+)(i\omega_n+\omega_{-\mathbf{k}}^+)(i\omega_n-\omega_{\mathbf{k}}^-)(i\omega_n+\omega_{-\mathbf{k}}^-)},\end{aligned} \tag{9.17}$$

where δ^+ is an infinite small positive number. $\omega_n=2\pi nT$ are the frequencies of bosons at a finite temperature where n is the integer number. $\mathbf{k}$ is the relative momentum against the condensed point $(-\lambda,0)$. We assume weak interactions such that the interaction-induced quantum depletions are small and $n_0\simeq\frac{N}{V}$.

The expression of $S(\mathbf{k})$ shows that it is symmetric with respect to the k_x and k_y axes. The result with $k_x>0$, $k_y>0$ is shown in Fig. 9.16. We see that $S(\mathbf{k})$ decreases as the momentum k becomes small, where the DW becomes soft. And it becomes 0 at the condensed point $\mathbf{k}=\mathbf{0}$. This means that the correlation goes to 0 where the quantum coherence is the largest. $S(\mathbf{k})$ remains finite and featureless as the momentum SW becomes soft.

When $\mathbf{k}$ is small, $S(\mathbf{k})$ is linear in k_x at $k_y=0$ and quadratic in k_y at $k_x=0$, as shown in Fig. 9.16. Like the behavior of the dispersion, in these two directions $S(\mathbf{k})$ at momenta near the condensation point are just the square of the dispersion for free

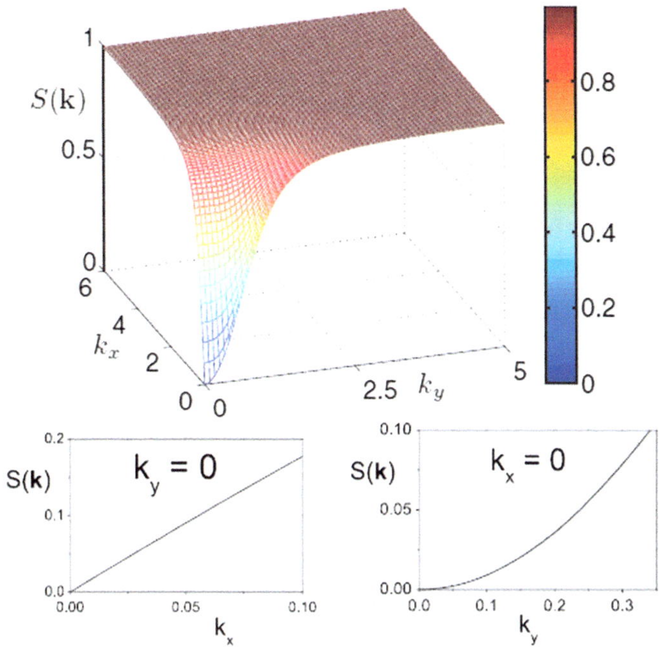

Fig. 9.16 The static structure factor $S(\mathbf{k})$. Momenta have been scaled in units of λ. Top: $S(\mathbf{k})$ goes to 0 at $\mathbf{k} = (0, 0)$, where the density-wave modes become soft. It becomes unity at a larger $\mathbf{k}$, which has the same behavior as free particles. Bottom: (Left) $S(\mathbf{k})$ versus k_x at $k_y = 0$; (right) $S(\mathbf{k})/k_y$ versus k_y at $k_x = 0$. We see that $S(\mathbf{k})$ is linear in the k_x direction and quadratic in the k_y direction. Reprint from Ref. [9], Copyright 2018, with permission from American Physical Society

bosons. For a high momentum, the static structure factor goes to unity just as for free bosons.

Now, we discuss the change in the static structure factor with a change in the SOC strength. The result is shown in Fig. 9.17 for a negative chemical potential and in Fig. 9.18 for a positive chemical potential. We find that enhancing λ reduces the static structure factor in a larger region of momenta. This is a reflection of the enhancement of the two processes of exciting opposite relative momenta; here, the relative momenta are momenta relative to the momentum of condensates.

In a one-component boson system, the Feynman relation, Eq. (9.6), holds true in the whole momentum space. It is a reflection of the f-sum rule. However, a multiple-band structure can make this simple relation exists only at a special momentum where the scattering processes between the lowest band and the condensates dominate. With SOC, the excitations consist of two bands. In addition, the dispersion is highly anisotropic and time-reversal symmetry is broken. For weak interactions, with $n_0 \simeq \frac{N}{V}$, by putting the dispersion for free bosons $\varepsilon_0(\mathbf{k}) = \left(\sqrt{(k_x - \lambda)^2 + k_y^2} - \lambda\right)^2$ in

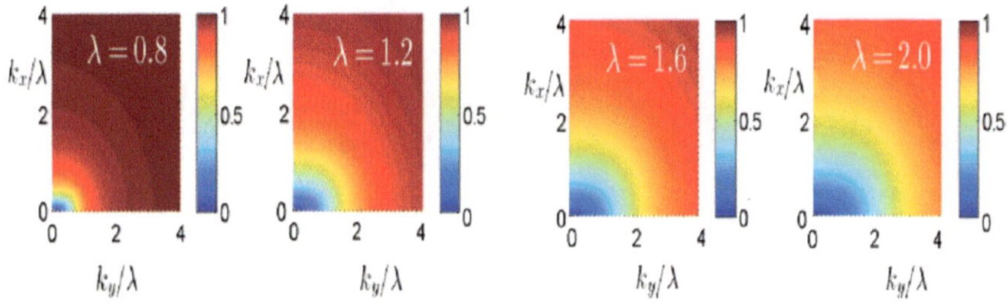

Fig. 9.17 The static structure factor $S(\mathbf{k})$ mapped onto the $k_x - k_y$ plane. The chemical potential $\mu = -0.5$ here. We see that increasing the spin–orbit coupling strength λ reduces the static structure factor and, hence, makes the bosons more coherent. Reprint from Ref. [9], Copyright 2018, with permission from American Physical Society

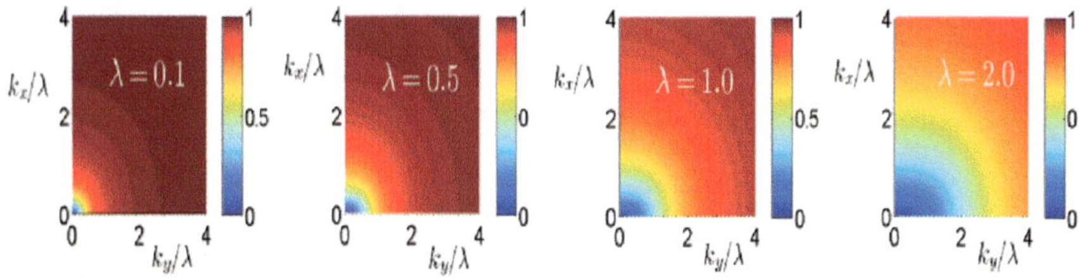

Fig. 9.18 The static structure factor $S(\mathbf{k})$ mapped onto the $k_x - k_y$ plane. The chemical potential $\mu = 0.1$ here. We see that increasing the spin–orbit coupling strength λ reduces the static structure factor and, hence, makes the bosons more coherent. Reprint from Ref. [9], Copyright 2018, with permission from American Physical Society

Eq. (9.6), we find that the Feynman relation does not apply in the whole momentum space (see Fig. 9.19).

In Fig. 9.19, the relative error of the Feynman relation, $\frac{\frac{\varepsilon_0(\mathbf{k})}{S(\mathbf{k})}}{\omega^-_{\mathbf{k}}} - \omega^-_{\mathbf{k}}$, is calculated and is mapped onto momentum space. There is symmetry for $k_y \leftrightarrow -k_y$, so only the $k_y > 0$ case is shown. We see that the Feynman relation is not satisfied around the "ring" in the momentum space. The energy of the free bosons is 0 on the ring in the momentum space. However, due to condensation of bosons onto a single point of the ring, the energy of excitations on the ring is increased except at the two points $(0, 0)$ and $(2\lambda, 0)$. And the static structure factor is 0 only at the condensed point $(0, 0)$. So the naive Feynman relation, Eq. (9.6), is clearly no longer applicable. In the same reason, we expect that the relation is not satisfied for general particle–particle interactions.

Furthermore, by setting the SOC strength λ to 0 in the Lagrangian, Eq. (9.9), it can be calculated easily that the static structure factor $S(\mathbf{k})$ and dispersion of excitations $\omega_{\mathbf{k}}$ satisfy the Feynman relation, Eq. (9.6). So the violation of the Feynman relation found in this book is not due to the fact that we are dealing with a two-component system. It is purely an effect of the existence of SOC.

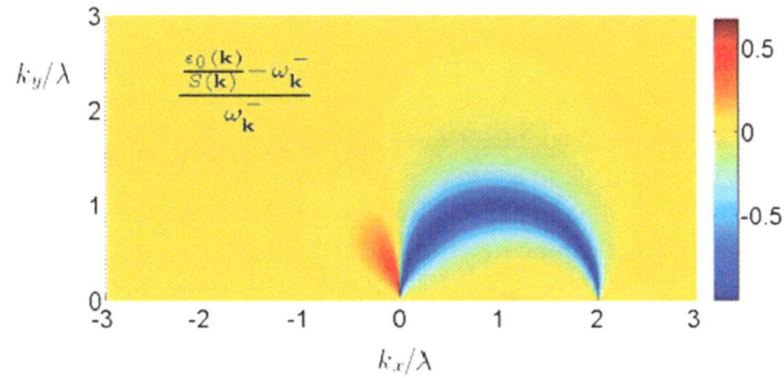

Fig. 9.19 $\frac{\frac{\varepsilon_0(\mathbf{k})}{S(\mathbf{k})}-\omega_{\mathbf{k}}^-}{\omega_{\mathbf{k}}^-}$ is the relative error of the Feynman relation $\omega_{\mathbf{k}}^- = \frac{\varepsilon_0(\mathbf{k})}{S(\mathbf{k})}$ in boson systems with spin–orbit coupling. Here, $\varepsilon_0(\mathbf{k}) = \left(\sqrt{(k_x-\lambda)^2+k_y^2}-\lambda\right)^2$ is the dispersion for free bosons. There is symmetry for $k_y \leftrightarrow -k_y$, so only the $k_y > 0$ case is shown in the diagram. We see that the Feynman relation is not satisfied around the "ring" in the momentum space. Reprint from Ref. [9], Copyright 2018, with permission from American Physical Society

9.5 Soliton Management in 1D Spin–Orbit-Coupled Bose–Einstein Condensates

We study the motion of matter-wave solitons in 1D BECs subjected to a spin–orbit coupling. The effect of the spin dynamics on the orbital motion of solitons is investigated.

9.5.1 *Presentation of the Model*

In the presence of SOC, the dynamics of the quasi-1D BECs, elongated in the direction of z, is modeled by the mean-field two-component NLS equations

$$i\frac{\partial}{\partial t}\begin{pmatrix}\psi_\uparrow\\ \psi_\downarrow\end{pmatrix} = \widehat{h}_0\begin{pmatrix}\psi_\uparrow\\ \psi_\downarrow\end{pmatrix} + \begin{pmatrix} g_{\uparrow\uparrow}|\psi_\uparrow|^2 & g_{\uparrow\downarrow}|\psi_\downarrow|^2\\ g_{\downarrow\uparrow}|\psi_\uparrow|^2 & g_{\downarrow\downarrow}|\psi_\downarrow|^2\end{pmatrix}\begin{pmatrix}\psi_\uparrow\\ \psi_\downarrow\end{pmatrix}, \tag{9.18}$$

$$\widehat{h}_0 = -\frac{1}{2}\frac{\partial^2}{\partial z^2} + V(z) + i\lambda\frac{\partial}{\partial z}\sigma_z + \Omega\sigma_x + \delta\sigma_z, \tag{9.19}$$

where ψ_σ are the pseudospin components of the BEC macroscopic wave function, with $\sigma = \uparrow, \downarrow$ labeling the spin states; these can represent, for instance, the hyperfine states $|1,-1\rangle$ and $|1,0\rangle$ of ^{87}Rb atoms. $\widehat{h}_0$ is a single-particle Hamiltonian which includes the Raman-induced SOC characterized by a strength λ, with and describing, respectively, the frequencies of the Raman coupling and the Zeeman detuning. $V(z) = \gamma^2 z^2/2$ is an effective 1D harmonic trap potential, and $\gamma \equiv \omega_z/\omega_\perp$ is the trap's aspect ratio, with ω_z and $\omega_\perp$ being the trapping frequencies along the longitudinal and transverse directions, respectively. The frequencies and lengths are measured in units $\omega_\perp$ and $a_\perp = \sqrt{\hbar/m\omega_\perp}$, respectively, and, as mentioned above, $\lambda = k_L a_\perp$ represents the SOC strength, with k_L being the momentum transfer. Note that, as the strengths of the inter- and intraspecies atomic interactions are very close in the experiment, it is reasonable to assume SU(2)-symmetric spin interactions, with all components $g_{\sigma\sigma'}$ taking a single value g. To focus on the SOC effects on the dynamics of solitons, we first consider the free space, while the external trap will be discussed afterward.

For $\lambda = \Omega = 0$, the system reduces to a normal binary BEC without the SOC. In this case, Eq. (9.18) is known as the integrable Manakov's system which gives rise to well-known exact soliton solutions. In particular, bright–bright solitons are $\psi_\sigma = (\eta\epsilon_\sigma/\sqrt{-g})\text{sech}[\eta z]\exp\left[i\eta^2 t/2\right]$ for the attractive sign of the nonlinearity, $g < 0$, where η^{-1} is the soliton's width and ϵ_σ satisfies the normalized condition $\left|\epsilon_\uparrow\right|^2 + \left|\epsilon_\downarrow\right|^2 = -g/2\eta$. We use such exact soliton solutions as an initially prepared wave function, and then study the soliton dynamics as the SOC is switched on. Note that Eq. (9.19) is an effective single-particle Hamiltonian in the frame transformed via the local pseudospin rotation by angle $\vartheta = 2\lambda z$ about the z-axis. The transformation adds opposite phase factors $\exp[\pm i\lambda z]$ to the two components of the input wave forms.

9.5.2 Dynamics of Soliton in the 1D SOC BECs

In the general case, the two-component NLS system (9.18) is no longer integrable. Therefore, we employ a variational approximation to investigate the soliton dynamics, based on the Lagrangian,

$$L(t) = \int_{-\infty}^{+\infty} \left\{ \frac{i}{2} \sum_{\sigma=\uparrow,\downarrow} \left[\psi_\sigma^* (\psi_\sigma)_t - \psi_\sigma (\psi_\sigma^*)_t\right] - \mathcal{H} \right\} dz,$$

where $\mathcal{H}$ is the Hamiltonian density of the system. We first consider the attractive nonlinearity, with $g < 0$. In this case, we introduce the following variational ansatz for bright-bright (BB) solitons, with the total norm fixed to be 1,

$$\begin{pmatrix} \psi_\uparrow \\ \psi_\downarrow \end{pmatrix} = \sqrt{\frac{\eta}{2}} \begin{pmatrix} \frac{\sin[\theta]}{\eta z + \xi_\uparrow} \exp\left[i\left(k_\uparrow z + \varphi_\uparrow\right)\right] \\ \frac{\cos[\theta]}{\cosh[\eta z + \xi_\downarrow]} \exp\left[i\left(k_\downarrow z + \varphi_\downarrow\right)\right] \end{pmatrix}, \tag{9.20}$$

where θ, η, ξ_σ, k_σ, and φ_σ are time-dependent variational parameters. Here, θ determines the population imbalance between the pseudospin components, η^{-1} defines their common width, k_σ is the wave number, and φ_σ the phase. For the SU(2) atomic interactions, two-component solitons favor the mixed phase. Therefore, the positions of spin-up and spin-down solitons will overlap, $\xi_\uparrow = \xi_\downarrow = \xi$, as confirmed by the numerical simulations below.

Inserting the ansatz (9.20) into the Lagrangian and performing the integration yield

$$
\begin{aligned}
L(t) = {} & \frac{\xi}{\eta}\frac{dk_+}{dt} - \frac{\xi}{\eta}\cos[2\theta]\frac{dk_-}{dt} - \frac{d\varphi_+}{dt} + \cos[2\theta]\frac{d\varphi_-}{dt} \\
& - \frac{1}{2}[k_+^2 - 2k_+k_-\cos[2\theta] + k_-^2] - \frac{1}{6}\eta^2 - \frac{1}{6}g\eta \\
& - \frac{\Omega\pi k_- \sin[2\theta]\cos\left[2\varphi_- - 2k_-\xi/\eta\right]}{\eta\sinh\left[\pi k_-/\eta\right]} + \delta\cos[2\theta] - \lambda[k_+\cos[2\theta] - k_-],
\end{aligned}
\tag{9.21}
$$

where $k_\pm \equiv (1/2)(k_\uparrow \pm k_\downarrow)$ and $\varphi_\pm \equiv (1/2)(\varphi_\uparrow \pm \varphi_\downarrow)$. The evolution of the variational parameters is governed by the corresponding Euler–Lagrangian (EL) equations.

Note that the EL equations produce simple results, $\eta \approx -g/2$ and $k- \approx \lambda$, in the case of a weak SOC, $\pi\lambda \gg \eta$. Indeed, in the absence of SOC, the relation $\eta = -g/2$ holds for the normalized wave function, indicating that the width of the solitons is determined by the nonlinearity, and k_- remains equal to the initial relative momentum λ between the components of the soliton. Thus, we arrive at a reduced system of the EL equations, in which η and k_- are considered as frozen quantities:

$$\dot{k}_+ = 2\lambda\widetilde{\Omega}\sin[2\theta]\sin[\phi], \tag{9.22a}$$

$$\dot{\phi} = -2\widetilde{\Omega}\cot[2\theta]\cos[\phi] + 2\lambda k_+ - 2\delta, \tag{9.22b}$$

$$\dot{\theta} = -\widetilde{\Omega}\sin[\phi], \tag{9.22c}$$

$$\langle\dot{z}\rangle = k_+. \tag{9.22d}$$

Here, $\langle z\rangle = \int_{-\infty}^{+\infty} z\left(|\psi_\uparrow|^2 + |\psi_\downarrow|^2\right)dz \equiv -\xi/\eta$ is the center-of-mass coordinate, $\widetilde{\Omega} = \Omega\pi\lambda/[\eta\sinh[\pi\lambda/\eta]]$, and $\phi \equiv 2\varphi_- + 2k_- z$ is the phase difference between the two components of the soliton. Thus, Eqs. (9.22a)–(9.22d) account for the nonlinear coupling of the center-of-mass momentum k_+, phase difference ϕ, population imbalance θ, and center-of-mass coordinate $\langle z\rangle$.

We introduce a normalized complex-valued spinor $\chi = (\chi_\uparrow, \chi_\downarrow)$ describing the two-component wave function $\psi_\sigma = \rho(z,t)\chi_\sigma$, where $\rho \equiv |\psi_\uparrow|^2 + |\psi_\downarrow|^2$ is the total density, with $|\chi_\uparrow|^2 + |\chi_\downarrow|^2 = 1$. Furthermore, we define the spin density $S = \chi T\sigma\chi$, where $\sigma \equiv \{\sigma x, \sigma y, \sigma z\}$ is the vector set of the Pauli matrices. Therefore, we have

$$\{S_x, S_y, S_z\} = \{\sin[2\theta]\cos(2\lambda z + 2\varphi-), -\sin[2\theta]\sin[2\lambda z + 2\varphi-], -\cos(2\theta)\}$$

for the ansatz given by Eq. (9.20). In what follows, we focus on the center-of-mass motion of the soliton by setting $z = \langle z\rangle = -\xi/\eta$. Consequently, the evolution of the atomic spin at the soliton's center-of-mass is governed by the following equations:

$$\dot{S}_x = 2(\lambda c_1 - \delta)S_y - 2\lambda^2 S_z S_y, \tag{9.23a}$$

$$\dot{S}_y = -2\widetilde{\Omega}S_z - 2(\lambda c_1 - \delta)S_x + 2\lambda^2 S_z S_x, \tag{9.23b}$$

$$\dot{S}_z = 2\widetilde{\Omega}S_y, \tag{9.23c}$$

where $c_1 \equiv \lambda S_{z,0}$, with $S_{\alpha,0}$ ($\alpha = x, y, z$) being the initial values of the components. These equations of motion for the soliton's spin can be rewritten as

$$\dot{S} = \mathbf{S} \times \mathbf{B}_{eff}, \quad \mathbf{B}_{eff} = \{2\tilde{\Omega}, 0, 2\lambda^2 S_z - 2(\lambda c_1 - \delta)\}. \tag{9.24}$$

This represents the Bloch equations for the spin precession under the action of the effective magnetic field $\mathbf{B}_{eff}$. The macroscopic SOC for the soliton as a quantum body is determined by the effect of the evolution of the spin on the soliton's longitudinal momentum, resulting in a coupled nonlinear dynamics of the soliton's spin and position. At first glance, nonlinear terms in Eqs. (9.23a)–(9.23c) arise essentially from the SOC strength λ. However, the atomic interactions also play a fundamental role, as in the no-interaction limit, $\tilde{\Omega} = 0$, these equations reduce to the linear Bloch precession under a fixed effective magnetic field.

To tackle solutions of the nonlinear Bloch equation, we first integrate Eq. (9.23a), dividing it by Eq. (9.23c). This yields $S_x = c_2 - (\lambda^2/2)\tilde{\Omega}^{-1} S_z^2 + (\lambda c_1 - \delta)\tilde{\Omega}^{-1} S_z$, where $c_2 \equiv S_{x,0} + (\delta - c_1\lambda)\tilde{\Omega}^{-1} S_{z,0} + (\lambda^2/2)\tilde{\Omega}^{-1} S_{z,0}^2$ is a constant determined by the initial conditions. Next, we focus on the case of $\delta = \lambda c_1$, which implies a particular relation between the strengths of the Zeeman splitting and SOC, making the analysis more explicit. By differentiating Eq. (9.23c), we then arrive at a standard equation of anharmonic oscillations for the single spin component S_z

$$\frac{d^2 S_z}{dt^2} + \Xi S_z + 2\lambda^4 S_z^3 = 0, \tag{9.25}$$

where $\Xi \equiv 4\tilde{\Omega}(\tilde{\Omega} - c_2\lambda^2)$ may be positive or negative. Equation (9.25) has a usual solution

$$S_z(t) = \frac{\sqrt{1 - \Xi\tau^2}}{2\lambda^2\tau} \mathrm{cn}\,[t/\tau, k], \quad k^2 = \frac{1}{2}\left(1 - \Xi\tau^2\right), \tag{9.26}$$

where τ is an arbitrary parameter taking values $\tau < 1/\sqrt{|\Xi|}$. In the case of $\Xi > 0$, the linearized version of Eq. (9.25), which corresponds to $\tau \to 1/\sqrt{|\Xi|}$ in Eq. (9.26), gives rise to free Rabi oscillations with frequency $\sqrt{\Xi}$. In the general case, the frequency given by solution (9.26), $\omega_{osc} = \pi/[2\tau K(k)]$, where $K(k)$ is the complete elliptic integral, exceeds $\sqrt{\Xi}$ due to the nonlinear shift. Note also that the nonlinearity may give rise to oscillations in the case of $\Xi < 0$, when the free Rabi oscillations are impossible. Further, in the spin representation, Eq. (9.22a) can be written as $\dot{k}_+ = -2\lambda\tilde{\Omega}S_y$, which accounts for the effect of the evolution of the spin on the center-of-mass momentum. This leads to the following equation of motion for the center-of-mass coordinate:

$$\frac{d^2\langle z\rangle}{dt^2} = -2\lambda\tilde{\Omega}S_y. \tag{9.27}$$

In other words, if we consider the soliton as a macroscopic quantum body carrying the intrinsic angular momentum, Eq. (9.27) represents the driving force, exerted by the intrinsic momentum and acting on the linear momentum, which is literally the

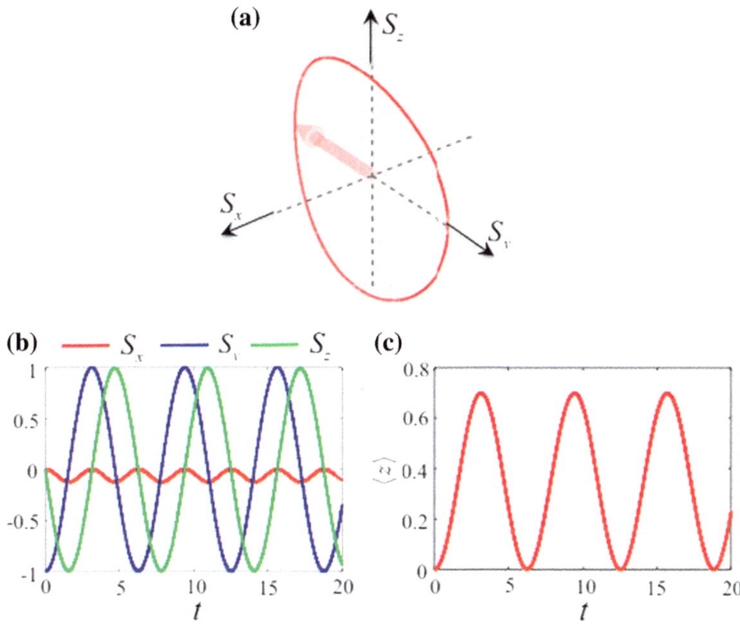

Fig. 9.20 The track of the spin density on the Bloch sphere (**a**), and the corresponding evolution of the spin components (**b**) and center-of-mass coordinate (**c**) of the soliton for the initially balanced state, with $\theta(t=0)=\pi/4$, and initial phase difference $\varphi_-(t=0)=\pi/4$ (see Eq. (9.20)). Other parameters are $\Omega=0.5$, $\lambda=0.5\sqrt{\Omega}$, $\delta=0$, and $g=-10$. z and t are measured in units of $a_\perp$ and $\omega_\perp^{-1}$. Reprint from Ref. [34], Copyright 2018, with permission from American Physical Society

macroscopic SOC. We stress that the mechanism of the soliton's motion is definitely different from the conventional collective dipole oscillations of BEC in a harmonic-oscillator trap. Indeed, in the present case, the force which drives the motion of solitons under the action of SOC is the spin precession, rather than the force induced by the external potential [34]. On the other hand, here we consider an effectively mechanical motion, which is produced by the interplay of the nonlinear self-trapping and SOC.

To illustrate the soliton dynamics in detail, we display, in Fig. 9.20, numerical solutions of Eqs. (9.23a)–(9.23c) and (9.27), with initially balanced populations in the two components, which correspond to $\theta(t=0)=\pi/4$ and $\varphi-(t=0)=\pi/4$. First, in Fig. 9.20a, we show that the soliton spin moves along a closed orbit on the Bloch sphere. Accordingly, perfect periodic oscillations of the spin can be identified in Fig. 9.20b, and perfectly periodic motion of the soliton's central coordinate is seen in Fig. 9.20c. In particular, for weaker SOC, the center-of-mass oscillations can be approximated as $\langle z\rangle \simeq \frac{\lambda}{\widetilde{\Omega}}\sin^2\left[\widetilde{\Omega}t\right]$ with amplitude $\lambda/\widetilde{\Omega}$ and a period $\pi/\widetilde{\Omega}$, which depend on the strengths of SOC, Raman coupling, and atomic interaction.

To test these findings obtained in the variational (i.e., effectively mechanical) approximation, we numerically solved the two-coupled NLS equations (9.18) with the input in the form of BB solitons, $\psi_\uparrow=\sqrt{\eta/2}(\sin[\theta_0])\text{sech}[\eta z]$

$\exp\left[i(\lambda z + \varphi_{\uparrow,0})\right]$ and $\psi_\downarrow = \sqrt{\eta/2}(\cos[\theta_0])\mathrm{sech}[\eta z]\exp\left[i(-\lambda z + \varphi_{\downarrow,0})\right]$. In Fig. 9.21a–c, the density of the soliton's components exhibits remarkable periodic oscillations, and the soliton maintains its initial hyperbolic secant profile for many oscillation cycles. Note that, although the initial momenta of the two components are opposite, the soliton does not split. In Fig. 9.21d and e, we show that the direct NLS simulations agree very well with the variational (mechanical) approximation. For weaker atomic interactions, the simulations show that solitons decay under the action of SOC for $\pi\lambda/\eta > (\pi\lambda/\eta)c \approx 0.4$, as shown in Fig. 9.21f and g. Furthermore, when $\delta \neq c_1\lambda$, we find that, besides the SOC-driven oscillation, the soliton would exhibit an additional linear motion, as found for the initially polarized case with $\theta(t = 0) = \pi/2$ and $\delta = c_1\lambda$.

Now, we proceed to the case of the repulsive BEC nonlinearity, which is more relevant to current experiments with SOC condensates [34]. In this case, we introduce the following variational ansatz for dark–dark (DD) solitons

$$\begin{pmatrix} \psi_\uparrow \\ \psi_\downarrow \end{pmatrix} = \sqrt{\frac{\eta}{2}} \begin{pmatrix} \sin[\theta]\tanh[\eta z + \xi]\exp\left[i(k_\uparrow z + \varphi_\uparrow)\right] \\ \cos[\theta]\tanh[\eta z + \xi]\exp\left[i(k_\downarrow z + \varphi_\downarrow)\right] \end{pmatrix}. \tag{9.28}$$

Inserting the ansatz (9.28) into the Lagrangian, one needs to renormalize the integrals to exclude divergent contributions of the non-vanishing background [3]. The analysis yields the same EL equations as Eqs. (9.22b)–(9.22d), but with $\eta = g/2$ for repulsive $g > 0$. We have also performed the respective direct simulations of the NLS system with the initial conditions corresponding to the DD solitons, $\psi_\uparrow = \sqrt{\eta/2}(\sin[\theta_0])\tanh[\eta z]\exp\left[i(\lambda z + \varphi_{\uparrow,0})\right]$ and $\psi_\downarrow = \sqrt{\eta/2}\cos[\theta_0]\tanh[\eta z]$ $\exp\left[i(-\lambda z + \varphi_{\downarrow,0})\right]$. The results are depicted in Fig. 9.22, where the DD soliton displays oscillations similar to those reported above for the BB configuration. Note that the two components of the background alternately disappear and revive, as shown in Fig. 9.22a and b.

In experiments, the condensate is usually trapped in a harmonic-oscillator potential, $V(z) = \gamma^2 z^2/2$, which affects the motion of solitons [5]. In this case, the EL equations for variational parameters k_+ and ξ are modified to $\dot{k}_+ = -2\lambda\widetilde{\Omega}Sy + (\gamma^2/\eta)\xi$ and $\dot{\xi} = -k_+\eta$, with η satisfying the condition $4\eta^4 + 2g\eta^3 = \pi^2\gamma^2$. Correspondingly, the center-of-mass motion would acquire an additional trap-dependent term $\frac{d^2z}{dt^2} + \gamma^2 z = -2\lambda\widetilde{\Omega}Sy$, which gives rise to an additional collective oscillation with a period $\simeq 2\pi/\gamma$, as shown in Fig. 9.23. For a weak trap, the period of trap-driven oscillation and the characteristic length of the trap are much larger than the period and amplitude of the SOC-driven oscillation, respectively. Hence, the SOC-driven oscillation of the soliton is not affected conspicuously by a weak trap and should be distinguishable in experiment. Here, we have used a bright soliton to illustrate the effect of a (weak) axial trap. Similar results are also identified numerically for a dark soliton in a harmonic trap.

Finally, we discuss some related experimental issues. So far, the Raman-induced SOC has been realized for the BEC in the ^{87}Rb gas with repulsive atomic interactions. We assume an elongated condensate of $\sim 10^4$ atoms is trapped in a weak

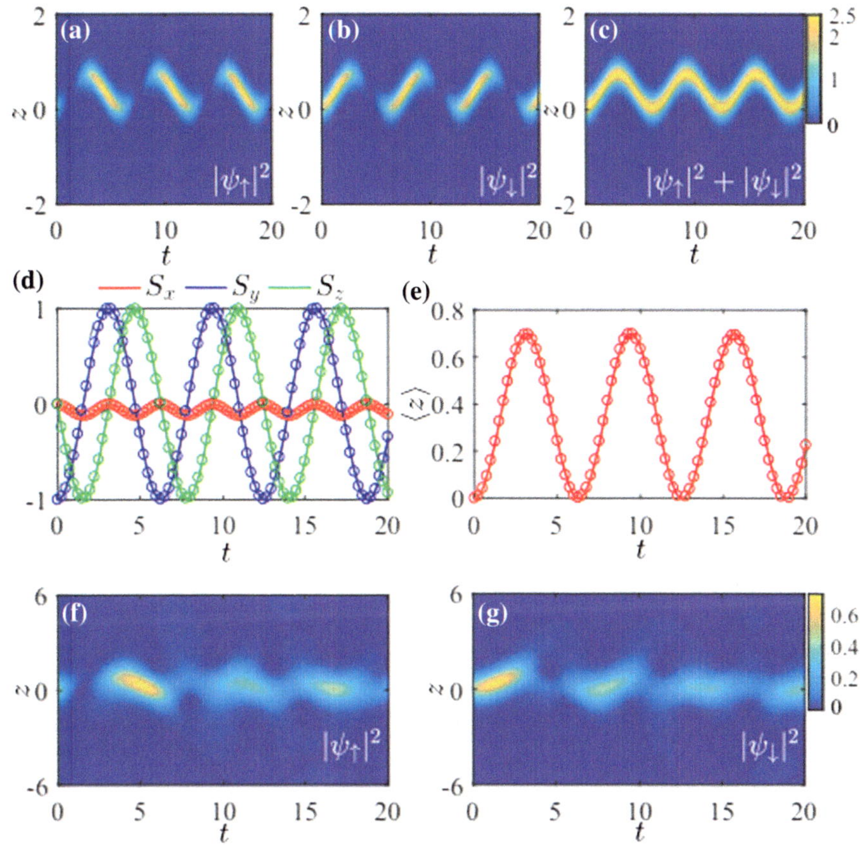

Fig. 9.21 **a–c** The evolution of the density in the two components of the soliton produced by the GP simulations for the initially balanced state with $\theta(t = 0) = \pi/4$ and $\varphi_-(t = 0) = \pi/4$. The parameters are $\Omega = 0.5$, $\lambda = 0.5\sqrt{\Omega}$, $\delta = 0$, and $g = -10$. The spin dynamics and center-of-mass motion of the soliton, generated by these simulations (circles), and by the variational approximation based on Eqs. (9.23a)–(9.23c) and (9.27) (solid lines), are depicted in **d** and **e**. **f** and **g** display decay of the soliton in the case of a weaker atomic interaction, $g = -3$. z and t are measured in units of $a_\perp$ and $\omega_\perp^{-1}$. Reprint from Ref. [34], Copyright 2018, with permission from American Physical Society

harmonic trap with frequencies $\omega_\perp = 2\pi \times 85$ Hz and $\omega_z = 2\pi \times 5.9$ Hz, and the corresponding ratio of the scattering lengths is $a_{\uparrow\uparrow} : a_{\uparrow\downarrow} : a_{\downarrow\downarrow} = 1 : 1 : 1.005$ [14]. In this case, at the center of the trap, we first create the stationary dark–dark solitons with width ~ 0.5 μm by means of phase and density-engineering techniques [34]. Then, two 804.1 nm Raman lasers with an intersection angle 20° are used to create a weak SOC. With the above parameters, the period and amplitude of SOC-driven oscillation of solitons are about 11 ms and 2 μm, respectively, which may be further adjusted by varying the SOC, Raman coupling, and atomic interaction. These results show that the SOC-driven oscillations can be experimentally observed.

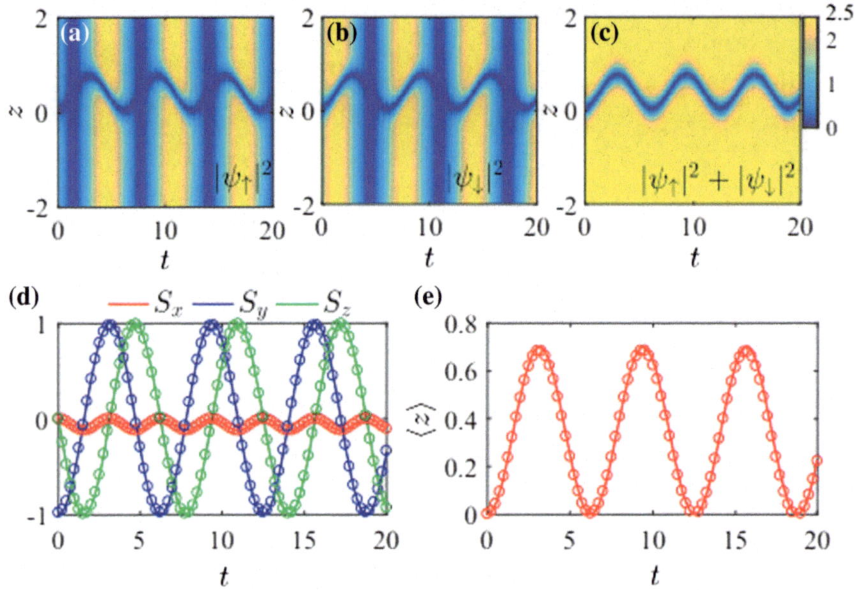

Fig. 9.22 **a–c** The same as in Fig. 9.21, but for the dark soliton, in the case of $g = 10$. z and t are measured in units of $a_\perp$ and $\omega_\perp^{-1}$. Reprint from Ref. [34], Copyright 2018, with permission from American Physical Society

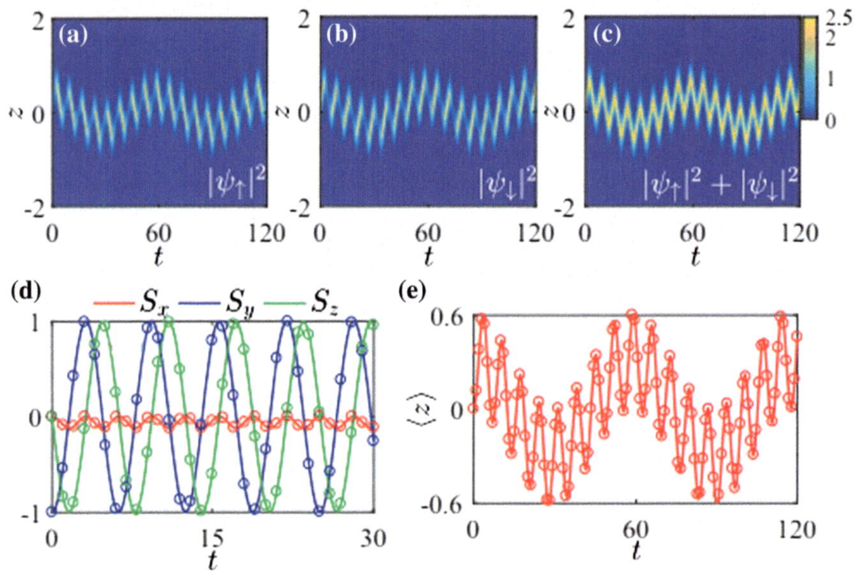

Fig. 9.23 The results for bright–bright solitons in the trap with $\gamma = 0.1$, the other parameters are the same as in Fig. 9.21. z and t are measured in units of $a_\perp$ and $\omega_\perp^{-1}$. Reprint from Ref. [34], Copyright 2018, with permission from American Physical Society

References

1. Aidelsburger, M., Lohse, M., Schweizer, C., Atala, M., Barreiro, J.T., Nascimbène, S., Cooper, N.R., Bloch, I., Goldman, N.: Measuring the Chern number of Hofstadter bands with ultracold bosonic atoms. Nat. Phys. **11**, 162 (2015)
2. Anderson, B.M., Spielman, I.B., Juzeliūnas, G.: Magnetically Generated Spin-Orbit Coupling for Ultracold Atoms. Phys. Rev. Lett. **111**, 125301 (2013)
3. Barashenkov, I.V., Harin, A.O.: Nonrelativistic Cherns-Simons theory for the repulsive Bose gas. Phys. Rev. Lett. **72**, 1575 (1994)
4. Barnett, R., Powell, S., Lewenstein, T.G., Das, Sarma S.: Order by disorder in spin-orbit-coupled Bose-Einstein condensates. Phys. Rev. A **85**, 023615 (2012)
5. Becker, C., Stellmer, S., Soltan-Panahi, P., Dorscher, S., Baumert, M., Richter, E.-M., Kronjäger, J., Bongs, K., Sengstock, K.: Oscillations and interactions of dark and dark-bright solitons in Bose-Einstein condensates. Nat. Phys. **4**, 496 (2008)
6. Chang, M.-S., Hamley, C.D., Barrett, M.D., Sauer, J.A., Fortier, K.M., Zhang, W., You, L., Chapman, M.S.: Observation of Spinor Dynamics in Optically Trapped ^{87}Rb Bose-Einstein Condensates. Phys. Rev. Lett. **92**, 140403 (2004)
7. Cheng, Y., Tang, G., Adhikari, S.K.: Localization of a spin -orbit coupled Bose -Einstein condensate in a bichromatic optical lattice. Phys. Rev. A **89**, 063602 (2014)
8. Han, W., Juzeliunas, G., Zhang, W., Liu, W.M.: Supersolid with nontrivial topological spin textures in spin-orbit-coupled Bose gases. Phys. Rev. A **91**, 013607 (2015)
9. He, P.-S., Liao, R., Liu, W.M.: Feynman relation of Bose-Einstein condensates with spin-orbit coupling. Phys. Rev. A **86**, 043632 (2012)
10. Ho, T.L.: Spinor Bose Condensates in Optical Traps. Phys. Rev. Lett. **81**, 742 (1998)
11. Kasamatsu, K., Tsubota, M., Ueda, M.: Vortex Molecules in Coherently Coupled Two-Component Bose-Einstein Condensates. Phys. Rev. Lett. **93**, 250406 (2004)
12. Kobayashi, J., Izumi, Y., Enomoto, K., Kumakura, M., Takahashi, Y.: Spinor molecule in atomic Bose-Einstein condensate. Appl. Phys. B **95**, 37 (2009)
13. Li, X., Liu, W.V., Balents, L.: Spirals and Skyrmions in Two Dimensional Oxide Heterostructures. Phys. Rev. Lett. **112**, 067202 (2014)
14. Lin, Y.J., Garcia, K.J., Spielman, I.B.: Spin-orbit-coupled Bose-Einstein condensates. Nature (London) **471**, 83 (2011)
15. Liu, C.F., Fan, H., Zhang, Y.C., Wang, D.S., Liu, W.M.: Circular-hyperbolic skyrmion in rotating pseudo-spin-1/2 Bose-Einstein condensates with spin-orbit coupling. Phys. Rev. A **86**, 053616 (2012)
16. Liu, G., Hao, N., Zhu, S.-L., Liu, W.M.: Topological superfluid transition induced by a periodically driven optical lattice. Phys. Rev. A **86**, 013639 (2012)
17. Liu, C.F., Juzeliunas, G., Liu, W.M.: Spin-orbit coupling manipulating composite topological spin textures in atomic-molecular Bose-Einstein condensates. Phys. Rev. A **95**, 023624 (2017)
18. Liu, C.F., Liu, W.M.: Spin-orbit-coupling-induced half-skyrmion excitations in rotating and rapidly quenched spin-1 Bose-Einstein condensates. Phys. Rev. A **86**, 033602 (2012)
19. Mandel, O., Greiner, M., Widera, A., Rom, T., Hänsch, T.W., Bloch, I.: Coherent Transport of Neutral Atoms in Spin-Dependent Optical Lattice Potentials. Phys. Rev. Lett. **91**, 010407 (2003)
20. Mazza, L., Bermudez, A., Goldman, N., Rizzi, M., Martin-Delgado, M.A., Lewenstein, M.: An optical-lattice-based quantum simulator for relativistic field theories and topological insulators. New. J. Phys. **14**, 015007 (2012)
21. Ohmi1 T., Machida K.: Bose-Einstein Condensation with Internal Degrees of Freedom in Alkali Atom Gases. J. Phys. Soc. Jpn. **67**, 1822 (1998)
22. Olson, A.J., Wang, S.-J., Niffenegger, R.J., Li, C.-H., Greene, C.H., Chen, Y.P.: Tunable Landau-Zener transitions in a spin-orbit-coupled Bose-Einstein condensate. Phys. Rev. A **90**, 013616 (2014)
23. Pethick, C.J., Smith, H.: Bose-Einstein Condensation in Dilute Gases. Cambridge University Press, Cambridge (2002)

24. Radić, J., Sedrakyan, T.A., Spielman, I.B., Galitski, V.: Vortices in spin-orbit-coupled Bose-Einstein condensates. Phys. Rev. A **84**, 063604 (2011)
25. Rooney, S.J., Bradley, A.S., Blakie, P.B.: Decay of a quantum vortex: Test of nonequilibrium theories for warm Bose-Einstein condensates. Phys. Rev. A **81**, 023630 (2010)
26. Sakaguchi, H., Ya, E.S., Malomed, B.A.: Vortex solitons in two dimensional spin orbit coupled Bose Einstein codensates: Effects of the Rashba -Dresselhaus coupling and Zeeman splitting. Phys. Rev. E **94**, 032202 (2016)
27. Schweikhard, V., Coddington, I., Engels, P., Tung, S., Cornell, E.A.: Vortex-Lattice Dynamics in Rotating Spinor Bose-Einstein Condensates. Phys. Rev. Lett. **93**, 210403 (2004)
28. See Supplemental Material at http://link.aps.org/supplemental/10.1103/PhysRevA.95.023624
29. Stamper-Kurn, D.M., Andrews, M.R., Chikkatur, A.P., Inouye, S., Miesner, H.-J., Stenger, J., Ketterle, W.: Optical Confinement of a Bose-Einstein Condensate. Phys. Rev. Lett. **80**, 2027 (1998)
30. Stamper-Kurn, Dan M., Ueda, M.: Spinor Bose gases: Symmetries, magnetism, and quantum dynamics. Rev. Mod. Phys. **85**, 1191 (2013)
31. Sun, K., Qu, C., Xu, Y., Zhang, Y., Zhang, C.: Interacting spin-orbit-coupled spin-1 Bose-Einstein Condensates. Phys. Rev. A **93**, 023615 (2016)
32. Szankowski, P., Trippenbach, M., Infeld, E., Rowlands, G.: Oscillating Solitons in a Three-Component Bose-Einstein Condensate. Phys. Rev. Lett. **105**, 125302 (2010)
33. Wang, D.-S., Hanc, W., Shid, Y., Lie, Z., Liu, W.M.: Dynamics and stability of stationary states for the spin-1 Bose-Einstein condensates in a standing light wave. Commun. Nonlinear Sci. Numer. Simulat. **36**, 45 (2016)
34. Wen, L., Sun, Q., Chen, Y., Wang, D.-S., Hu, J., Chen, H., Liu, W.M., Juzeliunas, G., Malomed, B.A., Ji, A.-C.: Motion of solitons in one-dimensional spin-orbit-coupled Bose-Einstein condensates. Phys. Rev. A **94**, 061602 (2016)
35. Woo, S.J., Park, Q.-H., Bigelow, N.P.: Phases of atom-molecule vortex matter. Phys. Rev. Lett. **100**, 120403 (2008)
36. Zhu, Q., Zhang, C., Wu, B.: Exotic superfluidity in spin-orbit coupled Bose-Einstein condensates. arXiv:1109.5811
37. Zinn-Justin, J.: Quantum Field Theory and Critical Phenomena. Clarendon Press, Oxford, UK (1989)

Chapter 10
Multicomponent Nonlinear Schrödinger Equations for Spin-1 and Spin-2 BECs

Abstract In this chapter, we use the multicomponent nonlinear Schrödinger equations to investigate the dynamics of spin–orbit coupling (SOC) in Bose–Einstein condensates. We will study the rotating spin-1 BECs with anisotropic SOC, the spin texture of the vortex chain for rotating spin-1 BECs, the spin singularity phenomenon in the spin-1 BECs, the topological defects in the ground states of SU(3) SO-coupled BECs, and the inhomogeneous spin patterns in the spatially homogeneous spin-1 BECs. Also, we investigate here the configuration of Dirac monopoles with a polar-core vortex induced by the SO coupling in ferromagnetic BECs, as well as the non-Abelian Josephson effect in $F = 2$ spinor Bose–Einstein condensates with double optical traps.

The present chapter deals with multicomponent nonlinear Schrödinger equations which describe the dynamics of spin–orbit coupling in Bose–Einstein condensates.

10.1 Vortex Chain in Anisotropic SOC Rotating Spin-1 Bose–Einstein Condensates

In this section, we study the rotating spin-1 BECs with anisotropic SOC by using the damped projected nonlinear Schrödinger (PNLS) equations with external potential. As we will see in what follows, the anisotropic SOC can cause some vortices to form the vortex chain, where the vortices link up with each other in a line. The formation of the vortex chain starts from very small rotation frequency and coexists with more remanent vortices with the increasing rotation frequency.

10.1.1 Damped PNLS Equations for Anisotropic SOC Rotating Spin-1 Condensates

The dynamics of $F = 1$ spinor BECs with SOC can be described by the following coupled NLS equations:

W.-M. Liu and E. Kengne, *Schrödinger Equations in Nonlinear Systems*,
https://doi.org/10.1007/978-981-13-6581-2_10

$$
\begin{aligned}
i\hbar\frac{\partial\Psi_j}{\partial t} &= \left[-\frac{\hbar^2}{2m}\nabla^2+\frac{m\omega^2\left(x^2+y^2\right)}{2}+g_n\left|\Psi\right|^2\right]\Psi_j \\
&+g_s\sum_{\alpha=x,y,z}\sum_{n,k,l=0,\pm1}\left(\widehat{F}_\alpha\right)_{jn}\left(\widehat{F}_\alpha\right)_{kl}\Psi_n\Psi_k^*\Psi_l \\
&-\Omega\widehat{L}_z\Psi_j+\sum_{\alpha=x,y}\sum_{n=0,\pm1}\kappa_\alpha\left(\widehat{F}_\alpha\right)_{jn}p_\alpha\Psi_n,
\end{aligned}
\tag{10.1}
$$

where Ψ_j ($j=0,\ \pm1$) denotes the macroscopic wave function of the atoms condensed in the spin state $|F=1,\ m_F=j\rangle$, m the mass of atom, ω the trapped frequency, $\widehat{F}_{\alpha=x,y,z}$ the spin-1 matrices, Ω the rotation frequency, $\widehat{L}_z=-i\hbar\left(x\frac{\partial}{\partial y}-y\frac{\partial}{\partial x}\right)$ the z component of the orbital angular momentum, $p_\alpha=-i\hbar\frac{\partial}{\partial\alpha}$ ($\alpha=x,\ y$) the momentum operator, and κ_α the strength of SOC which carries the unit of velocity. It should be noted that the coupling constants $g_n=\frac{4\pi\hbar^2(2a_2+a_0)}{3m}$ and $g_s=\frac{4\pi\hbar^2((a_2-a_0)}{3m}$ come from the density–density and spin–exchange interactions, respectively.

Generally, in real experiments, the ultracold Bose gases are only partially condensed with the noncondensed thermal cloud which provides a source of dissipation and causes the effect of damping. Meanwhile, it is a critical operation to obtain the BECs with the evaporative cooling which can be approximately viewed as a quenching process. In the present section, we use the damped PNLS equation with external potentials to describe the quenching process. Similarly to the stochastic projected NLS equation [3], the system is divided into the coherent region with the energies of the state below E_R and the noncoherent region with the energies of the states above E_R. The damped PNLS equation is written as

$$
d\psi_j=\mathcal{P}\left\{-\frac{i}{\hbar}\widehat{H}_j\psi_j dt+\frac{\gamma_j}{k_BT}\left(\mu-\widehat{H}_j\right)\psi_j dt\right\},
\tag{10.2}
$$

where $\widehat{H}_j\psi_j$ denotes the right-hand side part of Eq. (10.1), T the final temperature, k_B the Boltzmann constant, μ the chemical potential, and γ_j the growth rate for the jth component. The projection operator $\mathcal{P}$ is used to restrict the dynamics of the spin-1 BECs in the coherent region. The damped PNLS equation is equal to the limit case of zero temperature in the stochastic PNLS equation. In this scenario, all of the finite-temperature physics normally present in the PNLS equation theory is lost.

In numerical simulations, we use the spin-1 BECs of ^{87}Rb with $\frac{\gamma_j}{k_BT}=0.03$ and the trapped frequency $\omega=200\times2\pi$. The unit of length, time, energy, and strength of SOC correspond to $\sqrt{\hbar/(m\omega)}$ ($\approx0.76\,\mu$m), ω^{-1} ($\approx0.8\times10^{-3}$ s), $\hbar\omega$, and $\sqrt{\hbar\omega/m}$ ($\approx0.96\times10^{-3}$ m/s), respectively. The initial state of each ψ_j is generated by sampling the grand-canonical ensemble for a free ideal Bose gas with the chemical potential $\mu_{j,0}=3.6\hbar\omega$. This sampling would bring some initial thermal noise to break symmetry and give the true energetic ground state once the thermal fluctuations have been removed by the damping. Meanwhile, the condensate band must lie below the energy cutoff $E_R>E_k=\frac{\hbar^2|k|^2}{2m}$. Note that $k=2\pi(n_x,n_y)/L$, where n_x, n_y are integers and L is the size of the computation domain. Furthermore, the total number of modes is $n_x, n_y = 300$, and the energy cutoff is chosen at $n_{xc}, n_{yc} = 150$. To simulate the quenching process, the final chemical potential of the noncondensate band is altered to the new values $\mu = 25\hbar\omega > \mu_{j,0}$ $(j = 0, \pm 1)$.

10.1.2 Effect of Rotation Frequency Ω on the Creation of Vortex Chain in SOC Spin-1 BECs

First, with the fixed SOC strengths, $\kappa_x = 1$ and $\kappa_y = 0$, we aim to study the influence of the rotation frequency on the formation of the vortex chain in the spin-1 BECs of ^{87}Rb. We choose $\mu_{j,0}$ $(j = 0, \pm 1) = 3.6\hbar\omega$ and $\mu = 25\hbar\omega$. Figure 10.1 displays the densities and phases obtained under the equilibrium state with different rotation frequencies. The vortices can be identified in the phase image of each component of the BECs. For the case of the lowest rotation frequency ($\Omega = 0.2\omega$), all the vortices form a line along the x-axis and completely traverse the whole BECs [see Fig. 10.1a]. This new interesting configuration is referred to as the vortex chain, which essentially differs from the symmetric and equal distribution of vortices in the rotating BECs. With the rotation frequency $\Omega = 0.3\omega$, more vortices are excited, and there are also several vortices appearing on both sides of the center vortex chain, following a

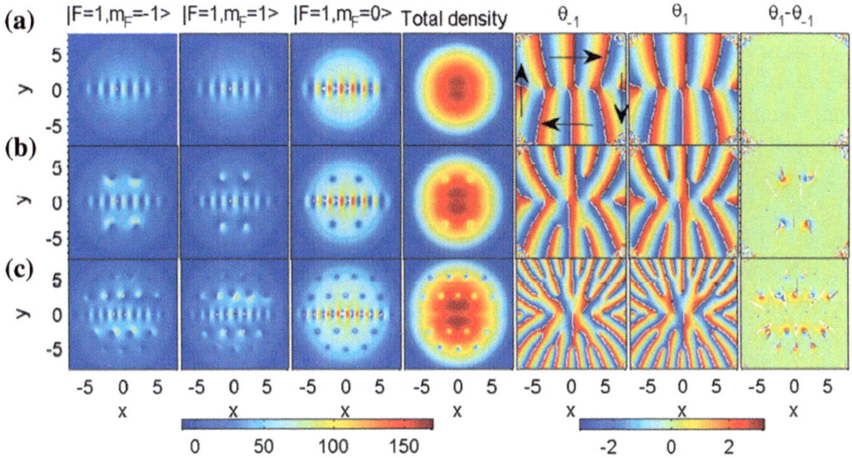

Fig. 10.1 Effect of rotation frequency on the densities and phases of the anisotropic SO-coupled spin-1 BECs of ^{87}Rb when the system reaches the equilibrium state. **a** $\Omega = 0.2\omega$, **b** $\Omega = 0.3\omega$, and **c** $\Omega = 0.5\omega$. Here, $\kappa_x = 1.0$, $\kappa_y = 0$, $a_0 = 101.8a_B$, and $a_2 = 100.4a_B$. Note that the fifth and sixth columns are the phases of $m_F = -1$ and $m_F = 1$ components, respectively. The atom numbers (N_{-1}, N_0, N_1) approximately are $(3.5 \times 10^3, 7.0 \times 10^3, 3.5 \times 10^3)$, $(4.1 \times 10^3, 7.3 \times 10^3, 3.6 \times 10^3)$, and $(4.8 \times 10^3, 9.1 \times 10^3, 4.9 \times 10^3)$, respectively. The length unit and strength of SOC are 0.76 μm and 0.96×10^{-3} m/s, respectively. Reprint from Ref. [26], Copyright 2018, with permission from American Physical Society

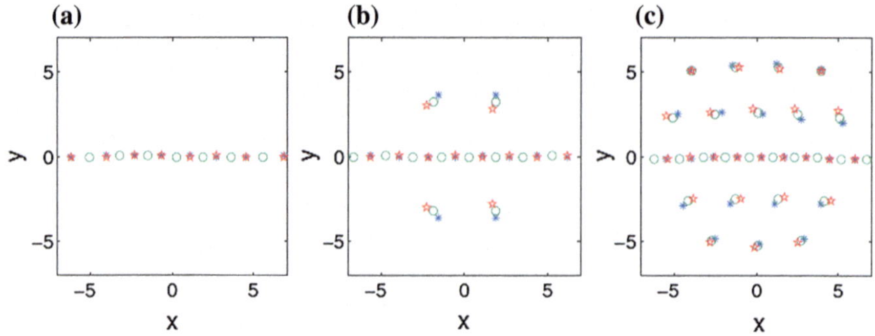

Fig. 10.2 Effect of rotation frequency on the position of vortices in the anisotropic SO-coupled spin-1 BECs of ^{87}Rb in Fig. 9.1. The asterisks, circles, and stars are the position of vortices formed by the $m_F = -1$, $m_F = 0$, and $m_F = +1$ components, respectively. **a** $\Omega = 0.2\omega$, **b** $\Omega = 0.3\omega$, and **c** $\Omega = 0.5\omega$. The length unit and strength of SOC are $0.76\,\mu$m and 0.96×10^{-3} m/s, respectively. Reprint from Ref. [26], Copyright 2018, with permission from American Physical Society

relatively symmetrical distribution [see Fig. 10.1b]. The number of vortices in the off-vortices-chain region increases with the rotation frequency [see Fig. 10.1c]. As compared with the coreless vortices that have been reported in Refs. [13, 14, 26], the maximum value of the total densities is not at the origin of the coordinate plane. Because of the vortex chain, some local minima of the total densities along the x-axis can be distinguished [see the fourth column].

Figure 10.2 illustrates the position of the vortices shown in Fig. 10.1. As shown in Fig. 10.2, the vortices of the $m_F = \pm 1$ components are overlapped in the vortex chain, and the vortices of the $m_F = \pm 1$ components and the vortices of $m_F = 0$ are arranged alternately in a line along the x-axis. The vortices of the three components are coupled in the regions off the x-axis, consisting of the three-vortex structure. The phase difference shown in Fig. 10.1a tends to be zero, which means that the $m_F = -1$ and $m_F = 1$ components are synchronous when the system reaches the equilibrium state. The phase differences shown in Fig. 10.1b and c also tend to be zero except in the region where the off-x-axis vortices locate. Figure 10.2 also indicates the patterns of the vortex distributions under the anisotropic SOC of $\kappa_x = 1$ and $\kappa_y = 0$. The vortices will first distribute along the x-axis as far as they can until the repulsive force between them is so strong that the remaining vortices prefer to appear on both sides of the x-axis in an approximately symmetric manner.

Figure 10.3 illustrates the profiles of the vortex chain shown in Fig. 10.1a–c. As shown in Fig. 10.3a, the $m_F = -1$ and $m_F = 1$ components are of the same density profiles along the x-axis. When the rotation frequency increases up to 0.3ω, the density profiles of the $m_F = -1$ and $m_F = 1$ components tend to differ. Meanwhile, the vortices continuously and homogeneously distribute along the x-axis so we have to view them as a whole: the vortex chain. It is obvious that the vortex chain expands over the whole BECs in the three cases in Fig. 10.1. Furthermore, the numbers of vortices in the three vortex chains increase as the rotation frequency increases. Note that the number of vortices in a chain is finite and intrinsic at the equilibrium state

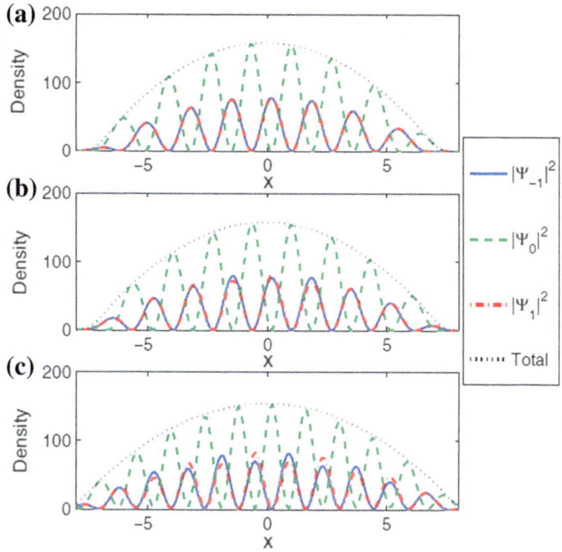

Fig. 10.3 Effect of rotation frequency on the profiles of the vortex chain in the anisotropic SO-coupled spin-1 BECs of ^{87}Rb in Fig. 1. **a** $\Omega = 0.2\omega$, **b** $\Omega = 0.3\omega$, and **c** $\Omega = 0.5\omega$. The length unit and strength of SO coupling are $0.76\,\mu$m and 0.96×10^{-3} m/s, respectively. Reprint from Ref. [26], Copyright 2018, with permission from American Physical Society

with all the considered parameters. We cannot change the intrinsic number of vortices in the chain artificially unless the parameters are changed. Although the fast rotation can cause larger BECs where more vortices would be stored along the x-axis to form the chain, we do not discuss the limit rotation frequency in this study.

The formation of the vortex chain derives from the anisotropic spin–orbit coupling. Similarly to the formation of vortex lattice in the spin-0 scenario, the rotation effect induces the creation of the vortices in the case of the SO-coupled spin-1 BECs. Thus, we observe some lines where the phases change discontinuously from red to blue in the fifth and sixth columns of Fig. 10.1, which corresponds to the branch cuts between the phases $-\pi$ and π. The ends represent phase defects, i.e., the vortex. All the lines extend to the outskirts of the BEC, where the density of the BEC is almost negligible, and end with another defect which offers neither the energy nor the angular momentum to the system. The additional anisotropic SOC terms cause the decrease in the system's energy. The κ_y term in the equation of motion does not affect the system's energy ($\kappa_y = 0$). The energy contributions of the additional anisotropic spin–orbit terms depend on κ_x, which is not zero in the equation of motion. In order to lower more energy, the gradient of phase should remain mainly on the x-direction. Therefore, it would cause the plane wave phase following the x-direction if there is no rotation and creation of vortices. In the presence of vortices, the plane wave phase is changed to form a circle [see the fifth column in Fig. 10.1a] and has to follow the x-direction mainly. Thus, the vortices are forced to distribute along the x-axis. Therefore, we observe the vortex chain along the x-axis with the anisotropic SOC ($\kappa_x = 1, \kappa_y = 0$) in Fig. 10.1. For the anisotropic SOC with $\kappa_x = 0$ and $\kappa_y = 1$, the vortices distribute along the y-axis.

We find that the vortex chain phenomenon occurs even when the rotation frequency is as small as $\Omega = 0.02\omega$. Without rotation, the BECs degenerate into the plane wave phase under the equilibrium state. We do not fix the ratio of the three components in the dynamical process, so the mixture ratio of the three components depends on the system itself. We now give a physical explanation about the vortex chain in the small rotation limit. The vortex chain results from two plane wave phases with opposite momenta. The phase gradient of the ground state with the plane wave phase is along the orientation of the SOC. The demarcation lines between the discontinuous $-\pi$ and π phases end at the boundary of the BECs, where the densities approach zero. However, when the BECs are rotated, the spatial distribution of phases changes. For the case of isotropic SOC, the gradients of the phase encircle the center, following the emergence of the vortices. Figure 10.4 shows the phase's gradients of the $m_F = -1$ component in Fig. 1a. The curves come from the interface between the values 0 and 2π. The value of $\partial\theta_{-1}/\partial x$ does not approach 0, and $\partial\theta_{-1}/\partial y$ approaches 0 in the off-x-axis region. Instead, the properties of $\partial\theta_{-1}/\partial x$ approach 0 and $\partial\theta_{-1}/\partial y$ does not approach 0 along the x-axis. This indicates that the gradients of the phase mainly continue in the same direction and the counterdirection of the anisotropic SOC. The phases form a whole circle as the arrows indicate in Fig. 10.1a (the fifth column). The demarcation lines between the discontinuous $-\pi$ and π phases then have to start from the domain wall of the two opposite plane wave phases and end at the outskirts of the BECs. Therefore, the vortex chain occurs although the rotation frequency is very low.

10.1.3 Effect of the Strength of Anisotropic SOC on the Vortex Chain

We now consider the influence of the strength of the anisotropic SOC on the vortex chain. The rotation frequency is fixed to be $\Omega = 0.5\omega$, and all other parameters are the

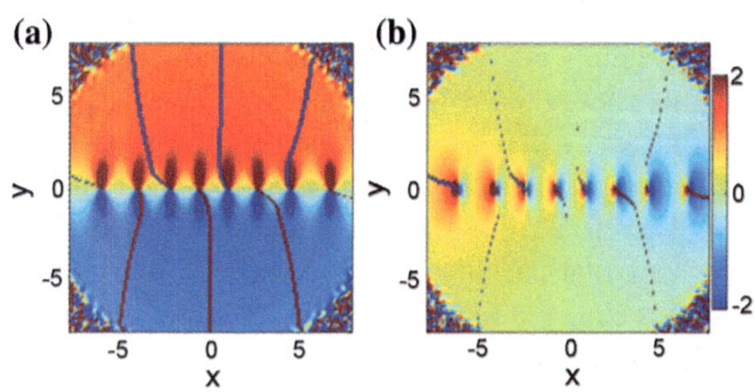

Fig. 10.4 The phase's gradient of $m_F = -1$ component in Fig. 9.1a. **a** $\partial\theta_{-1}/\partial x$ and **b** $\partial\theta_{-1}/\partial y$. The length unit and strength of SOC are $0.76\,\mu\text{m}$ and 0.96×10^{-3} m/s, respectively. Reprint from Ref. [26], Copyright 2018, with permission from American Physical Society

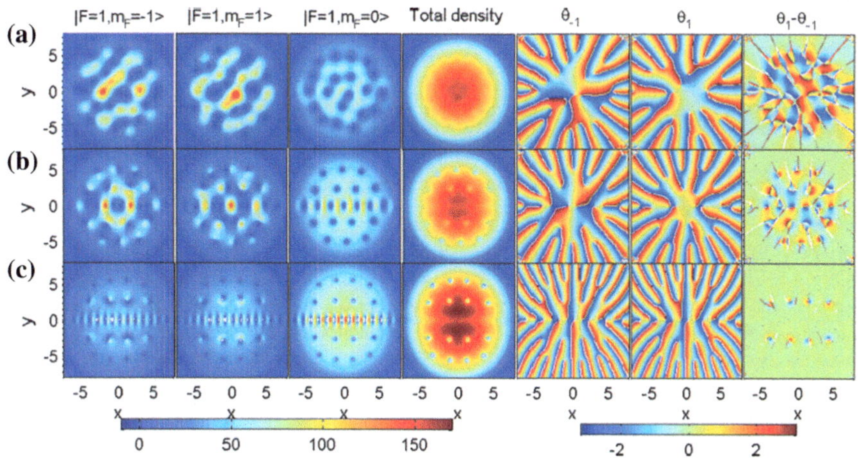

Fig. 10.5 Effect of anisotropic SOC on the densities and phases of the rotating spin-1 BECs of ^{87}Rb when the system reaches the equilibrium state. **a** $\kappa_x = 0.1$ and $\kappa_y = 0$. **b** $\kappa_x = 0.5$ and $\kappa_y = 0$. **c** $\kappa_x = 1.5$ and $\kappa_y = 0$. Here $\Omega = 0.5\omega$, $a_0 = 101.8a_B$, and $a_2 = 100.4a_B$. Note that the fifth and sixth columns are the phases of $m_F = -1$ and $m_F = 1$ components, respectively. The atom numbers (N_{-1}, N_0, N_1) approximately are $(5.4 \times 10^3, 5.3 \times 10^3, 5.5 \times 10^3)$, $(4.9 \times 10^3, 7.5 \times 10^3, 4.8 \times 10^3)$, and $(5.2 \times 10^3, 1.05 \times 10^4, 5.4 \times 10^3)$, respectively. The length unit and strength of SOC are $0.76\,\mu$m and 0.96×10^{-3} m/s, respectively. Reprint from Ref. [26], Copyright 2018, with permission from American Physical Society

same with those of Fig. 10.1 except the strength of the anisotropic SOC. Figure 10.5a indicates that no vortex chain is obtained for $\kappa_x = 0.1$ and $\kappa_y = 0$. Several vortices of the $m_F = -1, 0, 1$ components appear along the x-axis for $\kappa_x = 0.5$ and $\kappa_y = 0$, as shown in Fig. 10.5b. The phase difference in the seventh column shows that there is no synchronization of the phases. The vortex chains become clearer for each component for $\kappa_x = 1.5$ and $\kappa_y = 0$, as shown in Fig. 10.5c, and the phase difference along the x-axis is zero. In addition to the vortices on the x-axis, other vortices distribute approximately parallel to the x-axis. The phase difference of the off-x-axis vortices near $y = \pm 5$ approaches 0. This differs from those near the $y = \pm 2.5$ region. These results indicate that the stronger the anisotropic SOC, the more the regions of the $m_F = -1$ and $m_F = 1$ components tend to evolve synchronously.

Figure 10.6 shows the corresponding vortex positions in Fig. 10.5. It further displays the effect of the anisotropic SOC on the vortices. Figure 10.6a indicates that the relatively weak anisotropic SOC ($\kappa_x = 0.1$ and $\kappa_y = 0$) indeed cannot induce the vortex chain in Fig. 10.5a. Figure 10.6b shows a vortex chain where the vortices of the $m_F = -1$ component do not overlap with those of the $m_F = 1$ component in Fig. 10.5b. Furthermore, except the vortex chain along the x-axis, all other vortices form the three-vortex structure, just as that in Ref. [23].

Figure 10.6c indicates that the vortex chain in Fig. 10.5c is similar to that in Fig. 10.1. The anisotropic SO coupling can cause the vortices of different components to overlap with each other. The off-x-axis vortices near $y = \pm 5$ completely overlap,

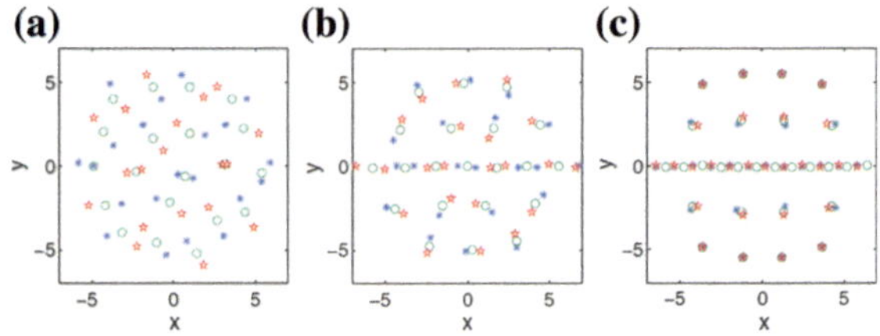

Fig. 10.6 Effect of anisotropic SOC on the position of vortices in the rotating spin-1 BECs of ^{87}Rb in Fig. 9.5. The asterisks, circles, and stars denote the position of vortices formed by the $m_F = -1$, $m_F = 0$, and $m_F = +1$ components, respectively. **a** $\kappa_x = 0.1$ and $\kappa_y = 0$. **b** $\kappa_x = 0.5$ and $\kappa_y = 0$. **c** $\kappa_x = 1.5$ and $\kappa_y = 0$. The length unit and strength of SOC are $0.76\,\mu$m and 0.96×10^{-3} m/s, respectively. Reprint from Ref. [26], Copyright 2018, with permission from American Physical Society

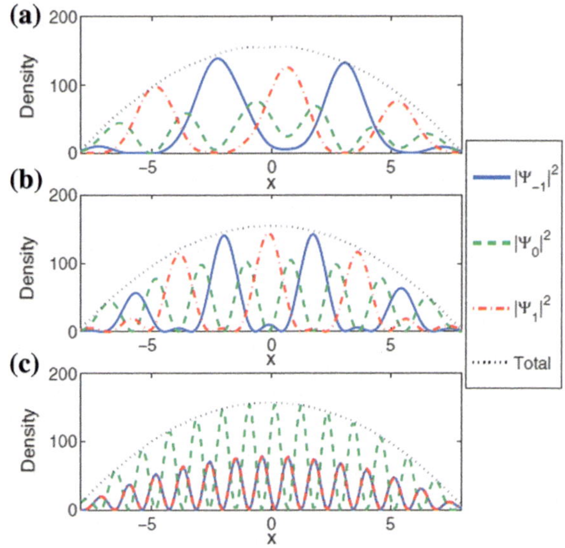

Fig. 10.7 Effect of anisotropic SOC on the profiles of vortex chain in the rotating spin-1 BECs of ^{87}Rb in Fig. 9.5. **a** $\kappa_x = 0.1$ and $\kappa_y = 0$. **b** $\kappa_x = 0.5$ and $\kappa_y = 0$. **c** $\kappa_x = 1.5$ and $\kappa_y = 0$. The length unit and strength of SOC are $0.76\,\mu$m and 0.96×10^{-3} m/s, respectively. Reprint from Ref. [26], Copyright 2018, with permission from American Physical Society

but the off-x-axis vortices near $y = \pm 2.5$ do not. Such a difference is because of the different densities at $y = \pm 5$ and $y = \pm 2.5$. In fact, this effect is caused by the trap potential, strength of SOC, and so on. Near the center, the density is bigger than that at the boundary. When the three vortices overlap, it will offer more energy and angular momentum to the system. Thus, the completely overlapped three vortices easily appear at the boundary of the BECs (low-density region).

The density profiles along the x-axis are plotted in Fig. 10.7. The vortex chain indeed does not occur along the x-axis in Fig. 10.5a, because some minimum values in Fig. 10.7a do not approach zero at all. Figure 10.7b indicates that the minimum values of the three components are separated from each other. When SOC of κ_x is up to 1.5, the property of the density profiles is similar to that in Fig. 10.3, where the density profiles of the $m_F = -1$ and $m_F = 1$ components are equal in Fig. 10.7c.

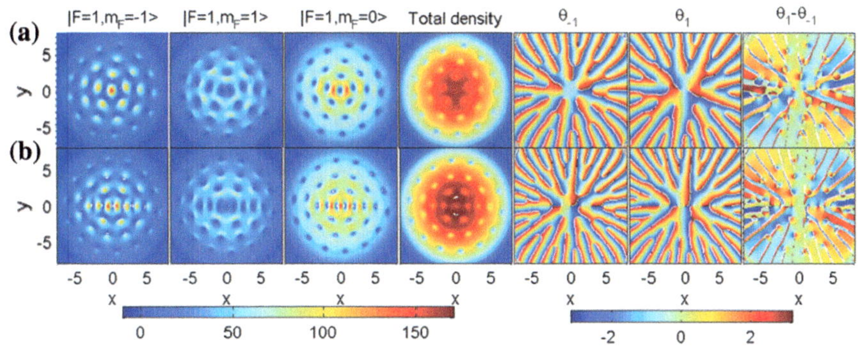

Fig. 10.8 Effect of anisotropic SOC on the densities and phases of the rotating spin-1 BECs of ^{87}Rb when the system reaches the equilibrium state. **a** $\kappa_x = 1.0$ and $\kappa_y = 0.7$. **b** $\kappa_x = 1.5$ and $\kappa_y = 1.2$. Here $\Omega = 0.5\omega$, $a_0 = 101.8a_B$, and $a_2 = 100.4a_B$. Note that the fifth and sixth columns are the phases of the $m_F = -1$ and $m_F = 1$ components, respectively. The atom numbers (N_{-1}, N_0, N_1) approximately are $(4.7 \times 10^3, 9.4 \times 10^3, 5.6 \times 10^3)$ and $(5.5 \times 10^3, 1.12 \times 10^4, 6.1 \times 10^3)$, respectively. The length unit and strength of SOC are 0.76 μm and 0.96×10^{-3} m/s, respectively. Reprint from Ref. [26], Copyright 2018, with permission from American Physical Society

According to the results of Figs. 10.3 and 10.7, we can conclude that the anisotropic SOC tends to equalize the density profiles of the $m_F = -1$ and $m_F = 1$ components in the vortex chain, but the fast rotation effect causes a slight deviation.

Further, the formation of the vortex configuration is investigated for different κ_y with the other parameters being the same with Figs. 10.1c and 10.5c, respectively, in order to explain the influence of κ_y on the vortex chains. As shown in Figs. 10.8a and 10.9a, the vortex chain is not observed for $\kappa_y = 0.7$, whereas in Figs. 10.8b and 10.9b, the vortex chain is observed for $\kappa_y = 1.2$. Figure 10.9c further plots the density profile of the vortex chain in Fig. 10.8b. By comparing Fig. 10.9c with Figs. 10.3 and 10.7, the total density profile is not smooth. Meanwhile, the vortex chain locates at the center of the BECs, not traversing the whole BECs. Therefore, not only κ_x but also κ_y of the anisotropic SOC can control the formation of the vortex chain.

We have also performed simulations with the antiferromagnetic BECs of ^{23}Na, where $g_s > 0$. We find that the vortex chain phenomenon can occur as long as the anisotropic SOC of κ_x ($\kappa_y = 0$) is strong enough. Compared with Ref. [25], which also uses the anisotropic SOC, the mechanism for obtaining the vortex chain is the same. However, the structure of the vortex chain in the spin-1 BECs is indeed more complicated than that in the pseudospin-1/2 BECs, where the vortices interlacedly array form a chain.

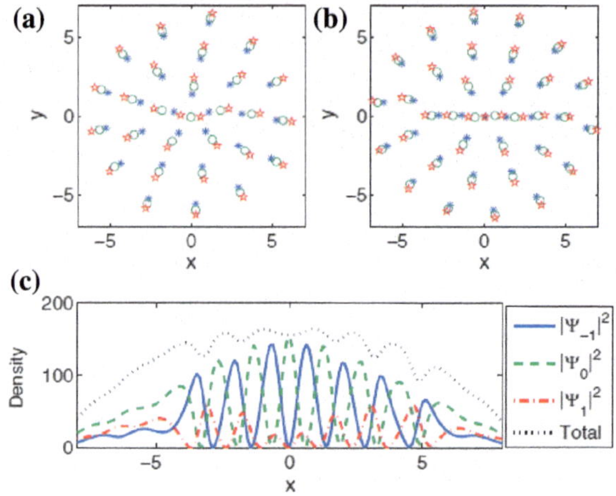

Fig. 10.9 The position of vortices in rotating spin-1 BECs of ^{87}Rb in Fig. 9.8a. The asterisks, circles, and stars are the position of vortices formed by the $mF = -1, mF = 0$, and $mF = +1$ components, respectively. **b** The position of vortices in rotating spin-1 BECs of ^{87} Rb in Fig. 9.8b. **c** The profile of vortex chain in Fig. 9.8b. The length unit and strength of SOC are 0.76 μm and 0.96×10^{-3} m/s, respectively. Reprint from Ref. [26], Copyright 2018, with permission from American Physical Society

10.1.4 Vortex Chains and the Related Vortex Configuration

Figures 10.1, 10.2, 10.3, 10.4, 10.5, 10.6, 10.7, 10.8 and 10.9 have shown the vortex chain phenomena induced by the anisotropic SOC in spin-1 BECs, where the vortex chain is always accompanied by various vortices as $(1_0, 2_{-1})_1$ or $(0_1, 1_0, 2_{-1})_1$ [12, 27], where, in the bracket, 0_1 denotes the $m_F = 1$ vortex with winding number 0, 1_0 denotes the $m_F = 0$ vortex with winding number 1, 2_{-1} denotes the $m_F = -1$ vortex with winding number 2, and the subscript 1 out the bracket means that the two vortices overlap. In addition, the nonaxisymmetric (1, 1, 1) vortices in Refs. [12, 27] can be presented as $(1_1, 1_0, 1_{-1})_3$, i.e., the three components contain a vortex with winding number 1, respectively, and they locate at three separated positions. Likewise, we characterize the three-vortex structure reported in Ref. [23] as $\langle 1_{-1}, 1_0, 1_1\rangle_3$. We use the angle brackets (parenthesis) to indicate the vortices, where the total density has (does not have) a local minimum at the region of the vortices.

With this description, the vortex chain in Figs. 10.1a–c and 10.3c can be characterized by $\langle\langle 1_{-1}, 1_1\rangle_1, 1_0\rangle_2 \times n$, where the overlapped vortices of the $m_F = -1$ and $m_F = 1$ components are denoted by $\langle 1_{-1}, 1_1\rangle_1$. The separation of the $\langle 1_{-1}, 1_1\rangle_1$ vortices and the 1_0 vortices is described by the subscript 2, and $\times n$ denotes that there are n $\langle\langle 1_{-1}, 1_1\rangle_1, 1_0\rangle_2$ vortex configurations linking one after another to form a chain. The vortex chain in Fig. 10.5b can be written as $[\langle 1_1, 1_0, 1_1\rangle_3 + \langle 1_{-1}, 1_0, 1_{-1}\rangle_3] \times n$.

The off-x-axis vortex configuration in Fig. 10.2b can be presented as the $\langle 1_{-1}, 1_0, 1_1\rangle_3$ vortices. In Fig. 10.6c, the vortex configuration near $y = \pm 5$ can be written as the $\langle 1_{-1}, 1_0, 1_1\rangle_1$ vortices, due to the minimum densities of the three components located at the same position. Generally, the total density profile always tends to be nonsingular and continuous although the vortices occur in the spin-1 BECs. However, the $\langle 1_{-1}, 1_0, 1_1\rangle_1$ vortices are an unusual case, behaving like a singular vortex in the single-component BEC.

Our results show that the phases of the $m_F = -1$ and $m_F = 1$ components always tend to be synchronous in the developed vortex chain structure, such as the $\langle\langle 1_{-1}, 1_1\rangle_1, 1_0\rangle_2 \times n$ vortex chain obtained by the combination of anisotropic SOC and rotation shown in Figs. 10.1a–c, 10.2a–c, 10.5c, and 10.6c. Even in the region near such developed vortex chains, where the vortex configuration can be described by the $\langle 1_{-1}, 1_0, 1_1\rangle_1$ vortices, the phases of the $m_F = -1$ and $m_F = 1$ components also tend to be synchronous, which is shown in Figs. 10.2c and 10.6c. As shown in Fig. 10.6, as κx increases from 0.5 to 1.5, the formation of the vortex chain is changed noticeably, where $[\langle 1_1, 1_0, 1_1\rangle_3 + \langle 1_{-1}, 1_0, 1_{-1}\rangle_3] \times n$ vortex structure develops into the $\langle\langle 1_{-1}, 1_1\rangle_1, 1_0\rangle_2 \times n$ vortex chains. As we have already known, the vortices of different components usually are not overlapped, which is energetically favorable, when the SOC interaction is absent. We find that κ_y also influences the formation of the vortex chains through comparing Figs. 10.6c and 10.9b. With the same κ_x as Fig. 10.6, κ_y is increased to 1.2 in Fig. 10.9b; therefore, the anisotropy in the x-direction becomes weaker in Fig. 10.9b as compared with Fig. 10.6c. Due to the decreased anisotropy, we find that the vortex chain is fragmented, which is described by $\langle 1_1, 1_0, 1_{-1}\rangle_3 \times 2 + \langle 1_1, 1_0\rangle_2 + \langle 1_{-1}, 1_1\rangle_1 + \langle 1_0, 1_1\rangle_2 + \langle 1_{-1}, 1_0, 1_1\rangle_3 \times 2$. Correspondingly, the $\langle 1_{-1}, 1_0, 1_1\rangle_1$ vortices configurations become the $\langle 1_{-1}, 1_0, 1_1\rangle_3$ configurations, where the phases of the $m_F = -1$ and $m_F = 1$ components around the vortices tend to be nonsynchronous. For clarity, we summarize the results in Table 10.1. The first column shows the strength of anisotropic SOC and the rotation frequency, the second column indicates the characterization of the vortex chain, the third column shows some other vortex configurations, and the fourth column shows the figures where the vortex chain appears.

Similarly, the vortex chain in the pseudospin-1/2 BECs can be approximately presented by $\langle 1_\uparrow, 1_\downarrow\rangle_2 \times n$. In Ref. [33], Radić et al. solve the dimensionless GP equations [see Eqs. (22) of Ref. [33]] and obtain the vortex chain (the line of vortices) along the x-axis, where the total density at minima is close to but not exactly equal to zero. This property implies the vortex formed by the two condensates should overlap with each other. So the chain can be written as $\langle 1_\uparrow, 1_\downarrow\rangle 1 \times n$. Our dynamical Eqs. (10.1) uses the anisotropic SOC on the $x - y$ plane, where the σ_z term and the spatially dependent $\delta(y)$ term have not been considered as in Ref. [33]. Therefore, our results concerning vortex chain differ from those in Ref. [33].

10.1.5 Spin Texture of the Vortex Chain

Usually, the vortex configuration can cause some topological excitation of the spin texture. The spin texture is defined by [28]

Table 10.1 A summary of the results in the above text. The first column indicates the parameters, the second column shows the characterization of the vortex chain, the third column shows some other vortex configurations, and the fourth column indicates the pictures where the results appear

	Vortex chain	Other vortex configuration	Where
$(\kappa_x = 1, \kappa_y = 0)_{\Omega=0.2\omega}$	$\langle\langle 1_{-1}, 1_1\rangle_1, 1_0\rangle_2 \times n$	No	Fig. 10.1a
$(\kappa_x = 1, \kappa_y = 0)_{\Omega=0.3\omega}$	$\langle\langle 1_{-1}, 1_1\rangle_1, 1_0\rangle_2 \times n$	$\langle 1_{-1}, 1_0, 1_1\rangle_3$	Fig. 10.1b
$(\kappa_x = 1, \kappa_y = 0)_{\Omega=0.5\omega}$	$\langle\langle 1_{-1}, 1_1\rangle_1, 1_0\rangle_2 \times n$	$\langle 1_{-1}, 1_0, 1_1\rangle_3$, $\langle 1_{-1}, 1_0, 1_1\rangle_1$	Fig. 10.1c
$(\kappa_x = 0.1, \kappa_y = 0)_{\Omega=0.5\omega}$	No vortex chain	$\langle 1_{-1}, 1_0, 1_1\rangle_3$, $\langle 1_1, 1_0, 1_1\rangle_3$, $\langle 1_{-1}, 1_0, 1_{-1}\rangle_3$	Fig. 10.5a
$(\kappa_x = 0.5, \kappa_y = 0)_{\Omega=0.5\omega}$	$[\langle 1_1, 1_0, 1_1\rangle_3 + \langle 1_{-1}, 1_0, 1_{-1}\rangle_3] \times n$	$\langle 1_{-1}, 1_0, 1_1\rangle_3$	Fig. 10.5b
$(\kappa_x = 1.5, \kappa_y = 0)_{\Omega=0.5\omega}$	$\langle\langle 1_{-1}, 1_1\rangle_1, 1_0\rangle_2 \times n$	$\langle 1_{-1}, 1_0, 1_1\rangle_3$, $\langle 1_{-1}, 1_0, 1_1\rangle_1$	Fig. 10.5c
$(\kappa_x = 1, \kappa_y = 0.7)_{\Omega=0.5\omega}$	No vortex chain	$\langle 1_{-1}, 1_0, 1_1\rangle_3$	Fig. 10.8a
$(\kappa_x = 1.5, \kappa_y = 1.2)_{\Omega=0.5\omega}$	$\langle 1_1, 1_0, 1_{-1}\rangle_3 \times 2$ $+\langle 1_1, 1_0\rangle_2$ $+\langle 1_{-1}, 1_1\rangle_1 +$ $\langle 1_0, 1_1\rangle_2$ $+\langle 1_{-1}, 1_0, 1_1\rangle_3 \times 2$	$\langle 1_{-1}, 1_0, 1_1\rangle_3$	Fig. 10.8b

$$\mathbf{S}_\alpha = \sum_{m,n=0,\pm 1} \Psi_m^* \left(\widehat{F}_\alpha\right)_{m,n} \Psi_m / |\Psi|^2 \quad (\alpha = x, y, z), \tag{10.3}$$

and the topological charge density is defined by $q(r) = \frac{1}{4\pi}\mathbf{s} \cdot \left(\frac{\partial \mathbf{s}}{\partial x} \times \frac{\partial \mathbf{s}}{\partial y}\right)$, where $\mathbf{s} = \mathbf{s}/|\mathbf{s}|^2$. Since we have introduced the vortex chain excitation, we want to further explore the corresponding spin texture. In order to investigate the topological excitation of the vortex chain, we check the spin texture of the vortex chains and vortex configurations obtained by various cases. As shown in Fig. 10.10a, which corresponds to Fig. 10.5b under a small anisotropy SOC ($\kappa_x = 0.5$, $\kappa_y = 0$), where the obtained vortex chain is described by $[\langle 1_1, 1_0, 1_1\rangle_3 + \langle 1_{-1}, 1_0, 1_{-1}\rangle_l] \times n$, we find that the $\langle 1_1, 1_0, 1_1\rangle_3$ vortex configuration relates to the circular skyrmion, while the $\langle 1_{-1}, 1_0, 1_{-1}\rangle_3$ vortex configuration relates to the hyperbolic skyrmion. Figure 10.10b corresponds to Fig. 10.5c under a large anisotropy ($\kappa_x = 1.5$, $\kappa_y = 0$), and Fig. 10.10c is the corresponding topological charge density. Being described by $\langle\langle 1_{-1}, 1_1\rangle_1, 1_0\rangle_2 \times n$, the obtained vortex chain is more developed and induces the domain of the two counterorientation arrows of the spin texture. In the regions of the $\langle 1_{-1}, 1_0, 1_1\rangle_3$ vortices, we find the half-skyrmion and the nonzero topological

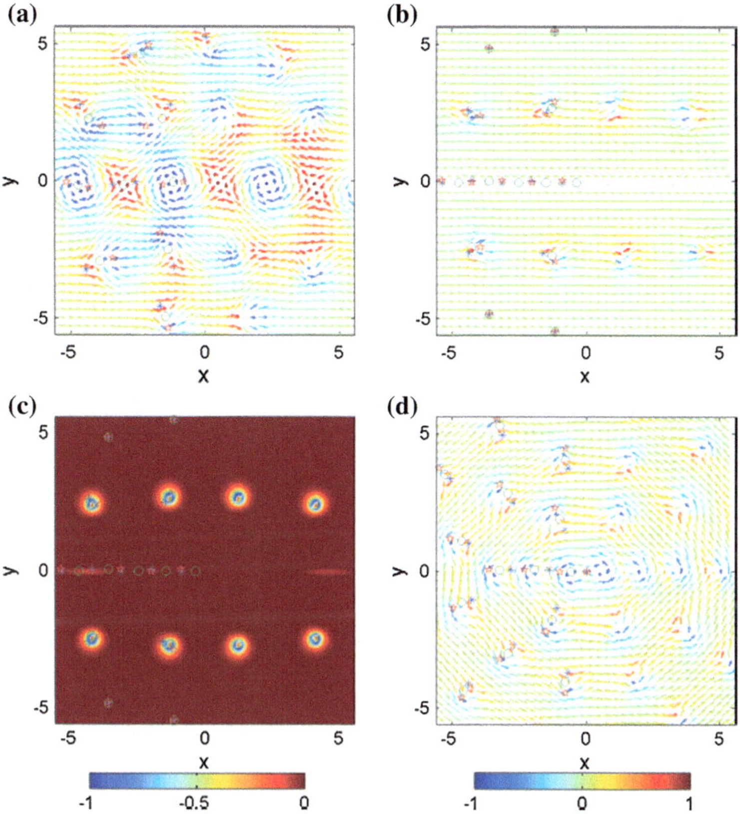

Fig. 10.10 **a** Vortex position and spin texture of spin-1 BECs of ^{87} Rb in Fig. 9.5b, where $\kappa_x = 0.5$, $\kappa_y = 0$, and $\Omega = 0.5\omega$. The color of each arrow indicates the magnitude of S_z. The asterisks, circles, and stars mark the position of vortices formed by the $m_F = -1$, $m_F = 0$, and $m_F = +1$ components, respectively. Noting that we only mark the vortices in $x < 0$ region in order to illuminate the spin texture and position of vortices clearly. **b** Vortex position and spin texture of spin-1 BEC of ^{87} Rb in Fig. 9.5c, where $\kappa_x = 1.5$, $\kappa_y = 0$, and $\Omega = 0.5\omega$. **c** Topological charge density and vortex position in Fig. 9.5c. **d** Vortex position and spin texture of spin-1 BEC of ^{87}Rb in Fig. 9.8b, where $\kappa_x = 1.5$, $\kappa_y = 1.2$, and $\Omega = 0.5\omega$. The length unit and strength of SOC are 0.76 μm and 0.96×10^{-3} m/s, respectively. Reprint from Ref. [26], Copyright 2018, with permission from American Physical Society

charge densities, which is consistent with the previously reported three-vortex configuration in the isotropic SO-coupled spin-1 BECs [23]. No spin texture is found for the $\langle\langle 1_{-1}, 1_1\rangle_1, 1_0\rangle_2 \times n$ vortex chain, and the topological charge density is around zero in the region of the vortex chain and the $\langle 1_{-1}, 1_0, 1_1\rangle_1$ vortices, which coincides with the general cases whereby $\langle 1_{-1}, 1_0, 1_1\rangle_1$ vortices usually do not cause nontrivial topological excitation. These properties indicate that the entire vortex configuration cannot induce the nontrivial spin texture. Figure 10.10d is the spin texture of vortex chain obtained in Fig. 10.8b ($\kappa x = 1.5$, $\kappa y = 1.2$), which corresponds to an interme-

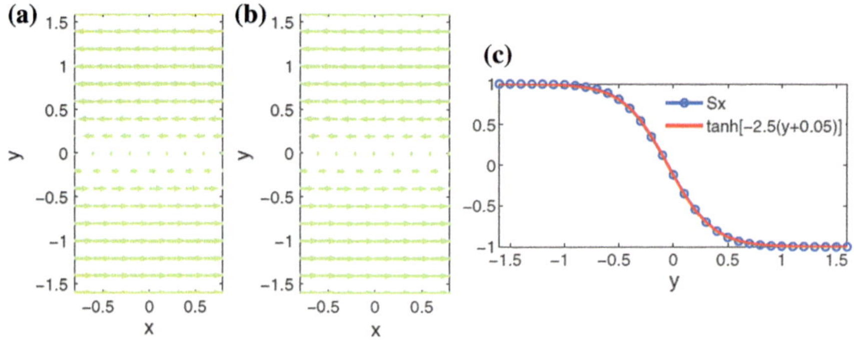

Fig. 10.11 **a** The enlarged spin texture in Fig. 9.10b. **b** The spin texture with Eqs. (10.3), where $c_1 = -2.5$ and $c_2 = 0.05$. **c** A comparison between the spin vector S_x and the function $\tanh[-2.5(y + 0.05)]$ along the y-axis. Reprint from Ref. [26], Copyright 2018, with permission from American Physical Society

diate strength of anisotropy, as compared with Fig. 10.10a and b. Some half-skyrmion structure caused by $\langle 1_{-1}, 1_0, 1_1\rangle_3$ vortices prevails, instead of the domain wall of the opposite arrows obtained in Fig. 10.10b for the larger anisotropy. The results shown in Fig. 10.10 indicate the relationship between the vortex configuration and the spin texture. When the phases of vortices formed by the $m_F = -1$ and $m_F = 1$ components are synchronous, the vortex configuration does not induce the nontrivial topological excitation of the spin texture. This means that we must ensure that the corresponding vortex excitation of the $m_F = -1$ and $m_F = +1$ components not to overlap in order to produce some topological excitation of spin texture such as skyrmion.

Figure 10.11a enlarges the spin texture which is induced by the vortex chain $\langle\langle 1_{-1}, 1_1\rangle_1, 1_0\rangle_2 \times n$ in Fig. 10.10b. It is a domain wall between two groups of opposite arrows. According to Fig. 10.7c, the density profiles of the $m_F = -1$ and $m_F = 1$ components are equal, and the density difference of the $m_F = -1$ and $m_F = 1$ components corresponds to the S_z vector of the spin texture. We find the spin texture can be presented by

$$\mathbf{S}_{xc} = \tanh\left[c_1(y + c_2)\right], \mathbf{S}_{yc} = 0, \mathbf{S}_{zc} = 0. \tag{10.4}$$

Figure 10.11b is the result with Eq. (10.3), where $c_1 = -2.5$ and $c_2 = 0.05$. We find that Eq. (10.3) can well describe the spin texture induced by the vortex chain $\langle\langle 1_{-1}, 1_1\rangle_1, 1_0\rangle_2 \times n$ through comparison of Fig. 10.11a and b and comparison of the spin vector S_x with the function $\tanh[-2.5(y + 0.05)]$ along the y-axis shown in Fig. 10.11c.

10.2 Spin Density Singularity Phenomenon in the Spin-1 BECs

In the BEC, the singular defect on the density is very common for producing the nontrivial quantum topological excitations, such as the monopole in the three dimensions and the quantized vortex in the two dimensions [7, 23, 24]. These products are the fundamental interest for most of the physical researches. Unlike the scalar BEC, the spinor BEC opens a new dimensionality on the spin.

This section deals with the investigation of spin singularity phenomenon in the spin-1 BECs. In what follows, we prove that the spin singularity is a common phenomenon in the spin-1 BECs. Meanwhile, by combining the SOC and the magnetic field, we show that a singular spin topological excitation, whose wave function possesses the property of the normal half-quantum vortex, can stably exist in the spin-1 BECs.

10.2.1 The Spin Density Singularity of the Spin-1 Bose–Einstein Condensates

For the spin-1 BECs, the spin vectors is defined by Eq. (10.3), that is,

$$\mathbf{S}_\alpha = \sum_{m,n=0,\pm 1} \Psi_m^* \left(\widehat{F}_\alpha\right)_{m,n} \Psi_n / \left|\Psi\right|^2 \quad (\alpha = x, y, z), \tag{10.5}$$

where Ψ_j $(j = 0, \pm 1)$ denotes the macroscopic wave function of the atoms condensed in the spin state $|j_F = 1, m_F = j\rangle$, and $\widehat{F}_{\alpha=x,y,z}$ is the spin-1 matrices. The total spin density is

$$|\mathbf{S}|^2 = \frac{2\Psi_0^{*2}\Psi_1\Psi_{-1}}{|\Psi|^4} + \frac{2\Psi_0^2\Psi_1\Psi_{-1}}{|\Psi|^4} + \frac{|\Psi_0|^2\left(|\Psi_1+\Psi_{-1}|^2 + |\Psi_1-\Psi_{-1}|^2\right)}{|\Psi|^4} + \frac{\left(|\Psi_1|^2 - |\Psi_{-1}|^2\right)^2}{|\Psi|^4}. \tag{10.6}$$

Providing $\Psi_0 \to 0$, we can get the expression $|\mathbf{S}|^2 = \left(|\Psi_1|^2 - |\Psi_{-1}|^2\right)^2 / |\Psi|^4 \le 1$. Providing $\Psi_{\pm 1} \to 0$ yields $|\mathbf{S}|^2 = \left[\left(|\Psi_0|^2 + |\Psi_{\pm 1}|^2\right)^2 - |\Psi_0|^4\right] / |\Psi|^4 \le 1$. Thus, the spin density is not conserved in the spin-1 BECs. We think about the creation of the spin singularity ($|\mathbf{S}|^2 \to 0$). For $\Psi_0 \to 0$, we must keep the condition $|\Psi_1| = |\Psi_{-1}| > 0$ to fulfill the spin singularity. And for $|\Psi_0| > 0$, we must keep $\Psi_{\pm 1} \to 0$.

In Fig. 10.12, we display four examples of the spin density and the spin singularity. In Fig. 10.12a, we assume the formation of the wave function to be

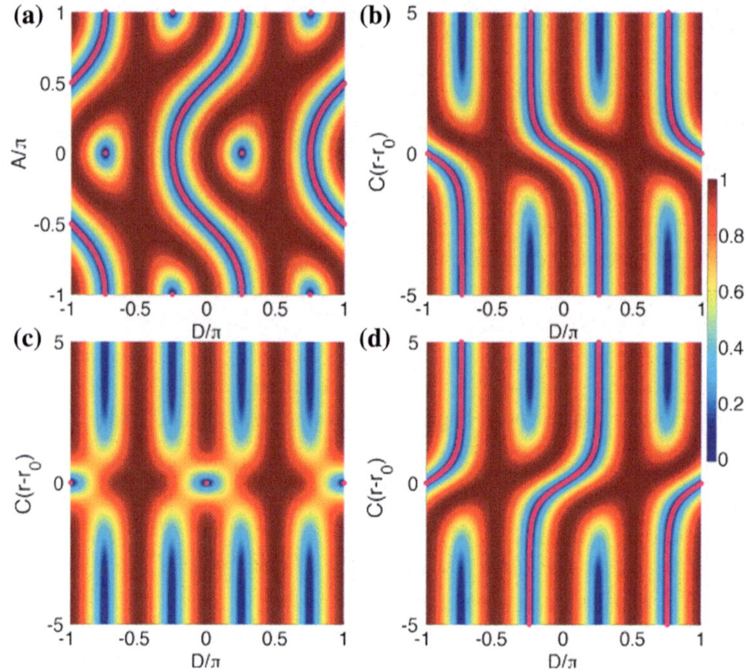

Fig. 10.12 Several schematics of the spin density and the spin singularity in the spin-1 BECs. **a** The formation of wave function is $\Psi = (\sin[D], \cos[D]\sin[A], \cos[D]\cos[A])^T$. In **b**, **c** and **d**, the formation of the wave function is $\Psi = (\sin[D], \cos[D]/\cosh[C|r-r_0|],$ $\cos[D]\tanh[C|r-r_0|]\exp[i\beta_0])^T$. **b** $\beta_0 = 0$, **c** $\beta_0 = \pi/2$, and **d** $\beta_0 = \pi$. The color denotes the value of $|S|$ and the mauve dots indicate the positions of the spin singularities [24]

$\Psi = (\sin[D], \cos[D]\sin[A], \cos[D]\cos[A])^T$, where the two angles D and A can well characterize all the ratio of the three components and keep $|\Psi|^2 = 1$. The mauve dots demonstrate the emergence of the spin singularities ($|\mathbf{S}| = 0$). Figure 10.12a shows that the spin singularities are very common in the spin-1 BEC even if there are no excitations. In Fig. 10.12b, c and d, we change the wave function to be $\Psi = (\sin[D], \cos[D]\,sech[C|r-r_0|], \cos[D]\tanh[C|r-r_0|]\exp[i\beta_0])^T$. For $\beta_0 = 0$ or π, the appearance of the spin singularities covers all value of $C|r-r_0|$. For $\beta_0 = \pi/2$, the spin singularities only occur when the parameters are $C|r-r_0| = 0$ and $D = n\pi$, where n is an integer. In fact, we find that $\sin[D] = 0$, $C|r-r_0| = 0$ are the common condition to create the spin singularities for different β_0. Therefore, the spin singularity is a common phenomenon in the spin-1 BECs.

10.2.2 Description of the Model

In the spin-1 BECs, to force one of the components approaching zero, i.e., Ψ_j ($j = 0, \pm 1$) $\rightarrow 0$, we can induce the vortices whose center is a singularity. Furthermore, we can also magnetize the BECs. This idea may produce a scalar BEC with only one component, while the SOC technology may change the result partly. In this study, we consider the combination of the SOC and the magnetic field. The dynamical equations of the system are written as

$$
\begin{aligned}
i\hbar\frac{\partial \Psi_j}{\partial t} &= \left[-\frac{\hbar^2}{2m}\nabla^2 + \frac{m\omega^2\left(x^2+y^2\right)}{2} + g_n\left|\Psi\right|^2\right]\Psi_j \\
&+ g_s \sum_{\alpha=x,y,z}\ \sum_{n,k,l=0,\pm 1} \left(\widehat{F}_\alpha\right)_{jn}\left(\widehat{F}_\alpha\right)_{kl}\Psi_n\Psi_k^*\Psi_l \\
&+ \sum_{n=01} \mathbf{B}_{jn}\Psi_n + \sum_{\alpha=x,y}\ \sum_{n=0,\pm 1} \kappa_\alpha\left(\widehat{F}_\alpha\right)_{jn} p_\alpha \Psi_n,
\end{aligned}
\tag{10.7}
$$

where m is the mass of atom, ω the trapped frequency, $\widehat{F}_{\alpha=x,y,z}$ the spin-1 matrices, $p_\alpha = -i\hbar\frac{\partial}{\partial\alpha}$ ($\alpha = x,\ y$) the momentum operator, and κ_α denotes the strength of SOC which carries the unit of velocity. In this section, we mainly consider an isotropic SOC with $\kappa_x = \kappa_y = \kappa$. Note that the coupling constants $g_n = \frac{4\pi\hbar^2(2a_2+a_0)}{3m}$ and $g_s = \frac{4\pi\hbar^2(a_2-a_0)}{3m}$ come from the density–density and spin–exchange interactions, respectively. The magnetic field can be described as

$$
\mathbf{B} = \begin{pmatrix} B_z & B_\perp \exp\left[-i\theta\right]/\sqrt{2} & 0 \\ B_\perp \exp\left[i\theta\right]/\sqrt{2} & 0 & B_\perp \exp\left[-i\theta\right]/\sqrt{2} \\ 0 & B_\perp \exp\left[i\theta\right]/\sqrt{2} & -B_z \end{pmatrix},
\tag{10.8}
$$

where B_z is the magnetic field in the z-direction and $B_\perp$ is the strength of magnetic field in the $x - y$ plane and its direction. The energy function of the SOC spin-1 BECs is given by

$$
\begin{aligned}
E = \int d^2r &\sum_{n=1,0,-1} \Psi_n^*\left(-\frac{\hbar^2}{2m}\nabla^2 + \frac{m\omega^2\left(x^2+y^2\right)}{2}\right)\Psi_n \\
&+ \frac{g_n}{2}\left(|\Psi_1|^2 + |\Psi_0|^2 + |\Psi_{-1}|^2\right)^2 + \frac{\hbar\kappa}{m}\left[\Psi_1^*\left(-i\partial/\partial x - \partial/\partial y\right)\right. \\
&\left.+ \Psi_0^*\left(-i\partial/\partial x - \partial/\partial y\right)\Psi_{-1} + H.c.\right] + B_z\left(|\Psi_1|^2 - |\Psi_{-1}|^2\right) \\
&+ \frac{g_s}{2}\left[\left(|\Psi_1|^2 - |\Psi_{-1}|^2\right)^2 + 2\left|\Psi_1^*\Psi_0 + \Psi_0\Psi_{-1}^*\right|^2\right] \\
&+ \frac{B_\perp}{\sqrt{2}}\left(\Psi_1^*\Psi_0\exp\left[-i\theta\right] + \Psi_0^*\Psi_{-1}\exp\left[-i\theta\right] + H.c.\right).
\end{aligned}
\tag{10.9}
$$

In a scalar BEC, we can set vortices by having the ground state wave function multiplied the ansatz function

$$\phi = \sqrt{n_0}\frac{|r - r_k|}{\sqrt{(r - r_k)^2 + \Lambda^{-2}}}\exp\left[\pm i\theta_k\right], \tag{10.10}$$

where $\Lambda^2 = 0.6805$, $r = (x, r)$ is the coordinate, $r_k = (x_k, y_k)$ is the position of the vortex, and $\theta_k(x, y) = \arctan\left[\frac{y-y_k}{x-x_k}\right]$. Similarly, in the spin-1 BECs, we can assume an approximate vortex solution as

$$\Psi = \begin{pmatrix} \Psi_1 \\ \Psi_0 \\ \Psi_{-1} \end{pmatrix} \sqrt{n_0(r)} \begin{pmatrix} \sin\left[D\right] \\ \frac{\cos[D]}{\cosh[C|r-r_0|]} \\ \cos\left[D\right]\tan\left[C\left|r - r_0\right|\right]\exp\left[i\beta\right] \end{pmatrix}, \tag{10.11}$$

where D is used to adjust the ratio of each component, C is a variational parameter for the vortex, the coordinate is $r = (x, y)$, $r_0 = (x_0, y_0)$ is the position of vortex, and $\beta(x, y) = \arctan\left[\frac{y-y_0}{x-x_0}\right]$ is the phase. Using Eq. (10.11), we have $\Psi^\dagger\Psi = \mathbf{n}_0(r)$, where $n_0(r)$ is the total density of the spin-1 BECs under the external potential $V(r)$.

10.2.3 Stability Analysis

In the numerical simulations, we will use here $\hbar = 1$, the atomic mass $m = 1$, and the chemical potential $\mu = 1\hbar\omega$. Meanwhile, we set the interaction coefficient $g_n = 1$ and $\omega = 0.1c/\xi$ for the external potentials $[V(r) = \frac{1}{2}m\omega^2 r^2]$, where $c = \sqrt{\mu/m}$ is the Bogoliubov speed of sound and $\xi = \hbar/\sqrt{\mu m}$ is the healing length. Thus, the length unit is ξ, the energy unit is μ and the time unit is ξ/c. The unit of $B_\perp$ and B_z is $\hbar\omega$. The unit of κ_x and κ_y is $\sqrt{\mu/m}$. And the unit of g_n and g_s is $\xi^3\mu$.

In the numerical calculations, we set the size of the system from -15ξ to 15ξ, where the size is much larger than that of the BECs. Then, we set the vortex with Eq. (10.11). We obtain n_0 through Thomas–Fermi (TF) approximation $n_0(r) = \left[\mu - V(r)\right]/g_n$. In two spatial dimensions, the wave function is discretized onto a linear grid with separation Δx (Δy) with index i (j), expressed as $\Psi_n(x_i, y_j), n = 0, \pm 1$. The central space-difference operator is given by $\frac{\partial\Psi_n(x_i,y_j)}{\partial x} = \frac{\Psi_n(x_{i+1},y_j)-\Psi_n(x_{i-1},y_j)}{2\Delta x}$, $\frac{\partial\Psi_n(x_i,y_j)}{\partial y} = \frac{\Psi_n(x_i,y_{j+1})-\Psi_n(x_{i-1},y_{j-1})}{2\Delta y}$. Then the double space derivative operator is given by

$$\frac{\partial^2\Psi_n(x_i, y_j)}{\partial x^2} = \frac{\Psi_n(x_{i+1}, y_j) - 2\Psi_n(x_i, y_j) + \Psi_n(x_{i-1}, y_j)}{\Delta x^2},$$
$$\frac{\partial^2\Psi_n(x_i, y_j)}{\partial y^2} = \frac{\Psi_n(x_i, y_{j+1}) - 2\Psi_n(x_i, y_j) + \Psi_n(x_{i-1}, y_{j-1})}{\Delta y^2}.$$

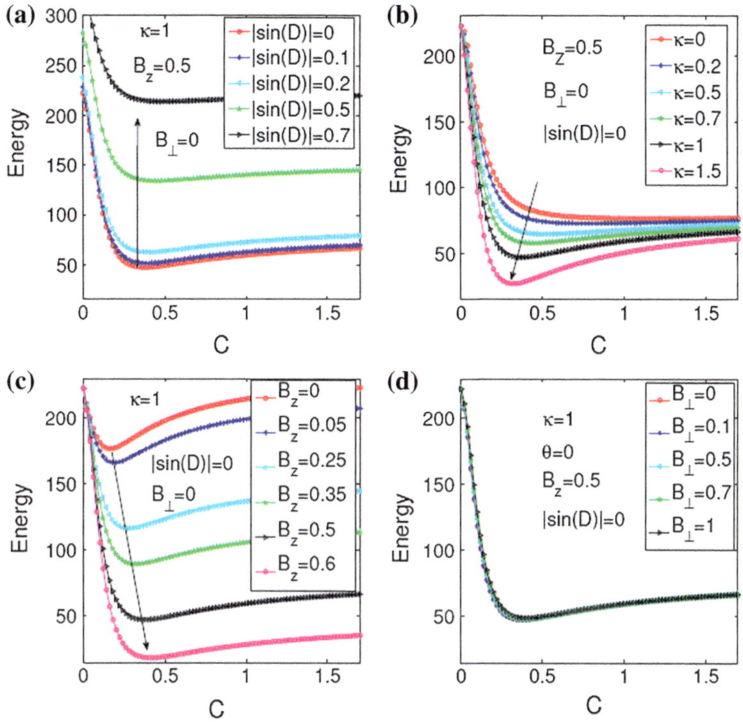

Fig. 10.13 Energy of the spin-1 BECs with a possible half-quantum vortex. **a** The effect of $|F = 1, m_F = 1\rangle$ component on the energy of the BECs as the variational parameter C changes. Other parameters are $g_n = 1$, $g_s = 0$, $\omega = 0.1$, $B_z = 0.5$, $B_\perp = 0$, $\kappa_x = \kappa_y = \kappa = 1$, and $(x_0, y_0) = (0, -0.5)$. **b** The effect of the SOC on the energy of the BECs as the variational parameter C varies. **c** The effect of the B_z on the energy of the BECs as the variational parameter C varies. **d** The effect of the $B_\perp$ on the energy of the BECs as the variational parameter C varies. The energy, magnetization, length, strength of the SOC, and strength of the interaction are in units of $\hbar\omega$, $\hbar\omega$, ξ, $\sqrt{\mu/m}$, and $\xi^3\mu$, respectively [24]

Using Eq. (10.9), we can obtain the total energy of the system.

The Stability of the Solution

Since Eq. (10.11) has assumed an approximate half-quantum vortex solution, we now check whether this solution is possible to occur in the spin-1 BECs. Using Eq. (10.9) of the dimensionless formation, we check the energy of the BECs. First, we consider the effect of ratio of the $|F = 1, m_F = 1\rangle$ component on the energy of the BECs. Figure 10.13a shows the curves of the energy of the BECs as the parameter C changes. When $|\sin[D]| = 0$, the energy curve locates at the bottom. As the $|F = 1, m_F = 1\rangle$ component increases, i.e., $|\sin[D]|$ grows, the energy curve shifts up. The black arrow indicates the trend. Thus, to minimize the energy of the BECs, we should set $|\sin[D]| = 0$, i.e., the $|_F = 1, m_F = 1\rangle$ component vanishes in the spin-1 BECs.

In Fig. 10.13b, we set the magnetic field $B_z = 0.5$. For $\kappa = 0$, the energy of BECs decreases monotonously as C increases. This implies that the energy of BECs will be the minimum when $C \rightarrow \infty$, i.e., the size of vortex approaches 0. Thus, we cannot get stable vortex with finite size when $\kappa = 0$. When is more than 0, there is a local dip of the energy of BECs. The larger the value of κ is, the more obvious the dip is. The dip indicates that we can obtain the minimum energy by using Eq. (10.11) with the corresponding $C^{E_{min}}$. Furthermore, the vortex is supposed to be stable. Figure 10.13b implies that the SOC is a necessary factor for the stable half-quantum vortex of Eq. (10.11).

In Fig. 10.13c, we further check the effect of the magnetic field B_z and C on the energy of the BECs. With $\kappa = 1$, the energy dip occurs in every case of B_z. And the large B_z causes a relatively low energy. This phenomenon could be easily understood. The main reason is that the magnetic field is effective on the $|F = 1, m_F = -1\rangle$ component. Figure 10.13d indicates the effect of $B_\perp$ on the energy of the BECs when the system possesses a half-quantum vortex of Eq. (10.11). Generally speaking, $B_\perp$ does not lead to an obvious change in the energy curves. However, $B_\perp$ can cause the transformation between the $|F = 1, m_F = 0\rangle$ component and the $|F = 1, m_F = 1\rangle$ component. This implies the increase of $|\sin[D]|$. Therefore, the magnetic field of $B_\perp$ should be set to be 0 for the stable half-quantum vortex.

Figure 10.14a–d shows an example of the half-quantum vortex of $C^{E_{min}} = 0.38$. Figure 10.14a and c shows the densities of the $|F = 1, m_F = 0\rangle$ and $|F = 1, m_F = -1\rangle$ components, respectively, and Fig. 10.14b and d shows the corresponding phases. Figure 10.14e is the density profile of the vortex. For purposes of comparison, we plot the density profile of a normal vortex of Eq. (10.10) with a homogenous background density. Figure 10.14f shows the corresponding spin density $|S|$, and Fig. 10.14g further indicates the profile of $|S|$. Obviously, the spin density $|S|$ decreases to be 0 at the center of vortex.

We plot the spin texture in Fig. 10.14h with the spin vector (S_x, S_y, S_z). Far away from the center of the spin vortex, all the arrows are perpendicular to the $x - y$ plane, pointing into the $-z$-direction. While approaching the coordinate (0, –0.5), the arrows become parallel to the $x - y$ plane and the arrows shorten to be zero (i.e., $|S| \rightarrow 0$). Meanwhile, the arrows form a circle around (0, –0.5). These properties show that it is a special topological excitation. Figure 10.14i indicates the topological charge density $q(x, y)$, which is defined by $q(x, y) = \frac{1}{4\pi}\mathbf{s} \cdot \left(\frac{\partial \mathbf{s}}{\partial x} \times \frac{\partial \mathbf{s}}{\partial r}\right)$, where $\mathbf{s} = \mathbf{S}/|\mathbf{S}|$. We find that there is a singular point at (0, –0.5) and that the sum of the topological density is $Q = \int\int q(x, y)dxdy = 0.5$. Therefore, we predict a half-quantum vortex in the spin-1 BECs and this vortex is related to a half-quantum topological spin texture with singularity.

The phase of spin defect can be written as $\theta_{spin} = \arctan\left[\frac{S_y}{S_x}\right]$. Using Eq. (10.11) with $\sin|D| = 0$, we can obtain $\theta_{spin} = \theta$.

We now check the effect of the position of the vortex on the energy of the BECs. In Fig. 10.15a, we use the variational method to examine the minimum energy $E_{min}(x, y)$. Note that we set the half-quantum vortex at position (x, y). Around $(0, 0^-)$, we can find a local energy minimum, which can be viewed as the most stable

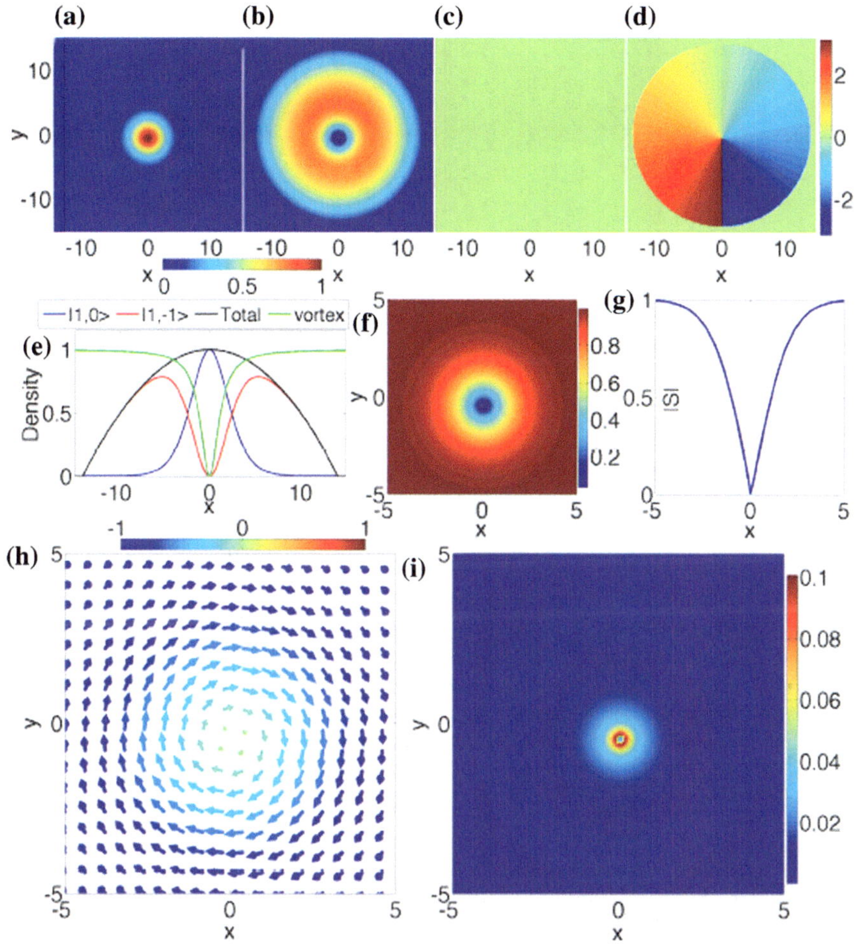

Fig. 10.14 Approximate half-quantum vortex solution in the spin-1 BEC and the corresponding singular spin texture. **a**, **b**, **c** and **d** show an example of the half-quantum vortex. The parameters are $g_n = 1$, $g_s = 0$, $\omega = 0.1$, $\sin[D] = 0$, $C = 0.38$, $B_z = 0.5$, $B_\perp = 0$, $\kappa_x = \kappa_y = \kappa = 1$, and $(x_0, y_0) = (0, -0.5)$. **a** and **b** are the densities of the $|F = 1, m_F = 0\rangle$ and $|F = 1, m_F = -1\rangle$ components, respectively. **c** and **d** are the corresponding phases. **e** shows the profile of the half-quantum vortex. The green curve is the profile of a normal vortex with Eq. (10.10). **f** Spin density $|S|$. **g** The profile of the spin density $|S|$. **h** Spin texture. **i** Topological charge density $q(x, y)$. The energy, magnetization, length, strength of the SOC, and strength of the interaction are in units of $\hbar\omega$, $\hbar\omega$, ξ, $\sqrt{\mu/m}$, and $\xi^3\mu$, respectively [24]

position (x_s, y_s). This implies that it is possible to obtain a stable vortex in the spin-1 BECs when we quench the BECs. If the local energy minimum is not around the center of BEC but approaching the boundary of the BECs, the vortex will be drifted out of the BECs though we first fix it at the center of the BECs. In this case, the half-quantum vortex of Eq. (10.11) is in an unstable state [see Fig. 10.15c, where the

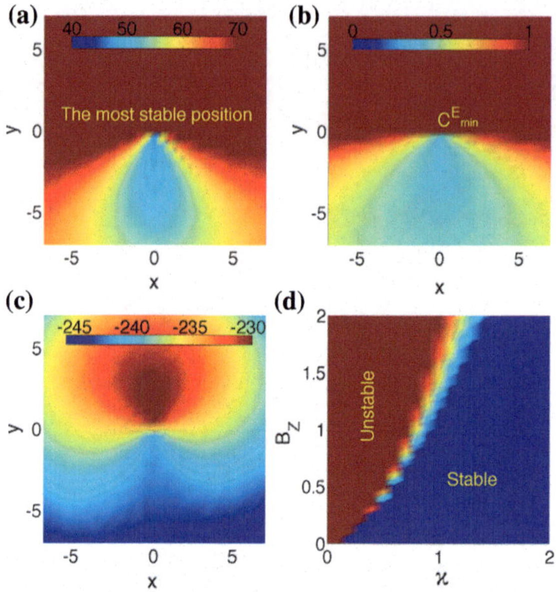

Fig. 10.15 **a** The stable minimum energy E_{min} of the BECs as a function of the position of the vortex (x, y). We set $B_z = 0.5$ and $\kappa_x = \kappa_y = \kappa = 1$. **b** The corresponding parameter $C^{E_{min}}$ as a function of the position of the vortex (x, y). **c** The unstable minimum energy E_{min} of the BECs as a function of the position of the vortex (x, y). We set $B_z = 1.5$ and $\kappa = 0.5$. **d** Stability of the half-quantum vortex under the magnetic field B_z and the SOC κ. The energy, magnetization, length, strength of the SOC, and strength of the interaction are in units of $\hbar\omega$, $\hbar\omega$, ξ, $\sqrt{\mu/m}$, and $\xi^3\mu$, respectively [24]

most stable position (x_s, y_s) approaches the boundary of the system]. Figure 10.15b indicates the corresponding parameter $C^{E_{min}}(x, y)$, in which the whole BECs with the half-quantum vortex have the minimum energy. Thus, the vortex of $C^{E_{min}}(x, y)$ possesses the minimum energy and should be the most possible solution in this system.

Obviously, each pair of parameters of (κ, B_z) can cause a most stable position (x_s, y_s) in the $x - y$ plane. If all the most stable positions (x_s, y_s) could be found out, we will be able to obtain an effective phase diagram, which shows the stable position and the unstable position in the $\kappa - B_z$ plane when the spin vortex is set. For reducing the amount of calculation, we can choose a finite region $r(x, y) = \sqrt{x^2 + y^2} < 10$, where the density $n_0(r = 10)$ is very low. The main work now is to check the formation of the minimum energy $E_{min}(x, y)$ of the BECs, where (x, y) is the position of the spin vortex. If $E_{min}(x, y)$ behaves like Fig. 10.15a, the spin vortex would be stable in the most stable position (x_s, y_s). If $E_{min}(x, y)$ behaves like Fig. 10.15c, the spin vortex would be unstable in the system. Figure 10.15d plots $\sqrt{x_s^2 + y_s^2}$ of the most stable position (x_s, y_s) for all the cases of (κ, B_z). On the other hand, Fig. 10.15d also shows the phase diagram of stability of the half-quantum

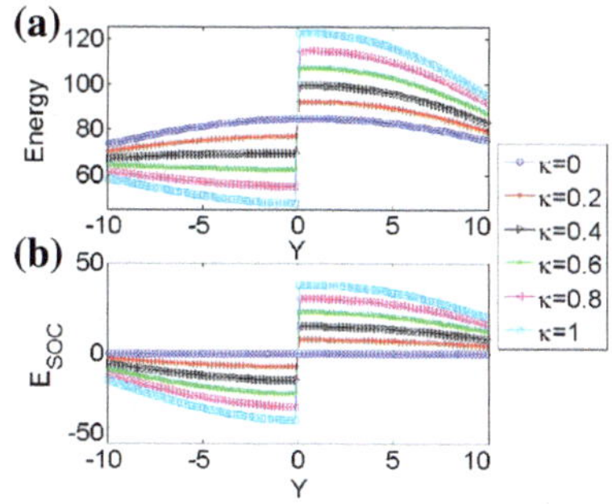

Fig. 10.16 The SOC causes the energy break as the position of the vortex changes. The strength of the SOC is $\kappa_x = \kappa_y = \kappa$. **a** The total energy of the system as the position of the vortex $(x_0 = 0, y_0 = Y)$ varies. **b** The corresponding energy of the SOC term, i.e., E_{SOC}. We set the half-quantum vortex with $C = 0.4$ in the BECs, and strength of the magnetic field $B_z = 0.5$. The energy, magnetization, length, strength of the SOC, and strength of the interaction are in units of $\hbar\omega$, $\hbar\omega$, ξ, $\sqrt{\mu/m}$, and $\xi^3\mu$, respectively [24]

vortex in the spin-1 BECs. If the value of $\sqrt{x_s^2 + y_s^2}$ is less than 10, we view the vortex to be stable in this system. Reversely, the vortex would be unstable and go to the boundary in this system. In fact, we cannot obtain the stable half-quantum vortex in the unstable region, because $E_{min}(x, y)$ has the formation like Fig. 10.15c. And the so-called most stable position (x_s, y_s) and the corresponding $\sqrt{x_s^2 + y_s^2} = 10$ is false. Thus, we do not give a scale in Fig. 10.15d to avoid the value to cause some misunderstanding.

Why does the position of the minimum energy approach $(0, 0^-)$ in Fig. 10.15a? To clarify this phenomenon, we examine the energy of the BECs when we set the position of the vortex to be $(x_0 = 0, y_0 = Y)$, and the size of the vortex is fixed with $C = 0.4$. Figure 10.16a shows the total energy. Obviously, when $\kappa = 0$, we can find that the total energy is continuous as Y varies. But the energy approaches the maximum value around $(0, 0)$, i.e., the vortex would be unstable when it locates at the center of BECs. With $\kappa > 0$, the total energy is uncontinuous as Y varies. There is a step at $Y = 0$. In Fig. 10.16b, we can find $E_{SOC} = 0$ when $\kappa = 0$. When $\kappa > 0$ and $Y < 0$, we have $E_{SOC} < 0$. When $\kappa >$ and $Y > 0$, we have $E_{SOC} > 0$. Thus, the SOC causes the energy break when the position of the vortex $(x_0 = 0, y_0 = Y)$ changes. However, the minimum of the total energy is around the center only when $\kappa > 0.6$ in Fig. 10.16a.

In the above text, the position of the minimum energy is $(0, 0^-)$, which is caused by the SOC. Note that $(0, 0)$ is a singular position for the half-quantum vortex of Eq. (10.11), and there is no corresponding image vortex. In the below discussion, we give a detail description of the image vortex. Thus, we cannot set the vortex at $(0, 0^-)$ directly but set it with a little deviation in the calculation. Here, the deviated position is $(0, -0.5)$.

Figure 10.17a further shows the most stable position $|r|$ ($|r| = \sqrt{x_s^2 = y_s^2}$) changes as the SOC κ varies. For the weak SOC, vortex tends to be drifted out of the BECs,

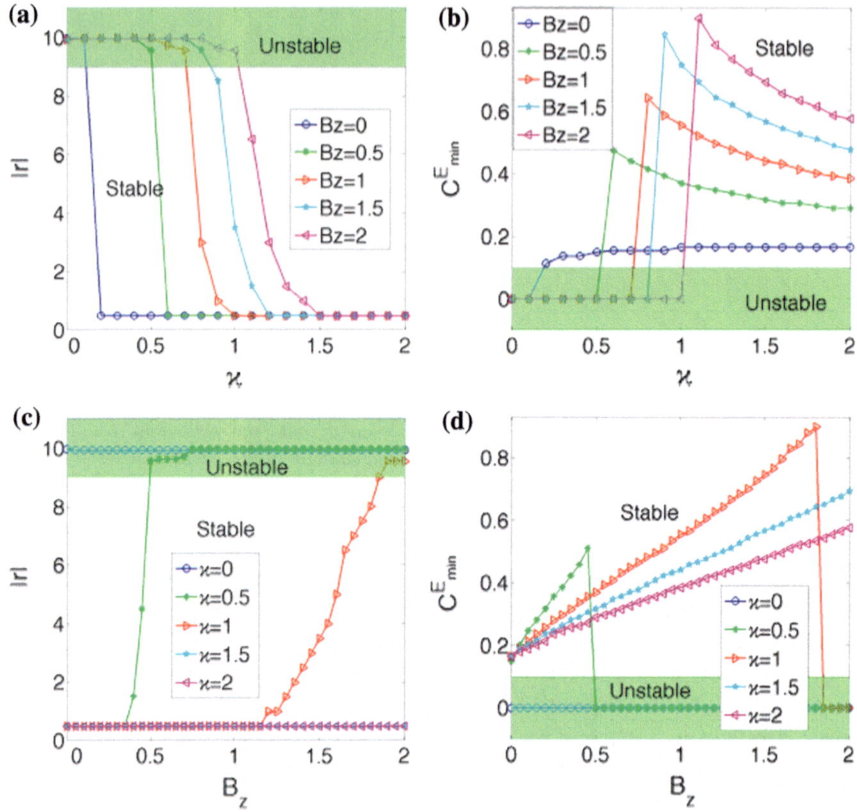

Fig. 10.17 **a** The most stable position of the half-quantum vortex with the parameter $C^{E_{min}}$ in the BECs as the SOC $\kappa(\kappa_x = \kappa_y = \kappa)$ varies. $|r| = \sqrt{x_s^2 + y_s^2}$. **b** The corresponding parameter $C^{E_{min}}$ as a function of κ. **c** The most stable position of the half-quantum vortex with $C^{E_{min}}$ in the BECs as the magnetic field B_z varies. **d** The corresponding parameter $C^{E_{min}}$ as a function of B_z. The energy, magnetization, length, strength of the SOC, and strength of the interaction are in units of $\hbar\omega$, $\hbar\omega$, ξ, $\sqrt{\mu/m}$, and $\xi^3\mu$, respectively [24]

i.e., the value of $|r|$ is up to 10 and we point out it is unstable, while the strong SOC can stabilize the vortex in the center of the BECs, i.e., the value of $|r|$ approaches 0. Figure 10.17b indicates the effect of the SOC on $C^{E_{min}}$ (x_s, y_s). Generally speaking, the strong SOC mainly decreases the values of $C^{E_{min}}$ (x_s, y_s) while the strong magnetic field increases the values of $C^{E_{min}}$ (x_s, y_s). Figure 10.17c further shows the most stable position $|r|$ $(|r| = \sqrt{x_s^2 = y_s^2})$ changes as the magnetic field B_z varies. In the absence of the SOC, the vortex would be drifted out of the BECs, and the magnetic field cannot cause an effective stable state. We find that the strong magnetic field Bz would cause the value of $|r|$ to increase. This point is obvious for the weak SOC ($\kappa = 0.5$ and 1). For the strong SOC ($\kappa = 1.5$ and 2), the vortex also should be drifted out of the BECs if Bz is strong enough. Figure 10.17d indicates that $C^{E_{min}}$

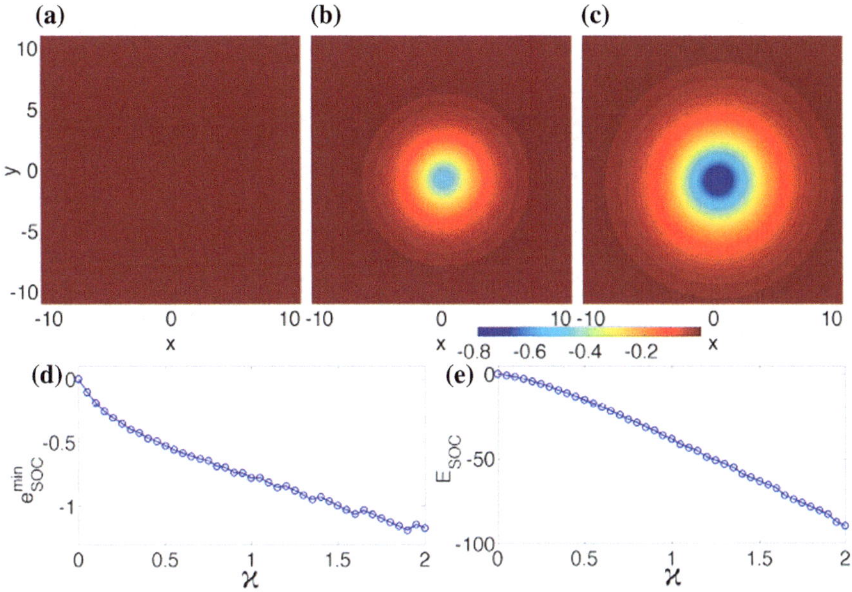

Fig. 10.18 The effect of the SOC on the SOC energy $e_{SOC}(x, y)$ of the BECs under the parameter $C^{E_{min}}(0, -0.5)$. **a** $\kappa_x = \kappa_y = \kappa = 0$. **b** $\kappa_x = \kappa_y = \kappa = 0.5$. **c** $\kappa_x = \kappa_y = \kappa = 1$. **d** The minimum values of e_{SOC} as a function of the SOC. **e** The sum of the SOC energy E_{SOC} [$E_{SOC} = \int\int e_{SOC}(x, y)\,dxdy$] as a function of the SOC. We set $B_z = 0.5$ and the position of the vortex to be $(0, -0.5)$. The energy, magnetization, length, strength of the SOC, and strength of the interaction are in units of $\hbar\omega$, $\hbar\omega$, ξ, $\sqrt{\mu/m}$, and $\xi^3\mu$, respectively [24]

(x_s, y_s) increases as B_z goes up. With the weak SOC, the value of $C^{E_{min}}(x_s, y_s)$ grows more quickly but the vortex is easy to reach the unstable state.

The Effect of the Spin–Orbit Coupling on the Stability

We find that the SOC plays a role in stabilizing the vortex in the BECs. In Fig. 10.18a, the strength of the SOC is 0, meaning that e_{SOC} is 0 naturally. If $\kappa > 0$, there is a local minimum value of $e_{SOC}(x, y)$, which is less than 0 and just at the position of the vortex [see Fig. 10.18b]. This local minimum value stabilizes the vortex in the BECs. The stronger the SOC is, the more obvious the local minimum of $e_{SOC}(x, y)$ is [see Fig. 10.18c]. In Fig. 10.18d, we further examine the minimum value of $e_{SOC}(x, y)$, i.e., e_{SOC}^{min}. It decreases as the SOC increases. Similarly, the whole energy E_{SOC} coming from the SOC also decreases as the SOC increases [see Fig. 10.18e].

The Effect of the Magnetic Field on the Stability

Figure 10.19 shows the effect of the magnetic field B_z on the SOC energy of the BECs. In Fig. 10.19a–c, we can find a local minimum value of $e_{SOC}(x, y)$. Meanwhile, comparing the three cases in Fig. 10.19a–c, we find that the magnetic field Bz would enhance the effect of the local minimum $e_{SOC}(x, y)$ and make it more obvious. Figure 10.19d further shows that the local minimum value of $e_{SOC}(x, y)$ is less than

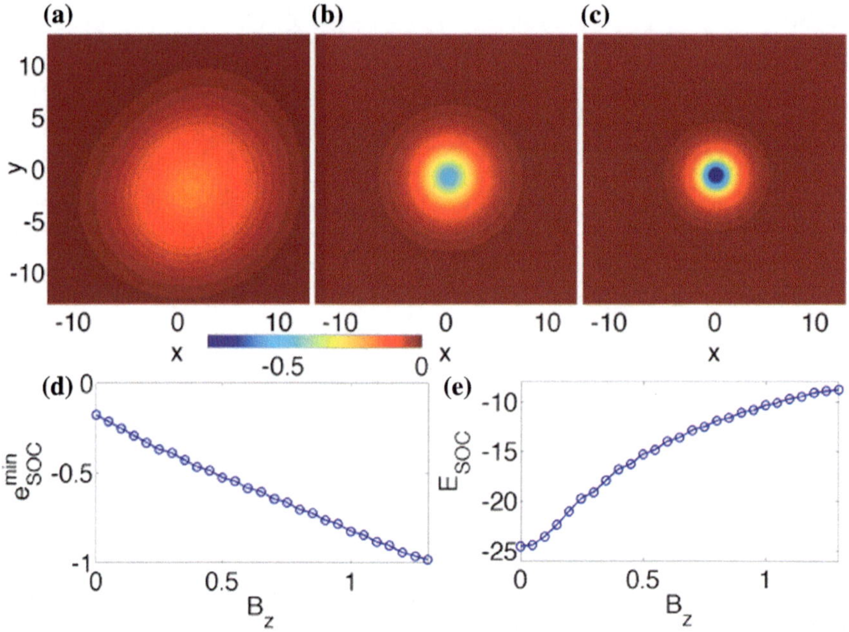

Fig. 10.19 The effect of the magnetic field B_z on the SOC energy $e_{SOC}(x, y)$ of the BECs under the parameter $C^{E_{min}}(0, -0.5)$. **a** $B_z = 0$. **b** $B_z = 0.5$. **c** $B_z = 1$. **d** The minimum values of e_{SOC} as a function of the magnetic field B_z. **e** The sum of the SOC energy E_{SOC} as a function of the magnetic field B_z. We set the SOC $\kappa_x = \kappa_y = \kappa = 0.5$ and the position of the vortex to be $(0, -0.5)$. The energy, magnetization, length, strength of the SOC, and strength of the interaction are in units of $\hbar\omega$, $\hbar\omega$, ξ, $\sqrt{\mu/m}$, and $\xi^3\mu$, respectively [24]

0, and the magnetic field B_z causes its value to decrease. In fact, if we examine the total SOC energy E_{SOC}, we find that it increases as the SOC goes up [see Fig. 10.19e].

The Effect of the Spin Interaction g_s on the Stability

The spin interaction g_s normally plays a key role in determining the properties of the spin-1 BECs. In the above text, we have assumed $g_s = 0$, i.e., $a_2 = a_0$ [$g_s = \frac{4\pi\hbar^2(a_2-a_0)}{3m}$]. We now consider the effect of g_s. According to the previous studies, we know $a_2 \approx a_0$. For example, the two parameters are $a_0 = 101.8a_B$ and $a_2 = 100.4a_B$ in the spin-1 BEC of ^{87}Rb, and are $a_0 = 50a_B$ and $a_2 = 55a_B$ in the spin-1 BEC of ^{23}Na. Thus, we can assume $g_s << g_n$.

Since the SOC and the magnetic field B_z are the two key points in our above analysis, we can assume that the spin interaction g_s would appear in the following four cases: (i) both the SOC and the magnetic field B_z are considered, (ii) only the SOC exists, (iii) both the SOC and the magnetic field B_z are turned off, and (iv) only the magnetic field B_z exists.

We have found that $(0, 0^-)$ is the stablest position if the half-quantum vortex can exist in the BEC stably. First, we test the energy of the BECs with the half-

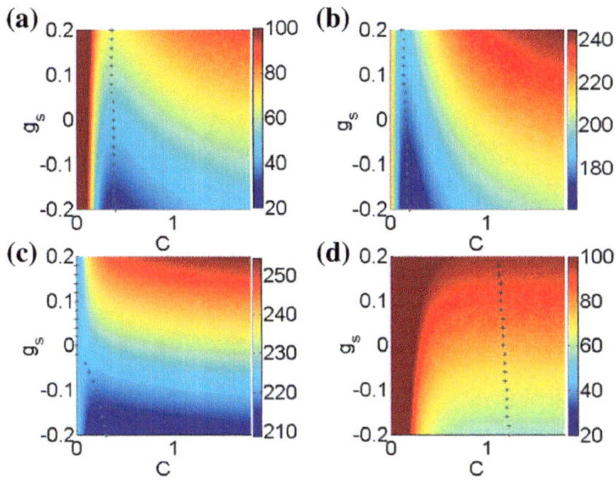

Fig. 10.20 The total energy of BECs as g_s and C change. We set the position of the vortex to be $(0, -0.5)$. **a** $\kappa_x = \kappa_y = \kappa = 1$ and $B_z = 0.5$. **b** $\kappa_x = \kappa_y = \kappa = 1$ and $B_z = 0$. **c** $\kappa_x = \kappa_y = \kappa = 0$ and $B_z = 0$. **d** $\kappa_x = \kappa_y = \kappa = 0$ and $B_z = 0.5$. The stars indicate the best C to create the lowest energy as g_s varies. The energy, magnetization, length, strength of the SOC, and strength of the interaction are in units of $\hbar\omega$, $\hbar\omega$, ξ, $\sqrt{\mu/m}$, and $\xi^3\mu$, respectively [24]

quantum vortex locating at $(0, -0.5)$. In Fig. 10.20a, we set $\kappa_x = \kappa_y = \kappa = 1$ and $B_z = 0.5$ and examine the total energy of the BECs. We can see the lowest energy with suitable C. We have marked them with stars. The lowest energy with $C > 0$ implies the possible stable solution of the half-quantum vortex in the BECs. Similarly, Fig. 10.20b is the result of the case (ii), in which we set $\kappa_x = \kappa_y = \kappa = 1$ and $B_z = 0$. The lowest energy with suitable C also exists (see the stars), no matter what the value of g_s is. Figure 10.20c shows the result of the case (iii) with $\kappa_x = \kappa_y = 0$ and $B_z = 0$. Obviously, when $g_s > 0$, the lowest energy approaches $C = 0$. When $g_s < 0$, we have the lowest energy with $C > 0$. This property indicates the half-quantum vortex is not stable in this system with $g_s > 0$. Thus, the ferromagnetic BECs ($g_s < 0$) are beneficial to keep the half-quantum vortex. For the case (iv), we only set the magnetic field $B_z = 0.5$ and turn off the SOC, i.e., $\kappa = 0$. In Fig. 10.20d, we can also find the lowest energy with $C > 0$.

Whether it is true that the lowest energy with $C > 0$ can keep the solution stable in Fig. 10.20? We now open the degree of freedom to let the vortex locate at different positions and compare the total energy of the BECs to examine whether the lowest energy can appear around $\left(0, 0^-\right)$. In Fig. 10.21, we show the total energy of the BECs when we change the position of the vortex to $(x_0 = 0, y_0 = Y)$. Figure 10.21a is the result of the case (i) with $g_s = 0.2$, which stands for a strong antiferromagnet. We indeed find that the lowest energy locates around $\left(0, 0^-\right)$. This result indicates that g_s does not affect the stability of the half-quantum vortex in Fig. 10.20a. In Fig. 10.21b, we also can find that the stablest region is around $\left(0, 0^-\right)$. However, we must point out that the half-quantum vortex of Eq. (10.11) cannot appear with $\sin[D] = 0$. Because the SOC would cause atoms to exchange among the $m_F = 0$

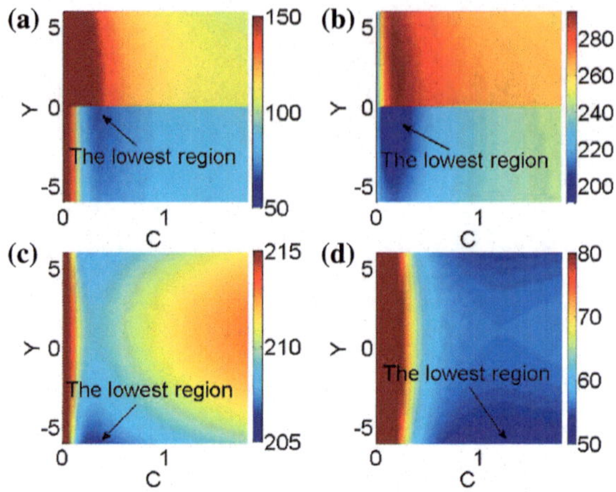

Fig. 10.21 The total energy of BECs as the position of the vortex ($x_0 = 0$, $y_0 = Y$) and C change. **a** $gs = 0.2$, $\kappa_x = \kappa_y = \kappa = 1$, and $B_z = 0.5$. **b** $g_s = 0 : 2$, $\kappa_x = \kappa_y = \kappa = 1$, and $B_z = 0$. **c** $g_s = -0.2$, $\kappa_x = \kappa_y = \kappa = 0$, and $B_z = 0$. **d** $g_s = -0.2$, $\kappa_x = \kappa_y = \kappa = 0$, and $B_z = 0.5$. The arrow indicates the stablest region, where the energy is the lowest. The energy, magnetization, length, strength of the SOC, and strength of the interaction are in units of $\hbar\omega$, $\hbar\omega$, ξ, $\sqrt{\mu/m}$, and $\xi^3\mu$, respectively [24]

and $m_F = +1$ component, it is hard to keep the $m_F = +1$ component approaching 0 without the magnet field B_z. In the previous studies, it has been proven that the pure SOC creates the plane wave phase and the stripe phase in BECs. Therefore, the stable vortex cannot be provided in the case (ii).

In Fig. 10.21c, we choose the strong ferromagnet BECs of $g_s = -0.2$. Our calculation indicates that the lowest energy region is not around $(0, 0^-)$ but around the boundary of the BECs (see the black arrow). Thus, we cannot obtain the stable half-quantum vortex without the SOC or the magnetic field B_z. Similarly, only with the magnetic field B_z, the lowest energy region also approaches the boundary of the BECs in Fig. 10.21d.

Generally speaking, with both the SOC and the magnetic field B_z, the half-quantum vortex can stably exist in the BECs. The adjustment of the spin interaction g_s could hardly affect the stability of the half-quantum vortex.

10.2.4 Discussion

We have studied the stability of the half-quantum spin vortex in the spin-1 SOC BECs, which are under the magnetic field. Especially in Fig. 10.15a, the minimum energy appears in the region of $y < 0$. In the discussion, we first discuss whether the half-quantum spin vortex is stable only in $y < 0$. In fact, the spin vortex can also

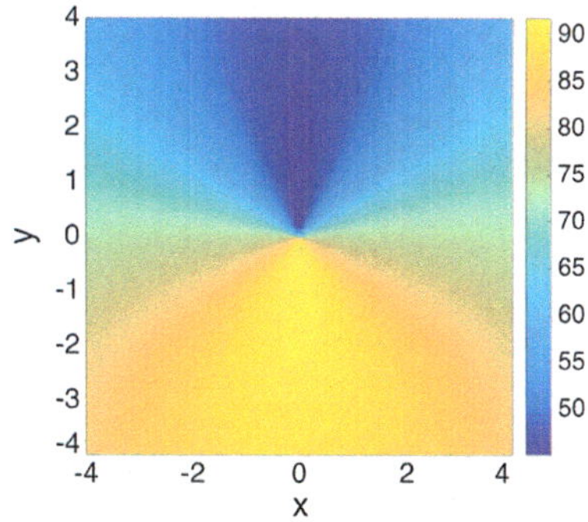

Fig. 10.22 The minimum energy E_{min} of the BECs as a function of the position of the vortex (x, y). We set $B_z = 0.5$ and $\kappa_x = \kappa_y = \kappa = -1$. The energy, magnetization, length, strength of the SOC, and strength of the interaction are in units of $\hbar\omega$, $\hbar\omega$, ξ, $\sqrt{\mu/m}$, and $\xi^3\mu$, respectively [24]

be stable in the region of $y > 0$ when $\kappa < 0$. In Fig. 10.22, we provide a case with $\kappa = -1$ and all other parameters are as the same as those in Fig. 10.15a. Obviously, the stable point, i.e., the position of the minimum energy, is in the region of $y > 0$. This case shows that the stable point of the spin vortex is affected by the SOC. And it is the SOC that causes the spin vortex to be deflected into a side of the $x - y$ plane.

Second, we illuminate a detail about setting the vortex in the BECs. Reference [6] provides a method to set vortex in the BEC. In fact, when applying the method, we need to set the corresponding image vortex, which has the reverse winding number outside of the BECs. Meanwhile, the vortex phase cut between π and $-\pi$ can extend fastest to the outskirts of the BECs when it connects the image vortex. The image vortex of the vortex is located at $r_0' = \frac{R^2 r_0}{|r_0|^2}$, where R is the radius of the BECs and $r_0' = (x_0', y_0')$. The corresponding phase is $\beta'(x, y) = \arctan\left[\frac{y-y'}{x-x'}\right]$. In our study, we set $R = 15$ which is a little larger than the size of the BECs and it can make the image vortex locate at the outside of the BECs. In Fig. 10.23a1 and b1, we display the phases with the vortex at the position of $(0, -0.5)$ and $(0, 0.5)$, respectively. Because the density outside of the BECs is set to be 0, the image vortex cannot be seen. But it changes the phases of the BECs. If we do not consider the image vortex, the phase cut between π and $-\pi$ would always parallel to the x-axis.

Third, we give a proof to further indicate how the SOC causes the vortex to deflect to one side of the $x - y$ plane. Take the two cases with spin vortex at the position of $(0, -0.5)$ and $(0, 0.5)$ as an example. Certainly, the size of the spin vortex in the two cases is the same, i.e., C is fixed and we only change the position of the spin vortex. In fact, we can obtain $\Psi_1^{(0,-0.5)}(x, y) = \Psi_1^{(0,0.5)}(x, y) = 0$, $\Psi_0^{(0,-0.5)}(x, y) = \Psi_0^{(0,0.5)}(-x, -y) \geq 0$, $\Psi_{-1}^{(0,-0.5)}(x, y) = \Psi_{-1}^{(0,0.5)}(-x, -y)$. The superscripts $(0, -0.5)$ and $(0, 0.5)$ denote the position of the vortex. Meanwhile, we have $\frac{\partial}{\partial x}\Psi_{-1}^{(0,-0.5)}(x, y) = -\frac{\partial}{\partial x}\Psi_{-1}^{(0,0.5)}(-x, -y)$ and $\frac{\partial}{\partial y}\Psi_{-1}^{(0,-0.5)}(x, y) = -\frac{\partial}{\partial y}\Psi_{-1}^{(0,0.5)}(-x, -y)$. For clearly displaying these features, Fig. 10.23a2 and a3 plots the real part and imaginary part of $\frac{\partial}{\partial x}\Psi_{-1}^{(0,-0.5)}(x, y)$, respectively. In addition,

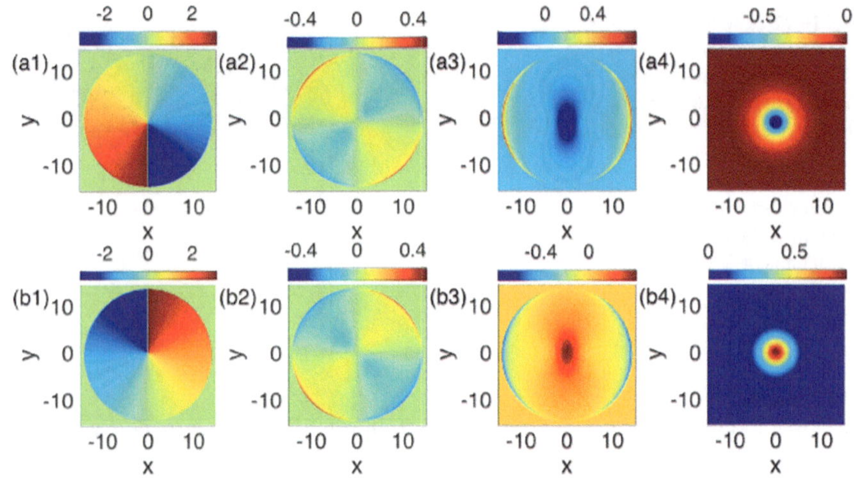

Fig. 10.23 **a1–a4** The results of the spin vortex with $C = 0.4$ at $(0, -0.5)$. **a1** Phase of $\Psi_{-1}^{(0,-0.5)}$. **a2** The real part of $\frac{\partial \Psi_{-1}^{(0,-0.5)}(x,y)}{\partial x}$. **a3** The imaginary part of $\frac{\partial \Psi_{-1}^{(0,-0.5)}(x,y)}{\partial x}$. **a4** $e_{SOC}^{(0,-0.5)}(x, y)$. **b1–b4** The results of the spin vortex with $C = 0.4$ at $(0, 0.5)$. **b1** Phase $\Psi_{-1}^{(0,0.5)}$. **b2** The real part of $\frac{\partial \Psi_{-1}^{(0,0.5)}(x,y)}{\partial x}$. **b3** The imaginary part of $\frac{\partial \Psi_{-1}^{(0,0.5)}(x,y)}{\partial x}$. **b4** $e_{SOC}^{(0,0.5)}(x, y)$. For all other parameters, we use $B_z = 0 : 5$, $B_\perp = 0$, and $\kappa_x = \kappa_y = \kappa = 1$. The energy, magnetization, length, strength of the SOC, and strength of the interaction are in units of $\hbar\omega$, $\hbar\omega$, ξ, $\sqrt{\mu/m}$, and $\xi^3\mu$, respectively [24]

Fig. 10.23b2 and b3 plots the real part and the imaginary part of $\frac{\partial}{\partial x}\Psi_{-1}^{(0,0.5)}(-x, -y)$, respectively. Thus, the SOC energy can be written as

$$\begin{aligned}
e_{SOC}^{(0,-0.5)}(x, y) &= \frac{\hbar\kappa}{m}\Psi_0^{(0,-0.5)}(x, y)\left[-i\frac{\partial \Psi_{-1}^{(0,-0.5)}(x, y)}{\partial x} - \frac{\partial \Psi_{-1}^{(0,-0.5)}(x, y)}{\partial y}\right] \\
&\quad + \frac{\hbar\kappa}{m}\left(\Psi_{-1}^{(0,-0.5)}(x, y)\right)^*\left[-i\frac{\partial \Psi_0^{(0,-0.5)}(x, y)}{\partial x} + \frac{\partial \Psi_0^{(0,-0.5)}(x, y)}{\partial y}\right] \\
&= \frac{\hbar\kappa}{m}\Psi_0^{(0,0.5)}(-x, -y)\left[i\frac{\partial \Psi_{-1}^{(0,0.5)}(-x, -y)}{\partial x} + \frac{\partial \Psi_{-1}^{(0,0.5)}(-x, -y)}{\partial y}\right] \\
&\quad + \frac{\hbar\kappa}{m}\left(\Psi_{-1}^{(0,0.5)}(-x, -y)\right)^*\left[i\frac{\partial \Psi_0^{(0,0.5)}(-x, -y)}{\partial x} + \frac{\partial \Psi_0^{(0,0.5)}(-x, -y)}{\partial y}\right] \\
&= -e_{SOC}^{(0,0.5)}(-x, -y).
\end{aligned} \tag{10.12}$$

In Fig. 10.23a4 and b4, we display the $e_{SOC}^{(0,-0.5)}(x, y)$ and $e_{SOC}^{(0,0.5)}(x, y)$, respectively. Obviously, the two figures well show the feature of Eq. (10.12). Thus, we can get

$$E_{SOC}^{(0,-0.5)}(x,y) = \int\int e_{SOC}^{(0,-0.5)}(x,y)dxdy = \int\int -e_{SOC}^{(0,0.5)}(-x,-y)dxdy$$
$$= \int\int -e_{SOC}^{(0,0.5)}(x,y)dxdy = -E_{SOC}^{(0,0.5)}(x,y).$$

Similarly, we can prove the identical relations $e_{SOC}^{(x_0,y_0)}(x,y) = -e_{SOC}^{(-x_0,-y_0)}(-x,-y)$ and $E_{SOC}^{(x_0,y_0)}(x,y) = -E_{SOC}^{(-x_0,-y_0)}(x,y)$, where (x_0, y_0) is the position of the spin vortex. That is the reason why the SOC inevitably causes the half-quantum spin vortex to deflect to one side of the plane.

10.3 Topological Defects in the Ground States of SU(3) SO-Coupled Bose–Einstein Condensates

The present section deals with the investigation of a type of topological defects, namely, the double-quantum spin vortices characterized by doubly quantized circulating spin currents and unmagnetized filled cores, in the ground states of SU(3) SO-coupled BECs. We mainly show that the SU(3) SOC and spin–exchange interaction play important roles in determining the ground state phase diagram.

10.3.1 SU(3) SOC

We consider the $F = 1$ spinor BECs with SU(3) SOC. Using the mean-field approximation, the Hamiltonian can be written in the GP form as

$$\mathcal{H} = \int d\mathbf{r}\left[\Psi^{\dagger}\left(-\frac{\hbar^2}{2m}\nabla^2 + \nu_{so}\right)\Psi + \frac{c_0}{2}n^2 + \frac{c_2}{2}|\mathbf{F}|^2\right], \tag{10.13}$$

where the order parameter $\Psi = (\Psi_1(\mathbf{r}), \Psi_0(\mathbf{r}), \Psi_{-1}(\mathbf{r}))^T$ is normalized with the total particle number $N = \int d\mathbf{r}\Psi^{\dagger}\Psi$. The particle density is $n = \sum_{j=0,\pm1}\Psi_j^*(\mathbf{r})\Psi_j(\mathbf{r})$, and the spin density vector $\mathbf{F} = (F_x, F_y, F_z)$ is defined by $F_{\nu}(\mathbf{r}) = \Psi^{\dagger} f_{\nu}\Psi$, with $\mathbf{f} = (f_x, f_y, f_z)$ being the vector of the spin-1 matrices given in the irreducible representation [9, 15, 41]. The SOC term is chosen as $\nu_{so} = \kappa\lambda \cdot \mathbf{p}$, where κ is the spin–orbit coupling strength, $\mathbf{p} = (p_x, p_y)$ represents 2D momentum, and $\lambda = (\lambda_x, \lambda_y)$ is expressed in terms of $\lambda_x = \lambda^{(1)} + \lambda^{(4)} + \lambda^{(6)}$ and $\lambda_y = \lambda^{(2)} - \lambda^{(5)} + \lambda^{(7)}$, with $\lambda(i)$ $(i = 1, \ldots, 8)$ being the Gell-Mann matrices, i.e., the generators of the SU(3) group [3]. Note that the SU(3) SO coupling term in the Hamiltonian involves all the pairwise couplings between the three states. This is distinct from the previously discussed SU(2) SO coupling in spinor BECs, where the states $\Psi_1(\mathbf{r})$ and $\Psi_{-1}(\mathbf{r})$ are coupled indirectly. The parameters c_0 and c_2 describe the strengths of density and spin–exchange interactions, respectively.

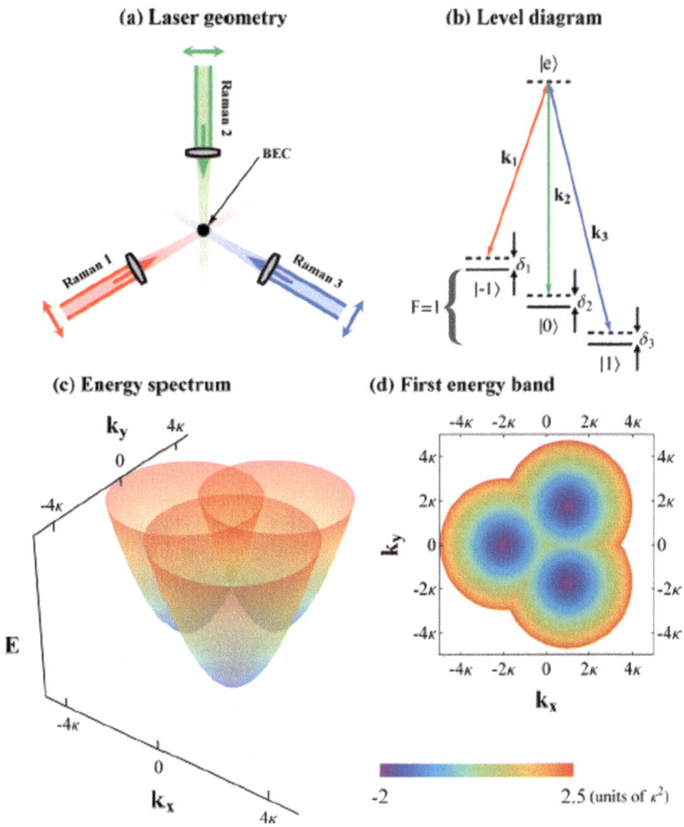

Fig. 10.24 Scheme for creating SU(3) spin–orbit coupling in spinor BECs. **a** Laser geometry. Three laser beams with different frequencies and polarizations, intersecting at an angle of $2\pi/3$, illuminate the cloud of atoms. **b** Level diagram. Each of the three Raman lasers dresses one hyperfine Zeeman level from $|F=1, m_F=1\rangle$, $|F=1, m_F=0\rangle$, and $|F=1, m_F=-1\rangle$ of the ^{87}Rb $5S_{1/2}$, $F=1$ ground state. δ_1, δ_2, and δ_3 correspond to the detuning in the Raman transitions. **c** Triple-well dispersion relation. The SU(3) spin–orbit coupling induces three discrete minima of the single-particle energy band on the vertices of an equilateral triangle in the $k_x - k_y$ plane. **d** Projection of the first energy band on a 2D plane. Units with $\hbar = m = 1$ are used for simplicity. Reprint from Ref. [9], Copyright 2018, with permission from American Physical Society

The Hamiltonian with SU(3) SOC can be realized using a similar method of Raman dressing as in the SU(2) case [21]. As shown in Fig. 10.24a, three laser beams with different polarizations and frequencies, intersecting at an angle of $2\pi/3$, are used for the Raman coupling. Each of the three Raman lasers dresses one hyperfine spin state from the $F=1$ manifold ($|F=1, m_F=1\rangle$, $|F=1, m_F=0\rangle$, and $|F=1, m_F=-1\rangle$) to the excited state $|e\rangle$ [see Fig. 10.24b]. When the standard rotating wave approximation is used and the excited state is adiabatically eliminated due to far detuning, one can obtain the effective Hamiltonian in Eq. (10.13) [9].

By diagonalizing the kinetic energy and SOC terms, we can obtain the single-particle energy spectrum, which can provide useful information about the ground state of Bose condensates. For the SU(2) case, it is known that the single-particle spectrum with the NIST-type SOC acquires either a single or two minima, depending on the strength of the Raman coupling [21], while for the case of Rashba-type, there exist an infinite number of minima locating on a continuous ring in momentum space. For the SU(3) SOC discussed here, we find that there are in general three discrete minima residing on the vertices of an equilateral triangle [see Fig. 10.24c, d]. This unique property of the energy band implies the possibility of a threefold-degenerate many-body magnetized state or a topologically nontrivial lattice state, depending on the choices among the three minima made by the many-body interactions.

10.3.2 Phase Diagram

Now, we discuss the phase diagram of the many-body ground states. For the case of SU(2) SOC, it is shown that two many-body ground states, the magnetized state and stripe state, can be stabilized in a homogeneous system [40]. Although the Rashba SOC provides infinite degenerate minima in the single-particle spectrum, a many-body ground state condensed in one or two points in momentum space is always energetically favorable due to the presence of spin–exchange interaction [40]. As a result, a lattice state with the condensates occupying three or more momentum points for SU(2) SOC is unstable, unless a strong harmonic trap is introduced [40].

For the present case of SU(3) SOC, we first analytically calculate the possible ground states using a variational approach with a trial wave function $\Psi = \alpha_1\Psi_1 + \alpha_2\Psi_2 + \alpha_3\Psi_3$, where

$$\Psi_1 = \frac{1}{\sqrt{3}}(1, 1, 1)^T \exp[-2i\kappa x], \tag{10.14a}$$

$$\Psi_2 = \frac{1}{\sqrt{3}}\left(\exp\left[-i\frac{\pi}{3}\right], \exp\left[i\frac{\pi}{3}\right], \exp[i\pi]\right)\exp\left[i\kappa(x-\sqrt{3}y)\right], \tag{10.14b}$$

$$\Psi_3 = \frac{1}{\sqrt{3}}\left(\exp\left[i\frac{\pi}{3}\right], \exp\left[-i\frac{\pi}{3}\right], \exp[i\pi]\right)\exp\left[i\kappa(x+\sqrt{3}y)\right], \tag{10.14c}$$

correspond to the many-body states with all particles condensing on one of the three minima of the single-particle spectrum, and $\alpha_{i=1,2,3}$ are expansion coefficients. Substituting Eqs. (10.14a)–(10.14c) into the interaction energy functional

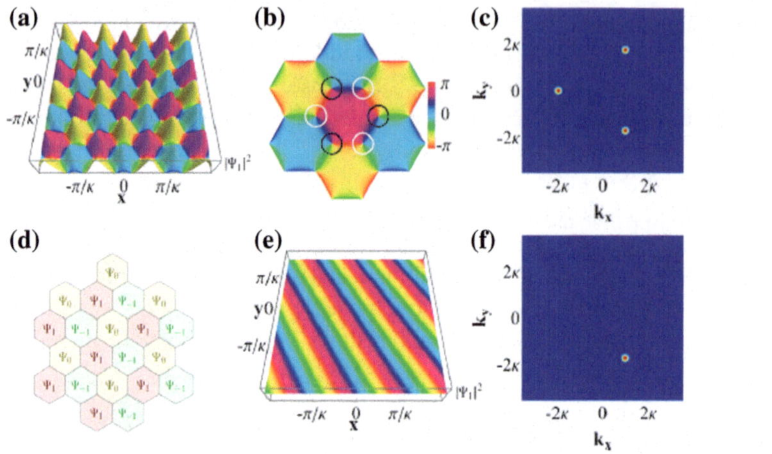

Fig. 10.25 Two distinct phases present in SU(3) spin–orbit-coupled BECs. **a–d** The topologically nontrivial lattice phase for antiferromagnetic spin interaction ($c_2 > 0$) with **a** the density and phase of 1 represented by heights and colors, **b** the phase within one unit cell showing the positions of vortices (white circles) and antivortices (black circles), **c** the corresponding momentum distributions, and **d** the structural schematic drawing of the phase separation. **e–f** The threefold-degenerate magnetized phase for ferromagnetic spin interaction ($c_2 < 0$) with **e** the density and phase distributions of Ψ_1 and **f** the corresponding momentum distributions. Reprint from Ref. [9], Copyright 2018, with permission from American Physical Society

$$E = \int d\mathbf{r} \left(\frac{c_0}{2} n^2 + \frac{c_2}{2} |\mathbf{F}|^2 \right), \tag{10.15}$$

one obtains

$$\frac{E}{N} = \left(\frac{c_0}{2} + \frac{4c_2}{9} \right) \widetilde{n} - \frac{7c_2}{9\widetilde{n}} \sum_{i \neq j} |\alpha_i| \left| \alpha_j \right|, \tag{10.16}$$

where $\widetilde{n} = |\alpha_1|^2 + |\alpha_2|^2 + |\alpha_3|^2$ is the mean particle density. By minimizing the interaction energy with respect to the variation of $|\alpha_i|^2$, one finds that the spin–exchange interaction plays an important role in determining the phase diagram.

When $c_2 > 0$, it favors $|\alpha_1|^2 = |\alpha_2|^2 = |\alpha_3|^2 = \widetilde{n}/3$, indicating that the ground state is a triangular lattice phase with an equally weighted superposition of the three single-particle minima. On the other hand, as $c_2 < 0$, the system prefers a state with either $|\alpha_1|^2 = \widetilde{n}$, $|\alpha_2|^2 = |\alpha_3|^2 = 0$, or $|\alpha_2|^2 = \widetilde{n}$, $|\alpha_1|^2 = |\alpha_3|^2 = 0$, or $|\alpha_3|^2 = \widetilde{n}$, $|\alpha_1|^2 = |\alpha_2|^2 = 0$, indicating that the ground state occupies one single minimum in the momentum space and corresponds to a threefold-degenerate magnetized phase.

Note that the variational wave function equations (10.14a)–(10.14c) is a good starting point as the SOC is strong enough to dominate the chemical potential. For the case with weak SOC, one must rely on numerical simulations to determine the many-body ground state. In such a situation, we find a stripe phase with two minima in momentum space occupied for $c_2 >> \kappa^2$, which will be discussed later.

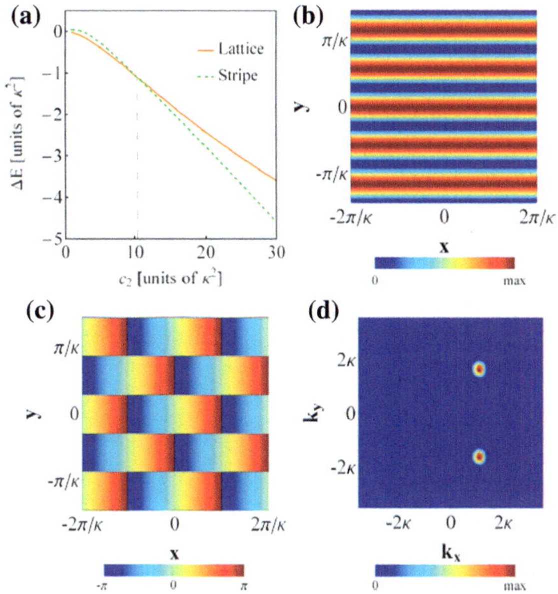

Fig. 10.26 **a** Energy comparison between the lattice and stripe phases. The energy difference ΔE between the numerical simulation and the variational calculation are shown by solid (lattice state) and dashed (stripe state) lines. **b–d** The ground state density, phase, and momentum distributions of the stripe phase with the parameters $c_2 = 20\kappa^2$ and $c_0 = 10c_2$. Reprint from Ref. [9], Copyright 2018, with permission from American Physical Society

The many-body ground states can be numerically obtained by minimizing the energy functional associated with the Hamiltonian Eq. (10.13) via the imaginary time evolution method. It is found that the numerical results are consistent with the analytical analysis discussed above for rather weak interaction with $c_2 \lesssim \kappa^2$. Figure 10.25 illustrates the two possible ground states of spinor BECs with SU(3) SOC. When $c_2 > 0$, the three components are immiscible and arranged as an interlaced triangular lattice with the spatial translational symmetry spontaneously broken [see Fig. 10.25a–d]. This lattice is topologically nontrivial and embedded by vortices and antivortices as shown in Fig. 10.25b. From this result, we conclude that a lattice phase can be stabilized in a uniform SU(3) SO-coupled BEC, which is in clear contrast to the SU(2) case where a strong harmonic trap is required [42]. More details on the structure of vortices as well as their unique spin configurations will be investigated later. On the other hand, as $c_2 < 0$, the three components are miscible, and the system forms a magnetized phase with the spatial transitional symmetry preserved but the time-reversal symmetry broken [see Fig. 10.25e, f]. This magnetized phase occupies one of the three minima of the single-particle spectrum by spontaneous symmetry breaking and hence is threefold degenerate instead of doubly degenerate in the SU(2) case [19].

For strong antiferromagnetic spin interaction with $c_2 >> \kappa^2$, however, a stripe phase is identified with two of three minima occupied in the momentum space. We take the states with two or three minima occupied in the momentum space as trial wave functions and perform an imaginary time evolution to find their respective optimized ground state energy. A typical set of results is summarized in Fig. 10.26a, showing the energy comparison with different values of interatomic interactions. Obviously, one finds that the stripe phase will have lower energy than the lattice phase when the interatomic interaction exceeds a critical value. Due to the finite momentum in vertical direction of the stripe [see Fig. 10.26d], both the spatial translational and time-reversal symmetries are broken [see Fig. 10.26b and c]. This is distinct from the stripe phase induced by SU(2) SOC, where the time-reversal symmetry is preserved [40].

10.3.3 Phase Requirement

The vortex configuration of spinor BECs depends on the phase relation between the three components. We next discuss the influence of SOC on the phase requirement of the vortex configuration. We first assume that the spinor order parameter of a vortex in the polar coordinate (r, θ) can be described as

$$\psi_j(r,\theta) = \phi_j \exp\left[i\omega_j\theta + \alpha_j\right], \tag{10.17}$$

where $\phi_j \geq 0$, $j = 0, \pm 1$.

Without Spin–Orbit Coupling

In the absence of SOC, the phase-dependent terms in the Hamiltonian are

$$H^{\text{phase}} = E_{kin}^{\text{phae}} + E_{int}^{\text{phae}} = -\frac{1}{2}\int \Psi^* \frac{1}{r^2}\frac{\partial^2}{\partial\theta^2}\Psi d\mathbf{r} + 2c_2 \int \text{Re}\left[\psi_1^*\psi_{-1}^*\psi_0^2\right] d\mathbf{r}, \tag{10.18}$$

where the first term results from the kinetic energy and the second from the spin–exchange interaction. Substituting Eq. (10.17) into Eq. (10.18), one obtains

$$E_{kin}^{\text{phae}} = \sum_{j=1,0,-1} \omega_j^2 \int \frac{\pi\phi_j^2}{r} dr, \tag{10.19a}$$

$$E_{int}^{\text{phae}} = 2c_2 \int \phi_1\phi_{-1}\phi_0^2 r dr \int \cos\left[(\omega_1 - 2\omega_0 + \omega_{-1})\,\theta + (\alpha_1 - 2\alpha_0 + \alpha_{-1})\right] d\theta. \tag{10.19b}$$

It is easy to read from Eq. (10.19a) that the system favors small winding numbers energetically. Moreover, from Eq. (10.19b) the energy minimization requires the winding number and phase to satisfy the following relations:

$$\omega_1 - 2\omega_0 + \omega_{-1} = 0, \tag{10.20a}$$

$$\alpha_1 - 2\alpha_0 + \alpha_{-1} = n\pi, \tag{10.20b}$$

where n is odd for $c_2 > 0$ and even for $c_2 < 0$. The phase requirement of Eq. (10.20a) indicates that the following types of winding combination, such as $\langle \pm 1, \times, 0 \rangle$, $\langle 0, \times, \pm 1 \rangle$, $\langle \pm 1, 0, \mp 1 \rangle$, $\langle \pm 1, \pm 1, \pm 1 \rangle$, $\langle \pm 2, \pm 1, 0 \rangle$, and $\langle 0, \pm 1, \pm 2 \rangle$, are allowed in a spinor BEC, where the symbol "$\times$" denotes the absence of the Ψ_0 component.

With SU(2) Spin–Orbit Coupling

For the case of SU(2) SOC, we take the Rashba-type as an example and write the Hamiltonian as

$$E_{SOC} = \int \kappa \psi^\dagger \begin{pmatrix} 0 & -i\frac{\partial}{\partial x} - \frac{\partial}{\partial y} & 0 \\ -i\frac{\partial}{\partial x} + \frac{\partial}{\partial y} & 0 & -i\frac{\partial}{\partial x} - \frac{\partial}{\partial y} \\ 0 & -i\frac{\partial}{\partial x} + \frac{\partial}{\partial y} & 0 \end{pmatrix} \psi d\mathbf{r}, \tag{10.21}$$

where $\psi = (\psi_1, \psi_0, \psi_{-1})^T$. Substituting Eq. (10.17) into Eq. (10.21) yields

$$\begin{aligned} E_{SOC} = \int dr d\theta \Big[&\left(\phi_0 r \frac{\partial \phi_1}{\partial r} - \omega_1 \phi_0 \phi_1\right) \exp\left[i\left\{(\omega_1 - \omega_0 + 1)\theta + \left(\alpha_1 - \alpha_0 - \frac{\pi}{2}\right)\right\}\right] \\ &- \left(\phi_1 r \frac{\partial \phi_0}{\partial r} + \omega_0 \phi_1 \phi_0\right) \exp\left[-i\left\{(\omega_1 - \omega_0 + 1)\theta + \left(\alpha_1 - \alpha_0 - \frac{\pi}{2}\right)\right\}\right] \\ &+ \left(\phi_0 r \frac{\partial \phi_{-1}}{\partial r} + \omega_{-1} \phi_0 \phi_{-1}\right) \exp\left[-i\left\{(\omega_{-1} - \omega_0 - 1)\theta + \left(\alpha_{-1} - \alpha_0 - \frac{\pi}{2}\right)\right\}\right] \\ &- \left(\phi_{-1} r \frac{\partial \phi_0}{\partial r} - \omega_0 \phi_{-1} \phi_0\right) \exp\left[-i\left\{(\omega_{-1} - \omega_0 - 1)\theta + \left(\alpha_{-1} - \alpha_0 - \frac{\pi}{2}\right)\right\}\right]\Big]. \end{aligned} \tag{10.22}$$

In order to minimize the SOC energy, it is preferred that

$$\omega_1 - \omega_0 + 1 = 0, \tag{10.23}$$

$$\omega_{-1} - \omega_0 - 1 = 0, \tag{10.24}$$

$$\alpha_1 - \alpha_0 - \frac{\pi}{2} = m\pi, \tag{10.25}$$

$$\alpha_{-1} - \alpha_0 - \frac{\pi}{2} = n\pi. \tag{10.26}$$

Then the SOC energy is rewritten as

$$\begin{aligned} E_{SOC} &= 2\pi \int \left[\phi_0 r \frac{\partial \phi_1}{\partial r} - \phi_1 r \frac{\partial \phi_0}{\partial r} - (\omega_1 + \omega_0)\, \phi_0 \phi_1 \right] dr \cos[m\pi] \\ &+ 2\pi \int \left[\phi_0 r \frac{\partial \phi_{-1}}{\partial r} - \phi_{-1} r \frac{\partial \phi_0}{\partial r} + (\omega_{-1} + \omega_0)\, \phi_0 \phi_{-1} \right] dr \cos[n\pi], \end{aligned} \tag{10.27}$$

where m and n are odd or even, which can be determined by minimizing the energy expressed in Eq. (10.27). It is found that the SU(2) SO coupling does not violate the ordinary requirement on the winding combination in Eq. (10.20a) but introduces further requirements in Eqs. (10.23) and (10.24). As a result, while $\langle -1, 0, 1\rangle$, $\langle -2, -1, 0\rangle$, and $\langle 0, 1, 2\rangle$ are still allowed, some winding combinations such as $\langle \pm 1, \pm 1, \pm 1\rangle$, $\langle \pm 1, \times, 0\rangle$, $\langle 0, \times, \pm 1\rangle$, $\langle 1, 0, -1\rangle$, $\langle 2, 1, 0\rangle$, and $\langle 0, -1, -2\rangle$ are forbidden. Obviously, one can see that the SOC breaks the chiral symmetry, thus may lead to chiral spin textures.

With SU(3) Spin–Orbit Coupling

For the case of SU(3) SOC, the effective Hamiltonian can be written as

$$E_{SOC} = \int \kappa \psi^{\dagger} \begin{pmatrix} 0 & -i\frac{\partial}{\partial x} - \frac{\partial}{\partial y} & -i\frac{\partial}{\partial x} + \frac{\partial}{\partial y} \\ -i\frac{\partial}{\partial x} + \frac{\partial}{\partial y} & 0 & -i\frac{\partial}{\partial x} - \frac{\partial}{\partial y} \\ -i\frac{\partial}{\partial x} - \frac{\partial}{\partial y} & -i\frac{\partial}{\partial x} + \frac{\partial}{\partial y} & 0 \end{pmatrix} \psi d\mathbf{r}. \tag{10.28}$$

Substituting Eq. (10.17) into Eq. (10.28), we get

$$\begin{aligned} E_{SOC} = \int dr d\theta \Bigg[&\left(\phi_0 r \frac{\partial \phi_1}{\partial r} - \omega_1 \phi_0 \phi_1 \right) \exp\left[i\left\{ [(\omega_1 - \omega_0 + 1)\theta + (\alpha_1 - \alpha_0 - \pi/2) \right\} \right] \\ &- \left(\phi_1 r \frac{\partial \phi_0}{\partial r} + \omega_0 \phi_1 \phi_0 \right) \exp\left[-i \left\{ (\omega_1 - \omega_0 + 1)\theta + (\alpha_1 - \alpha_0 - \pi/2) \right\} \right] \\ &+ \left(\phi_0 r \frac{\partial \phi_{-1}}{\partial r} + \omega_{-1} \phi_0 \phi_{-1} \right) \exp\left[i \left\{ (\omega_{-1} - \omega_0 - 1)\theta + (\alpha_{-1} - \alpha_0 - \pi/2) \right\} \right] \\ &- (\phi_{-1} r \frac{\partial \phi_0}{\partial r} - \omega_0 \phi_{-1} \phi_0) \exp\left[-i \left\{ (\omega_{-1} - \omega_0 - 1)\theta + (\alpha_{-1} - \alpha_0 - \pi/2) \right\} \right] \\ &+ \left(\phi_{-1} r \frac{\partial \phi_1}{\partial r} + \omega_1 \phi_{-1} \phi_1 \right) \exp\left[i \left\{ (\omega_1 - \omega_{-1} - 1)\theta + (\alpha_1 - \alpha_{-1} - \pi/2) \right\} \right] \\ &- \left(\phi_1 r \frac{\partial \phi_{-1}}{\partial r} - \omega_{-1} \phi_1 \phi_{-1} \right) \exp\left[-i \left\{ (\omega_1 - \omega_{-1} - 1)\theta + (\alpha_1 - \alpha_{-1} - \pi/2) \right\} \right] \Bigg]. \end{aligned} \tag{10.29}$$

By minimizing the SOC energy, one obtains the following relations:

$$\omega_1 - \omega_0 + 1 = 0, \tag{10.30a}$$

$$\omega_{-1} - \omega_0 - 1 = 0, \tag{10.30b}$$

$$\omega_1 - \omega_{-1} - 1 = 0, \tag{10.30c}$$

$$\alpha_1 - \alpha_0 - \pi/2 = m\pi, \tag{10.30d}$$

$$\alpha_{-1} - \alpha_0 - \pi/2 = n\pi, \tag{10.30e}$$

$$\alpha_1 - \alpha_{-1} - \pi/2 = l\pi. \tag{10.30f}$$

Then the SOC energy can be rewritten as

$$\begin{aligned} E_{soc} = 2\pi \int \left[\phi_0 r \frac{\partial \phi_1}{\partial r} - \phi_1 r \frac{\partial \phi_0}{\partial r} - (\omega_1 + \omega_0)\phi_0\phi_1]dr \cos[m\pi] \right] \\ +2\pi \int \left[\phi_0 r \frac{\partial \phi_{-1}}{\partial r} - \phi_{-1} r \frac{\partial \phi_0}{\partial r} + (\omega_{-1} + \omega_0)\phi_0\phi_{-1}]dr \cos[n\pi] \right] \\ +2\pi \int \left[\phi_{-1} r \frac{\partial \phi_1}{\partial r} - \phi_1 r \frac{\partial \phi_{-1}}{\partial r} + (\omega_1 + \omega_{-1})\phi_{-1}\phi_1]dr \cos[l\pi] \right], \end{aligned} \tag{10.31}$$

where m, n, and l are odd or even, which can be determined from Eq. (10.31). However, the three winding requirements Eqs. (10.30a)–(10.30c) cannot be satisfied simultaneously. Thus, the SU(3) SOC may choose two out of the three winding requirements for the following three cases:

Case I.

$$\omega_1 - \omega_0 + 1 = 0, \tag{10.32a}$$

$$\omega_{-1} - \omega_0 - 1 = 0, \tag{10.32b}$$

$$\alpha_1 - \alpha_0 - \pi/2 = m\pi, \tag{10.32c}$$

$$\alpha_{-1} - \alpha_0 - \pi/2 = n\pi. \tag{10.32d}$$

Case II.

$$\omega_1 - \omega_0 + 1 = 0, \tag{10.33a}$$

$$\omega_1 - \omega_{-1} - 1 = 0, \tag{10.33b}$$

$$\alpha_1 - \alpha_0 - \pi/2 = m\pi, \tag{10.33c}$$

$$\alpha_1 - \alpha_{-1} - \pi/2 = l\pi. \tag{10.33d}$$

Case III.

$$\omega_{-1} - \omega_0 - 1 = 0, \tag{10.34a}$$
$$\omega_1 - \omega_{-1} - 1 = 0, \tag{10.34b}$$

$$\alpha_{-1} - \alpha_0 - \pi/2 = n\pi, \tag{10.34c}$$
$$\alpha_1 - \alpha_{-1} - \pi/2 = l\pi. \tag{10.34d}$$

For **case I**, the winding combination $\langle -1, 0, 1\rangle$ is allowed, while $\langle 1, 0, -1\rangle$ is not allowed, indicating the chiral symmetry is broken. For **case II** and **case III**, one can find that the SU(3) SOC breaks the ordinary requirement on the winding combination in Eq. (10.20a); thus new winding combinations, such as $\langle 0, 1, -1\rangle$ and $\langle 1, -1, 0\rangle$, are possible.

10.3.4 Vortex Configurations

The vortex configurations of spinor BECs can be classified according to the combination of winding numbers and the magnetization of vortex core. For example, a Mermin–Ho vortex has winding combination $\langle \pm 2, \pm 1, 0\rangle$ with a ferromagnetic core, where the plus and minus signs represent different chirality of the vortices, and the expression of $\langle \omega_1, \omega_0, \omega_{-1}\rangle$ indicates that the components of Ψ_1, Ψ_0, and Ψ_{-1} in the wave function acquire winding numbers of ω_1, ω_0, and ω_{-1}, respectively. Using this notation, a polar-core vortex has winding combination $\langle \pm 1, 0, \mp 1\rangle$ with an antiferromagnetic core, and a half-quantum vortex has winding combination $\langle \pm 1, \times, 0\rangle$ with a ferromagnetic core, where the symbol "×" denotes the absence of the Ψ_0 component.

In the lattice phase induced by the SU(3) SOC with antiferromagnetic spin interaction, there exist three types of vortices: one is a polar-core vortex with winding combination $\langle -1, 0, 1\rangle$, and the other two are ferromagnetic-core vortices with winding combinations $\langle 1, -1, 0\rangle$ and $\langle 0, 1, -1\rangle$ [see Fig. 10.27a]. However, the vortex configurations with opposite chirality of each type, such as $\langle 1, 0, -1\rangle$, $\langle -1, 1, 0\rangle$, and $\langle 0, -1, 1\rangle$, are not allowed, because the chiral symmetry is intrinsically broken in SU(3) SO-coupled systems.

Surprisingly, one finds that the two types of ferromagnetic-core vortices $\langle 1, -1, 0\rangle$ and $\langle 0, 1, -1\rangle$ violate the conventional phase requirement $2\omega_0 = \omega_1 + \omega_{-1}$ for ordinary spinor BECs. This can be understood by noting that the relative phase among different wave function components is no longer uniquely determined by the spin–exchange interaction but also affected by the SU(3) SOC. Thus, the interlaced arrangement of the three types of vortices forms a class of vortex lattice which has no analog in systems without SOC.

The configurations of the three types of vortices induced by the SU(3) SO coupling with antiferromagnetic interaction are essentially different from those usually observed in ferromagnetic spinor BECs, as can be illustrated by the spherical har-

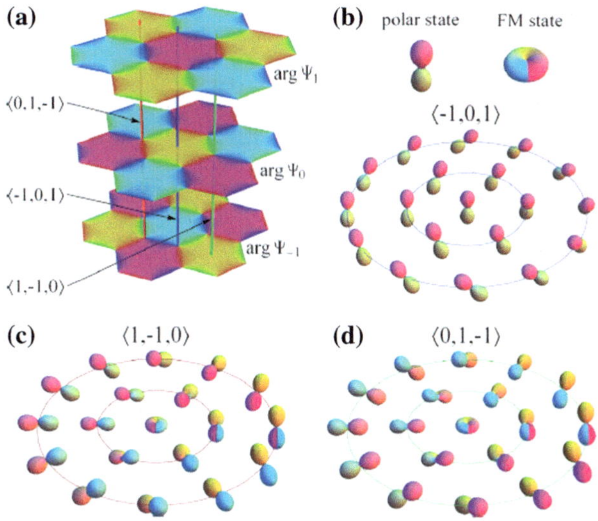

Fig. 10.27 Vortex configurations in antiferromagnetic spinor BECs with SU(3) spin–orbit coupling. **a** Vortex arrangement among the three components of the condensates. One can identify three types of vortices, including a polar-core vortex with winding combination $\langle -1, 0, 1\rangle$ (blue line) and two ferromagnetic-core vortices with winding combinations $\langle 1, -1, 0\rangle$ (green line) and $\langle 0, 1, -1\rangle$ (red line). **b**–**d** Spherical harmonic representation of the three types of vortices. The surface plots of $|\Phi(\theta, \phi)|^2$ for **b** the polar-core vortex $\langle -1, 0, 1\rangle$, **c** the ferromagnetic-core vortex $\langle 1, -1, 0\rangle$, and **d** the ferromagnetic-core vortex $\langle 0, 1, -1\rangle$ are shown with the colors representing the phase of $\Phi(\theta, \phi)$. Here, $\Phi(\theta, \phi) = \sum_{m=-1}^{m=1} Y_{1m}(\theta, \phi)\ \Psi_m$, and Y_{1m} is the rank-1 spherical harmonic function. Reprint from Ref. [9], Copyright 2018, with permission from American Physical Society

monic representation [15]. From Fig. 10.27b–d, one can find that for the polar-core vortex, the antiferromagnetic order parameter varies continuously everywhere, while for the ferromagnetic-core vortex, the magnetic order parameter acquires a singularity at the vortex core. In contrast, in the ordinary ferromagnetic spinor BECs, the ferromagnetic order parameter varies continuously everywhere for the ferromagnetic-core vortex but has a singularity at the core for the polar-core vortex [15].

10.3.5 Double-Quantum Spin Vortices

A spin vortex is a complex topological defect resulting from symmetry breaking, characterized by zero net mass current and quantized spin current around an unmagnetized core [15]. It is not only different from the magnetic vortex found in magnetic thin films but also from the 2D Skyrmion due to the existence of singularity in the spin textures. A single-quantum spin vortex with the spin current showing one quantum of circulation has been experimentally observed in ferromagnetic spinor BECs.

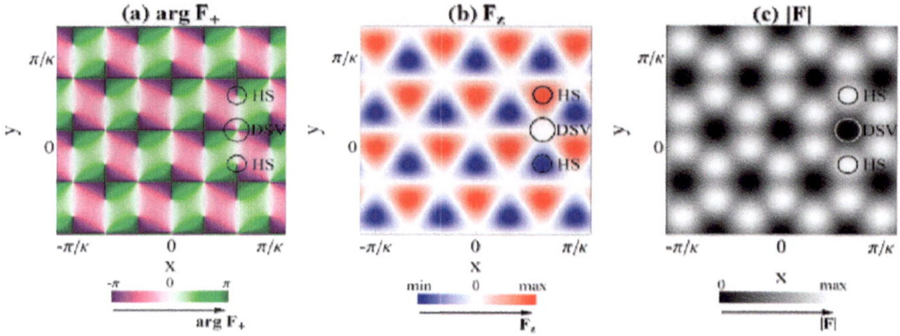

Fig. 10.28 Double-quantum spin vortex in antiferromagnetic spinor BECs with SU(3) spin–orbit coupling. **a** Spatial maps of the transverse magnetization with colors indicating the magnetization orientation. **b** Longitudinal magnetization. **c** Amplitude of the total magnetization $|\mathbf{F}|$. Two kinds of topological defects, double-quantum spin vortex (DSV) and half-Skyrmion (HS) are marked by big and small circles, respectively. The transverse magnetization orientation $\arg\left[F_+\right]$ along a closed path (indicated by big circles) surrounding the unmagnetized core shows a net winding of 4π, revealing the presence of a double-quantum spin vortex. Reprint from Ref. [9], Copyright 2018, with permission from American Physical Society

Multi-quantum spin vortices with l (l 2) quanta circulating spin current, however, are considered to be topologically unstable and have not been discovered yet [15].

A particularly important finding of our present work is that the polar-core vortex in the lattice phase has a spin current with two quanta of circulation around the unmagnetized core and hence can be identified as a double-quantum spin vortex. Figure 10.28 presents the transverse magnetization $F_+ = F_x + iF_y$, longitudinal magnetization F_z, and amplitude of the total magnetization $|\mathbf{F}|$ in the lattice phase, which are experimentally observable by a magnetization-sensitive phase-contrast imaging technique [Other experimental techniques, such as selective absorption and time-of-flight measurement, can also be used to observe the double-quantum spin vortex states. We suggest the magnetization-sensitive phase-contrast imaging method, just because it has an advantage in observing the phase information of the spin vortex structure]. From these results, one can find two distinct types of topological defects: the double-quantum spin vortex (DSV) and half-Skyrmion (HS), which correspond to the polar-core vortex with winding combinations $\langle -1, 0, 1\rangle$ and the ferromagnetic-core vortex with winding combinations $\langle 1, -1, 0\rangle$ or $\langle 0, 1, -1\rangle$, respectively. In particular, for the double-quantum spin vortex, the core is unmagnetized and the orientation of the magnetization along a closed path surrounding the core acquires a rotation of 4π. This finding indicates that a regular lattice of multi-quantum spin vortices can emerge spontaneously in antiferromagnetic spinor BECs with SU(3) SO coupling. By exploring the effect of a small but finite temperature, we confirm that the double-quantum spin vortices are robust against thermal fluctuations and hence are observable in experiments.

The emergence of spin current with two quanta of circulation can be analytically understood by expanding the wave function obtained by the variational methods around the center of a double-quantum spin vortex. We suppose that the wave function of the lattice phase is written as

$$\psi = \frac{1}{3}\begin{pmatrix}1\\1\\1\end{pmatrix}\exp[-2i\kappa x] + \frac{1}{3}\begin{pmatrix}\exp\left[-i\frac{\pi}{3}\right]\\ \exp\left[i\frac{\pi}{3}\right]\\ \exp[i\pi]\end{pmatrix}\exp\left[i\kappa(x-\sqrt{3}y)\right]$$
$$+\frac{1}{3}\begin{pmatrix}\exp\left[i\frac{\pi}{3}\right]\\ \exp\left[-i\frac{\pi}{3}\right]\\ \exp[i\pi]\end{pmatrix}\exp\left[i\kappa(x+\sqrt{3}y)\right]. \qquad (10.35)$$

Then one can expand ψ around the center of a vortex with winding number $\langle -1, 0, 1\rangle$, e.g., at the location of $(x, y) = [0, \pi/(3\sqrt{3}\kappa)]$. Substituting $x = \epsilon\cos\theta$ and $y = \pi/(3\sqrt{3}\kappa) + \epsilon\sin\theta$ into ψ and expanding with respect to the infinitesimal ϵ yield

$$\psi = \begin{pmatrix}-i\kappa\exp[-i\theta]\,\epsilon - \frac{1}{2}\kappa^2\exp[2i\theta]\,\epsilon^2\\ 1-\kappa^2\epsilon^2\\ -i\kappa\exp[i\theta]\,\epsilon - \frac{1}{2}\kappa^2\exp[-2i\theta]\,\epsilon^2\end{pmatrix} + O\left(\epsilon^3\right). \qquad (10.36)$$

Notice that the second-order terms with $\exp[\pm 2i\theta]$ have no essential influence on the phases; thus the winding number for each component can still be represented as $\langle -1, 0, 1\rangle$ [see Fig. 10.29a–c]. However, since the first-order terms are canceled out when calculating the transverse magnetization $F_+ = \sqrt{2}[\psi_1^*\psi_0 + \psi_0^*\psi_{-1}]$, the second-order terms play a dominant role, leading to the emergence of spin current with two quanta of circulation around an unmagnetized core,

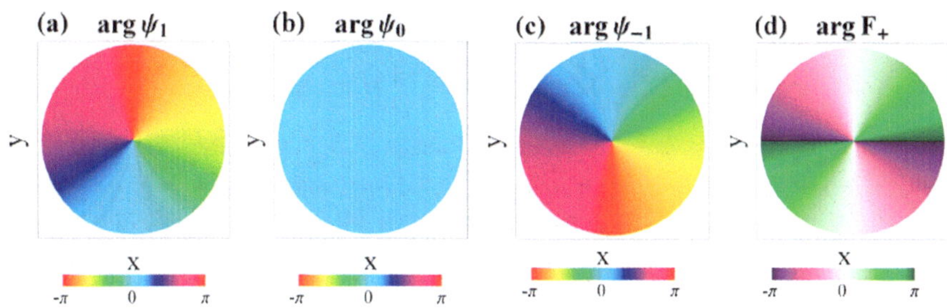

Fig. 10.29 **a–c** Phases of the polar-core vortex described by the wave function in Eq. (10.36), displaying the winding combination $\langle -1, 0, 1\rangle$. **d** Direction of the transverse magnetization, indicating the emergence of spin current with two quanta of circulation. Reprint from Ref. [9], Copyright 2018, with permission from American Physical Society

$$F_+ \propto \epsilon^2 \exp[-2i\theta],$$

as illustrated in Fig. 10.29d.

10.4 Inhomogeneous Spin Patterns in the Spatially Homogeneous Spin-1 BECs

In the presence of the positive and negative quadratic Zeeman effect, we obtain two exact solutions of broken-axisymmetry phase for spin-1 BEC. Two kinds of inhomogeneous spin patterns are displayed, the magnetization dynamics dominated by the quadratic Zeeman effect are discussed in detail, and the topological defects are described analytically.

10.4.1 Description of the Model

We consider the spatially homogeneous spin-1 BEC for which, under the mean-field theory, in consideration of the Zeeman effect, the quasi-one-dimensional dynamics is governed by the NLS equations in a dimensionless form [22, 29, 44]

$$\begin{aligned} i\frac{\partial \Phi_{\pm 1}}{\partial t} &= -\frac{1}{2}\frac{\partial^2 \Phi_{\pm 1}}{\partial x^2} + (c_0 + c_1)\,|\Phi_{\pm 1}|^2 + |\Phi_0|^2\,\Phi_{\pm 1} \\ &\quad + (c_0 - c_1)\,|\Phi_{\mp 1}|^2\,\Phi_{\pm 1} + c_1\Phi_{\mp 1}^*\Phi_0^2 + (q \mp p)\,\Phi_{\pm 1}, \qquad (10.37) \\ i\frac{\partial \Phi_0}{\partial t} &= -\frac{1}{2}\frac{\partial^2 \Phi_0}{\partial x^2} + (c_0 + c_1)\left(|\Phi_1|^2 + |\Phi_{-1}|^2\right)\Phi_0 \\ &\quad +2c_1\Phi_0^*\Phi_1\Phi_{-1} + c_0\,|\Phi_0|^2\,\Phi_{\pm 1}, \end{aligned}$$

where $c_0 = \overline{c}_0/\overline{c}_1$, $c_1 = \pm 1$. Here the constant $p = \overline{p}/\overline{c}_1 n$ denotes the linear Zeeman effect, the constant $q = \overline{q}/\overline{c}_1 n$ denotes the quadratic Zeeman effect, and $c_0 = (g_0 + 2g_2)/3$ and $\overline{c}_1 = (g_2 - g_0)/3$ denote effective constants of the mean-field and spin–exchange interaction, respectively [44], with the effective 1D couplings, $g_F = \frac{4\hbar^2 a_F}{Ma_\perp^2}\frac{1}{1-Ca_F/a_\perp}$. Here a_F are the s-wave scattering lengths in the total hyperfine spin $F = 0, 2$ channel; $a_\perp$ is the size of the transverse ground states; M is the atomic mass; $C = 1.4603, \ldots$, n is the average particle density; and $\overline{p} = -g\mu_B B$ is the sum of the linear Zeeman energy, with g the Landé factor, B the external magnetic field that is assumed to be applied in the z-direction, $\mu_B = e\hbar/2m_e$ (m_e is the electron mass, and $e > 0$ is the elementary charge) is the Bohr magneton, and $\overline{q} = \frac{(g\mu_B B)^2}{E_{hF}}$ is the quadratic Zeeman energy with the hyperfine splitting $E_{hF} = E_m - E_i$ given by the difference between the initial E_i and intermediate E_m energies. In addition, time t, position x, and $\Phi_{\pm1,0}$ are in units of $\hbar/(|\bar{c}|\,n)$, $\hbar/\sqrt{2M\,|\bar{c}|\,n}$ and $1/\sqrt{n}$, respectively. The interaction is ferromagnetic if $c_1 < 0$ (such as ^{87}Rb) or antiferromagnetic if $c_1 > 0$ (such as ^{23}Na). The value of q can be tuned in the negative direction by using a linearly polarized microwave field due to the ac Stark shift, and c_0 must be nonnegative to avoid the collapse. Specifically, for ^{87}Rb, $c_0 = 216$, $c_1 = -1$; for ^{23}Na, $c_0 = 32$, $c_1 = 1$. Typically, for ^{87}Rb and ^{23}Na, $n \sim 10^{20}\,\mathrm{m}^{-3}$, and $|\bar{c}_1|n \sim h \times 3.6\,\mathrm{Hz}$ or $h \times 24.1\,\mathrm{Hz}$. The time and length are measured in units of 10^{-2} s and 10^{-6} m, respectively.

In experiments, the linear Zeeman effect can be changed independently by applying a field gradient B along the axis of the trapped condensate, taken as $B'z_b$ where z_b is the gradient field range. The quadratic Zeeman effect can be changed independently by applying a weak external bias field B, which is given by $\overline{q} = \widehat{q}B^2$. It is found that $\widehat{q} = 71.65\,\mathrm{Hz\,G^{-2}}$ for ^{87}Rb and $278\,\mathrm{Hz\,G^{-2}}$ for ^{23}Na. In general, under a bias field, for spin-1 ^{23}Na or ^{87} Rb BEC, we have $q > 0$. However, by using ac Stark shift induced by a microwave dressing field, $q < 0$ is still accessible. The controllability of the sign and its magnitude of the quadratic Zeeman effect provides a powerful way to manipulate spinor BEC.

Our discussion is based on the analytical solutions of Eq. (10.37). As Eq. (10.37) is derived from mean-field theory with zero temperature, our results only provide spin configurations in the sense of a thermodynamic equilibrium state. Although some simplifications or approximations [44] have been made and some efficient analytical methods have been introduced [29], in the presence of the quadratic Zeeman effect, it seems that no analytical result has been reported. Even in the absence of the quadratic Zeeman effect, only some special cases of Eq. (10.37) have been studied analytically by a few authors. For example, in [38], the integrable models with the choice of the coupling constant $c_0 = c_1 = c$ (<0 or >0) have been investigated; in [17], the order parameters were presented analytically by assuming $|\Phi_1| = |\Phi_{-1}|$. But these kinds of solutions cannot describe the spontaneous magnetization. In what follows, we deal with a very general case, no assumption on c_0, c_1 is adopted, and in the solutions, $|\Phi_1| \neq |\Phi_{-1}|$, so these solutions can describe the spontaneous magnetization. We find that the uniform quadratic Zeeman effects can give rise to time-dependent topological defects which form the moving domain walls. It seems that such a phenomenon has not yet been reported in spinor BEC.

By a direct ansatz with Jacobian elliptic functions, we get two novel kinds of analytical solutions corresponding to a positive and negative quadratic Zeeman effect, respectively. To find explicit solutions of (10.37), we assume

$$\Phi_m(x,t) = r_m(x,t)\exp[i\chi m(x,t)] \quad (m = 1, 0, -1), \tag{10.38}$$

with the ansatz

$$\begin{aligned} r_m(x,t) &= A_m\mathrm{sn}(\xi,\lambda) + B_m\mathrm{cn}(\xi,\lambda) + C_m\mathrm{dn}(\xi,\lambda), \\ \chi_{\pm1}(x,t) &= k_{\pm1}x - \mu_{\pm1}t, \\ \chi_0(x,t) &= [\chi_1(x,t) + \chi_{-1}(x,t)]/2, \end{aligned}$$

where $\xi = kx + \omega t$, A_m, B_m, C_m, $k_{\pm 1}$, $\mu_{\pm 1}$, k, ω are parameters to be determined, $\mathrm{sn}(\cdot, \lambda)$, $\mathrm{cn}(\cdot, \lambda)$ and $\mathrm{dn}(\cdot, \lambda)$ are the real Jacobian elliptical functions, in which $\lambda^2 << 1$ gives the elliptic modulus, which determines the period of the functions. Denote by $K(\lambda) = \int_0^{\pi/2} d\theta/\sqrt{1-\lambda^2 sin^2[\theta]}$, then the periods of $\mathrm{sn}(\eta, \lambda)$, $\mathrm{cn}(\eta, \lambda)$, and $\mathrm{dn}(\eta, \lambda)$ are $4K(\lambda)$, $4K(\lambda)$, and $2K(\lambda)$, respectively. Furthermore, the definition of these functions can be extended for $\lambda^2 > 1$ by means of

$$\begin{aligned} \mathrm{sn}(\eta, \lambda) &= 1/\lambda \mathrm{sn}(\lambda\eta, 1/\lambda), \\ \mathrm{cn}(\eta, \lambda) &= \mathrm{dn}(\lambda\eta, 1/\lambda), \\ \mathrm{dn}(\eta, \lambda) &= \mathrm{cn}(\lambda\eta, 1/\lambda), \end{aligned} \tag{10.39}$$

so we think that λ is defined on the whole real line.

The Jacobian elliptical functions have the following properties:

$$\begin{aligned} \mathrm{sn}(x, 0) &= \sin[x], \;\; \mathrm{cn}(x, 0) = \cos[x], \\ \mathrm{dn}(x, 0) &= 1, \;\; \mathrm{sn}(x, 1) = \tanh[x], \\ \mathrm{cn}(x, 1) &= 1/\cosh[x], \;\; \mathrm{dn}(x, 1) = sech(x), \\ \mathrm{sn}^2(x, \lambda) + \mathrm{cn}^2(x, \lambda) &= 1, \\ \mathrm{dn}^2(x, \lambda) + \lambda^2 \mathrm{sn}^2(x, \lambda) &= 1. \end{aligned}$$

We see that the soliton form is a special case of a Jacobian elliptical function corresponding to $\lambda = 1$. In addition, the derivation operation is closed between the functions sn, cn, and dn as shown below:

$$\begin{aligned} \frac{d}{dx}\mathrm{sn}(x, \lambda) &= \mathrm{cn}(x, \lambda)\mathrm{dn}(x, \lambda), \\ \frac{d}{dx}\mathrm{cn}(x, \lambda) &= -\mathrm{sn}(x, \lambda)\mathrm{dn}(x, \lambda), \\ \frac{d}{dx}\mathrm{dn}(x, \lambda) &= -\lambda^2 \mathrm{sn}(x, \lambda)\mathrm{cn}(x, \lambda). \end{aligned}$$

Substituting the ansatz (10.38) into (10.37) and using the above identities, we obtain polynomials in sn, cn, and dn. Setting the coefficients of the polynomials to be zeros, we get a system of algebraic equations on the unknown parameters A_m, B_m, C_m and $k_{\pm 1}$, $\mu_{\pm 1}$, k, ω. This algebraic system can be solved, and we obtain our solutions. In fact, this procedure can provide many interesting solutions for the spin-1 system, but in what follows, we only care about the solutions with nonzero quadratic Zeeman effects.

10.4.2 Topological Defects and Inhomogeneous Spin by a Positive Quadratic Zeeman Effect

For the positive quadratic Zeeman effect, we get an exact solution of Eq. (10.37) only under the ferromagnetic interaction, i.e., $c_1 = -1$ (we still use the symbol c1 in the solution in order to emphasize its role) with the conditions $|c_1| < c_0$, which reads

$$\Phi_{\pm} = \left[\sqrt{-c_1/(c_0 - c_1)}\text{sn}(\xi, \lambda) \mp \frac{1}{2}\sqrt{-q/c_1}\text{dn}(\xi, \lambda)\right] \exp\left[i\chi_{\pm 1}\right],$$
$$\Phi_0 = \text{sn}(\xi, \lambda) \exp\left[i\chi_0\right], \tag{10.40}$$

where $\xi = \sqrt{2}q(x - k_1 t)$, $\lambda = \sqrt{-2c1/(q\delta)}$, and $\chi_{\pm 1} = k_1 x - \mu_{\pm 1} t$ with $\mu_{\pm 1} = k_1^2/2 - q(c_0 - 2c_1)/(2c_1) - 2c_1/\delta \mp p$, $2\chi_0 = \chi_1 + \chi_{-1}$, and $\delta = (c_0 - c_1)/(c_0 + c_1)$. Here k_1 is a free constant, which denotes the wave number of +1, and here is also the wave velocity of our traveling wave. For ^{87}Rb, $c_1 << c_0$, so we have $\Phi_{\pm} \approx \mp\frac{1}{2}\sqrt{-q/c_1}\text{dn}(\xi, \lambda) \exp\left[i\chi_{\pm 1}\right]$ which means that the density distribution of the $m_F = \pm 1$ components is almost the same. Obviously, in this solution, $q = 0$ is not permitted. This implies that such a solution can only exist in the presence of positive quadratic Zeeman effect; thus the initial components should be prepared under positive quadratic Zeeman effect. Theoretically, for Eq. (10.40), no initial conditions before the quadratic Zeeman effect exist. When $t = 0$, Eq. (10.40) gives the initial state

$$\Phi_{\pm} = \left[\sqrt{-c_1/(c_0 - c_1)}\text{sn}(\sqrt{2q}x, \lambda) \mp \frac{1}{2}\sqrt{-q/c_1}\text{dn}(\sqrt{2q}x, \lambda)\right] \exp\left[ik_1 x\right],$$
$$\Phi_0 = \text{sn}(\sqrt{2q}x, \lambda) \exp\left[ik_1 x\right], \tag{10.41}$$

which presents a periodic density distribution if $q = 2/\delta$ $(\lambda = 1)$, as shown in Fig. 10.30a. When $q = 2/\delta$ $(\lambda = 1)$, the density distribution becomes soliton type,

$$\Phi_{\pm} = \left[\sqrt{-c_1/(c_0 - c_1)} \tanh\left[\sqrt{2q}x\right] \mp \frac{\sqrt{-q/c_1}}{2\cosh\left[\sqrt{2q}x\right]}\right] \exp\left[ik_1 x\right],$$
$$\Phi_0 = \tanh\left[\sqrt{2q}x\right] \exp\left[ik_1 x\right], \tag{10.42}$$

as shown in Fig. 10.30b.

Solution (10.40) reveals many interesting features of the spin domain configuration induced by the positive quadratic Zeeman effect. It is known that the spin dynamics can be described by the spin density vector $F = \{F_x, F_y, F_z\}$, where F_z is the longitudinal magnetization parallel to the external magnetic field which describes the net magnetization, and $F_\perp = \{F_x, F_y\}$ indicates the transverse magnetization vector, which drives the atom exchange. The spin vector F is defined by

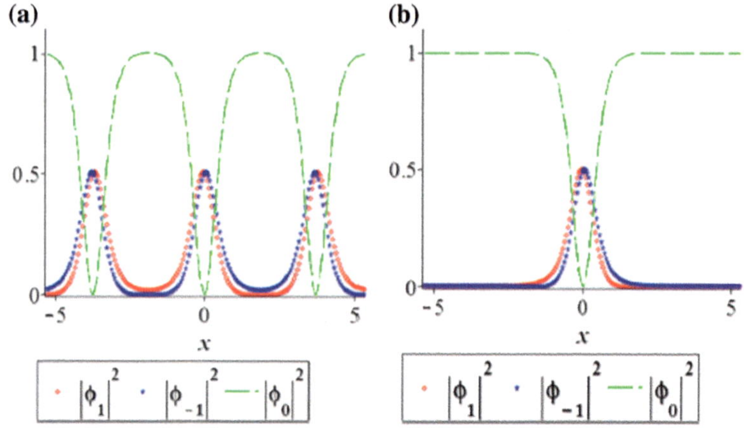

Fig. 10.30 Initial states of $\Phi_{\pm 1}$, Φ_0 given by (10.41) for ^{87}Rb: **a** $q = 2$, the periodic case; **b** $q = \frac{430}{217}$ ($\lambda = 1$), the soliton case. The other parameters are $p = 1$, $c_0 = 216$, $c_1 = -1$, $k_1 = 1$, $x \in [-5, 5]$. Such initial states should be prepared in the presence of positive quadratic Zeeman effect. Take $n = 2.8 \times 10^{20}\,\mathrm{m}^{-3}$, then $p = 1$ corresponds to $\overline{p} = 10.02\,\mathrm{Hz}$; $q = 2$ corresponds to $\overline{q} = 20.05\,\mathrm{Hz}$; and $q = \frac{430}{217}$ corresponds to $q = 19.86\,\mathrm{Hz}$. Reprint from Ref. [44], Copyright 2018, with permission from American Physical Society

$$F_x = \frac{1}{\sqrt{2}}\left[\Phi_1^*\Phi_0 + \Phi_0^*\left(\Phi_1 + \Phi_{-1}\right) + \Phi_{-1}^*\Phi_0\right],$$
$$F_y = \frac{i}{\sqrt{2}}\left[-\Phi_1^*\Phi_0 + \Phi_0^*\left(\Phi_1 - \Phi_{-1}\right) + \Phi_{-1}^*\Phi_0\right],$$
$$F_z = |\Phi_1|^2 - |\Phi_{-1}|^2 .$$

It is known that $|F_\perp| = \sqrt{F_x^2 + F_y^2}$ and $|F_z|$ determine the population exchange and the net magnetization, respectively.

For solution (10.40), we have

$$F_x = |F_\perp|\cos[pt],\ F_y = -|F_\perp|\sin[pt],\ F_y = 2\sqrt{q/(c_0 - c_1)}\mathrm{sn}(\xi,\lambda)\mathrm{dn}(\xi,\lambda), \tag{10.43}$$

where $|F_\perp| = \sqrt{-8c_1/(c_0 - c_1)}\mathrm{sn}^2(\xi,\lambda)$. Notice that when $\mathrm{sn}(\xi,\lambda) = 0$, we have $F_x = F_y = F_z = 0$, so all the zeros of $\mathrm{sn}(\xi,\lambda)$ provide the zeros of F. Usually, for system (10.37), in the absence of the quadratic Zeeman effect, F always has no zero and can be normalized, and thus the total magnetization density is conserved. In comparison, for solution (10.40), due to the zeros, F cannot be normalized, and the total magnetization density is not conserved.

It is seen from Eq. (10.43) that the Larmor frequency is determined uniquely by the linear Zeeman effect p, which is totally different from the case of negative quadratic Zeeman effect discussed later. The polar angle defined by Eq. (10.43) is $\arctan\left[|F\perp|/F_z\right] = \arctan\left[\sqrt{-\frac{2c_1}{q}\frac{\mathrm{sn}(\xi,\lambda)}{\mathrm{dn}(\xi,\lambda)}}\right]$; its evolution is shown in Fig. 10.31. It

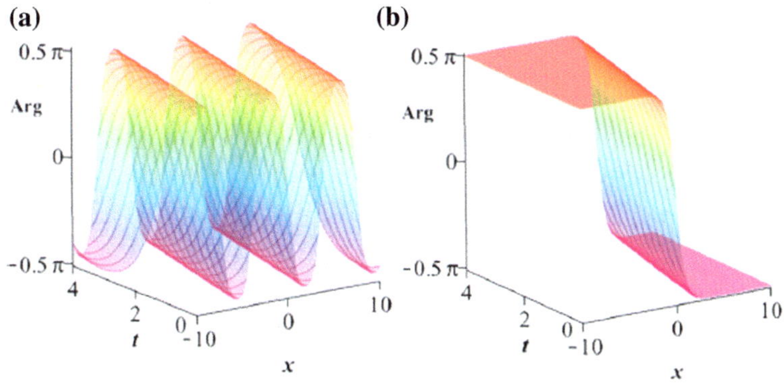

Fig. 10.31 Evolution of the polar angle determined by solution (10.40): **a** $q = 2$, the periodic case; **b** $q = \frac{430}{217}$ ($\lambda = 1$), the soliton case. The other parameters are $x \in [-10, 10]$, $t \in [0, 3]$, $p = 1, c_0 = 216, c_1 = -1, k_1 = 1$. Reprint from Ref. [44], Copyright 2018, with permission from American Physical Society

is clear that if $\mathrm{sn}(\xi, \lambda) = 0$, then $|F \perp| = 0$, and the solution is of axisymmetry-broken phase. However, the region such that $\mathrm{sn}(\xi, \lambda) = 0$ exactly gives the zeros of F, at which the local magnetization vanishes but the order parameter reads

$$\Phi_{\pm 1} = \mp \frac{1}{2}\sqrt{-q/c_1}\mathrm{dn}(\xi, \lambda) \exp\left[i\chi_{\pm 1}\right], \Phi_0 = 0, \tag{10.44}$$

which is of polar phase. Therefore, we can confirm that the zeros of $\mathrm{sn}(\xi, \lambda)$ are just the topological defects of the spinor gas, which corresponds to the unmagnetized areas that constitute the periodic DWs when $q = 2/\delta$ ($\lambda = 1$). Figure 10.31 displays the spatial distribution of F_x, F_y, F_z, where the intersection points of the three curves in Fig. 10.32 give the topological defects.

Figure 10.33 shows the spin configuration determined by Eq. (10.40). Figure 10.33a is the space–time surface of the spin density vector F (for ^{87}Rb). The color is assigned by the value of F_z, which distinguishes the different magnetization, where the point O corresponds to the zeros of F that indicate the topological defects. Figure 10.33b shows the space–time evolution of the transverse magnetization vector $F_\perp$, where the lengths of the arrows are $|F_\perp|$, and the color corresponds to that in Fig. 10.33a. The heavy yellow and pink regions correspond to the neighbor of the north pole and south pole in Fig. 10.33a, respectively. In such regions, the net magnetization $|F_z|$ reaches its maximum value. The regions between the light-green and light-blue lines between the yellow and pink regions are the unmagnetized areas, i.e., the topological defects, corresponding to the point O. The blue curve in Fig. 10.33a presents the spatial evolution of F at $t = 1$. When $q = 2/\delta$, along the curve, the spin vector F evolves periodically in the sequence $P \to O \to Q \to O \to P$, which shows that O actually corresponds to the periodic DWs. Correspondingly, in Fig. 10.33b, the color changes in the following sequence: green → yellow → green → blue → pink → blue → pink → blue → green →

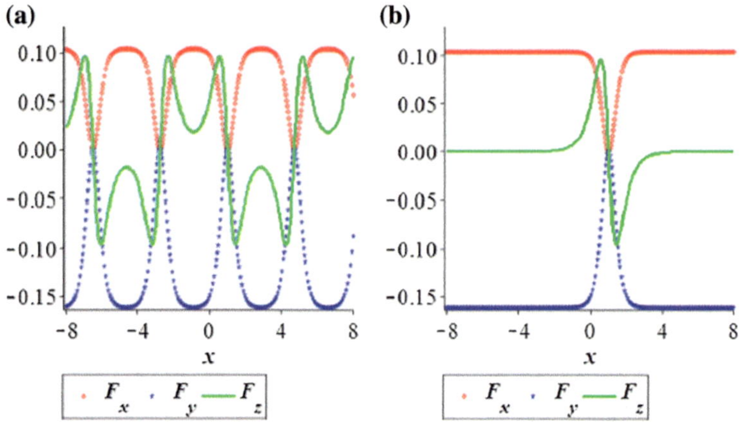

Fig. 10.32 The spatial distribution of F_x, F_y, F_z and topological defects: **a** the periodic case ($q = 2$), the common zeros of the three curves are the topological defects; **b** the soliton case ($q = 430/217$), the only common zero of the three curves indicates the topological defect. The other parameters are $p = 1, c_1 = -1,\ c_0 = 216, k1 = 1, t = 1, x \in [-8, 8]$. Reprint from Ref. [44], Copyright 2018, with permission from American Physical Society

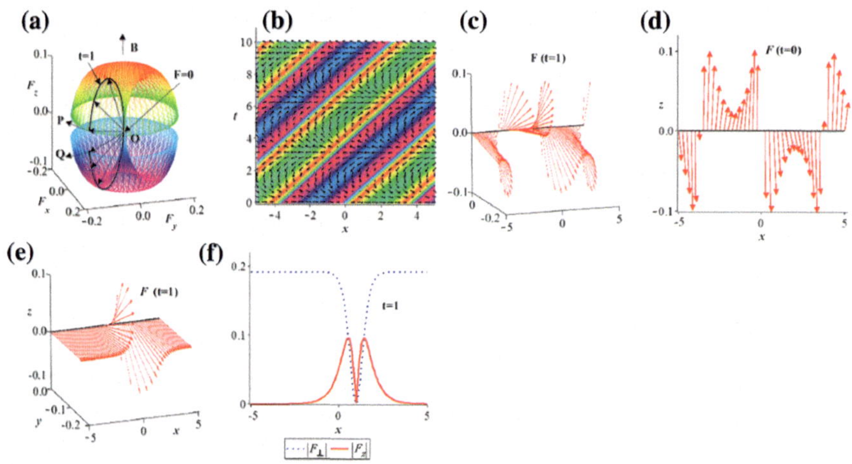

Fig. 10.33 Inhomogeneous spin structure and topological defects induced by positive quadratic Zeeman effect q. **a** The space–time evolution of the spin density vector F, the point O corresponds to topological defects. **b** The field plot of the transverse magnetization vector $F_\perp$, the green and blue lines between the yellow and pink regions correspond to the evolution of the topological defects. In **a** and **b**, $t \in [0, 10]$. **c** The spatial spin configuration at $t = 1$, and **d** the spatial spin configuration at t = 0. In **c** and **d**, the periodic domain walls are the topological defects. **e** The spatial spin configuration of local magnetism (the soliton case) at $t = 1$, where the unique domain wall is the topological defect. **f** The spatial distribution of $|F_\perp|$ and $|F_z|$ of local magnetism (the soliton case) at $t = 1$, where the only common zero of the two curves indicates the topological defect. In **a**–**d**, $q = 2$; in **e** and **f**, $q = 430/217$ ($\lambda = 1$). The other parameters are $x \in [-5, 5]$, $k_1 = 1, c_1 = -1, c_0 = 216$, $p = 1$. Reprint from Ref. [44], Copyright 2018, with permission from American Physical Society

yellow → green. Figure 10.33c gives the spatial formation of the spin vector F along the condensates at $t = 1$, indicating that the topological defects constitute the periodical Bloch-like DWs for the stripe domain. However, at time such that $\sin[pt] = 0$, all spin vectors lie in the $x - z$ plane. The solution presents helical domain formation with periodic Néel-like DWs formed by the unmagnetized atoms, as shown in Fig. 10.33d. Such two kinds of DWs in Fig. 10.33c and d can be converted to each other periodically. In this case, as $|F_\perp|$ and $|F_z|$ have common periodic zeros, the transverse ferromagnetic domains are divided by the topological defects periodically. These features agree with the experimental observation for a ^{87}Rb spinor BEC reported in [34]. However, in the experiment, the transverse ferromagnetic domains are formed from nearly pure spinor BEC prepared in the unmagnetized $|m_F = 0\rangle$ phase for $q >> 2/\delta$ at the beginning, and then rapidly quenched the gas to conditions $q >> 2/\delta$. Our discussion shows that for q around $2/\delta$, a similar phenomenon should also be observed. It is remarkable that when $q = 2/\delta$ ($\lambda = 1$), Eq. (10.40) gives a novel superposition soliton-like solution:

$$\Phi_\pm = \left[\sqrt{-c_1/(c_0 - c_1)}\tanh[\xi] \mp \frac{\sqrt{2/\delta}}{2\cosh[\xi]}\right]\exp\left[i\chi_{\pm 1}\right],$$
$$\Phi_0 = \tanh[\xi]\exp[i\chi_0].$$

The shape of such solutions is determined by the ratio $\gamma = |c_0/c_1|$. For ^{87}Rb, $\gamma >> 1$, each $|\Phi_{\pm 1}|^2$ behaves like a bright soliton. This fact implies that $q = 2/\delta$ is an unstable point of dynamical phase transition. As we usually have $|c_1| << c_0$, $q = 2/\delta$ can be compared with the quantum phase transition point $\overline{q} = 2|\overline{c}_1|n$ (i.e., $q = 2$) mentioned in [34]. In this case, we have

$$|F_\perp| = \sqrt{-8c_1/(c_0 - c_1)}\tanh^2[\xi], \quad F_z = 2\sqrt{q/(c_0 - c_1)}\tanh^2[\xi]/\cosh[\xi]. \tag{10.45}$$

The atom exchange between the sublevels is small near the soliton propagation direction, which permits only a unique spin DW. Far away from this region, the atom exchange of the component $m_F = \pm 1$ is almost uniform, which gives no net magnetization. Therefore, the net magnetization is concentrated in a local range, as shown in Fig. 10.33e and f.

When $|F_\perp| \neq 0$, the SO(2) symmetry is broken in the direction of the transverse magnetization. In our solution, the transverse magnetization is nonuniform, i.e., it is space- and time-dependent, which causes the inhomogeneous symmetry breaking in the spin evolution. To investigate the symmetry of the order parameters in spin space, we give the visualized realization of the order parameter for $q > 0$ in the spin space. Because an integer-spin state can be described in terms of the spherical harmonics $Y_f^m(\widehat{s})$ [$\widehat{s} = (\alpha, \beta)$, α is taken as the polar (colatitude) coordinate with $\alpha \in [0, \pi]$, and $\beta \in [0, 2\pi]$ is the azimuthal (longitudinal) coordinate], the above order parameter can be visualized by drawing the wave functions in spin space by the surface of $|\Psi(\widehat{s})|^2$, where $\Psi(\widehat{s}) = \Phi_{+1}(\widehat{s}) + \Phi_0(\widehat{s}) + \Phi_{-1}(\widehat{s})$.

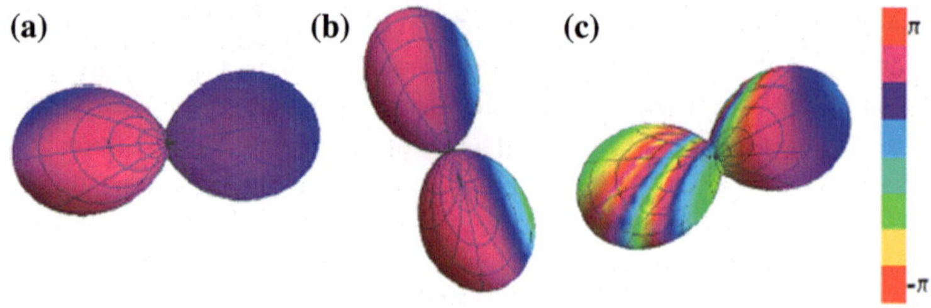

Fig. 10.34 Inhomogeneity of the order parameter $(\Phi_{-1}, \Phi_0, \Phi_1)$ in the spin space determined by the positive quadratic Zeeman effect. As a comparison for the spin state, **a** and **b** are for the same position ($x = 4$) and quadratic Zeeman effect ($q = 3$) but different time: $t = 2$ in **a**, $t = 4.5$ in **b**; **a** and **c** are for the same position ($x = 4$) and time ($t = 2$) but different quadratic Zeeman effect: $q = 3$ in **a**, $q = 9$ in **c**. The other parameters are taken as $p = 1, k_1 = 1, c_1 = -1, c_0 = 216$. Notice that the three figures have different morphologies near the centers, so they cannot be identified in the spin space via a rotation. Reprint from Ref. [44], Copyright 2018, with permission from American Physical Society

In mathematics, spherical harmonics are the angular portion of a set of solutions to Laplace's equation. Represented in a system of spherical coordinates, Laplace's spherical harmonics $Y_f^m(\widehat{s})$ are a specific set of spherical harmonics that form an orthogonal system.

Figure 10.34 shows the spatial variation of the order parameter in a unit cell visualized by plotting a surface with spherical coordinates, where the color represents the phase, i.e., the argument $\arg[\Psi(\widehat{s})]$ [39], which is defined on $[-\pi, \pi]$. It displays that the spin states of the order parameter in the spin space is space–time pointwise difference. As a comparison, Fig. 10.34a and b shows the spin states for the same position and quadratic Zeeman effect but different time, whereas Fig. 10.34a and c is for the same position and time but different quadratic Zeeman effect. One can see that for the order parameters for states (a) and (b), they not only differ in their orientations but also have different morphologies near the centers, so they cannot be identified in the spin space via a rotation, which is quite different from the case discussed in [39] for ground state. Figure 10.34a and c shows that the change of the quadratic Zeeman effect will change the spin state of the order parameter. This implies that the presence of the quadratic Zeeman effect can lead to different morphology for the space–time-dependent spin state.

We now turn to the population distribution determined by solution (10.40) and recall a conclusion in mathematics: f or a continuous function $f(t)$ with period T, we have $\int_0^T f(x)dx = \int_c^{T+c} f(x)dx = \int_0^T f(x+c)dx$ for any constant c. By this fact, we see that within a period, the atom number, $N_m = \int_0^T |\Phi_m[k(x+k_1t)]|^2 dx = \int_{kk_1t}^{T+kk_1t} |\Phi_m(kx)|^2 dx$ $(m = \pm 1, 0)$, of each component is time-independent; therefore, we confirm that the population exchange between the $m_F = \pm 1$ and $m_F = 0$ components is in an equilibrium state.

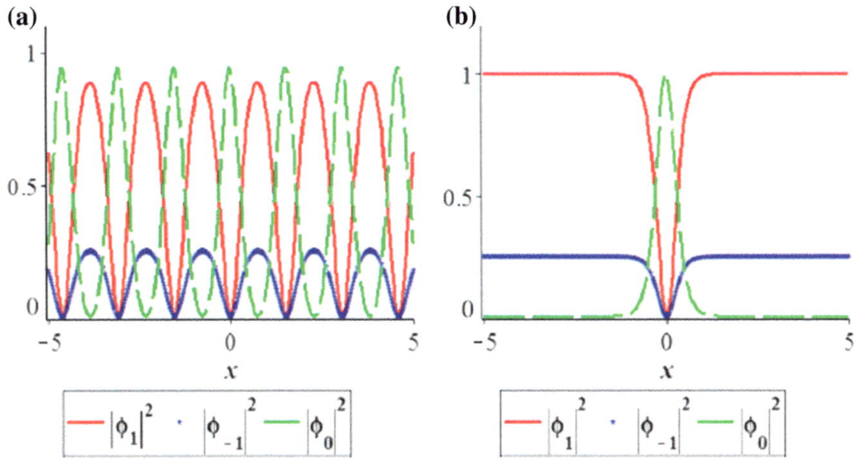

Fig. 10.35 Initial states of $\Phi_{\pm1}$, Φ_0 determined by solution (10): **a** $q = -3$, the periodic case; **b** $q = -3.875\,(\lambda = 1)$, the soliton case. The other parameters are $t = 0, x \in [-5, 5], p = 1, c_1 = 1$, $c_0 = 32$, $A_{-1} = -0.5$, $k_1 = 1$. Reprint from Ref. [44], Copyright 2018, with permission from American Physical Society

10.4.3 Topological Defects and Inhomogeneous Spin by Negative Quadratic Zeeman Effect

For a negative quadratic Zeeman effect, exact solutions are obtained for both ferromagnetic ($c_1 = -1$) and antiferromagnetic ($c_1 = 1$) interactions when $|c_1| < c_0$. Set $\nu = -2q/(c_0 - c_1)$ and $k = \sqrt{(c_0 + c_1)A_{-1}^2}$, and Eq. (10.37) has the solution

$$\Phi_{\pm1} = A_{\pm1}\mathrm{sn}(\xi, \lambda)\exp\left[i\chi_{\pm1}\right],\ \ \Phi_0 = \sqrt{-2A_1A_{-1}}\mathrm{cn}(\xi, \lambda)\exp\left[i\chi_0\right], \qquad (10.46)$$

where k_1, A_{-1} are free parameters, $A_1 = \sqrt{\nu} - A_{-1}$, $\lambda = \sqrt{(c_0 + c_1)\nu}/k$, $\xi = k(x - k_1 t)$, $\chi_{\pm1} = k_1 x - \mu_{\pm1} t$, $\chi_0 = (\chi_1 + \chi_{-1})/2$, with

$$\mu_{\pm1} = \left(k_1^2 + k^2\right)/2 + 2c_0 A_{-1}^2 - 2\sqrt{\nu}\,(c_0 \pm c_1)\,A_{-1} \pm c_1\nu \mp p.$$

Let $t = 0, q = 0$, and we get the initial conditions before the quadratic Zeeman effect as

$$\Phi_{\pm1} = \pm A_1 \sin\left[kx\right]\exp\left[ik_1 x\right],\ \ \Phi_0 = \sqrt{2}A_1 \cos\left[kx\right]\exp\left[ik_1 x\right], \qquad (10.47)$$

which is of a polar phase, where k, k_1 are arbitrary constants. The initial state ($t = 0$) of solution (10.46) is shown in Fig. 10.35.

In general, Eq. (10.46) indicates a solution of broken-axisymmetry phase, and the spin vector is given by

$$F_x = 2\sqrt{-A_1 A_{-1}\nu}\mathrm{sn}(\xi, \lambda)\mathrm{cn}(\xi, \lambda)\cos[\Omega t],$$
$$F_y = -2\sqrt{-A_1 A_{-1}\nu}\mathrm{sn}(\xi, \lambda)\mathrm{cn}(\xi, \lambda)\sin[\Omega t], \quad (10.48)$$
$$F_z = \left(\nu - 2\sqrt{\nu}A_{-1}\right)\mathrm{sn}^2(\xi, \lambda),$$

where $\Omega = (p - c_1\nu + 2c_1 A_{-1}\sqrt{\nu})$ gives the Larmor frequency. Different from the case of $q > 0$, the Larmor frequency is not only determined by the linear Zeeman effect but also by the quadratic Zeeman effect. Similar to the case of $q > 0$, the zeros of F determine the topological defects constituted by the unmagnetized atoms.

The expression of solution (10.46) requires $A_1 A_{-1} << 0$. The cases for $A_{-1} > 0$ and $A_{-1} < 0$ are different. One can see that if $A_{-1} > 0$, q should be restricted in the region $-12(c_0 - c_1) \leq A_{-1}^2 \leq q_0$, in this case, while $q = 0$ gives a periodic polar state

$$\Phi_{\pm 1} = \mp A_{-1}\sin[\xi]\exp\left[i\chi_{\pm 1}\right], \quad \Phi_0 = \sqrt{2}A_{-1}\cos[\xi]\exp[i\chi_0], \quad (10.49)$$

$q = -A_{-1}^2\left(c_0 - c_1\right)/2$ presents a ferromagnetic phase

$$\Phi_0 = \Phi_1 = 0, \ \Phi_{-1} = A_{-1}\tanh[\xi]\exp\left[i\chi_{\pm 1}\right]. \quad (10.50)$$

This suggests that by tuning q, it is possible to get polar phase or ferromagnetic phase from the broken-axisymmetry phase, which means that tuning q can change the population in each component.

If $A_{-1} < 0$, no further restriction on q is required. For simplicity, hereafter we only discuss the case for $A_{-1} < 0$. In such a situation, note that $q = -\frac{1}{2}(c_0 - c_1)A_{-1}^2$ gives the soliton state, the order parameter reads

$$\Phi_{\pm 1} = A_{\pm 1}\tanh[\xi]\exp\left[i\chi_{\pm 1}\right], \quad \Phi_0 = \frac{\sqrt{-2A_1 A_{-1}}}{\cosh[\xi]}\exp[i\chi_0],$$

and the spin vector becomes

$$F_x = 2\sqrt{-A_1 A_{-1}\nu}\frac{\tanh[\xi]}{\cosh[\xi]}\cos[\Omega t],$$
$$F_y = -2\sqrt{-A_1 A_{-1}\nu}\frac{\tanh[\xi]}{\cosh[\xi]}\sin[\Omega t],$$
$$F_z = \left(\nu - 2\sqrt{\nu}A_{-1}\right)\tanh^2[\xi].$$

Figure 10.36a–c display the spatial distribution of F_x, F_y, F_z and $|F_\perp|$, $|F_z|$, respectively. Similar to Fig. 10.32a and b, the intersection points of the three curves in Fig. 10.36a and b give the topological defects. In the soliton case, both the atom exchange between the sublevels and the net magnetization vanish on the topological defects. However, the transverse magnetization is concentrated on the region near the defect, and far away from this region, it becomes zero but F_z reaches its maximum,

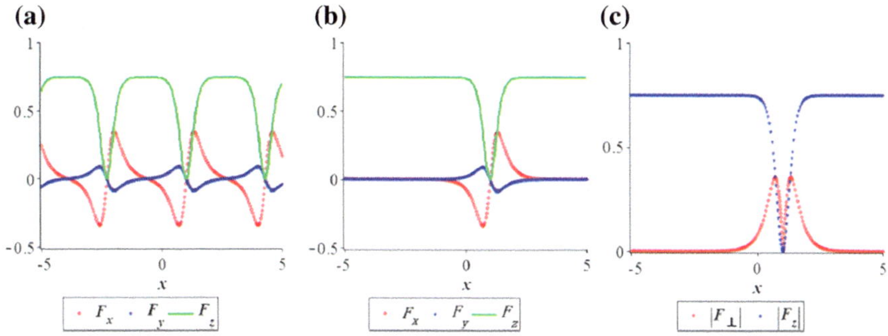

Fig. 10.36 **a** The spatial distribution of F_x, F_y, F_z for $q = -3.87$ (the periodic case), the common zeros of the three curves show the topological defects. **b** The spatial distribution of F_x, F_y, F_z for $q = -3.875$ ($\lambda = 1$) (the soliton case), the only common zero of the three curves indicates the topological defect. **c** The spatial distribution of $|F_\perp|$ and $|F_z|$ for $q = -3.875$ ($\lambda = 1$), the only common zero of the two curves is the topological defect. The other parameters are $p = 1$, $c_1 = 1$, $c_0 = 32$, $A_{-1} = -0.5$, $k_1 = 1$, $t = 1$, $x \in [-5, 5]$. Reprint from Ref. [44], Copyright 2018, with permission from American Physical Society

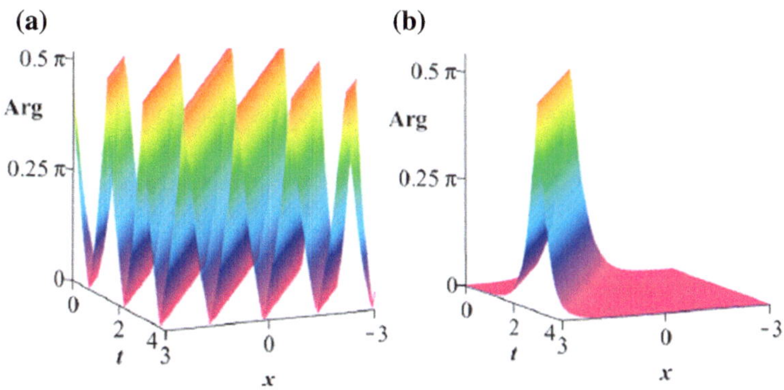

Fig. 10.37 Evolution of the polar angle determined by solution (10.46): **a** $q = -3$, the periodic case; **b** $q = -318$ ($\lambda = 1$), the soliton case. The other parameters are $x \in [-3, 3]$, $t \in [0, 3]$, $p = 1$, $c_1 = 1$, $c_0 = 32$, $k_1 = 1$, $A_{-1} = -0.5$. Reprint from Ref. [44], Copyright 2018, with permission from American Physical Society

which gives a uniform net magnetization with no population exchange, as Fig. 10.36c displays. This is contrary to the case of solution (10.40) shown in Fig. 10.33f. Roughly speaking, for the shapes of the curves in Figs. 10.36c and 10.33f, there is an exact interchange between the distribution of $|F_\perp|$ and $|F_z|$. On the other hand, it is the same as the case of $q > 0$, for solution (10.46). The zeros of $\mathrm{sn}(\xi, \lambda)$ present the topological defects, and at these defects, the order parameter becomes $\Phi_{\pm1} = 0$, $\Phi_0 = \sqrt{-2A_1 A_{-1}} \exp[i\chi_0]$, which is of polar phase, but besides them, the order parameter is of axisymmetry-broken phase.

The evolution of the polar angle determined by solution (10.46) is shown in Fig. 10.37. Similar to the case of $q > 0$, the population of each component is also

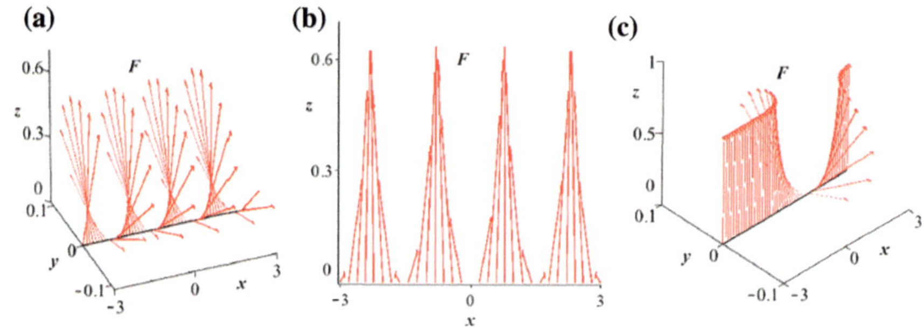

Fig. 10.38 Inhomogeneous spin structure and the topological defects determined by solution (10.46) for different t and q: **a** the spatial spin configuration at $t = 1$ for $q = -3$, the periodic case; **b** the spatial spin configuration at $t = 0$ for $q = -3$; and **c** the spatial spin configuration at $t = 1$ for $q = -31/8$ ($\lambda = 1$), the soliton case. The other parameters are: $x \in [-3, 3]$, $p = 1$, $c_1 = 1$, $c_0 = 32$, $k_1 = 1$, $A_{-1} = -0.5$. The domain walls are formed by the topological defects. Reprint from Ref. [44], Copyright 2018, with permission from American Physical Society

conserved within a period, and the atom exchange between the sublevels is also in an equilibrium state. However, the spin configuration for the case of $q < 0$ is rather different from the case of $q > 0$. First, the Larmor frequency is not only determined by the linear Zeeman effect but also by the quadratic Zeeman effect. In addition, the spin domain formation is completely different. We show the spin domain formation determined by solution (10.46) in Fig. 10.38 to compare with Fig. 10.33. One can find that Fig. 10.38a and c presents different spin patterns compared with that shown in Fig. 10.33c and e. Figure 10.38a is the spatial spin configuration, and its spin DWs are of Bloch-like type. Figure 10.38b shows the spin configuration at the time such that $\sin[t] = 0$. In this case, all spin vectors have almost the same direction and lie in the $x - z$ plane and the topological defect does not form domain walls, although they divided the domain into sub-domains periodically, which is very different from the case shown in Fig. 10.33d.

Taking ^{23}Na ($c_1 = 1$) as an example (for $c_1 = -1$, such as ^{87}Rb, the result is similar), Fig. 10.39a gives the spin density surface. Compared with the case of $q > 0$, it has a different topological structure. Figure 10.39b shows how the negative quadratic Zeeman effect affects the intensity of F, $F_\perp$, and F_z. The black curve in Fig. 10.39a shows the spatial evolution of F at $t = 1$. Along the curve, one cycle movement corresponds to one period of the spin domain. Similar to the case of $q > 0$, the nonuniform periodic population distribution and exchange bring on the periodic spin DWs formed by the topological defects, which is indicated by point O. In addition, the center areas of a spin domain correspond to the north poles in Fig. 10.39a, where $F_\perp = 0$, and no population exchange between the sublevels occurs. In such areas, the spin reaches its maximum net magnetization.

Experimentally, $q < 0$ can be reached by using a bias magnetic field B and a microwave dressing field M. Let q_B be the quadratic Zeeman shift given by B and q_M be the quadratic energy shift due to the ac Stark shift caused by M, q can be tuned

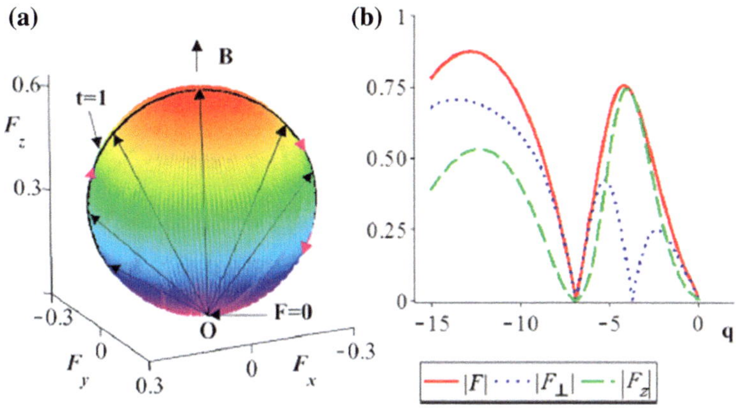

Fig. 10.39 Inhomogeneous spin structure induced by negative quadratic Zeeman effect: **a** the space–time evolution of the spin density vector F, the color indicates the intensity of F_z; **b** the influence of q on F, $F_\perp$ and F_z. In **a**, $p = 1, q = -3, x \in [-5, 5], t \in [0, 15]$; in **b**, $x = 0, t = 1, q \in [-15, 0]$. The other parameters are $A_{-1} = -0.5$, $p = 1$, $k_1 = 1$, $c_1 = 1$, $c_0 = 32$. Reprint from Ref. [44], Copyright 2018, with permission from American Physical Society

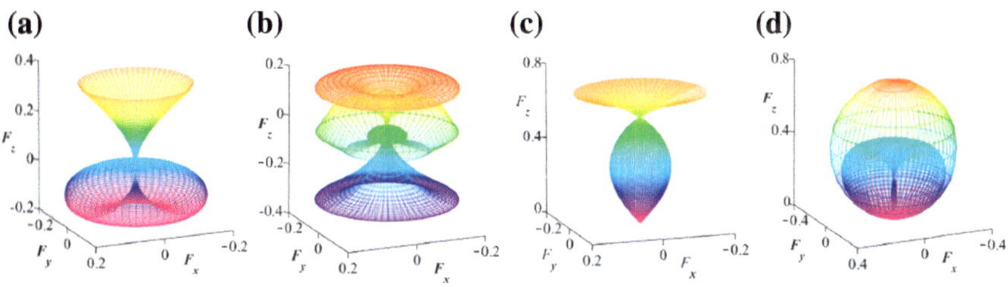

Fig. 10.40 Pointwise evolution of the spin density vector F in the linear and quadratic Zeeman effects p and q at the same time ($t = 1$) but different position, the color indicates the intensity of F_z: **a**, **b** for $q > 0$, where $x = 2$ in **a**, $x = 3.3$ in **b**, $q \in [0, 10]$; **c**, **d** for $q < 0$, where $x = 1.7$ in **c**, $x = 2.7$ in **d**, $q \in [-3.87, 0]$, $A_{-1} = -0.5$. In both cases, $p \in [0, 2\pi]$, $k_1 = 1$. Reprint from Ref. [44], Copyright 2018, with permission from American Physical Society

by virtue of $q = q_B + q_M$. In the recent experiment [5], in a quasi-1D case, with up to 3×10^6 sodium atoms with a peak atom density of $n_0 = 5.4 \times 10^{14}\,\mathrm{cm}^{-3}$, in an optical dipole trap, under a bias magnetic field $B = 97\,\mathrm{mG}$, by using microwave powers between 0 and 7.5 W, the quadratic Zeeman effect can be tuned between $q_B = h \times 2.5\,\mathrm{Hz}$ to $-18.5\,\mathrm{Hz}$. By rapidly switching the quadratic Zeeman effect from positive to negative values, the dynamics of an antiferromagnetic sodium BEC quenched across a quantum phase transition has been observed. However, the phase coherence between the dynamically created $m_F = \pm 1$ spin components in relation to topological defect formation was not explored. The above analytical solution for $q < 0$ provides a precise description of possible dynamics of this experiment and shows further that the same dynamics could be observed for the ^{87}Rb BEC. By probing the Larmor precession [10], the novel features are expected to be tested.

It is found that tuning the linear Zeeman effect and quadratic Zeeman effect will cause a pointwise-difference influence to the spin dynamics. Taking ^{87}Rb and ^{23}Na as an example, Fig. 10.40a and b shows the pointwise evolution of F in p and q for $q > 0$; Fig. 10.40c and d is for $q < 0$. These figures clearly show the inhomogeneous influence of the Zeeman effects on the spin structures.

10.5 Dirac Monopoles with a Polar-Core Vortex Induced by Spin–Orbit Coupling in Ferromagnetic BECs

We investigate a configuration of monopoles, Dirac monopoles with a polar-core vortex, induced by the SO coupling in ferromagnetic BECs. In our study, we assume that the monopoles in the SO-coupled spinor BECs are located at the endpoints of two nodal lines along the vertical axis. The nodal lines remain stable for a very long time, indicating that the monopoles with the polar-core vortex are a long-lived structure.

10.5.1 Dirac Monopoles with a Polar-Core Vortex

We consider the three-dimensional spin-1 BECs with a 2D SO coupling [16, 34] and a controllable magnetic field [2, 20]. In the mean-field approximation, the Hamiltonian is written as

$$\mathcal{H} = \int d^3\mathbf{r} \left(\Psi^\dagger \left[T + V_{\text{opt}}(\mathbf{r}) + \nu_{so} + g_F\mu_B\mathbf{B}(\mathbf{r}) \cdot \mathcal{F}\right] \Psi \right. \\ \left. + \left\{\frac{c_0}{2}n^2 + \frac{c_2}{2}\left[(n_1 - n_{-1})^2 \left|\Psi_1^*\Psi_0 + \Psi_0^*\Psi_{-1}\right|^2\right]\right\}\right), \tag{10.51}$$

where $\Psi = (\Psi_1(\mathbf{r}), \Psi_0(\mathbf{r}), \Psi_{-1}(\mathbf{r}))^T$ is the order parameter of the BECs with normalization $\int d^3\mathbf{r}\Psi^\dagger\Psi = \mathbf{N}$ and is the total particle number. The kinetic energy $T = -\frac{\hbar^2}{2m}\nabla^2$. The total particle density is defined by $n = \sum_m n_m$, wherein $n_m = |\Psi_m|^2$ $(m = 0, \pm 1)$. The optical trapping potential $V_{\text{opt}}(\mathbf{r}) = m\left[\omega_r^2\left(x^2 + y^2\right) + \omega_z^2 z^2\right]/2$, where ω_r and ω_z are the radial and axial trapping frequencies, and m is the mass of a ^{87}Rb atom. The vector of spin-1 matrices is defined by $\mathcal{F} = \left(\mathcal{F}_x, \mathcal{F}_y, \mathcal{F}_z\right)^T$ wherein $\mathcal{F}_x$, $\mathcal{F}_y$, and $\mathcal{F}_z$ are the 3×3 Pauli spin-1 matrices. The SO coupling term is written as $\nu_{so} = -i\hbar(\kappa_x\mathcal{F}_x\partial_x + \kappa_y\mathcal{F}_y\partial_y)$, where κ_x and κ_y are the strengths of the SO coupling. We define $\kappa_x = \kappa_y = \kappa$ for isotropic SO coupling (Rashba-type) and $\kappa_x \neq \kappa_y$ for anisotropic SO coupling. The external magnetic field is given by $\mathbf{B}(\mathbf{r}) = B_1'\left(x\widehat{e}_x + y\widehat{e}_y\right) + B_2' y\widehat{e}_z$, where the condition $2B_1' + B_2' = 0$ must be satisfied according to Maxwell's equation $\nabla \cdot \mathbf{B} = 0$. The Landé factor $g_F = -1/2$ and μ_B is the Bohr magneton. For the interaction terms, the coupling parameters are given by $c_0 = 4\pi\hbar^2(a_0 + 2a^2)/3m$ and $c_2 = 4\pi\hbar^2(a_2 - a_0)/3m$, where $\hbar$ is the Planck constant and $a_{0,2}$ are two-body s-wave scattering lengths for total spin 0, 2. The wave functions of spin-1 BECs are formulated as the dimensionless multicomponent NLS equations [31]

$$i\frac{\partial\psi_1}{\partial t} = \left[-\frac{1}{2}\nabla^2 + V + \lambda_0\rho + \lambda_2(\rho_1 + \rho_0 - \rho_{-1}) + B_2 z\right]\psi_1$$
$$+B_1(x - iy)\psi_0 + \kappa\left(-i\partial_x - \partial_y\right)\psi_0 + \lambda_2\psi_{-1}^*\psi_0, \qquad (10.52a)$$
$$i\frac{\partial\psi_0}{\partial t} = \left[-\frac{1}{2}\nabla^2 + V + \lambda_0\rho + \lambda_2(\rho_1 + \rho_{-1}) + B_2 z\right]\psi_0$$
$$+B_1(x + iy)\psi_1 + B_1(x - iy)\psi_{-1} + \kappa\left(-i\partial_x + \partial_y\right)\psi_1 \qquad (10.52b)$$
$$+\kappa\left(-i\partial_x - \partial_y\right)\psi_{-1} + 2\lambda_2\psi_1\psi_{-1}\psi_0^*,$$
$$i\frac{\partial\psi_{-1}}{\partial t} = \left[-\frac{1}{2}\nabla^2 + V + \lambda_0\rho + \lambda_2(\rho_0 + \rho_{-1} - \rho_1) - B_2 z\right]\psi_{-1}$$
$$+B_1(x + iy)\psi_0 + \kappa\left(-i\partial_x + \partial_y\right)\psi_0 + \lambda_2\psi_1^*\psi_0, \qquad (10.52c)$$

where $\psi_j = N^{-1/2}a_h^{3/2}\Psi_j$ is the dimensionless wave function and $\rho = \rho_1 + \rho_0 + \rho_{-1}$ with $\rho_j = \left|\psi_j\right|^2$ $(j = 0, \pm 1)$ is the total condensate density. The dimensionless optical trapping potential $V(\mathbf{r}) = \frac{1}{2}\left(\gamma_x x^2 + \gamma_y y^2 + \gamma_z z^2\right)$ with $\gamma_x = \omega_x/\omega$, $\gamma_y = \omega_y/\omega$, $\gamma_z = \omega_z/\omega$ and $\omega = \min(\omega_r, \omega_z)$. The dimensionless interaction strengths $\lambda_0 = 4\pi N(a_0 + 2a_2)/3a_h$ and $\lambda_2 = 4\pi N(a_2 - a_0)/3a_h$. We choose $a_2 = (100.4 \pm 0.1)a_B$ for total spin channel $F_{total} = 2$ and $a_0 = (101.8 \pm 0.2)a_B$ for total spin channel $F_{total} = 0$, where a_B is the Bohr radius. The characteristic length of the harmonic trap is defined as $a_h = \sqrt{\hbar/m\omega}$. The dimensionless strength of the magnetic field gradient complies the condition $2B_1 + B_2 = 0$. The time, energy, strength of the SO coupling, and strength of the magnetic field gradient are scaled by ω^{-1}, $\hbar\omega$, $\sqrt{\hbar\omega/m}$, and $\hbar\omega/(g_F\mu_B a_h)$, respectively.

The stationary state of the monopole is obtained by using the standard imaginary time propagation combined with finite-difference methods [43]. Equations (10.52a)–(10.52c) are solved by using the second-order centered finite difference for the spatial discretization and the backward and forward Euler schemes of the linear and nonlinear terms for the time discretization. The average energy decays monotonically with time until the steady states are reached. The computational grids are $120 \times 120 \times 120$, corresponding to the volume being $20 \times 20 \times 20$ (a_h^3) or $34.2 \times 34.2 \times 34.2$ (μm^3). The final steady state is independent of the initial trial wave function. The trial wave function is given by the normalized random number of complex values. The dynamic evolution is obtained by using the split-operator Crank–Nicolson method with the time stepping being $10^{-4}/\omega$.

We start by studying the structures of the monopoles in the absence of the SO coupling. To our knowledge, the homogeneous bias field is a necessary condition for observing the monopole creation process. However, our emphasis is to investigate the final steady state of the monopole, not the creation process of the monopole. There-

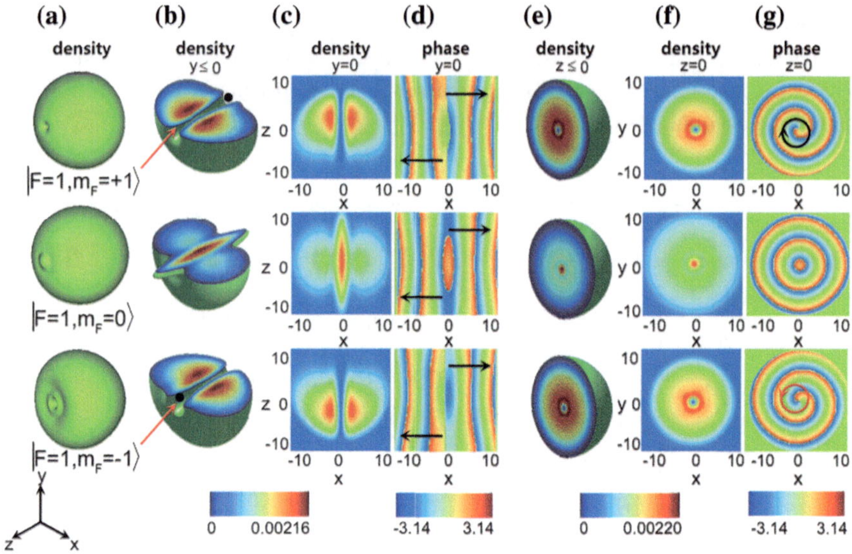

Fig. 10.41 The monopoles with the polar-core vortex. **a** Isosurface of particle densities. **b** Particle densities for $y \le 0$. The position of the nodal line (Dirac string) is highlighted by the red arrow. The position of the monopole is highlighted by the black dot. The result for $x \le 0$ is the same as that for $y \le 0$. **c** and **d** Densities and phase distributions in the $y = 0$ planes. The phases at both sides of the $y = 0$ planes are inverse, as highlighted by the black arrow in **d**. **e** Particle densities for $z \le 0$. **f** and **g** Densities and phase distributions in the $z = 0$ planes. The single vortex and antivortex are highlighted by the black and red circles, respectively. The simulation is conducted by using the dimensionless SO coupling strength $\kappa = 2$, spin-dependent interaction parameter $\lambda_2 = -75$, spin-independent interaction parameter $\lambda_0 = 7500$, strength of the magnetic field gradient $B_1 = 0.6$, and isotropic optical trap $\omega_r = \omega_z = 2\pi \times 250\,\text{Hz}$. Reprint from Ref. [20], Copyright 2018, with permission from American Physical Society

fore, the homogeneous bias field is not a necessary condition in the present work and therefore is not included in our model. In this case, the vortex line corresponding to the monopole will not be along the symmetry axis of the system. Therefore, the doubly quantized vortex line attached to the monopole shows a branchlike distribution, splitting into two singly quantized vortex lines partially in the $m_F = +1$ and -1 components.

When the SO coupling is weak, the monopoles with the polar-core vortex are found, as shown in Fig. 10.41. The structure of the monopoles with the polar-core vortex represents a single vortex line in the $m_F = +1$ component, a soliton in the $m_F = 0$ component, and a single antivortex line in the $m_F = -1$ component. Compared with the monopoles in BECs without the SO coupling, in the present system, there exist two monopoles that locate at the endpoints of two nodal lines along the vertical axis [as highlighted by the red arrows in Fig. 10.41b], which is caused by the coupling between the spin and the center-of-mass motion of the atom. In addition, we study the dynamics of the monopoles with the polar-core vortex and the dynam-

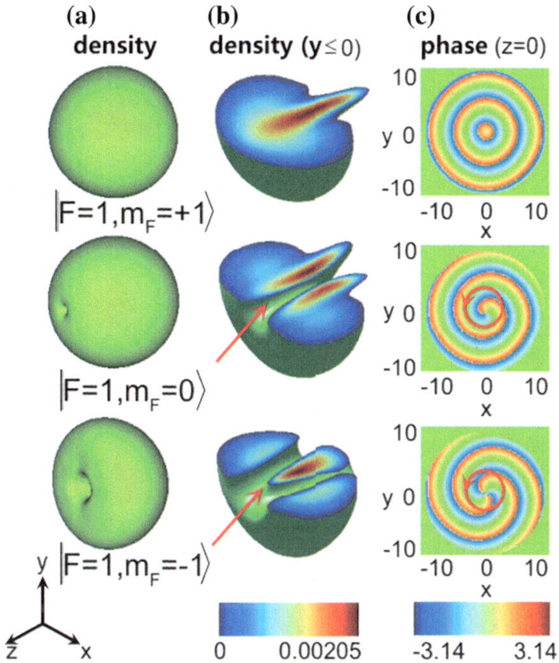

Fig. 10.42 The monopoles with the Mermin–Ho vortex. **a** Isosurface of particle densities. **b** Segments of isosurface of particle densities ($y \leq 0$). The position of the nodal line (Dirac string) is highlighted by the red arrow. **c** Phase distributions in the $z = 0$ planes. The single vortex ($m_F = 0$) and double vortex ($m_F = -1$) have the same circulations, as highlighted by the red circles. The simulation uses $\lambda_2 = -15$ and $\kappa = 2$ with the other parameters being the same as the ones in Fig. 10.41. Reprint from Ref. [20], Copyright 2018, with permission from American Physical Society

ics of the monopoles when the quadrupole field is turned off. We further discuss the influence of the anisotropic optical trap and the magnetic field gradient on the monopoles.

Next we investigate the influence of the interaction on the morphology of the monopoles. As shown in Fig. 10.42, the monopole with a Mermin–Ho vortex is obtained for a smaller λ_2 value. The monopole with a Mermin–Ho vortex represents a soliton in the $m_F = +1$ component, a single vortex line in the $m_F = 0$ component, and a double quantized vortex line in the $m_F = -1$ component. In the previous study, the monopole with a Mermin–Ho vortex has been observed in spinor BECs with a synthetic magnetic field. Figure 10.42 confirms that the monopole with a Mermin–Ho vortex can also exist in the condensates with the SO coupling.

10.5.2 Dirac Monopoles with Square Lattice

The monopoles with the square lattice are found as the SO coupling strength increases, as shown in Fig. 10.43. The square lattice is observed in the central region of the BECs. Meanwhile, the density distribution in the longitudinal direction is of the stripe structure. The phases at both sides of the $x = 0$ and $y = 0$ slice planes are inverse for all spin components. The morphology of the monopole with the square

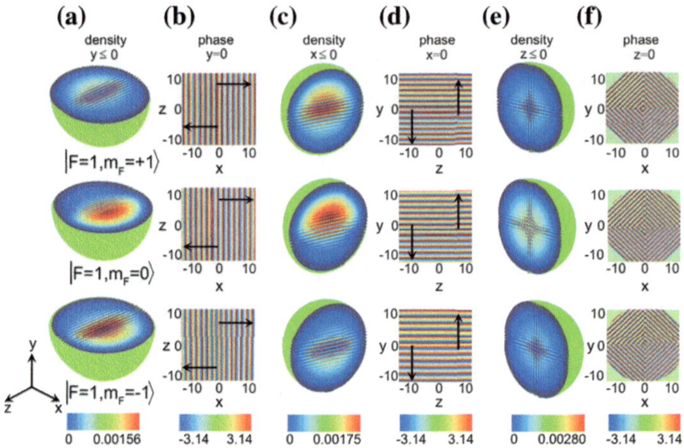

Fig. 10.43 The monopoles with the square lattice. **a**, **c**, and **e** Particle densities for $y \leq 0$, $x \leq 0$, and $z \leq 0$. **b**, **d**, and **f** Phase distributions in the $y = 0$, $x = 0$, and $z = 0$ planes. The phases at both sides of the $y = 0$ and $x = 0$ planes are inverse, as highlighted by the black arrows. The simulation uses $\kappa = 16$ with the other parameters being the same as the ones in Fig. 10.41. Reprint from Ref. [20], Copyright 2018, with permission from American Physical Society

lattice is not affected by the interactions, because the corresponding lowest energy state is actually not affected by the interactions when the SO coupling is strong. Finally, we also briefly discuss the influence of the anisotropic SO coupling on the structure of the monopole. The monopole may be canceled because the Dirac strings are removed in the case of anisotropic SO coupling.

The spinor BECs can be considered as a magnetic system, which can reflect the physical properties of topological defects. We study topological spin textures of the monopoles. The components of the spin vector are given by $F_x = \left[\psi_1^*\psi_0 + \psi_0^*(\psi_1 + \psi_{-1}) + \psi_{-1}^*\psi_0\right]/\sqrt{2}$, $F_y = i\left[-\psi_1^*\psi_0 + \psi_0^*(\psi_1 - \psi_{-1}) + \psi_{-1}^*\psi_0\right]/\sqrt{2}$, $F_z = |\psi_1|^2 - |\psi_{-1}|^2$. The spin textures of the monopoles with the polar-core vortex are shown in Fig. 10.44a. The spin aligns with the radially inward hedgehog distribution in the $x - y$ plane, representing the spin texture of a south magnetic pole, while the spin textures in the $x - z$ and $y - z$ planes show the cross-hyperbolic distribution. Compared with the case without the SO coupling, in our case, the SO coupling changes the spin direction, which causes the north magnetic pole to transform into the south magnetic pole. As shown in Fig. 10.44b, the spin textures of the antimonopoles represent the north magnetic poles. For the case of the strong SO coupling [see Fig. 10.44c], the spin textures are divided into four portions in the

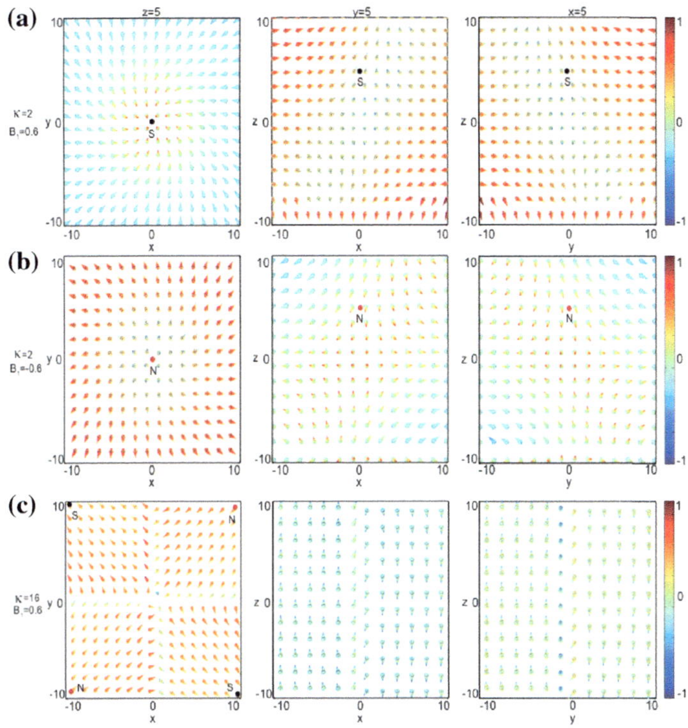

Fig. 10.44 **a** Spin textures of monopoles with the polar-core vortex describe the south magnetic poles (the black dot S). **b** Spin textures of the antimonopoles describe the north magnetic poles (the red dot N). **c** Spin textures of the monopoles with the square lattice. The other parameters are the same as the ones in Fig. 10.41. Reprint from Ref. [20], Copyright 2018, with permission from American Physical Society

$x - y$ plane, which behaves as the ferromagnetic distribution at each portion. The spin orientations at two diagonal portions are opposite, which indicates the structures of monopoles and antimonopoles that locate in the boundaries of the condensates. The spin textures in the $x - z$ and $y - z$ planes form the spin stripe. The dynamic evolution of the spin texture is shown in Fig. 10.45, which further proves that the monopoles with the polar-core vortex are stable.

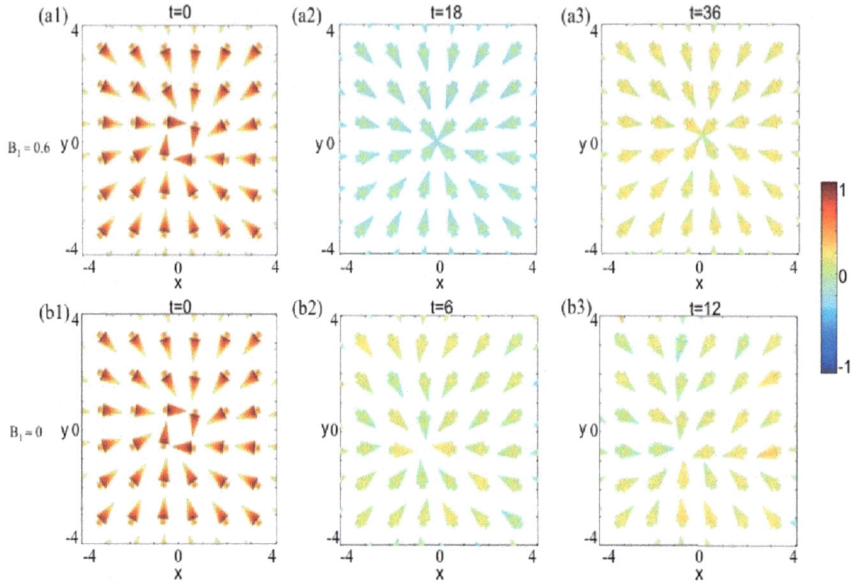

Fig. 10.45 **a1–a3** Dynamic evolution of the spin texture of the monopoles with the polar-core vortex for the $x - y$ plane in the presence of the quadrupole field, where $B_1 = 0.6$, $\kappa = 2$, $\lambda_2 = -75$, $\lambda_0 = 7500$, $\omega_r = \omega_z = 2\pi \times 250\,\text{Hz}$, and the real time $t' = 0.64t$ ms. **a1** $t = 0$, **a2** $t = 18$, and **a3** $t = 36$. **b1–b3** Dynamic evolution of the spin texture of the monopoles with the polar-core vortex for the $x - y$ plane when the quadrupole field is turned off, where $B_1 = 0$ and the other parameters are the same as those in **a1–a3**. **b1** $t = 0$, **b2** $t = 6$, and **b3** $t = 12$. Reprint from Ref. [20], Copyright 2018, with permission from American Physical Society

10.5.3 Phase Diagram and Experimental Protocol

The interplay between the SO coupling and the interaction leads to the rich phase diagrams of the monopoles. Figure 10.46a shows the phase diagram of the monopole structure as functions of κ and λ_0 for $\lambda_2 = -75$. The monopole is metastable when κ is less than a critical value $\kappa_c = 0.8$. The corresponding monopole structure is the same as the monopole structure when $\kappa = 0$, as shown in Fig. 10.47. For $0.8 < \kappa \leq 3$, the monopoles with the polar-core vortex exist when $0 < \lambda_0 \leq 3000$ and $5000 < \lambda_0 \leq 8000$, while the monopoles with the Mermin–Ho vortex emerge when $3000 < \lambda_0 \leq 5000$ and $8000 < \lambda_0 \leq 11000$. Because the spin-independent interaction causes the variation of the local magnetic order, a cyclic phase transition from Dirac monopoles with a polar-core vortex to those with a Mermin–Ho vortex can occur as the increasing λ_0. The monopoles with the square lattice appear when $\kappa > 3$, confirming that the monopoles are not affected by the interactions and are able to exist in the wide parameter region for the case of the strong SO coupling. Figure 10.46b shows the phase diagram of the monopole structure as functions of κ and λ_2 for $\lambda_0 = 7500$. The obtained monopole is metastable for $\kappa \leq 0.8$, where the

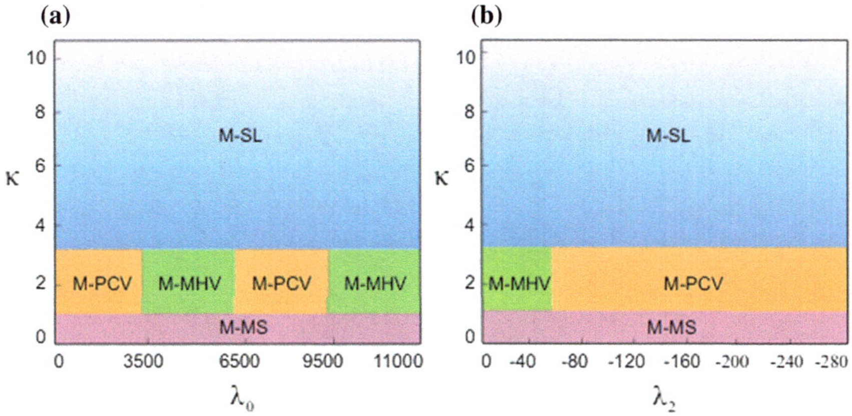

Fig. 10.46 Phase diagram of the monopole with the SO coupling strength κ and spin-independent interaction parameter λ_0 for a given spin-dependent interaction parameter λ_2. **b** Phase diagram of the monopole with κ and λ_2 for a given λ_0. Four phases are identified, including the metastable monopoles (M-MS), the monopoles with the polar-core vortex (M-PCV), the monopoles with the Mermin–Ho vortex (M-MHV), and the monopoles with the square lattice (M-SL). Reprint from Ref. [20], Copyright 2018, with permission from American Physical Society

monopole structure is the same as the structure when $\kappa = 0$. For $0.8 < \kappa \leq 3$, there exist the monopoles with the Mermin–Ho vortex when $\lambda_2 < 60$, while the monopoles with the polar-core vortex emerge when $\lambda_2 > 60$. Increasing λ_2 leads to the direct phase transition from the monopoles with the Mermin–Ho vortex to the monopoles with the polar-core vortex.

Finally, we discuss the experimental feasibility of creating the monopoles in SO-coupled BECs. We consider spin-1 BECs of alkali ^{87}Rb atoms, where particle number $N \approx 0.6(1.8) \times 10^5$. First, a quadrupole field realized by a pair of Helmholtz coils is turned on. The strength of the magnetic field is zero at the center of the quadrupole field, which corresponds to a point source of the superfluid flow. Afterward, when the formed point source of the superfluid flow is stable, the quadrupole field is turned off. At the same time, the pulsed magnetic fields are utilized to produce two-dimensional SO coupling [2]. We take some parameters from the recent experiments [2], including the isotropic optical trap $\omega_r = \omega_z = 2\pi \times 250\,\text{Hz}$, the anisotropic optical trap $\omega_r = 2\pi \times 160Z\,\text{Hz}$ and $\omega_z = 2\pi \times 220\,\text{Hz}$, the constant bias magnetic field $B^{(0)} = 20\,\text{G}$, and the quadrupole field gradient $B_1' = 0.03 - 0.1\,\text{T/m}$. In the case of the monopoles with the polar-core vortex, the SO coupling strength $\kappa \sim 0.8 - 3$. We find the dynamic period of the monopoles with the polar-core vortex $t' \sim 11\,\text{ms}$, which can be stably observed in experiments.

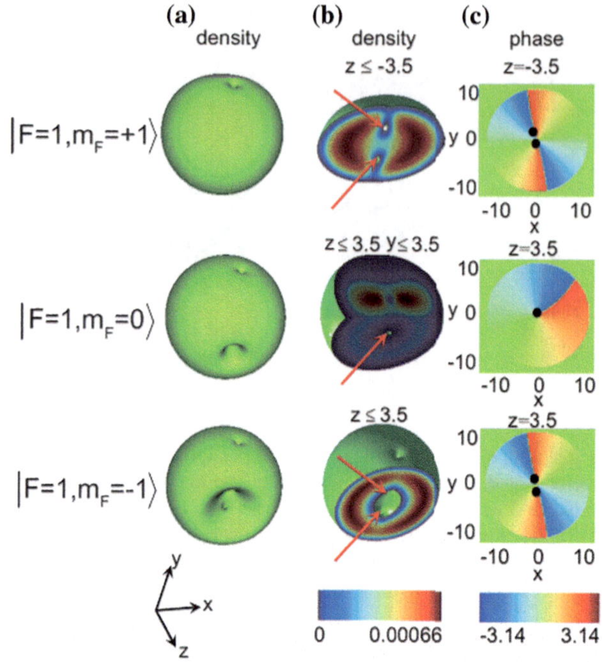

Fig. 10.47 **a** Isosurface of particle densities. **b** Segments of the isosurface of particle densities. **c** Phase distributions of slice planes. The black dots indicate the locations of vortices. The monopole is metastable, and a double vortex line has split into two single vortex lines partially, as highlighted by the red arrows. The simulation uses $\kappa = 0$, and the other parameters used in numerical calculation are the same as the ones in Fig. 10.41. Reprint from Ref. [20], Copyright 2018, with permission from American Physical Society

10.6 Non-Abelian Josephson Effect in $F = 2$ Spinor Bose–Einstein Condensates with Double Optical Traps

10.6.1 On the Five-Component NLS for Spin-2 BECs

In recent experiments, spin-2 BEC has been realized in ^{87}Rb and ^{23}Na systems with $F = 2$. The spin-2 condensates exhibit rich phenomenon, and many properties have been investigated including magnetic phases, charged vortices, and the phase separation, etc. In the mean-field theory, the $F = 2$ spinor BEC is characterized by the local order parameter (or the macroscopic wave function) with five components $\Psi(\mathbf{r},t) = (\psi_2, \psi_1, \psi_0, \psi_{-1}, \psi_{-2})^T$ (ψ_l for $m_F = l$ state, $l = -2, -1, 0, 1, 2$), reflecting the five spin degrees of freedom. At temperature T much smaller than the critical temperature T_c [15], the macroscopic wave function Ψ of the spin-2 BEC can be well described by the multicomponent NLS equations in the compact form as [11, 15] $i\hbar\frac{\partial\Psi}{\partial t} = [\widetilde{H} + c_0\rho - \widetilde{p} f_z + \widetilde{q} f_z^2 + c_1\mathbf{F}\cdot\mathbf{f}]\Psi + c_2 A_{00}\mathbf{A}\Psi$, where $\widetilde{H} = -\frac{\hbar^2}{2m}\nabla^2 + \widetilde{V}(x)$, is the single-particle Hamiltonian, $\widetilde{p}$ and $\widetilde{q}$ are the linear and quadratic Zeeman energy shifts, respectively. $\rho = |\Psi|^2 = \sum_{l=-2}^{2} |\psi_l|^2$ is the total density. $c_0 = \frac{4g_2+3g_4}{7}$, $c_1 = \frac{g_4-g_2}{7}$ and $c_2 = \frac{7g_0-10g_2+3g_4}{7}$ characterizes the spin-independent interaction, spin–exchange interaction and spin–singlet interaction, respectively, with $g_k = \frac{4\pi\hbar^2}{m} a_k$ $(k = 0, 2, 4)$ and a_k being the s-wave scattering length for scattering channel of total hyperfine spin k. The spin-2 matrices $\mathbf{f} = (f_x, f_y, f_z)^T$ are given as

$$f_x = \begin{pmatrix} 0 & 1 & 0 & 0 & 0 \\ 1 & 0 & \sqrt{\frac{3}{2}} & 0 & 0 \\ 0 & \sqrt{\frac{3}{2}} & 0 & \sqrt{\frac{3}{2}} & 0 \\ 0 & 0 & \sqrt{\frac{3}{2}} & 0 & 1 \\ 0 & 0 & 0 & 1 & 0 \end{pmatrix}, \quad f_y = i \begin{pmatrix} 0 & -1 & 0 & 0 & 0 \\ 1 & 0 & -\sqrt{\frac{3}{2}} & 0 & 0 \\ 0 & \sqrt{\frac{3}{2}} & 0 & -\sqrt{\frac{3}{2}} & 0 \\ 0 & 0 & \sqrt{\frac{3}{2}} & 0 & -1 \\ 0 & 0 & 0 & 1 & 0 \end{pmatrix},$$
$$f_z = diag\,(2, 1, 0, -1, -2)\,,$$

and the spin vector $\mathbf{F} = \mathbf{F}(\Psi) = (F_x(\mathbf{r}), F_y(\mathbf{r}), F_z(\mathbf{r}))^T$ with $F_\alpha(\mathbf{r}) = \Psi^* f_\alpha \Psi$ $(\alpha = x, y, z)$ is given as

$$F_x = \overline{\psi}_2\psi_1 + \overline{\psi}_1\psi_2 + \overline{\psi}_{-2}\psi_{-1} + \overline{\psi}_{-1}\psi_{-2} + \sqrt{\frac{3}{2}}\left(\overline{\psi}_1\psi_0 + \overline{\psi}_0\psi_1 + \overline{\psi}_0\psi_{-1} + \overline{\psi}_{-1}\psi_0\right)$$
$$F_y = i\left[\overline{\psi}_1\psi_2 - \overline{\psi}_2\psi_1 + \overline{\psi}_{-2}\psi_{-1} - \overline{\psi}_{-1}\psi_{-2} + \sqrt{\frac{3}{2}}\left(\overline{\psi}_0\psi_1 - \overline{\psi}_1\psi_0 + \overline{\psi}_{-1}\psi_0 - \overline{\psi}_0\psi_{-1}\right)\right]$$
$$F_z = 2|\psi_2|^2 + |\psi_1|^2 - |\psi_{-1}|^2 - 2|\psi_{-2}|^2,$$

with $F \cdot f = F_x f_x + F_y f_y + F_z f_z$. The matric $\mathbf{A}$ is given as

$$\mathbf{A} = \frac{\sqrt{5}}{5}\begin{pmatrix} 0 & 0 & 0 & 0 & 1 \\ 0 & 0 & 0 & -1 & 0 \\ 0 & 0 & 1 & 0 & 0 \\ 0 & -1 & 0 & 0 & 0 \\ 1 & 0 & 0 & 0 & 0 \end{pmatrix}$$

and $A_{00} := A_{00}(\Psi) = \Psi^T \mathbf{A} \Psi$ with $A_{00} = \frac{\sqrt{5}}{5}\,(2\psi_2(\mathbf{r})\psi_{-2}(\mathbf{r}) - 2\psi_1(\mathbf{r})\psi_{-1}(\mathbf{r}) + \psi_0^2(\mathbf{r}))$.

Following Kawaguchi and Ueda [15], the time-dependent multicomponent NLS equations for the spin-2 case is given as

$$
\begin{aligned}
i\hbar\frac{\partial\psi_{\pm 2}}{\partial t} &= \left[-\frac{\hbar^2}{2M}\nabla^2 + U_{\text{trap}(\mathbf{r})} \mp 2p + 4q + c_0 n \pm 2c_1 F_z - \mu\right]\psi_{\pm 2} \\
&\quad + c_1 F_{\mp}\psi_{\pm 1} + \frac{c_2}{\sqrt{5}} A_{00}\psi^*_{\mp 2}, \\
i\hbar\frac{\partial\psi_{\pm 1}}{\partial t} &= \left[-\frac{\hbar^2}{2M}\nabla^2 + U_{\text{trap}(\mathbf{r})} \mp p + q + c_0 n \pm c_1 F_z - \mu\right]\psi_{\pm 1} \\
&\quad + c_1\left(\sqrt{\frac{3}{2}} F_{\mp}\psi_0 + F_{\pm}\psi_{\pm 2}\right) - \frac{c_2}{\sqrt{5}} A_{00}\psi^*_{\mp 1}, \\
i\hbar\frac{\partial\psi_0}{\partial t} &= \left[-\frac{\hbar^2}{2M}\nabla^2 + U_{\text{trap}(\mathbf{r})} + c_0 n - \mu\right]\psi_0 \\
&\quad + \sqrt{\frac{3}{2}} c_1 \left(F_+\psi_1 + F_-\psi_{-1}\right) + \frac{c_2}{\sqrt{5}} A_{00}\psi^*_0,
\end{aligned}
$$

where $F_+ = F_-^* = 2\left(\psi_2^*\psi_1 + \psi_{-1}^*\psi_{-2}\right) + \sqrt{6}\left(\psi_1^*\psi_0 + \psi_0^*\psi_{-1}\right)$, $F_z = 2\left(|\psi_2|^2 - |\psi_{-2}|^2\right) + |\psi_1|^2 - |\psi_{-1}|^2$.

10.6.2 On the Josephson Effect

The Josephson effect is a quantum tunneling phenomenon which occurs when a pair of superconducting or superfluid systems are weakly linked by some kind of physical barrier. With the rapid experimental progress in cold atom physics, the Josephson junction has been realized for the trapped BECs of ^{87}Rb and ^{23}Na [1, 37]. However, most of the extensive studies about this effect focus on the Abelian case so far, in terms of a junction of two systems with spontaneously broken Abelian symmetry [4]. In the present section, we study the non-Abelian Josephson effect between two $F = 2$ spinor BECs with double optical traps. We propose a real physical system which contains non-Abelian Josephson effect and has very different density and spin tunneling characters compared with the Abelian case. We evaluate the frequencies of the pseudo-Goldstone modes in different phases between two traps, respectively, which are the crucial feature of the non-Abelian Josephson effect [32].

10.6.3 Ground State Structure

We consider a system of a homogenous spin-2 Bose gas with s-wave interaction. This system can be described by the following mean-field free energy:

$$
F(\psi) = \frac{1}{2}\left[c_0\left(\psi^\dagger\psi\right)^2 + c_1\langle\mathbf{f}\rangle^2 + \frac{c_2}{5}|\Theta|^2\right] - \mu\psi^\dagger\psi, \tag{10.53}
$$

where $\psi = (\psi_2, \psi_1, \psi_0, \psi_{-1}, \psi_{-2})^T$ is the order parameter of the system, c_0, c_1, and c_2 are interaction strengths related to the scattering length in different spin channels, $\langle \mathbf{f} \rangle = \psi^\dagger \mathbf{f} \psi$ is the mean value of the spin operator, and $\Theta = \sum_{a=-2}^{2} (-1)^a \psi_a \psi_{-a}$ represents a single pair of identical spin-2 particles. The ground state configuration can be determined by minimizing the free energy (10.53). There are several distinct phases in this system. We will discuss these phases under zero magnetic field and analyze the symmetry and low-lying excitation spectrum of each phase:

(I) Ferromagnetic phases: When $c_1 < 0$ and $c_1 - c_2/20 < 0$, two kinds of ferromagnetic phases are energetically favored. The corresponding ground state configurations are given by $\psi = \sqrt{n} \exp[i\theta] (1, 0, 0, 0, 0)$ or $\psi = \sqrt{n} \exp[i\theta]$ $(0, 1, 0, 0, 0)$, where $n = \mu/(c_0 + 4c_1)$ is the particle density and is an arbitrary global phase. It is obvious that these ground states have a $U(1)$ symmetry which leads to only one massless Goldstone mode. Therefore, two uncoupled systems have a $U(1) \otimes U(1)$ symmetry. This symmetry will softly break into a $U(1)$ diagonal symmetry when a weak coupling is applied. This pattern of symmetry breaking corresponds to an Abelian Josephson effect. The low-lying excitation spectrum of this state has been derived as $\omega_k = \sqrt{\epsilon_k (\epsilon_k + 2g_4 n)}$, where $\epsilon_k = k^2$ and $g_4 = c_0 + 4c_1$. It should be noted that this Goldstone mode will break into two modes when the coupling is applied: one zero energy mode and one pseudo-Goldstone mode. This pseudo-Goldstone mode has a finite but small gap and leads to a density mode fluctuation in dc Josephson current.

(II) Antiferromagnetic phase: In the absence of magnetic field, there is only one kind of antiferromagnetic phase when $c_2 < 0$ and $c_1 - c_2/20 > 0$ are satisfied. The corresponding

1. ground state configuration is degenerate with respect to five continuous variables which lead to five massless Goldstone modes. Four of them correspond to the $U(1) \times SO(3)$ symmetry of the free energy. The extra degeneracy besides the $U(1) \times SO(3)$ symmetry is correct only on the mean-field level and will be removed when the quantum fluctuation is considered. In what follows, we will work on the mean-field level and maintain all the five Goldstone modes.

(III) Cyclic phase: When $c_1 > 0$ and $c_2 > 0$, the cyclic phase is energetically favored. The ground state configuration is given by $\psi = \sqrt{n} \left(\frac{\exp[i\theta_2]}{2}, 0, \frac{\exp[i\theta_0]}{2}, 0, \frac{\exp[i\theta_{-2}]}{2} \right)$ where $n = \mu/c_0$ is the particle density and the global phase $\theta_{\pm 2}$ and θ_0 satisfy the condition $\theta_2 + \theta_{-2} - 2\theta_0 = \pi$ [32]. Using the Schwinger boson representation [17], one can see that the ground state of cyclic phase is mapped to a tetrahedron on the unit sphere. Therefore, the ground state of cyclic phase has a full $U(1) \times SO(3)$ degeneracy and leads to four Goldstone modes. We will show that these modes also lead to non-Abelian Josephson effect.

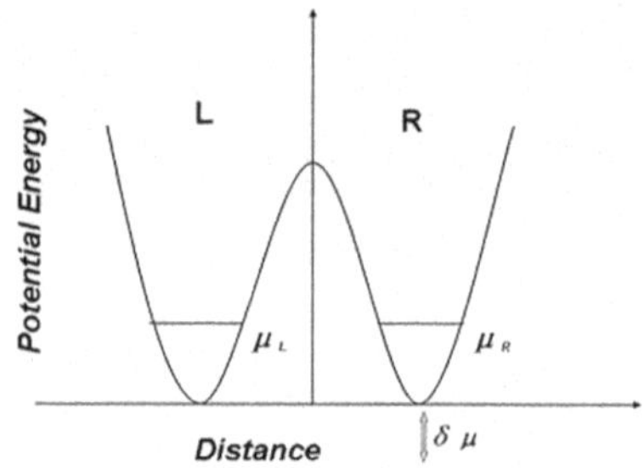

Fig. 10.48 The experimental schematic of a spin-2 Bose gas trapped in a double well with chemical potentials μ_L of left and μ_R of right trap, which initially satisfy $\mu_L = \mu_R$. To drive Josephson effect, we add a small distortion δ_μ to μ_R. Reprint from Ref. [32], Copyright 2018, with permission from American Physical Society

10.6.4 Non-Abelian Josephson Effect

Here, we analyze the Josephson effect of a spin-2 BEC system in a double-well optical trap, as shown in Fig. 10.48. We assume that the energy barrier between the two wells is strong enough so that the coupling between the Bose gas in each well is very weak and the overlap of the ground state wave functions in left and right well [which we denote as $\varphi_L(\mathbf{x})$ and $\varphi_R(\mathbf{x})$] can be safely neglected. We will also use the single mode approximation which means we take the same mode function for all five spin components; this is a widely used approximation, and it is valid when the spin interaction is symmetric. Under these assumptions, the system can be described by the following potential:

$$V_{\text{couple}} = \mathbf{F}(\psi_L) + \mathbf{F}(\psi_R) - \mathbf{J}\left(\psi_L^\dagger\psi_R + \psi_R^\dagger\psi_L\right), \tag{10.54}$$

where ψ_L and ψ_R are the order parameter of the Bose system in the left and right well, respectively, and $\mathbf{J}$ is the coupling parameter. It should be noted that we have taken the same chemical potential for the Bose gas in the right and left well, because we will be interested in only the dc Josephson effect which captures the essence of the non-Abelian symmetry breaking as simple as possible. Using the dynamic equations of the condensate $i\frac{d}{dt}\psi_{aL} = \frac{\delta V_{\text{couple}}}{\delta\psi_{aL}^*}$ and $i\frac{d}{dt}\psi_{aR} = \frac{\delta V_{\text{couple}}}{\delta\psi_{aR}^*}$, we can derive the equation of motion of the Josephson current in each phase and analyze the pseudo-Goldstone modes. However, as we have mentioned above, the symmetry of the ground state in ferromagnetic phase only leads to an Abelian Josephson effect, which is not of interest in the present book. Therefore, we will just analyze the antiferromagnetic phase and cyclic phase which are important realizations of non-Abelian Josephson effect.

(I) *Antiferromagnetic phase*: To obtain the pseudo-Goldstone modes, we need to linearize the equation of motion around the ground state $\psi_L = \psi_R = \left(\sqrt{n}/\sqrt{2}\right)$

$(1, 0, 0, 0, 1)$. By minimizing the potential V_{couple} under the above antiferromagnetic ground state, we get $n = (\mu + \mathbf{J}) / (c_0 + c_2/5)$. With this ground state configuration, we can obtain the linearized equations of the fluctuations in this phase and analyze the excitation spectrum.

(i) *The* $m = 0$ *mode*: The equation of motion of the $m = 0$ mode is given as

$$i\frac{d}{dt}\phi_{0L} = \left(-\frac{c_2}{5}n + \mathbf{J}\right)\phi_{0L} + \frac{c_2}{5}n\phi_{0L}^* - \mathbf{J}\phi_{0R}, \tag{10.55}$$

and a similar equation of ϕ_{0R}. Since this mode is decoupled from others, it corresponds to an Abelian Josephson current. By solving Eq. (10.55), we obtain the eigenenergies of this mode, one is zero corresponding to a massless Goldstone mode and the other is $\omega_0 = 2\sqrt{\mathbf{J}\left(\mathbf{J} + \frac{2|c_2|n}{5}\right)}$ corresponding to a pseudo-Goldstone mode.

(ii) *The* $m = \pm 1$ *coupled modes*: The $m = \pm 1$ modes are coupled in the following equations:

$$\begin{aligned} i\tfrac{d}{dt}\phi_{1L} &= \left[n\left(c_1 - \tfrac{c_2}{5}\right) + \mathbf{J}\right]\phi_{1L} + n\left(c_1 - \tfrac{c_2}{5}\right)\phi_{-1L}^* - \mathbf{J}\phi_{1R}, \\ i\tfrac{d}{dt}\phi_{-1L} &= \left[n\left(c_1 - \tfrac{c_2}{5}\right) + \mathbf{J}\right]\phi_{-1L} + n\left(c_1 - \tfrac{c_2}{5}\right)\phi_{1L}^* - \mathbf{J}\phi_{-1R}, \end{aligned} \tag{10.56}$$

and a similar set of equations of $\phi_{\pm 1R}$. The solution involves two pseudo-Goldstone modes with the same gap of $\omega_{\pm 1} = 2\sqrt{n\left(c_1 - \frac{c_2}{5}\right)\mathbf{J} + \mathbf{J}^2}$.

(iii) *The* $m = \pm 2$ *coupled modes*: The $m = \pm 2$ modes are coupled together in a set of equations similar to the ones of the $m = \pm 1$ case. We will not list the detailed equations of motion but just give the result here and after. We obtain two pseudo-Goldstone modes with energy gap $\omega_{\pm 2}^{(1)} = 2\sqrt{n\left(c_0 + \frac{c_2}{5}\right)\mathbf{J} + \mathbf{J}^2}$ and $\omega_{\pm 2}^{(2)} = 2\sqrt{n\left(c_1 - \frac{c_2}{20}\right)\mathbf{J} + \mathbf{J}^2}$. We can see that there are a total of five pseudo-Goldstone modes with four different gaps in this phase which consists with our analysis of the ground state degeneracy. In Fig. 10.49, we give the dependence of the above frequencies on coupling parameter $\mathbf{J}$. Recently, a polar behavior has been observed in the $F = 2$ ground state of ^{87}Rb condensate. We expect that the above modes of fluctuations can be observed in this system in future experiments. In the case of the ^{87}Rb system, the value of interacting strengths under typical experimental conditions is given as $c_1 n : 0 - 10$ nK, $c_2 n : 0 - 0.2$ nK, and $c_0 n$ about 150 nK. According to the weak coupling limit, we assume that the coupling parameter $\mathbf{J}$ is much smaller than the interaction energy of the condensate and given as about 0.1 nK. Under these conditions, we can obtain the frequencies of the fluctuation related to the antiferromagnetic phase, which is of order 100 Hz. The measurement of fluctuations on this characteristic timescale (about 10 ms) is accessible in current experiments.

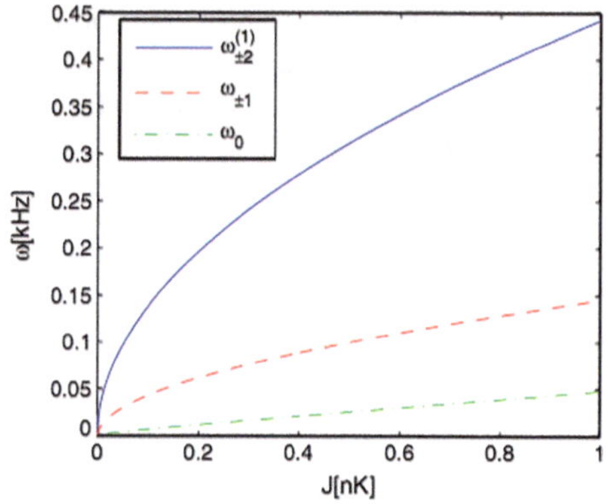

Fig. 10.49 The frequencies of the pseudo-Goldstone modes as a function of coupling parameter **J** in the case of antiferromagnetic phase. All of the frequencies are proportional to $\sqrt{\mathbf{J}}$ when **J** approaches zero and the dependence on **J** becomes linearizing when **J** is large compared to the interaction energy. Reprint from Ref. [32], Copyright 2018, with permission from American Physical Society

(II) *Cyclic phase*: Following the same procedure in the antiferromagnetic phase we can obtain the pseudo-Goldstone modes for the cyclic phase.

(i) *The $m = \pm 2, 0$ coupled modes.* We find that each Goldstone mode in the corresponding uncoupled system breaks into one massless mode and one pseudo-Goldstone mode. Since there are two Goldstone modes in the uncoupled system, we find two pseudo-Goldstone modes with energy $\omega^{(1)}_{0,\pm2} = 2\sqrt{\mathbf{J}^2 + c_0 n\mathbf{J}}$ and $\omega^{(2)}_{0,\pm2} = 2\sqrt{\mathbf{J}^2 + 2c_1 n\mathbf{J}}$.

(ii) *The $m = \pm 1$ coupled modes.* As we know, there are two massless Goldstone modes with the same energy in the uncoupled system. By solving the equation of motions for this mode, we find that each of them leads to a pseudo-Goldstone mode with a gap $\omega_{\pm1} = 2\sqrt{\mathbf{J}^2 + 2c_1 n\mathbf{J}}$. There are four pseudo-Goldstone modes in the cyclic phase which is consistent with our previous analysis on the symmetry of this phase. These kinds of fluctuations in cyclic phase are expected to be realized in a condensate of ^{85}Rb atoms [8]. Based on the current estimates of scattering lengths, the value of interacting strengths under typical experimental condition is given as: $c_1 n : 0 - 20\,\text{nK}$, $c_2 n : 0 - 0.6\,\text{nK}$, and $c_0 n$ about 600 nK. Under these conditions, we can also estimate the dc Josephson frequencies in cyclic phase, which is about 100–300 Hz. The dependence of the above frequencies on coupling parameter **J** is shown in Fig. 10.50.

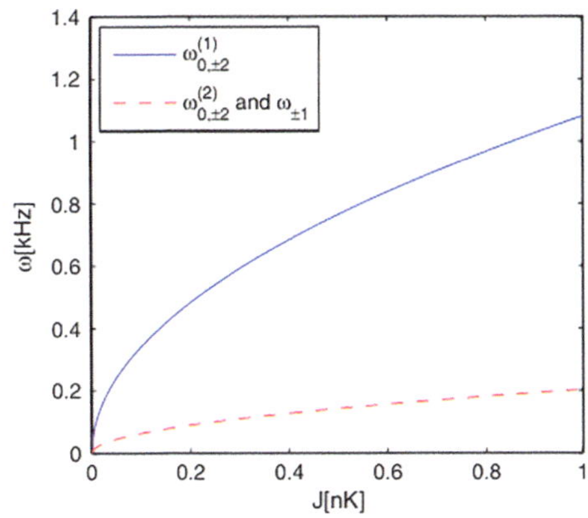

Fig. 10.50 The frequencies of pseudo-Goldstone modes as a function of coupling parameter **J** in the case of cyclic phase. The dependence on **J** is similar to the case in antiferromagnetic phase. However, there are two degenerate modes $\omega^{(2)}_{0,\pm2}$ and $\omega_{\pm1}$ in this case, which are originated from the two degenerate Bogoliubov modes in the uncoupled system. Reprint from Ref. [32], Copyright 2018, with permission from American Physical Society

10.6.5 Experimental Signatures of Non-Abelian Josephson Effect

The experimental setup of a spin-2 Bose gas trapped in a double well is illustrated in Fig. 10.48. The dc non-Abelian Josephson current can be detected with the following steps. The first step is to initiate a density oscillation in the system. This can be realized by slightly changing the depth of one well, which will cause a small imbalance in chemical potential ($\mu_R \to \mu_R + \delta\mu$) between the two wells, and then tuning it back. The next step is to detect the time dependence of the particle numbers in different spin components. Such kind of detection can be realized by first spatially separating different spin components with a Stern–Gerlach method during the time of flight after switching off the trapping potential. Then, the number of atoms in each spin component will be related to the respective spatial density distributions which can be evaluated by the absorption imaging method. Following the above steps, one can measure the density oscillation in each of the spin components which are coupled together due to the non-Abelian symmetry of the system. The measurement on the dependence of the oscillation frequencies on **J** can be realized by varying the barrier between the two wells and repeating the above measurement.

In mean-field theory, the condensates of ^{85}Rb atom in $F = 2$ state are predicted to be polar ($c_1 - c_2/20 > 0$ and $c_2 < 0$), but close to the border to cyclic phase ($c_1 > 0$ and $c_2 > 0$) [8, 36]. Furthermore, polar behavior in the $F = 2$ ground state of ^{85}Rb condensate has been observed in the recent experiment [8, 36]. As a result, we expect that the pseudo-Goldstone modes of the antiferromagnetic phase could be observed in experiments. As we have mentioned, the value of interacting strengths under typical experimental conditions is given as [35]: $c_1 n : 0 - 10\,\text{nK}$, $c_2 n : 0 - 0.2\,\text{nK}$, and $c_0 n$ about 150 nK, which leads to the fluctuation timescale of about 10 ms in this system. On this timescale, the measurement we proposed above is completely accessible within recent experimental technique in $F = 2$ spinor BECs of ^{85}Rb system [35]. In order to observe this dynamical oscillation clearly in the experiment, the temperature should be lower than the gap of the pseudo-Goldstone modes, which is about 1 – 10 nk. Although there is still no such kind of measurement performed in a system with cyclic phase, we expect that it will be realized in a condensate of ^{85}Rb atoms in the near future [8, 36].

References

1. Albiez, M., Gati, R., Fölling, J., Hunsmann, S., Cristiani, M., Oberthaler, M.K.: Direct observation of tunneling and nonlinear self-trapping in a single Bosonic Josephson junction. Phys. Rev. Lett. **95**, 010402 (2005)
2. Anderson, B.M., Spielman, I.B., Juzeliūnas, G.: Magnetically generated spin-orbit coupling for ultracold atoms. Phys. Rev. Lett. **111**, 125301 (2013)
3. Arfken, G.B., Weber, H.J., Harris, F.E.: Mathematical Methods for Physicists, 7th edn. Academic Press, New York (2000)
4. Barone, A., Paterno, G.: Physics and Applications of the Josephson Effect. Wiley, New York (1982)
5. Bookjans, E.M., Vinit, A., Raman, C.: Quantum phase transition in an antiferromagnetic spinor Bose-Einstein condensate. Phys. Rev. Lett. **107**, 195306 (2011)
6. Bradley, A.S., Anderson, B.P.: Energy spectra of vortex distributions in two-dimensional quantum turbulence. Phys. Rev. X **2**, 041001 (2012)
7. Busch, T., Anglin, J.R.: Wave-function monopoles in Bose-Einstein condensates. Phys. Rev. A **60**, R2669–R2672 (1999)
8. Ciobanu, C.V., Yip, S.-K., Ho, T.-L.: Phase diagrams of $F = 2$ spinor Bose-Einstein condensates. Phys. Rev. A **61**, 033607 (2000)
9. Han, W., Zhang, X.-F., Song, S.-W., Saito, H., Zhang, W., Liu, W.M., Zhang, S.-G.: Double-quantum spin vortices in SU(3) spin-orbit-coupled Bose gases. Phys. Rev. A **94**, 033629 (2016)
10. Higbie, J.M., Sadler, L.E., Inouye, S., Chikkatur, A.P., Leslie, S.R., Moore, K.L., Savalli, V., Stamper-Kurn, D.M.: Direct nondestructive imaging of magnetization in a spin-1 Bose-Einstein gas. Phys. Rev. Lett. **95**, 050401 (2005)
11. Ho, T.L.: Spinor Bose condensates in optical traps. Phys. Rev. Lett. **81**, 742 (1998)
12. Isoshima, T., Machida, K.: Axisymmetric vortices in spinor Bose-Einstein condensates under rotation. Phys. Rev. A **66**, 023602 (2002)
13. Kasamatsuv, K., Tsubota, M., Ueda, M.: Spin textures in rotating two-component Bose-Einstein condensates. Phys. Rev. A **71**, 043611 (2005)
14. Kasamatsu, K., Tsubota, M., Ueda, M.: Vortex molecules in coherently coupled two-component Bose-Einstein condensates. Phys. Rev. Lett. **93**, 250406 (2004)
15. Kawaguchi, Y., Ueda, M.: Spinor Bose-Einstein condensates. Phys. Rep. **520**, 253 (2012)
16. Kobayashi, J., Izumi, Y., Enomoto, K., Kumakura, M., Takahashi, Y.: Spinor molecule in atomic Bose-Einstein condensate. Appl. Phys. B **95**, 37 (2009)
17. Li, L., Li, Z.D., Malomed, B.A., Mihalache, D., Liu, W.M.: Exact soliton solutions and nonlinear modulation instability in spinor Bose-Einstein condensates. Phys. Rev. A **72**, 033611 (2005)
18. Li, X., Liu, W.V., Balents, L.: Spirals and skyrmions in two dimensional oxide heterostructures. Phys. Rev. Lett. **112**, 067202 (2014)
19. Li, Y., Pitaevskii, L.P., Stringari, S.: Quantum tricriticality and phase transitions in spin-orbit coupled Bose-Einstein condensates. Phys. Rev. Lett. **108**, 225301 (2012)
20. Li, J., Yu, Y.-M., Zhuang, L., Liu, W.M.: Dirac monopoles with a polar-core vortex induced by spin-orbit coupling in spinor Bose-Einstein condensates. Phys. Rev. A **95**, 043633 (2017)

21. Lin, Y.J., Jiménez-García, K., Spielman, I.B.: Spin-orbit-coupled Bose-Einstein condensates. Nature (London) **471**, 83 (2011)
22. Liu, C.F., Juzeliunas, G., Liu, W.M.: Spin-orbit coupling manipulating composite topological spin textures in atomic-molecular Bose-Einstein condensates. Phys. Rev. A **95**, 023624 (2017)
23. Liu, C.F., Liu, W.M.: Spin-orbit-coupling-induced half-skyrmion excitations in rotating and rapidly quenched spin-1 Bose-Einstein condensates. Phys. Rev. A **86**, 033602 (2012)
24. Liu, C.-F., Liu, W.M.: Spin-orbit coupling stabilizing half-quantum spin defect in spin-1 Bose-Einstein condensates. Opt. Express **25**, 307936 (2017)
25. Liu, C.F., Fan, H., Zhang, Y.C., Wang, D.S., Liu, W.M.: Circular-hyperbolic skyrmion in rotating pseudo-spin-1/2 Bose-Einstein condensates with spin-orbit coupling. Phys. Rev. A **86**, 053616 (2012)
26. Liu, C.F., Yu, Y.-M., Gou, S.-C., Liu, W.M.: Vortex chain in anisotropic spin-orbit-coupled spin-1 Bose-Einstein condensates. Phys. Rev. A **87**, 063630 (2013)
27. Mizushima, T., Machida, K., Kita, T.: Axisymmetric versus nonaxisymmetric vortices in spinor Bose-Einstein condensates. Phys. Rev. A **66**, 053610 (2002)
28. Mizushima, T., Kobayashi, N., Machida, K.: Coreless and singular vortex lattices in rotating spinor Bose-Einstein condensates. Phys. Rev. A **70**, 043613 (2004)
29. Nistazakis, H.E., Frantzeskakis, D.J., Kevrekidis, P.G., Malomed, B.A., Carretero-Gonzalez, R.: Bright-dark soliton complexes in spinor Bose-Einstein condensates. Phys. Rev. A **77**, 033612 (2008)
30. Olson, A.J., Wang, S.-J., Niffenegger, R.J., Li, C.-H., Greene, C.H., Chen, Y.P.: Tunable Landau-Zener transitions in a spin-orbit-coupled Bose-Einstein condensate. Phys. Rev. A **90**, 013616 (2014)
31. Pietilä, V., Möttönen, M.: Creation of Dirac monopoles in spinor Bose-Einstein condensates. Phys. Rev. Lett. **103**, 030401 (2009)
32. Qi, R., Yu, X.-L., Li, Z.B., Liu, W.M.: Non-Abelian Josephson effect between two F = 2 spinor Bose-Einstein condensates in double optical traps. Phys. Rev. Lett. **102**, 185301 (2009)
33. Radić, J., Sedrakyan, T.A., Spielman, I.B., Galitski, V.: Vortices in spin-orbit-coupled Bose-Einstein condensates. Phys. Rev. A **84**, 063604 (2011)
34. Sadler, L., Higbie, J.M., Leslie, S.R., Vengalattore, M., Stamper-Kurn, D.M.: Spontaneous symmetry breaking in a quenched ferromagnetic spinor Bose-Einstein condensate. Nature (London) **443**, 312 (2006)
35. Schmaljohann, H., Erhard, M., Kronjäger, J., Kottke, M., van Staa, S., Cacciapuoti, L., Arlt, J.J., Bongs, K., Sengstock, K.: Dynamics of F = 2 spinor Bose-Einstein condensates. Phys. Rev. Lett. **92**, 040402 (2004)
36. Schweikhard, V., Coddington, I., Engels, P., Tung, S., Cornell, E.A.: Vortex-lattice dynamics in rotating spinor Bose-Einstein condensates. Phys. Rev. Lett. **93**, 210403 (2004)
37. Shin, Y., Jo, G.-B., Saba, M., Pasquini, T.A., Ketterle, W., Pritchard, D.E.: Optical weak link between two spatially separated Bose-Einstein condensates. Phys. Rev. Lett. **95**, 170402 (2005)
38. Uchiyama, M., Ieda, J., Wadati, M., Low, J.: Soliton dynamics of F = 1 spinor Bose-Einstein condenstate with nonvanishing boundaries. Temp. Phys. **148**, 399 (2007)
39. Ueda, M., Kawaguchi, Y.: Topological aspects in spinor Bose-Einstein condensates. Phys. Rep. **520**, 253 (2012)
40. Wang, C., Gao, C., Jian, C.M., Zhai, H.: Spin-orbit coupled spinor Bose-Einstein condensates. Phys. Rev. Lett. **105**, 160403 (2010)
41. Wang, D.S., Shi, Y.R., Chow, K.W., Yu, Z.X., Li, X.G.: Matter-wave solitons in a spin-1 Bose-Einstein condensate. Eur. Phys. J. D **67**, 242 (2013)
42. Xu, Z.F., Kawaguchi, Y., You, L., Ueda, M.: Symmetry classification of spin-orbit-coupled spinor Bose-Einstein condensates. Phys. Rev. A **86**, 033628 (2012)
43. Zhang, X.-F., Dong, R.-F., Liu, T., Liu, W.M., Zhang, S.-G.: Spin-orbit-coupled Bose-Einstein condensates confined in concentrically coupled annular traps. Phys. Rev. A **86**, 063628 (2012)
44. Zhao, D., Song, S.-W., Wen, L., Li, Z.-D., Luo, H.-G., Liu, W.M.: Topological defects and inhomogeneous spin patterns induced by the quadratic Zeeman effect in spin-1 Bose-Einstein condensates. Phys. Rev. A **91**, 013619 (2015)

Chapter 11
Higher Dimensional Nonlinear Schrödinger Equations of Bose–Einstein Condensates

Abstract Based on the higher dimensional inhomogeneous nonlinear Schrödinger equations with external real/complex potential, we investigate in this chapter the dynamics of BECs in external potentials. A special attention will be paid on the both analytical and numerical studies of (i) BECs with both two- and three-bodies interactions, loaded into a 2D harmonic trapping potential, (ii) superfluid flows in inhomogeneous BECs, (iii) the ring dark solitons in 2D BECs with tunable interaction, (iv) the dynamics of 2D vortex quadrupoles in trapped BECs, and (v) the ring vortex solitons in 3D BECs under magnetic confinement.

11.1 Introduction

In the mean-field approximation, the dynamics of Bose–Einstein condensates in dilute ultracold bosonic gases are accurately described by the inhomogeneous nonlinear Schrödinger equation, where the spatial nonlinearity, being determined by the sign and magnitude of the s-wave scattering length, can be tuned by means of a spatially inhomogeneous magnetic and optical field in the vicinity of a Feshbach resonance (FR). A possibility of studying analytically the dynamics of nonlinear waves in BECs through NLS equations with inhomogeneous complex-valued coefficients have been reported. However, there is one main limitation of the presently available results. Most of the works on the analytical study of the dynamics of nonlinear waves in BECs were obtained for 1D or quasi-one-dimensional statements and only a few results on multidimensional problems were reported, so far.

In the mean-field approximation, the BEC dynamics at ultralow temperatures is accurately described by the inhomogeneous NLS equation with both two- and three-bodies interactions in n dimensions [24, 26, 32]

$$i\hbar\frac{\partial\Psi}{\partial t}=\left[-\frac{\hbar^2}{2m}\nabla^n+V(\mathbf{r},t)+G_3(\mathbf{r},t)\,|\Psi|^2+G_5(\mathbf{r},t)\,|\Psi|^4\right]\Psi, \qquad (11.1)$$

W.-M. Liu and E. Kengne, *Schrödinger Equations in Nonlinear Systems*,
https://doi.org/10.1007/978-981-13-6581-2_11

where t is the time, $\mathbf{r} = (x_1, x_2, \ldots, x_n)$ is the space coordinate, $\nabla^n = \partial^2/\partial x_1^2 + \partial^2/\partial x_2^2 + \cdots + \partial^2/\partial x_n^2$ is the n-dimensional Laplacian, and the wave function $\Psi(\mathbf{r}, t)$ is normalized by the particle number $N = \int_{\mathbf{R}^n} |\Psi|^2 \, d\mathbf{r}$; here, we focus our attention to $n = 3$. Function $G_3(\mathbf{r}, t) = 4\pi\hbar^2 a_s/m$ characterizes the strength of two-body interatomic interaction and is defined in terms of the s-wave scattering length a_s, which is allowed to be a function of $\mathbf{r}$ and t, the atomic mass m, and the Planck constant $\hbar$. $G_5(\mathbf{r}, t)$ represents the three-body interaction with the general form $G_5(\mathbf{r}, t) = 16\pi\hbar^2 a^4 (4\pi - 3\sqrt{3}) \ln(C\eta^2)/m$ for a Bose gas of hard spheres with radius a. The signs of strengths $G_3(\mathbf{r}, t)$ and $G_5(\mathbf{r}, t)$ can be positive or negative in the whole space, indicating that the interactions are repulsive or attractive, respectively. Throughout this chapter, the trapping potential is assumed to be harmonic and of the form $V(\mathbf{r}, t) = m\left(\omega_r^2 r^2 + \omega_z^2\right)/2$, where $r^2 = x^2 + y^2$, m is the atom mass, and ω_r and ω_z are the confinement frequencies in the radial and axial directions, respectively, with $\omega_r << \omega_z$. In this case, since $\frac{\omega_r}{\omega_z}$ is small, one can expect that the variation of the profile of the order parameter is slow in the radial direction. In order to control the dynamics of BECs in the trap, the trapping frequency in the radial axis ω_r can be a function of both time t and space x and y.

The present chapter deals with NLS equations (11.1) with inhomogeneous complex-valued coefficients, which describe the dynamics of BECs in external potentials.

11.2 2D Nonlinear Matter-Wave Functions in Bose–Einstein Condensates with Two- and Three-Bodies Interactions

Through the similarity transformation, we investigate analytically and numerically the nonlinear matter wave functions with odd and even parities in the framework of quasi-two-dimensional (quasi-2D) BECs with spatially modulated cubic–quintic nonlinearities in harmonic potential.

11.2.1 The Physical Model

In this section, we focus our attention on the situation that a BEC with both two- and three-bodies interactions is loaded into a 2D harmonic trapping potential. The dynamics of matter waves in such a BEC are governed by the inhomogeneous NLS equation (11.1) that we rewrite as

$$i\hbar\frac{\partial\Psi}{\partial t} = \left[-\frac{\hbar^2}{2m}\nabla^2 - \frac{\hbar^2}{2m}\frac{\partial^2}{\partial z^2} + V(\mathbf{r}, t) + G_3(\mathbf{r}, t)\,|\Psi|^2 + G_5(\mathbf{r}, t)\,|\Psi|^4\right]\Psi, \tag{11.2}$$

where $\nabla^2 = \partial^2/\partial x^2 + \partial^2/\partial y^2$ is the 2D Laplacian, and $V(\mathbf{r}, t) = m\left(\omega_r^2 r^2 + \omega_z^2\right)/2$. We assume that the motion of atoms in the axial direction is essentially frozen; that is, the trap is pancake shaped and $\omega_r << \omega_z$. Due to the strong confinement in the axial direction (z), the motion of the order parameter in the axial direction can be described by the ground-state wavefunction of a 1D quantum harmonic oscillator. Thus, by introducing the following transformation [9]:

$$\Psi(\mathbf{r}, t) = \frac{a_z}{\sqrt{\pi a_z^2}}\psi\left(\frac{x}{a_z}, \frac{y}{a_z}, \omega_z t\right)\exp\left[-i\frac{\omega_z}{2}t - \frac{1}{2a_z^2}z^2\right], \quad a_z = \sqrt{\hbar/\left[m\omega_z\right]},$$

one can obtain the following dimensionless quasi-2D spatially inhomogeneous cubic–quintic NLS equation

$$i\frac{\partial\psi}{\partial t} = \left[-\frac{1}{2}\nabla^2 + \frac{1}{2}\omega^2\left(x^2 + y^2\right) + g_3(x, y, t)\left|\psi\right|^2 + g_5(x, y, t)\left|\psi\right|^4\right]\psi, \quad (11.3)$$

where $\omega = \omega_r/\omega_z$, $g_3(x, y, t) = G_3(x, y, t)/\left[2\pi a_z^2\hbar\omega_z\right]$, $g_5(x, y, t) = G_5(x, y, t)/\left[3\pi^2 a_z^4\hbar\omega_z\right]$. This is the basic equation we wish to treat analytically and numerically.

11.2.2 Similarity Transformation and Analytical Solutions

We assume that the nonlinearity coefficients g_3 and g_5 are time independent, and seek exactly the spatially localized stationary solution of Eq. (11.3) in the form $\psi(x, y, t) = \phi(x, y)\exp[-i\lambda t]$, where λ is the chemical potential, and $\phi(x, y)$ is a real function for the localization and satisfies the conditions $\lim_{x,y\to\pm\infty}\phi(x, y) = 0$. Inserting this ansatz into the NLS equation (11.3) yields

$$\frac{1}{2}\left(\frac{\partial^2\phi}{\partial x^2} + \frac{\partial^2\phi}{\partial y^2}\right) - \frac{1}{2}\omega^2\left(x^2 + y^2\right)\phi - g_3(x, y)\phi^3 - g_5(x, y)\phi^5 + \lambda\phi = 0. \quad (11.4)$$

Seeking the exact solutions of Eq. (11.4) in the form $\phi(x, y) = \rho(x, y)U[R(x, y)]$ and imposing it on $U(R)$ to satisfy the condition

$$\frac{d^2U}{dR^2} + \delta_3 U^3 + \delta_5 U^5 + EU = 0 \quad (11.5)$$

with arbitrary real constants δ_3, δ_5, and E yield

$$\frac{\partial^2\rho}{\partial x^2} + \frac{\partial^2\rho}{\partial y^2} + 2\left[\lambda - \frac{1}{2}\omega^2\left(x^2 + y^2\right)\right]\rho = E\left[\left(\frac{\partial R}{\partial x}\right)^2 + \left(\frac{\partial R}{\partial y}\right)^2\right]\rho, \quad (11.6)$$

if the similarity variable $R(x, y)$ is chosen as $R(x, y) = \mathrm{erf}\left[\frac{\sqrt{2\omega}}{2}(x+y)\right]$ with the error function $\mathrm{erf}(X) = \frac{2}{\sqrt{\pi}}\int_0^X e^{-\zeta^2}d\zeta$.

Case $E = 0$ leads to the amplitude

$$\rho(x, y) = \exp[\omega xy]\left\{C_1 M\left[-\frac{\lambda}{2\omega}, \frac{1}{2}, \frac{\omega(x-y)^2}{2}\right] + c_2 U\left[-\frac{\lambda}{2\omega}, \frac{1}{2}, \frac{\omega(x-y)^2}{2}\right]\right\},$$

where $M(a, c, s)$ and $U(a, c, s)$ are respectively the Kummer M and U functions [1], which are two solutions of ordinary differential equation $s\frac{d^2\Theta}{ds^2} + (c-s)\frac{d\Theta}{ds} - a\Theta = 0$.

Exploiting a one-to-one correspondence between Eqs. (11.3) and (11.5), we obtain two families of exact nonlinear wave functions for Eq. (11.3) as

$$\psi_{1n}(x, y, t) = \frac{3\rho\,\mathrm{sn}[\eta K(m), m]}{\sqrt{3\mu[3-(m^2+1)\mathrm{sn}^2[\eta K(m), m]]}}\exp[-i\lambda t], \tag{11.7}$$

with $\eta = n\theta + 1$ for $n = 1, 3, 5, \ldots$, while, for $n = 2, 4, 6, \ldots$,

$$\psi_{2n}(x, y, t) = \frac{3\rho\,\mathrm{sn}[\eta K(m)\theta, m]}{\sqrt{3\mu[3-(m^2+1)\mathrm{sn}^2[\eta K(m)\theta, m]]}}\exp[-i\lambda t], \tag{11.8}$$

where μ is a real constant, $\theta = \mathrm{erf}\left[\frac{\sqrt{2\omega}}{2}(x+y)\right]$, sn is the Jacobian elliptic sine functions with modulus m, and $K(m) = \int_0^{\pi/2} 1/\sqrt{[1-m^2\sin^2\xi]}d\xi$ is the elliptic integral of the first kind. In what follows, we will see that integer n is associated with the matter wave order number. Moreover, nonlinear functions g_3 and g_5 are given by

$$g_j(x, y) = -\frac{2\omega}{\pi\rho^{j-1}}\exp\left[-\omega(x+y)^2\right], \quad j = 3, 5, \tag{11.9}$$

with $\delta_3 = \frac{2}{3}\mu n^2 K^2(m)(m^4 - m^2 + 1)$ and $\delta_5 = \frac{1}{9}\mu^2 n^2 K^2(m)(m^2-2)(2m^2-1)$ (m^2+1).

11.2.3 Quantum Properties of Exact Matter-Wave Functions with Attractive Two-Body Interactions

For the matter-wave functions (11.7) and (11.8) to be physical, from Eq. (11.9) it follows that $\rho(x, y)$ must not change its sign at any point (a sign-definite function), otherwise the local nonlinearity would diverge at points of $\rho = 0$. Therefore, this restriction requires that the chemical potential $\lambda < \lambda_0 = \omega$ and $c_1 c_2 > 0$. Interestingly,we find that λ_0 is exactly the minimum energy eigenvalue of the corresponding linear Schrödinger equation with harmonic potential $\frac{1}{2}\omega^2(x^2+y^2)$. Thus, these exact matter-wave functions of Eq. (11.3) exist in the semi-infinite band gap.

It is easy to prove the odd and even parities of the matter-wave functions (11.7) and (11.8). For the family of solutions (11.7), $|\text{sn}[(n\theta+1)K(m),m]| \sim |\text{cn}[nK(m)\theta,m]|$; thus, they are of even parity (see Figs. 11.2 and 11.4) since the matter-wave functions satisfy $\psi_n(-x,-y)=\psi_n(x,y)$. Similarly, the family of solutions (11.8) are of odd parity (see Figs. 11.2 and 11.4) and meet the criterion $\psi_n(-x,-y)=-\psi_n(x,y)$.

Generally, the modulus parameter m of Jacobian elliptic functions sn and cn varies from 0 to 1 and describes the degree of energy localization of the nonlinear matter waves. Solutions (11.7) and (11.8) require $\mu>0$, which thus gives $\delta_3>0$ for any m. Thus, the cubic nonlinearity can only be attractive, i.e., $g_3<0$ from Eq. (11.9). Similarly, one can find that δ_5 is positive for $0\le m<1/\sqrt{2}$ and negative for $1/\sqrt{2}<m<1$. Therefore, the quintic nonlinearity can be either attractive or repulsive, depending on m. Note that three special points of m influence solutions (11.7) and (11.8). The first point, for $m=0$, makes solutions (11.7) and (11.8) degenerate into trigonometrical sine and cosine function solutions. The second exceptional point occurs for $m=1/\sqrt{2}$, since it gives $g_5=\delta_5=0$, and thus the quintic nonlinearity identically vanishes [30]. For the last exceptional point, $m=1$, the elliptic integral of the first kind $K(m)\to\infty$, solutions (11.7) and (11.8) change into the trivial solutions. Next, we discuss the properties of exact nonlinear matter-wave functions in two cases, i.e., attractive two-body interaction atomic gases with attractive and repulsive three-body interactions, respectively.

Attractive Two-Body and Three-Body Interactions

Let us consider the case of attractive two-body and three-body interactions, that is, the case when, $g_3(x,y)<0$ and $g_5(x,y)<0$. Note that the energies of the localized states are

$$\begin{aligned} E_n &= \int\int dxdy\left[|\nabla\psi_n|^2+\frac{1}{2}\omega^2(x^2+y^2)|\psi_n|^2+\frac{1}{2}g_3(x,y)|\psi_n|^4+\frac{1}{3}g_5(x,y)|\psi_n|^6\right] \\ &\equiv \lambda N_n\int\int dxdy\left[\frac{1}{2}g_3|\psi_n|^4+\frac{2}{3}g_5|\psi_n|^6\right], \end{aligned} \tag{11.10}$$

with the number of atoms

$$N_n=\int\int dxdy|\psi_n|^2. \tag{11.11}$$

This expression, (11.10), suggests discrete energy levels from two aspects. On the one hand, for fixed chemical potential λ, the larger the matter-wave order number n, the larger the energy of the corresponding BEC state. Thus, one may conclude that ψ_1 corresponds to the ground state, and that the ψ_n are the *(n-1)*th excited states. On the other hand, for fixed n, the energy increases with respect to the increasing chemical potential.

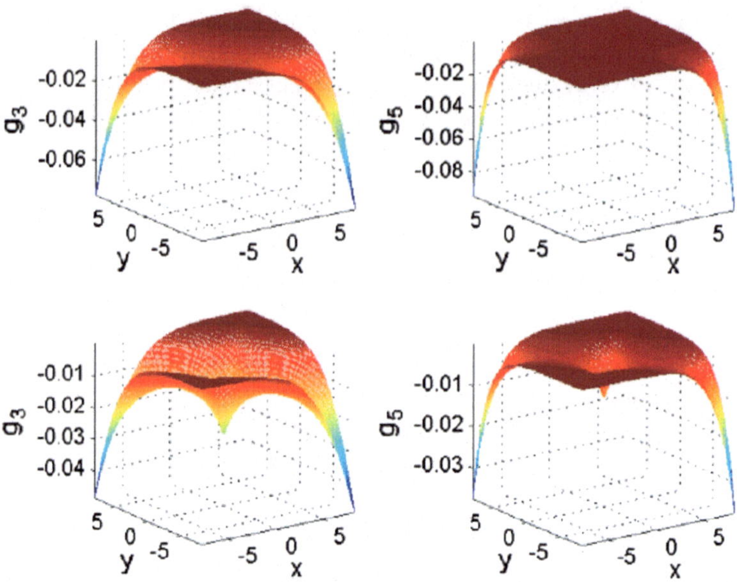

Fig. 11.1 The attractive two-body and three-body interactions parameters $g_3(x, y)$ and $g_5(x, y)$. The parameters are $\omega = 0.02$, $\mu = 0.1$, $m = 0.01$, $n = 1$, with $\lambda = 0.001$ for the first row and $\lambda = 0.01$ for the second row. Here, the unit of space length x, y is 1.69 μm. Reprint from Ref. [9], Copyright 2018, with permission from Annals of Physics

To translate the results into units relevant to experiments [9], we assume a ^{7}Li condensate containing $10^2 - 10^4$ atoms in a pancake-shaped trap with radial frequency $\omega_\perp = 2\pi \times 10$ Hz and axial frequency $\omega_z = 2\pi \times 500$ Hz. Thus, the ratio of trap frequency $\omega = \omega_\perp/\omega_z = 0.02$ in Eq. (11.3). The unit of length is 1.69 μm, the unit of time is 0.32 ms, and the unit of chemical potential is nK. Note that, from expression (11.9), for fixed matter-wave order number n and modulus parameter m, the chemical potential λ will modulate the strength of the cubic and quintic nonlinearities $g_3(x, y)$ and $g_5(x, y)$ via the amplitude $\rho(x, y)$ influenced by the Kummer M and U functions. For small chemical potential λ, the cubic and quintic nonlinearities $g_3(x, y)$ and $g_5(x, y)$ are smooth in space. When the chemical potential λ increases gradually and is close to the cutoff value λ_0, the cubic and quintic nonlinearities decrease and develop a valley (see Fig. 11.1). It is important to note that the value at the bottom of the valley does not develop a singularity, otherwise this case cannot be realized in physical experiments for infinite nonlinearities. Without loss of generality, here we choose the parameters $\omega = 0.02$, $\mu = 0.1$, $m = 0.01$, $n = 1$, $c_1 = 0$, $c_2 = 1$, with $\lambda = 0.001$ for the first row of Fig. 11.1 and $\lambda = 0.01$ for the second row of Fig. 11.1. For different chemical potential λ, the localized shapes of nonlinear matter-wave functions will exhibit different topological patterns (see Figs. 11.2 and 11.3).

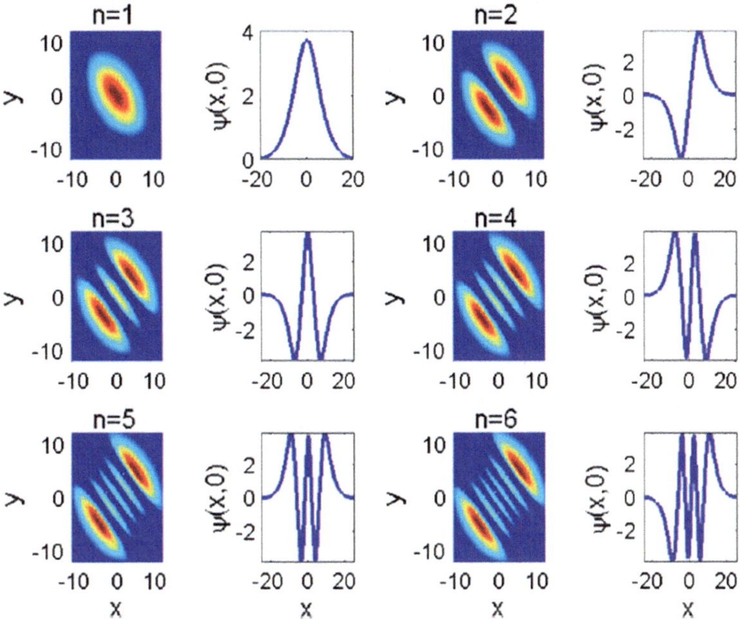

Fig. 11.2 The density distributions and wave functions of a quasi-2D BEC with attractive two-body and three-body interactions in the harmonic potential with the parameters $\omega = 0.02$, $\mu = 0.1$, $\lambda = 0.001$, $m = 0.01$. Here, the unit of space length x, y is 1.69 μm. Reprint from Ref. [9], Copyright 2018, with permission from Annals of Physics

In Fig. 11.2, we take $\omega = 0.02$, $\mu = 0.1$, $\lambda = 0.001$, $m = 0.01$, $c_1 = 0$, $c_2 = 1$. From the left subplots of Fig. 11.2, the wave functions ψ_n (n = 1, 3, 5, …) possess $n - 1$ (even number) zeros along line $y = x$, and display the property of even parity. Similarly, the right subplots of Fig. 11.2 demonstrate the density and wave function profiles of the odd-parity wave function (11.8) for $n = 2, 4, 6$ and odd-number density nodes along line $y = x$. From Eq. (11.11), the number of atoms for the ground state is $N_1 = 1.357 \times 10^3$, and for the lowest five excited states the numbers of atoms are $N_2 = 2.009 \times 10^3$, $N_3 = 2.482 \times 10^3$, $N_4 = 2.861 \times 10^3$, $N_5 = 3.181 \times 10^3$, $N_6 = 3.458 \times 10^3$, respectively. It is clear that the number of density packets increases successively along the line $y = x$ when the wave function order number n increases.

When the chemical potential λ is close to the cutoff value λ_0, the number of atoms will reduce. For example, if the chemical potential $\lambda = 0.01$ and the other parameters are the same as in Fig. 11.2, a novel quantum phenomenon in the quasi-2D BEC (see Fig. 11.3) appears: that is, all the density packets are symmetrical with respect to the line $y = \pm x$. A comparison of Figs. 11.2 and 11.3 indicates that the chemical potential can change the topological patterns of the atom packets. In this case, the number of atoms for nonlinear matter-wave functions (11.7) and (11.8) are $N_1 = 0.948 \times 10^3$, $N_2 = 1.345 \times 10^3$, $N_3 = 1.612 \times 10^3$, $N_4 = 1.812 \times 10^3$,

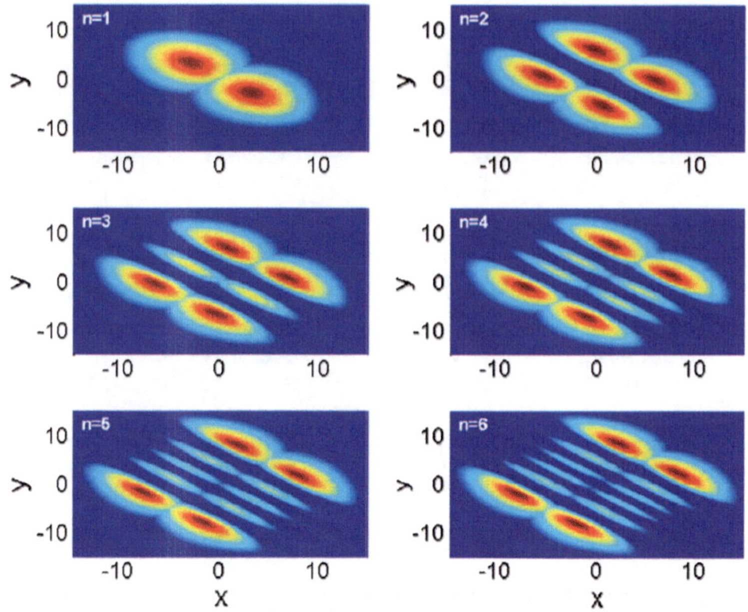

Fig. 11.3 The density distributions of a quasi-2D BEC with attractive two-body and three-bodies interactions in the harmonic potential with the parameters $\omega = 0.02$, $\mu = 0.1$, $\lambda = m = 0.01$. Here, the unit of space length x, y is 1.69 μm. Reprint from Ref. [9], Copyright 2018, with permission from Annals of Physics

$N_5 = 1.974 \times 10^3$, $N_6 = 2.107 \times 10^3$, respectively. The contrast of the number of atoms in Figs. 11.2 and 11.3 suggests that greater the chemical potential, the smaller the number of atoms of the matter wave. Clearly, this agrees with physical practice. The chemical potential means the energy required to add one more atom to the condensate for fixed entropy. Thus when the chemical potential increases, the number of atoms will decrease for a definite system.

Attractive Two-Body and Repulsive Three-Body Interactions

For $g_3(x, y) < 0$ and $g_5(x, y) > 0$, two-body and three-body interactions take effect in a competing way. From (11.10), for fixed chemical potential λ and states (i.e., fixed n), whether the energy is positive or negative is decided by the competition between two-body and three-body interactions. This competition involves matter waves balancing linear and nonlinear terms. Figure 11.4 displays the properties of odd and even parities for nonlinear matter-wave functions (11.7) and (11.8). Here, the parameters are chosen as $\omega = 0.02$, $\mu = 0.1$, $\lambda = 0.001$, $m = 0.99$. From Eq. (11.11), the number of atoms for the ground state is $N_1 = 3.942 \times 10^3$, and for the five excited states the numbers of atoms are $N_2 = 5.766 \times 10^3$, $N_3 = 7.076 \times 10^3$, $N_4 = 8.116 \times 10^3$, $N_5 = 9.011 \times 10^3$, $N_6 = 1.002 \times 10^4$, respectively. Thus, the number of density packets also increases successively along the line $y = x$ when the wave function order number n increases. The number of atoms with repulsive three-body interac-

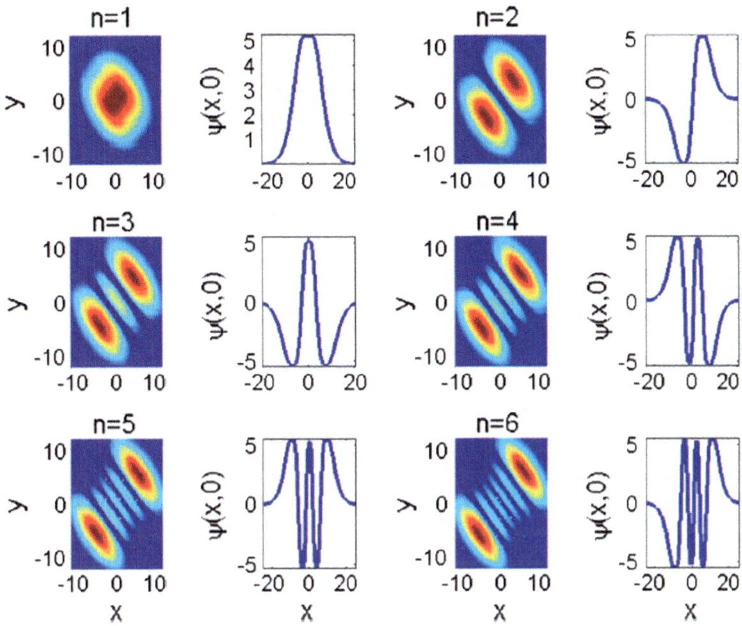

Fig. 11.4 The density distributions and wave functions of a quasi-2D BEC with attractive two-body and repulsive three-body interactions in the harmonic potential with the parameters $\omega = 0.02$, $\mu = 0.1, \lambda = 0.001, m = 0.99$. Here the unit of space length x, y is 1.69 μm. Reprint from Ref. [9], Copyright 2018, with permission from Annals of Physics

tion is bigger than the number with attractive three-body interaction, which can also be verified by Fig. 11.4. Similar to the discussion above, symmetry of the density packets with respect to the line $y = x$ will also emerge because the valley will also appear in the attractive two-body and repulsive three-body interactions.

11.2.4 Stability Analysis

It is worth mentioning that a corresponding and very important problem is the stability of solutions (11.7) and (11.8); that is, how they evolve in time when they are disturbed from their analytically given forms. This aspect of the problem must be addressed numerically (with the help of the linear stability analysis), and will be presented next.

The current situation for the propagation stability of solutions to higher dimensional NLS equations or GP equations is somewhat confusing and even controversial. Most workers study the stability of the radially symmetric structures without consideration of modulating the diffraction or dispersion coefficients. It is well known that 2D disk-shaped solitary waves are unstable due to transverse perturbations and collapse. Temporal periodic modulation of the scattering length in 2D disk-shaped BECs was reported to excite Faraday patterns in the atom density in the plane transverse to the tight confinement direction [9]. Moreover, the validity of the 1D description of the BEC by means of the 3D model was tested under the combined action of the time-dependent nonlinearity and confining potential, and the compression of the atom cloud from the initial cigar-shaped configuration into a quasi-spherical 3D configuration was achieved, leading to the collapse of the soliton. However, some authors have reported stable higher dimensional solitons. For example, the sign-alternating Kerr nonlinearity in a layered medium produces stable 2D solitons. In BECs, the alteration of atomic scattering length achieved by Feshbach resonance has been used to dynamically stabilize higher dimensional bright solitons [9]. Recently, Wang et al. [30] investigated 2D stable solitons and vortices for BECs with spatially modulated cubic nonlinearity and a harmonic trapping potential.

Our situation is different: the cubic and quintic nonlinearity strengths vary in time neither linearly nor periodically; they are expressed as Eq. (11.9), as shown in Fig. 11.1. Thus, it is vital to analyze the stability of exact solutions against finite perturbation, since experimental interest is in the stability of the stationary states (11.7) and (11.8) on physical time scales. To study the stability of the matter waves, we performed a linear stability analysis and direct simulations with initial white noise.

The linear stability of the solutions found in the previous section was explored by means of the method developed in [7]. To this end, we consider a perturbation of an exact solution $\psi(x, y, t) = [\phi_n(x, y) + \epsilon\psi_1(x, y, t)] \exp[-i\lambda t]$ of Eq. (11.3), where ϵ is an infinitesimal amplitude, $\phi_n(x, y)$ is a respective eigenmode [a solution of the stationary cubic–quintic NLS equation (11.4)], and $\psi_1(x, y, t) = [R(x, y) + I(x, y)] \exp[i\delta t]$ is decomposed into its real and imaginary parts. Inserting this expression into the quasi-2D NLS equation (11.3) and linearizing it around the unperturbed equation (the first-order term of ϵ), we arrive at the eigenvalue problem,

$$\begin{pmatrix} L_+ & 0 \\ 0 & L_- \end{pmatrix} \begin{pmatrix} R \\ I \end{pmatrix} = \delta \begin{pmatrix} I \\ R \end{pmatrix}, \tag{11.12}$$

where δ is the eigenvalue, and R and I are eigenfunctions with Hermitian operators

$$L_+ = -\frac{1}{2}\left(\frac{\partial^2}{\partial x^2} + \frac{\partial^2}{\partial y^2}\right) + 3g_3(x, y)\phi_n(x, y)^2 + 5g_5(x, y)\phi_n(x, y)^4 + \frac{1}{2}\omega^2(x^2 + y^2) - \lambda$$

and

$$L_- = -\frac{1}{2}\left(\frac{\partial^2}{\partial x^2} + \frac{\partial^2}{\partial y^2}\right) + g_3(x, y)\phi_n(x, y)^2 + g_5(x, y)\phi_n(x, y)^4 + \frac{1}{2}\omega^2(x^2 + y^2) - \lambda.$$

The wave function is unstable if any eigenvalue δ has an imaginary part.

The results of numerical calculations for the eigenvalue problem (11.12) reveal that only for the matter-wave order $n = 1$ is the eigenvalue δ of this eigenvalue problem real, which suggests that only for the ground-state wave order is the exact nonlinear matter-wave function (11.7) with $m = 0.01$ linearly stable, while higher order wave families are unstable (see Fig. 11.5). On the other hand, it is also easy to verify that, for all n, solution (11.8) with $m = 0.01$ is linearly unstable. Moreover, for matter waves with attractive two-body and repulsive three-body interactions, though not shown here, a similar case to that shown in Fig. 11.5 will appear (we calculate this case when the parameters are the same as in Fig. 11.4). The stability of the matter waves was further checked by direct numerical simulations with a substantial initial white-noise perturbation for Eq. (11.3). In particular, we analyze the ground state for $n = 1$ carefully for attractive and repulsive three-body interactions. The result exhibits an unstable evolution for attractive three-body interaction and then decays at time 19.2 ms, while other unstable waves for $n > 1$ quickly decay into noise (see Fig. 11.6). The competition between the attractive two-body and repulsive

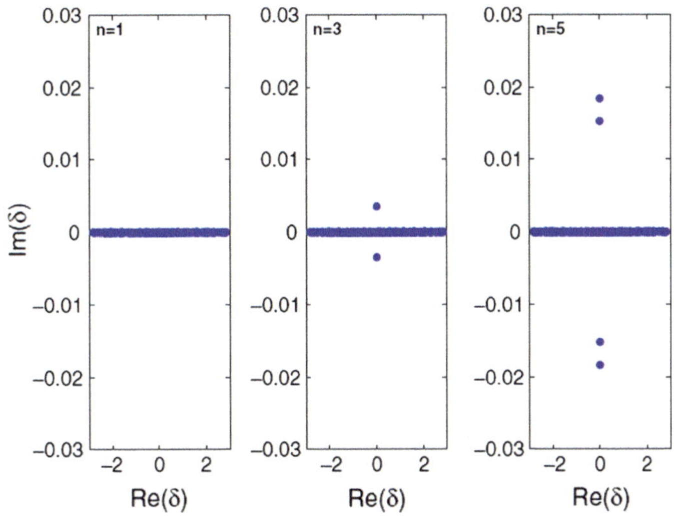

Fig. 11.5 Eigenvalues for different matter-wave order number n with parameters $\omega = 0.02$, $\mu = 0.1$, $m = 0.01$ and $\lambda = 0.001$. It is shown that only for $n = 1$ is the nonlinear matter-wave function (11.7) linearly stable. Reprint from Ref. [9], Copyright 2018, with permission from Annals of Physics

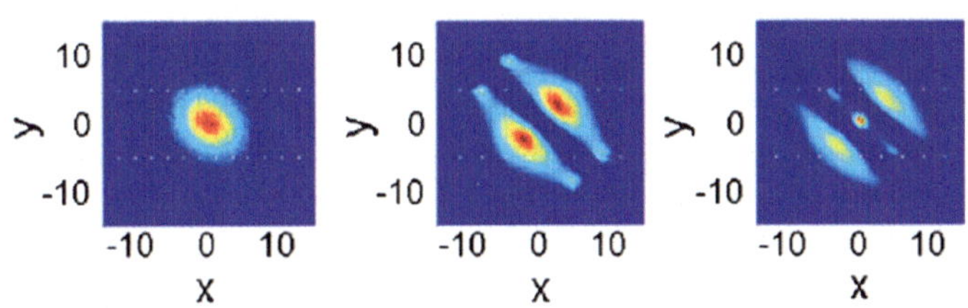

Fig. 11.6 Unstable matter waves with attractive two-body and three-body interactions, decaying at time 19.2 ms (left, for n =1), 6.72 ms (middle, for $n = 2$), and 3.68 ms (right, for $n = 3$). The other parameters are the same as in Fig. 11.1. The initial state is perturbed by an added 5% initial white noise. Reprint from Ref. [9], Copyright 2018, with permission from Annals of Physics

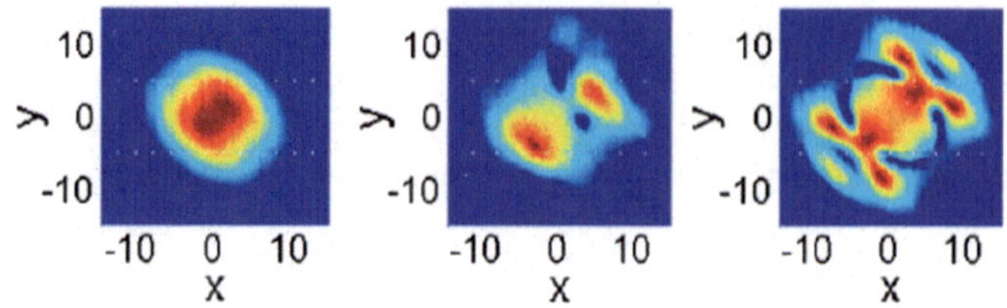

Fig. 11.7 Stable matter wave with attractive two-body and repulsive three-body interactions at time 160 ms (left, for $n = 1$) and unstable matter waves at 57.6 ms (middle, for $n = 2$) and 16 ms (right, for $n = 3$). The other parameters are the same as in Fig. 11.4. The initial state is perturbed by an added 5% initial white noise. Reprint from Ref. [9], Copyright 2018, with permission from Annals of Physics

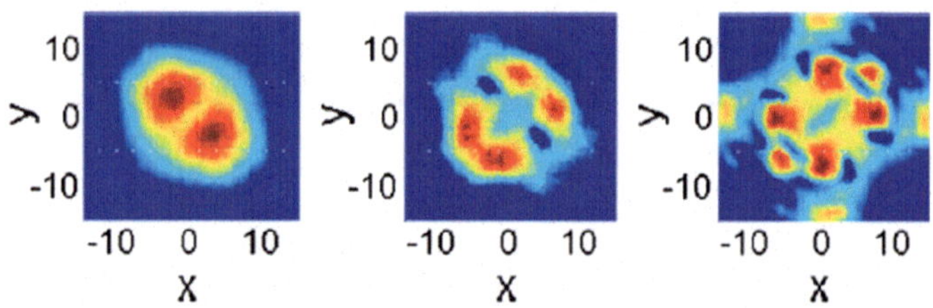

Fig. 11.8 Quasi-stable matter wave with attractive two-body and repulsive three-body interactions at time 160 ms (left, for $n = 1$) and unstable matter waves at 57.6 ms (middle, for $n = 2$) and 16 ms (right, for $n = 3$). The other parameters are the same as in Fig. 11.4 except that $\lambda = 0.01$. The initial state is perturbed by an added 5% initial white noise. Reprint from Ref. [9], Copyright 2018, with permission from Annals of Physics

three-body interactions makes the ground state ($n = 1$) stable and greatly prolongs the lifetime of the BEC for excited state waves ($n > 1$). In Fig. 11.7, the shape of the ground-state wave is almost unchanged and the lifetime exceeds 160 ms, which is considered stable because this lifetime greatly exceeds the value obtained from the experimental results. Similarly, in Fig. 11.8, the boundary of the ground-state aggregates, the shape is basically unchanged and the lifetime exceeds 160 ms, which is designated quasi-stable because this lifetime also greatly exceeds the value from the experimental results. However, for the excited state ($n > 1$), their humps become closer together with the increase of time and ultimately decay into noise. For the same state (fixed n), the greater the number of atoms the matter wave has, the better stability it possesses. Although some solutions are illustrated over short timescales for presentation purposes, all simulations were performed over timescales longer than the experimental lifetimes of the BECs.

11.2.5 Numerical Results for Repulsive Two-Body Interaction

In this section, we consider repulsive two-body interaction, $g_3 > 0$, and include attractive ($g_5 < 0$) and repulsive ($g_5 > 0$) three-body interactions with a harmonic trap to influence the system. In this case, the existence of elliptic-function solutions to Eq. (11.5) requires $E \neq 0$, and the wave function possess the following form:

$$\psi_n(x, y, t) = \frac{\rho \mathrm{sn}[nK(m)\theta + \epsilon K(m), m]}{\sqrt{\mu + \nu \mathrm{sn}^2[nK(m)\theta + \epsilon K(m), m]}} \exp[-i\lambda t], \quad (11.13)$$

with $\epsilon=1$ for $n=1, 3, 5, \ldots$ and $\epsilon = 0$ for $n=2, 4, 6, \ldots$ In solution (11.13), μ is a real constant, $\nu=\frac{[E-(1+m^2)n^2K^2(m)]\mu}{3n^2K^2(m)}$, $\theta=\mathrm{erf}[\frac{\sqrt{2\omega}}{2}(x+y)]$, $\delta_3 = \frac{-2E[\mu m^2(\mu+2\nu)+\nu(2\mu+3\nu)]}{\mu(m^2+1)+3\nu}$, and $\delta_5 = \frac{3E\nu(\mu+\nu)(\mu m^2+\nu)}{\mu(m^2+1)+3\nu}$, with a constraint $E > (m^2-2)n^2K^2(m)$. Of course, two exceptional points are remarkable for $E = (m^2+1)n^2K^2(m)$ and $E = (1-2m^2)n^2K^2(m)$, since they give $\delta_5 = 0$, and hence the quintic nonlinearity identically vanishes. Moreover, two exceptional points are also obvious for $\eta = \pm\sqrt{1-m^2+m^4}n^2K^2(m)$, since they give $\delta_3 = 0$, and thus the cubic nonlinearity identically vanishes. For $E \neq 0$, it is difficult to obtain an analytical solution of $\rho(x, y)$ from Eq. (11.6). Next, we discuss the properties of nonlinear matter-wave functions based on Eq. (11.13) by numerical simulations.

Figure 11.9 displays unstable matter waves with repulsive two-body and attractive three-body interactions. Matter waves with parameters $\omega = 0.02$, $\mu = 0.1$, $m = 0.99$, $E = 0.07$, $\lambda = 0.001$ quickly break down and decay into noise. For matter waves with repulsive two-body and three-body interactions, a quasi-stable matter wave (see Fig. 11.10) for the ground state is found with parameters $\omega = 0.02$, $\mu = 0.1$, $m = 0.99$, $E = 0.04$, $\lambda = 0.001$. This phenomenon can be understood as follows. The nonlinearities along $y = -x$ take effect in a matter wave in this direction, and then make the matter wave gradually rotate to this direction. Under the action of inertia, the matter wave eventually returns to the initial location. This process repeats periodically. For the excited state $n > 1$, matter waves with repulsive two-body and three-body interactions, though not shown here, are also unstable.

11.2.6 Experimental Realization

Finally, we discuss the experimental realization of quasi-2D localized nonlinear matter waves. We take the attractive ^{7}Li condensate [13], containing about $10^2 - 10^4$ atoms, confined in a pancake-shaped trap with radial frequency $\omega_\perp = 2\pi \times 10$ Hz and axial frequency $\omega_z = 2\pi \times 500$ Hz [25]. This trap can be obtained by a combination of spectroscopic observations, direct magnetic field measurement, and the observed spatial cylindrical symmetry of the trapped atom cloud [25].

The spatial variation of the scattering length can be realized via Fescbach resonance techniques. Near the Feshbach resonance [9], the scattering length $a_s(B)$ varies dispersively with magnetic field B in the form $a_s(B) = a[1 + \Delta/(B0 - B)]$, where a is the asymptotic value of the scattering length far from the resonance, B_0 is the resonant value of the magnetic field, and Δ is the width of the resonance. For

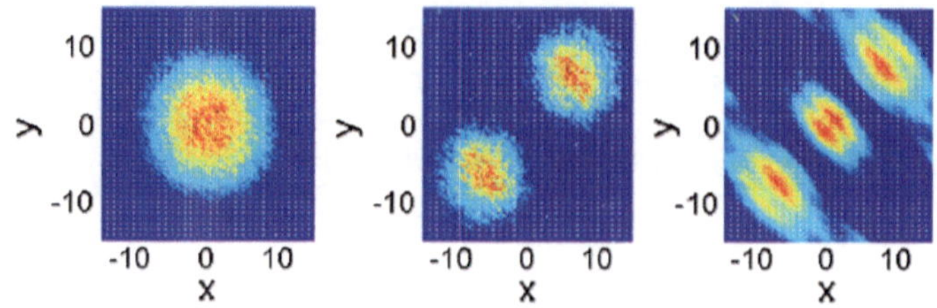

Fig. 11.9 Unstable matter waves with repulsive two-body and attractive three-body interactions at time 64 ms (left, for $n = 1$), 32 ms (middle, for $n = 2$), and 6.4 ms (right, for $n = 3$). The parameters are $\omega = 0.02$, $\mu = 0.1$, $m = 0.99$, $E = 0.07$, $\lambda = 0.001$. The initial state is perturbed by an added 5% initial white noise. Reprint from Ref. [9], Copyright 2018, with permission from Annals of Physics

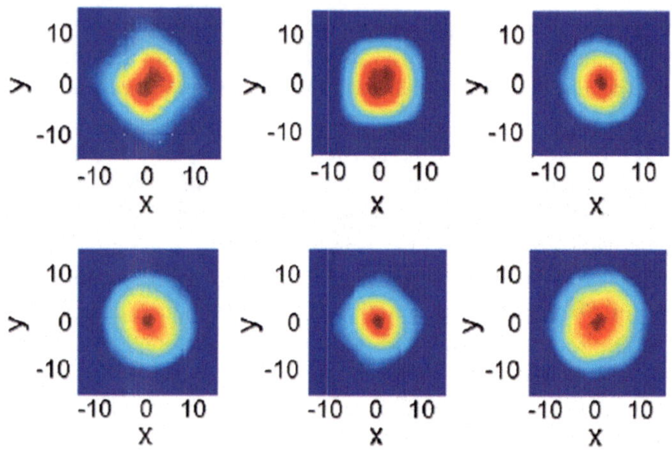

Fig. 11.10 Quasi-stable matter wave for $n = 1$ with repulsive two-body and three-body interactions at time 32,64, 96, 128, 160, 192 ms, respectively. The parameters are $\omega = 0.02$, $\mu = 0.1$, $m = 0.99$, $E = 0.04$,$\lambda = 0.001$. The initial state is perturbed by an added 5% initial white noise. Reprint from Ref. [9], Copyright 2018, with permission from Annals of Physics

the magnetic field in the z-direction with gradient α along the $x - y$ direction, one has $\overrightarrow{B} = [B_0 + \alpha B_1(x, y)]\overrightarrow{e}_z$. Thus, the scattering length depends on x and y.

In real experiments, the interactions in Fig. 11.1 can be realized by means of the Feshbach resonance technique. The corresponding nonuniform magnetic field may be created by a micro-fabricated ferromagnetic structure integrated into the matter-wave chip [29]. In order to observe the density distributions in Figs. 11.2, 11.13 and 11.4 clearly in experiments, the ^{7} Li atoms should be evaporatively cooled to low temperatures, say in the range $10 - 100$ nK. After the interaction parameter in Fig. 11.1 is realized by modulating the magnetic field properly, the density distributions in Fig. 11.2 can be observed for different numbers of atoms by evaporative cooling; for example, the numbers of atoms for three lowest excited states in Fig. 11.2 are

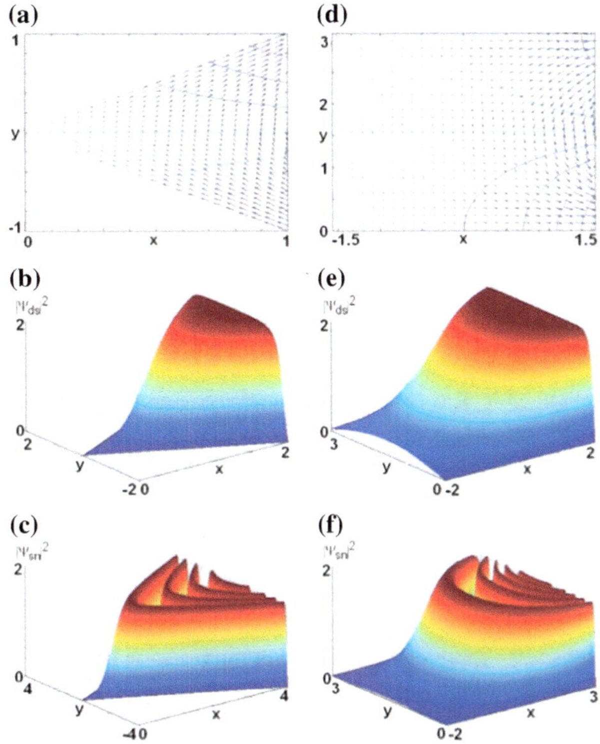

Fig. 11.11 The profiles for exact solutions given by Eqs. (11.25) and (11.26) and the corresponding velocity fields in the domains D_1 and D_2. Left column: Panels **a**–**c** show the velocity field $\nabla\varphi(\mathbf{r}) = (2y, 2x)$, and densities $|\Psi_{ds}(\mathbf{r})|^2$ and $|\Psi_{sn}(\mathbf{r})|^2$ for solutions given by Eqs. (11.25) and (11.26) in the domain D_1. Right column: Panels (d)–(f) show the velocity field $\nabla\varphi(\mathbf{r}) = (\exp[x]\cos[y], \exp[x]\sin[y])$, and densities $|\Psi_{ds}(\mathbf{r})|^2$ and $|\Psi_{sn}(\mathbf{r})|^2$ for solutions given by Eqs. (11.25) and (11.26) in the domain D_2. The parameters are $\varepsilon = 2$, $\nu = 0$, and $k = 0.8$. Reprint from Ref. [33], Copyright 2018, with permission from American Physical Society

1.357×10^3, 2.009×10^3, 2.482×10^3, respectively. When the chemical potential λ is close to the cutoff value λ_0, e.g., $\lambda = 0.01$, even-number humps of BEC in the domain of 642.6 μm^2 shown in Fig. 11.3 can be observed. Similarly, by detuning the scattering lengths via a magnetic field, the density distributions in Fig. 11.4 can also be observed.

11.3 Superfluid Flows in Inhomogeneous Bose–Einstein Condensates

This section deals with the algorithm of constructing the external potential and the nonlinearity of 2D inhomogeneous NLS equations in bounded subject to Dirichlet boundary conditions. The obtained solutions of the associated boundary problem represent superfluid flows in inhomogeneous BECs.

11.3.1 Physical Model and Useful Ansatz

We deal with the 2D cubic NLS equation with external potential

$$i\frac{\partial \Psi(\mathbf{r},t)}{\partial t} = \left[-\frac{1}{2}\nabla^2 + V_{\text{ext}}(\mathbf{r},t) + g(\mathbf{r})\,|\Psi(\mathbf{r},t)|^2\right]\Psi(\mathbf{r},t), \tag{11.14}$$

where $\mathbf{r} = (x, y) \in D \subset \mathbf{R}^2$ (D being an open domain with boundary ∂D) and $\nabla^2 = \partial^2/\partial x^2 + \partial^2/\partial y^2$. We are particularly interested in applications of our results to BEC flows with the macroscopic wave function $\Psi(\mathbf{r}, t)$. We explore the flexibility of potentials in the BEC applications, i.e., the possibility of manipulating them by external electric and/or magnetic fields for the sake of creation of desirable spatial configurations for the linear potential and for the scattering length of the two-body interactions (the latter performed through the Feshbach resonance technique). We notice that model (11.14) has also direct relevance to the mean field theory of exciton–polariton condensates. Applying the external pump from the free edges of a specimen one can create different kinds of nonzero conditions (nonzero currents, as required below).

We concentrate on a BEC in a domain D bounded by impenetrable walls. Respectively, Eq. (11.14) will be supplied by the Dirichlet boundary condition (the zero conditions given at the boundary of the domain D), i.e., we impose $\Psi(\mathbf{r}, t)|_{\mathbf{r}\in D} = 0$.

Our goal is to find an algorithm allowing for systematic constructions of the external potential $V_{\text{ext}}(\mathbf{r})$ and nonlinearity $g(\mathbf{r})$ so that the formulated Dirichlet problem allows for exact analytical solutions. To this aim, we assume that there exists a complex analytic function

$$\zeta(\mathbf{r}) = \eta(\mathbf{r}) + i\varphi(\mathbf{r}) = f(z) \tag{11.15}$$

of the complex variable $z = x + iy \in \mathbf{C}$, which provides the conformal mapping of the contour ∂D to the imaginary axis of $\zeta(\mathbf{r})$, i.e., to $\eta(\mathbf{r}) = 0$, such that the domain D is mapped into the right half-plane in terms of the new variables $(\eta, \varphi) : \eta(\mathbf{r}) > 0$. Due to analyticity of the mapping $\zeta(\mathbf{r})$ the Cauchy–Riemann equations on the pair of real-valued functions $\eta(\mathbf{r})$ and $\varphi(\mathbf{r})$ hold:

$$\frac{\partial \eta(\mathbf{r})}{\partial x} = \frac{\partial \varphi(\mathbf{r})}{\partial y}, \quad \frac{\partial \eta(\mathbf{r})}{\partial y} = -\frac{\partial \varphi(\mathbf{r})}{\partial x}. \tag{11.16}$$

Respectively, the following constraints are verified, as well

$$\nabla^2\eta(\mathbf{r}) = 0, \quad \nabla^2\varphi(\mathbf{r}) = 0, \nabla\eta(\mathbf{r})\cdot\varphi(\mathbf{r}) = 0. \tag{11.17}$$

In other words, $\eta(\mathbf{r})$ and $\varphi(\mathbf{r})$ are both 2D harmonic functions with orthogonal gradients. Moreover, on the basis of the Cauchy–Riemann equations we find the relation $|\nabla\eta(\mathbf{r})|^2 \equiv |\nabla\varphi(\mathbf{r})|^2$.

We restrict the consideration to the linear and nonlinear potentials,which can be represented in terms of $\eta(\mathbf{r})$ as follows:

$$V_{\text{ext}}(\mathbf{r}) = -\frac{\varepsilon}{2}\left|\nabla\eta(\mathbf{r})\right|^2, \quad g(\mathbf{r}) = \frac{\mathcal{G}}{2}\left|\nabla\eta(\mathbf{r})\right|^2, \tag{11.18}$$

where ε and $\mathcal{G}$ are real parameters. Without loss of generality, we choose $\mathcal{G} = \pm 1$. The parameter ε determines the proportionality coefficient between the potentials: $V_{\text{ext}}(\mathbf{r})/g(\mathbf{r}) = -\varepsilon\mathcal{G}$. Then, the change of the dependent variables $\Psi(\mathbf{r},t) \to \psi(\eta,\varphi,t)$, allows one to reduce Eq. (11.14) to the 2D form,

$$i\frac{\partial\psi}{\partial t} = \varrho(\eta,\varphi)\left(-\frac{\partial^2}{\partial\eta^2} - \frac{\partial^2}{\partial\varphi^2} - \varepsilon + \mathcal{G}\left|\psi\right|^2\right)\psi. \tag{11.19}$$

Here $\varrho(\eta,\varphi)$ is the positive-definite function, defined as $\varrho(\eta,\varphi) = \left|\nabla\eta\right|^2/2 \equiv \left|\nabla\varphi\right|^2/2$, where the gradients must be expressed in terms of η and φ. The obtained Eq. (11.19) is considered for $\eta > 0$ and has to be supplied with the zero boundary condition $\psi(\eta = 0, \varphi, t) = 0$.

First, we concentrate on time-independent solutions of Eq. (11.14): $\Psi(\mathbf{r},t) \equiv \psi(\eta,\varphi)$, where $\psi(\eta,\varphi)$ solves the 2D stationary NLS equation with constant coefficients $\varepsilon\psi = -(\frac{\partial^2}{\partial\eta^2} + \frac{\partial^2}{\partial\varphi^2})\psi + G|\psi|^2\psi$. Particular solutions of this equation can be represented as

$$\psi(\eta,\varphi) = \phi(\eta)\exp\left[i\nu\varphi\right], \tag{11.20}$$

where ν is a constant and the real-valued function $\phi(\eta)$ solves the problem

$$\mathcal{E}\phi(\eta) = -\frac{d^2\phi(\eta)}{d\eta^2} + \mathcal{G}|\phi(\eta)|^2\phi(\eta), \quad \phi(0) = 0, \tag{11.21}$$

with $\mathcal{E} = \varepsilon - \nu^2$ and $\eta > 0$.

Turning to the physical meaning of the obtained solutions, we observe that it follows from the ansatz, $\psi(\eta,\varphi) = \exp\left[i\nu\varphi\right]\phi(\eta)$, that $\nabla\varphi(\mathbf{r})$ can be identified as the superfluid velocity. Hence, the introduced analytic function $f(z)$ is nothing but the complex potential of the respective two-dimensional flow. As it is well known [18, 33] such a potential defines the current, J_C of the fluid through a given contour $C \subset D$, as well as the circulation Γ_C along C:

$$\int_C f'(z)dz = \Gamma_C + iJ_C \tag{11.22}$$

(the prime stands for the derivative with respect to z). Since in our case, $f(z)$ is analytic, this integral is zero for any closed contour C bounding a simply connected domain. Thus, the described flow has neither sources nor vorticity in D.

We also observe that if the change of variables implies growth of $|\nabla\varphi(\mathbf{r})|$ with $\mathbf{r}$, then the physical meaning might have only solutions with densities decaying at the infinity (thus ensuring decaying currents). Whenever one concerns with finite densities at the infinity, the physically meaningful solutions would correspond to $\nu = 0$. This last constraint is assumed in what follows.

Finally, we notice that if the contour ∂D included in the proposed scheme is closed, this implies that linear and (or) nonlinear potentials are divergent at some point(s) of the boundary. This is clear from the nature of the conformal mapping, since in this case there should be a point of the boundary, which is mapped into the infinity point. In its turn, such a point gives an origin to the singularity of the linear and nonlinear potentials. Such cases will be excluded in what follows, although they still may have physical relevance.

11.3.2 Exact Analytical Solutions

While large diversity of the domains can be considered, here we concentrate on the two simplest representative examples for the conformal mapping (11.15).

The Quadrant $x > |y|$ with the Boundary $x = \pm y$

The first domain considered below, is given by $D_1 = \{(x, y) \in \mathbf{R}^2 : x > |y|\}$ with the boundary $\partial D_1 = \{(x, y) \in \mathbf{R}^2 : x = y,\ y > 0\} \cup \{(x, y) \in \mathbf{R}^2 : x = -y,\ y < 0\}$ respectively. The conformal mapping is chosen as $f(z) = z^2$, and thus $\eta(\mathbf{r}) = x^2 - y^2$ and $\varphi(\mathbf{r}) = 2xy$. The linear and nonlinear potentials are now given by

$$V_{\text{ext}}(\mathbf{r}) = -2\varepsilon|\mathbf{r}|^2, \quad g(\mathbf{r}) = 2\mathcal{G}|\mathbf{r}|^2, \tag{11.23}$$

i.e., they are the linear expulsive parabolic potential and the parabolic nonlinearity. Now $\varrho(\eta, \varphi) = 2\sqrt{\eta^2 + \varphi^2}$.

The Strip $0 < y < \pi$ with the Boundary $y = 0, \pi$

The second domain explored below is a strip $D_2 = \{(x, y) \in \mathbf{R}^2 : 0 < y < \pi\}$ with the boundary $\partial D_2 = (\mathbf{R}, 0) \cup (\mathbf{R}, \pi)$. Now, the function performing conformal mapping to the upper half-plane is $f(z) = \exp[z]$ and the new variables are determined as $\eta(\mathbf{r}) = \exp[x]\sin[y]$ and $\varphi(\mathbf{r}) = \exp[x]\cos[y]$. The linear and nonlinear potentials allowing for the exact solutions are now given by

$$V_{\text{ext}}(\mathbf{r}) = -2\varepsilon \exp[2x], \quad g(\mathbf{r}) = 2\mathcal{G} \exp[2x], \tag{11.24}$$

and $\varrho(\eta, \varphi) = \frac{1}{2}(\eta^2 + \varphi^2)$.

Exact Solutions Associated with the Above Domains

Turning to the exact solutions, below we consider only the case of repulsive interactions $\mathcal{G} = 1$, as the most natural candidate to produce stable stationary flows. As the two simplest solutions of Eq. (11.14) [33] we study, the "dark soliton" shape

$$\Psi_{ds}(\mathbf{r}) = \psi_{ds}(\eta, \varphi) = \sqrt{\mathcal{E}}\tanh\left[\sqrt{\frac{\mathcal{E}}{2}}\eta(\mathbf{r})\right]\exp\left[i\nu\varphi(\mathbf{r})\right] \tag{11.25}$$

and the nonlinear periodic modulation

$$\Psi_{sn}(\mathbf{r}) = \psi_{sn}(\eta, \varphi) = \frac{k\sqrt{\mathcal{E}}}{\sqrt{1+k^2}}\text{sn}\left[\frac{\sqrt{\mathcal{E}}\eta(\mathbf{r})}{\sqrt{1+k^2}}, k\right]\exp\left[i\nu\varphi(\mathbf{r})\right], \tag{11.26}$$

with $\mathcal{E} = \varepsilon - \nu^2 > 0$ and $k \in (0, 1]$ being the modulus of the Jacobi elliptic sine function [notice that $\psi_{ds}(0, \varphi) = \psi_{sn}(0, \varphi) = 0$, in conformity with the imposed boundary conditions]. These exact solutions given by Eqs. (11.25) and (11.26) and the corresponding velocity fields in the domains D_1 and D_2 are illustrated in Fig. 11.11.

The obtained exact solutions, however, leave several open questions related to their practical feasibility. First, the stability of the flows was not investigated, so far. Second, in all considered cases infinitely growing potentials were used, while any cutoff (which exists in the real world) may strongly perturb, and even destroy the solutions. To address these issues, we now turn to direct numerical simulations.

In the case at hand, the potentials $V_{\text{ext}}(\mathbf{r})$ and $g(\mathbf{r})$ grow with x, while the density goes asymptotically to (or is bounded by) a certain constant level. Bearing this in mind, we construct a physical system with bounded potential in the following way. For $x < 0$, we consider shifted linear and nonlinear potentials $V_{\text{ext}}(\mathbf{r} - \mathbf{r}_0)$ and $g(\mathbf{r} - \mathbf{r}_0)$, where $\mathbf{r}_0 = (x_0, 0)$ with x_0 being a constant shift vector, while $V_{\text{ext}}(\mathbf{r})$ and $g(\mathbf{r})$ are given by the analytical formulas (11.18). To define the physical potentials at $x > 0$, we mirror them at $x = 0$, thus obtaining $V_{\text{ext}}(-\mathbf{r} + \mathbf{r}_0)$ and $g(-\mathbf{r} + \mathbf{r}_0)$. To confine the condensate along the y coordinate we introduce an additional linear tarp potential as follows: $v_{add} = 0$ for $\mathbf{r} \in D$ and $v_{add} = V_0$, where V_0 is large enough, for $\mathbf{r} \notin D$. For more details, see the next chapter where we discuss the numerical studies of the problem.

11.3.3 Direct Numerical Simulations

Now we can perform direct numerical simulations of Eq. (11.14) with the additional confining potential v_{add} taking the initial distribution of the field in the form given by the analytical formulas $\psi(\mathbf{r} - \mathbf{r}_0)$ for $x < 0$ and $\psi(\mathbf{r}_0 - \mathbf{r})$ for $x > 0$. As is clear, our Ansatz does not satisfy Eq. (11.14) only along the line $x = 0$ and in the areas $y < 0$ and $y > \pi$. Since, however, we impose strong confining potential, we expect that

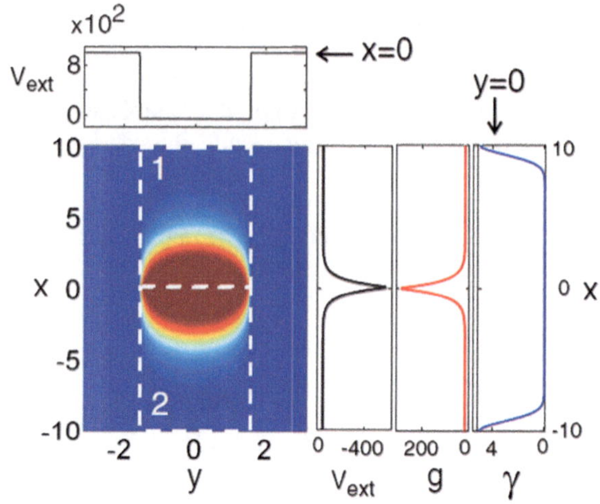

Fig. 11.12 The ansatz produced from the solutions with parameters $\varepsilon = 2, \nu = 0$. In the upper part of the figure, the distribution of the linear potential along y is shown for $x = 0$. On the right, the distributions of the linear and nonlinear potentials are shown alongside with the introduced loss for $y = 0$. In the areas marked by "1" and "2 " the ansatz exactly coincides with the analytical solution (neglecting exponentially weak loss). Reprint from Ref. [33], Copyright 2018, with permission from American Physical Society

the field is very weak outside the stripe $0 < y < \pi$ and our ansatz stays sufficiently close to the real stationary solution of Eq. (11.14) with the introduced physical potentials. The analytically found solution and the corresponding potentials are shown in Fig. 11.12. Within the areas "1 " and "2" the ansatz used in the numerics as the initial condition coincides with the analytical solutions. However, on the boundary between these areas the ansatz does not satisfy the equation for stationary fields. The ansatz does not satisfy the equation outside the areas either. However, the density of the condensate is low outside the areas and so there are some reasons to believe that the ansatz is close to the stationary solution. Let us remind here that in our numerical simulations to keep the condensate localized within the stripe we used strong linear potential; see the distribution of the linear potential along y.

We carried out numerical simulations for the solutions obtained in the domain D_2 and found out that some nonstationary excitations appear but the solution survive and stay rather close to the initial field distribution. To check the stability of the solution for a longer time we had to get rid of the propagating excitations. Since we perform numerical simulations in finite windows the only way to eliminate the propagating excitations is to introduce losses in the area where initially the density of the condensate is exponentially weak. To do this, we introduce linear losses in the form

$$\gamma = \gamma_0 \left(\exp\left[-\frac{(x - x_l)^2}{\omega_0^2} \right] + \exp\left[-\frac{(x + x_l)^2}{\omega_0^2} \right] \right).$$

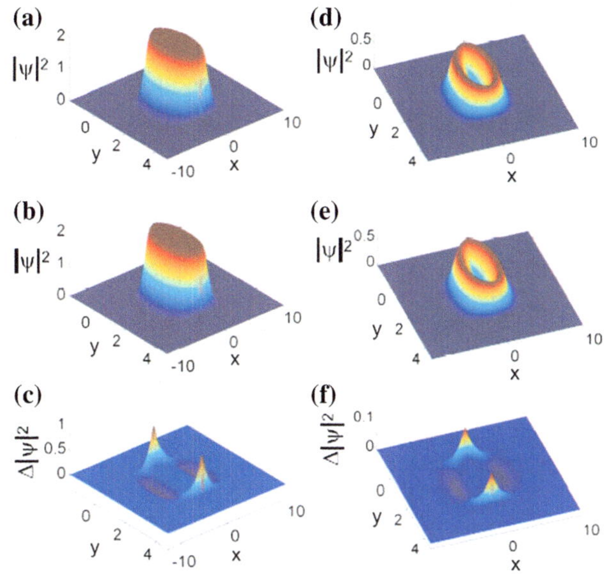

Fig. 11.13 Panels **a** and **d** show the initial distribution of the density of the condensates for cases 1 and 2; panels **b** and **e** show the distributions of the condensate densities at t $=$ 100; panels **c** and **f** show the differences between the initial density distributions and the density distributions at $t = 100$. The shifts of the initial conditions and the potentials are $x_0 = -2.5$ for **a**–**c** and $x_0 = -2$ for **d**–**f**. The solution parameters are $\varepsilon = 2$, $\nu = 0$. The field is kept within the stripe by strong repelling linear potential $va = 104$ for $y < 0$ and $y > \pi$. The parameters for the linear losses are $\gamma_0 = 5, \omega_0 = 1, x_l = 10$. Reprint from Ref. [33], Copyright 2018, with permission from American Physical Society

Then the losses are negligible in the area, where our analytical solution is big and so the solution is practically unaffected by the artificial losses used in the numerics. In the same time, all the perturbations propagating away are quickly absorbed. In this case, the excitations disappear very quickly and one can see in Fig. 11.13 that the surviving solution is very close to the initial distribution. As it is expected, the stationary solution deviates from the Ansatz most strongly around the points ($x = 0$, $y = 0$) and ($x = 0$, $y = \pi$).

In the case of attractive interactions (of a negative scattering length), the solution appears to be unstable against collapse.

11.3.4 The Generalized Conformal Mapping

Next, we address possibilities of getting more general types of the potentials, allowing for exact solutions. To this end, we introduce the generalized relations [cf. the Cauchy–Riemann equations (11.16)]:

$$\rho^2(\widehat{\eta})\frac{\partial}{\partial x}\widehat{\eta}(\mathbf{r}) = \frac{\partial}{\partial y}\widehat{\varphi}(\mathbf{r}), \quad \rho^2(\widehat{\eta})\frac{\partial}{\partial y}\widehat{\eta}(\mathbf{r}) = -\frac{\partial}{\partial x}\widehat{\varphi}(\mathbf{r}), \tag{11.27}$$

where $\rho^2(\widehat{\eta})$ is a positive-definite function of $\widehat{\eta}(\mathbf{r})$ only [as it is clear the arguments presented below can be also applied to the case where $\rho^2(\widehat{\varphi})$ is a function of $\widehat{\varphi}(\mathbf{r})$ only]. These equations still define transformation to the orthogonal coordinates $(\widehat{\eta}, \widehat{\varphi})$, however, now satisfying the relations [cf. Eq. (11.17)]

$$\nabla \cdot [\rho^2(\widehat{\eta})\nabla\widehat{\eta}] = 0, \quad \nabla \cdot [\rho^2(\widehat{\eta})\nabla\widehat{\varphi}] = 0, \quad \nabla\widehat{\eta} \cdot \nabla\widehat{\varphi} = 0. \tag{11.28}$$

Moreover, now we have that $|\nabla\widehat{\varphi}(\mathbf{r})|^2 \equiv \rho^4(\widehat{\eta})|\nabla\widehat{\eta}(\mathbf{r})|^2$. Notice that the generalized case [cf. Eq. (11.28)] can be reduced to the case mentioned above [cf. Eq. (11.17)] in the special case $\rho^2(\widehat{\eta}) \equiv 1$.

Next, we repeat the steps described above for the conformal mapping, and by the direct algebra show that the generalized Ansatz

$$\psi(\widehat{\eta}, \widehat{\varphi}) = \rho(\widehat{\eta}) \exp[i\nu\widehat{\varphi}(\mathbf{r})]\, \phi[\widehat{\eta}(\mathbf{r})] \tag{11.29}$$

solves the 2D stationary NLS equation (11.14) provided that $\phi(\mathbf{r})$ satisfy the stationary Eq. (11.21) with $\eta(\mathbf{r})$ substituted by $\widehat{\eta}(\mathbf{r})$ and the linear and nonlinear potentials are given by

$$V_{\text{ext}}(\mathbf{r}) \equiv \frac{\nabla^2\rho}{2\rho} + \frac{\nu^2(1-\rho^4) - \varepsilon}{2}|\nabla\widehat{\eta}|^2, \quad g(\mathbf{r}) \equiv \frac{\mathcal{G}|\nabla\widehat{\eta}|^2}{2\rho^2}. \tag{11.30}$$

Notice that now the linear and nonlinear potentials are not proportional to each other any more [cf. Eq. (11.18)].

To construct a particular example, we choose

$$\widehat{\eta}(\mathbf{r}) = 2\eta(\mathbf{r}) + \frac{\eta^2(\mathbf{r})}{2}, \quad \widehat{\varphi}(\mathbf{r}) = \varphi(\mathbf{r}), \quad \rho(\widehat{\eta}) = \frac{1}{\sqrt{2+\eta(\mathbf{r})}} \tag{11.31}$$

with $\eta(\mathbf{r})$ and $\varphi(\mathbf{r})$ solving Eq. (11.17). One can ensure that this choice also satisfies the conditions (11.28). Then, the simplest solution of Eq. (11.21) with the repulsive nonlinearity ($\mathcal{G} = 1$) is given by the study, the "dark soliton" shape (11.25) with $\eta(\mathbf{r})$ substituted by $\widehat{\eta}(\mathbf{r})$ and valid for the positive chemical potential $\mathcal{E} = \varepsilon - \nu 2 > 0$.

Then, according to the generalized ansatz with $\widehat{\eta}(\mathbf{r})$, $\widehat{\varphi}(\mathbf{r})$, and $\rho(\widehat{\eta})$ given by Eq. (11.31), as well as $\eta(\mathbf{r})$ and $\varphi(\mathbf{r})$ defined in the domain D_2, we obtain the "dark soliton"

$$\widehat{\Psi}_{ds}(\mathbf{r}) = \sqrt{\mathcal{E}}\rho(\widehat{\eta}) \tanh\left[\sqrt{\frac{\mathcal{E}}{2}}\widehat{\eta}(\mathbf{r})\right] \exp\left[i\nu \exp[x] \cos[y]\right], \tag{11.32}$$

with $\eta \equiv \eta(\mathbf{r}) = \exp[x]\sin[y]$, which solves Eq. (11.14) with the linear and nonlinear potentials given by [cf. Eq. (11.30)]

$$\begin{aligned} V_{\text{ext}}(\mathbf{r}) &= \frac{\exp[2x]}{8(2+\eta)^2}\{3 - 16(4\mathcal{E}+\nu^2) - 2\eta[\nu^2(3+2\eta) \\ &\quad +2\mathcal{E}(4+\eta)(\eta^2+4\eta+8)]\}, \\ g(r) &= \frac{1}{2}\exp[2x](2+\eta)^3. \end{aligned}$$

The presented solutions found in the domain D_2 behave in a way similar to the previous example; see Fig. 11.13. The numerical simulations prove that the found analytical solutions can indeed be used as a good approximation for the stable stationary solutions of Eq. (11.14) describing systems with physically relevant potentials.

11.4 Ring Dark Solitons in 2D Bose–Einstein Condensates with Tunable Interaction

This section deals with both the analytical and numerical investigation of the dynamics and modulation of ring dark solitons (RDS) in 2D BECs with tunable interaction.

11.4.1 The Physical Model

We consider a BEC trapped in an external potential and described by the macroscopic wave function $\Psi(\mathbf{r}, t)$, which is governed by the cubic NLS equation with external potential

$$i\hbar\frac{\partial\Psi}{\partial t} = \left[-\frac{\hbar^2}{2m}\nabla^2 + V(\mathbf{r}, t) + g_0(\mathbf{r}, t)|\Psi|^2\right]\Psi,$$

where the wave function is normalized by the particle number $N = \int_{\mathbf{R}^n}|\Psi|^2 d\mathbf{r}$ and $g_0(t) = 4\pi\hbar^2 a(t)/m$ m represents the strength of interatomic interaction characterized by the s-wave scattering length $a(t)$, which can be tuned by FR. The trapping potential is assumed to be $V(\mathbf{r}, z) = m\left(\omega_r^2 r^2 + \omega_z^2 z^2\right)/2$, where $r^2 = x^2 + y^2$, m is the atom mass, and $\omega_{r,z}$ are the confinement frequencies in the radial and axial directions, respectively. Further assuming that $\Omega = \omega_r/\omega_z << 1$ such that the motion of atoms in the z direction is essentially frozen in the ground state $[f(z)]$ of the axial harmonic trapping potential, the system can be regarded as quasi-2D. Then separating the degrees of freedom of the wave function and introducing the scales characterizing the trapping potential (the length, time, and wave function) as

$$\frac{\widetilde{\psi}\left(\widetilde{x},\widetilde{y},\widetilde{t}\right)}{2a_h\sqrt{\pi a_0 \eta}} f(z), \text{ with } \widetilde{x} = \frac{x}{a_h},\ \widetilde{y} = y,\ \widetilde{t} = \omega_z t,\ a_h = \sqrt{\hbar/\left[m\omega_z\right]},\ \eta = \int dz\,|f(z)|^4 \Big/ \int dz\,|f(z)|^2$$

lead to the dimensionless 2D NLS equation

$$i\frac{\partial\psi}{\partial t} = -\frac{1}{2}\nabla^2\psi + g(\mathbf{r},t)\,|\psi|^2\,\psi + \Omega^2 r^2\psi, \tag{11.33}$$

where $\nabla^2 = \frac{\partial^2}{\partial x^2} + \frac{\partial^2}{\partial y^2} = \frac{\partial^2}{\partial r^2} + \frac{1}{r}\frac{\partial}{\partial r} + \frac{\partial^2}{\partial\theta^2}$, $\Omega = \frac{\omega_r}{\omega_z}$, $g(t) = \frac{a(t)}{a_0}$, a_0 being a constant length that we choose to measure the time-dependent s-wave scattering length, and the tilde is omitted for simplicity. In what follows, we treat analytically and numerically the basic Eq. (11.33).

11.4.2 Basic Transformations and Exact Analytical Solutions

In order to study the dynamics of ring dark solitons, we consider the solution of Eq. (11.33) with circular symmetry, $\psi(r,t)$. In the case of $g(t) = C$, where C is a nonzero positive constant, the main difficulty in solving Eq. (11.33) is the existence of the last term, i.e., the trapping potential. Without a trap, i.e., $\Omega = 0$, the system is described by the standard NLS equation

$$i\frac{\partial Q(R,T)}{\partial T} + \frac{1}{2}\left(\frac{\partial^2 Q(R,T)}{\partial R^2} + \frac{1}{R}\frac{\partial Q(R,T)}{\partial R}\right) - C\,|Q(R,T)|^2\,Q(R,T) = 0. \tag{11.34}$$

Under the small-amplitude approximation, Eq. (11.34) has been transformed to the famous cylindrical Korteweg–de Vries CKdV equation by using the perturbation method [15]. The CKdV equation is known to be the basic nonlinear equation describing cylindrical and spherical pulse solitons in plasmas, electric lattices, and fluids see, e.g., [11] for a review and its exact solution has been derived [20]. Therefore, the soliton solutions of Eq. (11.34) are gained.

However, there is no effective method to solve Eq. (11.33) generally, especially when the last term is time dependent, $V(t)$. Now, we develop a method that can transform the general form of Eq. (11.33) to Eq. (11.34) by using the transformation

$$\psi(r,t) = Q(R(r,t),T(t))\exp\left[i\left(a(r,t) + c(t)\right)\right], \tag{11.35}$$

where $R(r,t)$, $T(t)$, $a(r,t)$, and $c(t)$ are assumed to be real functions and the transformation parameters read

$$
\begin{aligned}
R(r,t) &= \alpha(t)r, \\
T(t) &= \int \alpha^2(t)dt + C_0, \\
c(t) &= \frac{1}{2}\ln\frac{\alpha^2(t)}{C}, \\
a(r,t) &= -\frac{1}{2\alpha(t)}\frac{d\alpha}{dt}r^2,
\end{aligned}
\tag{11.36}
$$

where C_0 is a constant and $\alpha(t)$ is any real solution of the ODE

$$
\frac{1}{\alpha(t)}\frac{d^2\alpha(t)}{dt^2} - \frac{2}{\alpha^2(t)}\left(\frac{d\alpha(t)}{dt}\right)^2 - \Omega^2 = 0. \tag{11.37}
$$

Under the condition (11.37), all solutions of Eq. (11.34) can be recast into the corresponding solutions of Eq. (11.33). So we build a bridge between the extensive RDS study in nonlinear optics homogeneous system and the trapped BEC system. Furthermore, it is worth noting that the transformation method can be used to solve the general equation

$$
i\frac{\partial u(r,t)}{\partial t} + D(t)\left(\frac{\partial^2 u(r,t)}{\partial r^2} + \frac{1}{r}\frac{\partial u(r,t)}{\partial r}\right) + g(t)\left|u(r,t)\right|^2 u(r,t) + V(t)r^2 u(r,t) = 0, \tag{11.38}
$$

when $g(t)$ is proportional to $D(t)$. Equation (11.38) completely describes the dynamics and modulation of both the electric field in optical systems and the macroscopic-order parameter in atomic BECs in this case, $D(t)$ is a constant in quasi-2D systems with circular symmetry.

In order to get the transformation condition explicitly, we substitute $y(t) = d\alpha/[\alpha dt]$ into Eq. (11.37) and obtain

$$
\frac{dy}{dt} = y^2 + \Omega^2. \tag{11.39}
$$

This is a standard Riccati equation. We can solve it not only for $\Omega = const$, but also for various types of $\Omega^2(t)$, such as $a + b\sin[\lambda t]$, $a + b\cosh[t]$, $a\exp[\lambda t]$ (a, b, λ are arbitrary constants), and so on. Thus, we can study the dynamics of a system with a time-dependent external trap. In particular, when Ω is independent of t, the solution of Eq. (11.37) is

$$
\alpha(t) = \frac{C_1}{\cosh[\Omega t + C_0]}, \tag{11.40}
$$

where C_1 and C_2 are the integration constants.

Since under the small-amplitude approximation the solutions of Eq. (11.34) have been given, then by combining the solutions of Eqs. (11.34) and (11.40) we get the corresponding solutions for the RDS in a BEC in an external trap potential. They are exact solutions of the system when the depth of the RDS is infinitely small, which helps us to understand the dynamics of RDSs. However, when the RDSs get deeper, the small-amplitude approximation is invalid and we have to appeal to a numerical simulation. In the following sections, we study the dynamics and stability of RDSs numerically.

11.4.3 Stability Analysis

It is known that, starting from the initial configuration with strict circular symmetry, the RDS will oscillate up to a certain time until instabilities develop: a shallow RDS slowly decays into radiation and for a deep one snaking sets in, leading to formation of vortex–antivortex pairs arranged in a robust ring-shaped array (vortex cluster), because of transverse perturbations [10]. But the stability of the RDS to perturbation in the radial direction has not been analyzed. Here, we study numerically the stability of shallow RDSs against a small distortion of the ring shape, which is necessary for a practical experiment.

In the optical system, deformed RDSs on top of a constant background, called elliptic dark solitons, have been studied in [21]. As in the homogeneous case, our results show that deformed shallow RDSs in BECs exist and persist over long times for small distortions, and they become more circular when the radius increases. However, because of the inhomogeneous background in BECs, the evolution and instability of deformed RDSs change dramatically and the system displays diverse behaviors, which are illustrated in this section.

We study the stability of RDSs by solving Eq. (11.33) numerically with the parameters $g(t) = 1$, $\Omega = 0.028$, and the initial radius of the RDS $R_0 = 28.9$. Because of the large initial radius, a reasonable and good approximation is

$$\psi(x, y, 0) = \left(1 - \frac{\Omega^2}{4} r^2\right) [\cos[\phi(0)] \tanh[Z(r_1)] + i \sin[\phi(0)]], \qquad (11.41)$$

where $r = \sqrt{x^2 + y^2}$, $Z(r_1) = (r_1 - R_0) \cos[\phi(0)]$, $r_1 = \sqrt{(1 - e_c^2) x^2 + y^2}$, e_c is the eccentricity of the ring, and $\cos[\phi(0)]$ is proportional to the depth of the input soliton. When $e_c \neq 0$, R_0 represents the length of the semi-minor axis of the elliptical configuration. We take Eq. (11.41) as the initial configuration; the validity of this has been explained in detail and checked in the previous investigations [15].

We propagate the 2D time-dependent GP equation (11.33) using two distinct techniques: the alternating-direction implicit method [23] and the time-splitting Fourier spectral method [2]. The results from these two methods are cross checked and when $e_c = 0$, the properties of the results agree very well with those in [28].

To translate the results into units relevant to the experiment [10], we assume a ^{87}Rb ($a_0 = 5.7$ nm) condensate of radius 30 μm, containing 20000 atoms in a disk-shaped trap with $\omega_r = 218$ Hz and $\omega_z = 2628$ Hz. In this case, the RDS considered above has the radius $R_0 = 12.38$ μm, and the unit of time is 0.25 ms.

Our numerical results show that the shallow RDS refers to $\cos[\phi(0)] < 00.67$, where the soliton without distortion will not suffer from the snaking instability which is stable against a small distortion of the ring shape. There is a maximum eccentricity $e_{\max}$ for a given initial depth. When $e_c < e_{max}$, the RDSs are stable and oscillate, preserving their shape with the same period as the unperturbed RDSs until they decay. e_{max} becomes smaller with an increase of the initial depth see Fig. 11.14. This is because $2\pi/Qm_{ax}$ decreases as the RDS becomes deeper [15]; thus a shallower RDS can support a larger distortion.

When e_c exceeds e_{max}, snaking sets in and the dark soliton breaks into two vortex pairs, presenting a striking contrast to the multiples of four pairs reported previously. The evolution of the vortex pair is very different from the case of deep RDSs without distortion.

To illustrate the generic scenarios, we take a typical case with $\cos[\phi(0)] = 0.6$ and $e_c = 0.4$. The RDS initially shrinks and, when the minimum radius is reached in the short-axis direction, it starts snaking and forming two dark lump-like solitons in the horizontal direction; they move in opposite directions and then break into two vortex pairs see Fig. 11.15b. The vortex pairs arrange themselves in a ring configuration, which performs slow radial oscillations. Simultaneously, the vortices and antivortices move along the ring see Figs. 11.15c and 11.15d. The result of the motion is their collision in pairs in the vertical direction, followed by merging of the vortices and antivortices and formation of a pair of lump-like solitons see Fig. 11.16a; this process

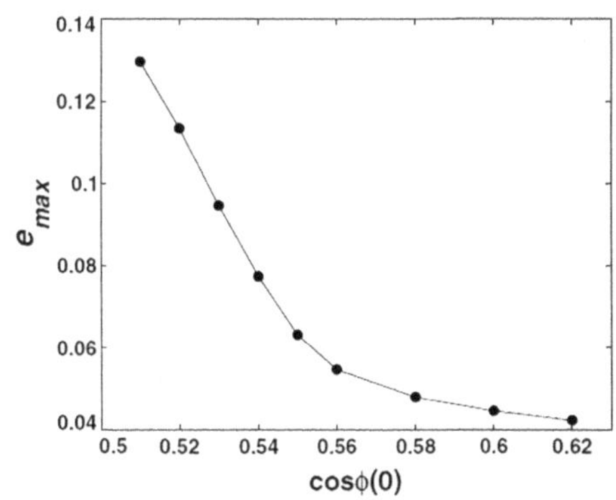

Fig. 11.14 Part of the stability diagram of a shallow ring dark soliton with $R_0 = 28.9$ and trap frequency $\Omega = 0.028$. The solid line is the instability threshold—below it the deformed ring dark solitons are stable, while above the line snaking sets in. The maximum eccentricity e_{max} of the ring in the initial configuration decreases as the initial depth $\cos[\phi(0)]$ of the ring dark soliton increases. Reprint from Ref. [10], Copyright 2018, with permission from American Physical Society

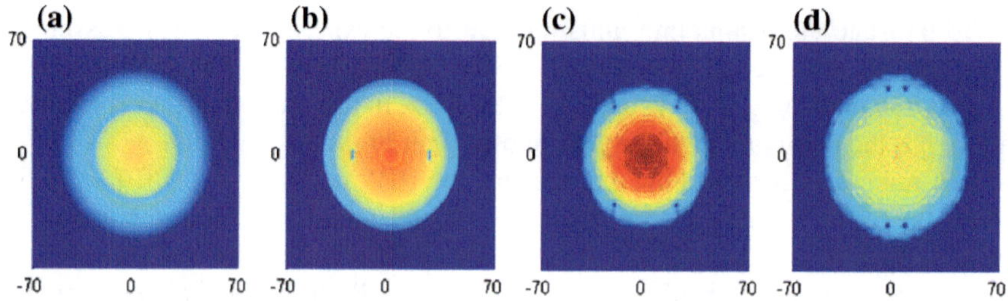

Fig. 11.15 Evolution of the ring dark soliton with initial depth $\cos[\phi(0)] = 0.6$, eccentricity $e_c = 0.4$, interaction strength $g(t) = 1$, and trap frequency $\Omega = 0.028$. **a** The initial profile shown by a color-scale density plot, where the ring dark soliton corresponds to the light-color ellipse with the length of the semi-minor axis $R_0 = 28.9$. **b**, **c**, and **d** correspond to $t = 80$, 400, and 540, respectively. The ring dark soliton with distortion first breaks into two lump solitons, and then two vortex pairs. See text for details. Reprint from Ref. [10], Copyright 2018, with permission from American Physical Society

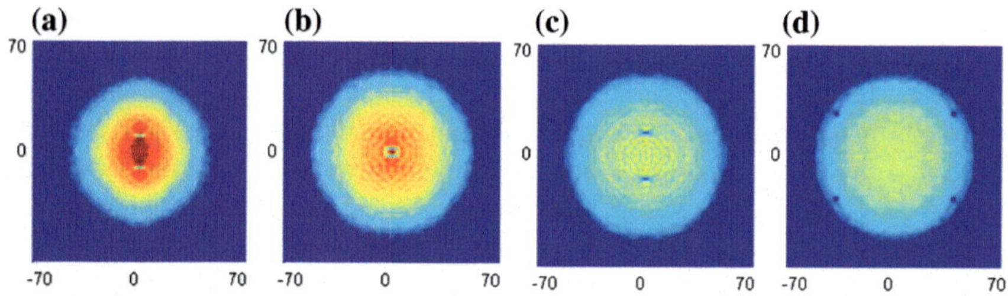

Fig. 11.16 Evolution of vortex pairs following Fig. 11.14 for $t =$ **a** 640, **b** 660, **c** 680, and **d** 1000. The vortex pairs collide and merge into dark lump-like solitons, which move, then form a ring dark soliton for a short time and revive. Finally the lump-like solitons again break into vortex pairs. See text for details. Reprint from Ref. [10], Copyright 2018, with permission from American Physical Society

is confirmed by the phase distribution of the system. The lump-like solitons move toward each other, i.e., to the center of the condensate, in the vertical direction. When they reach the minimum radius, a ring dark soliton forms see Fig. 11.16b. After a short time, the system returns to the state of two dark lump-like solitons. The lump-like solitons leave each other see Fig. 11.16c and along the way each lump-like soliton breaks into a vortex pair again. But the configuration is different from the initial one: the vortex–antivortex is substituted by an antivortex vortex, just as if the vortex and antivortex passed each other directly, following the motion before merging. Then the vortices and antivortices keep on moving see Fig. 11.16d, collide, and merge in the horizontal direction. The dynamical process repeats itself.

11.4.4 Control of Deep Ring Dark Solitons

The deep ring dark solitons [with $\cos[\phi(0)] > 00.67$] suffer from the snaking instability and can survive for only a short time typically 10 ms in numerical simulation before changing into are other soliton types. A longer lifetime is necessary for the practical observation of complex soliton physics such as oscillations or collisions. Our results show that the Feshbach resonance management technique can greatly extend the lifetime of the ring dark solitons, which makes it possible to study the long-time behavior of 2D dark solitons in experiments.

Feshbach resonance is a quite effective mechanism that can be used to manipulate interatomic interaction i.e., the magnitude and sign of the scattering length, which has been used in many important experimental investigations, such as the formation of bright solitons. Feshbach resonance management refers to the time-periodic changes of the magnitude and/or sign of the scattering length by Feshbach resonance, and has stimulated great interest in the theoretical and experimental study. Feshbach resonance management has been widely used to modulate and stabilize bright solitons, while its effect on the dynamics and stability of dark solitons is not very definite. Taking the ring dark soliton as an example, we demonstrate that the Feshbach resonance can dramatically affect the dynamics of dark solitons [10].

To show the effect of Feshbach resonance management (FRM), Eq. (11.33) is integrated numerically with eccentricity $e_c = 0$ and a time-dependent nonlinear coefficient $g(t)$; the other parameters and initial condition are the same as above. We consider the cases of $g(t) \sim a \exp[\pm\omega t]$ and $g(t) \sim a \sin[\omega]$, where a and ω are arbitrary constant parameters. Our results show that, for deep ring dark solitons, the Feshbach resonance management remarkably changes the evolution and the instability of solitons, as indicated by the number of vortex pairs which arise due to the snaking instability. Two typical examples are shown in Fig. 11.17: in the first, black ring dark solitons with $\cos[\phi(0)] = 1$ break into 20 vortex pairs, while the second has only four pairs. These phenomena are very different from results with constant scattering length, which have 16 vortex pairs. This makes it possible to study the dynamics of the snaking instability in detail. More interesting and more important are the cases in which the lifetime of the RDS can be extended, such as $g(t) = 1 - \sin[\omega t]$, as shown in Table 11.1, where the lifetime of RDSs corresponding to different modulation frequencies of the FRM (ω) is given.

In particular, when the system is subject to a modulation of $g(t) = 1 - \sin[\Omega t]$, the RDS can exist for a longer time before breakup. For example, when $\cos[\phi(0)] = 0.76$, the lifetime can be extended up to 45 ms from 10 ms; in this time, we can observe one complete cycle of oscillation of the RDS. If we just reduce the scattering length but keep it constant, the lifetime of the RDS will not change much. For example, if $g(t) = 0.2$, the RDS will start snaking at about 15 ms. So the lifetime extension effect is due to the FR management.

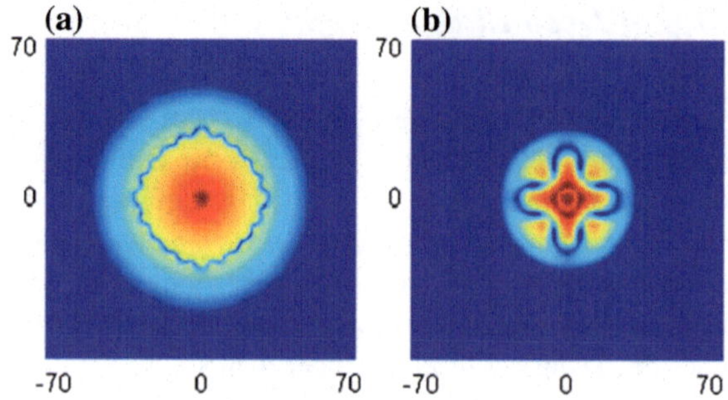

Fig. 11.17 Snapshots of condensate with a ring dark soliton under FR management. The parameters are **a** $e_c = 0, \cos[\phi(0)] = 1, g(t) = 2 - \sin[2\Omega t]$, and $t = 60\,(\approx 15\,\text{ms})$; **b** $e_c = 0, \cos[\phi(0)] = 1$, $g(t) = \exp[-\Omega t]$, and $t = 100$ (≈ 25 ms). Other parameters are the same as in Fig. 11.15. In **a**, the ring dark soliton will break into 20 vortex pairs, but only four vortex pairs in **b**. The total atom number is the same in these two plots, so for clarity we use different color scales. Reprint from Ref. [10], Copyright 2018, with permission from American Physical Society

Table 11.1 The lifetime of a RDS when it suffers the FRM $g(t) = 1 - \sin[\omega t]$. The parameters are $e_c = 0$, $\cos[\phi(0)] = 0.76$. The lifetime refers to the time interval between the start and the time point when snaking sets in

Frequency of FRM ω (units of Ω)	Lifetime of RDS (ms)
< 0.5	< 15
0.6	17
0.8	43
1.0	45
1.5	16
> 1.7	< 15

From Eq. (11.40), we can see that Ω is one intrinsic frequency of the system and the increase of the RDS lifetime is a resonance phenomenon, which is confirmed by Table 11.1. When the modulation frequency is tuned to be resonant with the intrinsic frequency of the system, Ω, on one hand, the RDS is less affected by the inhomogeneous background because of the approximately synchronous movement of the ring dark soliton and the background condensate resulting from FRM; on the other hand, the deep RDSs become effectively shallower (see Fig. 11.18), prolonging the time needed for the onset of snaking. So, the lifetime of the RDS increases.

Furthermore, this effect is even valid for shallow RDSs suffering from the instability due to the distortion of the ring shape. It has been checked that under Feshbach resonance management $g(t) = 1 - \sin[\Omega t]$, the RDS with $\cos[\phi(0)] = 0.6$ and $e_c = 0.4$ has a lifetime of 50 ms, which is much larger than the original 10 ms. Thus, under large perturbation the RDS can exist long enough to be observed, which makes the experimental study of RDSs easier.

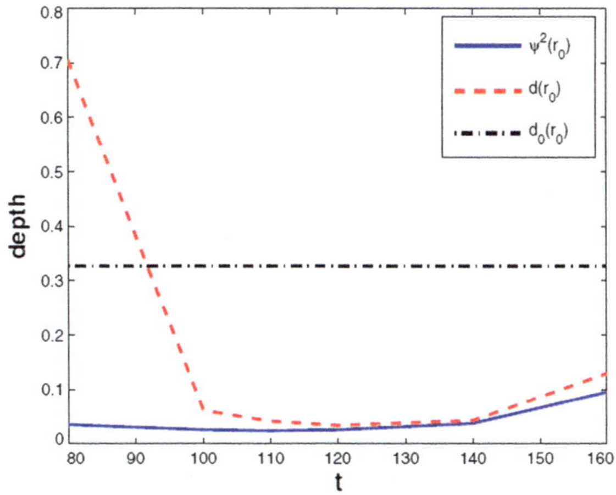

Fig. 11.18 Depth of the ring dark soliton as a function of time. The parameters are $e_c = 0, \cos[\phi(0)] = 0.76$, and $g(t) = 1 - \sin[\Omega t]$. $d_0(r_0)$ (black dot-dashed line) is the initial depth of the ring dark soliton, and $d(r_0)$ red dashed line is the depth of the ring dark soliton, where r_0 is the position of the center of the ring dark soliton. The depth of the RDS is much smaller than its initial depth at long times due to FRM. $\psi^2(r_0)$ (blue solid line) is the density at the center of the ring dark soliton, which changes very little. This fact shows that the soliton senses a more homogeneous local environment. Reprint from Ref. [10], Copyright 2018, with permission from American Physical Society

11.5 Dynamics of 2D Vortex Quadrupoles in Trapped BECs

A vortex quadrupole is a vortex cluster with zero total vorticity constructed as the product of four singly charged vortices. The aim of this section is to study, by numerical simulations, the detailed dynamics of 2D vortex quadrupoles with different initial configurations in trapped BECs. Also, we identify the parameter ranges which give the distinct regimes for the revival and disappearance of a quadrupole structure.

11.5.1 Description of the Physical Model

In the frame of zero-temperature mean-field theory, a trapped BEC system with only two-body interatomic interactions is well parameterized by the wave function $\psi(x, y, z, t)$ which satisfies the NLS equation with external potential

$$i\hbar\frac{\partial\psi}{\partial t} = \left[-\frac{\hbar^2}{2m}\nabla^2 + \frac{1}{m}\left(\omega_x^2x^2 + \omega_y^2y^2 + \omega_z^2z^2\right) + g_{3D}N\,|\psi|^2\right]\psi, \qquad (11.42)$$

with ω_j being the trap frequency and $g_{3D} = 4\pi\hbar^2 a_s/m$ being the three-dimensional coupling constant, where the wave function ψ is normalized to 1. For a large aspect ratio of the trap potential $\omega_z >> \omega_x = \omega_y = \omega_\perp$, the condensate can be regarded as a 2D disk-shaped cloud, where the motion in the z-direction is squeezed. The 2D order parameter $\psi(x, y, t)$ is governed by the corresponding 2D NLS equation, where the coupling constant g_{3D} in Eq. (11.42) is substituted by $g_{2D} = g_{3D}/\sqrt{2\pi}a_z$ with $a_z = \sqrt{\hbar/[m\omega_z]}$. In our simulations of a ^{87}Rb BEC containing $N = 1.5 \times 10^5$ atoms, the bulk s-wave scattering length is $a_s = 5.4$ nm. For the trap frequencies, we used a 2D disk-shaped configuration, where $\omega_x = \omega_y = 2\pi \times 5\text{rad} \cdot \text{s}^{-1}$ and $\omega_z = 2\pi \times 100\text{rad} \cdot \text{s}^{-1}$. It is convenient to turn all quantities dimensionless by introducing a time scale $t_0 = 1/\omega_\perp$ and a length scale $a_0 = \sqrt{\hbar/[m\omega_\perp]}$, which can provide better precision in simulations. For these parameters, the 2D Thomas–Fermi radius of the condensate is $R_{TF} = (8\sqrt{2}a_s N_0/\sqrt{\pi}a_z)^{1/4} \approx 8.32a_0$.

The order parameter of the system can be well described by a single- particle wave function $\psi(\mathbf{r}, t) = \sqrt{\rho(\mathbf{r}, t)}\exp[i\theta(\mathbf{r}, t)]$, where $\rho(\mathbf{r}, t)$ and $\theta(\mathbf{r}, t)$ are the density and phase distributions of atoms, respectively, with $\mathbf{r} = (x, y)$. Vortices can be seeded into the system by employing the phase imprinting techniques [31] experimentally. Numerically, the quadrupole states of interest in our simulations are obtained by using the method in Refs. [12, 22, 34] and 65. First, the vortex-free ground state of the condensate, ψ_g, is obtained by the imaginary time evolution. The initial phase factor of the order parameter of the quadrupole state is

$$\theta(\mathbf{r}, t = 0) = \prod_j s_j \arctan \frac{y - y_j}{x - x_j}, \tag{11.43}$$

where $s_j = \pm 1$ is the topological charge of the *jth* vortex, imprinting the velocity field of a vortex located at (x_j, y_j). The radial position of each vortex is then $r_j = \sqrt{x_j^2 + y_j^2}$. With new initial condition $\psi_g \exp[i(r, t = 0)]$, the quadrupole vorticity state of a BEC system can be obtained by evolving the NLS equation again in imaginary time while forcing the phase distribution during the calculation to be $\theta(r, t = 0)$ until the solution is numerically well converged, i.e., the relative change in norm of number of atoms and in energy of the system is smaller than 10^{-6}. The converged quasi-equilibrium state is then the desired quadrupole configuration ready for a real-time evolution. In this book, we focus on a subset of possible configurations, namely, the initial vortex quadrupole structures are axisymmetric. An example density profile and the corresponding phase diagram are shown in Fig. 11.19e, where the vortices with opposite charges are nested in the condensate alternately forming dipole structures along each axis. The thermal and quantum fluctuations are experimentally only important for temperatures which are larger or comparable to the mean-field interactions. Since we are studying systems in the frame of zero-temperature mean field theory, it is reasonable to think that these fluctuations will not affect the dynamics significantly, especially during the short-time evolution. However, taking the symmetry of systems into account, quantum fluctuations might have obvious effects on the long-time evolution of the systems. To introduce quantum fluctuations into our systems, the coordinate space-order parameter equivalent to our choice of stochastic initial state is easily shown to be

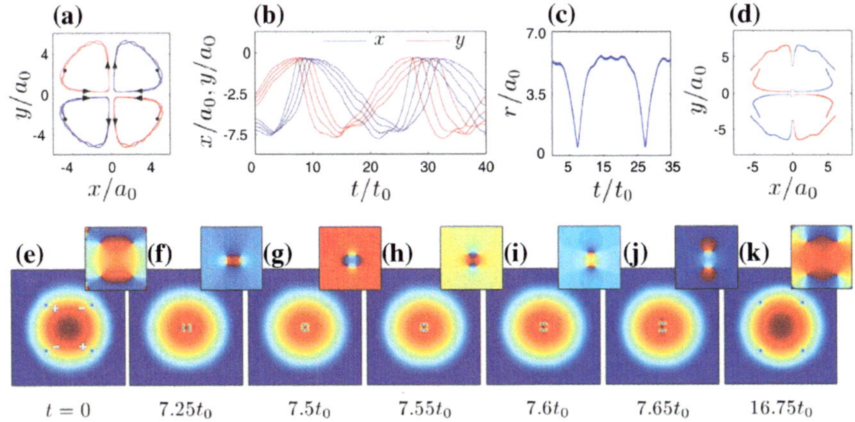

Fig. 11.19 Dynamics of quadrupoles in the "**recombination**" **regime**. **a** The trajectory of the quadrupole (black dots) initially seeded at ($\pm 4.75, \pm 2.5$) in the condensate. Arrows show the directions of their movements. **b** The time series of the x (blue) and y (red) coordinates of the down-left vortex for four initial configurations, ($-4, -3.25$), ($-4.25, -3$), ($-4.5, -2.75$), and ($-4.75, -2.5$), respectively. **c** The time series of the radial positions of the initially seeded vortices with charge $s = -1$ (blue dots). The initial configuration is the same as that used in **a**. **d** The trajectory of the quadrupole initially seeded at ($\pm 4.75, \pm 3.35$) in the condensate. (**e**–**k**) Typical snapshots for the density distribution and corresponding phase diagram of the condensate. The vortices are initial placed at ($\pm 4, \pm 4$) [34]

$$\Psi(\mathbf{r}, t = 0) = \psi_0(\mathbf{r}) + \xi(\mathbf{r}) \tag{11.44}$$

where $\psi_0(\mathbf{r})$ is chosen to be the initial state containing the quadrupole structure used in the situations without fluctuations, i.e., wave function of the condensed atoms. $\xi(\mathbf{r})$ is a complex Gaussian random field, which represents the uncorrelated quantum fluctuations

$$\langle \xi(\mathbf{r}) \rangle = \langle \xi^*(\mathbf{r}) \rangle = 0, \tag{11.45a}$$

$$\langle \xi(\mathbf{r})\xi(\mathbf{r}') \rangle = \langle \xi^*(\mathbf{r})\xi^*(\mathbf{r}') \rangle = 0, \tag{11.45b}$$

$$\langle \xi^*(\mathbf{r})\xi'(\mathbf{r}) \rangle = \frac{1}{2}\delta(\mathbf{r} - \mathbf{r}'). \tag{11.45c}$$

In practice, one can implement this initial condition by utilizing a complete set of plane-wave basis functions $\{\phi_j(\mathbf{r})\} = \frac{1}{\sqrt{V}} \exp\left[i\mathbf{k}_j \cdot \mathbf{r}\right]$, such that the random field can be written as

$$\xi(\mathbf{r}) = \sum_j \xi_j \phi_j(\mathbf{r}), \tag{11.46}$$

where ξ_j are complex gaussian random variables:

$$\langle \xi_j \rangle = \langle \xi_j^* \rangle = 0, \tag{11.47a}$$

$$\langle \xi_j \xi_{j'} \rangle = \langle \xi_j^* \xi_{j'}^* \rangle = 0, \tag{11.47b}$$

$$\langle \xi_j^* \xi_{j'} \rangle = \frac{1}{2}\delta_{jj'}. \tag{11.47c}$$

By using the truncated Wigner method [27, 34], and integrating the fully coordinate space stochastic differential equation of a 2D system

$$i\hbar\frac{\partial \Psi(\mathbf{r},t)}{\partial t} = -\frac{\hbar^2}{2m}\nabla^2\Psi(\mathbf{r},t) + \left[V_{\text{trap}}(\mathbf{r}) + g_{2D}\,|\Psi(\mathbf{r},t)|^2\right]\Psi(\mathbf{r},t) \tag{11.48}$$

in real time, the dynamics of systems under the influence of quantum fluctuations can be explored. The normalization condition is defined as $N_T = \int d\mathbf{r}\,|\Psi(\mathbf{r},t)|^2$ which includes both the condensed atoms and the virtual particles introduced into the field. It is obvious that even though Eq. (11.48) has a form similar to that of the NLS equation (11.42), the important difference is that quantum fluctuations (virtual particles) are included in this stochastic differential equation while the NLS equation is only valid for the condensed atoms.

In our calculations, only the "low-energy" modes in momentum space contribute to the fluctuations, which means the noise is just added into the modes with amplitude $\xi_j, j \in L$, where $L = [-25, 25]$ is a subspace containing 51×51 modes, only about 1% of the 501×501 momentum space used for our simulations. All other ξ_j, $j \notin L$ are set to be 0.

11.5.2 Numerical Analysis

To get an insight into the dynamics of vortex quadrupoles in a disk-shaped BEC, we first review the motion of a single vortex and a vortex dipole in such a system. A 2D disk-shaped condensate in an axisymmetric harmonic trap is a nonuniform system. In the TF limit, for a single vortex generated in this nonrotating system the precession velocity is a function of the vortex displacement r from the center of the condensate as $v = \omega_m r/\left(1 - r^2/R_\perp^2\right)$, where $\omega_m = \left(3\hbar^2/\left[2mR_\perp^2\right]\right)\ln\left(R_\perp/R_c\right)$ is the metastable frequency for the appearance of a central vortex in a disk-shaped condensate, with $R_\perp$ being the mean condensate radius and R_c being the vortex core radius. It is obvious that a central vortex does not undergo any evolution over time. At zero temperature, an off-center vortex with positive (negative) charge will perform a circular motion around the center of the condensate in a anticlockwise (clockwise) direction in a constant radius with a constant energy without considering the interaction between the vortex and the excited phonon mode. This pattern of behavior repeats as the density wave oscillations induced by the vortex motion are all confined within the condensate, and no net phonon radiation is expected. However, due to the interaction of the vortex with the collective phonon mode, the vortex trajectory is not a perfect circle. At finite temperatures, it will spiral out of the condensate due to its interaction with the thermal component.

For a dipole structure of vortex–antivortex pair in a trap, the dynamics is the result of the competition between the density inhomogeneity driven and the dipole interaction driven (mutually driven of each vortex of the pair). In the case where the mutually driven and inhomogeneity-driven balance with each other, a stationary dipole structure generates. For configurations, where a dipole is placed symmetrically about the trap center, vortices make close orbits about fixed points. Nearly the same as the motion in a single-vortex system, each of the vortices begins its orbit along the equipotential line as if they move independently, but once the vortices begin to approach each other they begin to interact. Instead of annihilating each other, both vortices alter their paths and begin to travel down the condensate from one side to the other side, parallel to each other. This continues until the vortices begin to reach the boundary of the condensate, then both of them move back on to their original orbits.

The situation gets more complicated when a quadrupole structure is taken into account. By setting different initial configurations, i.e., varying the radial position of the vortices, we find that the dynamics of the vortices shows three distinct regimes, which are named as "recombination", "annihilation" and "exchange".

In the recombination regime as shown in Fig. 11.19a, the vortices take quarter-circular motion and near-linear motion repeatedly, where the solid dots are the initial positions of the vortices, and the arrows show the direction of movement of each vortex. We can see clearly that, when they approach each other, two vortices in the left half-plane move as a dipole towards the trap center along the x-axis, while the other two in the right half-plane make a dipole movement as well facing to the left pair. However, after the four vortices reach their minimal separation, two of them in the upper half-plane combine into a new dipole structure and move away the trap center toward the edge of the condensate. So do the remains in the lower half-plane. The trajectories of the upper and lower dipoles are mirror symmetric to the x-axis. During the movement of vortices, the density wave oscillations arise across the condensate, and are severe around the edge of the condensate where the density is low. Therefore, the oscillation of the quarter-circular parts of the vortex trajectories in the low-density region is obvious as seen in Fig. 11.19a–d. For different initial positions, the time evolution of the $x-$ and $y-$coordinates of the down-left vortex is plotted in Fig. 11.19b. Typical density and phase distributions in this regime are plotted in Fig. 11.19e–k in time series. We note that for given parameters in Fig. 11.19a–c and e-k four vortices never contact. After the first period, the vortices will move away from their initial position a little bit due to the interactions with the density wave oscillations arising in the system, unlike a single vortex or a vortex dipole movement which generates close orbits. In Fig. 11.19c, we plot time series for the radial positions of the $s = -1$ vortices (blue dots). If we increase the initial radial displacement of vortices, the dynamics of the quadrupole can still stay within the recombination regime for the first period. But the dipole annihilation and revival can occur, which makes the deviation of the vortex positions from the initial positions sizable as seen in Fig. 11.19d $[(x, y) = (\pm 4.75, \pm 3.35)]$. During the following time evolution, this may induce a transition of the dynamical regime from recombination to exchange (annihilation).

Figure 11.20 shows the trajectory of the down-left vortex ($s = -1$) for the first period in the exchange regime. From the symmetry of the system, we know that the $s = 1$ vortices will give the same trajectories but with different starting position and in opposite direction (anticlockwise). Therefore, the left dipole and the right dipole move toward each other along the x-axis to the trap center at the beginning, which is the same as the situation in the recombination regime. As they approach each other, the distance between the two vortices of the left (right) dipole decreases gradually in the y-direction as shown in Fig. 11.20f. However, instead of separate in the y-direction after getting their minimal separation in the recombination regime, the two dipoles join together, and finally form a soliton ring as shown in Fig. 11.20g–i. This can also be distinguished from the fact that, for some certain time intervals, no vortex trajectories are plotted in Fig. 11.20b–d. The dark soliton ring in our system is formed by the connection of four moving vortices during the dynamical process. The soliton ring then breaks into two dipole structures as the original left and right dipoles exchange their positions and then keep moving in the x-direction toward the edge of the condensate. After that the vortices in the dipoles separate in the y-direction and take the place of their counterpart. The charge flipping occurs as seen in Fig. 11.20k as if two dipoles go through each other. The vortices continue to move in a quarter-circular orbit until they approach each other again. But at this time two vortices in

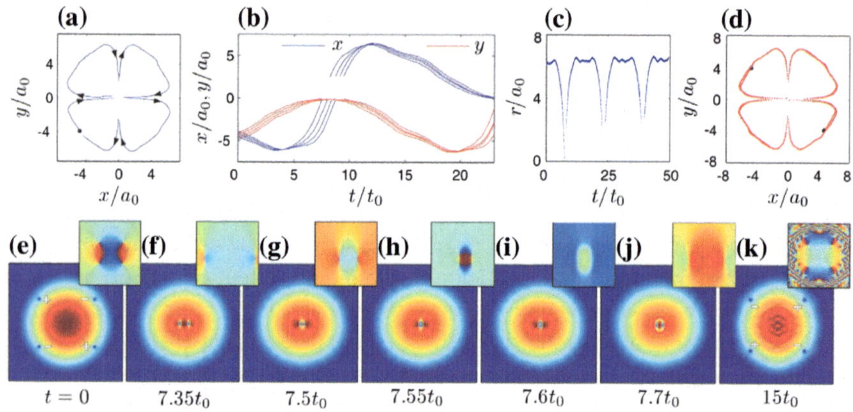

Fig. 11.20 Dynamics of quadrupoles in the "exchange" regime. **a** The trajectory of the vortex initially seeded at (−4.75, −4) in the condensate. Arrows show the directions of its movement. The initial positions of four vortices are denoted by the black dots for reference. **b** The time series of the x (blue) and y (red) coordinates of the down-left vortex for four initial configurations, (−4, −4.75), (−4.25, −4.5), (−4.5, −4.25), and (−4.75, −4), respectively. **c** The time series of the radial positions of the initially seeded vortices with charge $s = -1$ (blue dots). **d** The trajectories of the initially seeded vortices with charge $s = 1$ (red dots) in the condensate. **e**–**k** Typical snapshots for the density distribution and corresponding phase diagram of the condensate. The initial configurations of **c**–**k** are the same as that used in **a** [34]

the upper half-plane and the other two in the lower half-plane combine into two dipoles, respectively. They move in the y-direction and repeat what happened in the x-direction. After four times charge flipping, the system will restore to its original configuration.

The dynamics in the annihilation regime is the most simplest, where the vortices vanish permanently as long as they collide in the trap center for the first time as shown in Fig. 11.21a. In this process, there is also a ring structure formed. However, different from the situation in the exchange regime, the ring expands across the condensate quickly and smear at the edge of the condensate, which induces severe density wave oscillations as shown in Fig. 11.21k.

As is well known, the energy and angular momentum associated with a vortex increase as the vortex is moved to higher density region, which coincides with the fact that the energy cost to create a vortex is high when the local density is high. So we note that in all the three regimes, there are two timescales in the motion: the slow quarter-circular precession around the trap center and the fast motion nearly parallel to the axis. We note that the density wave oscillations are greater after the annihilation of vortex dipoles and the decay of the soliton ring, which indicates that the decay of vortex dipoles or soliton rings can make much severe excitations in dynamical systems. The revival of vortex dipoles can depress the excitations in the system.

In Fig. 11.22 we identify the coordinate parameters which separate regions of "recombination" (circles), "annihilation" (dots) and "exchange" (crosses) dynamics.

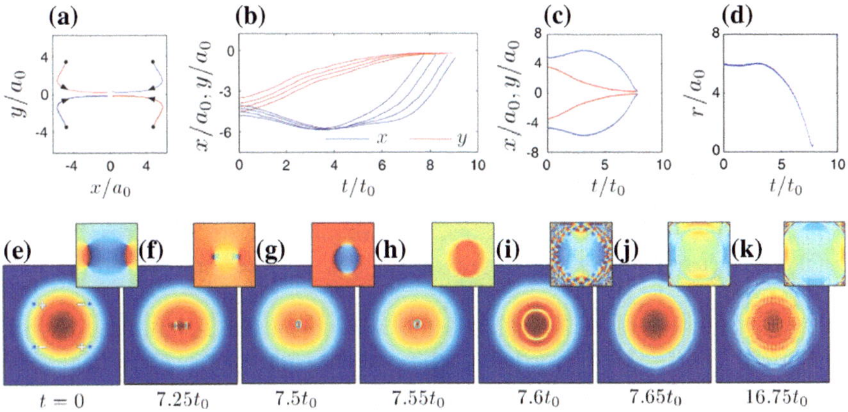

Fig. 11.21 Dynamics of quadrupoles in the "annihilation" regime. **a** The trajectories of the vortices initially seeded at (±4.75, ±3.5) (black dot) in the condensate. Arrows show the directions of their movements. **b** The time series of the x (blue) and y (red) coordinates of the down-left vortex for four initial configurations, (−4, −4.45), (−4.25, −4.15), (−4.5, −3.9), and (−4.75, −3.5), respectively. **c** The time series of the coordinates of the depicted vortex positions with charge $s = -1$ (initially seeded). Blue and red dots are x and y coordinates versus time, respectively. **d** The trajectories of the vortices with charge $s = \pm 1$ initially seeded at (4.75, −3.5) and (−4.75, 3.5) in the condensate. **e**–**k** Typical snapshots for the density distribution and corresponding phase diagram of the condensate. The initial configuration is the same as that used in **a** [34]

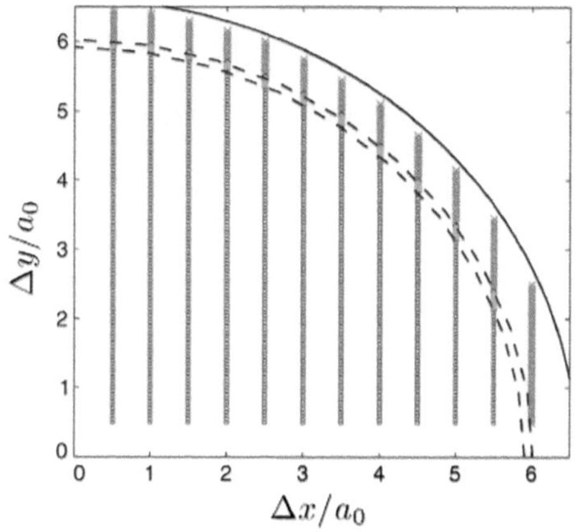

Fig. 11.22 Distinguish three dynamical regimes in coordinate space. Three regions in coordinate space where the dynamics of vortex quadrupoles show whether the vortices will recombine into dipoles (circles), annihilate each other (dots), or exchange their positions (crosses). Δx, and Δy are the initial displacement of the vortices in $x-$ and $y-$directions, respectively. The dashed-lines show the boundaries between these regions. The solid line is the arc with radius $6.6a_0$. The number of atoms is 1.5×10^5 [34]

The boundaries between these regions are both nearly quarter-circles, which are denoted by the dashed-lines. For a relatively small radial position $r_j < 5.91a_0$, the dynamics of the quadrupole is characterized by the combination of the vortices into dipoles in x- and y-directions alternately, while for a large radial position $r_j > 6.02a_0$ exchanging of positions of the vortices and charge flipping occur during the dynamical process. In between these two regions both the quadrupole and dipole structures vanish due to the annihilation of the vortices with opposite charges. For $r_j > 6.6a_0$ (black solid curve in Fig. 11.22, the density of the condensate is too low to identify a complete vortex structure. In Fig. 11.23, we give the radial position indicating the boundaries of three dynamical regimes with respect to varying TF-radius, which is applicable at any time step during the dynamical process. We find that the fitting curves of the boundaries (the blue and red lines) are both proportional to R_{TF}.

Through the analysis of isotropic systems, we know that the quarter-circular precessions are nearly along the equipotential lines, which are concentric ellipses in anisotropic traps (with different trapping frequencies in $x-$ and $y-$ axis, $\omega_x/\omega_y = 1.25$). We find that the dynamics of vortices still depends on the configuration of the quadrupole structure in anisotropic systems. The parameter ranges can also be determined according to Fig. 11.23. However, r which is chosen to identify the dynamical regimes is not the radial position of the vortices as in the isotropic systems, but the major radius or minor radius of the ellipse decided by the vortex positions, since the TF-radius of the condensate in each direction is different. By releasing the trapping

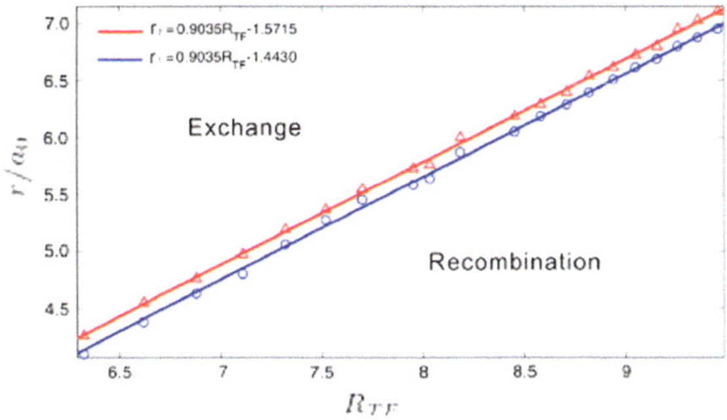

Fig. 11.23 Distinguish three dynamical regimes with radial position of vortices. The boundaries of the three dynamical regimes of vortex quadrupoles as linear functions of the TF-radius. r is the radial position of the vortices in isotropic systems, and the major (minor) radius of the elliptical equipotential lines decided by the vortex positions in anisotropic systems. The circles and triangles are data obtained from the numerical simulations, while the red and blue lines are plotted by the equations shown in the figure legend. The area between the blue and red lines is the annihilation regime [34]

potential in one direction (both directions) suddenly or adiabatically (the trapping frequencies are decreased linearly within $10t_0$) to a smaller value (form $2\pi \times 5$ Hz to $2\pi \times 3$ Hz or $2\pi \times 4$ Hz in our calculations), we confirm that the dynamics of quadrupole structures still obeys the same rule, which supports the conclusion in Ref. [16], where they showed that the sudden reduction in the number of atoms following expansion is expected to excite a small-amplitude (a few percent) breathing mode, with only slight alteration of vortex trajectories, which means that the dynamics of vortices is robust under the expansion of condensates.

To examine the robustness of the quadrupole dynamics, we introduce quantum fluctuations into the systems we studied above. We find that in the recombination regime with fluctuations the trajectories of vortices are nearly the same as those in the unperturbed systems during the first few periods. However, after that the quadrupole structure is eventually destroyed and the vortices take chaotic trajectories as shown in Fig. 11.24a. In the exchange regime with fluctuations, the dynamics of the vortices still obeys the rule of the unperturbed systems but with an additional rotation of the quadrupole structure as shown in Fig. 11.24b. This rotation also occurs if a dipole is placed asymmetrically about the origin [34]. Moreover, the vortices will not maintain a rectangular configuration anymore, leading to the destruction of the quadrupole structure and chaotic trajectories of vortices eventually. In the annihilation regime, the quantum fluctuations introduced will not change the dynamics of the system since the vortices annihilate permanently as soon as they meet at the trap center during the first period.

In the frame of zero-temperature mean field theory, we expect that quantum fluctuations will not change the dynamics of the system dramatically especially for the

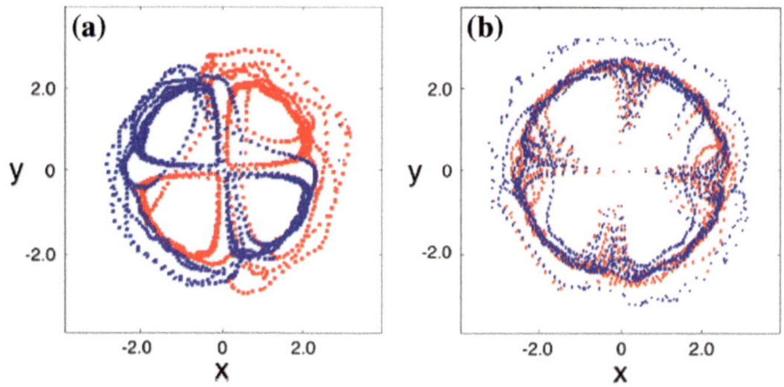

Fig. 11.24 Dynamics of quadrupoles in the systems with quantum fluctuations. The blue and red dots show the trajectories of vortices with charge $s = 1$ and $s = -1$, respectively. **a** The trajectories of the vortices initially seeded at (± 4.75, ± 2.5) in the condensate. The total evolution time is $125t_0$. The chaotic motion of vortices starts from about $86t_0$. **b** The trajectories of the vortices initially seeded at (± 4.75, ± 4.0) in the condensate. The total evolution time is $167t_0$. The chaotic motion of vortices starts from about $128t_0$

short-time evolution. However, symmetries of the system are affected by quantum fluctuations, which leads to chaotic motion of vortices for the long-time evolution. As we focused on the first few periods of quadrupole dynamics in our discussions, it shows that our results are robust, i.e., quantum fluctuations do not change the dynamical regimes of the system we studied at least for the short-time evolution. This also demonstrates that vortices have very stable topological structures.

We note that even though our calculations are limited at zero temperature, the vortices of the quadrupole cluster do deviate from their original locations during the periodical movement, which is similar to the situation in finite temperatures. The dynamics of the next period is fully decided by the fact that which region in Fig. 11.23 the radial positions (major radius or minor radius) of the vortices will fall into. Transition from one regime to another of the system dynamics may occur during the time evolution if the vortices do not annihilate permanently, which means that the initial configurations whose dynamics is ruled by "recombination" in the first period may transit to the other two regimes in the following period. And this may finally result in expulsion of the vortices from the condensate.

11.6 Ring Vortex Solitons and Their Stabilities in 3D Bose–Einstein Condensates Under Magnetic Confinement

The present section deals with the derivation of ring vortex solitons for both attractive and repulsive 3D Bose–Einstein condensates subject to harmonic potential confinement.

11.6.1 Physical Model and Stationary Exact Soliton Solutions

We consider the BECs in an external harmonic trapping potential $V_{\text{ext}}(r,z) = m(\omega_r^2 r^2 + \omega_z^2 z^2)/2$, where $r^2 = x^2 + y^2$, m is the atom mass, and ω_r and ω_z are the radial and axial trapping frequencies. The wave function ψ of the BECs satisfies the dimensionless NLS equation

$$\frac{\partial\psi}{\partial t} = -\frac{1}{2}\left[\frac{1}{r}\frac{\partial}{\partial r}\left(r\frac{\partial}{\partial r}\right) + \frac{1}{r^2}\frac{\partial^2}{\partial\theta^2} + \frac{\partial^2}{\partial z^2}\right]\psi + V_{\text{ext}}(r,t)\psi + g\left|\psi\right|^2\psi + i\Omega\frac{\partial\psi}{\partial\theta}, \tag{11.49}$$

where θ is the azimuthal angle, $g = 4\pi N a_s/a_0$ is the interaction strength, as determined by the total number of particles N in the condensate, the s-scattering wavelength a_s, and the harmonic oscillator length $a_0 = \sqrt{\hbar/[m\omega]}$, and Ω is the rotation angular frequency. The dimensionless $V_{\text{ext}}(r,t) = \frac{1}{2}\left(\gamma_r^2 r^2 + \gamma_z^2 z^2\right)$ with $\gamma_r = \omega_r/\omega$, $\gamma_z = \omega_z/\omega$, and $\omega = \min\{\omega_r, \omega_z\}$. Equation (11.49) is obtained by rescaling the length by a_0, the time by ω^{-1}, and the energy by $\hbar\omega$.

We seek the solution ψ of Eq. (11.49) in the form

$$\psi(r,z,\theta,t) = \phi(r,z)\exp\left[iS\theta - i\mu t\right], \tag{11.50}$$

where S is azimuthal quantum number, i.e., the intrinsic vorticity, and μ is the chemical potential. Asking that Eq. (11.50) satisfies Eq. (11.49) yields

$$-\frac{1}{2}\left[\frac{1}{r}\frac{\partial}{\partial r}\left(r\frac{\partial}{\partial r}\right) + \frac{\partial^2}{\partial z^2} - \frac{S^2}{r^2}\right]\phi + V_{\text{ext}}(r,t)\phi - S\Omega\phi + g\phi^3 = \mu\phi \tag{11.51}$$

for the function $\phi(r,z)$. We impose on $\phi(r,z)$ the zero boundary conditions $\lim_{r\to 0,|z|\to 0}\phi(r,z) = 0$ and $\lim_{r\to\infty,|z|\to\infty}\phi(r,z) = 0$. The linear-limit solution of Eq. (11.51) is written as the scaled linear combinations of products of the wave functions of the harmonic oscillator [3, 17],

$$\phi(r,z) = \frac{\gamma_r^{(S+1)/2}\gamma_z^{1/4}}{\sqrt{\pi C_n^S}\pi^{1/4}} r^S \widehat{L}_n^S\left(\gamma_r r^2\right)\exp\left[-\frac{1}{2}(\gamma_r r^2 + \gamma_z z^2)\right], \tag{11.52}$$

where $C_n^S = \Pi_{j=1}^n (n+j)$, $\widehat{L}_n^S$ is the generalized Laguerre polynomials, and n is the quantum number in the r direction (the quantum number in the z direction has set to be zero). The chemical potential corresponding to Eqs. (11.51) and (11.52) is written as

$$\mu_n S = \gamma_r[2(n-1) + S + 1] - S\Omega + \frac{1}{2}\gamma_z. \tag{11.53}$$

The stationary soliton solutions can be obtained as numerical solutions of Eq. (11.51) using the Newton continuation method [3] with $\mu_n S$ input, starting with the linear-limit solutions, under the constraint that the normalization of the wave function $\widetilde{N}$ and the energy E are conserved with numerical iteration time, where

$$\widetilde{N}(\psi) \equiv \int_{\mathbf{R}} |\psi(\mathbf{R}, t)|^2 \, d\mathbf{R} = \widetilde{N}(\psi_0) = 1, \quad t \geq 0, \tag{11.54}$$

and the energy

$$E_{g,\Omega}(\psi) = \int_{\mathbf{R}} \left[\frac{1}{2} |\psi(\mathbf{R}, t)|^2 + V_{\text{ext}}(\mathbf{R}, z) \, |\psi(\mathbf{R}, t)|^2 + \frac{g}{2} |\psi(\mathbf{R}, t)|^4 - \Omega S \right] d\mathbf{R}$$
$$\equiv E_{g,\Omega}(\psi_0), \quad t \geq 0. \tag{11.55}$$

Shown in Fig. 11.25 are various ring vortex solitons obtained in the case of the attractive interaction under different shaped harmonic potentials. For the isotropic harmonic potential of $\gamma r = \gamma_z = 1$, $g = -0.0083N$ refers to a ^{7}Li system both with $\omega_r = \omega_z = 20\pi$ Hz, for the cigar-shaped harmonic potential of $\gamma_r = 1.44$ and $\gamma_z = 1$, $g = -0.0117N$ referring to ^{7}Li with $\omega_r = 178$ Hz and $\omega_z = 123$ Hz [5], and for the pancaked harmonic potential of $\gamma_r = 1$ and $\gamma_z = 20$, $g = -0.0083N$ referring to ^{7}Li with $\omega_r = 20\pi$ Hz and $\omega_z = 400\pi$ Hz [25]. In the case of isotropic harmonic potential, we also study the ring vortex solitons for the repulsive interactions, where $g = 0.0188N$ refers to ^{87}Rb BECs with $\omega_r = \omega_z = 20\pi$ Hz. In this book, $\Omega = 0.7$ is fixed, except that is mentioned specifically. The vortex solitons obtained for the repulsive interactions (not shown here for the sake of brevity) resemble those shown in Fig. 11.25 for the attractive interaction.

The ring vortex solitons are further illustrated according to the radial profile of the wave function. Figure 11.26a shows that the influence of the quantum number n on the ring vortex solitons, i.e., the radially excited level is increasing with n, corresponding to more nodes of the wave function in the radial direction. Figure 11.26b shows the influence of the quantum number S on the ring vortex solitons. The ring vortex

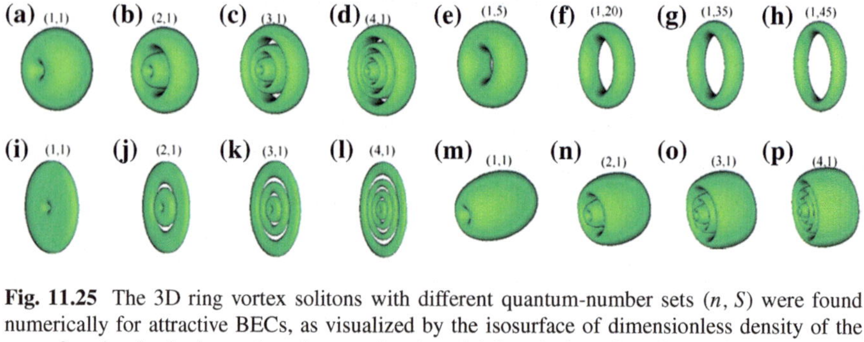

Fig. 11.25 The 3D ring vortex solitons with different quantum-number sets (n, S) were found numerically for attractive BECs, as visualized by the isosurface of dimensionless density of the wave function in the isotropic **a–h**, pancake-shaped **i–l**, and cigar-shaped **m–p** harmonic traps, respectively. Reprint from Ref. [17], Copyright 2018, with permission from American Physical Society

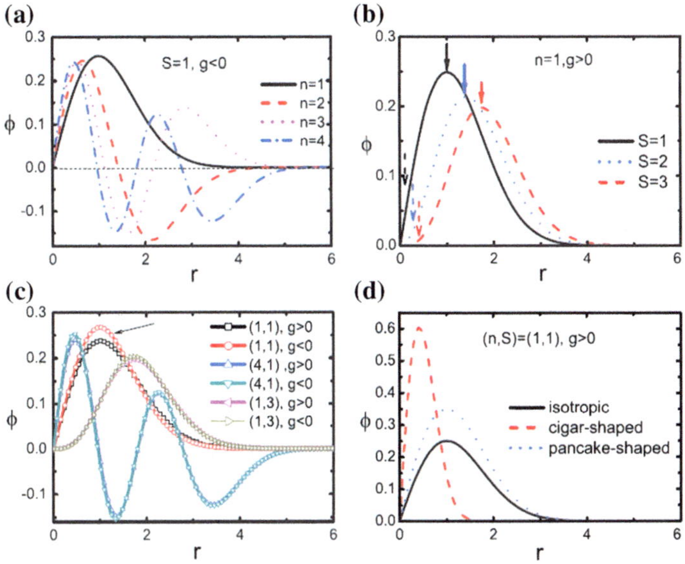

Fig. 11.26 The radial density profile of the ring vortex solitons characterized by the quantum numbers (n, S) for the attractive $(g < 0)$ or repulsive $(g > 0)$ interactions under the isotropic harmonic potential in **a–c** and also the cigar-shaped and pancake-shaped harmonic potentials in **d**. Reprint from Ref. [17], Copyright 2018, with permission from American Physical Society

solitons with the larger S demonstrate the deeper depletion of density at the core area (denoted by dashed arrows) and the lower density peak (denoted by solid arrows). Figure 11.26c shows the influence of the sign of the interaction nonlinearity on the shape of the ring vortex solitons for a given interaction strength. The density of the wave function at the first peak [denoted by the arrow in Fig. 11.26c] in the case of the attractive interaction is higher a little than that in the case of the repulsive interaction. Such a difference is the most prominent in the case of $(n, S) = (1, 1)$ but is weak in the cases of large n or large S. Figure 11.26d illustrates variation of the radial profile of the ring vortex solitons under the different shaped harmonic potentials.

Shown in Fig. 11.26 is the $\mu(N)$ curves of the ring vortex solitons. For a given N, as compared to the SRVS family with $n = 1$, the MRVS families with $n = 2, 3, 4, \ldots$ are of increasing μ. Within a SRVS or MRVS family, μ is increasing with S. Each data shown in Fig. 11.27 corresponds to a convergent stationary state, which is further investigated for the stability and dynamic evolution.

11.6.2 Stability Analysis

The stability of the ring vortex solitons is analyzed using the linear stability analysis. The wave function that deviates slightly from the stationary solutions is constructed as

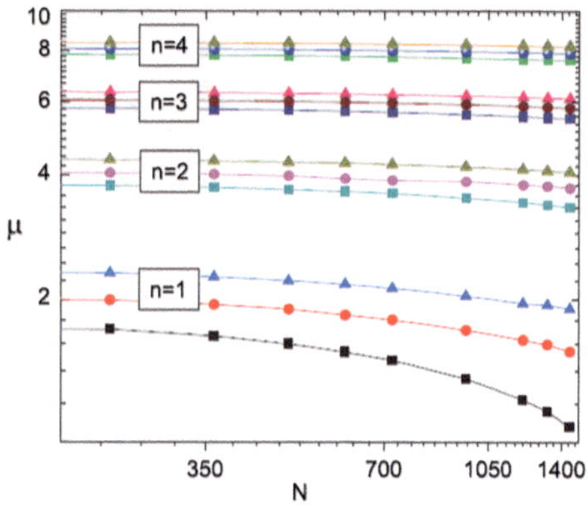

Fig. 11.27 The $\mu(N)$ curves of the ring vortex solitons of $n = 1, 2, 3, 4$ with $S = 1$ (square), $S = 2$ (circle), and $S = 3$ (triangular), as calculated with $\gamma_r = \gamma_z = 1$ and $g = -0.0083N$, and $\Omega = 0.7$, referring the ^{7}Li condensate under the isotropic harmonic potential with $\omega_r = \omega_z = 20\pi$ Hz. Reprint from Ref. [17], Copyright 2018, with permission from American Physical Society

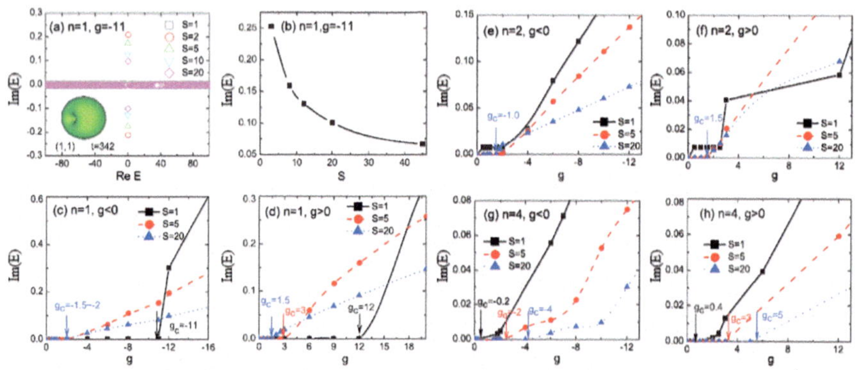

Fig. 11.28 The imaginary part in the energy spectrums [Im (E)] of the stability analysis, **a**, **b**: Im (E) vs S obtained for SRVS under a given attractive interaction $g = -11$. The inset in **a** displays a robust SRVS wit $S = 1$ at dimensionless time $t = 342$ against noise perturbation, and IIm (E) vs g obtained for SRVS [(**c**), (**d**)] and MRVS of $n = 2$ [(**e**) and (**h**)] and $n = 4$ [(**g**), (**h**)] with different S. Reprint from Ref. [17], Copyright 2018, with permission from American Physical Society

$$\psi = \left\{\phi(r, z) + u \exp\left[iEt\right] + \omega^* \exp\left[-iE^*t\right]\right\} \exp\left[iS\theta - i\mu t\right], \qquad (11.56)$$

where $|u|, |\omega| << 1$ are eigenmodes. Substituting Eq. (11.56) into Eq. (11.49) gives

$$\begin{pmatrix} L & -g\phi^2 \\ g\phi^2 & -L \end{pmatrix} \begin{pmatrix} u \\ \omega \end{pmatrix} = E \begin{pmatrix} u \\ \omega \end{pmatrix}, \qquad (11.57)$$

where $L \equiv \frac{1}{2}\left(\frac{\partial^2}{\partial r^2} + \frac{1}{r}\frac{\partial}{\partial r} + \frac{\partial^2}{\partial z^2} - \frac{S^2}{r^2}\right) - V(r, z) - 2g\phi^2 + S\Omega + \mu$, and E is the eigenvalue related to u and ω, which is obtained by diagonalization of Eq. (11.57) under the boundary conditions demanding that $u(r, z), \omega(r, z) \to 0$ at r as $|z| \to \infty$ and $r, |z| \to 0$. Applying Eq. (11.57), we analyze the ring vortex solitons of different radial-excited states, not only the SRVS, i.e., $n = 1$, but also the MRVS, from the first radially excited state, i.e., $n = 2$, to the higher radially excited states.

Figure 11.28a shows the real and imaginary parts of eigenvalues of the SRVS obtained for a given interaction strength in an isotropic harmonic potential. When $S = 1$, all eigenvalues are real, indicating that the corresponding state is stable. When $S = 2$, complex eigenvalues emerge, indicating that the corresponding states are dynamically unstable. The inverse of the imaginary part of the complex eigenvalue, $\mathrm{Im}\,(E)$, gives the time scale of such a dynamic instability. As shown in Fig. 11.28a, b, it is found that $\mathrm{Im}\,(E)$ decreases with S, which indicates that the ring vortex soliton with the larger intrinsic vorticity is of the longer lifetime before collapse, i.e., the better experimental stability.

Moreover, Fig. 11.28c, d show the g dependence of $\mathrm{Im}\,(E)$ obtained for the SRVS with different S. The eigenvalues become complex above the critical interaction strength g_c, showing the onset of dynamic instability. The g_c value is dependent on S. For example, our results show that g_c is around $-11$when $S = 1$ but decreases to -1.5 and -2 when $S = 5$ and 20 in the case of the attractive interaction. The corresponding value of g_c becomes a little larger in the case of the repulsive interaction, being around 12, 3, or 1.5 for the SRVS with $S = 1, 5$, and 20, respectively. The case of the SRVS, i.e., the general ring vortex soliton without radial excitation, has been studied previously [19]. It was found that for sufficiently weak interaction strength the SRVS with $S = 1$ is stable, while for $S \geq 2$ the SRVS is unstable to quadrupole oscillations. For instability times much longer than the timescale $2\pi/\omega$ of the BECs, the SRVS with $S \geq 2$ is said to be experimentally stable for small interaction strength. Our results shown in Fig. 11.28a–c are consistent with the previous study for the general ring soliton without radial excitation [17].

Figure 11.28e, f show the curve of $\mathrm{Im}(E)$ vs g obtained for the MRVS of $n = 2$, i.e., the first excited state. The stability of the ring vortex soliton of the first radial-excited state has been studied based on the 2D solutions [8], which shows that the first radial-excited state with $S = 1$ is stable for the sufficiently small interaction strength below a critical value. In our 3D study, for $S = 1$, no stable regime is found for the first excited state. This result means that the 3D MRVS of the first excited state with $S = 1$ could be different from its counterpart in 2D. The critical dimensionality for the GP equation is 2D. Further, we find that for large S, the 3D MRVS can be made stable. For example, we give two $\mathrm{Im}\,E$ vs g curves obtained for MRVS with $S = 5$ and $S = 20$ in Fig. 11.28e, f, where the stable regime, though being very small, is found up to $g_c = -1.0$ in the case of the attractive interactions and $g_c = 1.5$ in the case of the repulsive interactions. Such results indicate that the stability of the MRVS is dependent on the intrinsic vorticity. The MRVS with large S is expected to be stable.

Shown in Fig. 11.28g, h are results for the MRVS of $n = 4$. The S dependence on the stability becomes more prominent when the MRVS is of the higher radial excitation. The stable region, though being very small, below $g_c = -0.2$ and $g_c = 0.4$, is found for the $n = 4$ MRVS, with $S = 1$ under the attractive and repulsive interactions. The stable region increases for large S. When $S = 5$, the stable regions are up to $g_c = -2$ and $g_c = 3$ in the cases of the attractive and repulsive interactions, respectively. Furthermore, when $S = 20$, the stable regions increase up to $g_c = -4$ and $g_c = 5$ in the cases of the attractive and repulsive interactions, respectively.

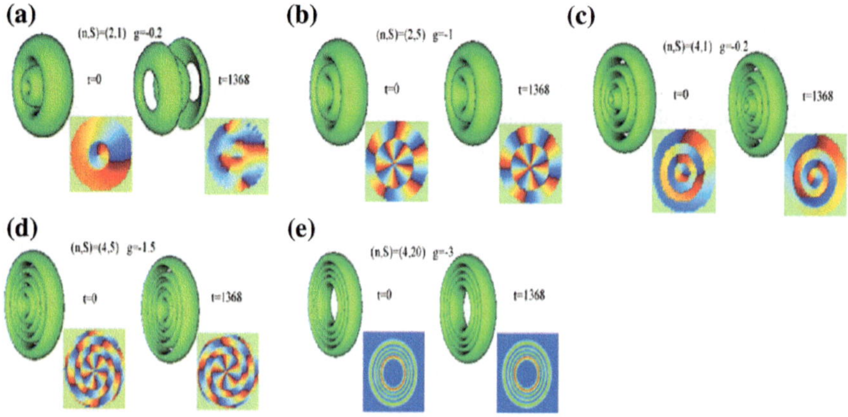

Fig. 11.29 The dynamic evolution of the MRVS in unstable and stable regions in Fig. 11.22e, f, where **a** is dynamically unstable and **b–e** are dynamically stable states which are consistent with linear stability analysis. Referring to a ^{7}Li system confined by $\omega_r = \omega_z = 20\pi$ Hz, the size is about 120 μm and the real time $t' = 0.0159t$ s, and $t = 0$ corresponds to the initial state of a 1% random perturbation in density. Reprint from Ref. [17], Copyright 2018, with permission from American Physical Society

Through comparing two cases of $n = 2$ [Fig. 11.28e, f] and $n = 4$ [Fig. 11.28g, h], we can find that the intrinsic vorticity tends to improve the stability of the MRVS, and such an effect becomes more significant when the radial excitation of MRVS is high. The MRVS of large S has a prominent stable region.

The linear stability analytic results for the S dependence of the MRVS stability is further confirmed by the directed simulation based on Eq. (11.49). Corresponding to Fig. 11.28e, Fig. 11.29a, b prove that the radially first excited state of MRVS of n = 2 is unstable against random perturbation when $S = 1$ but becomes stable when S is large, such as $S = 5$, under a small g. Next, corresponding to Figs. 11.28g, 11.29c–e show the stable MRVS of $n = 4$ with different S values under the given interaction strength below an increasingly large g_c.

11.6.3 Dynamics of Evolution

The ring vortex solitons are unstable against dynamic instability when the interaction strength is larger than g_c. Such dynamic instability is simulated through numerically integrating Eq. (11.49) using the time-splitting-spectral technique [4]. In our simulation, the stationary solutions of the ring vortex solitons are used as initial states after adding a random perturbation of a 1% relative amplitude of density.

We first simulate the dynamic evolution of the SRVS. As shown in Fig. 11.30, the SRVS evolves into polygonal rings, then splits into small fragments, and finally collapses. The multiple symmetry is revealed during such evolution. As S increases,we

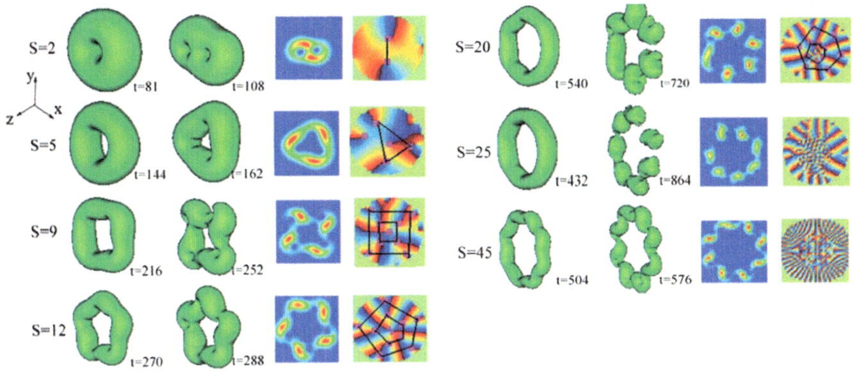

Fig. 11.30 Noise-induced splitting of the ring vortex solitons of different S at dimensionless time t, and the corresponding density and phase (the third and forth columns) plots in the $x-y$ plane. To translate the results into the experiment-related units, we assume a ^{7}Licondensate containing about 1500 atoms in an isotropic harmonic potential with $\omega_r = \omega_z = 20\pi$ Hz. The radii of the ring vortex solitons here are about 120 μm, decaying at the real time $t' = 0.0159t$ s. Reprint from Ref. [17], Copyright 2018, with permission from American Physical Society

obtain the two-, three-, four-, five-, six-, seven-, and eight fold symmetry, as shown in Figs. 11.30 and 11.31. As seen in the phase profile of the wave function, the multiple quantized vorticity splits into many singly quantized vortices. Such a singly quantized vortex cannot be seen in the density plot, and hence they are called hidden (or ghost) vortices [17]. The singly quantized vortices are self-organized into the regular lattice, contributing to the symmetry in splitting. Such symmetry in instability can be regarded as simultaneous, because the random initial perturbation we add is of no azimuthal mode. The multiple symmetric splitting phenomena vebeen demonstrated for giant vortices [14]. The symmetry in splitting and the number of the fragments are considered to be equal to the added angular momentum quantum number of the Bogoliubov excitation mode responsible for the splitting [14]. One finds a total of three types of splitting patterns—twofold, threefold, and fourfold symmetries—when the initial perturbation is random noise in the density of the wave function [17]. Figures 11.30 and 11.31 indicate that there could be more types of symmetries in splitting of the ring vortex soliton. As a matter of fact, the number of symmetry can be manipulated by the interaction strength g. As shown in Fig. 11.32, the number in symmetry (or the number of the fragments) decreases with g in the case of the repulsive interaction and increases with g in the case of the attractive interaction.

Second, we simulate the dynamic evolution of the MRVS. As shown in Fig. 11.33, the MRVS demonstrates expand–merge cycles, then collapses. The similar phenomenon, expand–shrink cycles occurring in the transverse direction, was predicted for the dark SRVS [19]. As compared with the dark SRVS, a principal difference is that the expand–merge cycles of the 3DMRVS occurs in the longitudinal direction. The condensate expands along the axial direction. The outer rings expand faster than the internal rings. At $t = 54$, the outmost ring splits into two pieces, and the internal

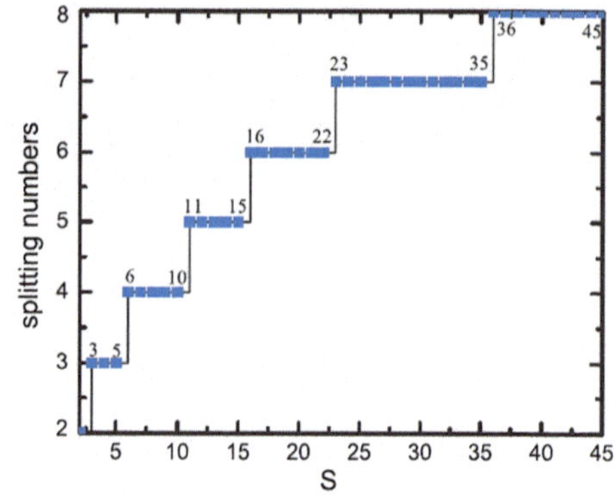

Fig. 11.31 Noise-induced splitting of giant vortices. The splitting numbers as a function of different S for the attractive interaction are shown. Reprint from Ref. [17], Copyright 2018, with permission from American Physical Society

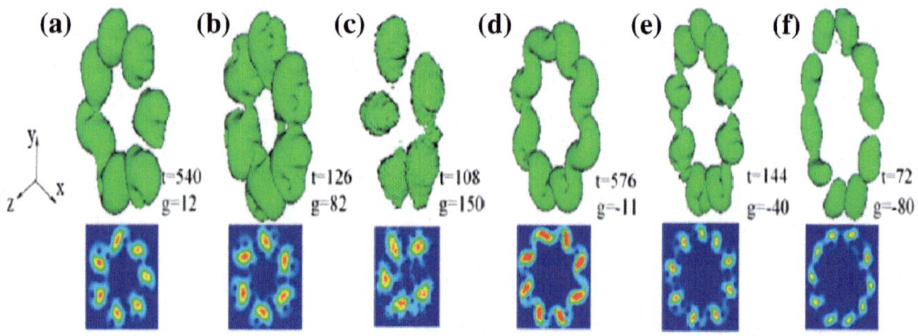

Fig. 11.32 Noise-induced splitting of the 3D vortex solitons of large vorticity in **a**–**c** with $S = 25$ for the repulsive interaction and **d**–**f** with $S = 45$ for the attractive interaction. To translate the results into the experiment-related units, we assume the ^{87}Rb condensate containing about 640 **a**, 4360 **b**, and 8000 **c** atoms and the ^{7}Li condensate containing 1325 **d**, 4820 **e**, and 9638 **f** atoms in a isotropic harmonic potential with $\omega_r = \omega_z = 20\pi$ Hz. The radii of the ring vortex solitons here are about 31 μm for **a**–**c** and 120 μm, being snapshots at the real time $t' = 0.0159t$ s. Reprint from Ref. [17], Copyright 2018, with permission from American Physical Society

rings expand but remain united. In contrast, the inmost ring shrinks, as seen in the density profile in the longitudinal plane shown in Fig. 11.33, middle row.

The expanded condensates and pieces subsequently unite to restore the original shape at $t = 162$. Such cycles repeat two times, then collapse finally. The expand–merge cycles reflect the expansion in the axial direction and the oscillation in the radial direction of the density of the wave function. In addition, during the dynamic evolution of the MRVS, as seen in the phase profile of the wave function in the transverse plane shown in Fig. 11.33, lower row, the multiple quantized vorticity is maintained well during the expand–shrink cycles. No vortex splitting is observed until the final collapse, which indicates that for the MRVS, the instability of the multiple radial waves prevails over the azimuthal instability. The similar expand–merge evolutions are demonstrated in the case of the repulsive interaction, where the cycles repeat three times before collapse, indicating the longer lifetime.

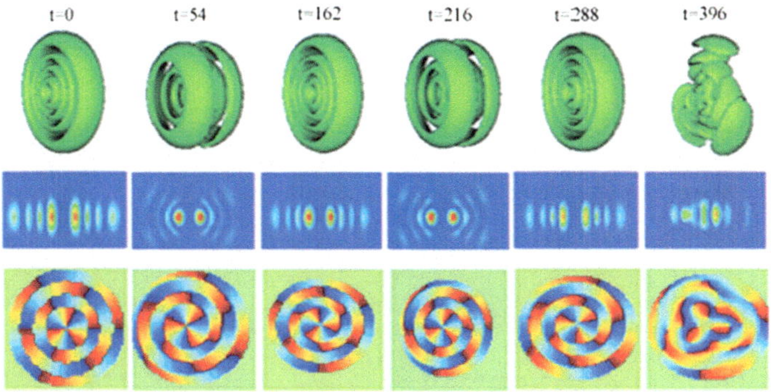

Fig. 11.33 Noise-induced splitting of a vortex soliton of $(n, S) = (4, 3)$ at the dimensionless time t and the corresponding density plot in the $x - z$ plane (middle row) and phase plot in the $x - y$ plane (lower row). To translate the results into experiment-related units, we assume a ^{7}Li condensate containing about 1500 atoms in an isotropic harmonic potential with $\omega_r = \omega_z = 20\pi$ Hz. The radii of the ring vortex solitons here are about 83.68 μm, decaying at the real time $t' = 0.0159t$ s. Reprint from Ref. [17], Copyright 2018, with permission from American Physical Society

We also simulate the dynamic instability under the pancake-shaped and cigar-shaped harmonic potentials. The expand–merge cycle occurring for the isotropic harmonic potential is suppressed under the pancake confinement potential. Under the cigar-shaped harmonic potential, being elongated in the axial direction, the MRVS does not demonstrate expand–merge cycles either. Besides, another interesting dynamics phenomenon, intertwining of vortices in the cigar-shaped BECs [17], is also not observed in our simulations.

The present section is conducted under the rotating NLS equation frame, which facilitates the future further study that considers the possible dependence of the form and properties of the ring vortex soliton on the rotation angular frequency. The parameter $\Omega = 0.7$ used for Figs. 11.25, 11.26, 11.27, 11.28, 11.29, 11.30, 11.31, 11.32, and 11.33 is chosen arbitrarily, referring to the rotating BEC experiment condition [6]. However, the results shown in Figs. 11.25, 11.26, 11.27, 11.28, 11.29, 11.30, 11.31, 11.32, and 11.33 are of general interest, and the stable ring vortex solitons do not necessarily rely on such a specific value, as found in our additional calculations with other values. For example, the dimensionless density profiles of two kinds of stable stationary ring vortex solitons, the SRVS with $S = 1$ and the MRVS of $n = 4$ with $S = 20$ subject to the sufficient small interaction strength $g = g_c$, obtained for different values overlap together, as shown in Fig. 11.34a, b. The results indicate that the form of such ring vortex solitons hardly changes with variations of. On the other hand, for a sufficiently small interaction strength $g = g_c$, all $\text{Im}(E)$ obtained for $\Omega = 0$, 0.4, 0.75, and 0.95 are equal to zero, as shown in Fig. 11.34c, d, which indicates that the SRVS with $S = 1$ and the MRVS with $S = 20$ maintain robust stability for different rotating frequencies. For an interaction strength larger than g_c, as shown in the insets in Figs. 11.34c and 11.10d, $\text{Im}(E)$ decreases with,

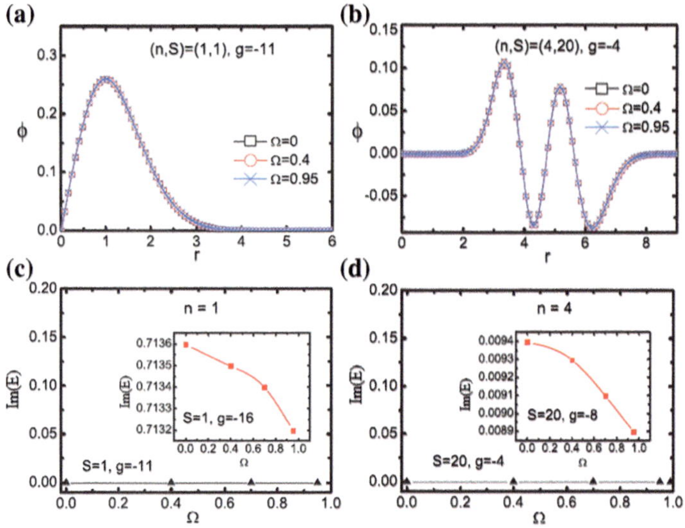

Fig. 11.34 The ring vortex solitons obtained for different rotation frequencies : **a** and **b** the radial density profile of the wave function, **c** and **d** the imaginary part in energy spectrums ($\mathrm{Im}(E)$) of the stability analysis. Reprint from Ref. [17], Copyright 2018, with permission from American Physical Society

corresponding to longer lifetime in the dynamic unstable evolution; however, such influence is in the minor quantity range.

References

1. Abramowitz, M., Stegun, I.: Handbook of Mathematical Functions. Dover, New York (1965)
2. Bao, W., Jaksch, D., Markowich, P.A.: Numerical solution of the Gross-Pitaevskii equation for Bose-Einstein condensation. J. Comput. Phys. **187**, 318 (2003)
3. Bao, W., Li, H., Shen, J.: A generalized-Laguerre-Fourier-Hermite pseudospectral method for computing the dynamics of rotating Bose-Einstein condensates. SIAM J. Sci. Comput. **31**, 3685 (2009)
4. Bao, W., Wang, H.: An efficient and spectrally accurate numerical method for computing dynamics of rotating Bose-Einstein condensates. J. Comput. Phys. **217**, 612 (2006)
5. Bradley, C.C., Sackett, C.A., Tollett, J.J., Hulet, R.G.: Evidence of Bose-Einstein condensation in an atomic gas with attractive interactions. Phys. Rev. Lett. **75**, 1687 (1995)
6. Bretin, V., Stock, S., Seurin, Y., Dalibard, J.: Fast rotation of a Bose-Einstein condensate. Phys. Rev. Lett. **92**, 050403 (2004)
7. Bronski, J.C., Carr, L.D., Deconinck, B., Kutz, J.N.: Bose-Einstein condensates in standing waves: the cubic nonlinear Schrödinger equation with a periodic potential. Phys. Rev. Lett. **86**, 1402 (2001)
8. Carr, L.D., Clark, C.W.: Vortices in attractive Bose-Einstein condensates in two dimensions. Phys. Rev. Lett. **97**, 010403 (2006)

9. Dai, C.-Q., Wang, D.-S., Wang, L.-L., Zhang, J.-F., Liu, W.M.: Quasi-two-dimensional Bose-Einstein condensates with spatially modulated cubic-quintic nonlinearities. Ann. Phys. **326**, 2356 (2011)
10. Hu, X.H., Zhang, X.F., Zhao, D., Luo, H.G., Liu, W.M.: Dynamics and modulation of ring dark solitons in two-dimensional Bose-Einstein condensates with tunable interaction. Phys. Rev. A **79**, 023619 (2009)
11. Infeld, E., Rowlands, G.: Nonlinear Waves, Solitons and Chaos. Cambridge University Press, Cambridge (1990)
12. Jackson, B., Proukakis, N.P., Barenghi, C.F., Zaremba, E.: Finite-temperature vortex dynamics in Bose-Einstein condensates. Phys. Rev. A **79**, 053615 (2009)
13. Junker, M., Dries, D., Welford, C., Hitchcock, J., Chen, Y.P., Hulet, R.G.: Photoassociation of a Bose-Einstein condensate near a Feshbach resonance. Phys. Rev. Lett. **101**, 060406 (2008)
14. Kawaguchi, Y., Ohmi, T.: Splitting instability of a multiply charged vortex in a Bose-Einstein condensate. Phys. Rev. A **70**, 043610 (2004)
15. Kivshar, Y.S., Yang, X.P.: Ring dark solitons. Phys. Rev. E **50**, R40 (1994)
16. Kuopanportti, P., Huhtamäki, J.A.M., Möttönen, M.: Size and dynamics of vortex dipoles in dilute Bose-Einstein condensates. Phys. Rev. A **83**, 011603 (2011)
17. Li, J., Wang, D.-S., Wu, Z.-Y., Yu, Y.-M., Liu, W.M.: Three-dimensional ring vortex solitons and their stabilities in Bose-Einstein condensates under magnetic confinement. Phys. Rev. A **86**, 023628 (2012)
18. Majda, A.J., Bertozzi, A.L.: Vorticity and Incompressible Flow. Cambridge University Press, Cambridge (2002)
19. Mihalache, D., Mazilu, D., Malomed, B.A., Lederer, F.: Vortex stability in nearly-two-dimensional Bose-Einstein condensates with attraction. Phys. Rev. A **73**, 043615 (2006)
20. Nakamura, A., Chen, H.H.: Soliton solutions of the cylindrical KdV equation. J. Phys. Soc. Jpn. **50**, 711 (1981)
21. Papacharlampous, I.E., Kevrekidis, P.G., Nistazalis, H.E., Frantzeskakis, D.J., Malomed, B.A.: Elliptic dark solitons. Phys. Scr. **69**, 7 (2004)
22. Parker, N.G., Allen, A.J., Barenghi, C.F., Proukakis, N.P.: Coherent cross talk and parametric driving of matter-wave vortices. Phys. Rev. A **86**, 013631 (2012)
23. Press, W.H., Teukolsky, S.A., Vetterling, W.T., Flannery, B.P.: Numerical Recipes in Fortran 77. Cambridge University Press, Cambridge (1992)
24. Rodas-Verde, M.I., Michinel, H., Pérez-García, V.M.: Controllable soliton emission from a Bose-Einstein condensate. Phys. Rev. Lett. **95**, 153903 (2005)
25. Rychtarik, D., Engeser, B., Nägerl, H.-C., Grimm, R.: Two-dimensional Bose-Einstein condensate in an optical surface trap. Phys. Rev. Lett. **92**, 173003 (2004)
26. Salerno, M., Konotop, V.V., Bludov, Y.V.: Long-living bloch oscillations of matter waves in periodic potentials. Phys. Rev. Lett. **101**, 030405 (2008)
27. Sinatra, A., Lobo, C., Castin, Y.: The truncated Wigner method for Bose-condensed gases: limits of validity and applications. J. Phys. B **35**, 3599 (2002)
28. Theocharis, G., Frantzeskakis, D.J., Kevrekidis, P.G., Malomed, B.A., Kivshar, Y.S.: Ring dark solitons and vortex necklaces in Bose-Einstein condensates. Phys. Rev. Lett. **90**, 120403 (2003)
29. Vengalattore, M., Conroy, R.S., Rooijakkers, W., Prentiss, M.: Ferromagnets for integrated atom optics. J. Appl. Phys. **95**, 4404 (2004)
30. Wang, D.S., Hu, X.H., Hu, J.P., Liu, W.M.: Quantized quasi-two-dimensional Bose-Einstein condensates with spatially modulated nonlinearity. Phys. Rev. A **81**, 025604 (2010)
31. Williams, J.E., Holland, M.J.: Preparing topological states of a Bose-Einstein condensate. Nature **401**, 568 (1999)
32. Xiong, H.W., Liu, S.J., Zhang, W.P., Zhan, M.S.: Ultracold two-component fermionic gases with a magnetic field gradient near a Feshbach resonance. Phys. Rev. Lett. **95**, 120401 (2005)
33. Yan, Z., Konotop, V.V., Yulin, A.V., Liu, W.M.: Two-dimensional superfluid flows in inhomogeneous Bose-Einstein condensates. Phys. Rev. E **85**, 016601 (2012)
34. Yang, T., Hu, Z.-Q., Zou, S., Liu, W.M.: Dynamics of vortex quadrupoles in nonrotating trapped Bose-Einstein condensates. Sci. Rep. **6**, 29066 (2016)

Index

W.-M. Liu and E. Kengne, *Schrödinger Equations in Nonlinear Systems*,
https://doi.org/10.1007/978-981-13-6581-2

Printed by Printforce, the Netherlands